ISBN 978-1-330-45388-9
PIBN 10064476

This book is a reproduction of an important historical work. Forgotten Books uses
state-of-the-art technology to digitally reconstruct the work, preserving the original format
whilst repairing imperfections present in the aged copy. In rare cases, an imperfection in
the original, such as a blemish or missing page, may be replicated in our edition. We do,
however, repair the vast majority of imperfections successfully; any imperfections that
remain are intentionally left to preserve the state of such historical works.

1 MONTH OF
FREE
READING

at
www.ForgottenBooks.com

By purchasing this book you are eligible for one month membership to ForgottenBooks.com, giving you unlimited access to our entire collection of over 1,000,000 titles via our web site and mobile apps.

To claim your free month visit:
www.forgottenbooks.com/free64476

English
Français
Deutsche
Italiano
Español
Português

www.forgottenbooks.com

Mythology Photography **Fiction**
Fishing Christianity **Art** Cooking
Essays Buddhism Freemasonry
Medicine **Biology** Music **Ancient**
Egypt Evolution Carpentry Physics
Dance Geology **Mathematics** Fitness
Shakespeare **Folklore** Yoga Marketing
Confidence Immortality Biographies
Poetry **Psychology** Witchcraft
Electronics Chemistry History **Law**
Accounting **Philosophy** Anthropology
Alchemy Drama Quantum Mechanics
Atheism Sexual Health **Ancient History**
Entrepreneurship Languages Sport
Paleontology Needlework Islam
Metaphysics Investment Archaeology
Parenting Statistics Criminology
Motivational

NOTICE TO MEMBERS.

THE publications of the New Zealand Institute consist of—

1. **Transactions,** a yearly volume of scientific papers read before the local Institutes. This volume is of royal-octavo size, and is issued about the 1st June of each year.

2. **Proceedings,** containing reports of the meetings of the Board of Governors of the New Zealand Institute and of the local Institutes, abstracts of papers read before them and of papers dealing with New Zealand scientific matters and published elsewhere, list of members, &c. The Proceedings are of the same size as the Transactions, and are bound up with the yearly volume of Transactions supplied to members.

3. **Bulletins.** Under the title of "Bulletins" the Board of Governors hopes to be able to issue from time to time important papers which for any reason it may not be possible to include in the yearly volume of the Transactions. The bulletins are of the same size and style as the Transactions, but appear at irregular intervals, and each bulletin is complete in itself and separately paged. The bulletins are not issued free to members, but may be obtained by them at a reduction on the published price.

LIBRARY PRIVILEGES OF MEMBERS.—Upon application by any member to the Librarian of the New Zealand Institute or of any of the affiliated Societies such works as he desires to consult which are in those libraries will be forwarded to him, subject to the rules under which they are issued by the Institute or the Societies. The borrower will be required to pay for the carriage of the books.

MEMORANDUM FOR AUTHORS OF PAPERS.

1. All papers must be typewritten, unless special permission to send in written papers has been granted by the Editor for the time being.

2. The author should read over and correct the copy before sending it to the Secretary of the society before which it was read.

3. A badly arranged or carelessly composed paper will be sent back to the author for amendment. It is not the duty of an editor to amend either bad arrangement or defective composition.

4. In regard to underlining of words, it is advisable, as a rule, to underline only specific or generic names, or foreign words.

5. In regard to specific names, the International Rules of Zoological Nomenclature in relation to zoological names, and the International Rules for Botanical Nomenclature, must be adhered to.

6. Titles of papers should give a clear indication of the scope of the paper, and such indefinite titles as, *e.g.*, "Additions to the New Zealand Fauna" should be avoided.

7. Papers should be as concise as possible.

8. Photographs intended for reproduction should be the best procurable prints, unmounted and sent flat.

9. *Line Drawings.*—Drawings and diagrams may be executed in line or wash. If drawn in line—*i.e.*, with pen and ink—the best results are to be obtained only from good, firm, black lines, using such an ink as Higgin's liquid India ink, or a freshly mixed Chinese ink of good quality, drawn on a smooth surface, such as Bristol or London board. Thin, scratchy, or faint lines must be avoided. Bold work, drawn to about twice the size (linear) of the plate, will give the best results. Tints or washes may not be used on line drawings, the object being to get the greatest contrast from a densely black line (which may be fine if required), drawn on a smooth, white surface.

10. *Wash Drawings.*—If drawing in wash is preferred, the washes should be made in such water-colour as lamp-black, ivory black, or India ink. These reproduce better than neutral tint, which inclines too much to blue in its light tones. High lights are better left free from colour, although they may be stopped out with Chinese white. As in line drawings, a fine surface should be used (the grain of most drawing-papers reproduces in the print with bad effect), and well-modelled contrasted work will give satisfactory results.

11. *Size of Drawings.*—The printed plate will not exceed $7\frac{1}{4}$ in. by $4\frac{1}{2}$ in., and drawings may be to this size, or preferably a multiple thereof, maintaining the same proportion of height to width of plate. When a number of drawings are to appear on one plate they should be neatly arranged, and if numbered or lettered in soft pencil the printer will mark them permanently before reproduction. In plates of wash drawings, all the subjects comprising one plate should be grouped on the same sheet of paper or cardboard, as any joining-up shows in the print.

12. In accordance with a resolution of the Board of Governors, authors are warned that previous publication of a paper may militate against its acceptance for the Transactions.

13. In ordinary cases twenty-five copies of each paper are supplied gratis to the author, and in cases approved of by the Publication Committee fifty copies may be supplied without charge. Additional copies may be obtained at cost-price.

14. *Citation.*—Authors are recommended to adopt the following method of citation : In the text write the date of the paper referred to after the author's name, with also a page reference if necessary, enclosing these in parentheses, thus : "Smith (1910, p. 6)." The reference may then be placed in a list at the end of the article in alphabetical order. Form of reference :—

TRANSACTIONS

AND

PROCEEDINGS

OF THE

NEW ZEALAND INSTITUTE

FOR THE YEAR 1915

VOL. XLVIII

(New Issue)

EDITED AND PUBLISHED UNDER THE AUTHORITY OF THE BOARD
OF GOVERNORS OF THE INSTITUTE

ISSUED 16th OCTOBER, 1916

Wellington, N.Z.
MARCUS F. MARKS, GOVERNMENT PRINTING OFFICE
WILLIAM WESLEY AND SON, 28 ESSEX STREET, STRAND, LONDON W.C.

CONTENTS.

I. BOTANY.

II. ZOOLOGY.

32410

III. GEOLOGY.

IV. CHEMISTRY AND PHYSICS.

V. MISCELLANEOUS.

PROCEEDINGS.

APPENDIX.

LIST OF PLATES.

(Text figures not included.)

ii—Trans.

List of Plates.

TRANSACTIONS.

TRANSACTIONS

OF THE

NEW ZEALAND INSTITUTE,

1915.

ART. I.—Records of Unconformities from Late Cretaceous to Early Miocene in New Zealand.

By P. G. MORGAN, M.A., F.G.S., Director of the Geological Survey of New Zealand.

[Read before the Wellington Philosophical Society, 29th September, 1915.]

INTRODUCTION.

VARIOUS geologists, notably Hutton, have striven, mainly on palaeontological grounds, to show that unconformity exists between the Cretaceous and Tertiary rocks of New Zealand. On the other hand, Hector and the various workers associated with him as members of the staff of the Geological Survey stoutly supported what is commonly known as the Cretaceo-tertiary theory, and consistently denied the existence of any stratigraphical break in the horizon postulated by Hutton. Being willing, however, to admit unconformities between Cretaceo-tertiary and supposed Eocene strata, and again between the latter and Miocene rocks, in the course of field-work they frequently recorded breaks as present in one or other of these positions. Park, though at one time a supporter, in 1900 (34,* p. 350) definitely dissociated himself from the Cretaceo-tertiary hypothesis, and since then, except in minor details, has been in essential agreement with Hutton. In 1911 Marshall, Speight, and Cotton, in a joint paper (47), expressed their opinion that the Cretaceous and Tertiary rocks of New Zealand form a single series, unbroken by any marked unconformity. Their views must therefore be regarded as an extension of the old Cretaceo-tertiary hypothesis, the main point of difference being that the two unconformities supposed by Hector and his colleagues to exist in the Lower and Middle Tertiary sequence were eliminated.

Since the Cretaceous and Tertiary rocks of New Zealand contain practically all the workable coal, the interpretation of their stratigraphical relations becomes one of great importance. As an example, the problem

---

* This and other numbers similarly enclosed in parentheses refer to the bibliography at the end of this paper.

1.—Trans.

of discovering hidden coalfields may be mentioned. The geologist's estimates of the probability of buried coal and of its depth in a district where
only younger Tertiary rocks are exposed at the surface will be strongly
influenced by his preconceived views of what to the layman appear to be
purely scientific and therefore economically immaterial questions. Progress in stratigraphical geology has till recently been greatly impeded by
the lack of detailed surveys and of reliable .determinations of fossils.
Advance in these matters is now being made, but for some years to come
the imperfect data available will afford scope for rival hypotheses such as
those that were so keenly debated in past years.

The object of the present paper is to review the stratigraphical evidence
that has been advanced for unconformities in the succession of rocks em-
braced by Hector's Cretaceo-tertiary, Eocene, and Miocene formations,
or, alternately, by Hutton's Waipara, Oamaru, and Pareora systems. A
perusal of the paper by Marshall, Speight, and Cotton already mentioned
will give the reader a good idea of the confusion introduced into New
Zealand literature by the uncoördinated efforts of independent observers,
each of whom was endeavouring to add to the sum of human knowledge.
On this occasion the writer, taking a somewhat different standpoint from
that of Marshall and his colleagues, hopes to be able to clear a portion of
the field. In order to do so the various localities where stratigraphical
breaks have been suspected will be separately discussed. Most attention will
be given to an unconformity that seems to be invariably present at the
base of the strata included in Hutton's Oamaru System. The palaeontological evidence for this may here be summarized by the statement that
it is nowhere inconsistent with this view, and, as a rule, strongly supports it.

The following classification of the Cretaceous and Tertiary formations
roughly represents the writer's views :—

Age.	Extent.
Pliocene	Ormond beds, Petane Series, upper part of Wanganui System, &c.
Local unconformities ..	(Gisborne district, Reefton district.)
Upper Miocene	Waitemata Series ; Pareora System of Hutton ; Eocene and Miocene of Hector and McKay.
Probable local unconformities	(Reefton district, North Canterbury, &c.)
Middle and Lower Miocene ..	Papakura Series, &c. ; Oamaru System of Hutton ; Cretaceo-tertiary System of Hector, in part.
Unconformity	(Not fully proved in all districts.)
Eocene	Bituminous-coal measures on west coast of South Island ; Cretaceo-tertiary Sys- tem, in part.
Unconformity (?).	
Cretaceous (with possibly some early Tertiary strata) ..	Waipara System of Hutton ; Cretaceo- tertiary System, in part.

Local unconformities such as those shown in the above table between
Middle and Upper Miocene, and again between Miocene and Pliocene,
almost certainly do not belong only to the horizons mentioned. It is
highly probable that detailed field-work will demonstrate that from the
Middle Miocene to the Pleistocene differential movements in some part

or another of New Zealand were almost constantly in progress, and consequently that a series of local stratigraphical breaks exists, no two of which are exactly synchronous.

Owing to the close relationship between Hutton's Oamaru and Pareora systems, and to the fact that in many localities no trace of unconformity can be found between them, they may well be included in one series or system, as is done in various Geological Survey bulletins.

The localities to be discussed so far as conveniently possible will be taken in order from north to south, and will be arranged under the headings of—(I) North Auckland ; (II) Waitemata, Drury, and Waikato Districts ; (III) Gisborne – East Cape District; (IV) Hawke's Bay ; (V) East Wellington ; (VI) Marlborough and South-east Nelson ; (VII) North Canterbury ; (VIII) West Coast of South Island ; and (IX) Otago.

I. NORTH AUCKLAND.

1. *Whangaroa District.*—In 1877 McKay (4, p. 57) states that on the east•side of Whangaroa Harbour Secondary rocks are overlain unconformably by green and brown sandstones. The latter rocks may be definitely identified with the upper part of Bell and Clarke's Kaeo Series (42, p. 46 *et seq.*), which contains Tertiary fossils; and, since McKay records *Inoceramus* in the Secondary rocks, it is probable that he refers to a contact between the lower and upper portions of the Kaeo Series. In 1892 McKay (16, pp. 68–71) somewhat fully discusses the relations of the green and brown sandstones to underlying " hydraulic limestone," and doubtfully decides in favour of conformity.

In 1909 Bell and Clarke, as the result of detailed field-work, state that the lower part of the Kaeo Series contains Mesozoic fossils (42, pp. 16, 49, 58). They sought for a stratigraphical break, but finding evidence of this at one point only (p. 49), in the form of a conglomerate, remain doubtful as to whether there really is a physical unconformity in the Kaeo Series or not. They add, however, that " it is quite probable that more extended observation will justify the division on stratigraphical grounds of the Kaeo rocks into two distinct series of Late Mesozoic and Early Tertiary age respectively—a division which is certainly warranted by the palaeontological evidence " (p. 49). A similar statement appears on a later page of the Whangaroa Bulletin, and is quoted by Park in a paper published in 1911 (48, p. 549).

2. *Kawakawa District.*—Cox reported in 1877 (5, pp. 135–38) on the Kawakawa Colliery, but nothing very definite can be gathered from his remarks or from the accompanying bore-logs, though the latter, if trustworthy, indicate unusual variation in the beds overlying the coal. In 1884 McKay (11, pp. 122–34) gives a fuller account of the geology of the district. His statement that the clays found in diamond-drill bores above a hard crystalline limestone are equivalent to the Amuri limestone is open to adverse criticism (39, p. 410). Both the upper and the lower surfaces of the limestone encountered in the bores are irregular—in the former case probably owing to Quaternary erosion ; in the latter case, according to McKay, through replacement of the greensand overlying the coal by limestone, or *vice versa*. In 1894 Hector (17, pp. ix *et seq.*) and McKay (17, pp. 60–61) discuss the Kawakawa Coalfield, the former giving several plans and sections. The latter, as drawn, induce suspicion of an unconformity between the hard limestone (" Whangarei limestone ") and the coal-

1*

measures. This view has been advanced by Cox (8, pp. 17–18), but the available evidence is inconclusive, and for the present one must regard the supposed discordance as doubtful.

3. *Hikurangi, Kamo, and Whangarei Districts.*—Cox, reporting on the Whangarei district in 1877 (5, pp. 95–106) and 1881 (8, pp. 15 *et seq.*), finds the geology of the coal-measures and associated beds difficult to explain. He places unconformities both above and below the Whangarei limestone (8, pp. 17–18). In 1884 McKay (11, pp. 110–22) refers all the sedimentary strata associated with or overlying the coal-measures to the Cretaceo-tertiary formation, and recognizes no unconformities. In 1894 the same geologist states that at Hikurangi the apparently unconformable relation of the Whangarei limestone to the beds above and below " has to be accounted for by dislocation and faulting of the beds along certain lines " (17, p. 61), which, however, he does not definitely indicate. He further mentions that the fossils from the beds immediately overlying the coal at Hikurangi are found also in corresponding beds at Kamo and Kawakawa, and occur in the coalfields of Canterbury and the west coast of the South Island. The fossils referred to are presumably those mentioned by Cox in his reports—namely, *Cardium brunneri* and " *Ostrea carbonacea.*" A brief examination of the Whangarei and Kawakawa districts made by the writer a few months ago has not afforded any fresh evidence bearing on the question of conformity or unconformity in the so-called Cretaceo-tertiary sequence. Flows of volcanic rock and deep surface clays offer impediments to the collection of geological data. Faulting, too, seems to be prevalent. Thus, with our present knowledge, no decided conclusion regarding the unconformities suggested by Cox can be reached. A suspicion that there is an unconformity below the Whangarei limestone seems justified.

4. *Waipu.*—Hector, in 1877 (4, p. vi of Progress Report), states that at Morrison's Caves, behind Waipu, limestone rests unconformably on greensand.

5. *Pahi and Paparoa.*—In 1881 and 1882 Cox reports that near Pahi the Whangarei limestone of Eocene age rests unconformably upon the chalk-marls of Cretaceo-tertiary age, and is also unconformably overlain by Miocene greensand (8, pp. 18–19, 33–34*; 9, p. 23). His section (9, p. 19) intended to illustrate these unconformities is to some extent unsatisfactory, for although it clearly shows the Miocene rocks resting on the upturned edges of the hydraulic limestone, yet it brings the Whangarei limestone into a position of which faulting is the most probable explanation, thus leaving the question of unconformity between it and the hydraulic limestone entirely open. In 1887 Park confirms the presence of an unconformity between the Miocene greensand and the underlying rocks, which he considers to be of Jurassic age, and illustrates his views by sections (13, pp. 222, 224; see also remarks on Komiti Point, *postea*). In an earlier report, however, he describes the Pahi rocks without mentioning any unconformity (12, pp. 168–69), and gives the generic names of a large number of fossils collected from a steeply dipping greensand which both he and Cox state underlies the hydraulic limestone. The list certainly has a Miocene facies, and unless it can be proved that the apparently inferior position of the greensand is due to faulting or to overfolding the age of the hydraulic limestone becomes very doubtful. The Pahi section

---

* Presumably the word " unconformably " ought to be inserted in the bottom line of page 33 of Cox's report.

has been discussed by Hutton, who, reasoning from the available evidence, decides that the hydraulic limestone is of Upper Oamaru age (30, pp. 380, 381, 382), a view that the writer is by no means inclined to endorse.

6. *Komiti (Kumete) Point.*—As early as 1877 Hector notes an unconformity at Komiti Point between Miocene and Cretaceo-tertiary strata (4, p. v of Progress Report). In 1881 Cox records that here, and at other localities in the Arapaoa Arm of the Kaipara Harbour, strata containing a varied Tertiary fauna, and referable to the Waitemata Series, rest unconformably upon chalky clay or hydraulic limestone supposed to belong to the Cretaceo-tertiary Series (8, pp. 17, 33, 37). The next year he renews the statement, and illustrates it by a section (9, pp. 23, 24). In 1886 Park after a brief visit to Komiti Point states that the junction of the Komiti beds and the chalky marls is obscured, but omits to give any definite opinion concerning the unconformity reported by Cox (12, pp. 165–66). In 1887, however, after making a close examination of Komiti Peninsula, he describes a sharp discordance, which he shows in section, between the Komiti beds and blue shaly clays (13, p. 221). The lower beds, for no very convincing reasons, he regards as Jurassic (p. 229), and therefore not part of the Cretaceo-tertiary sequence. He also comes to the conclusion that there is only a limited amount of unconformity between the Komiti beds and the hydraulic limestone, and that the obtainable evidence strongly favours a stratigraphical break between the latter rock and the underlying shaly clays (pp. 228, 229).

The sections and descriptions given by Cox and Park may be accepted as proof of an unconformity between the Komiti beds of Miocene age and probable Cretaceous strata. Recently (September, 1915) an unconformable junction—doubtless that seen by Cox and Park—has been observed in the south-eastern part of Komiti Peninsula by Mr. J. A. Bartrum and the writer, but owing to adverse weather-conditions was not closely examined. The angular discordance was clear, but apparently not very great. How far there is justification for extending this unconformity to neighbouring areas where less clear sections are observable is open to discussion. Two points of view have been presented, by Park (48, p. 546) and by Marshall (49, p. 319).

7. *Kaipara Flats, Wellsford, &c.*—According to Cox, the Cretaceo-tertiary chalk-marl of the Kaipara Flats district is unconformably overlain by Miocene concretionary greensand (8, p. 19). In the same report it is stated that between Wellsford and Mahurangi, greensand unconformably caps hydraulic limestone. In 1914 J. Henderson notes the occurrence of pebbles derived from the hydraulic limestone and associated rocks in the overlying sandstones as evidence of unconformity (60, p. 157).

8. *Mahurangi (Warkworth) District.*—In 1881 Cox gives a section showing strong unconformity at Wilson's (site of lime and cement works) between Cretaceo-tertiary hydraulic limestone and Lower Miocene greensand (8, p. 19). He further states that at Matakana South Head firestone is unconformably overlain by greensand, &c., and that " outside the Mahurangi North Head breccia conglomerate occurs containing fragments of slate, quartz, and white granular limestone, which gives additional proof, if it were needed, of the unconformity of these beds to the hydraulic limestone " (p. 29). He also gives a section showing strong unconformity between the hydraulic limestone and Miocene strata west of Little Omaha, near Cape Rodney (p. 31). In 1882 Cox again describes an unconformity in the Mahurangi district, shown by horizontal limestones resting on steeply dipping

chalk-marl and hydraulic limestone (9, p. 24). Two years later McKay confirms the existence of a pre-Miocene unconformity near Warkworth, and gives a section similar to Cox's illustrating the unconformity at Wilson's (11, pp. 105, 106). In 1914 J. Henderson reports unconformity between probable Miocene and Cretaceous rocks, in the following terms : " Although the relationships of the hydraulic limestone and the green sandstones are nowhere within the district shown clearly in section, the discordance of the formations is proved by the numerous coarser bands which occur in the sandstones, containing waterworn pebbles of hydraulic limestone, calcareous marl, and, rarely, carbonaceous shale derived from the denudation of the hydraulic limestone and associated beds " (60, p. 157).

9. *Wade District.*—In 1884 McKay reports that on the banks of the Orewa River " Cretaceo-tertiary rocks are unconformably overlain by soft dirty greensandstones " (11, p. 104). Some years later Park confirms the presence of an unconformity in this locality by a section given without any comment (13, p. 226). This, however, he supplies in 1911 and 1912 (48, p. 545, and 50, p. 493).

Summary of North Auckland Data.—Cox, McKay, and Park agree in observing angular unconformity between Miocene and Cretaceo-tertiary strata, and show it as very sharp in several sections—for example, at Komiti Point (Cox and Park), at Wilson's near Warkworth (Cox and McKay), and Orewa River (McKay and Park). The existence of a basal Miocene conglomerate containing pebbles derived from the hydraulic limestone and other rocks of probable Cretaceous age is demonstrated by Hector, Cox, and Henderson. Again the wide overlap of the Miocene strata on the pre-Cretaceous (Trias-Jura) rocks here as elsewhere in New Zealand is highly significant. As a whole, the evidence for unconformity between Miocene and all older rocks appears to be conclusive. Cox and Park agree in finding an unconformity between the Whangarei limestone and the hydraulic or chalky limestone, but the supporting evidence is unsatisfactory, and at a later date Park takes quite a different view (45, p. 95).

II. WAITEMATA, DRURY, AND WAIKATO DISTRICTS.

In 1871 Hutton contends that the Tertiary rocks south of Auckland exhibit two unconformities—one between the brown-coal measures of Eocene age and the Aotea Series of Oligocene age, the other between the Aotea rocks and the Waitemata Series of Upper Miocene age (19 ; and see also 22). In 1882 Cox reports a probable unconformity near Howick (9, pp. 95, 96). In 1884 McKay, supported by Hector, suggests an unconformity at the horizon of the Parnell grit (11, pp. 106, xviii ; 12, p. xxxviii) — that is, in the Waitemata Series — and assigns the beds below the grit to the Cretaceo-tertiary system. Park rejects the supposed unconformity at the Parnell-grit horizon, but believes that there is one at the base of the Waite-mata Series (12, p. 158 ; 31, pp. 391–99). In 1892, however, he maintains that there is conformity to the base of the Cretaceo-tertiary (32, p. 381). Subsequent papers by E. K. Mulgan (37) and C. E. Fox (38) deal very hardly with the Parnell-grit unconformity, an attitude which is in agreement with the writer's observations, though it must be admitted that at St. George's Bay there is a minor local unconformity, indicated by the wedging-out of several feet of the beds underlying the grit. Neither Mulgan nor Fox seems to support unconformity in any part of the Tertiary sequence as developed near Auckland, and Clarke's paper of 1905 (40), together with his latest statement, contributed to Marshall, Speight, and Cotton's paper of 1911 (47, pp. 396–99), appears to dispose of the unconformity believed by Hutton to exist between the Aotea (Papakura) Series and the Waitemata rocks. No evidence other than Hutton's concerning the supposed discordance between the Aotea Series and the brown-coal measures is available, but apparently it was observed by him at a single point only, in fault-broken country (19, pp. 246, 247), and therefore additional data are necessary before it can be accepted. On the other hand, it cannot be rejected without further investigation.

III. GISBORNE – EAST CAPE DISTRICT.

In 1877 Hector distinguished the Tawhiti beds of supposed Lower Tertiary age from the Turanganui Series of Cretaceo-tertiary age, and gave sections showing unconformities between the two sets of rocks (3, pp. xvi and xvii; map and sections opposite p. xvi). His sections also show a strong discordance between Upper Tertiary and Cretaceo-tertiary rocks in localities where the Tawhiti beds are absent. McKay, in the same year, described evidence of unconformity at several points in a post-Cretaceous horizon, his strongest point being that at Whareponga, "behind the church, an unconformable junction between the marls with concretionary masses and the younger rocks is seen" (3, p. 121). In 1887 McKay perceived unconformity between the Tawhiti beds and the Turanganui Series at Akuakua (Akuaku), and between the former rocks and the Waipiro Series at Riparua (Reporua) (13, section on p. 218). Inland he discovered thick beds of conglomerate containing igneous boulders in the Lower Miocene rocks of the Ihungia and Mata Rivers, but so far as can be judged from his report did not regard the occurrence as significant of stratigraphical break. In 1901, however, he placed an unconformity in the Gisborne district between rocks of supposed Cretaceous age and undoubted Miocene strata. Conglomerates similar to those just mentioned, and containing boulders of limestone derived from the underlying strata, he considered to mark the base of the Miocene (36, p. 24). J. H. Adams, however, as the result of a detailed survey over the limited area included in the Whatatutu Subdivision, came to the conclusion that the so-called Cretaceous rocks of that part of the Gisborne district were conformable to the overlying Miocene strata, and, although confirming McKay's description of the conglomerate, did not recognize it as evidence of unconformity (44, pp. 12, 18–20). Marshall, as the result of field examinations and of palaeontological work, confirmed Adams's statements (44, pp. 21–23; see also 47, p. 393). Neither Marshall nor Adams found any fossils of Cretaceous aspect in the Whatatutu district, but McKay's statement that *Inoceramus* occurs in concretions in the bed of the Waipaoa River has since been confirmed by at least two observers— one a geologist of repute—and, in addition, a belemnite has been found on Waitangi Hill by Mr. M. P. Poole, of Tuparoa. During a visit to the Gisborne district in January, 1914, the writer, guided by Mr. John Mouat, saw numerous specimens of *Inoceramus* near Motu Falls, at a point about five miles from the north-western corner of the Whatatutu Subdivision. The fossils occur in a dark mudstone, probably corresponding in horizon to the clay shale forming the lowest portion of Adams's Whatatutu Series (44, pp. 12–13).

Summary.—The available evidence strongly supports the presence of an unconformity below the Miocene strata of the Gisborne – East Cape district.

IV. HAWKE'S BAY.

In 1877 McKay records that Cretaceo-tertiary (probably Miocene) rocks unconformably overlie Secondary rocks four miles north of Waimirima (Waimarama) and at Paonui Point (4, p. 47). In his sections (opposite p. 50) various apparently unconformable contacts of the two sets of rocks are shown. Some of these, however, are probably merely discordant juxtapositions due to faulting. A few years later McKay states that the Te Aute limestone rests unconformably on the upturned edges of Cretaceo-tertiary strata at Mount Vernon (near Waipukurau) and elsewhere (7, pp. 71, 72, and sections opposite p. 72). In 1887 he redescribes the so-called Cretaceo-tertiary rocks north of Waimarama as Tertiary beds surrounding an isolated outcrop of Cretaceo-tertiary or young Secondary rocks (13, p. 191), and in a section (p. 192) shows strong unconformity between Upper Miocene and Cretaceo-tertiary strata near the Tukituki River.

Summary.—The available evidence leads to the conclusion that the lower part of the Oamaru Series* is probably not developed in the Hawke's Bay district, and that there is a strong unconformity between the Upper Miocene and late Mesozoic rocks.

V. EAST WELLINGTON.

In 1877 Hector found that the Cretaceo-tertiary rocks of eastern Wellington were " overlain unconformably by the Hawke's Bay Series, forming the Taipos " (3, p. ix). Next year McKay reported that there was an unconformity in the Wairarapa district between the Upper Miocene and the Taipo beds, regarded by him as of Lower Miocene age (6,·p. 20). The writer must here remark that, although Miocene fossils have been collected from supposed Taipo beds near Tinui, there can be little doubt but that the true Taipo beds are of pre-Miocene if not pre-Tertiary age. In 1888 Park reported an unconformity at the Blairlogie gas-spring between Miocene and probable Cretaceous rocks (14, p. 21) which, according to the writer's observations in 1910, overlie the Taipo sandstones (43A, p. 2). There may, however, be a fault at this point, so that until further examination has been made the field evidence for physical unconformity remains inconclusive. Park's statement of 1904 (39, p. 412) that in the Mangapakeha Valley there is an unconformity between sandy clays with Tertiary fossils and soft shaly claystones with probable *Inoceramus* presumably has reference to the Blairlogie section. McKay considers that Hutchinson Quarry beds of reputed Eocene (*i.e.*, probably Miocene) age occur near Cape Palliser, and apparently regards them as unconformable to the Cretaceo-tertiary strata of the neighbourhood (7, pp. 80, 81).

Summary.—As in Hawke's Bay, the lower part of the Oamaru Series appears to be absent from the East Wellington district. Though everything points to strong unconformity between the Upper Miocene. and the late Mesozoic strata, more field-work is required in order to establish it as a fact of observation.

VI. MARLBOROUGH AND SOUTH-EAST NELSON.

Von Haast in 1871 records an unconformity between the "Amuri Bluff beds " and the "*Scalaria* beds " at Kaikoura and near the Jed River

---

* Unless otherwise stated, the Oamaru Series of this paper includes Hutton's Pareora System.

(1A, p. 39, and sections vii, viii, and xvi). According to his section xv, the Weka Pass stone rests discordantly upon Saurian conglomerate " one mile above junction of Eden River with Waiau-ua " (Dillon River).

The sections accompanying Hutton's report of 1877 on the north-east part of the South Island (3, opposite p. 56) show physical unconformity between Amuri limestone and overlying Tertiary beds at the Conway River, Amuri South Bluff, east head of Kaikoura Peninsula, and Flaxbourne (Ward). In the same volume, however, Hector states that at Kaikoura " a corrugated concretionary disturbance of the calcareous beds has given rise to an apparent unconformity " (3, p. xi), and that he is unable to satisfy himself " of any stratigraphical break between the Amuri limestone and the overlying Grey Marls " (p. x). McKay, reporting on the Cape Campbell district in the same year, states that the Awatere beds rest with high unconformity on the greensand and Amuri groups (4, p. 186). He also mentions conglomerates containing boulders both of Awatere and of Amuri rocks (p. 190). Similar conglomerates exposed in the Mead River and elsewhere are described by Hector and McKay as resting discordantly on Grey Marl, Weka Pass stone, or Amuri limestone (12, pp. xvi, xxxiv *et seq.*, 113 *et seq.*; sections on pp. 81, 82, 85, 90, 94, 95, 96, 103, &c.). The age of this conglomerate they consider to be post-Miocene. McKay has sections on pages 88 and 89 showing probable unconformity between Tertiary strata and Amuri limestone near Flaxbourne (Ward). On page 83, after discussing the apparent discontinuity at Kaikoura Peninsula between the Amuri limestone and the overlying Grey Marl, he comes to the conclusion that their relations are conformable. In his Progress Report of 1887 (13, pp. ix *et seq.*) Hector again mentions that there is no unconformity immediately above the Amuri limestone at Kaikoura, as Hutton supposes. In 1890 McKay again describes what he calls the " Great Post-Miocene Conglomerate " as unconformably younger than the Awatere beds. In several sections he shows it resting discordantly on the Grey Marl (15, pp. 170, 171, 175), and in one as in contact with the Amuri limestone (p. 173 ; see also 12, pp. 81, 82). C. A. Cotton, however, considers that the conglomerate in localities examined by him conformably succeeds the Grey Marl, and ingeniously accounts for included masses of Grey Marl, Amuri limestone, &c., by supposing that they are derived from a block of adjacent territory which was faulted and uplifted whilst conformable deposition proceeded at its base (57, p. 360). Although Cotton observed several facts supporting his " hypothesis of block-faulting with the restriction that the faulted block alone moved," the very strongest evidence is required in order to establish the correctness of such a startling explanation, which involves a differential elevation " of perhaps as much as 12,000 ft." (57, p. 359).

Summary.—Since Hector and McKay disagree with Hutton concerning the horizon of a supposed unconformity separating Miocene from older rocks, and since Cotton finds no evidence of a stratigraphical break (other than that involved in block-faulting of part of the area), disconformity must be considered unproved. Hence detailed field-work is necessary to establish the truth or otherwise of any hypothesis.*

---

* In December, 1915, the writer visited Kaikoura Peninsula and Amuri Bluff, and there obtained clear evidence of an unconformity between the Amuri limestone proper and an upper band of limestone corresponding to the Weka Pass stone. At Kaikoura appearances also support the view of a local stratigraphical break or period of non-deposition with slight erosion between the upper limestone and the Grey Marl.

VII. NORTH CANTERBURY.

1. *Waipara and Weka Pass District.*—The Waipara–Weka Pass district has been frequently examined, and much has been written concerning supposed unconformities at various horizons from early Tertiary upwards. In 1869 Hector published some brief remarks on the geology of the Waipara district (1, pp. x–xiii). He mentions an unconformity which on Hutton's authority is to be understood as occurring between the Grey Marl and the Mount Brown beds, though the description might almost equally well apply to the contact of the Amuri limestone with the Weka Pass stone (see also 39, p. 413). McKay and Park (in early reports) also place a stratigraphical break at this horizon. The evidence, however, appears to be very slender, and probably at most justifies the opinion that only slight local unconformity is present.* This seems to be the opinion of von Haast (1A, pp. 14, 16 ; 21, p. 306), who also in his earliest report suggests a local unconformity above the Grey Marl (1A, p. 17).†

In 1877 Hutton describes in clear terms what he believes to be the unconformable junction of the Weka Pass stone with the underlying Amuri limestone (3, pp. 43, 44). In 1885 he repeats this statement, with additional data in its favour (24, pp. 269–70). Hector, in his Progress Report of 1877 (3), expresses his inability to convince himself of the unconformity. Von Haast in 1879 discusses the question, and decides that there is no break of any consequence, but, somewhat strangely, does not refer to the views of other observers (21, pp. 297–98). In 1887 (13, pp. 78 *et seq.*) and in 1892 (16, pp. 98, 102) McKay unequivocally opposes the supposed unconformity. Park in 1888 perceives evidence of change in the conditions of deposition, perhaps accompanied by a degree of elevation, but not of true unconformity (14, pp. 28, 31, &c.). In 1904 Park agrees with Hutton (39, p. 413), but in 1905 shifts the unconformity to the upper surface of the Weka Pass stone (41, pp. 542, 546). Marshall, Speight, and Cotton cannot see any evidence of unconformity in the younger rock-series of the Waipara and Weka Pass districts. In 1912 J. A. Thomson observes " apparent conformity in section throughout the Waipara district " (52, p. 8). Owing, however, to palaeontological evidence of the Tertiary age of the Weka Pass stone having been discovered—or, rather, rediscovered—by Thomson and Cotton, Park in 1912 returns to Hutton's view (50, pp. 496–97). An examination lately made by the writer has convinced him that the upper surface of the Amuri limestone has been eroded, and that local unconformity (disconformity) at least is present (63, p. 92). On palaeontological grounds there is reason for believing that the unconformity represents a considerable time interval. Further evidence, however, is required before one can assert with any degree of confidence that the break extends from the Cretaceous to the Oligocene or Miocene, as inferred by Hutton and Park.

2. *Motunau (Stonyhurst) District.* — In 1877 Hutton shows an unconformity at Motunau between Amuri limestone and Tertiary rocks (3, section vi, opposite p. 56), and in 1885 gives further data (24, pp. 270–71).

---

* Since this was written the writer has seen clear evidence of at least a local unconformity between the Grey Marl and the Mount Brown beds in the valley of the Weka Creek.

† Von Haast's section of 1871 (1A, opposite p. 18) shows violent unconformity near Boby's Creek between the Grey Marl and underlying beds. A fault, however, is the cause of the structure interpreted by von Haast as an unconformity. (See 14, section on p. 30, and map opposite same page.)

McKay in 1881 confirms this by a section, three miles north of Motunau, which shows the Amuri limestone followed unconformably by Mount Brown and Pareora beds (8, p. 115). He also reports that the Teredo limestone (a part of the " Cretaceo-tertiary " sequence) is in unconformable contact with Tertiary rocks. Marshall, Speight, and Cotton, however, state that a careful examination of the creek section fails to reveal any discordance (47, p. 392).

3. *Trelissick Basin.*—According to McKay's report of 1881, the Trelissick basin contains a fairly complete succession of Cretaceo-tertiary, Upper Eocene, and Lower Miocene rocks, with unconformities between the main divisions (8, p. 60 *et seq.*). The map and sections in this report are stated to be the work of Hector in 1872 (p. 54). In 1887 Hutton also considers that there are two unconformities in the Cretaceous and Tertiary rocks, but apparently disagrees with McKay as to the horizon of the upper one (26, p. 408). In 1905 Park finds himself unable to reach any definite conclusion concerning unconformity (41, p. 534). Marshall, Speight, and Cotton state that the interpretation is very difficult owing to disturbances caused by volcanic action, but no undoubted unconformity can be seen (47, p. 392). In a recent paper Speight again takes a similar view (61, p. 342).

Summary. — Although palaeontological data strongly support unconformity between Cretaceous and Tertiary rocks in North Canterbury, the recorded stratigraphical evidence is inconclusive, the weight of authority being possibly in favour of conformity. The writer, however, has observed what is at least local unconformity at Weka Pass, and can see little or no reason, other than the opposition of Hector, McKay, and Marshall, why Hutton's view should not be at least provisionally accepted for this important section.

VIII. West Coast of South Island.

1. *Collingwood and West Wanganui (Westhaven).*—In this district bituminous and brown coals exist in two horizons, separated by a considerable thickness of strata, which, so far as known, consist of shale, sandstone, and conglomerate. In 1883 Cox places the bituminous coal in the Lower Greensand (Cretaceous) Series, and the brown coal in the Cretaceo-tertiary sequence (10, pp. 71–72). His sections, however, show the two coal-bearing formations as conformable. On the other hand, he states that the calcareous rocks at a higher horizon are probably unconformable to the coal-bearing series, the evidence for this consisting in overlap on the eastern side of the Whakamarama Mountains (p. 73). He admits, however, that in the West Wanganui section no unconformity can be traced.* A few years later Park reports that the bituminous and brown coals both belong to one conformable series, the Cretaceo-tertiary (15, p. 238). In 1910 he places both coal horizons in his Waimangaroa Series of Upper Eocene age (45, pp. 310–12). From the data at present available, it seems that in the Collingwood district there is at least 2,000 ft. of fresh-water strata containing a number of thin coal-seams, the uppermost of which are pitch, and the lowest bituminous in character. A bore recently drilled at Rakopi, on the east side of West Wanganui Inlet (now known as Westhaven), passed through no fewer than seventy-three seams of coal from 3 in. to 1 ft. 9 in.

---

* As a matter of fact, overlap does exist south of West Wanganui Inlet.

in thickness, and finally reached granite at a depth of 1,421 ft.   The
brown coal of the district is at a considerably higher horizon, and may or
may not be separated from the pitch and bituminous-coal series by an
unconformity.   At present no evidence for discordance other than overlap
and analogy with areas farther south can be brought forward.

2. *Westport District.*—The writer has lately published evidence in favour
of an unconformity (disconformity) between the Oamaru Series of approximate Miocene age and the bituminous-coal measures of probable Eocene
age (58, pp. 271 *et seq.* ; 62, pp. 58–88).   Angular discordance (clino-
unconformity) between the two sets of strata has not been observed, and
the most striking evidence of a stratigraphical break consists in the presence
of numerous pebbles of coal and carbonaceous shale, derived from the bituminous-coal measures, in the basal and occasionally in the higher beds
of the Oamaru.   The latter series widely trangresses the Eocene strata,
in such a manner as might well be considered proof of unconformity were
it not that New Zealand conditions are so peculiar that overlap alone can
hardly ever be regarded as a decisive criterion.   In at least two localities,
however, upper Oamaru beds rest on the Eocene beds to the exclusion of
the lower portion of the Oamaru formation (58, p. 275 ; 62, pp. 92–93),
and at the mouth of the Fox River Miocene strata unconformably overlie
a breccia conglomerate correlated with the Hawk's Crag breccia, the lowest
member of the bituminous-coal measures (58, pp. 274–75).

3. *Greymouth District.*—Von Haast in 1861 evidently thought that the
series of rocks having the Cobden limestone as its upper member rested
unconformably on bituminous-coal measures (18, p. 109), and this view
was quoted with approval by Hutton in 1887 (29, pp. 268–69).   In 1873
McKay discovered strata containing detrital coal, which at that period he
thought were below the Cobden limestone and were indicative of unconformity with the coal-measures (3, pp. 77 *et seq.* ; 35, p. 7).. In 1901,
however, he placed the detrital-coal beds above the Cobden limestone
(35, p. 8), an horizon where he considered there was independent evidence of
a stratigraphical hiatus.   In 1909 and 1911 the writer showed that McKay's
original view was the correct one (43, p. 13 ; 46, pp. 63, 66, &c.), and in
the course of field-work found that no break of any kind existed above
the Cobden limestone in the Greymouth district.*   As at Westport, angular
discordance has not been detected, and the chief evidence for unconformity
between the writer's Greymouth Series (equivalent to Oamaru and Pareora
beds) and the bituminous-coal measures of Eocene age consists in the presence of immense quantities of detrital coal in the lowest, or Omotumotu,
beds of the Greymouth Series, as originally announced by McKay.   In
1911 Marshall, Speight, and Cotton regard the evidence as inconclusive
(47, pp. 392–93), and in 1912 Marshall, writing alone, takes the same view
(55, p. 68).

4. *South Westland.*—Rocks of Tertiary or possibly in part of Cretaceous
age have been described by Cox and von Haast as occurring near the
mouth of the Paringa River.   Owing to the presence of a conglomerate con-

---

* McKay at one locality observed a difference of strike between the Cobden limestone and so-called nummulitic limestone (3, p. 75).   The writer examined the same
locality, and came to the conclusion that a fault or other disturbance accounts for
the discordance of strike, which is not very great.   The "nummulitic" limestone is
really an *Amphistegina* limestone (46, pp. 68, 71).   At Greymouth a clear and continuous section shows it in perfect conformity with the Cobden limestone.

taining boulders from an underlying lithographic limestone, Cox suspected a stratigraphical break, but was not certain (4, p. 83). Von Haast spent some time in the locality (21, pp. 160 *et seq.*, 299–300), but made no definite statement regarding the conglomerate or other evidence of unconformity. The Paringa section is complicated by the presence of volcanic lavas and tuffs. For this reason it is not likely to furnish clear evidence of the presence or absence of unconformity in the purely sedimentary strata of the district.

Summary. — The stratigraphical evidence for a hiatus between the Miocene and Eocene (bituminous-coal measures) rocks of the Westport and Greymouth districts appears conclusive to the writer. The palaeontological data (see 46, 62), though not strongly in favour of unconformity, do not oppose it. The difference between the somewhat scanty Eocene fauna and the Miocene fauna is probably quite as great as that between the latter and the present-day fauna.

IX. OTAGO.

1. *Oamaru District.*—The presence of stratigraphical breaks in the Tertiary rocks of the Oamaru district has been affirmed at various times. Hector and McKay believe that an unconformity separates the Hutchinson Quarry beds from the underlying Oamaru limestone (5, pp. ix, 57–58). In 1885 Hutton discusses the matter, and reaches the conclusion that no definite discontinuity exists (25, pp. 560–64). With this view most other observers are practically in accord. In the writer's opinion, slight local unconformities connected with the volcanic eruptions during Miocene times at Oamaru Cape and elsewhere may be present, as indicated by Park (41, p. 502) and again by Marshall and his colleagues (47, p. 405), but these cannot invalidate the substantial unity of the Tertiary rocks of the district.

2. *Shag Point.*—In 1872 and 1877 von Haast reports that between the boat-harbour and the mouth of Shag River are sections showing discordance between the coal-bearing rocks and a Tertiary series (2, p. 150; 3, p. 25). From later reports, especially one by McKay (13, p. 11), it would appear that von Haast was mistaken with regard to the existence of a break in these sections. Hutton in 1875 states that in the Shag Point district there is a complete unconformity between the Waipara coal-measures and the Tertiary rocks (20, pp. 50, 103), and in various later papers retains this view (*e.g.*, 30, p. 379, &c.). Cox, in his report of 1883, does not mention any stratigraphical break, and maps von Haast's Tertiary rocks with the Cretaceo-tertiary Series. He describes a fault which obviates the necessity of assuming an erosional unconformity (10, pp. 55–56). In various reports McKay consistently supports Cox's view (11, pp. 58, 63; 13, pp. 10 *et seq.*). In 1904 Park is positive that the Miocene rocks of the Shag Point district are unconformable to the underlying coal-bearing series (39, p. 414), and reaffirms this statement in 1910, 1911, and 1912 (45, p. 116; 48, pp. 541–45; 50, pp. 493–95). Marshall, Speight, and Cotton, however, can find no evidence of discontinuity, and accordingly support McKay (47, p. 393). On the other hand, according to A. G. Macdonald, there is "no room for doubt as to the unconformity of the beds" (53, p. 1037). Unquestionably, at Shag Point both Cretaceous and Miocene rocks are present, apparently in unconformable relations; but, since no actual contact has been observed, it is possible that, as suggested by Cox,

a fault is the true explanation of the supposed discordance. It is also possible (1) that no fault is present, or (2) that unconformity as well as a fault exists.

3. *Green Island and Brighton.*—The Cretaceo-tertiary age of the Green Island and Brighton coal-seams was naturally assumed by the Geological Survey of Sir James Hector's time. This view was supported by the discovery of *Belemnites lindsayi* in an impure pebbly limestone almost immediately overlying the Brighton coal. Near Green Island the occurrence of the clearly Tertiary and apparently conformable Caversham sandstone in an horizon high above the coal has led most observers, including the writer, to consider the coal as of Tertiary age, the presence of a belemnite at Brighton being regarded as anomalous. Park, however, explains the situation as due to unconformity between the Brighton and Green Island coals (45, pp. 90, 315–17 ; 50, p. 496). The writer still believes the coals to be of one age, but is now inclined to suppose that the Green Island coal is late Cretaceous, and that there is an unconformity in some upper horizon. Since the Brighton limestone is a beach deposit, its absence from the Green Island district is not hard to explain. The possible stratigraphical break, owing to the amount of clay burdening the surface almost everywhere in the critical area, will be hard to discover. What appears to be an unconformity, but only a very slight one, is visible between marl and overlying greensand at the Burnside marl-pit. The greensand contains a few poorly phosphatic pebbles. Since, however, phosphatic concretions or beds, and to a smaller extent glauconite, are frequently associated with unconformities (56, pp. 46, 71, 215–17), a careful examination of the phosphatic horizon, which is exposed elsewhere in the Kaikorai Valley, may be recommended as likely to yield evidence of value.

4. *Kaitangata.*—Hutton observed that the upper surfaces of the Kaitangata coal-seams are in places eroded, and covered by a conglomerate containing pieces of coal (20, p. 106). He does not, however, regard this as evidence of other than local unconformity, of no importance. In 1911 Park gives reasons, partly founded on field-work by A. G. Macdonald, for believing that an unconformity similar to that at Shag Point exists between Tertiary and Cretaceous coal-bearing strata, but is not able to report actual contact of the two sets of beds (48, pp. 544–45). Shortly after he reaffirms his position, in order to meet criticism by Marshall (49, p. 318). The writer's own observations show that coal-seams belonging to two distinct horizons occur in the Kaitangata and adjoining districts, as previously stated by A. G. Macdonald, who places both horizons in the Waipara Series (53, p. 1089). Unconformity between these is probable, but cannot be regarded as demonstrated

General Summary and Conclusions.

In sifting the evidence presented by the various writers who have been cited on the previous pages, the chief difficulty arises in separating what was probably seen in the field from what was inferred. While the actual observations demand credit, the inferences made in the literature are by no means of equal value. Thus we find sections drawn without any indication of the blanks filled in by the ideal extension of actual outcrops. In some cases parallel bedded strata are shown as gently diverging immediately below the outcrops, and thus a possibly unwarranted inference of unconformity is given support. In one or two instances faults seem to be a much more likely explanation of discordant contacts than unconformity as supposed by the observer. The irreconcilably opposing interpretations of critical sections offered by the several groups of workers add much to the difficulties of the student, and, combined with the knowledge that various errors have been made, induce a spirit of scepticism, and incline him to reject the observations made by those with whom he happens not to be in sympathy. Until, however, the field exposures and other data upon which past observers relied have been critically re-examined it would be a mistake to follow such an inclination. Various examples of older work being cast aside by a later investigator (who in some cases was also the original observer), and subsequently found to be practically correct, could be cited.

The following conclusions have been reached by the writer :—

(1.) That there are various local unconformities in the rock-successions discussed. Some investigators have hastily assumed one or more of these to be major unconformities, whilst others have denied their existence altogether.

(2.) That, since in no locality are marine Cretaceous rocks known to be in contact with marine Eocene rocks, no definite opinion as to the existence or non-existence of unconformity between Cretaceous and Eocene can be formed. The absence of such contacts, however, favours unconformity.

(3.) That in all parts of New Zealand there is a decided unconformity between the Miocene and any Eocene or older rocks present. The evidence for this is most unmistakable in North Auckland and on the west coast of the South Island.

From these opinions it also follows that—

(4.) The Cretaceo-tertiary theory of Hector is untenable as a working hypothesis applicable to all New Zealand; for, although proof of unconformity between the marine Cretaceous and the marine Eocene is still wanting, evidence of conformity is equally far to seek. In any case, the inclusion of the whole or the greater part of Hutton's Oamaru System in the Cretaceo-tertiary is an error which has led to great confusion.

(5.) In like manner the modification of the Cretaceo-tertiary hypothesis advanced by Marshall, Speight, and Cotton is unsustainable, for unconformities do exist between Cretaceous and Pliocene, and, moreover, the investigators named have made various extremely questionable correlations. On the other hand, they show considerable respect for palaeontological evidence, have disproved various erroneous correlations of other writers, and have clearly demonstrated the essential unity of the Oamaru and Pareora. systems of Hutton.

(6.) In those localities where the unconformity reported by Hector and his colleagues between the Cretaceo-tertiary and the Eocene happens to coincide with the lower limit of the Miocene rocks it has a real existence, but elsewhere is either non-existent, or local, or requires further investigation.

(7.) The unconformity frequently reported by the old Geological Survey between the Eocene and the Miocene generally coincides with the boundary between Hutton's Oamaru and Pareora systems, and therefore is either nonexistent or is represented only by a slight local unconformity.

(8.) Hutton may or may not have been right in placing an unconformity at the top of the Cretaceous. He was right, however, in placing an unconformity at the base of the Oamaru System or Series.

BIBLIOGRAPHY.

(A.) *Reports of the Geological Survey under Sir J. Hector.*

1. Hector, J., in Rep. Geol. Expl. during 1868–69, No. 5, 1869.

1A. Haast, J. von, in Rep. Geol. Expl. during 1870–71, No. 6, 1871.

2· —— in Rep. Geol. Expl. during 1871–72, No. 7, 1872.

3. Hector, J.; Hutton, F. W.; Haast, J. von; and McKay, Alex., in Rep. Geol. Expl. during 1873–74, No. 8, 1877.

4. Hector, J.; Cox, S. H.; and McKay, Alex., in Rep. Geol. Expl. during 1874–76, No. 9, 1877.

5. —— in Rep. Geol. Expl. during 1876–77, No. 10, 1877.

6. McKay, Alex., in Rep. Geol. Expl. during 1877–78, No. 11, 1878.

7. —— in Rep. Geol. Expl. during 1878–79, No. 12, 1879.

8. Cox, S. H., and McKay, Alex., in Rep. Geol. Expl. during 1879–80, No. 13, 1881.

9. Cox, S. H., in Rep. Geol. Expl. during 1881, No. 14, 1882.

10. —— in Rep. Geol. Expl. during 1882, No. 15, 1883.

11. Hector, J.; Cox, S. H.; and McKay, Alex., in Rep. Geol. Expl. during 1883–84, No. 16, 1884.

12. Hector, J.; McKay, Alex.; and Park, James, in Rep. Geol. Expl. during 1885, No. 17, 1886.

13. —— in Rep. Geol. Expl. during 1886–87, No. 18, 1887.

14. McKay, Alex., and Park, James, in Rep. Geol. Expl. during 1887–88, No. 19, 1888.

15. —— in Rep. Geol. Expl. during 1888–89, No. 20, 1890.

16. McKay, Alex., in Rep. Geol. Expl. during 1890–91, No. 21, 1892.

17. Hector, J., and McKay, Alex., in Rep. Geol. Expl. during 1892–93, No. 22, 1894.

(B.) *Other Publications.*

18. Haast, J. von. " Report of a Topographical and Geological Exploration of the Western Districts of the Nelson Province." 1861.

19. Hutton, F. W. " On the Relative Ages of the Waitemata Series and of the Brown-coal Series of Drury and Waikato." Trans. N.Z. Inst., vol. 3, 1871, pp. 244–49.

20. —— " Geology of Otago," 1875.

21. Haast, J. von. " The Geology of Canterbury and Westland," 1879.

22. Hutton, F. W. " On the Age of the Orakei Bay Beds near Auckland." Trans. N.Z. Inst., vol. 17, 1885, pp. 307–13.

23. —— " Sketch of the Geology of New Zealand." Q.J.G.S., vol. 41, 1885, pp. 191–220.

24. —— " On the Geological Position of the Weka Pass Stone." *Ibid.*, pp. 266–78.

25. —— " On the Correlations of the ' Curiosity Shop Beds ' in Canterbury, New Zealand." *Ibid.*, pp. 547–64.

26. —— " On the Geology of the Trelissick or Broken River Basin." Trans. N.Z. Inst., vol. 19, 1887, pp. 392–412.

27. —— " On the Geology of the Country between Oamaru and Moeraki." *Ibid.*, pp. 415–30·

28. Hutton, F. W. " On some Railway Cuttings in the Weka Pass." Trans. N.Z. Inst., vol. 20, 1888, pp. 257–63.

29. —— " On some Fossils lately obtained from the Cobden Limestone at Greymouth." *Ibid.*, pp. 267–69.

30. —— " On the Relative Ages of the New Zealand Coalfields." Trans. N.Z. Inst., vol. 22, 1890, pp. 377–87.

31. Park, James. " On the Conformable Relations of the different Members of the Waitemata Series." *Ibid.*, pp. 391–99.

32. —— " On the Prospects of finding Workable Coal on the Shores of the Waitemata." Trans. N.Z. Inst., vol. 24, 1892, pp. 380–84.

33. Hutton, F. W. " The Geological History of New Zealand." Trans. N.Z. Inst., vol. 32, 1900, pp. 159–83.

34. Park, James. " Notes on the Coalfields of New Zealand." *New Zealand Mines Record*, vol. 3, 1899–1900, pp. 349–52. See also Proc. Inst. Min. and Metall., vol. 8, p. 146.

35. McKay, Alex. " Report on Supposed Coal-seams in Kaiata Range, Greymouth." Parl. Paper C.–10 (in Mines Report), 1901, pp. 2, 7–8. See also *New Zealand Mines Record*, vol. 4, 1900–1, pp. 203–4.

36. —— " Report on the Petroleum-bearing Rocks of Poverty Bay and East Cape Districts." Parl. Paper C.–10 (in Mines Report), 1901, pp. 21–25.

37. Mulgan, E. K. " On the Volcanic Grits and Ash-beds in the Waitemata Series." Trans. N.Z. Inst., vol. 34, 1902, pp. 414–35.

38. Fox, C. E. " The Volcanic Beds of the Waitemata Series." *Ibid.*, pp. 452–93.

39. Park, James. " On the Age and Relations of the New Zealand Coalfields." Trans. N.Z. Inst., vol. 36, 1904, pp. 405–18.

40. Clarke, E. de C. " The Fossils of the Waitemata and Papakura Series." Trans. N.Z. Inst., vol. 37, 1905, pp. 413–21.

41. Park, James. " On the Marine Tertiaries of Otago and Canterbury, with Special Reference to the Relations existing between the Pareora and Oamaru Series." *Ibid.*, pp. 489–551.

42. Bell, J. M., and Clarke, E. de C. " The Geology of the Whangaroa Subdivision, Hokianga Division." N.Z. Geol. Surv. Bull. No. 8 (n.s.), 1909.

43. Morgan, P. G. " Field-work in the Greymouth Subdivision." Third Ann. Rep. (n.s.) of N.Z. Geol. Surv., 1909, pp. 11–17.

43A. —— " Petroleum and other Minerals in Eastern Wairarapa District." Parl. Paper C.–16, 1910.

44. Adams, J. H. " The Geology of the Whatatutu Subdivision, Raukumara Division, Poverty Bay." N.Z. Geol. Surv. Bull. No. 9 (n.s.), 1910.

45. Park, James. " The Geology of New Zealand," 1910.

46. Morgan, P. G. " The Geology of the Greymouth Subdivision, North Westland." N.Z. Geol. Surv. Bull. No. 13 (n.s.), 1911.

47. Marshall, P.; Speight, R.; and Cotton, C. A. " The Younger Rock-series of New Zealand." Trans. N.Z. Inst., vol. 43, 1911, pp. 378–407.

48. Park, James. " The Unconformable Relationship of the Lower Tertiaries and Upper Cretaceous of New Zealand." Geol. Mag. (n.s.), dec. v, vol. 8, Dec., 1911, pp. 539–49.

49. Marshall, P. " The Younger Rock-series of New Zealand." Geol. Mag. (n.s.), dec. v, vol. 9, July, 1912, pp. 314–20.

50. Park, James. " The Supposed Cretaceo-tertiary Succession of New Zealand." *Ibid.*, Nov., 1912, pp. 491–98.

51. Speight, R. " A Preliminary Account of the Lower Waipara Gorge." Trans. N.Z. Inst., vol. 44, 1912, pp. 221–33.

52. Thomson, J. Allan. " Field-work in East Marlborough and North Canterbury." Sixth Ann. Rep. (n.s.), N.Z. Geol. Surv. 1912, pp. 7–9.

53. Macdonald, A. G. " The Brown Coals of Otago." *Colliery Guardian,* 22nd and 29th Nov., 6th and 13th Dec., 1912; pp. 1036–38, 1089–91, 1140–41, and 1190.

54. Marshall, P. " Geology of New Zealand," 1912.

55. —— " New Zealand and Adjacent Islands." Handbuch der regionalen Geologie, 5 Heft, (Band vii, Abteilung 1), 1912.

56. Hatch, F. H., and Rastall, R. H. " The Petrology of the Sedimentary Rocks," 1913, pp. 46, 71, 215–17.

57. Cotton, C. A. " On the Relations of the Great Marlborough Conglomerate to the Underlying Formations in the Middle Clarence Valley, New Zealand." Journal of Geology, vol. 22, 1914, pp. 346–63.

58. Morgan, P. G. " Unconformities in the Stratified Rocks of the West Coast of the South Island." Trans. N.Z. Inst., vol. 46, 1914, pp. 270–78.

59. —— " Preliminary Report on the Gisborne – East Cape District." 8th Ann. Rep. (n.s.) N.Z. Geol. Surv., in Parl. Paper C.-2, 1914, pp. 124–28.

60. Henderson, J. " Coal Possibilities of the Warkworth District." *Ibid.*, pp. 157–58.

61. Speight, R. " The Intermontane Basins of Canterbury." Trans. N.Z. Inst., vol. 47, 1915, pp. 336–53.

62. Morgan, P. G. " The Geology and Mineral Resources of the Buller–Mokihinui Subdivision, Westport Division." N.Z. Geol. Surv. Bull. No. 17 (n.s.), 1915.

63. —— " Weka Pass District, North Canterbury." 9th Ann. Rep. (n.s.) N.Z. Geol. Surv., in Parl. Paper C.-2, 1915, pp. 90–93.

Art. II.—*The Geology of the Neighbourhood of Kakanui.*

By G. H. Uttley, M.A., M.Sc.

Communicated by Dr. J. Allan Thomson.

[*Read before the Wellington Philosophical Society, 27th October, 1915.*]

Introduction.

Previous geological work in the Oamaru coastal district has resulted in somewhat discordant conclusions. McKay (1877) has assumed an unconformity to exist between the Cretaceo-tertiary and Tertiary series, but he relies on differences of dip and on volcanic action as evidencing the existence of a land surface. It is pretty well agreed now, however, that the greater part, if not all, of the volcanic rocks in the district were ejected beneath the sea. Difference of dip at widely separated points carries no weight where the rocks have been folded and faulted as at Oamaru Cape, where McKay seeks to establish his unconformity. The supposed stratigraphical break, therefore, between a Tertiary and a Cretaceo-tertiary system does not exist, and there is no palaeontological ground for the recognition of the latter system so far as the Oamaru district is concerned.

Hutton (1887) placed a break between the Oamaru limestone and the overlying greensand in certain localities where a nodular band, described in detail below, is found to occur—viz., at All Day Bay and Deborah—but other observers except McKay have refused to recognize the unconformity. Hutton did not rely on palaeontological evidence, and included both the greensand and limestone in his Oamaru system, but inferred an unconformity from the waterworn surface of the limestone and the difference of dip. The evidence will be discussed in detail below.

Park (1905) introduced an altogether new interpretation of the succession when he stated that two limestones existed in the Oamaru district, separated by the Hutchinson Quarry and Awamoa beds, the lower being termed the Oamaru stone and the upper the Waitaki stone. It is certain that the limestone of the Devil's Bridge, which he calls the Waitaki stone, is not underlain, as he states, but overlain by glauconitic sandstone—the Hutchinson Quarry beds. The rocks underlying the limestone may be seen to the north at Brockman's Hill, and consist of tuff and a sill of dolerite dipping beneath the limestone—that is to say, the limestone of the Devil's Bridge has the usual stratigraphical position of the Ototara limestone, forming the middle member of the Oamaru system. It is only in the neighbourhood of Kakanui that Park shows the two limestones in superposition in the same section, and I shall endeavour to show below that he has misread the sequence. There are two limestones represented in the same section on the south side of the Kakanui River, but they are separated by volcanic rocks, and the upper limestone is followed by the Hutchinson

Quarry beds and the Awamoa blue clay.   The complete sequence in the
Oamaru coastal district, east of the Waiareka Valley, is as follows :—

Top. Blue clay   ..   ..   ..   } Awamoa beds.
      Shell-bed   ..   ..   ..   }
      Greensand with *Pachymagas parki*   )
      Greensand with *Aetheia gaulteri* .. |
      Nodular bed with *Isis dactyla* and   } Hutchinson Quarry beds.
        Mopsea hamiltoni   ..   ..   )
      Limestone   ..   ..   ..)
      Brecciated pillow lava and tuff   .. |
      Mineral breccia..   ..   ..   } Ototara limestone.
      Marl ..   .. |
      Limestone   ..   ..   ..)
      Fine brown tuff   ..   ..   )
      Tachylite tuff (fine)   ..   ..   } Waiareka tuffs.
      Diatomaceous earth   ..   .. |
      Tuff ..   ..)

In the following detailed accounts of sections in the neighbourhood
of Kakanui the lists of fossils are the result of determinations of specimens
collected afresh by myself, with a few additional brachiopods collected by
Dr. Thomson.   The species marked with an asterisk have been determined
by Mr. H. Suter in the case of *Mollusca* and by Dr. Thomson in the case
of *Brachiopoda*, the remainder being determined by myself.   The lists take
no account of earlier determinations, and must not be regarded as complete
lists, but rather as illustrations of the fauna.   Much promising work still
remains to be done before the distribution of the fossils can be accurately
known.

DETAILED ACCOUNT OF SECTIONS.

(1.) *All Day Bay.*

This bay is situated in the extreme north-east of the Otepopo Survey
District, and the section to be described is exposed on the southern side

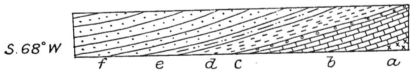

S. 68° W

f   e   d   c   b   a

Fig. 1.—Section north end of All Day Bay. *a*, mineral breccia† ; *b*, limestone, 15 ft.
thick ; *c*, concretionary greensand band, 18 in. thick ; *d*, greensand ; *e*, greensand ;
f, blue clay.

of Kakanui Point.   The point consists of a mass of volcanic breccia, and
has been described by Thomson (1906).   The section exposed on the coast
immediately south of the point is shown in fig. 1.‡

---

† The "mineral breccia" was so termed by Thomson on account of the abundance
of fragments of hornblende, feldspar, olivine, augite, garnets, &c., which it contains.
It is easily distinguished on this account from the other basaltic breccias and tuffs of
the Oamaru district.
‡ The nature of this section has altered since my visit in 1903, probably by heavy
seas removing gravel fallen down from the cliffs on to the beach, and it is now much
clearer.   This explains the difference between Mr. Uttley's account, with which I agree
entirely, and my earlier account, and probably also explains how Hutton could think
there was an unconformity.—J. A. T.

The breccia dips S. 60° W. at an angle of 8°, but the upper beds gradually flatten and the blue clay becomes almost horizontal.

The limestone is rather fine in texture, and contains cleavage fragments of minerals similar to those in the underlying breccia, but the junction appears quite conformable. In its upper portion the limestone becomes glauconitic, and gradually assumes the character of bed *c*. The fossils obtained from the base of the limestone were : *Aetheia gaulteri* (Morris), *Hemithyris* cf. *H. squamosa* (Hutt.), *Terebratulina suessi* (Hutt.), *Epitonium rugulosum lyratum* (Zittel). Sharks' teeth occur in abundance at the base.

The concretionary greensand (*c*) marks the change of conditions on the sea-bottom which brought about the deposition of the more glauconitic beds above. It is essentially a mixture of limestone and greensand, and where it forms the present sea-beach the waves have removed the looser sands and the surface is irregular and nodular. Minerals that occur in the breccia are still present here, though sparingly. Fossils are abundant, but, unfortunately, mainly as casts, and this has rendered specific identification difficult. The following were obtained here : *Turbo* sp., *Struthiolaria* sp., *Polinices ovatus* (Hutt.) ?, *Cypraea ovulatella* Tate ?, *Epitonium rugulosum lyratum* (Zittel), *Siphonalia* sp. nov., *Lapparia corrugata* (Hutt.) ?, *Euthria media* (Hutt.), *Pecten polymorphoides* Zittel, *Lima lima* (L.), *Aetheia gaulteri* (Morris), *Hemithyris* cf. *H. squamosa* (Hutt.), *Terebratulina suessi* (Hutt.), *Terebratula oamarutica* Boehm, *Terebratula* sp. nov.

Beds *d* and *e* are glauconitic greensands 10 ft. thick, but they have been separated, as the upper band (*e*) is very much indurated, and does not contain the same variety of fossils as bed *d*. I obtained the following fossils from the lower greensand : *Epitonium browni* (Zittel), *Aetheia gaulteri* (Morris), *Hemithyris* cf. *H. squamosa* (Hutt.), *Terebratula* sp.

Two forms which are very abundant both in the concretionary greensand and in the lower portion of the greensand (*d*) are *Mopsea hamiltoni* (Thomson) and *Isis dactyla* Ten.-Woods.

The hardened band (*e*) contains *Pachymagas parki* (Hutt.).

The blue clay (*f*) is not very fossiliferous in its lower portions, but higher in the section a number of fossils were obtained which would seem to correlate the bed with the Awamoa horizon. Although the change from the greensands below is very gradual, the exposure of *f* farther along the beach clearly shows that the bed is lithologically similar to the blue clay of the Awamoa Creek deposits. A list of fossils from bed *f* collected by Professor Marshall and myself has been published by Professor Marshall in the last volume of the " Transactions of the New Zealand Institute " (vol. 47, p. 384).

It will be seen from the above description that there is a gradual transition from the limestone through greensands to the typical Awamoa blue clay, and that the beds are conformable throughout. The horizons *b*, *c*, *d*, *e*, and *f* are clearly recognizable in many other parts of the district. The limestone is probably only the upper portion of the Ototara limestone, the continuous deposition of which was interrupted by volcanic action which resulted in the accumulation of the mineral breccias at Kakanui, the volcanic rocks of Oamaru Creek, and much of Oamaru Cape.

(2.) *Kakanui River (Right Bank).*

Thomson (1906, pp. 485, 486, fig. 2) has given a section of the beds exposed here. In the bed of the river at very low water a small isolated outcrop of volcanic rock is exposed, of which he makes no mention. The exact locality is where the line of road bounding the townships of Riverview and Kakanui South would strike the river if produced. It is about 300 yards west of the outcrop of the lower limestone of his section, and dips in the same direction and at the same angle. The rock is breccia, but of a different nature from the mineral breccia higher in the section. It has a fine tufaceous matrix with vesicular masses of basalt, together with tachy-lyte tuff similar to the glass tuffs that are closely associated with the deposits of diatomaceous earth in the Waiareka Valley, which in the latter locality always lie below the limestone and never above it. As much confusion has arisen in the past through the erroneous correlation of the various volcanic rocks of the Oamaru district, it has been thought advisable to mention this isolated outcrop, for the dip of the beds is exactly the same as the more easterly beds, and, although no actual junction is seen, they undoubtedly form part of the same series.

The lower limestone is impure and tufaceous, but similar in texture to the typical Oamaru building-stone. It is 35 ft. thick, but the base is not seen. A few brachiopods were collected from the upper 6 ft. of this rock, and the following were identified : *Terebratula oamarutica* Boehm and *Terebratulina suessi* (Hutt.).

The limestone is followed by a pure-white foraminiferal marl 24 ft. thick, which contains occasional lines of rounded volcanic pebbles. This bed is capped by 6 in. of limestone, which is followed by 56 ft. of fragmental volcanic rocks. The lowest 16 ft. is a fine volcanic ash, brown to black in colour, but it does not show any minerals. The upper 40 ft. is very conspicuous from the great abundance of minerals that it contains. Overlying these tufaceous beds is a limestone of unknown thickness, but the thickness exposed is about 35 ft. The limestone contains in its lower part much very fine tufaceous matter, and is very friable. The highest 5 ft. of the limestone is more like the building-stone, and it is much harder and more compact than the more tufaceous portion immediately below. From the tufaceous limestone the following species were obtained : *Epitonium rugulosum lyratum* (Zittel), *Pecten aldingensis* Tate, *P. dendyi* Hutt., *P. delicatulus* Hutt., *Aetheia gaulteri* (Morris), *Hemithyris* sp. nov., *Terebratula oamarutica* Boehm, *Terebratulina suessi* (Hutt.), *Neothyris* sp. nov., *Magella carinata* Thomson.

This assemblage of fossils clearly correlates this limestone with the Kakanui Quarry limestone, to be described in the next section.

(3.) *Sea-coast near Kakanui Township.*

The section to be described is seen on the coast for about a mile north of Kakanui Quarry, and has been described and figured by Hutton (1887, p. 420, pl. xxvi, sec. iv.), Park (1905, pp. 509, 510, figs. 3, 4), and Thomson (1906, p. 484, fig. 1). Hutton and Thomson agree in their interpretation of the stratigraphy, but Park differs from both these observers. The present writer is satisfied that the latter's interpretation of the succession is due to some error, and as Park in his section (p. 510, fig. 4) shows two separate limestones (Oamaru and Waitaki stones), which he asserts are separated by the Hutchinson Quarry and Awamoa beds—an altogether new interpretation of the Tertiary sequence at Oamaru—a rather detailed description will be necessary.

Park's first section (1905, p. 509, fig. 3) represents the southern limb of a syncline (see fig. 2), and the second (*loc. cit.*, p. 510, fig. 4) cuts the northern limb of the same syncline at right angles. I have given my own interpretation of this section in fig. 3, which represents the same line of section as Park's section at Trig. T.

The reference numbers of the beds in figs. 2 and 3 are the same as those given by Park in his sections. This, perhaps, will facilitate a comparison of the sections.

Fig. 2.—Section along sea-cliff north of Kakanui. Distance = ¾ mile. 1, calcareous tuff, fossiliferous; 2, blue micaceous tuff bed; 3, limestone; 3a, nodular band; 4, sand; 5, greensand; 6, breccia.

The lowest bed exposed near the quarry is a very calcareous tuff (1), which contains a considerable variety of fossils, but the latter are difficult to extricate owing to the great hardness of the rock. The following forms were collected : **Turbo marshalli* Thomson, **Turritella carlottae* Watson, **Epitonium browni* (Zittel), **E. rugulosum lyratum* (Zittel), **Siphonalia turrita* Suter, **S. conoidea* (Zittel), *S. costata* (Hutt.), *Lapparia* sp., **Dentalium solidum* Hutt., **Pecten polymorphoides* Zittel, **P. aldingensis* Tate, **P. delicatulus* Hutt., *Lima angulata* Sow., *Venericardia difficilis* var. *benhami* Thomson, **Chione meridionalis* (Sow.), *Aturia australis* McCoy, *Aetheia gaulteri* (Morris), **Hemithyris* sp. nov., *Terebratula oamarutica* Boehm, **Terebratula* sp. nov., *Terebratulina suessi* (Hutt.), **Neothyris* sp. nov.

Overlying the tuff is a less calcareous blue micaceous tuff bed (2) about 14 ft. in thickness, from which I obtained the following fossils : *Turbo marshalli* Thomson, *Turritella carlottae* Watson, *Dentalium solidum* Hutt., *Aetheia gaulteri* (Morris), *Terebratulina oamarutica* (Hutt.), **Terebratula* sp. nov., *Pentacrinus* sp.

The limestone (3) is about 20 ft. thick, and very pure. It makes excellent material for the lime-kiln, but it is too hard for building purposes. It is very fossiliferous in parts, and from the quarry near the road (Everett's Quarry) the following species were identified : *Aturia australis* McCoy, *Pecten aldingensis* Tate, *Venericardia* sp., **Thecidellina hedleyi* Thomson, **Aetheia gaulteri* (Morris), **Hemithyris* sp. nov., **Terebratula oamarutica* Boehm, *Terebratulina suessi* (Hutt.), **Rhizothyris rhizoida* (Hutt.), **Neothyris* sp. nov., *Magella carinata* Thomson, **Terebratella oamarutica* Boehm, **Mopsea hamiltoni* Thomson.

Towards the top the limestone becomes glauconitic and much less pure, and at the surface the glauconite sand and the limestone are so much intermingled that the bed assumes the concretionary—or, rather, nodular—structure (3a) similar to the beds at All Day Bay. Some of the nodules are brown, and are invariably covered with a thin, much darker, shining brown veneer.† Fragments of minerals and small pieces of volcanic rock occur in this nodular bed (3a). Fossils are abundant, but mainly as

---

† Dr. Thomson informs me that Mr. B. C. Aston has determined this veneer to be phosphatic.

casts, and identification has been rendered difficult. The following forms occurred : *Trochus conicus* Hutt. ?, *Astraea* sp. aff. *sulcata* (Martyn), *Turbo* sp., *Cypraea ovulatella* Tate, *Euthria media* (Hutt.), *Olivella neozelanica* (Hutt.), *Lapparia corrugata* (Hutt.) ?, *Cardium* sp. aff. *C. brachytomum* Suter, *Cardium* sp., *Arca decussata* (Smith), *Pecten polymorphoides* Zittel, *Lima jeffreysiana* Tate, *L. angulata* Sow., *L. lima* (L.), *Aetheia gaulteri* (Morris), *Hemithyris* cf. *H. squamosa* Hutt., *Terebratula* sp. *Mopsea hamiltoni* (Thomson) and *Isis dactyla* Ten.-Woods are very abundant here, and they pass up into the overlying greensand (5).

Bed 5 is of unknown thickness, as it is unconformably overlain by quartz sand. It is a glauconitic foraminiferal greensand, and is fossiliferous. The species were : *Epitonium browni* (Zittel), *Siphonalia nodosa* (Martyn) ?, *Teredo heaphyi* Zittel, *Pecten delicatulus* Hutt., *P. polymorphoides* Zittel, *Lima angulata* Sow., *L. bullata* Born, *Aetheia gaulteri* (Morris), *Terebratulina suessi* (Hutt.), *Pachymagas parki* (Hutt.), *Isis dactyla* Ten.-Woods, *Mopsea hamiltoni* (Thomson).

Bed 6, which underlies the limestone at the northern end of the beach about due east of Trig. T, is a somewhat coarse calcareous mineral breccia. In its upper portions it is interstratified with limestone bands, and limestone occupies vertical cracks in the breccia forming dykes. These limestone bands contain *Terebratula oamarutica*. The breccia itself in its upper 20 ft. is very fossiliferous, but there is a noticeable absence of brachiopods. A collection included the following forms : *Turbo marshalli* Thomson, *Turritella carlottae* Watson, *Crepidula* sp., *Polinices laevis* (Hutt.) ?, *Cypraea ovulatella* Tate, *Siphonalia turrita* Suter ?, *Siphonalia costata* Hutt. ?, *Arca* sp., *Glycimeris laticostata* Q. & G., *Pecten delicatulus* Hutt., *P. williamsoni* Zittel ?, *P. hutchinsoni* Hutt., *Lima* sp. aff. *angulata* Sow., *L. jeffreysiana* Tate, *L. bullata* (Born), *Venericardia australis* Lamk., *V. difficilis* Desh. var. *benhami* Thomson, *Diplodonta zelandica* (Gray), *Chione chiloensis truncata* Suter, *C. crassa* Q. & G., *Dosinia caerulea* (Reeve), *Cardium* sp., *Mesodesma subtriangulatum* (Gray), *Siphonium planatum* Suter.

Fig. 3 represents the same line of section as that shown by Professor Park (*loc. cit.*, fig. 4, p. 510). The reference numbers of the beds are the same as those used in Park's figures.

FIG. 3.—Trig T to the sea.   3, limestone ; 3*a*, nodular band ; 5, greensand ; 6, mineral breccia ; 7, limestone.

The section runs E.-W., with a flattening dip to the sea. Bed 6 is the mineral breccia which is exposed in the fields to the west of Trig T. The limestone (7) on which the trig. station has been erected is the Waitaki stone of Professor Park. This calcareous band can be traced as a continuous ridge from the station to Kakanui Quarry in a southerly direction, and from the quarry to the sea-beach.

Professor Park in his fig. 3 (p. 509) shows this beach outcrop (bed 3). Now, it has been shown above (see fig. 2) that this outcrop of limestone is continuous along the coast, and covered by greensand (5) as far as a point directly east of Trig. T; and Professor Park evidently interprets the coastal section in the same way, for he shows 3 and 5 in their proper position. How, then, can bed 7 (Waitaki stone) be at a different horizon from bed 3 (Oamaru stone), as his section (fig. 4) shows? Palaeontological evidence is not wanting to support the contention that there is but one limestone horizon, for in the road-cutting near the farmhouse near Trig. T I found the following fossils in the limestone, and these are certainly the fossils that occur at Kakanui Quarry : *Terebratula oamarutica* Boehm, *Aetheia gaulteri* (Morris), **Hemithyris* sp. nov., *Terebratella oamarutica* Boehm, *Terebratulina suessi* (Hutt.), *Magella carinata* Thomson.

I searched carefully for the basalt flow (bed 6), but could find none, although the mineral breccia is present beneath the limestone everywhere in this locality.

If the evidence detailed above is accepted, it is impossible that Professor Park's classification of the Tertiaries in the Oamaru district can stand (see "Geology of New Zealand," 1910, Whitcombe and Tombs, p. 120). The sequence in the present locality, at all events, does not show two limestones separated by greensands (Hutchinson Quarry and Awamoa beds).

(4.) Three Roads.

I have given this name to a locality one mile north-east of Trig. T. It is just where the mineral breccia crops out on the coast for the last time going north.

The section given above (fig. 2) showed the beds which formed the northern limb of the syncline dipping south. The mineral breccia then extends about a mile along the coast to the present locality. Its strike-lines are clearly traceable on the beach at low tide, swinging round in sweeping curves from a N.E.–S.W. to a N.W.–S.E. strike at the present locality. The mineral breccia here has much the same nature as that described in the last section, though it is finer in texture. Thin veins of limestone penetrate the breccia near the surface. From the highest portion exposed I collected the following fossils : *Turbo marshalli* Thomson ?, *Lima* sp. aff. *angulata* Sow., *Ostrea* sp., *Venericardia australis* Lamk., *Diplodonta zelandica* (Gray), *Chione chiloensis truncata* Suter, **Dosinia caerulea* (Reeve), *Mesodesma subtriangulatum* (Gray).

Although the fossils are not numerous here, it is pretty evident that the above are similar to those found near Trig. T; in the present locality, therefore, the breccia collected from represents the upper portion. About 200 yards from the outcrop of breccia on the beach, in a cutting in the road that runs east and west, another small section is exposed, showing a glauconitic nodular band 6 in. thick. Cleavage fragments of hornblende were scattered sporadically throughout the limestone, and the nodules are characterized by the peculiar brown sheen that usually covers the exterior. I collected the following species from this bed : **Pecten polymorphoides* Zittel, *Dentalium solidum* Hutt., *Aetheia gaulteri* (Morris), **Terebratula oamarutica* Boehm, *Terebratulina suessi* (Hutt.), *Isis dactyla* Ten.-Woods.

Overlying this bed is a calcareous glauconitic greensand, but the thickness could not be estimated. In the highest part of the exposure I collected *Pecten delicatulus* Hutt. and *Pachymagas parki* (Hutt.).

The beds dip N. 40° E. at an angle of 18°. As shown above, the nodular band is in this locality again followed by greensands. In the lower layers of the latter, for some distance above the junction, I obtained no fossils, and possibly the species detailed here represent the horizon of band *e* in the All Day Bay section, the lower unfossiliferous portion representing band *d*. About 15 chains north-east of this locality a channel sunk for drainage purposes reveals very fine glauconitic sand, and, although an area of only two or three square yards has been exposed, I collected the following species : *Malletia australis* Q. & G., *Nucula hartvigiana* Phil., *Pecten (Pseudamussium) huttoni* Park, *Lima colorata* Hutt., *Limopsis aurita* (Brocchi), *Venericardia australis* Lamk., *Macrocallista assimilis* (Hutt.), *Crassatellites obesus* A. Ad.

This bed represents the Awamoa horizon, and the dip of the beds at Three Roads is such as to take them beneath these beds.

ANALYSIS OF FOSSIL LISTS.

I have to thank Dr. Thomson for drawing up the following table showing the range of the various species within the Kakanui district.

1. Tuffs, &c., below the Kakanui limestone (Lower Ototaran).
2. Kakanui limestone (Upper Ototaran).
3. Hutchinson Quarry beds of Kakanui district (Hutchinsonian).
4. Awamoa beds of All Day Bay (see Marshall, 1915, p. 384) and Three Roads (Awamoan).

---	1.	2.	3	4.
CORALS.				
Isis dactyla Ten.-Woods			×	
Mopsea hamiltoni (Thomson)		×	×	
BRACHIOPODA.				
Aethia gaulteri (Morris)	×	×	×	
Hemithyris cf. *H. squamosa* (Hutt.)		×	×	
Hemithyris sp. nov.	×	×		
Magella carinata Thomson		×		
Neothyris sp. nov.	×	×		
Pachymagas parki (Hutt.)			×	
Rhizothyris rhizoida (Hutt.)		×		
Terebratella oamarutica Boehm		×		
Terebratula oamarutica Boehm	×	×	×	
Terebratula sp. nov.	×		×	
Terebratulina suessi (Hutt.)	×	×	×	
Thecidellina hedleyi Thomson		×		
MOLLUSCA.				
Ancilla novaezelandiae (Sow.)			...	×
Arca novae-zealandiae Smith			?	
Astraea aff. *sulcata* (Martyn)			×	
Aturia australis McCoy	×	×		
Cardium aff. *brachytonum* Sut.			×	
Chione chiloensis truncata Sut.	×			
Chione crassa (Q. & G.)	×			
Chione meridionalis (Sow.)	×			
Corbula pumila Hutt.				×
Crassatellites obesus A. Ad.				×
Cypraea ovulatella Tate	×			×
Dentalium mantelli Zitt.				×
Dentalium solidum Hutt.	×		×	
Diplodonta zelandica (Gray)	×			
Dosinia caerulea (Reeve)	×			
Epitonium browni (Zitt.)	×		×	×

---	1.	2.	3.	4.
MOLLUSCA—*continued.*				
Epitonium rugulosum lyratum (Zitt.)	×	×	×	..
Euthria media (Hutt.)	..	..	×	..
Glycymeris laticostata (Q. & G.)	×	..	..	..
Lapparia corrugata (Hutt.)	..	..	?	×
Lima angulata Sow.	×	..	×	..
Lima bullata (Born)	×	..	×	..
Lima colorata Hutt.	..	..	..	×
Lima jeffreysiana Tate	×	..	×	..
Lima lima (L.)	..	..	×	..
Limopsis aurita (Brocchi)	..	..	..	×
Limopsis zitteli Iher.	..	..	..	×
Macrocallista assimilis (Hutt.)	..	..	..	×
Mangilia rudis (Hutt.)	..	..	..	×
Marginella conica Harris	..	..	..	×
Marginella harrisi Cossman	..	..	..	×
Mesodesma subtriangulatum (Gray)	×	..	..	..
Murex octogonus Q. & G.	..	..	..	×
Nucula hartvigiana Pfr.	..	..	..	×
Olivella neozelanica (Hutt.)	..	..	×	..
Pecten aldingensis Tate	×	×	..	..
Pecten delicatulus Hutt.	×	×	×	..
Pecten dendyi Hutt.	..	×	..	..
Pecten hutchinsoni Hutt.	×	..	..	..
Pecten huttoni Park	..	..	..	×
Pecten polymorphoides Zitt.	×	..	×	..
Pecten williamsoni Zitt.	?	..	..	..
Phalium achatinum pyrum (Lamk.)	..	..	..	×
Placunanomia zelandica (Gray)	..	..	..	×
Polinices gibbosus (Hutt.)	..	..	..	×
Polinices laevis (Hutt.)	?	..	..	..
Siphonalia conoidea (Zitt.)	×	..	..	..
Siphonalia costata (Hutt.)	×	..	..	..
Siphonalia nodosa (Mart.)	..	..	?	..
Siphonalia turrita Sut.	×	..	..	..
Siphonium planatum Sut.	×	..	..	..
Teredo heaphyi Zitt.	..	..	×	..
Trochus conicus (Hutt.)	..	?	?	..
Turbo marshalli Thomson	×	..	..	..
Turbonilla oamarutica Sut.	..	..	..	×
Turris altus (Harris)	..	..	..	×
Turritella carlottae Wats.	×	..	..	..
Turritella cavershamensis Harris	..	..	..	×
Typhis McCoyi Ten.-Woods	..	..	..	×
Venericardia australis Lamk.	×	..	..	×
Venericardia difficilis benhami (Thomson)	×	..	..	..
Verillum apicale (Hutt.)	..	..	..	×

LIST OF PAPERS REFERRED TO.

Hutton, F. W. (1887). "On the Geology of the Country between Oamaru and Moeraki." Trans. N.Z. Inst., vol. 19, pp. 415–30.

Marshall, P. (1915). "Cainozoic Fossils from Oamaru." Trans. N.Z. Inst., vol. 47, pp. 377–87.

McKay, A. (1877). "Oamaru and Waitaki Districts." Rep. Geol. Explor. during 1876–77, pp. 41–66.

Park, J. (1905). "On the Marine Tertiaries of Otago and Canterbury, with Special Reference to the Relations existing between the Pareroa and Oamaru Series." Trans. N.Z. Inst., vol. 37, pp. 489–551.

Thomson, J. A. (1906). "The Gem Gravels of Kakanui, with Remarks on the Geology of the District." Trans. N.Z. Inst., vol. 38, pp. 482–95.

Art. III.—*On Stage Names applicable to the Divisions of the Tertiary in New Zealand.*

By J. Allan Thomson, M.A., D.Sc., F.G.S., Director of the Dominion Museum, Wellington.

[Read before the Wellington Philosophical Society, 20th October, 1915.]

Contents.

[Note.—References are indicated by the date after the author's name, and will be found in the list of papers at the end of this article.]

I. Introduction.

There are two objects to be aimed at in framing a classification of the younger rocks of New Zealand, and it is important to distinguish them. The first is to set up a standard of reference by which rocks from different parts of the country may be correlated with one another; the second is to correlate the various divisions of the classification thus established with their equivalents in the classifications of other parts of the world, and particularly in the accepted time-scale based on the rocks of Europe. The need for attacking the problem of classification in this order is imposed by the differentiation of the world's fauna into geographical provinces, a differentiation that has been, on the whole, accentuated as the present day is approached. It may be possible to correlate our Cretaceous rocks in individual districts directly with the European equivalents, but for the divisions of the Tertiary such a procedure is impossible in the present state of our knowledge.

It is undesirable, therefore, to speak of Eocene or Miocene rocks of New Zealand if what is meant is really rocks of the age of the Ototara limestone, the age of which has not yet been firmly established. For this reason the classification of the Upper Cretaceous and Tertiary rocks by Hector and McKay as Cretaceo-tertiary, Upper Eocene, Lower Miocene, Upper Miocene, and Pliocene must be abandoned, at least temporarily, so far as the names are concerned, whether or not one agrees with the distinctness of the groups of rocks on which it is based. Instead we must adopt a classification on the lines followed by Hutton, Haast, Park, and Marshall, in which names for the various series recognized are derived from New

Zealand place-names. The subdivisions of Hector and McKay's classification followed the latter principle.

The difficulties in the way of framing a classification that shall receive general support are twofold. New Zealand geologists are not agreed as to the presence or absence of unconformities between certain rocks, and they are not agreed on the correlations that should be made between the rocks of different localities. As a matter of fact, neither of these difficulties need stand in the way of a classification by stages, provided that there is agreement as to the order of superposition in the localities on which they are based, and as to the order of preference of localities from which they are named.

It may be pointed out that the European time - scale (Cambrian, Ordovician, &c.) is now entirely independent of unconformities, although these were formerly used in drawing it up. There is no gap between Jurassic and Cretaceous time, although in many parts of England there are unconformities between Cretaceous and Jurassic, and again between different divisions of the Jurassic. In other parts of England the missing stages are represented. What we should aim at in New Zealand, then, is a similar series of stage names corresponding to all the divisions of geological time represented in our rocks. If we can agree so far, a great advance will be made in several directions. In the first place, although the stages may be variously combined by different authors into geological systems, the subdivisions of their systems will remain the same, and the points at issue will be easily grasped by those who have not followed the whole perplexing controversies in New Zealand literature. In the second place, attention will be drawn to the lack of a satisfactory knowledge of the fauna of many of the stages, a lack which largely explains the divergent views hitherto held as to correlation. When this lack is supplied it will be possible to compare the rocks of other localities with those chosen as types for the various stages, and if important unconformities exist they will either be detected by the absence of known stages in the new localities or by the discovery of new stages unknown in the type localities. Such a palaeontological solution, of course, cannot be expected in cases like the Weka Pass, where one of the rocks—the Amuri limestone—is practically without fossils, unless the microscopic fauna can be brought into use. The subdivision of the Tertiary into stages, then, is not likely to advance greatly the solution of the Cretaceo-tertiary problem in areas where the Amuri limestone is developed, but should clear up our views on the stratigraphy of the Tertiary rocks.

In choosing localities and rocks to serve as types, and to give names to the various stages to be recognized, it would be desirable to adopt the principle of priority of mention in geological literature if this were practicable.* Names might thus be chosen from widely separated localities, and the order of the stages then determined by a comparison of the faunas. This method, adopted in Europe, South America, and Australia for the stages of the Tertiary, has frequently created controversies as to the equivalence or relative position of the various stages, and is undesirable

---

* The following is the rule adopted by the United States Geological Survey on this point (24th Ann. Rep., 1903, p. 24) : " In the application of names to members, formations, and larger aggregates of strata, the law of priority shall generally be observed, but a name that has become well established in use shall not be displaced by a term not well known merely on account of priority. In general, a newly defined formation shall not receive a name that has been previously used in a different sense."

in New Zealand, where so many localities exist in which numerous stages are in superposition, the younger on the older. It is, therefore, preferable to choose one or more such localities and base the stage names on these alone, applying priority of nomenclature wherever possible, but not insisting on its rigid application. As, however, I propose to use stage names with the termination "ian,"* on a strict reading of priority it is necessary to consider only previous uses of similar names, such as Park's use of "Oamaruian" and "Wanganuian." In the selection of the localities the abundance of fossils in the various stages represented must be a leading consideration, in order that correlations may prove possible. Thus, although on the assumption of the conformity of the Amuri limestone and Weka Pass stone the Waipara section offers the most complete sequence of Tertiary rocks known in New Zealand, the paucity of fossils in the Amuri limestone, Weka Pass stone, and grey marls is such that correlations on purely palaeontological grounds would be extremely difficult for stages based on these rocks.

For the youngest Tertiary stages, apparently unrepresented in the South Island by marine rocks, unless in the Awatere Valley, we must choose between the rocks of Wanganui and of Hawke's Bay. Both have been used already for names of series or systems, but Wanganui has come to be recognized as the type locality, and on grounds of priority may also be justified as such, since Mantell described the rocks of that district long before Hochstetter obtained younger Tertiary fossils collected by Triphook from the neighbourhood of Napier. For the middle and old Tertiary stages the Oamaru district is indicated alike from priority (Mantell's descriptions), general usage, and abundance of fossils. For still older Tertiary stages, apparently not represented at Oamaru, we must go to the West Coast of the South Island.

I propose further to consider these localities in the following order— Oamaru, Wanganui, West Coast—and to exhaust the possible stages of each in turn. This is necessary, because probably all the stages found at Oamaru occur also in the West Coast, and some at least may underlie the younger beds at Wanganui as they do in the Awatere district, although in none of these cases are they so suitably developed as their correlatives at Oamaru to form the type occurrences of stages.

The geographical names at present applied to geological systems involve considerations not only of geographical occurrence, but of conformity and unconformity. Thus the term "Oamaru system" has been used in three distinct senses :—

(1.) By Hutton for part only of the Tertiary sequence at Oamaru and its correlatives. Hutton excluded the Awamoa beds because of a supposed unconformity and faunal difference between them and the uppermost member of his Oamaru system—viz., the Hutchinson Quarry beds.

(2.) By Park (1905) for the whole of the Tertiary sequence at Oamaru and its correlatives. Park included the Awamoa beds in his Oamaru system.

---

* Abbreviated to "an" where the word from which it is derived ends in "i," and to "n" where it ends in "a." Hybrids between Maori words and anglicized Latin terminations are doubtless unfortunate, but they can hardly be avoided in view of the retention of the Maori place-names throughout New Zealand. They are less objectionable as technical geological terms than as words in the vernacular, and we find that such terms as "Tahitian," "Fijian," and "Samoan" have been easily enough assimilated into our everyday speech.

(3.) By Marshall (1911) for the whole of the Tertiary sequence at Oamaru and its correlatives, and also for all other beds which in other parts of New Zealand are conformable with those correlatives. Marshall included the Cretaceous beds of the Waipara and Amuri Bluff districts in the Oamaru system because he considered that they lay conformably beneath Tertiary beds correlative with those at Oamaru.

Since I am not here concerned with the validity or otherwise of an Oamaru system, these three usages need not be discussed in that light; but it may be suggested that for a system such as Marshall believes to exist the term " Waipara system " would be preferable on geographical grounds, for all the beds he places in it, including those at Oamaru, have correlatives at Waipara. Park used such a geographical principle in 1910 when he replaced his Oamaru system of 1905 by a Karamea system, which included not only his former Oamaru system (now Oamaru series), but also lower beds without correlatives at Oamaru (Waimangaroa series). This is, I believe, a sound principle to follow wherever possible in the use of geological terms derived from place-names. To me, then, " Oamaruian " denotes the whole Tertiary succession as developed at Oamaru, and its correlatives elsewhere, but no other beds. This use of the term fortunately coincides in extent with its former application by Park.

There is, however, a subtle but important difference between his usage and that which I now propose. In speaking, for instance, of the Weka Pass stone, the grey marls, and the Mount Brown beds of the Waipara section as Oamaruian we both mean that these beds are the correlatives of the beds at Oamaru; but Park means also that they are unconformable with the underlying and overlying beds, whereas I express by the term no opinion on the point. In other words, Park's use of the term is systematic, while mine is geographical. I make the plea for such a geographical use of stage names terminating in " ian " as tending to bring about a greater uniformity of nomenclature.

II. STAGE NAMES DERIVED FROM THE OAMARU DISTRICT.

The Oamaruian, as above defined, consists of a series of rocks to the members of which we may apply stage names with advantage. Thus the European Miocene, with which the Oamaruian is sometimes correlated, has been divided into six stages. It is not contended that the stages here proposed are of the same value as those of the European Miocene : that will be an interesting point to discuss when we are in a position to do so. All that is claimed is that these divisions can be recognized at Oamaru and the neighbouring districts, and probably will be capable of identification throughout New Zealand when our knowledge of the faunas is more complete. Incidentally, the recognition of distinct stages will throw into relief our lack of knowledge of some of these faunas.

The succession of rocks at Oamaru is somewhat varied in different localities within the district owing to the unequal development of volcanic rocks at different points, and some difference of opinion exists as to the actual order of succession. The difficulty is in reality palaeontological more than stratigraphical, and is explained by the fact that the fauna of the beds underlying the Ototara limestone and its correlative the Waihao limestone resembles that of the beds overlying the limestone more closely than that of the limestone itself. This is due, no doubt, to the similarity of conditions in the formation of the two first-named sets of beds and the difference of conditions in the latter, combined with a relatively slow rate of evolution throughout the whole period of deposition. Hutton (1887) thus placed certain beds near Enfield and Windsor in his Pareora system, correlating them on palaeontological grounds with the Awamoa beds, and supposing them to occupy a valley of erosion in the Oamaru system, whereas McKay (1884), from an examination of the field evidence, had placed them under the Ototara limestone. Park (1887) came to a similar conclusion to McKay; and my own observations leave me in no doubt that Hutton was mistaken in his account of the sequence.

Park, in 1905, suggested an entirely new reading of the field evidence at Oamaru, designed to get over the problem, which he clearly recognized, of the " Pareora fauna " above and below the limestone. Thus he correlated, as Hutton did, the Waihao greensands, which lie below the Waihao limestone, with the Awamoa beds, which lie above the Ototara limestone. In consequence, he considered the Waihao limestone to represent a higher horizon than the Ototara limestone. My study of the brachiopods occurring above the limestone of Landon Creek and below the Ngapara and Maerewhenua limestones could be construed in the same way to support Park's explanation of two distinct limestones, but the stratigraphical facts do not seem to support it, and suggest rather that the similarity of the upper and lower faunas is due to the slow rate of evolution. In the Waihao district the existence of the Mount Harris beds with a " Pareora fauna " above the Waihao limestone is otherwise inexplicable. Park found these beds resting on Lower Mesozoic beds, and, as they have a wide distribution, this may be locally the case through overlap. Between Mount Harris and the Waihao River, however, they rest on greensands resembling those of the Hutchinson Quarry beds, and these in turn rest on the limestone (Thomson, 1914). This suggests strongly that the Waihao limestone is the correlative of the Ototara limestone, and the Waihao greensands the correlatives of the Enfield-Windsor beds.

The only locality where Park described the two limestones developed in the same section is near Kakanui. I am in full accord with Uttley's interpretation of this section, contained in another paper in this volume (p. 19)—viz., that there is only one limestone, which is sharply flexed from the beach to the top of the hill. It is easy, however, to see how any one starting off with the view that there were two limestones could interpret it in the manner done by Park in a brief visit.

Park's interpretation of the Oamaru and Waihao districts was an ingenious attempt to meet a real difficulty which had long ago been stated by Haast (1879), and had never been boldly faced by other geologists. It had the merit of suppressing Hutton's ambiguous Pareora system, and thus initiated a real advance in the classification of the Tertiary rocks. I hope, therefore, that Professor Park will retire with honour from his insistence on the presence of two limestones at Oamaru, and clear the way for a complete accord as to the stages of the Oamaruian. Such a step will not at all affect his general position as to the relationship of the Tertiary and Cretaceous in New Zealand.

With these prefatory remarks, I may now give a short generalized account of the Oamaru district, based mainly on my own observations, but supplemented for the coastal district by information received from Mr. G. H. Uttley.

There is a great escarpment of limestone, first described by W. Mantell (1850), and by him named the Ototara limestone, which runs from the neighbourhood of Kakanui for several miles in a northerly and then north-westerly direction towards the Awamoko River. The escarpment overlooks the Waiareka Valley opposite Alma and at Enfield, and the rock dips in an easterly direction towards Weston and Totara, where the Oamaru lime-stone quarries are situated. The Oamaru building-stone, the Oamaru limestone of Park (1905), is thus the Ototara stone, although Park in 1910 calls his supposed upper limestone the Ototara stone and the lower lime-stone the Oamaru stone. The dip of the beds does not remain long constant to the east, however, for the limestone reappears at numerous points in the triangle of country between the main escarpment, the sea-coast, and the Waitaki Valley. In all clear exposures the overlying beds are found to consist of calcareous greensand, often crowded with brachiopods, and generally known as the Hutchinson Quarry beds. The junction between the greensand and the limestone is not a simple one, for the upper surface of the limestone is often very irregular, and in many places covered with a brown varnish which has proved to be phosphatic (*e.g.*, All Day Bay, Kakanui, Deborah, Devil's Bridge). In the upper layers of the limestone there are frequently rounded calcareous concretion-like masses, varying in size up to 3 in. or 4 in. in diameter, and containing tissue-like com-partments within, suggesting an algal growth. The basal layers of the greensand are crowded with segments of *Mopsea hamiltoni* (Thomson), often in a phosphatized condition, while both unaltered and phosphatized brachiopods may be obtained. This peculiar junction has some analogies with that between the Weka Pass stone and the Amuri limestone, and seems to correspond to an abrupt change of conditions. It almost cer-tainly denotes a non-sequence, though not in most places physical uncon-formity ; but there is no reason at present for believing that the non-sequence is of any great extent. At Hutchinson's Quarry the greensands rest on a concretionary conglomerate consisting of rounded pieces of basalt and concretions similar to those described above, set in a calcareous cement ; this bed, several feet thick, rests conformably on a thin band of hard white limestone, which has recently been determined by Morgan to be phos-phatic, and this latter rests quite unconformably on a series of tuffs containing thin limestone bands. The formation of the limestone at this and other points seems to have been nearly prevented by the amount of volcanic material strewed into the sea, so that the limestone is represented by calcareous tuffs and thin bands of limestone.

Wherever the greensands of the Hutchinson Quarry horizon can be traced upwards they reveal less and less glauconite, and finally pass into shell-bearing sands and blue sandy mudstones—the Awamoa beds. No higher Tertiary rocks are seen in the Oamaru district.

In a few localities to the east of the main outcrop of the Ototara lime-stone, notably in the neighbourhood of Kakanui, Deborah, and Cape Wanbrow, exposures of older beds are seen. In all cases these are of volcanic origin, and it is evident at Kakanui, at any rate, that the upper-most of these must represent a replacement of the lower part of the Ototara limestone, since the Kakanui limestone is much thinner than the former and shows no characters which would lead one to suppose it formed at a slower rate, but rather the reverse.

To the west and north of the main outcrop of the Ototara limestone lies the broad low valley of the Waiareka River, on the far side of which again rises the Ngapara tableland. The latter is a well-preserved part of an elevated plain of denudation which was formerly continuous to the coast-line, but has been more dissected in the neighbourhood of Oamaru, though it is still recognizable there. This old plain truncates the Tertiary rocks around Ngapara at a low angle, and it is probable that, under the loess deposits which cover it, remnants of the Hutchinson Quarry and Awamoa beds are preserved, though they have not so far been described. Near Ngapara itself the plateau is cut across the Ngapara limestone, which runs thence to the Awamoko River and the Waitaki Valley. There can be no doubt that the Ngapara limestone was formerly continuous across the Waiareka Valley with the Ototara limestone. The rocks underlying the two limestones are not, however, the same. The Ngapara limestone is underlain by greensands of considerable thickness, and these pass down into quartz sands without glauconite, interstratified with grits and conglomerates and a seam of brown coal. The succession thus exactly resembles that at Waihao. The Ototara limestone, on the other hand, is underlain opposite Alma by volcanic breccias of no great thickness, and these again by a deposit of diatomaceous earth, which in turn rests upon a considerable thickness of volcanic tuffs and lavas. The latter at Enfield rest on greensands, which can be traced thence at various points through Windsor to Ngapara. The volcanic series underlying the Ototara limestone in the Waiareka Valley has long been known as the Waiareka series, which, strictly speaking, must include the diatomaceous earth. The difference between the succession at Enfield and that at Ngapara is due mainly to the volcanic conditions existing during deposition at the former place, though a local subsidence, perhaps correlated with the volcanic activity, seems necessary to explain the formation of the diatomaceous earth, regarded by Hinde and Holmes as a deep-water deposit. The Ngapara succession represents more normal conditions of marine deposition, such as prevailed in the Waitaki Valley, Waihao, and other South Canterbury localities.

For the above-described succession at Oamaru the following stage names are suggested :—

Top.			
Awamoa beds	Awamoan	⎫	Upper Oamaruian.
Hutchinson Quarry beds and concretionary band	Hutchinsonian	⎬	
Ototara limestone	Ototaran	..	Middle Oamaruian.
Waiareka tuffs and Enfield-Windsor greensands = Ngapara greensands ..	Waiarekan	⎫	Lower Oamaruian.
Coal-measures, sands, conglomerates, and coal-seams	Ngaparan	⎬	

All ambiguity caused by the use of the terms "Pareora series," "Pareora fauna," &c., may be avoided by the adoption of "Awamoan" for the uppermost stage of the Oamaruian. On grounds of strict priority, the name for this stage should perhaps be founded on the Onekakara formation of Mantell (1850); but although the blue clay of All Day Bay, which Mantell included in that formation, and from which he described its fauna, is undoubtedly Awamoan, there is considerable doubt whether the Onekakara (Hampden) beds are not really Waiarekan, as McKay (1884) supposed, since their stratigraphical relationship to the Ototara limestone is not clear, and all that is known about them palaeontologically is that they contain the "Pareora fauna."

No exception can be taken to "Hutchinsonian" on the grounds of priority, for no other earlier name has been used for this horizon in the Oamaru district. Park speaks of the Hutchinson Quarry beds as the Mount Brown beds; but the latter, in the Weka Pass district, include certainly the Awamoan as well as the Hutchinsonian stage, and possibly also the Ototaran. In the Hutchinsonian I would place all beds between the Ototara limestone and the shell-bed of Target Gully, described by Marshall and Uttley (1913), the latter bed forming the base of the Awamoan.

"Ototaran" is alike indicated by priority and subsequent usage as the most suitable name for the limestone member of the Oamaruian, and should be accorded general acceptance.

The Waiarekan stage has been studied in most detail as the Waihao greensands, but as the Waihao formation of Haast refers to much older rocks no stage name can be based on these greensands without ambiguity. The Waiareka tuffs are frequently included in older classifications, and a name based on them should find ready acceptance. Should this stage be too large in comparison with the others, the Waiarekan may be restricted to the tuffs themselves, and the underlying Windsor and Enfield beds may be made the type of a new stage.

The Ngaparan is the only one of the proposed stages that is not based on a well-established name for a geological series. Park, in 1905, called the coal-beds at the base of his Oamaru system the Awamoko beds. In 1910, however, he called them the Ngapara beds when speaking of the Oamaru district only, but for New Zealand generally he designated this horizon the Kaikorai coal-measures. For a subdivision of an Oamaru series a name derived from the Oamaru district is obviously most suitable. I select Ngapara as the type locality because it is nearer to the outcrop of the Ototara limestone than Awamoko, and the coal-beds there are demonstrably the basal member of the Tertiary series represented at Oamaru. The limits between the Ngaparan and Waiarekan cannot be exactly drawn without a more detailed investigation of the Ngapara district.

The Ngaparan in the North Otago and South Canterbury district does not contain a marine fauna, and cannot be correlated directly with marine beds of the same stage, such as probably occur in North Canterbury, and possibly on the West Coast. I suggest that "Ngaparan" should be restricted to coal-beds, and that a different stage name should be used for the normal marine beds of the same horizon. Applying the same principle throughout, we shall get a double set of stage names for the normal marine and for the littoral or terrestrial beds.

III. Stage Names derived from the Wanganui District.

As in the Oamaru district, there has not been complete agreement amongst geologists regarding the order of succession of the rocks in the Wanganui district. The principal point in dispute relates to the beds of Kaimatera Cliff, considered by Hutton (1886) as an upper series of Pleistocene beds unconformable with the Pliocene beds of the district, and by Park (1887) as an integral part of the "Newer Pliocene." As Hutton's view was founded on a brief inspection of the cliff without a full knowledge of the geological structure of the district, while Park's was based on an extensive reconnaissance survey, the latter may be accepted.

2*

Park described a continuous section on the sea-coast between Wanganui and Patea, the beds of which are shown in the following table along with his later groupings :—

Park, 1887.		Park, 1905.	Park, 1910.

(1.) Wanganui beds—
    Upper sandy beds
    Lower blue clays
(2.) Kai-iwi blue clays
(3.) Okehu pumice beds
(4.) Okehu sandy shell beds
(5.) Nukumaru Rotella beds
	} Newer Pliocene ..	} Wanganui series	} Petane series.

(6.) Nukumaru limestone .. Older Pliocene			

(7.) Waitotara Coralline series—
    Brown micaceous sand-
      stone
    Coralline beds
    Yellowish-blue sand-clays
    Whenuakura blue clays
	} Upper Miocene	} Te Aute or Wai-
		totara series
			} Waitotara series.

(8.) Patea blue clays and brown sands			} Awatere series.

Amongst the fossils of the Waitotara Coralline series given by Park in 1887 there are some extinct species not found in the higher beds, but known from the Oamaruian; some extinct species apparently not found either in the Oamaruian or in the higher beds—*e.g., Ostrea ingens;* and some Recent species not known from the Oamaruian, but found in the higher beds—*e.g., Terebratella rubicunda.* This shows, if Park's identifications are correct, that these beds occupy an intermediate position faunally between the Awamoan and the beds around Wanganui as seen at Shakespeare Cliff, Castlecliff, &c. There are, then, at least two stages represented in the above succession which are clearly superior to the Awamoan, and may be safely named. For the lower stage I propose "Waitotaran," based on Park's "Waitotara Coralline series." For the upper, "Wanganuian" is not available, as that name has been already used by Park to embrace the whole of the succession between Wanganui and Patea, and therefore includes the Waitotaran. The upper stage, to which Hutton's Wanganui system was confined, is best known by the often-described sections at Shakespeare Cliff and Languard's Bluff (Putiki), on the latter of which Hutton based his Putiki series. In neither of these localities, however, is the base of the upper stage represented, for it must include at least the Kai-iwi blue clays, which are inseparable from the blue clays of Park's Wanganui beds of 1887. I propose, therefore, to base its name on Castlecliff, from the point at which Park's section along the coast commences. The limits between the Castlecliffian and the Waitotaran must be left vague until the fauna of the beds between the Kai-iwi blue clays and the Waitotara Coralline series is better known. It is possible, but not probable, that a further stage may be found necessary between the Waitotaran and Castlecliffian.

It must also be left to further research to decide whether any further stages are necessary between the Waitotaran and the Awamoan. Park's Awatere series of 1910, which occupies such a position, is not founded on any definite set of beds in the Awatere district, and apparently presupposes that only one stage occurs there. As I pointed out in 1913, the

Awatere district contains a sequence of beds, and includes Oamaruian as well as Wanganuian stages. What is probably the uppermost stage, represented in the Starborough Creek beds described by Park in 1905, contains far fewer Oamaruian species than the Waitotaran list given by Park in 1887 ; so that, even allowing for some incorrect identifications in the latter, the former can hardly be earlier than Waitotaran. The same applies to the " Motunau beds " of the Waipara section, which rest directly, and apparently without unconformity, on the Awamoan. It does not appear probable, therefore, that any stage between the Waitotaran and the Awamoan will be necessary ; but, as said before, further research, including a detailed investigation of the Waitotara beds, is necessary before the matter can be settled.

IV. STAGE NAMES DERIVED FROM THE WEST COAST OF THE SOUTH ISLAND.

It has been generally held that the rocks of the West Coast coalfields contain in their lower members horizons lower than those represented at Oamaru. At one time these horizons were considered Cretaceous, but as no definitely Cretaceous marine fossils have been described it is now generally agreed that the whole marine sequence is Tertiary. It is, therefore, necessary to examine the evidence available with a view to discovering whether infra-Oamaruian stages may be safely named.

The fullest sequence, showing not only all the upper horizons represented on the West Coast, but also lower horizons than exist in any other district, is contained in the Greymouth district, recently surveyed in detail by the Geological Survey (Morgan, 1911). So far as the development and the relationships of the rocks are concerned, this area is one of the best known in the Dominion, and Morgan's account, based on a detailed survey, must be accepted in preference to all earlier accounts. The Tertiary rocks are grouped by him as follows :—

Pliocene beds	Soft sandstones, lignite, and somewhat consolidated gravels.
Greymouth series, Miocene ..	Blue Bottom formation. Cobden limestone. Port Elizabeth beds. Lower Kotuku conglomerate. Omotumotu beds.
Coal-measures (Mawheranui series), Eocene	Kaiata mudstone. Island sandstone. Brunner beds. Paparoa beds.

The so-called Pliocene beds are fluviatile, and contain no marine fossils. The whole of the Greymouth series and the Kaiata mudstone and Island sandstone of the Mawheranui series are marine, but the lower beds are sparingly fossiliferous. The Brunner and Paparoa beds consist of conglomerates, sandstones, grits, and shales, with coal-seams, and contain no marine fossils, but well-preserved leaf-impressions at several horizons.

Morgan correlated the middle and lower beds of the Greymouth series with the Oamaru formation of Hutton, suggesting the following individual correlations :—

Cobden limestone = Ototara stone.
Port Elizabeth beds = Middle Oamaru.
Omotumotu beds = Possibly the coal horizon of the Oamaru
  formation.

This view of the correlation was influenced by Park's interpretation of the Oamaruian, in which the Ototara stone (*i.e.*, the Waitaki stone) was considered as the closing member. In consequence, Morgan suggested that the Blue Bottom formation must, if Park's views were correct, be the equivalent of the Wanganui series. As a matter of fact, the Blue Bottom contains an Oamaruian fauna, and, if the Cobden limestone be really Ototaran, must represent the Hutchinsonian or the Awamoan, or both stages.

Morgan considered that the Kaiata mudstone and the Island sandstone represent a lower horizon than is developed at Oamaru ; but the fossils he enumerates contain a large number of well-known Oamaruian forms, and only three—viz., *Cardium brunneri, Kleinia conjuncta,* and *Schizaster exoletus*—which are unknown from the Oamaruian. On the evidence of the marine fossils, then, it is unsafe to conclude that these beds are not Oamaruian ; they may well be Waiarekan, so far as our present knowledge goes. The fact that an unconformity exists between these beds and the Greymouth series hardly affects the question, for there is no *a priori* reason why an unconformity should not be present between certain stages of the Oamaruian in some parts of New Zealand, although there is conformity in others. On the other hand, as we do not know what was the marine fauna of the times immediately preceding the Waiarekan, it is quite possible that it contained a large number of forms which survived into the Oamaruian, and that the Kaiata mudstone and Island sandstone do represent such an horizon. In the present state of our knowledge it is unsafe to base any stage names on these rocks.

The plant-fossils contained in the Brunner and Paparoa beds may yet yield important evidence as to the position of these horizons relative to the Oamaruian. Von Ettingshausen considered the flora of the Brunner beds Cretaceous and that of Shag Point Tertiary, while Morgan has shown that the flora of the Paparoa beds is more closely allied than that of the Brunner beds to the flora of Shag Point. The following explanation of this tangle may be suggested : Von Ettingshausen's correlation of the Shag Point beds as Tertiary, and of the Pakawau, Wangapeka, Grey, and Brunner beds as Cretaceous, was based on botanical comparisons with the floras of Europe, and cannot be held to have weight against the evidence yielded by the correlation of the overlying marine beds with those of other parts of New Zealand ; but his correlation of the Pakawau, Wangapeka, Grey, and Brunner plant-beds with one another may be accepted. The marine beds overlying the Shag Point plant-beds must be correlated with the Cretaceous beds of the Waipara and Amuri Bluff district, and these plant-beds are therefore Upper Cretaceous. The marine beds overlying the various West Coast localities contain a marine fauna agreeing with or closely related to the Oamaruian, and certainly Tertiary. The West Coast plant-beds are therefore either Tertiary or Danian, and most probably Tertiary, but certainly higher in horizon than the Shag Point beds. The flora of the Paparoa beds as determined by Morgan shows them to be intermediate between the Brunner beds and the Shag Point beds, as might be expected from their position. They contain four species (viz., *Podocarpus Parkeri, P. Hochstetteri, Dacrydium prae-cupressinum,* and *Aralia Tasmani*) hitherto known only from Shag Point, and three others (viz., *Quercus lonchitoides, Fagus ninnissiana,* and *Cinnamomum intermedium*) which have been recorded both from Shag Point and from Tertiary localities, but only two

(*Dacrydium cupressinum* and *Ulmophyllum latifolium*) which are known from the Pakawau horizon.

Although relatively very little is known of the fossil floras of undoubtedly Oamaruian horizons, the absence of any common species in the few species known and in the relatively well-known Pakawau-Brunner flora suggests that the latter is older than the Oamaruian; but until either the flora has been shown to be distinct from the Ngaparan flora or the fauna of the Kaiata mudstone and Island sandstone has been shown to be distinct from the Waiarekan fauna it would be unsafe to base a stage name on the Brunner series. It is certain, however, that even if the latter is Ngaparan the Paparoa series is older, and a stage may be safely based on it, and may be termed the Paparoan.

V. Summary and Conclusions.

Care has been taken in the above discussion to admit no stages that are likely to overlap on others. The following table gives the complete sequence of those proposed, and indicates the possible gaps :—

		Marine Stages.	Coal-beds.
Wanganuian	..	{ Castlecliffian. { Waitotaran. (Other stages possible.) Awamoan.	
Oamaruian	..	Hutchinsonian. Ototaran. Waiarekan.	
			Ngaparan.
		(Other stages possible.)	
			Paparoan.

Other stages for coal-beds are clearly necessary. Thus at Maharahara the coal-measures are immediately followed by marine beds with a Wanganuian and probably a Waitotaran fauna. The naming of these is best postponed until the correlation of the overlying marine beds with the named marine stages is rendered more definite.

The determination of the molluscan and brachiopodan fauna of the above marine stages has occupied the attention of Mr. Suter and myself for some considerable time, and it is hoped to publish the results before long. Fresh collections from any of the localities chosen as the types for stages will be gladly welcomed, for, although large collections have been examined, it is certain that much further collecting is still necessary before the complete range of the species can be stated, and until that is known correlations based on purely palaeontological grounds cannot have much value. The determinations so far made show that there a number of species which have a wide range in horizon and are also widespread in occurrence and abundant in the beds in which they occur. Such fossils are of little use as indications of horizon when taken singly. Fossils that are rare, and known from few localities, are obviously also of little use, for their rarity prevents their range being adequately known, while their absence in any given locality cannot be taken as evidence that they did not exist elsewhere at the time the beds of the said locality were being deposited. What must be sought are fossils which are at least moderately abundant and yet restricted in range. If such species can be found—and the experience of other countries suggests that they can—correlation will be enormously simplified, and it will be then possible to test the value of diastrophic considerations in correlation, and to apply these if they stand the test. Until the last few years the published lists of fossils and the analyses of the same were quite inadequate to support the classifications put forward. The three immediate needs of Tertiary geology in New Zealand may be stated as collecting, more collecting, still more collecting.

VI. List of Papers referred to.

Haast, J. von, 1879. "Geology of the Provinces of Canterbury and Westland," Christchurch, p. 315.

Hutton, F. W., 1886. "The Wanganui System," Trans. N.Z. Inst., vol. 18, pp. 336–67.

—— 1887. "On the Geology of the Country between Oamaru and Moeraki," *ibid.*, vol. 19, pp. 415–30.

Mantell, G. A., 1848. "Additional Remarks on the Geological Position of the Deposits in New Zealand which contain Bones of Birds," Quart. Journ. Geol. Soc., vol. 4, pp. 238–41.

—— 1850. "Notice of the Remains of the *Dinornis* and other Birds, and of Fossils and Rock-specimens, recently collected by Mr. Walter Mantell in the Middle Island of New Zealand; with Additional Notes on the Northern Island," *ibid.*, vol. 6, pp. 319–42, pl. 28, 29.

Marshall, P., 1911. "New Zealand and Adjacent Islands," Handbuch der regionalen Geologie, Bd. vii, Abt. 1.

—— 1912. "Geology of New Zealand," Wellington.

Marshall, P., and Uttley, G. H., 1913. "Some Localities for Fossils at Oamaru," Trans. N.Z. Inst., vol. 45, pp. 297–307.

McKay, A., 1884. "On the North-eastern District of Otago," Rep. Geol. Explor., 1883–84, pp. 45–66.

Morgan, P. G., 1911. "The Geology of the Greymouth Subdivision, North Westland," Bull. No. 13, N.Z. Geol. Surv.

Park, J., 1887. "On the Geology of the Western Part of Wellington Provincial District, and Part of Taranaki," Rep. Geol. Explor., 1886–87, pp. 24–73.

—— 1887. "On the Age of the Waireka Tufas, Quartz-grits, and Coal at Teaneraki and Ngapara, Oamaru," Rep. Geol. Explor., 1886–87, pp. 137–41.

—— 1905. "On the Marine Tertiaries of Otago and Canterbury, with Special Reference to the Relations existing between the Pareora and Oamaru Series," Trans. N.Z. Inst., vol. 37, pp. 489–551.

—— 1910. "The Geology of New Zealand," Christchurch.

Thomson, J. A., 1913. "The Tertiary Beds of the Lower Awatere Valley," 7th Ann. Rep. N.Z. Geol. Surv., Mines Statement, 1913, p. 123.

—— 1914. "Classification and Correlation of the Tertiary Rocks," 8th Ann. Rep. N.Z. Geol. Surv., Mines Statement, 1914, pp. 123–24.

—— 1914. "Coal Prospects of the Waimate District, South Canterbury," *ibid.*, pp. 158–62.

ART. IV.—*Additions to the Knowledge of the Recent and Tertiary* Brachiopoda *of New Zealand and Australia.*

By J. ALLAN THOMSON, M.A., D.Sc., F.G.S., Director of the Dominion Museum, Wellington.

[*Read before the Wellington Philosophical Society, 27th October, 1915.*]

Plate I.

CONTENTS.

(1.) A NEW SPECIES OF CRANIA FROM NEW ZEALAND WATERS.

Crania huttoni sp. nov. Plate I, figs. 1, 2.

1873. *Crania* sp. ind. Hutton, Cat. Mar. Moll. N.Z., p. 87.
1906. *Crania* sp. Hamilton, Bull. Col. Mus., No. 1, p. 41.

" Dorsal valve rugose, with a few radiating lines in places; ventral valve smooth. Light brown. Diameter about 0·5 (inch)."—Hutton.

There is in the Dominion Museum a tablet holding three valves from an unknown locality which purport to be the specimens upon which Hutton based the above brief description. All three valves, however, appear to be dorsal valves, and the apparent absence of radiating lines on two of them is due in one case to attrition on the sea-bottom and in the other to an encrusting organism. There is also in the Museum another tablet of seven valves- which are those mentioned by Hamilton as "Trail; Whangaroa, Cook Strait." Of these seven, the two smallest are the young of *Anomia* sp., but the other five are dorsal valves of the same species of *Crania* as Hutton's specimens. In view of the locality record attaching to the latter specimens, the best preserved of these is chosen as the holotype.

In outline the shell tends to be nearly square, with rounded angles, but there is considerable irregularity among the eight specimens, and in two the posterior margin is slightly embayed in the middle. In the nearly square shells the commissures are practically in one plane, but in the more irregularly shaped specimens they are sinuous. The dorsal valve is conical or limpet-shaped, with the vertex generally in the middle line of the shell, although in the holotype it is slightly to one side; it is directed obliquely backward, and is situated at from one-third to one-fifth of the length of

the shell from the posterior margin. The surface of the valve is ornamented by fine radial ribs, which are obsolete, or may never have been
developed, near the apex, and are crossed by irregularly developed growth-
lines. The shell is thick and solid in the older specimens, and the fine
punctuation can be seen in the interior by means of a lens. The colour
varies from light brown to almost colourless.

In the interior of the dorsal valve there is a narrow margin encircling
the valve, separated from the rest of the interior by a rounded shoulder.
and not granulated. The muscular impressions are white and porcellanous,
and afford an easy means of distinguishing the species from the molluscan
limpets. The posterior adductor-scars are large, nearly round, and situated
near the posterior angles of the shell. The unpaired (posterior) muscle has
left practically no impression in any of the specimens. The anterior
adductor-scars are smaller than those of the posterior adductors, are
crescentic in shape with the convex sides directed posteriorly inwards, and
approach one another on their inner ends. On their concave sides there
are well-marked depressions in the floor of the valve. The impressions
of the dorsal protractors lie close to the outer and hinder sides of the
posterior adductor-scars, and are small and oval in shape. The impressions
of the retractors of the arms lie close to the outer ends of the anterior
adductor-scars, and are small and rounded in shape. The protractors of
the arms have left a single minute impression situated mesially in front of
the anterior adductors, and from it a well-marked groove extends nearly
to the front margin. This groove appears to be one of the pallial-sinus
impressions, and on each side of it lie four similar grooves of irregular
length.

The dimensions of the holotype are : Length, 12 mm. ; breadth, 11 mm. ;
height of dorsal valve, 4·5 mm.

C. huttoni differs from most living species of *Crania* by the absence of
a granulated rim in the interior of the dorsal valve, and in this respect, as
well as in shape, agrees with *Crania japonica* Adams. From this species,
however, it is easily distinguished by the radiating ornament as well as
by internal characters. The only other Recent species with radiating
ornament is the Australian *Crania suessi* Reeve, which is described as sub-
orbicular, and therefore differs in shape. The genus is not yet known.
fossil in the Tertiary of the Southern Hemisphere.

(2.) On the Generic Position of the Tertiary Terebratulids of the Southern Hemisphere.

The correct generic assignment of the New Zealand, Australian, and
Western Antarctic Tertiary fossils formerly known as species of *Terebratula*
is a matter of no little difficulty in view of the close restriction of that
genus by Buckman (1907). *Terebratula terebratula*, the genotype, is a biplicate Pliocene shell from Italy, and only those species in actual genetic
connection with it may be admitted to the genus. The biplicate Cretaceous and Jurassic species may be excluded easily enough, since the
great difference in time is alone sufficient to prove that these species have
attained biplication from different ancestral forms ; but a difficulty arises
in connection with the uniplicate and biplicate older and middle Tertiary
species, for there is no *a priori* reason why these should not be in actual
genetic connection with *Terebratula terebratula*. Some of them, however,
must belong to *Liothyrina*, which is believed to have existed since the

Cretaceous, and there may be a number of other stocks equally worthy of generic rank. Owing to the simplicity in external form of Terebratulids, and the narrow limits of possible variation, the chances of homoeomorphy are very considerable, and, unless genera are closely restricted and homoeomorphy thus excluded, zoological comparisons between the fossils of different countries cannot carry much weight for purposes of correlation.

Since the geological record is too imperfect and the differences in external form are too slight to enable the species to be grouped in linear series, thus allowing their phylogenetic relationships to be traced, it is necessary, if the genera are to be closely restricted, to find some anatomical characters common to groups of related species by which they may be distinguished from other groups of species. The characters chosen may be such as have hitherto been considered unimportant, even in specific differentia, provided that they are persistent within the group. The chances are that a group thus defined will prove to be a good genetic series.

Buckman (1910) has taken the first step, so far as fossils are concerned, towards the discovery of such characters in his treatment of the Tertiary Brachiopods from the islands of the Weddell Sea collected by the Swedish Antarctic Expedition. His words are worth quoting in full :—

" Among the material brought from the Antarctic are several specimens which belong to various species of *Terebratula*, using that term in its wide sense ; but it is probable that none of them really belongs to the typical series which would be grouped around the genotype *Terebratula terebratula* Linné sp. There are two series which differ conspicuously in the character of their test. The first series shows coarse and distant punctae associated with a rather thick test. . . . The second series shows a finely and closely punctate test . . . which is also thin, as if it were a deep-water series. Further, in the older specimens particularly, there is an outer layer of test which is undoubtedly grooved—the grooves waved and irregular . . . very suggestive of the ornament seen in certain species of Lower Jurassic *Lima* (*Plagiostoma*). This finely punctate series is not punctate so finely and minutely as *Terebratula variabilis* of the English Tertiary, which is presumably a true *Terebratula :* the punctae of the Antarctic species are larger and therefore seem more approximate. There is evidently much yet to be learnt concerning these differences of punctuation. . . . These [finely punctate] species with their thin test have much the appearance of species of *Liothyrina ;* but it has not seemed desirable to place them in that genus, of which the characters and limits are none too fully known. Of these species the loop cannot be seen, and they show no indication of the four internal radiating furrows which serve for the attachment of the pallial sinuses ; and these furrows so marked in the type species *Liothyrina vitrea* would serve in fossil species as an outward index where interior details could not be seen : thus they show well through the test of the Chalk species assigned to *Liothyrina* — namely, *T. carnea, T. subrotunda, T. semiglobosa,* &c."

There are at least two Australasian Tertiary species which agree closely with those of Buckman's finely punctate species—namely, *Terebralula tateana* Tenison-Woods* and *Terebratula concentrica* (Hutton). Both these shells have a thin and finely punctate test, and *T. tateana* shows in addition a

---

* It may be remarked that Buckman identified one of his species as " *Terebratula vitreoides* Tate (1880) ? of Woods," a name which Tate in 1899 corrected to *Terebratula tateana* T. Woods.

fine but irregular radial ribbing similar to that described by Buckman. Another character which they have in common is the possession of a short, low, thin, mesial septum in the dorsal valve, perhaps hardly strong enough to be worthy of the name of septum, but sufficiently marked to leave a distinct groove on well-preserved internal casts. This character is additional to that mentioned by Buckman as separating this series from typical *Liothyrinae*. Apparently, also, it serves to distinguish it from *Terebratula*, for, although we are in ignorance of the internal characters of *Terebratula terebratula*, the English Crag species, *T. spondylodes*, *T. variabilis*, &c., considered by Buckman (1908) as probably true *Terebratulae*, possess very strong and widely separated adductor muscular impressions, between and behind which is a broad, raised, nearly flat platform, apparently corresponding to the thin septum mentioned above. (Plate I, fig. 5.)

There is, moreover, a series of Recent Terebratulids commonly ascribed to *Liothyrina* which possess all the above-mentioned peculiarities. Clearly, then, we are in a position to recognize a distinct genetic series.

Liothyrella gen. nov.   Genotype, *Terebratula uva* Broderip.

Type of folding from non-plicate, through dorsally uniplicate, to dorsally biplicate. Loop short, terebratuloid. Muscular impressions of the dorsal valve separated by a thin, low, short, mesial septum. Test thin and hyaline, but becoming thick and opaque in old age, finely punctate, with grooves for the reception of the pallial sinus, which are, however, only occasionally visible through the shell in fossil examples. Surface ornamented with a fine radial ribbing, very irregularly distributed, and absent on some individuals.

The following species may safely be included in the genus : *Liothyrina uva* var. *notocardensis* Jackson, *Terebratula tateana* Tenison-Woods, *Waldheimia concentrica* Hutton, and a new species dredged off Tasmania by the Mawson Expedition. Probably also many of the other southern species ascribed to *Liothyrina* will be included here, but the descriptions do not state whether or not a mesial septum is present.

Liothyrella uva has been described in great detail by Blochmann (1908 and 1912) under the name of *Liothyrina*. He divides the *Liothyrinae* into two groups, according to the presence or absence of certain spicules at the base of the cirrhi (Cirrensockeln). The group in which they are absent, which includes *Liothyrina vitrea*, occurs chiefly in the Atlantic Ocean and the Mediterranean Sea, with the exception of one species in Japan and one off Madagascar. The group in which they are present, which includes *Liothyrella uva*, is restricted to the Antarctic and South Temperate seas, with the exception of one species in the Mediterranean, while *L. uva* itself ranges from the Argentine coast, around the Horn, and up the western American coast as far as the Gulf of Tehuantepec in Mexico. The spicules of several species, however, are as yet unknown. It does not seem probable that the presence or absence of these basal spicules will, taken by itself, separate genetic groups, but it is quite probable that their presence will be found to be a constant character of Recent *Liothyrellae*. For fossils, however, this criterion cannot be used.

In the thickness of the shell, *Liothyrella* stands between *Liothyrina* and *Terebratula*, while all three genera are finely punctate. In view of the known occurrence of *Liothyrella* in the Recent seas both of South America and Tasmania, and in the Tertiaries of these countries and also of New Zealand, it will be preferable to include the finely punctate fossils from these countries of which the interiors are unknown in *Liothyrella* rather than in *Terebratula senso latu.* Buckman's coarsely punctate series, however, must be regarded as a distinct genetic stock.

(3.) THE GENERIC POSITION OF TEREBRATULINA DAVIDSONI *Etheridge* AND MAGASELLA EXARATA *Verco.*

Terebratulina davidsoni Etheridge, an Australian Tertiary fossil, differs from typical *Terebratulinae* in its type of folding. The dorsal valve is almost flat, but there is a faint medial sinus revealed by the course of the anterior commissure. In *Terebratulina caput-serpentis*, on the other hand, the dorsal valve shows a fold. It becomes, therefore, a matter of considerable interest to know the type of the loop of *Terebratulina davidsoni*. Tate (1880, p. 159) makes the following statement : " The founder of this species, being unacquainted with its internal portions, placed it with a doubt in the genus *Terebratulina*, but having seen the loop, which offers no special character, I can confidently refer it to that genus." This statement does not help us much, for at the time at which Tate wrote the genus *Terebratulina* was made to include such divergent loop forms as *Dyscolia wyvillei* and *Eucalathis murrayi*, while in 1886 Davidson also referred the shell subsequently known as *Chlidonophora incerta* to *Terebratulina* (?). It is quite possible, then, that the loop of *Terebratulina davidsoni* is not that of a typical *Terebratulina*. I have endeavoured to expose the loop of some specimens in the Dominion Museum from the River Murray, but without success, owing to the hardness of the matrix. My preparations, however, reveal several important characters. In the first place, the internal margin of both valves is strongly crenate, a character not found in typical *Terebratulina*. There is no median septum on either valve, so that the loop is probably short and simple as in *Terebratula* and *Terebratulina*. The socket-ridges and cardinal process, however, are much stronger than in typical *Terebratulina*, and recall those of primitive *Pachymagas* and of *Campages*, except for the absence of the bifurcating septum. The origin of the crura from small processes below the inner anterior ends of the socket-ridges is also distinct from the conditions in *Terebratulina*, where the crural bases are united to the inner sides of the socket-ridges at a slightly higher level.

All these peculiarities exist also in *Magasella exarata* Verco, a Recent shell from South Australia of which also the loop is imperfectly known. It cannot belong to any Magaselloid genus, since it does not possess a high septum, but only a low median ridge. We have, therefore, a new genus, which we may designate as follows :—

Murravia gen. nov. Genotype, *Terebratulina davidsoni* Etheridge. Plate I, fig. 4.

Shell with incipient ventral uniplication, the dorsal valve flattened. Delthyrium partially closed by two lateral deltidial plates, leaving a sub-mesothyrid foramen which is margined anteriorly by the dorsal valve. Surface of valves ornamented with radiating ribs, which are continued as crenulations on the inner margins of the valves. Dorsal valve without a median septum, and with strong socket-ridges and a stout pyramidal cardinal process. Loop unknown, but crura springing from swollen bases below the inner anterior ends of the socket-ridges.

Besides the genotype, *Murravia exarata* is the only other species which may be included in the genus with certainty. *Terebratulina ornata* Giebel from the Oligocene of Latdorf, Germany, and Hoesfelt, Belguim (*cf.* Davidson, 1874), agrees in possessing the internal crenulation, but has, according to Davidson's figure, a different type of cardinal·process. It also certainly cannot be a *Terebratulina*.

A genus nearly related in shape is *Disculina* Deslongchamps, 1884, based on *Terebratula hemisphaerica* Sowerby. This species is radially costate, and has a flat dorsal valve and a foramen similar to that of *Murravia*, but there are no internal crenulations in the dorsal valve.

(4.) On a New Form of Terebratella from New Zealand Waters.
Plate I, fig. 3.

A small collection of Brachiopods dredged off Cape Colville, Auckland, in 20 fathoms, by Mr. Anderton, of the Portobello Marine Fish-hatchery, and presented through Mr. G. M. Thomson to the Dominion Museum, consists of a form closely allied to *Terebratella sanguinea* (Leach), but presenting some interesting differences. As it is possibly only a local race of that species, and may be linked to it by a series of intermediate forms, I do not propose to burden the literature with a new name, that may prove synonymous, until collections from other northern localities have dispelled the doubt.

The collection consists of one adult shell and seven half-grown or still smaller specimens. In shape the adult, which is slightly unsymmetrical, agrees fairly well with moderately elongate examples of *T. sanguinea*, but the beak is suberect instead of erect, and in consequence the foramen is farther removed from the dorsal umbo and the pseudo-deltidium is higher. The beak-ridges also are less pronounced, and the sinus on the dorsal valve is broader, shallower, and commences more imperceptibly, so that the anterior commissure presents a regular ventrally directed curve instead of a nearly rectilinear angled trough, and the front margin is rounded instead of straight. The most striking difference, however, lies in the almost complete reduction of the radial multicostation. The radial ornament that still exists is irregularly distributed on all the specimens, and varies in its distribution in different individuals, in some being more marked on the sides, and in others along the fold and sinus. One shell shows an area corresponding to the middle period of growth in which the ribbing is entirely absent on the sides, although it was present at an earlier stage and appears again at a later stage. The ribbing in no case commences near the umbo, the earliest stage for about 5 mm. being completely smooth, but it is much better developed in shells of about 15 mm. than it is in the adult, which measures 31 mm. in length.

A somewhat analogous case in the same genus is furnished by *T. dorsata* var. *submutica* Fischer and Oehlert (1892), in which the shells start with a smooth stage, develop multicostation for a short period, and finally become smooth again before adolescence.

List of Papers referred to.

Blochmann, F. (1908). "Zur Systematik und geographischen Verbreitung der Brachiopoden." Zeitschr. f. wiss. Zool., Bd. 90, pp. 596–641.

—— (1912). "Die Brachiopoden der Schwedischen Südpolarexpedition." Wissensch. Ergebn. Schwed. Südpolar-Exped. 1901–3, Bd. 6, Lief. 7.

Buckman, S. S. (1907). "Brachiopod Nomenclature: the Genotype of *Terebratula*." Ann. Mag. Nat. Hist., ser. 7, vol. 9, pp. 525–31.

Plate I.

Buckman, S. S (1908).   " Brachiopod Nomenclature : the *Terebratulae* of the Crag."   Ann. Mag. Nat. Hist., ser. 8, vol. 1, pp. 444–47.

—— (1910).   " Antarctic Fossil *Brachiopoda* collected by the Swedish South Polar Expedition."   Wissensch. Ergebn. Schwed. Südpolar-Exped. 1901–3, Bd. 3, Lief. 7, pp. 23–28.

Dall, W. H. (1903).   " Contributions to the Tertiary Fauna of Florida, &c. : *Brachiopoda*."   Trans. Wagner Free Inst. Sci., vol. 3, pt. 6, pp. 1533–40 (for *Chlidonophora*, pp. 1538–39).

Davidson, T. (1874).   " On the Tertiary *Brachiopoda* of Belguim, &c."   Geol. Mag., dec. 2, vol. 1, pp. 150–59.

Deslongchamps, E. (1884).   " Notes sur les modifications à apporter à la classification des *Terebratulidae*."   Bull. Soc. Linn. Normandie, ser. 3, vol. 8, pp. 161–297 (for *Disculina*, p. 241).

Fischer, P., and Oehlert, D. P. (1891).   " Expédition Scientifiques du Travailleur et du Talisman, &c. : Brachiopodes " (*Dyscolia*, pp. 18–29 ; *Eucalathis*, pp. 40–51).

—— (1892).   " Mission Scientifique du Cap Horn (1882–83) : Brachiopodes." Bull. Soc. Hist. Nat. d'Autun, t. 5, pp. 254–334 (*Terebratella dorsata submutica*, pp. 280–81, pl. xi, figs. 1–6).

Tate, R. (1880).   " On the Australian Tertiary Palliobranchs."   Trans. Roy. Soc. S. Austral., vol. 3, pp. 140–69.

—— (1889).   " A Revision of the Older Tertiary *Mollusca* of Australia : Part I."   Trans. Roy. Soc. S. Austral., vol. 23, pp. 249–77 (*Terebratula tateana*, pp. 250–51).

———

EXPLANATION OF PLATE I.

Fig. 1. *Crania huttoni* Thomson, Whangaroa, Cook Strait.   Dorsal valve viewed dorsally.   Enlarged about 2 diameters.

Fig. 2. The same.   Dorsal valve viewed ventrally.

Fig. 3. *Terebratella sanguinea* var., off Cape Colville.   Enlarged 2 diameters.

Fig. 4. *Terebratella davidsoni* (Etheridge), River Murray, South Australia.   Enlarged 5 diameters.

Fig. 5. *Terebratula spondylodes* W. Smith, English Crag, exact locality unknown.   Dorsal valve viewed ventrally.   Natural size.   (From specimen in the Dominion Museum).

ART. V.—*The Flint-beds associated with the Amuri Limestone of Marlborough.*

By J. ALLAN THOMSON, M.A., D.Sc., F.G.S., Director of the Dominion Museum, Wellington.

[*Read before the Wellington Philosophical Society, 27th October, 1915.*]

Plates II, III.

CONTENTS.

I. THE AMURI LIMESTONE.

The Amuri limestone, which takes its name from Amuri Bluff, seventeen miles south of Kaikoura, is a great limestone formation composed of a variety of rocks, of which the most characteristic and the most abundant is a much-jointed, thin-bedded limestone of chalky appearance, but considerably indurated and much harder than chalk. Typical soft chalk is known from the formation only at Oxford, in Canterbury, near the southernmost part of its range. In North Canterbury the lower beds of the limestone are generally softer and more argillaceous, with a coarser and more conchoidal fracture, and pass down gradually in the Waipara district into a dark-coloured mudstone. In East Marlborough, where the formation is very much thicker, there are several coarse alternations of hard chalky limestone with more argillaceous bands, and in places there are fine alternations of hard white limestone with greenish argillaceous limestone, in all cases fine-grained. The base of the formation in this district is formed by the beds of flint which form the subject of this paper. By its fine grain and chalky appearance, together with the manner in which it breaks into small cuboidal blocks, the Amuri limestone is easily distinguished from all other limestones in New Zealand,* and it has evidently had a different mode of origin from the Oamaruian (Ototaran) limestones of Otago and South Canterbury, which are largely composed of fragments of *Bryozoa*. It contains a fair proportion of unbroken tests of *Foraminifera*, especially *Globigerina*, but the greater part of its mass is made up of an exceedingly fine-grained calcareous mud, with little terrigenous matter. Analyses, however, reveal a considerable amount of silica, which must also be present in a state of minute subdivision, while there are occasionally grains of glauconite.

Although the Amuri limestone is now distributed in an irregular and discontinuous manner throughout the east part of North Canterbury and Marlborough, there can be little doubt from its lithological characters that it once spread over the greater part of these districts, a conclusion previously

---

* *Cf.* Hutton, 1877, pp. 27–58.

reached by Hector (1890), McKay (1891), and Cotton (1913). In thickness, however, it is very variable, being less than 300 ft. in the Waipara district, and, together with the flint-beds, over 2,500 ft. in the Middle Clarence Valley. At Amuri Bluff, roughly half-way between these places, the thickness is estimated by Hector (1877) at 780 ft., and it probably increases gradually between this point and the Middle Clarence Valley. Near Keke-rangu the limestone appears to be divided into two or three parts by mud-stone intercalations, but it is possible that this appearance is deceptive, and is due to faulting or sharp folding, since a mudstone overlies the lime-stone. A somewhat similar phenomenon occurs in the hills north of the Lower Ure River. The thickness of the formation decreases again in the Cape Campbell district, but it is still considerable. Apparently the typical Amuri limestone reappears on the north side of Cook Strait in the Cape Palliser district, but it has not been personally investigated in that area by the writer.

The age of the Amuri limestone is approximately fixed by the fact that it always overlies beds containing Cretaceous fossils, and is always followed by beds containing a Tertiary (Oamaruian) fauna. In all localities where a continuous section can be traced the limestone follows the Cretaceous beds with complete conformity of bedding; and in localities south of Amuri Bluff there is also a lithological gradation between these two sets of rocks, proving that the apparent conformity is there a real one. The Amuri limestone is thus much older at its base, even where it is thinnest, than any beds in the Oamaru district; and the diastrophic correlation of the Amuri and Ototara limestones by Marshall, Speight, and Cotton (1911) is in conflict with the palaeontological evidence. The Cretaceo-tertiary problem in New Zealand owes much of its complexity to an earlier false correlation of these rocks, which, it may be noted, was never accepted by Hutton. Within the central part of New Zealand, where the Amuri lime-stone is developed, Cretaceous beds always underlie the limestone con-formably, while it is followed by physically conformable beds containing an Oamaruian fauna. Outside this area—at any rate, to the south of it—the sequence generally commences with Oamaruian coal-beds, and, in the few areas where Cretaceous rocks are developed, there is probably unconformity.

In the Waipara district the Amuri limestone is followed by a calcareous sandstone, the Weka Pass stone, and this in turn is followed by more or less calcareous mudstones, the "grey marls." In places the Weka Pass stone is glauconitic at the base, and in such places there is the peculiar and often-described junction which gives the appearance of an erosion of the lime-stone prior to the deposition of the glauconitic bed. Whether or not this is a correct explanation of this peculiar contact, it is certain that no such contact is present at places within the same district where the base of the Weka Pass stone is not glauconitic, but that there is a passage so gradual between the two rocks that one cannot say within a foot of rock where the one ends and the other begins. It is, therefore, evident that even if there has been some erosion of the Amuri limestone prior to the deposition of the Weka Pass stone, it is a purely local phenomenon, and not indicative of a non-sequence of any extent. The argument that an unconformity between the two limestones may well be present from the analogy of similar well-proved unconformities in other parts of the world entirely overlooks the facts that in these other places the apparent conformity of the two limestones in question is local, and that in New

Zealand the parallelism in dip and strike of the Amuri limestone and the succeeding Tertiary rocks is found throughout ' some hundreds of miles of outcrop.

In the Middle Clarence area the uppermost band of the Amuri limestone is argillaceous, and passes quite gradually into fossiliferous Oamaruian mudstones similar to the grey marls. The use of the term " grey marls " by Marshall, Speight, and Cotton (1911) for mudstones following the Ototara limestone, and, in particular, for the Wanganuian mudstones of the Wanganui River, has robbed this term of any geological significance, and made it practically synonymous with Tertiary mudstone of any age or position. Some of McKay's uses of the term in South Canterbury and elsewhere are also equally unfortunate; but as originally used by him within the area occupied by the Amuri limestone it has a perfectly definite significance—viz., for mudstones following the Amuri limestone—and if confined to this usage it may continue as a most useful geological term. It does not follow that by terming certain rocks the grey marls of the Clarence Valley one necessarily considers them the correlatives of the grey marls of the Waipara district, for the greater thickness of the Amuri lime-stone of the Clarence Valley may be due not only to an earlier beginning, but also to a later cessation of deposition than in the case of the lime-stone at Amuri Bluff and the Waipara district. All that the term implies is that they are mudstones following the Amuri limestone.

On the theory of conformity between the Weka Pass stone and the Amuri limestone, the age of the top beds of the latter can be fixed if the age of the former is known; but, although a considerable number of fossils are known from the Weka Pass stone and the grey marls, the range of these species within the Oamaruian is not yet well enough ascertained to allow it to be stated whether these beds are Middle or Lower Oamaruian. The Weka Pass stone certainly does not correlate with the Hutchinson Quarry beds, as supposed by Marshall, since the main band of the Mount Brown beds, containing *Rhizothyris rhizoida* and *Pachymagas parki*, occupies this horizon in the Waipara district. The probabilities are that it will be found that the Weka Pass stone correlates with the Waiarekan.

It has apparently been assumed by the opponents of a conformable Cretaceo-tertiary succession in any part of New Zealand that the whole of the Amuri limestone is Cretaceous in age. Actually, however, the only determinable fossils obtained from the limestone, excluding fish-teeth and crinoid stems, are of Tertiary aspect. McKay has recorded *Pecten zitteli* Hutt. and *Rhynconella squamosa* Hutt. from the limestone of Amuri Bluff. The former species I have also collected from fallen blocks of the limestone in the same locality, but the specimen identified as *Rhynconella squamosa* is, in my opinion, not specifically determinable. In addition, Mr. H. Suter has identified the following forms from the collections made by Mr. A. McKay and myself from Amuri Bluff :—

> *Pecten zelandiae* Gray.
> *Pecten delicatulus* Hutt. (?)
> *Pecten* sp. nov. cf. *P. chathamensis.*
> *Pecten* sp. nov.

Pecten zelandiae is a Recent species, and its occurrence strongly suggests that the rock from which it comes is of Tertiary age. Un-fortunately, the exact horizon within the limestone of any of the above forms is unknown.

In the green argillaceous limestone of the Ure River, a little above the junction of the Isolated Hill Creek, *Teredo heaphyi* occurs. Again the exact horizon within the limestone cannot be stated, owing to the complex faulting in the vicinity.

In the Coleridge Creek section of the Trelissick Basin, Mr. R. Speight and I recently discovered a band of fossiliferous tuff, about 25 ft. thick, and interbedded 10 ft. below the top of the Amuri limestone. This tuff yielded twenty-one species of *Mollusca*, determined by Mr. Suter as follows :—

> *Calliostoma aucklandicum* E. A. Smith.
> *Seila huttoni* Suter. .
> *Siphonum planatum* Suter.
> *Polinices huttoni* Ihering.
> *Polinices ovatus* (Hutton).
> *Ampullina miocoenica* Suter.
> *Ampullina suturalis* (Hutton).
> *Epitonium zelebori* (Dunker) var.
> *Epitonium rugulosum lyratum* (Zittel).
> *Fusinus bicarinatus* Suter.
> *Hemifusus goniodes* Suter.
> *Siphonalia turrita* Suter.
> *Cominella intermedia* Suter (?)
> *Admete trailli* (Hutton).
> *Ancilla papillata* (Tate).
> *Ancilla subgradata* (Tate).
> *Marginella harrisi* Cossman.
> *Surcula seminuda* Suter.
> *Terebra costata* Hutton.
> *Limopsis catenata* Suter.
> *Chione chiloensis truncata* Suter.

Of the above, two species are new, four (or 19 per cent.) are Recent, while the others are all well-known Tertiary, and mostly Oamaruian, species.

It seems clear from these facts that the top at least of the Amuri limestone is of Tertiary age, and that the Amuri limestone is in itself a Cretaceo-tertiary rock, Cretaceous at the base and Tertiary at the top. In view of this, the conformity or unconformity of the Weka Pass stone sinks to a question of purely local Tertiary geology without significance for the relationship of the Cretaceous and Tertiary in New Zealand.

The Cretaceous fossils of the rocks underlying the Amuri limestone have recently been studied by Mr. H. Woods, of Cambridge, and, as stated by Morgan (1915), he has come to the conclusion that two distinct faunas occur. "The older of these, found at Coverham, is considered to correspond to the Lower Utatúr (approximately Upper Greensand and Gault) fauna. The younger, of approximately Senonian age, occurs at Amuri Bluff and other points to the south."

It would appear at first sight from this statement that the Amuri limestone of the Clarence Valley (Coverham) must be separated from the underlying Cenomanian beds by an unconformity. All the natural sections at Coverham and in the neighbourhood, however, show a perfectly conformable junction between the flint-beds at the base of the limestone and the underlying Cenomanian mudstones. One would expect, in any case, that the base of the limestone in this locality, where it is over 2,500 ft. thick (including the flint-beds), would be lower in horizon than the base of the limestone at Amuri Bluff, where it is less than 800 ft. thick, in view of the lithological similarity of the limestones in the two localities. It appears that depression and deposition commenced in the Middle Clarence area at an earlier date than farther south, and continued in this area while it gradually spread farther south (and probably north). A confirmation of this view will ensue if the Amuri limestone in the Puhipuhi Mountains, between Kaikoura and Clarence Mouth, is found to be intermediate in thickness between that at Amuri Bluff and that in the Middle Clarence area, and if fossils from the underlying Cretaceous beds, none of which have been available to send to Mr. Woods, prove to be intermediate in age between Senonian and Cenomanian—*i.e.,* Turonian.

II. Distribution of the Flints.

Small rounded flints similar in manner of occurrence to those of the English Chalk occur fairly abundantly embedded in the limestone at Amuri Bluff.* They are generally of a light-brown or flesh colour, but are sometimes opaque white and hardly distinguishable from the limestone in which they occur. Unlike the English flints, they are never hollow, thus destroying any hopes of discovering sponge-spicules or other fossils in their interiors; nor does the surrounding limestone exhibit the hollow casts of sponge-spicules which are often so well displayed in the Chalk. In the Middle Clarence and the Ure River area similar flints are found in the upper part of the limestone, but in the lower part an entirely different mode of occurrence prevails. The base of the limestone is entirely replaced by beds which are composed of large lenticules of black flint in the centre, with a variable amount of grey external matter, very often composed of euhedral crystals of a carbonate set in a flint matrix.

These flint-beds were first described by McKay (1877) from the Hapuka River, a little to the north of Kaikoura, where he found " 20 ft. of a peculiar rock consisting of flinty nodules in a matrix of dark shale " between sandy micaceous (Cretaceous) clays and the base of the limestone. This appears to be the southernmost point of their range. At Waipapa boat-harbour, farther to the north, they are 50 ft. or more in thickness, and are described by McKay (1886) as " formed by black flint in layers averaging about 6 in. in thickness, which are parted by a lesser thickness of white decomposed flint, or fine-grained sandstone." They have not been studied in detail between Clarence Mouth and Kekerangu, but are described by McKay (1886) as of " very considerable thickness " in the Benmore Stream, and here he notices the light-coloured exterior to the flints in their upper part. From this point the Amuri limestone swings round in a great curve through Benmore, the Isolated Hill, and Brian Boru, leaving its former course approximately parallel to the coast to enter the Middle Clarence Valley (*cf.* Cotton, 1913; Thomson, 1915). In the Isolated Hill Creek, a tributary of the Ure River which separates Benmore from the Isolated Hill, a very good section of the flint-beds is displayed. They are about 400 ft. thick, and are perfectly conformable to the underlying mudstones, into which they pass almost imperceptibly within a thickness of 2 ft. of rock. Forty feet below the main mass of flints there is a thin intercalation of flint and flint-carbonate rock within the Cretaceous mudstones. At Coverham, in the Middle Clarence Valley, a few miles to the south-west,

---

* *Cf.* Hutton (1877), Hector (1887), Liversidge (1877). I have not seen any flint in the Amuri limestone of the Waipara district.

[*C. A. Cotton, photo.*

Section across flint-beds, Mead Gorge, Clarence Valley, showing the dark lenticules of black flint surrounded by a grey or white exterior.

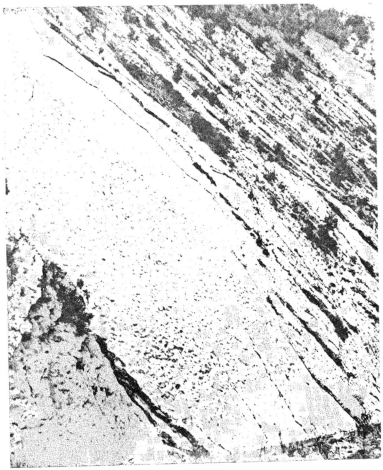

[*C. A. Cotton, photo.*

View of dip slope of flint-beds, Mead Gorge, Clarence Valley, showing the
irregular surface of the bedding-planes.

the flint-beds are estimated by McKay (1886) as over 500 ft. in thickness.
In Sawpit Gully, a tributary of the Nidd Stream, at Coverham, the junction
between the flint-beds and the underlying sandy mudstones is sharp, and
not gradual. The mudstones contain numerous pyritic concretions in their
topmost 20 ft., and at the junction itself there is a strong yellow efflorescence.
The lowest flint lenticules are light-coloured on the exterior, but do not
show the carbonate rhombohedra that are so characteristic of the lenticular
flint-beds in general. In the valley of the Nidd Stream, above the junction
of Sawpit Gully there is a thick band of dark, but not black, flint or chert
within the Cretaceous mudstones, over 1,000 ft. below the base of the main
flint-beds. This band of flint is not lenticular in character, and does not
exhibit the flint-carbonate rock associated with the flints at the base of the
limestone series.

Between Coverham and the Dee River, still farther to the south-west,
the flint-beds at the base of the limestone, as well as the limestone itself,
attain their maximum thickness. In the Mead Gorge there are massive
beds of flint at the base of the limestone series, and then alternations of
limestone and flint-beds for some distance higher up. By measuring the
dip and pacing the width of the outcrops I made the following rough
estimate of the thickness at this point, neglecting some small faults which
reduce the apparent thickness :—

					Ft.
Top.	Grey marls (fossiliferous Oamaruian mudstone).				
	Hard argillaceous limestone (Weka Pass stone of McKay)				150
	Marly limestone				400
	Hard chalky limestone				280
	Marly limestone				420
	Hard chalky limestone				90
	Flint-beds with limestone intercalations				1,410

2,750

To the south-west of the Dee River the flint-beds again decrease in thickness, and, according to McKay (1886), they disappear altogether before
the limestone reaches the Dart River. In the Upper Awatere Valley the
Amuri limestone is much thinner than in the Clarence Valley, and the
flint-beds are mentioned by McKay (1890) as occurring at its base. Finally,
a considerable development of these beds with the Amuri limestone is
reported by McKay (1877) near the mouth of the Flaxbourne River, in the
Cape Campbell district.

III. CHARACTER OF THE FLINTS.

What have been termed " flint-beds " in the above account are layers
composed either wholly of black flint or of large lenticules of flint surrounded
by a semi-crystalline material. The layers are distinct from one another
above and below, and are in close contact along surfaces resembling bedding-
planes. Where the layers are composed of flint they are generally about
6 in. to 8 in. in thickness, and the surfaces of contact are approximately
parallel. Where the layers consist of dark flint only in certain lenticules,
the bounding surfaces are more irregular, as the layers exhibit numerous
swellings following in a reduced degree the shape of the lenticules, and
reaching in some cases a thickness of 18 in. or more. Plates II and III
will give a better idea of the phenomena than any lengthy description. It
seems reasonable to suppose that the surfaces of separation of the layers

are bedding-planes, possibly in some cases distorted by the chemical re-arrangement that has taken place.

The flint-beds are in most outcrops much shattered, and it is difficult to collect hand specimens free from flaws, although more compact specimens may be obtained from the gravels of the streams cutting through the series. In general the boundary between the light-coloured exterior and the inner dark flint is sharp, but there is no surface of easy separation between the two, and flints of the Chalk type preserving their original surface cannot be obtained either naturally or artificially. The amount of the light-coloured exterior varies considerably from place to place, and there is also considerable variation in its composition. It may consist of dark flint containing more and more rhombohedra of carbonate as the exterior is approached, until finally a rock is obtained which consists only of carbonate, and resembles marble. Such a gradation from dark flint has been observed in boulders obtained from the Ure River, and in this case there is no sharp boundary between the inner and outer layers. In the usual case there is a sudden change from a flint practically without carbonate crystals to one containing them abundantly, the outermost layer again consisting almost entirely of carbonate crystals without any flint matrix. Occasionally thin lenticules of flint occur within beds of limestone, and there may or may not be an intermediate rock with crystals of carbonate set in either a flint

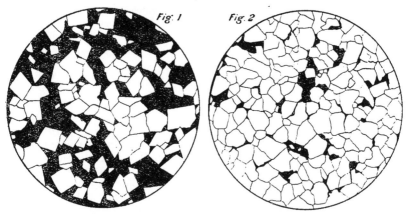

Fig. 1.—Flint with relatively few euhedral crystals of dolomite, Ure River gravels. Magnified 15 diameters.
Fig. 2.—Dolomite rock with a little flint in the interstices of the crystals, Mead Gorge. Magnified 15 diameters.

or a partly calcareous matrix. Text fig. 1 shows the appearance in micro-scopic section of a flint with relatively few crystals of the carbonate, which are euhedral towards the flint, but anhedral towards one another. Text fig. 2 shows the carbonate rock with only a little flint in the interstices of the crystals. Where larger crystals of carbonate occur in a fine-grained calcareous matrix they are less regularly euhedral than when they occur in a flint matrix.

The size of the carbonate crystals is approximately constant in any one specimen, but varies considerably in different specimens, a common size being a little over $\frac{1}{2}$ mm. in the greatest diameter. One exceptional speci-men, obtained from the gravels of the Isolated Hill Creek, a tributary of the Ure River, contains rhombs with a largest diameter of as much as 11 mm.

There is reason to suspect that this specimen came from the band of flint-carbonate rock intercalated near the top of the mudstones in this section. for this band exhibits much larger rhombs than usual.

A qualitative chemical test, carried out by Mr. B. C. Aston, showed that the carbonate of the coarsely crystalline specimen from the Isolated Hill Creek was strongly magnesian. A specimen from the Ure River gravels of the normal type of flint-carbonate rock, containing little flint matrix, was then submitted to the Dominion Analyst, Dr. Maclaurin, for analysis, with the following result :—

SiO_2	..	..	26·18	
$Al_2O_3 + Fe_2O_3$	..	..	3·12	$CaCO_3$ .. 41·60
MgO	..	..	13·57	$MgCO_3$ .. 28·50
CaO	..	..	23·30	
$H_2O +$	..	..	0·31	or
$H_2O -$	..	..	0·15	$CaMg(CO_3)_2$ 62·44
CO_2	..	..	33·23	$CaCO_3$ .. 7·66
Undetermined	..	..	0·14	

100·00

This analysis reveals two interesting facts. In the first place, it shows that the carbonate crystals have approximately the composition of dolomite. It seems most probable that they are pure dolomite, in view of the known power of crystallization of that mineral, and that the excess of calcium carbonate over the dolomite molecule is contained in the matrix. In the second place, it shows that the silica of the flint is little hydrated, and is not opal. In microscopic section the flint matrix is nearly dark between crossed nicols, but often shows numerous very small laths with moderate birefringence, straight extinction, and positive elongation, probably one of the orthorhombic forms of silica. They are altogether too minute for study in convergent light with the apparatus available.

The flint-beds at the base of the flint series in Sawpit Gully, Coverham, differ from those higher up in the series in that they contain a large amount of clastic quartz and other minerals in a matrix similar to that of the more normal flint-beds. The quartz is the most plentiful of the included minerals, and is in small angular grains. Muscovite in thin flakes appears in considerable amount, while there are scattered grains of glauconite and numerous minute crystals of rutile and pyrite with some iron-hydroxide staining. There are small rectangular areas of finely crystalline silica, often containing sericite, which possibly represent a replacement of clastic grains of feldspar. The light-coloured exteriors of these flints have a similar composition, except that there is less of the flint matrix and a greater proportion of subparallel plates of muscovite. There seems little doubt that in this case the flint-beds are replacing an impure fine-grained sandstone containing a little glauconite.

The flinty band intercalated in the Cenomanian mudstones of the Nidd Stream, above Sawpit Gully, is similar in composition and structure to the last described, with the difference that the minerals enclosed in the flint matrix are about half as large, and include very little glauconite.

IV. ORIGIN OF THE FLINT-BEDS.

Chert-beds not occurring in association with limestones or dolomites are usually the result of the molecular rearrangement *in situ* of silica derived from skeletons of such organisms as *Radiolaria*, sponges, or diatoms. The absence of such skeletons in any of the numerous microscopic sections examined removes any ground for accepting such an explanation for the origin of the silica in the present case.

Chert-beds associated with dolomites are of frequent occurrence in the pre-Cambrian and Palaeozoic, and in these cases it is generally held that the silica has replaced limestone or dolomite, whether its origin lies within the mass of the carbonate rock or not. The Chalk flints are believed to result from the replacement of the chalk by silica, the latter being dissolved from sponge-spicules throughout the mass of the chalk and collected into concretionary bodies by some process not well understood. The absence of the casts of sponge-spicules in the Amuri limestone militates against such an explanation for the origin of the silica in the present case, although it is not a fatal objection, since the limestone has practically everywhere been subjected to such pressure as might be competent to close up any spaces of dissolution so left. The unbroken state of the tests of *Foraminifera*, however, suggests that little internal movement of the limestone has taken place. The fact that the flint-beds are thickest where the whole series is thickest is in accord with the derivation of the silica of the flints from the overlying limestone series, but again the total absence of the flint-beds at Amuri Bluff and farther south is, on this assumption, difficult of explanation. It seems most probable, therefore, that the silica was chemically deposited along with the dolomite or that it is of extraneous origin. If the flint-beds were of strictly local occurrence, an extraneous origin would be not improbable, though it would be difficult to explain the silicification of only the lowest beds of the limestone series; but in view of the widespread occurrence of the flint-beds, apparently at a definite horizon, the theory of original deposition seems most acceptable. The chemical deposition of silica along with dolomite is believed to have taken place in other parts of the world. Writing of the magnesian limestone of Durham, England, Trechmann (1914, p. 258) states, "At certain periods, it seems, a deposit of siliceous material took place. Silica, in the form of compact or friable nodules of chert, occurs in several beds, but chiefly in the middle bedded rocks on the eastern side of the reef. In many cases the nodules merge with every gradation into the surrounding dolomite."

The conditions that permit the precipitation of silica are intimately related to those that permit also the precipitation of dolomite, calcium carbonate, and sulphides (Daly, 1907). They are found best displayed at the present day in the Black Sea (Andrussow, 1897), and are believed to have been operative also in the closed sea of Permian times in western Europe, in which were deposited the Kupferschiefer and the Zechstein dolomite (Schuchert, 1915). The fundamental cause of these conditions is the absence of bottom-scavengers, the lack of which allows the sea-bottom to become foul with accumulated remains of the swimming animals of the upper levels of the sea. The absence of bottom-dwelling life is caused by the lack of oxygen, a condition which can exist only in a more or less enclosed or sheltered sea not accessible to the creep of heavier and colder polar waters which oxygenate the bottoms of the open oceans. In the foul bottoms sulphur bacteria thrive, and decompose the organic matter, precipitating sulphides and liberating ammonium carbonate. The latter acts on the calcium and magnesian salts in solution in the sea, precipitating calcium and magnesium carbonates, and probably also on dissolved colloidal silica, which is precipitated as flint or chert.

There is no direct evidence that such conditions caused the formation of the flint-dolomite rocks at the base of the Amuri limestone, but it is difficult to formulate any other hypothesis for their origin, while there are other groups of rocks the formation of which can also most easily be explained by the adoption of such an hypothesis. The Amuri limestone itself appears to be in large part a chemical deposit, and its silica-content and its poverty in fossils becomes then easily explicable. The Senonian sulphur sands and mudstones of Amuri Bluff and the Waipara district are certainly the kind of deposits to be expected not far from the margins of a sea with a foul bottom.

It is quite possible that the sea in which the Amuri limestone and the underlying rocks were deposited was more or less enclosed, for although Marshall, Speight, and Cotton (1911) considered that after the close of the post-Jurassic folding a shore-line ran in a north-east direction from New Zealand to New Caledonia, recent work by Cotton and by the officers of the Geological Survey on the West Coast has abundantly demonstrated that the present trend of the mountains of the north part of the South Island is due not to original folding in the directions of their present trends, mainly north-east and south-west, but to post-Oamaruian block-faulting. Older rocks similar to those forming these ranges outcrop in the Chatham Islands, and the presence of Oamaruian sediments in these islands suggests that there was a land surface close at hand during Oamaruian times to furnish material for the sediments. Upper Cretaceous rocks have not been recorded from the Chatham Islands, and, should their absence under the Oamaruian rocks be proved, it will establish a considerable degree of probability that there was a land surface there during the Upper Cretaceous. Should this be established, east and west coasts to the Amuri sea will be known to exist. The sediments that mark the northern and southern coasts, if any, are unfortunately below the waters of the ocean. It is not necessary, however, to have an entirely closed sea for the existence of sulphur bacteria in quantity, for these are known in foul bottoms in sheltered fiords in Norway and in the Bay of Kiel (Schuchert, 1915). It is, indeed, quite possible that foul bottoms exist in deeps of the open ocean that are cut off from the currents of oxygenating cold polar waters by submarine ridges, and on such bottoms, if they exist, rocks resembling the Amuri limestone must be in course of formation.

According to the hypothesis here put forward, the sequence of events was somewhat as follows : The Amuri sea was at first restricted in extent so far as the present New Zealand area is concerned, and on its margins it received terrigenous deposits (conglomerates, sandstones, and mudstones) in Cenomanian times. One point of the margin at this epoch is fixed by the coal-beds and fresh-water fossils of Quail Flat, at the upper end of the Middle Clarence Valley. As depression extended the margins, the area of original deposition passed beyond the range of the terrigenous deposits, and the precipitation of flint and dolomite, together with the shower of tests of *Foraminifera*, built up the lower layers of the Amuri limestone in the area of which Coverham is the centre. During this period—possibly the Turonian—the terrigenous sediments presumably continued on the sea-margins, and should be found, as indicated above, in the Puhipuhi Mountains. With further depression the area covered by the limestone increased, but for some reason the deposition of flint, and probably also of dolomite, ceased. During the Senonian the area south of Kaikoura Peninsula apparently came under similar conditions, during which the peculiar sulphur sands and mudstones were deposited. Still further depression removed this area also beyond the range of terrigenous deposits, and the limestone between Kaikoura and Oxford was formed between the Senonian and some stage of the Oamaruian, together with the upper part of the limestone of the northern area. The marginal terrigenous deposits of this period are unknown unless they are represented in the Malvern Hills beds, though the presence of *Conchothyra parasitica* in these renders this unlikely. The area occupied by the Amuri sea now appears to have been elevated, or at least to have remained stationary, allowing it to shoal with sediments, while at the same time its margins became very widely extended north and south, forming the eastern Oamaruian sea, which was probably only a part of the Pacific Ocean. It will be noticed that the above account postulates that the Amuri sea was not entirely closed, for an extension of its margins with depression is dependent on a connection with the open ocean.

Much of the above is speculative, but will serve a useful purpose if it calls attention to the peculiar nature of the Amuri limestone, the flint-beds, and the sulphur sands, and provokes an alternative explanation. During most of the field-work on which this paper is based I was accompanied by Dr. C. A. Cotton, to whose kindly criticism I desire to express my indebtedness.

List of Papers referred to.

Andrussow, N. 1897. "La Mer Noire," Guides des Excursions, VII⁰ Cong. Géol. Internat., St. Pétersbourg, 1897, art. xxix.

Cotton, C. A. 1913. "The Physiography of the Middle Clarence Valley, New Zealand." Geogr. Journal, vol. 42, pp. 225–45.

Daly, R. A. 1907. "The Limeless Ocean of Pre-Cambrian Times." Am. Jour. Sci., vol. 23, pp. 93–115.

Hector, Sir J. 1877. "Marlborough and Amuri Districts." Rep. Geol. Explor. during 1873–74, pp. ix–xiii.

—— 1890. "Marlborough District." Rep. Geol. Explor. during 1888–89, No. 20, pp. xxxvi–xxxviii.

Hutton, F. W. 1877. "Report on the Geology of the North-east Portion of the South Island, from Cook Strait to the Rakaia." Rep. Geol. Explor. during 1873–74, pp. 27–58.

Liversidge, A. 1877. "Notes on some of the New Zealand Minerals belonging to the Otago Museum." Trans. N.Z. Inst., vol. 10, pp. 490–505.

Marshall, P. ; Speight, R. ; and Cotton, C. A. 1911. "The Younger Rock-series of New Zealand." Trans. N.Z. Inst., vol. 43, pp. 378–407.

McKay, A. 1877. "Report on Kaikoura Peninsula and Amuri Bluff." Rep. Geol. Explor. during 1874–76, pp. 172–184.

—— 1886. "On the Geology of the Eastern Part of Marlborough Provincial District." Rep. Geol. Explor. during 1885, No. 17, pp. 27–136.

—— 1887. "Report on Cape Campbell District." *Ibid.*, pp. 185–91.

—— 1890. "On the Geology of Marlborough and the Amuri District of Nelson." Rep. Geol. Explor. during 1888–89, No. 20, pp. 85–185.

—— 1891. "On the Geology of Marlborough and South-east Nelson: Part II." Rep. Geol. Explor. during 1890–91, No. 21, pp. 1–28.

Schnohert, C. 1915. "The Conditions of Black Shale Deposition as illustrated by the Kupferschiefer and Lias of Germany." Proc. Am. Phil. Soc., vol. 44, pp. 260–69.

Thomson, J. A. 1915. "Oil-indications in the Benmore District, East Marlborough." 9th Ann. Rep. (n.s.) N.Z. Geol. Surv., Parl. Paper C.–2, pp. 100–1.

Trechmann, C. T. 1914. "On the Lithology and Composition of Durham Magnesian Limestone." Quart. Jour. Geol. Soc., vol. 70, pp. 232–65.

Art. VI.—*Block Mountains and a " Fossil " Denudation Plain in Northern Nelson.*

By C. A. Cotton, D.Sc., F.G.S., Victoria University College, Wellington.

[*Read before the Wellington Philosophical Society, 27th October, 1915.*]

Plates IV, V.

Contents.

Introduction.

For some years the writer has been interested in the geomorphogeny of northern Nelson, and in 1913, though at that time without personal knowledge of the district, he presented a brief note on the subject before the Geological Section of the Wellington Philosophical Society, questioning in some measure the interpretation of the relief given by Bell in the Parapara bulletin of the Geological Survey.* The views then expressed having been favourably received by geologists acquainted with the district, and Professor W. M. Davis having in the meantime advised him to take up the subject of block mountains in New Zealand, the writer paid two visits to northern Nelson in 1915, and is now able to give a detailed description of a small portion of the district, and to express some general opinions as to the remainder. In this paper a condensed and generalized description of northern Nelson is given, and that is followed by a more detailed description of some features of the Aorere Valley and of the Gouland Downs.

" Block " Features throughout New Zealand.

As a result of observations made in many parts of New Zealand at various times, and confirmed by special visits recently made to a number of critical localities, the writer has come to the conclusion that the present relief is very largely—almost entirely—due to recent differential movement of crust blocks, both large and small. Though this explanation of the relief is not to be found in any general work on the geology

---

* J. M. Bell, " The Geology of the Parapara Subdivision," N.Z. Geol. Surv., Bull. 3, 1907.

or geography of New Zealand, it is by no means altogether new. It was clearly in the mind of McKay as early as 1883,* and it is implied in the descriptions of various areas in Nelson and Westland made during the last decade by Henderson, Morgan, Webb, and others.

The structure of the younger rock formations, where these have been preserved, affords ample confirmation of this explanation of the relief. Most of the larger relief features are tectonic forms—of course modified by erosion to a greater or less extent—while the river-courses are very largely consequent, still following very closely the courses taken upon the tumbled and irregular surface produced by a late disorderly uplift.

In many cases the blocks or units of the disorderly tumbled crust are bounded by faults. This is, however, by no means universally true, and where either formerly horizontal strata—that is to say, horizontally deposited strata which are so young that they must be supposed · to have lain in their original horizontal attitude in the period immediately preceding that of uplift—or formerly horizontal, planed surfaces are present, these frequently exhibit evidence of considerable deformation by folding. In some parts of the country, indeed, the intense folding and mashing that the young strata have undergone point to strong compression accompanying—perhaps initiating—the movements. In other parts, however, such evidence is lacking, formerly horizontal strata and planed surfaces remaining flat, though frequently tilted, over considerable areas. In the present state of our knowledge it cannot be stated whether normal or reverse faults predominate. No attempt will be made in this paper to explain the cause of any earth-movements, but attention will be directed to their effects as seen in the present form of the land-surface.

New Zealand may be described as a concourse of earth-blocks of varying size and shape, in places compressed, the highest blocks lying in the north-east and south-west axis of the land-mass, so that the whole structure may be termed a geanticline; the blocks initially consisting of an undermass of generally complex structure much denuded and largely planed, and covered over most of the area by an overmass which had not been disturbed before the "blocking" movements took place; the whole since the movements considerably modified by both degradation and aggradation.

In the present paper, a sketch of but a small portion of the New Zealand area, the writer cannot hope to demonstrate the accuracy of the foregoing general description, but it is offered as a working hypothesis which gives much assistance in the interpretation of this particular district.

Northern Nelson.

In the north-western corner of the South Island two fault-angle depressions, the Aorere and Takaka Valleys (see fig. 1), open out broadly towards the north-east and north, separating three composite upland blocks which, owing to diminishing throw of the boundary faults and consequent dwindling of the fault-angle depressions towards the south-west, coalesce in that direction. The north-western, or Wakamarama, block presents a fault-scarp front, but little dissected, towards the Aorere Valley, while north-westward its back slope, much dissected by conse-

---

* A. McKay, " On the Origin of the Old Lake Basins of Central Otago," Geol. Surv. of N.Z., Rep. Geol. Expl. dur. 1883–84, pp. 76–81, 1884 ; see also " On the Geology of Marlborough and South-east Nelson," *ibid.*, 1890–91, pp. 1–28, 1892.

quent streams, descends towards the Tasman Sea. The mountain-ranges of the middle, or Haupiri, block—that between the Aorere and Takaka Valleys—appear to have been carved from a mass which had initially a rough anticlinal or domed form, its present surface descending towards the north-west, north-east, and east from heights of over 5,000 ft. at the south-western end. The block is perhaps composed throughout of a number of smaller or secondary blocks separated from one another by faults and flexures. This is certainly the case towards the west, where the Haupiri block coalesces with the Wakamarama block in a region of flexed and broken plateaux known as the Gouland Downs. These plateau surfaces are remants of a stripped, " fossil " denudation plain

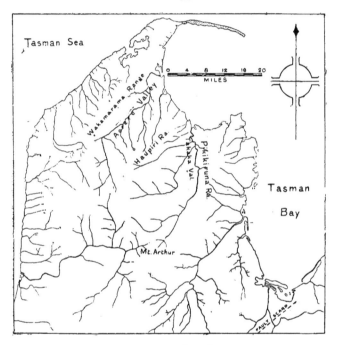

FIG. 1.—Locality map of northern Nelson.

which formed the floor underlying a series of marine strata since removed over large areas, but here and there preserved.

These marine covering strata are weak in comparison with the resistant undermass, and so the planed surface of the latter, which truncates indifferently indurated and metamorphosed clastic sediments and crystalline limestone of complex structure and also granitic intrusions, and forms the floor upon which the covering strata have lain, survives in favourable situations as a plateau long after the cover has been removed. It has, however, been destroyed, as might be expected, where its attitude favours deep dissection.

The Takaka Valley fault-angle depression is bounded on the east for twelve miles by an almost undissected fault-scarp nearly 3,000 ft. in height, with a north-and-south trend, which is the western edge of the block forming the Pikikiruna Range, and which may, therefore, be

appropriately named the Pikikiruna fault-scarp (see fig. 2). The Pikikiruna block, which is tilted towards the east, is much dissected on its
eastern or back slope, the recently drowned margin of which forms the
indented western shore of Tasman Bay.

FIG. 2.—View looking south along the Pikikiruna fault-scarp, which bounds the Takaka
Valley on the east.. In the centre is seen a hog-back of Tertiary limestone, which
is turned up along the fault.

Tasman Bay itself no doubt had its origin in the subsidence of an
earth-block, while the lowlands forming its southerly continuation have
the appearance of a fault-angle depression, now much modified by
erosion, bounded on the east by the scarp of the Richmond or Waimea
fault* and on the west by the eastern boundary, probably in great part
a dissected back slope, of the great block or complex of blocks constituting the highlands of Mount Arthur, the Mount Arthur tableland, and
the neighbouring ranges, which are more or less continuous with the
Haupiri and Pikikiruna blocks towards the north. The structure of
the covering strata as interpreted in a series of sections by McKay,†
however, indicates a considerable complication of the block movements
in this neighbourhood by folding.

THE AORERE-GOULAND DEPRESSION.

The Aorere River in the lower seventeen miles of its course, in
which it flows north-eastward, is guided by the fault-angle depression
previously referred to as the Aorere Valley, a name which it is convenient
to restrict arbitrarily to this obviously consequent portion of the whole
river-valley. At the head, or south-western end, of the Aorere Valley
the river enters it from the south, emerging from a deep, narrow, and
steep-sided valley between high mountains—a valley which, unlike the
other, appears to owe the whole of its depth and width to erosion, which
is perhaps consequent but possibly insequent, and which may be conveniently designated the " Upper Aorere Valley."

The Aorere Valley depression is bounded on the north-western side
by the fault-scarp front of the Wakamarama block, and on the south-
eastern side by the tilted surface of a portion of the Haupiri block.
To the north-east it is open to the sea, while at the south-western end,

---

* See Bell, Clarke, and Marshall, "The Geology of the Dun Mountain Subdivision,
Nelson," N.Z. Geol. Surv., Bull. No. 12, p. 12, 1911.
† A. McKay, "The Baton River and Wangapeka Districts and Mount Arthur Range,"
Geol. Surv. of N.Z., Reports of Geol. Expl. dur. 1878-79, p. 122, 1879.

just beyond the point at which the Upper Aorere Valley opens to the south, the Aorere Valley is terminated by a maturely dissected scarp, certainly of tectonic origin and probably a fault-scarp, the streams dissecting which supply the Aorere with a tributary of considerable size, Brown's River. Though this scarp forms the boundary of the Aorere Valley as here defined, and though its crest is a divide between the streams of the Aorere system and those flowing westward to the Tasman Sea, it does not terminate the tectonic depression of which the Aorere Valley forms a part, for, beyond the scarp, which constitutes a step upwards of 2,000 ft. in height, the depression is continued in a south-westerly direction towards the western coast, the floor of this portion being the plateau known as the Gouland Downs. The composite depression as a whole may, therefore, be conveniently termed the Aorere-Gouland depression.

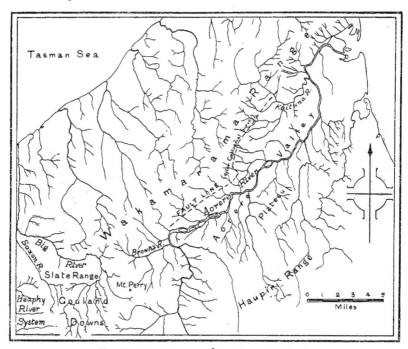

Fig. 3.—Locality map of the Aorere-Gouland area.

The continuation of the Aorere Valley depression in the Gouland Downs was recognized by Bell, who regarded the depression as an ancient strait in which the marine Tertiary strata found in the Aorere Valley and on the Gouland Downs were deposited. Bell's explanation will be referred to again on a later page.

The geological evidence as to the fault-angle origin of the Aorere Valley has been clearly stated by McKay[*] (p. 10), who summarizes his views in the following words: " The disposition of the rocks in every

---

[*] A. McKay, "The Geology of the Aorere Valley, Collingwood County, Nelson," Papers and Reports rel. to Minerals and Mining, Parl. Paper, C.–11, 1896, pp. 4–27.

part of the Aorere Valley indicates that it was first formed along a line of earth-fracture, trending in the general direction of the present valley, and having its downthrow on the south-east side of the line of rupture" (p. 9).

The Wakamarama Fault-scarp.

The scarp front of the Wakamarama Range, which forms the north-western side of the Aorere Valley depression, belongs to a class of composite fault-line- and fault-scarps common in New Zealand, its upper part being a true fault-scarp, and its lower part, from which the covering strata of the downthrown block have been in part stripped by erosion since faulting took place, being a fault-line scarp. This "composite fault-scarp," as it may be called, rises to a height of about 3,000 ft. above the valley lowland. Only one stream—the Kaituna—breaks through the scarp and has its head far back in the range, and, with the exception of this break, the divide between the heads of the numerous steep-grade consequent ravines of the scarp and the north-westward-flowing rivers of the back slope of the Wakamarama block lies at no great distance back from the fault-line. The edges of the facets into which the scarp is divided by the ravines which dissect it are rounded off, and no doubt the steepness of the facets has been much reduced by weathering, slipping, and soil-creep. In spite of a covering of forest which obscures the details, however, the alignment of the blunt-ended spurs is still very striking; and, even in the absence of the geological evidence of the existence of a fault along this line, which is afforded by the presence of the covering strata in the valley lowlands along the base of the scarp, the morphological evidence would indicate faulting. In that case it would be necessary to give very careful consideration to an alternate hypothesis that the scarp is the result of lateral cutting in the course of normal erosion by the Aorere River, which flows at its base, or perhaps that it has been formed by glacial erosion. The most convincing argument against either of these hypotheses is the absence of a similar scarp on the opposite side of the Aorere Valley. Though the whole scarp cannot by any stretch of the imagination be regarded as the work of fluviatile erosion, the streams of the Aorere system, while they were engaged in eroding away the initial floor of the depression, undoubtedly swung occasionally against the base of the scarp, which may thus be expected to be now a little way back from the fault-line.

For the greater part of the length of the scarp the undermass rocks alone occur in the upthrown block in the neighbourhood of the fault-line, and the crest appears from the valley as a line of rounded prominences of roughly accordant height, while, according to Bell, there are on the highland surface patches of *pakihi* (flat open spaces) (p. 24), suggesting that the range is a dissected block of an uplifted denudation plain. From the crest of the range, which, as stated above, is but a short distance back from the fault-line, the summit-levels descend towards the north-west: and the north-western slope is dissected by large sub-parallel streams of apparently consequent (more strictly, probably, superposed consequent) origin. The lower ground to the north-west is formed of a continuous sheet of the covering strata, maturely dissected, and there are outliers of the same towards the crest of the range.* All

---

* Cox, S. H., "On the District between Collingwood and Big River," Geol. Surv. of N.Z., Rep. Geol. Expl. dur. 1882, pp. 62–74, 1883 ; Park, J., "On the Geology of Collingwood County, Nelson," *ibid.*, 1888–89, pp. 186–243, 1890 ; Bell, J. M., *loc. cit.*, map opp. p. 89.

these facts give strong support to the view that the Wakamarama Range is the dissected remnant of a tilted block (or possibly a complex of minor folded and faulted blocks which may in a general description be conveniently considered as a single block) presenting its front or scarp to the Aorere Valley and its back slope to the Tasman Sea, a view involving the not unreasonable assumption that the covering strata, lying on a planed surface of the undermass, were continuous across the site of the range prior to the uplift of the block, and that since the uplift they have been more completely removed from the higher than from the lower ground.

As the foregoing assumption is contrary to the interpretation of the geological history and physiography of the district given by Bell in the Parapara bulletin to which reference has already been made, the conclusions there arrived at may be also stated. According to Bell's interpretation, maturely dissected mountains occupied the area in the period immediately preceding that in which the covering strata were laid down, and the period of deposition was one of only partial submergence. The unsubmerged mountains are regarded as still surviving in a form but little altered, and, on account of their supposed relation to the younger deposits, they are termed the "old land." While the occurrence of some faulting and therefore of some differential movements of later date than the period of submergence is recognized, the uplift of the already mountainous "old land" to its present height is ascribed mainly to "bodily secular movement since the Miocene era" (pp. 21, 23–24).

" The old land represents physiographically an ancient mountain-range which had probably been maturely dissected prior to Miocene times. . . . One sees generally the rounded outlines so characteristic of elevated land-surfaces subjected to long-continued subaerial erosion " (p. 23).

The North-eastern Portion of the Scarp.

Towards the north-east, as on the north and north-west, the surface of the Wakamarama block descends towards the sea, and maturely dissected covering strata survive on it. The fault-scarp facing the Aorère Valley decreases in height towards the north-east, therefore, and towards the mouth of the Aorere Valley the covering strata make their appearance on the crest of the range. The line of unconformable contact between the undermass and the cover runs obliquely up the face of the fault-scarp towards the south-west, but, unfortunately, owing to the covering of forest, details of the contact are not easily seen. At first sight it appears as though the even crest of the range on the undermass is continued on the bevelled edges of the cover as in fig. 4, *a*. As there can be little doubt that the even crest on the undermass is determined by the resurrection of a denudation plain, this would mean the presence of intersecting denudation plains, and would involve strong unconformity between a lower series of conglomerates and coals* (those exposed on the face of the scarp) and an upper series (also coal-bearing)† on the back slope of the Wakamarama Range farther to the south-west. As, however, the beds are all regarded by Park as belonging to one conformable series, it is probable that the appearance of intersecting denudation plains is false, arising from an increasing inclination

---

* Described by Cox and Park in the papers previously referred to.
† Bell, *loc. cit.*, p. 55.

towards the north-east of the covering strata and of the floor on which they lie, as shown in fig. 4, *b*, combined with resistance to erosion offered by indurated conglomerate at the base of the cover.

Tilted and Stripped Plateau of the South-eastern Side of the Aorere Valley.

Turning now to the south-eastern side of the Aorere Valley, one's attention is arrested by a tilted denudation plain which rises from the valley lowland to a height of about 1,500 ft. in a distance of three miles and a half. This plain from some points of view appears very well preserved, but closer inspection reveals the fact that it is deeply divided by the steep-walled gorges of numerous streams tributary to the Aorere. The form of this south-eastern slope was noted by Hochstetter,[*] who estimated its inclination as 8°, but did not remark upon its genesis. It was remarked upon later by Park (p. 198), McKay (1896, p. 23), and Bell (pp. 24–25), McKay explaining it as a plain of marine denudation and Bell referring to it as an "old sea-shelf."

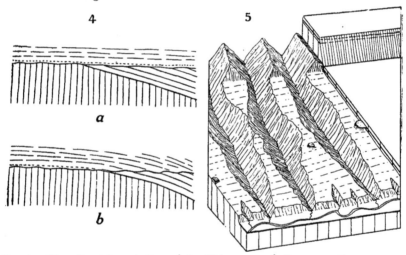

Fig. 4.—Alternative interpretations of the Wakamarama fault-scarp section.
Fig. 5.—Dissection of a sloping surface, such as that of the south-eastern side of the Aorere Valley, by streams from a higher block behind it. The initial form is shown in the right-hand block.

McKay describes it as a " feature characterizing the east side of the Aorere Valley," and adds, " This is an uniform slope of the country to the north-west and the low grounds of the valley, from heights 1,200 ft. to 1,500 ft. above the sea, which slope, as seen from a distance, appears to be remarkably uniform, both as regards its dip towards the low grounds and as regards its extension along this side of the Aorere Valley, and suggests at once the idea of a plane [*sic*] of marine denudation, which, by the elevation of the mountain region to the south, has acquired a steeper slope than it had when first formed " (1896, p. 23). Illustrations are given by Hochstetter (woodcut on p. 103) and by Bell (pl. i, lower view, and pl. ii, upper view).

---

[*] F. v. Hochstetter, "New Zealand," Stuttgart, 1867, p. 105.

This sloping plateau is certainly, as McKay noted, a plain of marine denudation—a plain of denudation, that is to say, to the formation of which the finishing touches at least were given by marine erosion as the sea advanced over it and deposited upon it the covering beds since largely removed. The occurrence, however, of terrestrial formations at the base of the covering strata in neighbouring areas, notably on the north-western and northern parts of the Wakamarama Range, is perhaps an indication that the surface of the undermass had there been reduced to small relief by subaerial agencies in the period preceding submergence. It is, therefore, reasonable to suppose that here also a peneplain was in existence, and that the plain of marine erosion is one of those produced by the stipping and removal of the waste mantle from a peneplain, accompanied by a minimum of erosion of the fresh underlying rock. The large extent of the planed surface—which, indeed, probably extended formerly over a much larger area than that in which the plain is now preserved—thus receives a simple and probably correct explanation.

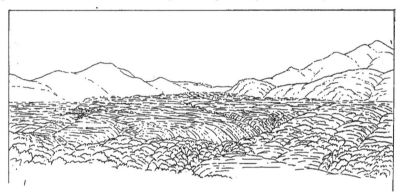

Fig. 6.—View looking south-east up the sloping plateau of the Aorere Valley. After a photograph by the Geological Survey.

The tilted plateau of the Aorere Valley has been revealed owing to the removal from its surface of the covering strata, a few residual areas of which testify to their former wide extension. Where seen the floor beneath the covering strata proved to be a cleanly eroded surface of fresh rock, and the basal beds in this locality have always been described as conglomerate of marine origin, generally a few feet in thickness and thoroughly sorted, only the hardest of the materials of the undermass, as a rule, surviving. Within a few feet vertically this quartz conglomerate passes into pure limestone, the presence of which indicates prevailing open water, the shore-line being some distance away and any neighbouring land being of small relief and supplying a negligible quantity of mechanical waste.

Bell regards the formation of the denudation plain as having taken place in a strait necessarily initially narrow and afterwards opened by marine erosion to the width of the sloping plateau of the present day, the shores of which strait constituted an "old land" of mature mountains. There are, however, several rather serious objections to this explanation.

First, if the land-forms of the region prior to submergence were mature mountains, and if submergence of such a region of strong relief

3*

led to the formation of a strait, it is difficult to believe that the feeble waves of the waters of a strait, which was necessarily initially narrow and, at least in parts, almost or completely landlocked, can have reduced mature mountains of resistant rocks to a plain of marine erosion, over a width of several miles, in a period so short that the mountainous relief of the "old land" was not destroyed in neighbouring areas by subaerial denudation.

Secondly, submergence of a region of mature mountains must result in the drowning of many valleys. Thus not only would a single narrow strait be formed, but also many branching bays of considerable depth. If we suppose planation of the partially drowned ridges, the peninsulas, and the islands of such a region to take place, it is apparent that the result cannot be a continuous denudation plain. While a more or less perfect plain will have been produced, and this end will have been achieved by the cutting-down of the salient features and the filling-up of the drowned valleys to a common level. The sloping plateau is, however, a plain of denudation throughout. Had it been otherwise the plateau in its present form could not have survived, but the ancient filled valleys would have been re-excavated by modern erosion, revealing again the postulated maturely opened valley forms and the ancient drainage pattern.

Thirdly, a strait the floor of which is a platform of marine erosion must have been much widened by wave-action, and if it has been so widened in a region of mature mountains it must be bordered by wave-cut cliffs of great height. No such cliffs, however, have been pointed out by Bell. It is true that a scarp, previously referred to as the Wakamarama fault-scarp, bounds the Aorere Valley on the north-west side, and it is true also that steep mountain-slopes ascend from the sloping plateau on the south-east side; but recourse has never been had to marine erosion to account for these, and other and more satisfactory explanations are not difficult to find.

Fourthly, the nature of the sediments of the covering strata of the Aorere Valley has been previously referred to. The limestone which follows the basal conglomerate is free from admixture of terrigenous material, and is not the kind of deposit that might be expected to occur in a strait between mountains. It has, moreover, not been shown to pass into a littoral facies towards either side of the supposed strait.

Erosional, Sedimentary, and Deformational History.

It has been necessary to state the objections to Bell's theory of the genesis of the physical features of the Aorere Valley somewhat fully, since they are all arguments in favour of the hypothesis which is here offered in its place. It would appear that the strong relief which the deformed undermass presumably had in some earlier period had been almost completely destroyed prior to the deposition of the covering strata. It is reasonable to suppose that this reduction of the ancient mountains was effected largely by subaerial erosion, though planation was completed, over at least the area of the Aorere Valley, by the advancing sea at the commencement of the period of deposition of the covering strait. Next followed a period of deposition over a wide area, and, later, the episode of strong differential movements, which sketched out the broad outlines of the land-forms of the present day, led to the formation of many consequent rivers, and inaugurated the cycle of erosion in which the majority of the details of the surface were developed.

Dissection.

The sloping plateau is crossed, as noted above, by a number of young gorges. These are evidently superposed consequent ravines, for they descend the slope of the tilted plateau, and the courses of the streams which cut them were evidently guided by the slope of the surface of the former cover. They are quite indifferent to both dip and strike of the undermass rocks upon which they now flow. These gorges are, in general, narrow-floored and steep-walled, and some distance back from their debouchures the larger of them are incised to a depth of several hundred feet below the sloping plateau. The larger streams head in a range of mountains to the south-east, some peaks of which rise to heights of 2,000 ft. and more above the plane of the sloping plateau produced in that direction. These mountains, the Haupiri Range, may be satisfactorily explained as the dissected remains of a higher block separated from the sloping plateau block by a fault or flexure. The streams rising in the Haupiri Mountains have already considerable volume when they begin to cross the sloping plateau in extended, superposed consequent courses, and it is to the action of these vigorous streams that the somewhat advanced dissection of the plateau is to be ascribed. The diagram, fig. 5, represents the dissection of a sloping block surface such as that of the south-east side of the Aorere Valley by streams from a higher block behind it. A moderate area of the stripped plateau is represented as surviving, and also a few remnants of the cover. On the sloping plateau of the Aorere Valley a few such residuals. small limestone mesas and buttes, remain (Bell, pp. 24–25).

The Actual Valley of the Lower Aorere.

In the foregoing sections the name " Aorere Valley " has been used with reference to a tectonic depression (which, as will have been gathered from the description, is strictly neither a *Graben* nor a fault-angle depression, but partakes of the nature of both), modified as it is at the present day by erosion. Nothing has been said of the Aorere River beyond the fact that it occupies the depression.

It must be noted, however, that the present open valley is largely the work of the river, the amount of waste which it has transported seaward since the episode of differential movements initiating the depression being truly enormous, and consisting of almost the whole of the covering strata (of unknown thickness, but probably several hundred feet at least), as well as a very large contribution from the dissection of the undermass by its tributaries and its own upper course. The actual erosive work performed by the river itself has been considerable, as is evidenced by the occurrence of terrace after terrace up to a height of many hundred feet on the sloping plateau. As these terraces occur on the lower portion only of the sloping plateau, and as the form of the latter has been modified by them only to an inconsiderable extent, they have not been mentioned in the account of the plateau given on earlier pages. When, however, the valley-side is examined closely, the terraces and their thick covering of gravel are readily recognizable. As the gravel is auriferous, it has been largely excavated, and the workings reveal some former channels of the river refilled when from time to time degradation gave place to aggradation.

The terraces occur up to a height of perhaps 600 ft.* above the present level of the river. and there are also residuals of gravel-covered

---

* Accurate data are not available.

floors at various heights forming flat-topped and terraced hills, which have been noted by many observers, throughout the three- to four-mile-wide valley lowland, indicating that the Aorere has wandered widely on a broad flood-plain more than once during the excavation of its valley in the covering strata of the fault-angle. It would appear that, after the movements of deformation by which the major topographic features were blocked out, the whole region stood some hundreds of feet lower than at present, and that while the land was in this attitude there occurred the great denudation which resulted in the removal of a great part of the covering strata from the upland areas and the carving of the blocks into forms approximating to those of the present day. It is possible that the sea entered the north-eastern end of the fault-angle; but, if such was the case, the re-entrant so formed was no doubt rapidly filled up, and at the same time a considerable amount of aggradation must have taken place throughout the lower parts of the depression. The traces of any such filling have been since removed by erosion. It is a safe assumption that towards the end of the great denudation the Aorere River was not only graded, but had developed a flood-plain with a width of about four miles, of which the highest terraces are remnants. The lower terraces and the present valley of the river, with its discontinuous narrow flood-plain, are to be ascribed to excavation by the Aorere and its tributaries during recent intermittent movements of uplift.

The Gouland Downs Depression.

As stated on an earlier page, the Aorere-Gouland depression is continued beyond the divide between the Aorere Valley system and the streams flowing westward to the Tasman Sea. The south-westward continuation, though obviously cognate with the Aorere Valley portion, differs

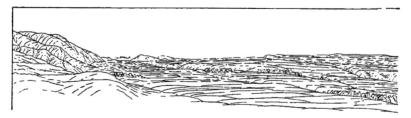

Fig. 7.—The Gouland Downs, showing the Slate Range on the

from it so much in certain particulars that it demands separate description. The floor, known as the Gouland Downs, is in most respects homologous with the sloping plateau of the Aorere Valley, while the true boundary of the depression on the north-western side is a scarp which, after an interruption, continues the line of the Wakamarama fault-scarp. A subsidiary tilted block, the Slate Range, however, at the base of and parallel with the Wakamarama Range, separates the latter from the nearly level floor of the depression, which is bounded on the north by the fault-scarp front of the Slate Range block.

The Floor.

The surface of the " downs " plateau (with an area of about twelve square miles) is a plain of erosion similar to that forming the sloping plateau of the Aorere Valley, with here and there small mesas of covering strata—pure limestone passing downward into a quartz grit with a calcareous cement, with a thin layer of conglomerate at the base. The limestone mesas are riddled with caves, and it is obvious that the removal of this lowest stratum of the cover is being effected mainly by solution. As the limestone areas are forested, they show out conspicuously in contrast with the rest of the plateau, which is bare of vegetation, with the exception of rushes and a few tussocks of coarse grass struggling for existence in a " sour " and slimy soil (see figs. 7 and 8).

There are areas many acres in extent which are quite flat and nearly level; but the surface of the plateau as a whole is by no means uniform. Besides a number of narrow gorges cut recently below the general surface, but collectively not affecting a large area of the plateau, to which reference will be made in a later section dealing with the drainage and dissection of the "downs," the principal irregularities are such as may be ascribed to deformation of the denudation plain—of course, along with its cover—at the period when the larger differential movements were also taking place. Owing to the presence of a system of small faults, or possibly flexures, generally transverse in direction to the general elongation of the depression, the surface of the "downs" descends towards the middle of the northern boundary in a series of broad irregular steps, each differing in height from its neighbours by a few tens of feet. These may be seen in Plate IV, fig. 1, which is a view looking south-westward across the "downs." A wider panorama from about the same point of view is shown in fig. 7. The foregoing appears to be the most satisfactory explanation of the irregularities in the floor of the depression, but it must be remembered that the surface has been long subject to erosion, that an unknown thickness of cover has been removed from it, and that

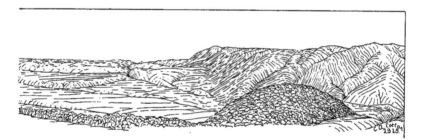

right. Angle of view, from south to west-north-west.

the initial forms of the small fault-line scarps, if such there be, have been much modified. Moreover, the stream which has effected the removal of the debris of the covering strata has wandered rather widely over the area, and as a result there are some more or less definite fluviatile terraces cut on the undermass in the lower central part of the "downs." A layer of river-gravel occurs on, and proves the origin of, the more definite terraces bordering for some distance the stream (the head of the Big River) which now drains the "downs," and a sprinkling of gravel over a much wider area—perhaps sporadically over the whole floor—may have a similar origin; but it is probable that most of the surface gravel is a residuum of the conglomerate at the base of the former cover. The presence of scattered knolls of the relatively weak limestone of the cover proves that the higher flat areas are not to any appreciable extent the result of recent lateral planation by streams on the resistant underlying rocks, and indicates quite clearly that they are portions of a stripped floor. The dissection of the floor by streams, though not far advanced, obscures the minor tectonic features of the relief to some extent, for the streams can in no case be consequent on the form of the floor. Theoretically we may expect to find, in addition to insequent, and possibly subsequent, streams, a drainage pattern superposed from the cover, the streams of which, even if wholly consequent on the form of the cover, may be to a great extent indifferent to minor breaks in the floor.

The general form of the floor of the Gouland Downs depression may, if we neglect the minor irregularities referred to above, be described as sloping gently in the form of a half-basin from the east, south, and west against the scarp of the Slate Range, which forms the northern boundary. The lowest part of the basin is at a height of about 2,000 ft. above sea-level. On the western and southern sides the plateau slopes gently up to an even sky-line at a height of about 3,000 ft., beyond which lies the valley system of the Heaphy River.

The Eastern Boundary.

Towards the eastern boundary the slope of the surface becomes steeper, as though passing in an anticlinal form, if produced, over Mount Perry and other peaks at the northern end of a range about 4,000 ft. in

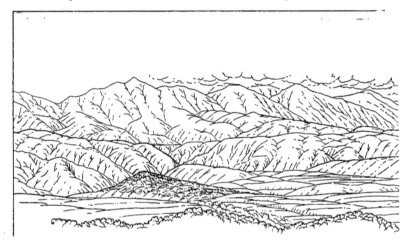

Fig. 8.—View looking north-east across the Gouland Downs and the "catenary" distance in the centre.

height which bounds the depression on the east. Mount Perry is seen to the left of the centre in Plate IV, fig. 2, and on the right in text fig. 8. In both figures the rise of the plateau surface with increasing steepness towards the range may be noted. The preservation of the surface on slopes of considerable steepness is explained by the fact that the rocks are indurated shale and quartzite, offering great resistance to erosion. Naturally, the stage of dissection of the slopes rapidly approaches maturity as the steepness increases, until, on the flank of the range, the denudation plain is completely dissected away, and only fully mature forms developed in the current cycle are seen. So far as the writer is aware, no remants of the denudation plain or of the covering strata are preserved on the higher parts of this range. There can be little doubt, however, as to the general truth of the foregoing explanation of Mount Perry and neighbouring peaks—namely, that they have been carved by erosion, in the cycle still current there, from an upfolded mass of the older rocks, the upper surface of which was, in all probability, prior to the deformation, a denudation plain continuous with that preserved on the Gouland Downs, and carried, like it, a cover of younger strata. Several miles farther south, however, where the same range still forms the boundary of the " downs," the fold structure of the mountain-flank may be replaced by a fault; but as the range is there composed of granite which is somewhat easily decomposed, and as the slopes are forest-clad, the interpretation of the scarp is not a simple matter.

A " Catenary " Saddle.

The southward-facing scarp of the Slate Range was referred to above as forming the northern boundary of the plateau-floor of the Gouland Downs. As the floor rises towards the east, however, its level approaches that of the crest of the Slate Range, the scarp of the latter diminishing in height and finally dying away. The north-eastern part of the " downs " surface would apparently be continuous, therefore, with that

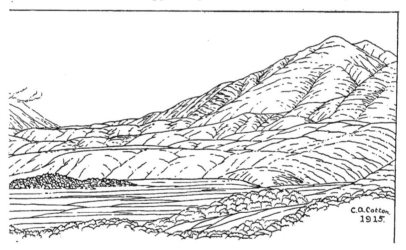

saddle. Mount Perry is on the right, and the Wakamarama fault-scarp in the Angle of view, about 75°.

of the summit of the range were it not for the fact that it is separated from it by a deeply eroded gorge—that of the Big River. In the neighbourhood of this gorge dissection of the surface is naturally in a somewhat advanced stage, but there is still an accordance of the levels of the interfluves indicating the initial shape of the warped denudation plain. In this—north-easterly—direction the Gouland Downs surface rises gradually to the saddle which separates the Gouland depression from the Aorere Valley.

The saddle is one of the most striking features in the whole district. Viewed either from the south-west—*i.e.*, from the Gouland Downs (see fig. 8)—or from the north-east—*i.e.*, from the Aorere Valley—it appears against the sky as a perfect catenary curve four miles in length. It sags from a height of about 4,000 ft. at the south-eastern end (Mount Perry) and about 3,750 ft. at the north-western end (the Wakamarama Range) to a height of about 2,500 ft. above sea-level in the centre. The catenary form of the curve is so striking that the feature was pointed out to the writer by a resident of the Aorere Valley as an indication of profound glacial erosion. There can, however, be no doubt that it is of tectonic origin. It is as though, during the episode of uplift, while the Aorere-Gouland depression as a whole lagged behind the blocks which now form the ranges to north-west and south-east, a strip here had failed to break away from either side, but had sagged in the middle so as to assume the true catenary form.

The Slate Range.

Mention has already been made of the Slate Range subsidiary block. For three miles—the full length of the block—its scarp forms the northern boundary of the Gouland Downs, while its width in a north-south direction cannot exceed one-third of its length. Unfortunately, a complete description of this interesting little block cannot be given, as the writer saw only the southern side.

At the eastern end, as previously noted, the upper surface of the Slate Range would but for the ravages of erosion be continuous with the higher north-eastern part of the Gouland Downs; but on all other sides the block appears to be bounded by dislocations. The scarp facing the Gouland Downs has been referred to above as the "front" of the range. An assumption has thus been made that the Slate Range is a tilted block, and that opposite to the fault-scarp front facing south there is a back slope to the north. This must be so, because the evenness of the crest-line and the small dissection of the front show that in the vicinity of the crest the upper surface is nearly flat and slopes back so as to lead the drainage northward. A glimpse of the top of the range caught from the slope of Mount Perry confirms the above view. The flat surface of the top of the range can only be a portion of the dislocated denudation plain found throughout the district; and it seems probable that initially the northward slope of the surface and its probable cover formed, with the scarp of the western Wakamarama Range, a fault-angle depression determining a consequent east-west reach of the Big River (see fig. 3), now, no doubt, superposed on and deeply sunk in the undermass.

The southward-facing fault-scarp of the Slate Range, which has an average height of about 700 ft., presents the usual appearance of blunt spurs ending in line (see fig. 7, and Plate V, fig. 1). The spurs and the intervening steep-grade gullies are forested, with the exception of one spur, which stands out as a bare and also sharp-edged facet because its surface is veneered with a thick vein of quartz. The quartz vein evidently filled an ancient fissure which has guided the more modern fault.

The Drainage and Dissection of the Gouland Downs.

The drainage of the Gouland Downs is collected in the fault-angle depression between the gently sloping "downs" surface and the scarp

[C. A. Cotton, photo.

Fig. 1.—View looking south-westward across the Gouland Downs.

[C. A. Cotton, photo.

Fig. 2.—Mount Perry and the eastern margin of the Gouland Downs.

[C. A. Cotton, photo.

FIG. 1.—The scarp of the Slate Range. View looking north-north-west.

[C. A. Cotton, photo.

FIG. 2.—The gorge of the Blue Duck Creek, incised below the plateau surface of the Gouland Downs.

of the Slate Range, the greater part making its way out eastward as the Big River, and the remainder westward as the Saxon River, which joins the Big River before reaching the sea. These east-and-west stream-courses are obviously consequent. The eastward-flowing reach of the Big River is of great interest, as the stream meanders in full maturity upon a wide flood-plain cut but little below the surface of the denuda-tion plain of the "downs," indicating that the cycle initiated by the great differential movements is still current. Rejuvenation of the valley of the Big River, due to the later movements of regional uplift which have affected this part of New Zealand, has not yet proceeded so far up-stream as to modify the form of the Gouland Downs.

A number of streams which cross the western "downs" in a north-ward and north-eastward direction to join the streams in the fault-angle are obviously superposed consequents as they follow the general slope of the surface. They cross obliquely the outcrops of the strata of the oldermass which have a uniform north-north-west strike. The streams are roughly graded, though still in narrow, steep-walled gorges, and hence are sunk most deeply beneath the plateau in their middle courses. Plate V, fig. 2, illustrates the type of features thus produced. The sharp contrast between the steep walls of the gorges and the level plateau above is very striking. As the floors of the gorges are occupied to their full width by the streams, the latter rise rapidly, and become impassable after a shower of rain.

Farther to the east the streams flowing towards the Big River cross the "downs" in a north-north-westerly direction—a direction more northerly than that of the general slope of the surface. They have perhaps been guided by irregularities of the initial surface, but it is noticeable that they are parallel with the strike of the strata of the oldermass. Dissection is here more advanced than it is farthest west, and the numerous longitudinal gullies are separated by rounded quartzite ridges suggesting a subsequent origin.

The eastward-flowing consequent reach of the Big River is connected with a westward-flowing consequent reach farther down-stream (see fig. 3) by a northward-flowing reach, where the stream makes its way in a gorge around the eastern end of the Slate Range. There can be little doubt that this is of (superposed) consequent origin also. The south-western rim of the "downs" basin is high, and the probable former extension of the surface to the south-westward would be still higher. Thus the present outlet of the Big River may well mark the position of the lowest gap in the rim of the basin-shaped initial surface of the covering strata, some hundreds of feet above the site of the Gouland Downs.

The outlet gorge has been cut to a depth of many hundreds of feet in the extremely resistant rocks of the oldermass. During the process of gorge-cutting the local base-levels on the Gouland Downs have been very slowly lowered, and thus perfect conditions have been afforded for the stripping of the cover from the plateau.

Art. VII.—*The "Red Rocks" and Associated Beds of Wellington Peninsula.*

By F. K. Broadgate, M.Sc.

[*Read before the Wellington Philosophical Society, 27th October, 1915.*]

Plates VI, VII.

Introduction.

The area examined in connection with this paper is the south-west corner of the Wellington Land District, and more particularly is defined by Port Nicholson on the east, and a line drawn west from the head of that inlet to the coast; Cook Strait forms the remaining boundary. The area is conveniently termed the Wellington Peninsula (fig. 1).

Red and green argillites, often with associated tuff-beds, have been noted in various parts of New Zealand. The present paper gives the results of an examination of these rocks as they are represented in the Wellington Peninsula. Such examination must take account of the series of greywackes and dark-coloured argillites, as interbedded members of which the red and green argillites occur; some general notes on this series precede the main problem.

The conclusion reached is that the red and green argillites were originally green argillites not differing, save in colour, from the ordinary dark-coloured argillites. An attempt is made to explain the changes undergone by the green argillites subsequently to their deposition, and a hypothesis to account for the formation of such argillites is put forward.

For convenience, the well-known name "Maitai" is here adopted for the whole of the rock-series under review. No age significance is to be attached to its use in this connection; the only fresh evidence bearing on this question, as related in the text, is of a destructive rather than constructive kind.

Topography.

The mountainous area of the Wellington Peninsula, together with the Rimutaka Mountains to the east, form the southern extremity of the main structural axis of the North Island. This axial line is continued in the South Island by the Kaikoura Mountains, of Marlborough.

The existing topography is that of a recently uplifted block which had already suffered close folding and distortion, and, according to J. M. Bell (1, p. 535), had been reduced to a state of peneplanation before uplift. In a report on the Maharahara district, Woodville, J. A. Thomson states that the second period of general uplift began in late Pliocene times (2, p. 165); while A. McKay believes that "the Kaikoura Mountains . . . have been elevated to their present height from a moderately elevated plateau solely by earthquake action, and this since the commencement of Pliocene times" (3, p. 11). Elevation commencing in the Pliocene, and, with intervals of standstill, continuing to the present day (4, p. 246), may be fairly assumed to be the case in the Wellington Peninsula, lying midway between the districts mentioned.

Compressional forces acting on an area already much disturbed have resulted in tilting and differential movements, with the institution of faults along planes of weakness, and probably a renewal of faulting along old lines.

Fig. 1.

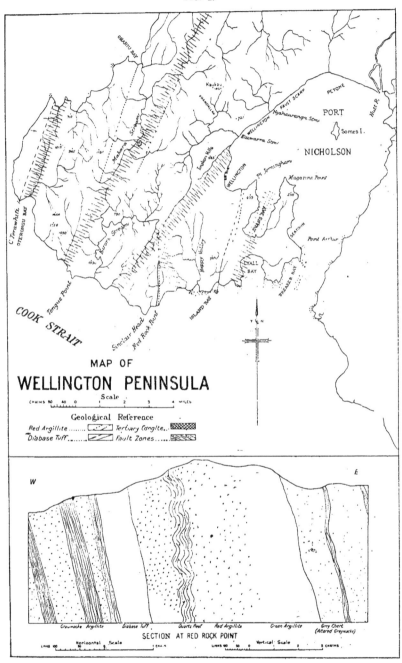

MAP OF
WELLINGTON PENINSULA
Scale

Geological Reference

Red Argillite .... Tertiary Conglte..
Diabase Tuff........ Fault Zones.....

SECTION AT RED ROCK POINT

Fig. 2.

OUTLINE OF THE GEOLOGY.

The Maitai rocks of Wellington Peninsula are a series of argillites and greywackes with a general strike N. 12° E., and having an average dip of 70°, the direction of dip varying from east to west. The rocks to be described under the heading of " Red Rocks " are members of this series.

In the Makara Valley, four miles west of Port Nicholson, is a small patch of marine conglomerate which is considered by A. McKay (5) to be of Miocene age. From the physiographic evidence it appears that the deposit cannot be younger than Pliocene. The conglomerate is 150 ft. above sea-level, and rests on a fault-zone. Its total areal distribution does not exceed 10 acres, and the deposit has no parallel elsewhere in the peninsula. If, as stated by C. A. Cotton (4, p. 246), the Wellington block was uniformly uplifted to a height of 800–1,000 ft., this conglomerate owes its position, and probably its preservation, to down-faulting since that time.

The gravel veneers, relics of former base levels, as described by C. A. Cotton (4, p. 250), are found at varying heights. Save where complications have arisen through faulting, the highest gravels are the oldest.

THE MAITAI SERIES.

Stratigraphy.

These rocks are recognized throughout the length of New Zealand. The complex foldings they have undergone and their general scarcity of fossils, except at a few widely separated localities, have caused much confusion as to their age and correlation.

The various classifications proposed by A. McKay for the rocks about Wellington were seemingly based not on evidence from the stratigraphy of the Wellington development of the series so much as from their supposed relations in other places. In the first scheme proposed (6, p. 132) the rocks were divided into—

Trias and Permian	..	(*a.*) Sandstones and slaty shales, Magazine Point.
Carboniferous	..	(*a.*) Red and green slates, Sinclair Head.
,,	..	(*b.*) Sandstones and earthy slates.
Devonian ..	..	(*a.*) Sandstones and drossy slates, with numerous veinlets of quartz.

The Devonian rocks are spoken of by Dr. Hector as Lower Carboniferous (7, p. 30); and this age, he says, is assigned them on account of lithological resemblances to the rocks of the Rimutaka Range. These latter, he considers, resemble rocks underlying the fossiliferous limestone at the base of the Maitai slate near Nelson (7, pp. 28, 29).

Omitting the red and green slates, the rocks mentioned in this classification do not differ beyond the changes induced by weathering and faulting. In fact, the description of Devonian rocks receives significance when considered as applied to a fault-zone ; and a study of the "Devonian" outcrops mentioned—parts of the south coast, Tinakori Valley, and Makara Valley—make it clear that the chief fault-zones of the district were mistaken for an older series of rocks. Thus, in his Progress Report for 1878–79, Dr. Hector says, "Before leaving these rocks [the ' Devonian ' rocks of A. McKay] it may be as well to point out that generally when rocks of this age and character occur in the South Island traces of gold are found." This is easily understood, seeing that the quartz veins of Wellington and Marlborough are silicified fault-zones.

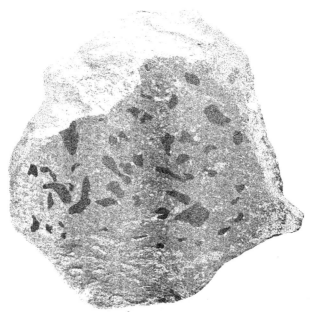

FIG. 1.—Greywacke with argillite inclusions, Makara Valley: polished slab.

FIG. 2.—Green argillite, partly reddened: polished slab.

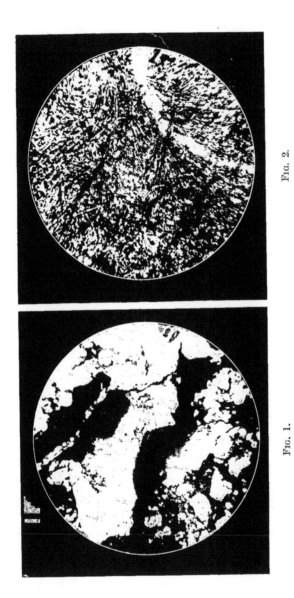

FIG. 2.

FIG. 1.

FIG. 1.—Greywacke with argillite inclusions, Makara Valley; ordinary light. Magnified 34 diameters.
FIG. 2.—Diabase tuff, Red Rock Point; ordinary light. Magnified 34 diameters.

In the argillites of the Maitai Series of Wellington Peninsula are found a few obscure animal-remains of small size not yet identified, and some indistinct plant-remains. The only fossil so far determined is the annelid *Torlessia mackayi* Bather ; this fossil has as yet no definite stratigraphic value.

Petrography.

Microscopic study of the greywacke shows the presence of quartz, feldspar, hornblende, augite, biotite, muscovite, and epidote.

The quartz which makes up most of the recognizable material is always angular, with no appearance of secondary growth. Many of the pieces show undulose extinction. Feldspars are both orthoclase and plagioclase. The orthoclase is always considerably altered. The plagioclase feldspars identified range from albite to medium labradorite. Hornblende and augite are in very small amount. The biotite, muscovite, and epidote are probably secondary in some cases. Needles of rutile are sometimes seen in the quartz, and rods of apatite in the hornblende. The greywacke cement is siliceous ; much of the groundmass is of indeterminable brown material.

The material of the argillites is too fine to reveal much under the microscope. Feldspar and quartz are recognizable under high power. Small veinlets are generally present, these being probably of secondary silica. The common argillite is dark-coloured ; prolonged heating leaves a dull red-coloured product. The dark colouring-matter of the rock is most likely carbonaceous material (9, p. 44).

Conditions of Deposition.

The criteria for recognition of estuarine deposits as given by Hatch and Rastall (10, p. 13) apply well to the rocks of Wellington Peninsula. The rapid alternation of different types of sediments, often in thin layers, is very marked. The greywackes vary in texture from fine conglomerates to fine-grained sandstones, though the change is not gradual in any one band ; layers of different degrees of fineness are separated by bands of argillite, the argillite itself being probably only an extreme case of a fine-grained greywacke.

The presence of feldspar and the persistent angularity of even the large quartz grains both point to deposition not far from the source of supply. The facies is, indeed, that common to deposition in shallow water of the waste resulting from denudation of an igneous mass—the greywacke being a quartz product from the siliceous content, and the argillite representing the product of decay of the more aluminous constituents.

A peculiar type of greywacke is that from Makara Valley (Plate VI, fig. 1, and Plate VII, fig. 1). Besides the ordinary constituents, this has numerous pieces of argillite up to half an inch in length. These argillite inclusions have an irregular outline, and appear to be invaded by the quartz, as though they were still unhardened at the time of inclusion. This type has been noted in Westland (11, p. 45 ; 12, p. 85). Such a rock could originate only in shallow water.

FAULTS.

The chief fault-zones are indicated on the map. Most likely this represents only a proportion of the total number, but it probably includes the more important lines of dislocation.

Generally the argillite has afforded planes of weakness, so that the fault-lines have often a N.N.E. direction, following approximately the general strike of the rocks. Only in the case of the Wellington fault is there good physiographic evidence of faulting ; in other cases, however, the fault-lines are readily recognized by their zones of crushed and slickensided rock.

A secondary effect of the faulting has been the production of silicified fault-zones. In the western part of the area these have been worked for their gold-content.

The ·"Red Rocks."

Tuff-beds, cherts, and red and green slates (with associated quartz veins) have been recorded together at various points, and referred to as the "Red Rocks." They may conveniently be described under this heading.

Tuff-beds.

Thin beds of diabase tuff occur at two, and probably three, points, inter-bedded with the argillite and greywacke. Two of these occurrences are indicated on the map ; the third was noted by A. McKay as a dyke rock encountered in the gold-mines of Terawhiti, the term "dyke" having been used by that writer for rocks later called "tuff" (8, pp. 66–7).

The exposure at Red Rock Point is associated with red and green slates. The rock is chocolate-coloured, but this colour is largely masked by green stainings of epidote and chlorite ; it shows also irregular veins and nests of calcite. Under the microscope the rock is seen to be composed largely of colourless needles arranged in sheaf-like bundles ; these are embedded in an irresolvable brown paste. The needles polarize in low colours—whitish-grey of the first order—and show negative elongation. Calcite veins and smaller veins of haematite are present. Plate VII, fig. 2, shows the microscopic appearance of this rock. Dr. J. A. Thomson has kindly shown me sections of a rock from Western Australia, which he terms "fine-grained greenstone" (15, p. 634). These are very similar to sections of the diabase tuff. The greenstones, according to Dr. Thomson, are altered diabase lavas or tuffs (15, p. 670) ; and the colourless needles, an alteration from the secondary amphibolite, he considers to be hornblende which has in some way had its birefringence lowered.

Notwithstanding the evident alteration which the diabase tuff has undergone, its chemical composition still approximates to that of an average diabase rock as given by R. A. Daly (16)—compare analyses 6 and 7. In comparison the tuff shows a deficiency of magnesia, differs in the proportions of ferrous and ferric oxide, and has a higher percentage of loss on ignition. The formation of epidote and chlorite, as noted above, would increase the hydrous content of the rock. The relative proportions of ferrous and ferric iron have probably altered in a way similar to that of the red and green slates.

Grey Cherts.

The rocks named "grey cherts" by A. McKay are abundant in the area. They are well seen on the coast between Lyall Bay and Island Bay, east of Red Rock Point, and at several places between there and Oterongu Bay. They form most of the block west of Oterongu and Ohau Bays (17, p. 3). The rock is quite evidently a greywacke altered by secondary silicification ; small veins form a closely anastomosing mesh-work in the rock, so it is not possible in parts to select even a small sample free from veining.

Under the microscope the grey chert shows quartz and feldspars as in greywacke, and small amounts of the other minerals noted in that rock. An analysis of a sample of grey chert, as free as possible from quartz veining, is appended ; comparison with the analysis of a greywacke from Breaker Bay shows how little the grey chert differs from greywacke.

Argillite bands in the grey chert are generally distorted. The grey cherts are readily altered by weathering, the effects of which are well seen on the coast between Lyall Bay and Island Bay. It is these rocks which A. McKay classed as the Otapiri series, and described as "gritty grey sandstones decomposing to a light-brown colour " (8, p. 61).

Table : Rock Analyses.

—	1.	2.	3.	4.	5.	6.	7.
SiO_2	70·75	70·20	57·89	58·70	61·10	48·62	50·12
TiO_2	0·62	0·66	1·00	0·93	0·95	1·66	1·41
Al_2O_3	12·30	13·53	19·03	18·29	17·60	13·98	15·68
Fe_2O_3	2·72	1·68	4·48	1·99	3·84	7·68	4·55
FeO	2·74	3·24	2·88	6·03	3·85	2·16	6·73
MnO	0·06	n.d.	0·04	n.d.	n.d.	n.d.	0·23
MgO	1·22	1·48	1·47	2·96	2·07	1·36	5·85
CaO	1·90	1·80	0·16	1·30	1·07	10·27	8·80
Na_2O	3·18	2·04	2·02	2·63	2·78	·3·56	1·38
K_2O	2·48	3·18	4·21	3·30	3·45	1·69	2·95
CO_2	Nil	n.d.	Nil	n.d.	n.d.	n.d.	n.d.
P_2O_5	0·16	n.d.	0·26	n.d.	n.d.	n.d.	0·37
SO_3	0·18	n.d.	0·11	n.d.	n.d.	n.d.	n.d.
Moisture at 100° C.	0·16	} 2·03	{ 1·15	} 4·11	3·42	8·22	1·93
Organic matter and combined water	1·79		{ 5·65				
	100·26	99·84	100·35	100·24	100·13	99·20	100·00

(1.) Greywacke, Breaker Bay, Cook Strait.
(2.) Grey chert (= altered greywacke), Red Rock Point, Cook Strait.
(3.) Argillite, Point Arthur, Wellington Harbour.
(4.) Green argillite, Red Rock Point, Cook Strait.
(5.) Red argillite, Red Rock Point, Cook Strait.
(6.) Diabase tuff, Red Rock Point, Cook Strait.
(7.) Diabase, average analysis (from R. A. Daly, " Igneous Rocks and their Origin ").

(Analyses 1–6 by Dominion Laboratory, Wellington.)

Red and Green Argillites.

In the reports of the old Geological Survey the term " slates " has been used for these rocks as also for those now called " argillites." The red and green slates do not differ, except in colour, from the common argillites ; they show parting parallel to the bedding-planes, and, save where weathering has been active, slaty cleavage is no more developed than in the argillites (9, p. 44 ; 18, p. 43). Their chemical similarity to a typical argillite is seen by comparing analyses Nos. 3, 4, and 5. The name " argillite " is here used for these rocks. The exposures of red argillites in the Wellington Peninsula are indicated on the map (fig. 1). The outcrop at Red Rock Point, illustrated by the section (fig. 2), is the clearest. Only in the case of this outcrop is the development of green argillite in connection with red argillite plain. Outcrops on hillsides are conspicuous by reason of their red colour, but the green rock is hardly to be distinguished from the common darker green or black argillite.

The material of the green argillites shows, under the microscope, quartz, magnetite and pyrite, and sericite. The green colouring-matter is indeterminable; in the case of green argillites examined elsewhere this colouring has been considered to be epidote (9, p. 47) or amphibole (12, p. 99). Haematite is the chief recognizable mineral of the red argillites.

The so-called red and green slates are green argillites partly reddened. Sometimes a gradual change of tint from deep red through light red to green is observable. In other cases the colour changes abruptly from red to green. Sometimes the red argillite shows veins of deeper red (Plate VI, fig. 2). The appearance under the microscope with reflected light is of a roughly equidimensional mass of grains, each coated with red, while in places red colouring is gathered in veins which show a deeper tone.

The description of the red clays and shales of Nova Scotia as given by J. W. Dawson (19, p. 26) applies equally to the microscopic appearance of these argillites: "[the colouring-matter] having indeed the aspect of a chemical precipitate rather than of a substance triturated mechanically. In addition to oxide of iron distributed through the beds, there is, in fissures traversing them, a considerable quantity of the same substance in the state of brown haematite and red ochre, as if the colouring-matter had been subperabundant and had been in part removed and accumulated in these veins."

Where the red argillites have been subjected to weathering, cleavage is more pronounced; and often the red colour has been leached out, leaving a light-grey to white product. A similar result of weathering of argillite has been noted by C. Fraser (18, p. 47).

Writing in "The Geological History of New Zealand" (20, p. 164), F. W. Hutton says, when speaking of the Maitai system, "In several localities in both Islands red-jasperoid slates occur, sometimes associated with manganese oxide, and this, together with the paucity of fossils and the general absence of plant-remains, points perhaps to a deep-sea origin." The manganese oxide which occurs in the rocks of the Wellington Peninsula is found in small nests or stringers in the softer strata. In most cases it forms no more than an incrustation sufficient for blowpipe testing: a sample from Duck Creek, Porirua, yielded 5·4 per cent. of MnO_2. It is never found as concentric shells around a nucleus, nor exhibiting mammillated structure, nor yet impregnating a mass of palagonite or forming layers alternate with any such substance—these being the more general modes of occurrence of manganese oxide in deep-sea deposits (21 : 22). Here its occurrence seems to have no more significance than that of iron oxides, and in habit it appears to parallel closely the latter oxides.

Further, the Maitai rocks of Wellington Peninsula, as mentioned before, yield plant-remains, these having been collected on both sides of the argillites at Sinclair Head and elsewhere, while the black colouring-matter of the argillites is presumably carbonaceous. Paucity of fossils is no criterion of deep-sea deposits. The absence of radiolarian cherts and glauconitic sands anywhere in the rocks of the Maitai series, as well as the nearness in the series of conglomerate bands both above and below the red and green slates, are evidence that the Maitai rocks are not of deep-sea origin.

Analyses Nos. 4 and 5 of the accompanying list are of green and red argillite respectively. Save for the proportions of ferrous and ferric iron, the two analyses are almost identical. The total content of iron does not differ by more than 0·4 per cent., but, while in the green argillite the proportions of ferrous and ferric oxide are 3 : 1, in the red argillite the proportions are equal. This shows that the difference in colour is due to the presence of more ferric iron in the red argillite, and indicates that the change of colour has been brought about by the production of ferric iron from the ferrous iron present in the green argillite.

In discussing the formation of red sediments it seems a common premise that the iron-content in the sediment was in the higher state of oxidation at the time of its deposition, either as a hydrated sesquioxide, as supposed by Joseph Barrell (23, p. 286), or already in the form of anhydrous red haematite, in which state I. C. Russell concludes all red sediments were deposited (24, p. 56).

J. D. Dana has attributed the red colour of certain shales to the oxidation of their iron-content by the action of heat resulting from orogenic movements (25). How the oxidation has been brought about is not stated. Some such hypothesis as Dana's seems best suited to the case of the red argillites of Wellington Peninsula.

The connection between the red argillites and the strike-faults of the Maitai rocks is indicated by the map. That the areas of red argillites have suffered from faulting-effects is shown by the fact that a quartz lode is developed in connection with each band. As stated above, these lodes are silicified fault or shear zones (26, p. 135). Genetically, however, they are segregated veins, as distinct from true fissure-veins, and as such have been described by J. Park (27, p. 64). Siliceous solutions, circulating mostly in a downward or lateral direction, may be considered efficient agents in supplying the oxygen necessary to convert the ferrous iron to the ferric state.

In the field the appearance of the green argillites is consistent with the idea of leaching ; the argillites are of a dull greyish-green colour, and, although quite compact as distinct from weather-rotted, they are without the sheen commonly noticed in light-coloured argillites (11, p. 47 ; 29, p. 50).

The effects of vein solutions on country rock composed of ' 'clay slates, greywacke slate, and similar rocks " have been investigated by A. von Groddeck (28). He finds that the result of such action on " variegated slates " will be a leaching-out of iron and magnesia, a loss in sericite, and a gain in quartz. The final product, however, is a slate composed of " quartz and sericite with a little rutile and considerable specular iron."

On comparing the analysis of a typical dark-coloured argillite with that of the red argillite it will be seen that the small differences in the analyses vary in accordance with von Groddeck's results ; while the percentage of silica shows a slight increase in the red argillite, the amounts of alumina and potash are slightly less. The total loss on ignition is high in the case of the dark-coloured argillite, due presumably to the carbonaceous matter present. In the case of the red argillite, water-content is probably responsible for the ignition-loss of 3·42 per cent.

The leached-out products of iron and magnesia are not necessarily lost to the rock ; in the case of the slates investigated by A. von Groddeck the resulting product has " considerable specular iron." Probably the leached iron, as in the case of the Wellington argillites, has been oxidized and redeposited in the rock.

Discussing the origin of red formations, Joseph Barrell (23, p. 290) concludes that the chief factors operating in the production of red shales from those of lighter colour are—(1) Dehydration of iron oxides under great pressure and moderate temperature ; (2) diffusion operating under conditions of warmth and moisture. Oxidation of the ferrous iron might be accompanied by hydration ; if so, the conditions postulated by Barrell for its dehydration would obtain in the neighbourhood of a fault-zone. That some diffusion of the dehydrated iron oxide has taken place is evident from the appearance of the reddened argillite, and is also shown by the fact that in most cases the silica of the associated quartz veins is also slightly reddened. The necessary conditions for this diffusion would also be found as an accompaniment of the faulting.

The changes mentioned above have no doubt been induced as the result of more than one movement of faulting. The folded quartz vein at Sinclair Head points to movement after its formation. In this connection it is interesting to recall the position of the Tertiary conglomerate at 150 ft. above sea-level. That there has been revival of faulting along old fault-lines has been pointed out by C. A. Cotton (13, p. 295).

Origin of the Green Argillites.

It seems clear that the land from the waste of which the Maitai series was derived supported a considerable vegetation, as evidenced by the plant-remains and the general dark colouring of the argillites. The absence of organic matter from some argillites, and its general absence from or less proportion in the greywackes, are points to be explained.

The theory of Joseph Barrell (31, p. 428) to account for the sharp demarcation of shales and sands seems the best so far put forward. He assumes that the waste from a land area is laid down in shallow water, the coarser material in general nearer the shore. A violent storm will be effective in churning up the material and carrying the sand-product to a greater depth, where it is deposited. The silt of this depth is also stirred by the storm, and is in part worked farther seaward, in part settles back in place. This latter portion, being lighter, is held longer in suspension, and is deposited after the coarser sands. Repetition of this process results in the development of such finely demarcated series of sands and clays as are common the world over. With the lighter material held in suspension will be the vegetable matter contained in the rock-waste.

Lack of organic matter in quantity sufficient to impart a dark colour to the argillites may have been due to an actual scarcity of vegetation on the parent land-mass of the time. To assign causes for such scarcity is rather speculative. It is suggestive, however, to note that in three cases the red argillites of Wellington Peninsula are accompanied by bands of diabase tuffs which in the two cases examined appear to underlie the green argillites; and similar association of tuffs and green argillites has been noted elsewhere (14 ; 17, p. 2 ; 35 ; 36). The tuff-beds are proof of energetic volcanic action on the land surface of that time, a condition which would be inimical to plant-growth during and immediately succeeding the time the tuff-beds were laid down.

The presence of carbonaceous matter explains why the argillites, where other conditions are favourable, are not more generally red-coloured. That no oxidation of ferrous iron can take place in the presence of organic matter has been pointed out by J. S. Newberry (32, pp. 7 and 8) and H. Newton (33). These writers refer to oxidation of ferrous iron at the time of deposition of the sediments ; the reasoning may equally be applied to the case of later oxidation.

" RED ROCKS " ELSEWHERE.

The term " jasperoid rock " has frequently been used in describing these developments (14 ; 17, p. 2 ; 35 ; 36). On examining samples from Marlborough and Hawke's Bay I conclude that these rocks did not differ originally from those described above. In the jasperoid rocks silicification has taken place on a microscopic scale throughout the rock, while in the case of the red argillites of Wellington Peninsula the introduction of silica has resulted in the formation of quartz veins. In all cases it seems likely that the various metalliferous ores reported in connection with the red rocks have been introduced as an accompaniment of the silicification.

REFERENCES.

(1.) J. M. Bell. " The Physiography of Wellington Harbour." Trans. N.Z. Inst., vol. 42 (1910), pp. 534–40.

(2.) J. A. Thomson. " Mineral Prospects of the Maharahara District, Hawke's Bay." Mines Statement for 1913, N.Z. Parl. Paper C.–2 (1914), pp. 162–70.

(3.) A. McKay. " Report on the Recent Seismic Disturbances within Cheviot County in Northern Canterbury and the Amuri District of Nelson." Wellington, 1902.

(4.) C. A. Cotton. " Notes on Wellington Physiography." Trans. N.Z. Inst., vol. 44 (1912), pp. 245–65.

(5.) A. McKay. " Report on the Tertiary Rocks at Makara." N.Z. Geol. Surv., Rep. Geol. Explor. during 1874–76 (1877), p. 54.

(6.) A. McKay. " The Geology of the Neighbourhood of Wellington." N.Z. Geol. Surv., Rep. Geol. Explor. during 1878–79 (1879), pp. 131–35.

(7.) J. Hector. Progress Report. N.Z. Geol. Surv., Rep. Geol. Explor. during 1878–79 (1879), pp. 1–41.

(8.) A. McKay. " On the Tauherenikau and Waiohine Valleys, Tararua Range." N.Z. Geol. Surv., Rep. Geol. Explor. during 1887–88 (1888), pp. 58–67.

(9.) J. M. Bell. " The Geology of the Parapara Subdivision, Karamea, Nelson." N.Z. Geol. Surv. Bull. 3 (n.s.), 1907.

(10.) Hatch and Rastall. " Text-book of Petrology : The Sedimentary Rocks." London, 1913.

(11.) J. M. Bell. " The Geology of the Hokitika Sheet, North Westland Quadrangle." N.Z. Geol. Surv. Bull. 1 (n.s.), 1906.

(12.) P. G. Morgan. " The Geology of the Mikonui Subdivision, North Westland." N.Z. Geol. Surv. Bull. 6 (n.s.), 1908.

(13.) C. A. Cotton. " Supplementary Notes on Wellington Physiography." Trans. N.Z. Inst., vol. 46 (1914), pp. 294–98.

(14.) A. McKay. " On the Copper-ore at Maharahara, near Woodville." N.Z. Geol. Surv., Rep. Geol. Explor. during 1892–93 (1894), p. 3.

(15.) J. A. Thomson. " The Petrology of the Kalgoorlie Goldfield." Quart. Journ. Geol. Soc., vol. 69 (1913), pp. 621–77.

(16.) R. A. Daly. " Igneous Rocks and their Origin," New York, 1914, p. 27.

(17.) A. McKay. " On Mineral Deposits in the Tararua and Ruahine Mountains." N.Z. Geol. Surv., Rep. Geol. Explor. during 1887-88, (1888).

(18.) C. Fraser. " The Geology of the Coromandel Subdivision, Hauraki, Auckland." N.Z. Geol. Surv. Bull. 4 (n.s.), 1907.

Transactions.

(19.) J. W. Dawson. "On the Colouring of Red Sandstones and of Greyish and White Beds associated with them." Quart. Journ. Geol. Soc., vol. 5 (1849), pp. 25–30.

(20.) F. W. Hutton. "The Geological History of New Zealand." Trans. N.Z. Inst., vol. 32 (1900), pp. 159–83.

(21.) Murray and Lee. "The Depths and Marine Deposits of the Pacific." Memoirs of the Museum of Comparative Zoology at Harvard College, vol. 38, No. 1, Cambridge, U.S.A., 1909, pp. 28–35, and 143–48.

(22.) Murray and Renard. "Deep-sea Deposits," Report of the Exploring Voyage of H.M.S. "Challenger," 1875–76, 1891, pp. 341–78.

(23.) J. Barrell. "Studies for Students—Relations between Climate and Terrestrial Deposits." Journ. Geol., vol. 16 (1908), pp. 255–95.

(24.) I. C. Russell. "Subaerial Decay of Rocks, and Origin of the Red Colour of certain Formations." U.S.G.S. Bull. 52. Washington, 1889.

(25.) J. D. Dana in Amer. Journ. Sci., 3rd series, vol. 5 (1873), p. 431. Quoted from reference 24, p. 49.

(26.) A. McKay. "On the Gold-mines at Terawhiti." N.Z. Geol. Surv., Rep. Geol. Explor. during 1883–84 (1884), pp. 135–40.

(27.) J. Park. "On the Proposed Low-level Drive, Perseverance Mine, Terawhiti Goldfield." N.Z. Geol. Surv., Rep. Geol. Explor. during 1888–89 (1890), pp. 63 and 64.

(28.) R. Beck. "The Nature of Ore-deposits." Trans. and revised by W. H. Weed. New York, 1909, p. 397.

(29.) P. G. Morgan. "The Geology of the Greymouth Subdivision, North Westland. N.Z. Geol. Surv. Bull. 13 (n.s.), 1911.

(30.) A. McKay. "On the Geology of Marlborough and South-east Nelson." N.Z. Geol. Surv., Rep. Geol. Explor. during 1890–91 (1892), pp. 1–28.

(31.) J. Barrell. "Criteria for the Recognition of Ancient Delta Deposits." Bull. Geol. Soc. of America, vol. 23 (1912), pp. 377–446.

(32.) J. S. Newberry. "Fossil Fishes and Fossil Plants of the Triassic Rocks of New Jersey and the Connecticut Valley." Monograph U.S.G.S., vol. 14. Washington, 1888.

(33.) H. Newton. "Report on the Geology and Mineral Resources of the Black Hills of Dakota." Washington, 1888, pp. 138–39. Quoted from reference 24, p. 52.

(34.) W. H. Hobbs. "Earth Features and their Meaning." New York, 1912.

(35.) A. McKay. "On the Geology of Marlborough and the Amuri District of Nelson." N.Z. Geol. Surv., Rep. Geol. Explor. during 1888–89 (1890), p. 118.

(36.) S. H. Cox. "On Mount Somers and Malvern Hills District." N.Z. Geol. Surv., Rep. Geol. Explor. during 1883–84 (1884), pp. 27–28.

ART. VIII.—*The Younger Limestones of New Zealand.*

By Professor P. MARSHALL, M.A., D.Sc., F.G.S., Otago University.

[*Read before the Otago Institute, 7th December, 1915.*]

Plates VIII–X.

IN previous papers attention has been drawn to the great differences of opinion which have been expressed and published in regard to the age of the younger rocks of New Zealand. It has also been stated that it is very generally the fact that in all those districts where the younger rock-series is well developed there is a conspicuous bed of limestone forming one of its members. In various parts of the country this limestone has been placed in different chronological divisions by geologists. In some localities it has been classed as Cretaceous, in others as Oligocene or Miocene. Identical beds of limestone which have by some authors been classed as Cretaceous have been placed in the Oligocene or Miocene by others. The statements about the age of the limestones have frequently not been based upon the internal evidence found in the limestones themselves, but upon external evidence found in their stratigraphical relations, or in the palaeontological remains found in the beds above or below the limestones in various localities. In some instances both of these methods have failed, and reliance has then been placed upon the lithological resemblance of some outcrop of limestone to another, often situated in some distant part of the country. Up to the present time no attempt has been made to describe with any degree of accuracy the minute structure of these rocks, or to state even generally the nature of the organisms of which they are composed.

It is intended in this paper to make a general statement of this nature, though it is recognized that a full description of the rocks, with a specific identification of the organic remains which occur in them, would require the work of a specialist in each particular group of organisms. In the case of the Echinoderms, which are very abundantly represented, the remains are merely spines or isolated plates, which even in the hands of an expert would not admit of accurate identification. The *Polyzoa* also require very expert knowledge in order that they should be identified in chance sections. In the case of the *Foraminifera* generic classification alone is attempted. The Sponges and *Radiolaria*, which occur more rarely, are referred to in a general way only. Notes are also given on the classification of these limestones by various geologists who have examined them in the laboratory or in the field.

1. WHANGAREI – KAIPARA – BAY OF ISLANDS REGION OF NORTH AUCKLAND.

Within this district limestones cover a large area of country, and two distinct types have been generally recognized — the so-called hydraulic limestone and a semi-crystalline type generally called the Whangarei limestone. Near Whangarei the latter occurs at Horabora, Kamo, and Waro. Near the Bay of Islands the same type is found at Waiomio, and in the Kaipara area at the so-called Gibraltar Rocks at the end of the Pahi Arm. The hydraulic limestone has a much wider occurrence. Typical localities from which specimens have been obtained and are here described are Mahurangi, Port Albert, Wellsford, Kaiwaka, and Limestone Island, in the Whangarei Harbour.

The earliest attempt to classify the limestones of this district was made by Cox.* He calls the hydraulic limestone of Limestone Island the " lower

SKETCH-MAP OF NEW ZEALAND, SHOWING THE LOCALITIES WHERE THE LIMESTONE ROCKS DESCRIBED HERE WERE OBTAINED.

limestone," and correlates it with the Lower Greensand formation of England. The other type—Whangarei limestone—is classed in the Cretaceo-tertiary, and is correlated with the Amuri limestone of North Canterbury.

* S. H. Cox, N.Z. Geol. Rep., 1876–77, pp. 101–4.

A further description of the district was written by McKay.* He correlates the hydraulic limestone with the Amuri limestone in the Cretaceo-tertiary division, but he places the Whangarei limestone in a lower horizon both because of stratigraphical relations in the Whangarei Harbour and because of a clear sequence in a section at Waiomio, where he describes the superposition of the Whangarei as undounted. He classes all the limestones in the Cretaceo-tertiary division.

In the Kaipara district the first description again was written by Cox,† who this time classes the hydraulic limestone as Cretaceo-tertiary. The Whangarei limestone of the Gibraltar Rocks is said to overlie the Cretaceo-tertiary unconformably, and it is therefore placed in the Upper Eocene. Park‡ was the next geologist to describe this area. He makes the hydraulic limestone an equivalent of the Amuri limestone of Cretaceo-tertiary age, and the Whangarei limestone is placed below it conformably and interstratified with the greensands.

The next geologist to visit the district was McKay,§ and in his report he entirely agrees with Park, and also gives a diagram showing the inferior position of the Whangarei limestone to the hydraulic limestone at Waiomio. This opinion has been strengthened by J. A. Bartrum, Lecturer on Geology at the Auckland University College, who assures me in correspondence that the semi-crystalline limestone (Whangarei limestone) lies interstratified in the greensands below the hydraulic limestone near Pahi. Hector and Hutton do not appear to have made any specific references to the stratification or age of the limestones of this North Auckland district.

Park's latest statements in regard to these limestones show a considerable change of view.|| The Whangarei limestone, including the stone of the Gibraltar Rocks, is said to rest conformably on the hydraulic limestone which is still correlated with the Amuri limestone, but both the hydraulic and the Whangarei limestones are now placed in the Cretaceous system.

2. Cabbage Bay.

A limestone similar to the Whangarei stone comes from Cabbage Bay, on the Coromandel Peninsula. McKay, in 1885 and 1897, correlated it with the Whangarei limestone of Cretaceo-tertiary age. Park, in 1897, placed it in the Lower Eocene. McLaren (Parliamentary paper, C.-9, "Geology of Coromandel Goldfield"), in 1900, classed it with Hector's Cretaceo-tertiary or Hutton's Oligocene. In Bull. N.Z. Geol. Surv. No. 4, 1907, p. 56, Fraser and Adams placed it in the Lower Eocene, and described it as a hard compact limestone consisting largely of bryozoan corals, among which *Foraminifera* are sparsely scattered. Park, in 1910, placed this limestone as an equivalent of the Ototara stone (Oamaru) of Miocene age.

3. Kawhia.

Another similar limestone comes from the Wiwiku Island, Kawhia Harbour. McKay, in 1883, describes it as a hard subcrystalline limestone of Cretaceo-tertiary age. He says that it lies below the ordinary limestone of this locality. At Morant Island, half a mile away, there is a greensand

---

* A. McKay, N.Z. Geol. Rep., 1883–84, pp. 115–28.
† S. H. Cox, N.Z. Geol. Rep., 1879–80, p. 17.
‡ J. Park, N.Z. Geol. Rep. 1886–87, pp. 222–28.
§ A. McKay, N.Z. Geol. Rep. 1887–88, pp. 43–53.
|| J. Park, "Geology of New Zealand," 1910, p. 96.

which lies below a flaggy limestone that is apparently identical with that of Wiwiku Island. The greensand contains a variety of fossil *Mollusca*, and of these all that were collected belong to Miocene species. Morant Island is not mentioned by McKay. The ordinary limestone of this locality is well developed on both sides of the entrance of the Rakaunui River.

4. Tata Islands, Nelson.

The limestone found here is also a coarse-grained type, somewhat less crystalline and less compact than the Whangarei limestone. Hutton, in 1885, placed this stone in his Oligocene system. Park, in 1889, correlated it with the Ototara (Oamaru) stone of Cretaceo-tertiary age. This stone contains a number of fossil *Mollusca* which indicate a Miocene age.

5. Mount Somers.

From this South Canterbury locality a coarse limestone has been procured for use as a building-stone. It is still less crystalline and compact than the stone from the Tata Islands. Haast, in 1873, said that this limestone was the equivalent of the Weka Pass stone—the highest member of the Saurian formation (Cretaceous). Cox, in 1884, placed the limestone in the Eocene.

6. Oamaru.

The limestone formation here has generally been called the Ototara stone by the Geological Survey. It has been largely used as a building-stone, with great success. It is not conspicuously crystalline, and is far from compact, but in practice it is found to possess satisfactory resisting-powers to weathering influences. The references to this stone are numerous. Hector and the officers of his Geological Survey consistently placed it in the Cretaceo-tertiary; Hutton always included it in his Oligocene system; and Park has of recent years always considered it of Miocene age. The last observer has described two beds of this stone separated by a bed of fossiliferous greensand.

7. Raglan, Mokau, Te Kuiti.

Limestone occurs in a thick stratum at all of these localities. It appears to have been correlated with the Ototara stone by all observers, and it has therefore been placed in the Cretaceo-tertiary by Hector, in the Oligocene by Hutton, and in the Miocene by Park.

8. Cobden Limestone of Greymouth.

This, again, has on all hands been correlated with the Ototara stone.

9. Amuri Stone.

This is the limestone which occurs in beds of great thickness over a large part of Marlborough and North Canterbury. Very generally it has been correlated with the hydraulic limestone of the North of Auckland. This correlation appears to have been based on its fine-grained lithological nature, for, like the hydraulic limestone, it is practically destitute of molluscan fossils. Hector and the officers of his Geological Survey consistently classed it in the Cretaceo-tertiary, but Hutton always considered it a Cretaceous horizon. Park also has classed it in the Cretaceous system.

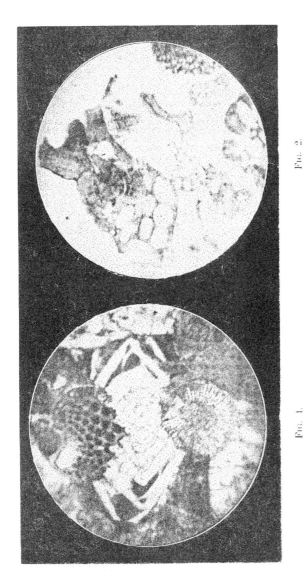

FIG. 1.

FIG. 2.

FIG. 1.—Micro-photograph of limestone from Waiomio, Bay of Islands. × 20. A polyzoan above. *Amphistegina* in the middle, spine of echinoid below.
FIG. 2.—Micro-photograph of Oamaru stone. *Amphistegina* above, with fragments of *Polyzoa* and echinoids. × 20.

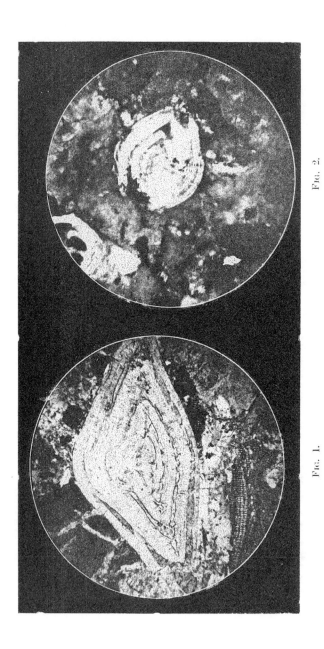

FIG. 1.

FIG. 2.

FIG. 1.—Mokau limestone. Large *Amphistegina*, with *Lithothamnium* below.   × 20.
FIG. 2.—Hoeohora (Whangarei) limestone. A nearly round *Amphistegina*.   × 20.

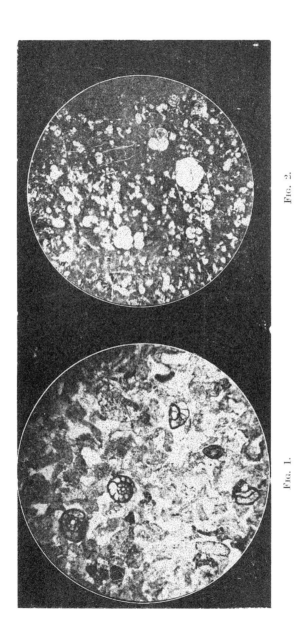

Fig. 2.

Fig. 1.

Fig. 1.—Foraminiferal limestone, Milburn, showing *Rotalia.* × 20.
Fig. 2.—Hydraulic limestone, Kaiwaka, showing *Globigerina.*

DESCRIPTIONS OF LIMESTONES FROM THE LOCALITIES NAMED ABOVE.

A. POLYZOAL LIMESTONES.

(*a.*) *Waiomio Limestone, near Kawakawa, Bay of Islands.* Plate VIII, fig. 1.

Foraminifera of relatively large size frequent: *Truncatulina, Nodosaria,* a thick-walled *Carpenteria, Globigerina,* and a moderate-sized nearly round form of *Amphistegina.* Spines and plates of echinoderms in a perfect state of preservation are very frequent. *Polyzoa* are abundant. There is much glauconite in separate rounded grains. Occasional small grains of fine sandstone. Mr. F. Chapman, of Melbourne, writes, "The Waiomio limestone is of the typical Miocene character."

(*b.*) *Horahora, Whangarei.* Plate IX, fig. 2.

Very similar to the rock from Waiomio, but it contains much less glauconite. *Polyzoa* are somewhat more abundant at the expense of the *Foraminifera.* In particular, *Globigerina* and *Carpenteria* are not in my sections, but *Amphistegina* is rather more abundant.

(*c.*) *Waro, near Whangarei.*

Echinoderm fragments are very plentiful, and, with *Polyzoa,* constitute nearly the whole material of the rock. There are occasional *Foraminifera,* including *Amphistegina* and *Rotalia.* Some secondary calcite has crystallized round the echinoderm fragments. A little glauconite is present, but there are no grains of foreign rock matter. The subcrystalline nature often mentioned in this and similar limestones is entirely due to the frequent occurrence of plates and fragments of echinoderms.

(*d.*) *Gibraltar Rocks, Pahi Arm, Kaipara Harbour.*

A coarse-grained type, mainly composed of *Polyzoa.* Echinoderm spines and plates are large and perfect, with a little secondary calcite in the interstices. *Foraminifera* are not numerous, but *Carpenteria, Globigerina,* and *Rotalia* occur, and the small round *Amphistegina* is fairly common. No *Lithothamnium* can be seen. A little glauconite, but no grains of foreign rock matter. Mr. Chapman says, "The *Amphisteginae* of this rock are of a varietal form denoting rather deep water."

(*e.*) *Limestone Band in Greensand, Pahi, North Auckland.*

Polyzoa and echinoderm fragments are here of less importance, and *Foraminifera* and *Lithothamnium* are in considerable quantity. Of the *Foraminifera, Spiroplecta, Truncatulina,* and *Globigerina* are the most common; but there are many individuals of a smaller type of *Amphistegina* than in the preceding rocks. Very little quartz, but a considerable quantity of glauconite.

(*f.*) *Wiwiku Island, Kawhia Harbour.*

In hand-specimens a coarse-grained type. In section remains of *Polyzoa* and echinoderms are seen to form the greater part of the material. *Amphistegina,* apparently the same species as at Waiomio and Horahora, is quite plentiful. Angular quartz grains are quite frequent. Several specimens of *Rotalia calcar* in the section. The ordinary limestone of the mainland varies considerably in different horizons. *Amphistegina* occurs in a band 100 ft. above the base.

(g.) *Tata Island, Golden Bay, West Nelson.*

Again *Polyzoa* and echinoderms make up the greater part of the rock. *Foraminifera* rather numerous, including *Truncatulina*, *Rotalia*, and a somewhat doubtful specimen of *Amphistegina*. A little *Lithothamnium* is present. There are a great many quartz grains.

(h.) *Mount Somers, South Canterbury.*

This rock consists mainly of *Polyzoa*, though echinoderm remains are frequent. *Foraminifera* are less abundant, and only *Globigerina*, *Operculina*, and the small round species of *Amphistegina* could be distinguished. Grains of glauconite and occasional grains of quartz are to be seen.

(i.) *Castle Hill, Trelissick Basin, Canterbury.*

Polyzoa are again the most frequent organisms in this rock, though echinoderm plates are common. *Foraminifera* are less common, and only *Cristellaria* could be distinguished. There is some *Lithothamnium*.

(j.) *Pahau River, Culverden, North Canterbury.*

In some specimens shell fragments are very abundant, but in others the material is mainly polyzoal. The echinoderm plates are plentiful, and generally in a good state of preservation. *Foraminifera* are generally rather few in number, but a *Truncatulina* and a small round *Amphistegina*, apparently the same species as in the northern limestones, are quite distinct. There is some *Lithothamnium*. A little glauconite. Much secondary calcite and a single quartz grain.

(k.) *Oamaru, North Otago.* Plate VIII, fig. 2.

This is the Ototara stone of the Geological Survey, which has been largely used as a building-stone throughout New Zealand. The stone is less compact than the majority of the limestones, and during the preparation of the sections the organisms become separated to some extent. *Polyzoa*, echinoderm plates, and *Foraminifera* are present in about equal proportions. *Rotalia* is the most common of the *Foraminifera*. *Truncatulina* is not infrequent, and there is an occasional *Amphistegina*. No *Lithothamnium*, glauconite, or quartz are to be seen in the half-dozen rock-slices that were prepared.

Beneath the main outcrop on the north side of the Kakanui Valley, near Clark's mills, there are several narrow bands of limestone interbedded in the soft marls. The lowest of these has a small quantity of *Lithothamnium*, but in other respects these bands are not to be distinguished from the ordinary limestones of the main outcrop.

(l.) *Limehills, Winton, Southland.*

Composed mainly of *Polyzoa*, though *Foraminifera* are rather more frequent than usual. Shell fragments and plates of echinoderms are common. Among the *Foraminifera* there are *Globigerina*, *Rotalia*, *Truncatulina*, and a very small *Amphistegina*. Some small grains of quartz and a few round grains of glauconite.

(m.) *Cabbage Bay, Coromandel Peninsula.*

A great preponderance of *Polyzoa* and relatively little echinoderm material. *Foraminifera* consist of *Globigerina*, *Carpenteria*, and *Amphi-*

stegina. A small quantity of glauconite and one or two grains of foreign rock. This rock closely resembles the limestones from Waiomio and Whangarei described above.

(n.) *Mokau, Taranaki.* Plate IX, fig. 1.

The specimen that I have consists mainly of *Amphistegina* of relatively large size, 2·5 mm. in diameter. In many of the individuals the canal system is most distinctly seen, because all the minute canals are filled with glauconite, as well as some of the chambers. A few *Operculina* are also present, as well as an occasional *Spiroplecta.* Very few *Polyzoa* and echinoderm plates. *Lithothamnium* is quite abundant. A little quartz and some shell fragments.

B. Foraminiferal Limestones.

(a.) *Raglan, West Auckland.*

Globigerina abundant, but larger individuals than in the true oozes of New Zealand. One large *Cristellaria.* *Rotalia* and *Truncatulina* not frequent. No glauconite and no detrital grains. Much of the material of the rock is extremely fine-grained calcite that shows no distinct characters. Mr. Chapman says, " This is a deeper-water type than the Milburn rock, and was formed at a fair distance from the shore-line."

(b.) *Waitetuna, near Raglan.*

Not greatly different from the above, but it contains *Bolivina* as well as *Globigerina,* a fragment of a shell, and a small polyzoan. A small piece of brown hydrous mica. Many slender rods, probably calcified sponge-spicules.

(c.) *Cobden, Greymouth, Westland.*

A relatively fine-grained type, which shows an approach to *Globigerina* ooze. Most of the tests of *Globigerina* are broken. Small *Truncatulina* and *Rotalia* Some carbonaceous matter and brown hydrous mica. A few small quartz grains and some glauconite. Numerous slender rods that may be calcified sponge-spicules.

(d.) *Te Kuiti, South Auckland.*

Very similar to the Raglan types, but containing several small fragments of *Polyzoa.* *Globigerina, Truncatulina,* and *Rotalia* are abundant. The internal cavities of the tests are filled with calcite.

(e.) *Sandymount, Dunedin.*

Globigerina is the most abundant organism, and the cavities of the tests are filled with crystalline calcite. *Truncatulina* and *Spiroplecta* occur as well. A small amount of carbonaceous matter is present, and there are some quartz grains.

(f.) *Milburn, South Otago.* Plate X, fig. 1.

Remains of *Polyzoa* few and inconspicuous. Echinoderm plates fairly frequent, but thin and small material. *Foraminifera* rather abundant, but small. *Globigerina, Rotalia,* and *Spiroplecta* are the most frequent. A few grains of quartz and some glauconite. Mr. Chapman says, " The Milburn limestone is of moderately deep-water origin, and resembles some of our Miocene foraminiferal marls."

C. Globigerina Ooze.

The first six limestones described here are generally called hydraulic limestones. The others are generally classed as Amuri limestone.

(a.) *Kaiwaka, Kaipara Harbour.* Plate X, fig. 2.

The only distinct organism is *Globigerina*, which is present in great abundance. The chambers are generally filled with secondary calcite, though here and there with glauconite. Several sponge-spicules of perfectly isotropic silica. A few radiolarians. Some brown hydrous mica and a very few extremely minute quartz grains. This limestone is associated with a stratum of diatomaceous and radiolarian ooze.

(b.) *Port Albert, Kaipara Harbour.*

Globigerina very abundant, and a small *Dentalina* ? Chambers generally filled with calcite, but occasionally with pyrite. One radiolarian and many calcified sponge-spicules. A small amount of glauconite and brown mica.

(c.) *Port Albert Wharf.*

Much glauconite and a few minute grains of quartz. No *Foraminifera* except *Globigerina* can be seen. There is a great deal of finely granular calcite and some of the brown mica.

(d.) *Wellsford, North Auckland.*

Slice much broken by small veins of crystalline calcite. *Globigerina* is the only organism, but the greater part of the rock is formed of finely granular calcite. There is a little glauconite and some of the brown mica.

(e.) *Limestone Island, Whangarei Harbour.*

A true *Globigerina* ooze, with crystallized calcite occupying the interior of the chambers. Very few minute quartz grains, and some glauconite and some pyrite.

(f.) *Mahurangi, North Auckland.*

Many small veins of secondary calcite. *Globigerina* the only recognizable organism. The larger chambers are crushed. A little glauconite, but no detrital grains.

(g.) *Kaikoura.*

A pure *Globigerina* ooze, with a very occasional grain of glauconite. Some sponge-spicules and several examples of a radiolarian. A little brown mica.

(h.) *Otaio, Canterbury.*

A fine-grained type, with many minute quartz grains and a good deal of glauconite. Mainly *Globigerina*, but one specimen of *Amphistegina*.

(i.) *Amuri, Bluff, Marlborough.*

The organic remains are nearly entirely *Globigerina*, the chambers of which are separated. A few grains o glauconite and sponge-spicules. *Radiolaria* are not infrequent. There is a little of the brown mica.

(j.) Esk River, North Canterbury.

A very fine-grained type, consisting mainly of fine-grained calcite, which does not appear to show any trace of organic origin. One specimen of *Rotalia*. Many *Globigerinae* of small size.

(k.) Amuri Limestone, Weka Pass, North Canterbury.

The chambers of *Globigerina*, which are generally isolated, are fairly numerous. By far the greater part of the rock consists of very finely grained calcite. This appears to be the general structure of the Amuri limestone over the whole of North Canterbury and Marlborough, and even in the highly siliceous and flinty varieties of Ward and the Ure River no remains of siliceous organisms can be distinguished, probably because they have been dissolved, for their presence is clearly shown in small number in the limestones of Amuri Bluff and Kaikoura.

A type of this rock has been found near Oxford, and in this locality it is soft and chalky. Hutton stated that this was the remains of an old coral reef, a most unlikely origin in the light of these descriptions of New Zealand limestones, for in none of them has the remains of any coral been found.

At the Weka Pass a glauconitic quartzose limestone rests on the Amuri limestone. A number of rock-slices from the immediate vicinity of this junction have been examined. It is found that near the junction the Amuri limestone contains a considerable number of grains of quartz sand and some glauconite, as well as some brown mica. The presence of these minerals is associated with the appearance of different and larger species of *Foraminifera*, including *Cristellaria* and *Rotalia*. These characters emphasized to a greater extent are the features that distinguish the overlying Weka Pass stone from normal Amuri limestone. There are small isolated nodules of the Weka Pass stone embedded in the Amuri stone without a sharp boundary between them, In addition, there are also inclusions of the Amuri limestone embedded in the Weka Pass stone. The microscopic structure and relations of these limestones, therefore, serves to indicate that there is a strong resemblance between these stones near their contact, and that such differences as there are would be a natural result of the shallowing of the water and of an increase in the velocity of the ocean-currents. There is independent evidence of the shallowing of the water at this time in the nature of the deposits that everywhere rest on the limestone. From a geological standpoint it is a most remarkable suggestion that a pure *Globigerina* ooze should be overlaid uncomformably, though without change of strike or dip, by another relatively deep-water deposit containing the same microscopic minerals and organisms as the underlying rock, especially when the one rock lies directly on the other continuously over a wide area of country.

OPINIONS SUGGESTED BY A STUDY OF THESE LIMESTONES.

The examination of the structure of these limestones and the recognition of the general nature of their component organisms would naturally be expected to throw some considerable light upon the conditions under which they were formed, and on their age.

So far as the conditions under which they were formed are concerned, we gain a decided amount of information. The limestones are, as a rule, remarkably free from contaminating sediment, and this fact alone indicates that they were formed at some distance from land, or that the land area was small and did not support large rivers that could supply any important quantity of sediment. This idea is further supported by the actual position of some of the limestone outcrops, which may be far inland, and are even in some cases situated within the recesses of the mountain-ranges. The sediment that does occur in the limestones is either extremely fine quartz sand or small fragments of volcanic rock produced by volcanic eruptions of a submarine nature. There is thus strong evidence that at the time or times at which these limestones were formed the land was reduced to comparatively small dimensions.

It is very noticeable that in the microscopic examination of these limestones no remains of any kind of coral has been found. This is the more remarkable when it is stated that in the greensands which often occur above and below the limestones there are frequently a large number of species of *Flabellum* and related species of coral. It is thus evident that the statements which have frequently been made that the Tertiary limestones are wholly or in part the remains of former coral reefs are quite incorrect, so far at least as those localities are concerned from which the specimens described here were obtained. This is the more important when it is realized that the limestones from which these specimens have been described were collected in all parts of the country, and that they include the outcrops which have been referred to by other geologists as of coral origin.

The actual depth of water in which the deposit of these limestones took place is also indicated by the organic remains that occur in the rocks.

Those limestones that consist of *Polyzoa* and echinoid fragments and of the larger types of *Foraminifera* with occasional *Lithothamnium* were deposited in water of no great depth. The soundings round the New Zealand coast show the presence of so-called coral and coral sand at depths of 105–772 fathoms to the west of Cape Maria van Diemen, in places as much as 100 miles from the coast-line. Similar material occurs at a distance of twenty-five miles east of the Bay of Islands in water 305 and 325 fathoms deep. The only other locality in which there is a record of a similar bottom is near the Snares, in water 77–81 fathoms deep. It is probable that this material closely resembles the deposit that developed into the polyzoal type of limestones, such as Whangarei, Cabbage Bay, Wiwiku Island, Tata Islands, Mount Somers, Oamaru, and Winton. The echinoid remains are certainly more abundant in the rock than they are likely to be in the "coral sand," but this may indicate warmer climatic conditions. Thus in the Funafuti report (Roy. Soc. Lon. Rep. of Coral Reef Com.), 1904, p. 329, it is stated, "Detached spines of echinids are in many cores so abundant as appreciably to contribute to the mass of the rock." In those localities where "coral sand" has been found on the sea-floor in the New Zealand area the land is of small dimensions, there are no rivers near at hand to supply sediment, and the ocean-currents flow with greater swiftness than at other places on the coast-line. Here, then, we appear to have important evidence in favour of the idea that the land was of much smaller size at the time when this limestone material was deposited.

In those instances where the limestone contains much *Lithothamnium*, as at Mokau and the Kaipara, it is probable that the depth of water was much less, though it is noticeable that at Funafuti this alga grows at a depth of 200 fathoms. The glauconite, which occurs not uncommonly in the limestone, especially in the polyzoan types, also indicates clear oceanic water; but it also suggests a relatively steep coast-line and peculiar conditions of ocean-currents.

In those types of limestone such as occur at Milburn, Raglan, and Cobden, where the *Polyzoa* are much fewer and the *Foraminifera* more numerous and smaller, it is probable that the depth of water was much more considerable—perhaps some 500 fathoms. The fossil remains found in this type of limestone are relatively few, and they belong to genera which may occur at considerable depths. *Pecten huttoni, Lima laevigata, Pericosmus compressus*, and *Magellania* sp. are those that are most usually found. In these rocks, again, glauconite is usually found in small quantity.

The third type of limestone is represented by the hydraulic limestone of Auckland and the Amuri limestone of Canterbury. It is a pure *Globigerina* ooze, and may have been deposited in water of any depth between 600 and 2,500 fathoms, though it is not very frequent in water of a less depth than 1,000 fathoms. The soundings round the New Zealand coast are most numerous in the extreme north, and here it is found that such an ooze covers the sea-floor at a depth of 1,095 fathoms about twenty miles east of the North Cape; generally, however, it does not occur at distances less than 100 miles from the coast. Probably the hydraulic limestone, which contains a notable quantity of the most finely divided detrital material, is analogous to the "*Globigerina* w. cl." of the charts—*Globigerina* ooze and white clay (?)—which has been found at depths of 600 fathoms and more near the northern part of New Zealand, and at 400 fathoms seventy miles to the east of the Bay of Islands.

The occurrence of siliceous organisms in this limestone at Amuri Bluff, where the deposit is 630 ft. thick, and the highly siliceous nature of the limestone near Kaikoura, Ward, and the Clarence Valley, indicate that in these areas at least the water was of great depth—perhaps more than 2,500 fathoms—for the siliceous organic remains begin to displace the calcareous types. This is the same in the North of Auckland, where much of the hydraulic limestone is free from glauconitic matter, and it is often highly flinty or even siliceous throughout. At Kaiwaka this limestone is associated with radiolarian and diatomaceous ooze.

The nature of the organic remains in the limestones has not been studied with any exactness, though such as the results are they certainly throw some light on their age; but we still have to rely in the main on the nature of the organisms contained in the strata associated with the limestones. It is a matter of very general agreement that the limestones at Te Kuiti, Raglan, Wiwiku Island, Mokau, Tata Islands, Mount Somers, Oamaru, Milburn, and Winton are of Miocene age. In nearly all of these localities these limestones are associated with strata that contain a variety of Miocene fossils. These fossils are mainly molluscan, and they have been regarded as Miocene chiefly because of the high percentage of Recent species that are represented—20 per cent. or more.

The Whangarei limestone is not so easily dismissed. No molluscan fossils have yet been recorded in strata that are associated with it, and it becomes necessary to consider again the nature of the organisms that occur in it. The presence of *Amphistegina* is of great importance in this connection. This genus appears to be .regarded as a typical Miocene genus. Thus Chapman says, "The inequilateral *Amphistegina* took the place of the equilateral *Nummulites* towards the close of the Oligocene, and was the predominant form in many foraminiferal deposits of Miocene age."[*] The frequent occurrence of *Amphistegina* in the Whangarei limestone thus points decisively to a Miocene age for this rock, and this

---

[*] F. Chapman, Mem. Nat. Mus. Melbourne, No. 5, 1914, p. 23.

4—Trans.

organism may be used to correlate all those limestones in which it occurs, for it appears to be the same species in all of them. The stones from the following localities contain this species: Wiwiku Island, Kawhia, Tata Islands, Culverden, Otaio, Mount Somers, Oamaru, and Winton.

It appears to be the general opinion of geologists that the Whangarei limestone is a lower horizon than the hydraulic limestone in the North of Auckland, or at least that the two limestones belong to the same series. If this is the case, the hydraulic limestone also cannot be older than the Miocene. The suggestion is in part supported by the occurrence of *Cucullaea* and *Dentalium* in this rock at Limestone Island, Whangarei Harbour, specimens of which were on view in the Auckland Exhibition of 1913. It is further supported by the occurrence of a varied Miocene molluscan fauna in the greensands below the hydraulic limestone at Pahi.*

The Amuri limestone still remains to be considered. It has already been stated that microscopic examination of the Amuri limestone at Weka Pass, and of the Weka Pass stone which rests on it, has shown that the limestones have an identical composition in kind both mineralogically and organically, and that the differences between these two strata, so far as specimens taken from the immediate neighbourhood of the junction are concerned, are due solely to the difference of proportions of the various constituents. Very few molluscan fossils have as yet been obtained from the Amuri stone. *Pecten zitteli* Hutton is practically the only species. In the Trelissick basin, however, Thomson and Speight have discovered a molluscan fauna in the beds beneath the Amuri limestone, and at the Weka Pass the Weka Pass stone, which rests on the Amuri limestone, also contains a Miocene fauna. Again, the Otaio limestone, which is always regarded as an outcrop of the Amuri limestone, and which is a very fine-grained type of rock with an abundance of *Globigerina*, also contains this *Amphistegina*, which is apparently the same species as in other parts of the country.

This Amuri limestone has always been correlated with the hydraulic limestone of North Auckland. It is quite reasonable to suppose that these two rocks are of the same age. Under the microscope the two cannot be distinguished, and their composition indicates that they were deposited on an oceanic floor in deep water. This implies great depression, which is likely to have taken place during the same earth-movements in localities that are relatively close to one another. If this correlation, which has been so very generally adopted, is correct, the Amuri limestone, like the Whangarei limestone and the hydraulic limestone, must be of approximately the same age as the Oamaru and other limestones that are so largely composed of remains of *Foraminifera*.

The presence of *Amphistegina* in so many of these limestones both north and south supports the correlation of the limestones from another and quite a different point of view. So far as our knowledge of the New Zealand *Foraminifera* goes at present, the genus *Amphistegina* does not occur in our waters. There is no mention of it in Hutton's "Index Faunae Novae Zelandiae," nor in Haeusler's account of the *Foraminifera* in the Hauraki Gulf, nor in Chapman's description of the *Foraminifera* dredged in 110 fathoms west of the Great Barrier. The genus is, however, well represented at the present day in the warmer waters of the Pacific Ocean. Thus Brady† states, "In the living condition *Amphistegina* is distinctly a tropical genus; its home is among the shallow-water sands of warm seas." He

---

* J. Park, N.Z. Geol Surv., 1885, p. 168.
† "Foraminifera," Chall. Rep., vol. 9, p. 741.

further states that it commences at 30 fathoms and continues with some frequency to 300 or 400 fathoms. At Funafuti it occurs at all depths down to 200 fathoms.

The formation of these limestones must, therefore, have taken place during a period of warm climatic conditions. Such a condition of relatively warm climate must be regarded as exceptional in the New Zealand latitudes, and is not likely to have been repeated within any but relatively long-time periods. There is independent evidence that the Miocene period was one of relatively warm conditions in New Zealand, as in many other regions of temperate climate. Within New Zealand such evidence is to be found in the relatively large size of many species of *Mollusca*, of which mention has been made by several geologists. To this may be added the fact that such genera as *Cypraea*, *Trigonia*, and *Chama* then existed in New Zealand, and, with *Trivia*, occurred as far south as Oamaru.

That the younger limestones of New Zealand are all of the same age, and represent approximately the same horizon, has already been suggested by Marshall, Speight, and Cotton.* It was, however, stated in that paper that the probable age of the limestone was Oligocene, this opinion being based upon the classification of various groups of organisms that had been collected from the limestone and submitted to experts for description. The Miocene age, however, is supported not only by the nature of the molluscan fossils that have been found in the limestone horizon itself at Otiake, but in the molluscan fauna below the limestone at Pahi and in the Trelissick basin, and in a similar fauna in the beds resting on the limestone in many localities. The percentage of Recent species in all of these cases amounts to rather more than twenty, a fact that suggests that the age is really the Upper Miocene. As has been stated on many occasions, however, it is probable that too much importance has been attached to this percentage.

The very general occurrence of *Amphistegina* supports the idea that the limestones are of Miocene age. This genus, however, existed in the Oligocene, though it is extremely characteristic of the Miocene. Thus Chapman† says, "The inequilateral *Amphistegina* took the place of the equilateral *Nummulites* towards the end of the Oligocene, and was the predominant form in many foraminiferal deposits of Miocene age."‡

I am greatly indebted to Mr. J. A. Bartrum, Lecturer on Geology at Auckland University College, and to Mr. G. S. Thomson, of the Whangarei High School, for many specimens of limestone from the North of Auckland; to Mr. R. Speight, Curator of the Canterbury Museum, for micro slices of many Canterbury limestones. I am particularly grateful to Mr. F. Chapman, of the National Museum, Melbourne, for the identification of the *Foraminifera* and for some comparative notes.

---

* Marshall, Speight, and Cotton, Trans. N.Z. Inst., vol. 43, 1910, p. 407.

† F. Chapman, Mem. Nat. Mus. Melbourne, No. 5, 1914, p. 23.

‡ Since the geological range of *Amphistegina* is a matter of prime importance in connection with this paper, I have made inquiries of Mr. F. Chapman, of the National Museum, Melbourne, as to the occurrence of the genus. He has kindly sent me the following statement : "*Amphistegina* is (leaving out doubtful forms) Cainozoic and Recent. It occurs very sparingly in one or two localities in the Middle or Upper Eocene, and perhaps Oligocene, in Europe; but is increasingly abundant towards the Burdigalian (Middle Miocene) ; then moderately rare till the Recent formations, when it occurs in all tropical and subtropical seas. In India *Amphistegina* appears to be confined to the Miocene. In Australia it is fairly abundant in the Oligocene, excessively abundant in the Miocene, and thence to Recent more sparingly met with, when it is found living at or near the coasts in low latitudes. Whenever it is found abundantly in fossil deposits one may conclude we are dealing with Miocene strata."

4*

ART. IX.—*Relations between Cretaceous and Tertiary Rocks.*

By Professor P. MARSHALL, M.A., D.Sc., F.G.S., Otago University.

[*Read before the Otago Institute, 7th December, 1915.*]

I. CLASSIFICATIONS OF THE YOUNGER ROCK-SERIES OF NEW ZEALAND.

(*a.*) *General.*

IT is well known that much discussion has taken place in New Zealand in regard to the relation between the Cretaceous and the Tertiary rocks from the stratigraphical and palaeontological points of view. Much difference of opinion has been expressed, and, while some of the views which have been advanced have now been allowed to lapse for want of support, it would be foolish to suggest that any unanimity of opinion exists in regard to this vexed question.

The matter is not one of mere academic importance, but is actually the geological question that carries in its train the most important conclusions in regard to the extent and occurrence of the productive coal-measures in all parts of the country. So far as the coal-measures themselves are concerned, the difference of opinion amounts to this : Some geologists maintain that there are at least two coal-bearing horizons, while others think that there is only one horizon of any importance at any locality, and that this horizon is always at the base of that development of the younger rocks which is found in the locality.

The following table summarizes the classifications of the younger rock-series which have been proposed by various geologists :—

Hector, 1886.	Hutton, 1885.	Park.	Morgan.	Marshall.
Pliocene ..	Wanganui, lignite at base	Wanganui ..	..	Wanganui.
Miocene { Upper } { Lower }	Pareora, coal at base	Oamaru, coal at base	Oamaru, coal at base	The whole series classed
Oligocene ..	Oamaru, coal at base	..	..	as Oamaru series. Coal
Eocene—Upper ..	..	Waimangaroa, coal at base	Waimangaroa, coal at base	at the base of that de-
Cretaceo-tertiary, coal at base	..	..	..	velopment of the system
Cretaceous, Lower Greensand, coal	Waipara, coal at base	Amuri, coal at at base	..	present in any locality.

The rather surprising variety of opinion shown by this table is partly accounted for by the isolated position of many of the coal-basins, and by the somewhat confusing nature of the palaeontological collections which have been made in some of the coal-bearing localities. From the purely stratigraphical standpoint the perfect conformity of the members of the whole series does not give room for much difference of opinion—in those places, at least, where the complete series is developed. This matter has, however, been dealt with at some length in a previous paper,* in which it was shown that at the Waipara and at Amuri Bluff there occurs a complete conformable series of sediments. The lowest beds of this sequence contain distinct Cretaceous fossils, whilst the nature of the abundant fossils in the highest member makes it necessary to regard this as of Upper Miocene age.

More recently sections have been examined in the Trelissick basin by Speight which demonstrate the same fact more clearly than ever.

In the paper just referred to emphasis was laid on the fact that the series in all those three districts in which it is well developed shows in its lower members a sequence of sediments that clearly indicates progressive deepening of the water in which the sediments were deposited. In the upper members of the series the reverse is found, for these strata show that a gradual shallowing of the water was in progress.

The order of superposition of the sediments throughout the country is so similar in its general bearings that it is thought possible, and advantageous, and sufficiently exact, to state the succession of the strata in the following order :—

(8.) Gravels.
(7.) Sands.
(6.) Mudstones.
(5.) Greensands.
(4.) Limestone.
(3.) Greensands.
(2.) Sands.
(1.) Gravels (often with coal).

In those localities, such as Waipara, where the series is complete, fossils which indicate a Cretaceous age are found as high in the formation as the

---

* Marshall, Speight, and Cotton, "The Younger Rock-series of New Zealand." Trans. N.Z. Inst., vol. 43, 1911, p. 378.

lower part of the greensands, which are often of a concretionary nature. In other localities a Tertiary fauna has been found in the upper part of the greensands, though it usually happens that these strata are destitute of fossils.

The order of succession of the strata which is given here is, of course, subject to no small variation. In particular, the greensands, which it is well known require particular and special conditions of temperature and currents for their formation, are not developed in some localities. In others it may happen that thick beds of mudstone occur between the sands and the limestone, and these are sometimes oil-bearing. In all localities as far as known at present there is a considerable thickness of unfossiliferous strata between the Cretaceous horizon and the lowest Tertiary horizon in which fossils occur.

The highly diverse classifications of the younger rocks of New Zealand which have already been quoted clearly show that there is considerable difficulty in harmonizing those details of stratigraphical and palaeontological facts that have been discovered. All of the classifications appear to be ultimately due to the recognition of the fact that the coals of New Zealand are clearly of different age in the various localities in which they occur, for they are associated with different molluscan faunas. It is perhaps as well to state here some of the merits and demerits of each of the classifications.

(b.) *Hector's Work.*

The classification of Hector had its first appearance in the New Zealand Geological Reports, 1876–77, pp. iii–iv, but its first comprehensive description was in 1886, in the " Outline of the Geology of New Zealand," prepared for the Indian and Colonial Exhibition, held that year in London. On page 55 it is stated, " These strata constitute the Cretaceo-tertiary formation, being stratigraphically associated, and containing many fossils in common throughout ; while at the same time, though none are existing species, many present a strong Tertiary facies from both the highest and lowest part of the formation, but even in the upper part a few are decidedly Secondary forms."

His position is, however, more clearly stated in the New Zealand Geological Reports for 1890–91, p. 1. He there states, " Successive efforts to prove the age of the beds by the fossils were disappointing, and led to attempts to correlate the beds with their supposed equivalents by means of stratigraphy and the sequence of the prominent lithological characters of the various divisions of the strata constituting the Waipara formation. Over wide areas a correspondence of the sequence in regular and varying order was traced, and at very many places in both Islands these beds proving richly fossiliferous were largely collected from. But the palaeontological evidence over the whole of South Canterbury and Otago and on the west coast of the South Island was apparently wholly in favour of the Tertiary age of these beds ; while in the north-east, or Marlborough, district of the South Island, and along the east coast of the North Island from Cape Palliser to Cape Kidnappers, and from Poverty Bay to East Cape, the evidence obtained from beds stratigraphically and lithologically the same demonstrated their Cretaceous age."

Other sentences on page li show clearly that Hector thought that no unconformity separated his Cretaceo-tertiary from his Lower Greensand formation. This extract clearly shows that an identical stratigraphical sequence was found which contained in the basal beds fossils of different ages in various localities; but at the same time the faunas were so related in all the localities that Hector considered it advisable to place them in the same series.

This hardly seems to me$_{rit}$ the remarks of Wilckens,* "Hector's whole Cretaceo-tertiary formation, which contains pell-mell a motley collection of Cretaceous and Tertiary fossils alongside of Jurassic and Recent plant-remains, is at least doubtful."

Hector's work clearly shows the essential nature of the problem, which is admitted by all observers: (1.) That the younger series of rocks rests unconformably on an older series, whether Jurassic or Triassic sediments or schists, and on these older rocks the younger series was deposited after a prolonged period of erosion. (2.) The basal members of this younger series of rocks are of Cretaceous age at Amuri Bluff, Waipara, Trelissick basin, East Wellington, Hawke's Bay, North Auckland, and elsewhere. This is amply proved by the fossils being in some cases reptilian types—*Plesiosaurus, Mauisaurus,* &c.; though in others merely *Mollusca,* such as *Trigonia, Inoceramus, Conchothyra, Ammonites,* and *Belemnites.* (3.) In other localities the basal fauna is as clearly Tertiary, which is generally the case in South Canterbury and Otago. (4.) In those places where the basal beds are of Cretaceous age they pass up into strata that are distinctly Tertiary; but, unfortunately, in any particular section, with a considerable thickness of intervening beds which contain no fossil remains.

Actually the only point at issue is whether this change of fauna is associated with a stratigraphical break or whether there is a complete stratigraphical conformity. Opinion based upon geological knowledge and experience obtained in other countries or from reading would at once decide on the former of these alternatives. It is only work in the field in this country that has caused various geologists to adopt the latter alternative, and it may at once be said that this attitude is adopted not only after a close inspection of the stratigraphical sequence, but in accordance with the lithological character of the beds, which demands a gradually increasing depth of water as one rises through the series from the Cretaceous to the Tertiary strata.

(c.) Hutton's Classification.

Hutton's position is perfectly clear, and it is certainly one that must strongly commend itself to those geologists who have not visited the country, or who have not become personally acquainted with the problem in the field. He maintained that the Cretaceous beds are separated from those of Tertiary age by a distinct stratigraphical break.† In the typical districts of North Canterbury, at the Weka Pass, and at the Waipara this break is placed between a *Globigerina* limestone (Amuri limestone) and a glauconitic arenaceous limestone (Weka Pass stone). The former is in these localities the highest member of the unfossiliferous beds which separate the beds containing Cretaceous fossils from those with a Tertiary fauna.

---

* O. Wilckens, "Revision der Fauna der Quiriquina Schichten," Neues Jahrb. für Min., &c., Beil.-Band xviii, p. 279.

† F. W. Hutton, "Geological Position of the Weka Pass Stone," Quart. Journ. Geol. Soc., 1885, p. 266; also, "Sketch of the Geology of New Zealand," Quart. Journ. Geol. Soc., 1885, p. 207.

In regard to this matter it can only be stated that the great majority of geologists who have seen this junction regard it as a conformable one. These include Hector, McKay, Skeats, Marshall, Speight, and Cotton. So far as published statements are concerned, Park is the only observer who has supported Hutton. This support, which is in total opposition to all previous statements of this author, has been accorded within the last few years only. The change of opinion is due to the recent discovery of Tertiary fossils in the Weka Pass stone, which formation Park had for many years placed in the Cretaceous System, and had separated by an unconformity from the well-known Tertiary beds above. Recent examination of the two limestones under the microscope has satisfied me as to their essential identity. As the junction is approached closely the Amuri limestone loses its most characteristic pure *Globigerina* nature and acquires a notable amount of minute quartz grains, as well as some glauconite and particles of a hydrous brown micaceous mineral. These are the characteristic minerals of the glauconitic Weka Pass stone, which rests on it. This surely proves that there is no stratigraphic break.

Hutton claimed that palaeontological evidence supported his position; but this statement must be accepted with caution, as the Amuri limestone is almost wholly unfossiliferous, and those fossils which have been collected in it, such as *Pecten zitteli* Hutton, are certainly Tertiary forms. In addition, the Amuri limestone, both at Waipara and at Amuri Bluff, rests on other beds of considerable thickness which also contain no fossils.

(d.) Park's Classification.

Park's latest position appears to be identical with that of Hutton so far as the relationship of the Cretaceous to the Tertiary beds of North Canterbury is concerned. This position, however, has not yet been fully stated, while there are long statements by this author showing—(1) The conformable relations of the whole series; (2) the conformity between the Amuri limestone and the Weka Pass stone.* Until full reasons are given for this change of front it must be assumed that there is no further information beyond that given by Hutton.

(e.) Morgan's Classification.

Morgan's position is at present based upon coal-occurrences in Westland. He states,† that there is in that district a stratigraphic break between the so-called Eocene and the Miocene; but the evidence of this is of a lithological nature mainly. Insistence in particular is laid upon the occurrence of pebbles of coal derived from the lower members of the series embedded in the upper members. This, however, is not an uncommon occurrence in coal-measures. Thus in Arber's book on "The Natural History of Coal," p. 131, instances are quoted of the occurrence of pebbles of coal in the sandstones of the coal-measures of the South Wales coalfield and that of Bristol, as well as in the Midlands. Further instances are quoted from the coalfields of France. Recently Prouty has recorded the occurrence of large and small pebbles of coal in the detrital beds of the coalfields of Alabama.‡

---

* J. Park, N.Z. Geol. Rep., 1883, p. 33; Trans. N.Z. Inst., vol. 37, 1905, p. 542; "Geology of New Zealand," 1910, p. 88; Geol. Mag., dec. v., vol. viii, 1911, p. 541.
† P. G. Morgan, Trans. N.Z. Inst., vol. 46, 1914, p. 271.
‡ W. F. Prouty, Jour. of Geol., vol. 20, 1912, p. 769.

In New Zealand, pebbles of coal occur in the grit near the Selwyn Rapids, in Canterbury. These beds are stratigraphically lower than the marine fossiliferous beds of Cretaceous age, yet there is no doubt that their coal is derived from the neighbouring coal-seams of Cretaceous age. This occurrence of detrital coal derived from beds of the same series may therefore be regarded as by no means abnormal.

Further attention is called by Morgan to a discordance between strike and dip, and to possibly unconformable contacts; but it is stated by him that such appearances may be due to faulting, which has been pronounced in that district. An overlap of the Miocene over the Eocene is also referred to by him as an evidence of a stratigraphical break, though this does not seem to imply such a structure when it is realized that during the deposition of the series of younger rocks there is throughout the country strong evidence of rapid depression. No palaeontological evidence has been brought forward by Morgan. Bartrum, however, says of this district, "It is a significant fact that wherever both are developed the Miocene series has been found to rest with perfect conformity on the Eocene."[*]

Morgan has recently given a brief statement of the palaeontology of his Eocene.[†] Among the *Foraminfera* is *Amphistegina* or a closely allied genus. This appears to suggest the Miocene, or perhaps Oligocene, rather than the Eocene age. Ten species of *Mollusca* have been identified, and nine of these species are well known to occur elsewhere in New Zealand in beds that have always been regarded as of Miocene age. The tenth species is undescribed. Two species—*i.e.*, 20 per cent. of this small collection—are Recent species. It is difficult to see any reason for supposing that such a fauna could be characteristic of the Eocene as compared with the New Zealand Miocene. Morgan has also reviewed the opinions held in regard to the stratigraphical relation between the Amuri limestone and Weka Pass stone at the Weka Pass.[‡] He considers that there is some discordance, and, in opposition to all other observers, suggests that this discordance may mark the plane of separation between the Eocene and Miocene. He offers no palaeontological evidence.

(*f.*) Marshall, Speight, and Cotton.

The position taken by Marshall, Speight, and Cotton is this: It was recognized that the base of the younger series of rocks is of Cretaceous age in several localities, notably at the Weka Pass, at Amuri Bluff, and at Waipara; and it was maintained that there was a clearly conformable stratigraphical sequence from beds with Cretaceous fossils, through a thick series of unfossiliferous strata, to beds apparently of Miocene age. Since this paper was published Speight has discovered a section of a similar nature in the Trelissick basin where the thickness of the unfossiliferous strata is less, but the conformity of the rock-series is equally evident.

It is a matter of common agreement among all those authors that have been quoted that there are many localities where the fossiliferous Miocene rocks, often with coal at the base, rest directly upon the basement of the older rock-series. This occurrence of Cretaceous rocks at the base in some places and of Miocene rocks at others is the point which caused Hector to establish the Cretaceo-tertiary series, Hutton to insist upon a Cretaceous (Waipara) and an Oligocene (Oamaru) formation, and Park to describe a

---

[*] J. A. Bartrum, Trans. N.Z. Inst., vol. 46, 1914, p. 257.
[†] P. G, Morgan, Bull. N.Z. Geol. Surv. No. 17, 1915, pp. 80, 81.
[‡] P. G. Morgan, Ninth Ann. Rep. Geol. Surv. N.Z., 1915, p. 92.

Cretaceous (Amuri) and a Miocene (Oamaru) division. A similar difference in the age of the basement rocks in Westland caused Morgan to separate the Eocene (Waimangaroa) and Oamaru (Miocene) formations in Westland.

It was stated by Marshall, Speight, and Cotton, that this difference in the age of the basement beds in various sections was the necessary overlap of the younger over the older members of a conformable series deposited during a prolonged period of depression during which the downward movement was more rapid than the building-up by the accumulation of sediment. This rapid depression caused the series to be deposited in the following order of succession :—

> (4.) Limestones.
> (3.) Greensands.
> (2.) Sands.
> (1.) Conglomerates.

It was held that the stratigraphical and the lithological evidence were perfectly definite, but it was admitted that at that time there was no palaeontological evidence in support of the position taken up by the authors. So far as palaeontological work had gone, it appeared that the Cretaceous fossils at the base of the series were quite distinct from the Miocene fossils in the higher members of the series, and, so far as known, could not even be regarded as an ancestral fauna of the latter.

At that time, however, the Cretaceous fauna had never been described; but that great gap in our knowledge has now been partly filled, for Mr. H. Woods has recently described the collections of lamellibranchs and cephalopods from Amuri Bluff and other typical Cretaceous localities. His work has not yet been published, but it is understood that he classes the Amuri Bluff, Waipara, and Selwyn Rapids beds in the Senonian.*

The knowledge of the Tertiary fauna was also in an unsatisfactory state, because the types had not been figured, and for the most part they were not available for study, and the collections had not been closely defined as to locality and horizon. Specific identification was therefore inexact, and it was impossible to find in many places in what portion of a series the fossils had been collected—whether from sands, greensands, limestones, or marls.

Within the last few years extensive collections have been made in definite horizons in various districts—Clarke in the Waitemata beds of Auckland, Speight in the lower Waipara Gorge, Gudex at the Blue Cliffs in South Canterbury, Marshall and Uttley at many localities near Oamaru. These collections have shown that the division of these Tertiary rocks into Upper Miocene, Lower Miocene, Eocene, and Cretaceo-tertiary by Hector, and into Oligocene and Miocene by Hutton, and into various divisions of the Miocene by Park need considerable revision.

Hutton had previously drawn attention to the close relationship between his Miocene (Pareora) and Oligocene (Oamaru) formations by stating that, of 268 species of *Mollusca* found in the Tertiary series, thirty-three species were restricted to the Oamaru, and 184 species to the Pareora, while fifty-one species (a percentage of 19) were common to the two formations; or, of the eighty-four species of the Oamaru, as much as 60 per cent. occur in the Pareora. Marshall, in fairly complete collections near Oamaru, finds in the greensands of Wharekuri, below the limestone, fourteen Recent

* N.Z. Parl. Paper C.–2, Geol. Surv., Ninth Ann. Rep. (n.s.), 1915, p. 76.

species in a collection of sixty, a percentage of 23·3 ; in the horizon of the limestones at Otiake, fifteen Recent species in a collection of sixty-one, a percentage of 24 ; in the upper greensands directly above the limestone, forty-seven Recent species in a collection of 155, a percentage of 33 ; in the next horizon—that of the marls at Awamoa—twenty-one Recent species in a collection of sixty-four, a percentage of 35. Of these strata, it appears that the last two would have been placed in Hutton's Pareora (Miocene) system, and the two former in the Oamaru (Oligocene) system ; yet some 75 per cent. of the species of the limestone horizon occur also in the upper greensands. So far as it goes, this collection of fossils from Oamaru shows a constant increase in the numbers of Recent species. No allowance had been made for the increase in the depth of water in which the sediments were deposited, nor for our lack of knowledge of the fauna of the deeper water off the New Zealand coasts. In this case, however, it is probable that the differences in the depth of water was not very great.

It is noticeable that *Murex* and *Arca* have not been found in the lower beds, while *Exilia* and *Niso* have been found in the lower only. This, of course, may be due to incomplete collecting, and the dominant fact emerges that the species are so similar throughout that the beds obviously all belong to one series; and the palaeontological results confirm the statement of the order of succession of the strata that was based on stratigraphical considerations. Even in the lowest of the strata, however, the fauna is distinctly of a middle Tertiary type when judged by European equivalents or by the percentage of Recent species. As previously pointed out, this latter criterion may be misleading, for it is quite possible that the rate of faunal change in New Zealand, owing to its complete isolation, may be extremely slow. So far as the results that have been obtained at Oamaru up to the present time are concerned, it may be said that there is no indication of a transition from Cretaceous to Tertiary types. Collections have not yet been made from the Black Point beds, where McKay collected some Cretaceous fossils, though Park subsequently collected Tertiary types only.

II. RELATIONS BETWEEN CRETACEOUS AND TERTIARY ROCKS IN OTHER COUNTRIES.

It appears to be generally thought that there is a great stratigraphical break between the Cretaceous and the Tertiary strata throughout the world. Wilckens, in his work on the younger sediments of Patagonia, has expressed this opinion in its most extreme form. He says, "The division between the Cretaceous and the Tertiary is one of the sharpest known in the whole earth's record. Here occurs a break in our knowledge of our planet which has up to the present been maintained complete. . . . We know of no marine sediments which correspond in age to the interval between the two periods."[*] In such sweeping statements it appears that the palaeontological side of our knowledge is given great emphasis at the expense of our stratigraphical information.

In regard to this point, it may be of value to quote the statements made in standard works, since the detailed researches in which the actual original observations of geologists have been recorded are not available. Thus Chamberlin and Salisbury, in discussing the age of some American strata, state, "If the presence of saurian fossils demonstrates the Cretaceous

---

* O. Wilckens, "Die Meeresablagerungen der Kreide- und Tertiärformation in Patagonien," Neues Jahrb. Min., &c., Beil.-Band xxi, 1905, p. 147.

age of the beds containing them, the Arapahoe and Denver beds are Cretaceous, but every other consideration seems to point rather to their inclusion to the early Tertiary. . . . The invertebrate fauna of the Denver beds is little known, and the identified species are common to both the Cretaceous and Eocene."* On page 216, in connection with the relation between these formations in Europe, the authors say, " The break between the Cretaceous and the Eocene was long regarded as one of the great breaks in the geological record, but the hiatus is partially and imperfectly bridged by the estuarine, lacustrine, and other deposits of the early Eocene. It is not to be lost sight of that the one period merged insensibly into the next, even though the strata which recorded the transition are not to be found in every region. In southern Europe the separation of Cretaceous from Eocene is much less sharp, showing that the notable geographic changes of the western region did not affect the southern and south-eastern parts of the continent, or, at least, not to the same extent."

(a.) *Western and Southern Europe.*

The most detailed treatise that is available in regard to the stratigraphy of the geological formations of Europe is the " Traité de Géologie " by de Lapparent. In discussing the Danian, on page 1469 of vol. iii (5th ed.), he says, " Divers auteurs, notamment M. de Grossouvre, se sont fondés sur l'extinctions des ammonites pour faire du danien le premier terme du groupe tertiaire. Ce qui nous détermine à le laisser dans la série néocrétacée, c'est le phénomène de régression qui a marqué la période danienne, et qui semble mieux convenir à la fin d'une époque qu'a l'inauguration d'une ère nouvelle." Here at least it is clear that the demands of stratigraphy are given a place of prior importance to the requirements of palaeontology. Later, where the same author speaks of the Montian formation, he says, " Plusieurs auteurs font du calcaire du Mons le terme inférieur du terrain tertiaire. Pour le maintenir dans la série néocrétacée, nous nous fondons, non seulement sur l'affinité paléontologique de cette assise avec le calcaire pisolithique de Paris, mais aussi sur ce fait que son dépôt correspond à un maximum de régression, précédant la grande invasion marine du début des temps tertiaires " (pp. 1470–71). Here again great importance is attached to the stratigraphical aspect of the question as a criterion for deciding upon the dividing plane between the Cretaceous and the Eocene. On the following pages further statements are made showing that the passage from the Cretaceous to the Tertiary is to be found in a continuous series of conformable sediments in Istria, in the Peloponnesus, in the Deccan of India, and in Tunisia.

(b.) *Pacific Region.*

It is, however, with the Pacific region that we are mostly concerned, and here it will be found that in many localities there is the greatest difficulty in deciding where the Cretaceous ends and where the Tertiary begins. There are at least five regions where marine fossiliferous beds show in the same series both Cretaceous and Tertiary strata. These regions are California, Chile, Patagonia, Seymour Island, and New Zealand. Since the descriptive and critical literature in regard to the geological features of these districts is scattered and not readily obtainable in this country, it is advisable to summarize the main points, as they have a great importance in connection with the geology of New Zealand.

---

* T. C. Chamberlin and R. D. Salisbury, " Geology," vol iii, 1906, p. 158.

(1.) *California.*

The localities are in the great valley of California, near Chico and Martinez. The fossils found there were first described by Conrad,[*] who regarded them as Tertiary, while Gabb[†] thought them to be of Cretaceous age. Later on White[‡] gave a more detailed account of the stratigraphy. He says, " The strata which constitute the Tejon, Martinez, and Chico groups of Gabb form one unbroken series, which rests unconformably on all the rocks beneath it, and on which the Miocene rests conformably. The Tejon portion of the series represents the Eocene ; the Chico portion the closing epoch of the Cretaceous. But there is an alternate mingling of types throughout the whole series, so that no horizon can be distinguished that will separate all the Cretaceous types on the one hand and all the Tertiary types on the other. In other words, there is an unbroken faunal and stratigraphic continuity from the Cretaceous to the Tertiary part of the series."

The same author says elsewhere, " In the case of the Tejon Chico series unbroken marine conditions existed. . . . It is true that on the western border of the continent we find the marine Cretaceous merging into the marine Eocene."[§]

Fairbanks remarks of this series, " The Chico-Tejon has a thickness of at least 20,000 ft. in several places. We have no knowledge at present of a stratigraphic break in the series."[||]

The latest worker on this series is Stanton, who states,[¶] " (1.) In all known sections which contain both Chico and Tejon [faunas] the strata are apparently conformable. . . . (5.) The Chico is characteristically Cretaceous, its so-called Tertiary types being persistent or modern types that have changed but little from the Cretaceous to the present day. (6.) An examination of the species supposed to occur in both the Chico and Tejon reduces the number to not more than six, and with one exception these are all persistent types that cannot be classed as Mesozoic. The one exception is *Ammonites jugalis.* It is held that the Tejon fauna is essentially Eocene, and very distinct from the Chico, even though this ammonite should prove to belong to it. (7.) The time interval indicated by the decided change in faunas cannot now be estimated. In fact, there is little evidence that the later fauna is directly derived from the earlier except in a few species, and it is possible that all the changes took place by extinction and migration of species during the period in which the barren beds between the latest Chico and the earliest Tejon were laid down."

In the Martinez group, which constituted the transition series between the Cretaceous and Eocene of Gabb, fifty-two species of *Mollusca* occur. Of these, Stanton admits four species as occurring certainly in both, and an additional six species as occurring doubtfully in both. Gabb had previously stated that sixteen species occurred generally in both the Chico and Tejon series. Stanton reduces this number to six. He places the division between the Chico and the Tejon in the middle of the Martinez. He gives no stratigraphical diagrams or maps. He further states (p. 1033), " Excepting the

---

[*] Am. Journ. Sci., 44, 1867, p. 376.
[†] " Palaeontology of California," vol. 1, ii.
[‡] C. A. White, Bull. U.S. Geol. Surv. 15, 1885, p. 3.
[§] U.S. Geol. Surv. Bull. 82, 1891, pp. 200, 201.
[||] Jour. of Geol., vol. 3, 1895, p. 433.
[¶] U.S. Geol. Surv., 17th Annual Rep., 1896, p. 1035.

meagre evidence of the occurrence of ammonites in the Tejon, it cannot be held that the fauna of any part of the Tejon contains important Mesozoic elements."

Wilckens has discussed Stanton's work, and states that Stanton " Die vollige zeitliche Verschiedenheit der Chico und der Tejon Gruppe nachgewiesen hat wird. Die Chico Gruppe ist obercretasich, die Tejon eocän."* This statement appears to imply that there is a well-marked division between the beds, a conclusion that is certainly not supported by the quotations from Stanton that have been given above. One must, however, agree with Wilckens that definite lists of the .Chico and Tejon species are urgently required. He appears to regard the Chico as the highest division of the Senonian, and with it he places the Nanaimo division of the Canadian Pacific. He also draws attention to the fact that a detailed description of the Chico is still wanting, and that it is not possible at present to find out from the literature what forms belong to the Chico and what forms are of Tejon age.

Finally, he says that, so far as the literature enables him to form an opinion, it appears that the highest member of the Cretaceous rests in places on the Lower Cretaceous, and in the east on the crystalline rocks at the foot of the Sierra Nevada. The fauna of these beds reveals their age as Upper Cretaceous by means of the *Ammonites* and *Baculites,* and still more by the typical occurrence of gastropods such as *Pugnellus* and *Gyrodes;* and it has a Pacific character. It is more closely related to the Indian and Quiriquina fauna than to that of other American localities. He draws more comparisons between the Chico, Nanaimo, and Quiriquina beds, and shows that the Nanaimo transgressed as did the Chico on the older crystalline rocks.

Even in the Gulf of Mexico region the plane of separation from the Cretaceous is not definitely decided. The latest suggestion is to include in the Cretaceous a stratum which contains a molluscan fauna without cephalopods, and which has, with one exception, only Tertiary genera.†

(2.) *Chile.*

The work of d'Orbigny‡ and of Darwin§ first gave us information of the occurrence of fossiliferous beds in this part of the world. One of the most important localities is the Island of Quiriquina, in the Bay of Concepcion. Of the fossils found here, Darwin remarks, " Although the generic character of the Quiriquina fossils naturally led M. d'Orbigny to conceive that they were of Tertiary origin, yet as we now find them associated with the *Baculites vagina* and an ammonite we must, in the opinion of M. d'Orbigny, if we are guided by the analogy of the Northern Hemisphere, rank them in the Cretaceous system." On page 131 he further says, " From these [stratigraphical] facts, and from the generic resemblance of the fossils from the different localities, I cannot avoid the suspicion that they all belong to nearly the same epoch, which epoch, as we shall immediately see, must be a very ancient Tertiary one." Included in this general statement were the beds of Navidad, 160 miles north of Concepcion and sixty miles south of Valparaiso.

---

    * O. Wilckens, "Revision der Fauna der Quiriquina Schichten," Neues Jahrb. für Min., &c., Beil.-Band xiii, 1904, p. 281.
    † Journ. of Geol., xxiii, 1915, p. 523.
    ‡ " Voyage dans l'Amérique méridionale," 1842, Parties iii, iv.
    § " Geological Observations in South America," 1851, p. 126.

The first really detailed description of these beds and of those at Coquimbó, near Valparaiso, was written by Steinmann and Möricke,* who classed the former as Miocene and the latter as youngest Miocene or Pliocene.

The Tertiary beds of New Zealand are quoted as a parallel because of the occurrence of *Natica solida* Sow. and of *Limopsis insolita* Sow. in both countries, and of the closely related forms in New Zealand to *Scalaria rugulosa* Sow. and *Crepidula gregaria* Sow. which occur in the South American beds.

The comparison by Wilckens of the Quiriquina beds of South America with the New Zealand strata is, however, of greater importance. These Quiriquina beds were first fully described by Steinmann and Möricke, but later, with the aid of more abundant material, by Wilckens. The collections were obtained from two neighbouring localities on the mainland, as well as the Island of Quiriquina itself. The fauna from the three localities is practically identical, and is also of very similar age to that of Algarrobo, 210 miles farther north, near Valparaiso. Wilckens compares the gastropods of these beds with those of other Cretaceous strata in the Pacific region. *Pugnellus uncatus*, however, is not found at Quiriquina, which has the species *Pugnellus tumidus* Gabb. *Eriptycha chilensis* d'Orb. comes near to *Cinulia obliqua* Gabb, and *Pyropsis hombriana* d'Orb. is related to *Pyropsis* species from the Foxhill beds of Chico. *Pugnellus* is here stated to be very characteristic of the Senonian of the Pacific coast. It ranges from the Turonian to the Senonian. Emphasis is laid on the absence of knowledge of the Waipara fauna of New Zealand.

Wilckens continues with the statement that it is obvious that a sea united all those regions with a similar Upper Cretaceous fauna on the margin of the Pacific. This sea overflowed its coasts in later Senonian time, and laid down such deposits as those at Quiriquina. Finally he states that " The descendants of the Quiriquina fauna are found in the Patagonian Tertiary. An important problem lies in their relationship to the Eocene fauna of central Europe, as shown by the many similar species that occur in both.".

The Indian Cretaceous contains in the Ariyalur group a formation generally similar in its fauna to the Quiriquina. The gastropod genera *Pugnellus*, *Gyrodes*, *Pyropsis*, and *Eriptycha* are again represented. Kossmat classes with these Indian strata the Cretaceous beds in Natal, Madagascar, Assam, Borneo, Yesso, Vancouver, and Quiriquina.

(3.) *Patagonia.*

A highly important formation in respect of the relationship of the Upper Cretaceous of the Pacific region to the Tertiary is found in Patagonia. For much of our knowledge of this we are again indebted to Wilckens.† He at once states that this horizon is distinctly higher than that of the Cretaceous of Columbia, Peru, and the Chilian and Argentine Cordillera. Though on the east side of the Cordillera, and resting on a great thickness of Older Cretaceous sediments, instead of crystalline rocks as at Quiriquina, it is still the case that the nearest affinities of the fauna are with the Quiriquina, though there is also a distinct resemblance to the Patagonian Miocene. In

---

* G. Steinmann and W. Möricke, "Die Tertiärbildungen des nördlichen Chili und ihre Fauna," Neues Jahrb. für Min., &c., Beil.-Band x, 1896, pp. 533–612.

† O. Wilckens, "Die Lamellibranchiaten, Gastropoden, &c., der oberen Kreide Südpatagoniens," Ber. der. nat. Ges., Freiburg, Bd. xv, 1907.

many respects the fauna is intermediate between that of the Quiriquina and that of the Patagonian Miocene, for it has many relationships to both. Nevertheless, he regards the Patagonian Cretaceous as the equivalent of the Quiriquina ; the differences that are noticed between the faunas are regarded as due to the 1,000 miles of distance which separates the localities. The Patagonian formation is considered by Wilckens to be equivalent to the Navidad, of Miocene age, and the differences between these two series, though somewhat greater than those between the Quiriquina and the Patagonian Cretaceous, are still of the same order of magnitude.

At first sight, Wilckens says, the Patagonian Upper Cretaceous is apparently Tertiary, especially since, with the exception of *Baculites,* no cephalopods have been found. The bivalve and gastropod fauna, however, contains so many very characteristic Cretaceous genera that no doubt can be entertained that it is of Cretaceous age. *Pugnellus* and *Cinulia* are wholly absent from the Tertiary. *Trigonia, Pyropsis,* and *Struthiolariopsis* also indicate the Cretaceous. Then come also the very close relationships to the Quiriquina (previously described by him as Senonian). The Patagonian Cretaceous contains many elements of the Miocene, and the following eighteen genera are quoted as occurring in both : *Schizaster, Nucula, Leda, Malletia, Cucullaea, Pecten, Mytilus, Amathusia, Corbula, Panopea, Martesia, Dentalium, Scalaria, Galerus, Natica, Turritella, Aporrhais, Bulla.*

(4.) *Magellan Region.*

Here again the exact stratigraphical relations between the Patagonian formation (Miocene) and the Senonian beds has given rise to a large amount of discussion, and a great divergence of opinion has been expressed. The latest description of the fossils in this part of South America is by Steinmann and Wilckens.* This work, however, gives little stratigraphical information on the question, except that the statement is made that fossils indicating the Patagonian molasse (Miocene) were obtained from the deep part of a river-valley where the mapping indicates Cretaceous rocks.

A full discussion of the stratigraphical relations of the various members of the Patagonian region was written by Wilckens† in 1905. He revises the opinions of those geologists who have given descriptive accounts of the stratigraphical relations of the strata. Ameghino speaks of the gradual transition from the Cretaceous to the Tertiary which is to be seen in Patagonia. Ihering speaks of a gradual passage from the Cretaceous fauna to that of the Patagonian formation (Miocene). Hauthal says, in opposition to the statements of Steinmann, that he has found no discordance between the Cretaceous and the Tertiary in south-west Patagonia.

Wilckens, after discussing the statements of these various geologists, and basing his opinion solely on their researches, concludes not only that there is no gradual transition, but also that there was inserted a period of elevation lasting throughout the Eocene and the Oligocene. This period of elevation separates the San Jorge (Upper Senonian or Danian) from the Patagonian formation (Miocene). The reasons that he adduces for this opinion seem rather slender. His classification appears to be mainly based on the statements of Tournouer and Ameghino in regard to the relation

---

* G. Steinmann and O. Wilckens, "Kreide- und Tertiarfossilien aus den Magellanslandern," Archiv for Zoologi, Stockholm, Band iv, No. 6, 1908.

† O. Wilckens, " Die Meeresablungerungen der Kreide- und Tertiärformation in Patagonien," Neues Jahrb. für Min., &c., Beil.-Band xxi, 1905, pp. 98–195.

between the beds with mammalian and reptilian remains (*Notostylops, Pyrotherium,* and Dinosaurs) to the marine beds of *Ostrea pyrotherorium.* His arguments do not appear to be convincing, especially as they are opposed to the observations and opinions of those geologists who actually saw the formations in the field.

To one who merely reads the statements of the various authors it appears quite possible that the whole series was deposited during a period of continuous marine transgression. At any rate, the close relationship between the Senonian and Miocene faunas in various parts of Patagonia and in other countries of South America seems to be a matter of considerable importance in this connection. This close affinity is admitted and is even emphasized by Wilckens, and it appears wholly contradictory to the idea of an elevation and palaeontological break extending throughout the Eocene and Oligocene periods.

(5.) *Antarctica.*

Another locality where Upper Cretaceous and Miocene sediments are found in association is East Antarctica, at Seymour Island. Here there is a distinct Senonian formation, which contains a considerable cephalopod as well as other molluscan fauna. The latter was described by Wilckens,* who classes it as distinctly Upper Senonian, and states that it has an Indo-Pacific character, together with certain elements of its own. It includes such genera as *Lima, Nucula, Malletia, Limopsis, Turritella, Fusus,* and *Cassidaria,* all of which have a large occurrence in the Miocene formation of South America.

Again, there are on the same small island fossil-bearing Tertiary rocks correlated with the Patagonian molasse of Miocene age. The junction of these beds with the Senonian appears to be indistinct. Andersson, in his paper on the geology of Grahamland,† says that no discordance can be seen. Apparently the junction is to be placed in a poorly fossiliferous horizon which shows some false bedding. It is obvious that here the relationships in the Chile, Patagonia, and Magellan district are repeated ; in other words, there is a distinct Senonian fauna clearly ancestral to the Miocene fauna, while from a stratigraphical standpoint no break has been found.

(6.) *New Zealand.*

In this country no lists of fossils which have been collected from those localities generally admitted to be rightly classed as of Cretaceous age have yet been published. It is well known that species of *Inoceramus, Trigonia, Aporrhais,* and *Conchothyra,* as well as *Belemnites* and *Ammonites,* occur in them ; but no species of *Mollusca* similar to those of recognized Tertiary rocks have yet been recorded from them. Collections of the Cretaceous fossils from several New Zealand localities have recently been classified by Mr. H. Woods, but the results of his work are not yet available.

On the other hand, the fossils from the various Tertiary localities have now been moderately well studied, and no fewer than 751 species of *Mollusca*, excluding cephalopods, are classified in Suter's " Hand-list of New Zealand Tertiary Mollusca," 1915. Amongst this large number of species the only two which may be regarded as possessing Mesozoic affinities belong to the

---

* O. Wilckens, "Die Mollusken der antarktischen Tertiärformation," Wiss. Erg. schwed. Südpolarexp., Bd. iii, Lief. 13 (1911) ; "Handbuch der regionalen Geologie," 15 Heft, Band vii, 6, pp. 7, 8.
 † J. G. Andersson, Bull. Geol. Inst. Univ. Upsala, vol. 7, p. 60.

genus *Trigonia*. There is, however, at Brighton, twelve miles south of Dunedin, a pebbly, hard shell-bed containing a belemnite, which rests almost directly on the coal. This shell-bed has been classified by Hector, Hutton, Park, and others as of Tertiary age. The belemnite has not yet been accurately classified. Hector called it *Belemnites lindsayi*, though he afterwards suggested that it might be classed with *Acanthocamax* (*Actinocamax*) of Miller. Specimens sent by me to Otto Wilckens, of Jena, were submitted to specialists in Europe, who stated that they were too much rolled for accurate description, but they were certainly specimens of a true belemnite. The two *Trigoniae* and this belemnite are the only Cretaceous types hitherto admitted to occur in Tertiary rocks in New Zealand. It is, however, true that the collections of Tertiary fossils have been almost entirely made from the upper beds, as the lower are generally quite destitute of fossils.

(v.) Wangaloa.

In those localities where these lower beds are fossiliferous very small collections have been made. One of these localities is Wangaloa, situated on the east coast, about seven miles to the north of the mouth of the Clutha River. These beds were first classed by Hector in 1872 as Upper Tertiary. Hutton in 1875 classed them as Pareora (Miocene). In 1910* Suter described a fossil from this bed as *Turritella semiconcava*. He states that Park, from whom the fossil came, has classed the bed as Cretaceous because of the occurrence of *Conchothyra, Belemnites,* and *Aporrhais*. Park, in the " Geology of New Zealand," 1910, classes the Kaitangata coal-measures, in which the Wangaloa beds occur, as Cretaceous. Hector referred to them more fully in the Geological Survey Reports, 1890–91, p. lviii, and placed them in his Cretaceo-tertiary as a lower horizon than the Ototara stone.

No lists of species have yet been recorded from this locality. The one given here does not pretend to completeness, though some two days were spent in collecting. The fossils occur in concretionary masses in a calcareous sandstone with some glauconite. These concretionary masses often unite and form a continuous stratum in the quartz grits. There is a general agreement that the horizon is a little higher than the Kaitangata coal.

In the collections that were made the following fifty-two species were found. They have been classified by Mr. H. Suter, to whom I am deeply indebted, for without his aid the precision of the identifications would be much less complete, and the lists would have a much smaller value. I give the list as I received it from Mr. Suter :—

> *Gibbula* n. sp., near *G. strangei* A. Ad.
> *Minolia* sp.
> *Bittium* n. sp. ?
> *Cerithiopsis* n. sp. ?
> *Turritella symmetrica* Hutton.
> *Struthiolaria* (*Pelicaria*) n. sp.
> *Struthiolaria* n. sp. Perhaps young shells of the above.
> *Natica australis* Hutton.
> *Polinices gibbosus* Hutton.
> *Ampullina* n. sp.
> *Architectonica* n. sp.
> *Niso neozelanica* Suter.

---

* H. Suter, Trans. N.Z. Inst., vol. 43, 1911, p. 595.

Euthriofusus n. sp.
—— n. sp.
Latirus (*Mazzalina*) n. sp.
—— n. sp., near *Leucozonia straminea* Tate.
Siphonalia compacta Suter ?
Cominella n. sp., near *C. lurida* Phil.
Phos n. sp.
—— n. sp., near *P. liraecostatus* T.-Woods.
Turris sp. ind.
—— n. sp.
—— n. sp.
Surcula fusiformis Hutton.
Daphnella n. sp.
—— n. sp.
Actaeon n. sp.
—— n. sp., near *A. ovalis* Hutton.
Pupa n. sp. ?
Avellana n. sp.
Cylichnella enysi Hutton.
Roxania n. sp.
Haminea n. sp.
Dentalium mantelli Zittel.
—— *pareorense* Suter.
Nucula sagittata Suter.
Malletia n. sp.
Glycymeris n. sp.
Cucullaea alta Sow.
Limopsis aurita Brocchi, juv.
Venericardia difficilis Desh.
—— *zelandica* Desh. ?
—— *patagonica* Sow.
Mactra crassa Hutton ?   Twice the size of the type.
Dosinia greyi Zittel.
Protocardia pulchella Gray.
Corbula zelandica Q. & G.
Panopea orbita Hutton.
Teredo heaphyi Zittel.

The species of *Avellana* sent to Mr. Suter is described in this volume (p. 120) under the name *Avellana paucistriata* Marshall. In addition to the species in this list, *Turritella semiconcava* Suter is abundant, and I also found *Pugnellus australis* Marshall and *Avellana curta* Marshall. The last two of these are described elsewhere in this volume (pp. 120, 121). Suter has stated on Park's authority that *Conchothyra, Aporrhais,* and *Belemnites* occur as well, but I could find none of them, though the occurrence of *Belemnites* in strata of the same age at Brighton makes it quite possible that this genus at least is represented.

The molluscan species in this collection thus total fifty-two, and of these the three species *Pugnellus australis* Marshall, *Avellana paucistriata* Marshall, and *A. curta* Marshall belong to genera which are not known to have representatives in strata higher than the Cretaceous. Of these two genera *Pugnellus* is usually Senonian, but Cossmann mentions no species of *Avellana* in strata higher than the Cenomanian. There are twenty-six other extinct species, and of these one—*Roxania* n. sp.—belongs to a genus not previously recorded among the extinct or Recent fauna of New Zealand. The same remark applies to the two species of the subgenus *Mazzalina* of the genus *Latirus*. The genera *Gibbula, Bittium,* and *Haminea* occur among the Recent fauna, but no extinct species have previously been recorded in New Zealand.

Thus, of thirty-nine genera, as many as four, or 10 per cent., are extinct in New Zealand. Of the fifty-two species, twenty-nine, or 56 per cent., are not represented among the 751 Tertiary species previously known. Only twenty-one species, or 40 per cent., occur among the previously described Tertiary species, and, of these, seven, or 13·5 per cent., are Recent. The Recent species are: *Turritella symmetrica, Natica australis, Venericardia difficilis, V. zelandica, Dosinia greyi,* and *Corbula zelandica.*

This analysis of the Wangaloa fossils shows that the fauna is a very peculiar one, and it is difficult to state the age of the strata in which such an association of organisms occurs. On the one hand, the occurrence of *Pugnellus* and *Avellana* shows an affinity with the Chico and Quiriquina fauna and with the Patagonian fauna of Senonian age. It is true that amongst the fauna of these localities *Eriptycha* or *Cinulia* occurs with the *Pugnellus,* but the replacement of these genera by *Avellana* merely emphasizes its Cretaceous affinities, for there is no previous record of this genus in strata higher than the Cenomanian.

The suggestion of an earlier age than the ordinary Tertiary of New Zealand is further supported by the very high number (56 per cent.) of species which had apparently become extinct before the ordinary Tertiary strata of New Zealand were deposited.

The fact that *Roxania* and *Mazzalina* also occur, though unknown in the Tertiary or Recent of New Zealand, lends further support, as well as the fact that *Niso* has hitherto been recorded only from the low Tertiary greensands of Wharekuri.

On the other hand, the 40 per cent. of New Zealand Miocene species closely relates the Wangaloa beds to that age; and this point is further emphasized by the occurrence of 13·5 per cent. of Recent species.

(w.) Hampden.

The strata at the north end of the Onekakara Beach, near Hampden, have long been known to be fossiliferous. Mantell, in 1851, classed them as Upper Tertiary. Hutton, in 1875 and in 1885, placed them in his Pareora formation, of Miocene age; and the same was done by Park in 1905 and in 1910. McKay alone, in 1887, classed them as Cretaceo-tertiary, basing his opinion on purely lithological and stratigraphical grounds.

Both Hutton and Park have given lists of fossils from this locality. That of Hutton included *Trigonia pectinata* Lam.; but the specimen obtained by him has been recently examined by Suter, and he has shown that it is distinct from the Australian species, and it is now known as *Trigonia neozelanica* Suter. In 1905 Park failed to find this species; and even Hutton appears to have been doubtful about it, for he does not mention it in his list of New Zealand Tertiary *Mollusca* in 1886, and failed to find any more specimens when he visited the locality in that year.

In October, 1915, a visit was paid to Hampden, and no fewer than three specimens of a *Trigonia* distinct from *T. neozelanica* were found. In addition, a species of *Avellana* was obtained. This species is described elsewhere in this volume (p. 121) as *Avellana tertiaria* Marshall. The stratification of the Tertiary rocks of this coast is correctly described by McKay, as follows :—

    (4.) Onekakara sands (*Trigonia, Avellana*).
    (3.) Moeraki boulder beds.
    (2.) Katiki Beach beds.
    (1.) Sandstones and conglomerates with coal.

These strata McKay thought to be conformable, and there does not seem to be any reason to doubt the truth of this statement. In the upper division of No. 1 McKay found fossils that were not named, but were said to be typical species of the New Zealand Cretaceous. No fossils have yet been found in this district in the beds between (1) and (4), in the latter of which there are the two Cretaceous genera mentioned, but also a considerable number of Miocene and even Recent forms. The latter in Hutton and Park's lists amount to as much as 48 and 41 per cent. respectively. Here, then, two genera usually decidedly Cretaceous survived until the fauna had become of a definite Miocene type. This statement is made in accordance with the work of Hutton and Park, as my collections have not yet been worked out. It is, however, noticeable that *Surcula hamiltoni* Hutton, which has hitherto only been found in the greensands below the limestone at Wharekuri and Waihao, is well represented in my collections, though it is not mentioned in the other lists.

It is, of course, the case that *Trigonia* is found in the Australian Tertiaries, especially in the Janjukian, and it is also a Recent genus on the Australian coast. The Janjukian formation has been placed by Pritchard and Hall in the Eocene, and by Chapman in the Miocene period.

This collection at Hampden is particularly interesting in view of the results stated by Wilckens in Chile and Patagonia ; in fact, the occurrence of these forms would probably cause him to class the Hampden formation as Cretaceous.

(x.) Brighton.

This locality lies twelve miles to the south of Dunedin. Here the quartz gravels at the base of the Tertiary rocks with coal strata rest on an eroded surface of schist. For many years the belemnite previously mentioned has been known to occur in these beds, but no other fossils in a state of preservation that allowed of identification had been obtained from it. A new species of *Pecten* has now been found. There are, in addition, many remains of *Ostrea* and a fragment of *Venericardia*. There can be little doubt that this bed is of the same age as the Wangaloa strata, from which it is thirty miles distant. Every geologist has up to the present time admitted its Tertiary age, and it is certainly almost at the base of the Tertiary series.

Thus in these three localities—Wangaloa, Hampden, and Brighton—all close to the base of the younger series of rocks, there are faunas which together include the following species with decided Cretaceous affinities : *Pugnellus australis* Marshall, *Avellana paucistriata* Marshall, *Avellana curta* Marshall, *Avellana tertiaria* Marshall, *Trigonia neozelanica* Suter, *Trigonia* n. sp., *Belemnites lindsayi* Hector.

The occurrence of these fossils would apparently be sufficient to cause Wilckens to classify the beds in the Senonian, judging by his classification of the Quiriquina and Patagonian strata (see p. 112). He has practically used the genera *Pugnellus*, *Cinulia*, *Trigonia*, and *Baculites* for assigning the Cretaceous age to these formations.

The presence of a strong Miocene element in the fauna, on the other hand, is not destructive of this position, as in the South American localities Wilckens emphasizes the fact that the Navidad (Miocene) fauna is the descendant of the Quiriquina (Senonian) fauna, and that the Patagonian Senonian contains many elements of the Patagonian Miocene.

There is in the Wangaloa fauna, however, an important and notable fact in the occurrence of 13·5 per cent. of Recent species, and, according to Hutton and to Park, a much greater percentage of Recent species at Hampden. There appears to be no parallel for this in South America— apparently *Cardium acuticostatum* is the only Recent species in the Senonian.

(y.) Selwyn Rapids.

The first mention of the fossils of this locality was made by Haast.* On page 68 he gives a list of species obtained from these beds. The list includes *Inoceramus, Conchothyra parasitica, Struthiolaria?* as well as many other genera well represented in the Tertiary strata of New Zealand.

Cox, in 1876–77, included these beds in the Cretaceo-Tertiary, a classification that was also adopted by Hector in the same year. Haast, in the " Geology of Canterbury and Westland," in 1879 still includes these Selwyn Rapids beds in the Waipara system, of Cretaceous age. This arrangement was also followed by Hutton in 1885, and Park, in the " Geology of New Zealand," 1910, also groups them as Cretaceous.

The collection made for the Geological Survey by McKay, who recorded *Trigonia* and *Aporrhais*, has been submitted to Professor Woods, who, it is understood, classes them as Senonian.

A small collection was made from these beds in January, 1915, but in nearly all cases the species are too imperfect for exact identification. Mr. Suter, however, says that the following genera are represented : *Panopea, Glycymeris, Myodora, Paphia curta?, Struthiolaria (Pelicaria)* (good specimen), *Conchothyra parasitica.*

The occurrence of *Struthiolaria*, which was doubtfully recorded by Haast but not found by McKay, is important. This genus has not elsewhere been recorded in rocks older than the Miocene. It must consequently be concluded either that New Zealand is the home of this genus, which developed here at an earlier period than in other countries, or that in New Zealand the Cretaceous genera *Inoceramus* and *Conchothyra* lingered on until some Miocene genera had appeared. The other Tertiary genera, though very much more numerous than those of definite Cretaceous age, appear to be represented in the Tertiary rocks of other countries.

III. Conclusions.

The following conclusions are suggested by the considerations detailed in the foregoing pages :—

In New Zealand there is as yet no agreement—(1) as to the division-line between Cretaceous and Tertiary rocks ; (2) as to the age of the oldest series of Tertiary strata. Various formations in Europe and America have given rise to a similar difference of opinion.

In California the Eocene rocks appear to be conformable to the Cretaceous.

Various localities in the south and west of South America have formations which have been assigned by Wilckens to the Cretaceous and to the Miocene,

---

* N.Z. Geol. Rep., 1871–72.

though other authorities have considered the two formations thus classified to be conformable.

The rocks in South America described as Cretaceous contain a number of fossils which belong to genera which are well represented in the Miocene rocks of the neighbourhood.

The Cretaceous found in South America thus contains a distinct Miocene element, which may be regarded as ancestral of the Miocene fauna.

Cephalopods are either very scarce or are absent from these Cretaceous strata.

The strata at Wangaloa, Hampden, and Brighton, admitted to be near the base of the New Zealand younger series of rocks, together contain species of *Mollusca* similar to those which have caused the Chico, Quiriquina, and some Patagonian strata to be classed as Senonian. The Wangaloa beds in particular contain also a large number of extinct species of *Mollusca* belonging usually to genera well represented in the New Zealand Tertiary strata. These species include 50 per cent. of the Wangaloa fauna. There are also at Wangaloa 40 per cent. of species found elsewhere in the Miocene rocks of New Zealand, and of these some 13·5 per cent. are Recent species.

If these strata are not of Senonian age, they clearly indicate that there were several highly important Cretaceous survivals when the deposition of the Tertiary beds began in New Zealand.

If they are of Senonian age, they indicate that the Miocene fauna was already partly developed before the close of the Cretaceous.

At the Selwyn Rapids, where the strata have always been classed as Cretaceous, the distinctly Miocene genus *Struthiolaria* has now been definitely found.

The fauna clearly suggests that a palaeontological break between the Cretaceous and the New Zealand Tertiary either does not occur or that it is of little importance. This conclusion agrees with the result previously stated, that in the opinion of the author and of several other New Zealand geologists no stratigraphical break has been found between these formations.

These results appear to agree satisfactorily with those obtained in western California, Chile, Patagonia, the Magellan area, and Seymour Island in Antarctica.

POSTSCRIPT.

The whole question as to whether the fossils and formations referred to in this paper as Miocene are rightly classed in this way is a matter that will probably be much discussed in the future. The classification as Miocene is based almost wholly on the high percentage of Recent species. It would obviously be better to use New Zealand local names for the horizons of the Tertiary rocks of this country. Thus the Wangaloa series might include all those rocks that contain from 5 to 20 per cent. of Recent species, the Waitaki series from 20 to 40 per cent., and the Wanganui series any higher percentage of Recent species.

Art. X.—*Some New Fossil Gastropods.*

By Professor P. Marshall, M.A., D.Sc., F.G.S., Otago University.

[*Read before the Otago Institute, 7th December, 1915.*]

Plate XI.

The following new species, belonging to the genera *Pugnellus* and *Avellana*, which appear to occur in Cretaceous strata only in other parts of the world, were obtained from the younger strata of Otago during the past year.

Pugnellus australis n. sp. Plate XI, figs. 1, 2, 3.

Shell of moderately large size, turreted, carinate. Sculpture consists of a conspicuous nodulose spiral rib on the carina, showing 9 nodes on the part of the body-whorl that is not hidden by the very large callosity of the lips. Two other less conspicuous nodulous spiral ribs on the body-whorl. A large number of fine spiral striae on the body-whorl—in all, about 50. Nodules on the body-carina extend backward as ridges, gradually decreasing in size, and disappearing near the suture. Nodules also extended forward and bent round towards the aperture. Nodules near the callosity of the outer lip less pronounced. Spire of 4 whorls, each whorl convex, with a nearly flat base. Suture linear, slightly-canaliculate. Aperture narrowly -oval, rounded at the top, but with 2 rudimentary canals at the base. Columella straight. Callosity of the outer lip very large—18 mm. wide. It extends over the top of the spire, and unites with the callosity of the inner lip, and together they cover about half of the body-whorl. The middle part of the outer lip is produced into a prominent rounded claw, with a scaly structure. Columella straight.

Height, 47 mm.; width, 40 mm.

Type and two paratypes in the Otago University Museum, Dunedin.

Locality.—Wangaloa, South Otago.

The genus *Pugnellus* has hitherto been recorded from Cretaceous rocks only. It has rather a wide occurrence, which is, however, almost wholly circum-Pacific. It has been found in the Ripley and the Chico beds of California, the Ariyalur of India, the Quiriquina of Chile, the Cazador of south Patagonia, in the Cretaceous of Borneo and of Libya. It has not yet been found in Australia.

The genus is closely allied to *Conchothyra;* in fact, Wilckens considers the genera may be identical.* *Conchothyra* is well known to occur in beds at the Selwyn Rapids, Waipara, and the Trelissick basin. It is always associated with a Cretaceous fauna.

Avellana paucistriata n. sp. Plate XI, figs. 4, 5, 6, 7.

Shell small, nearly globular. Sculpture a series of about 17 small spiral lines on the body-whorl. Near the base these are narrow, with nearly equal interstices. Towards the middle of the body-whorl they are much wider and flat, with narrow interstices. Those near the suture can hardly be distinguished. In one well-preserved specimen 7 of these spiral ribs are distinct near the base, but there are none on the middle of the whorl, and an additional 2 are developed close to the suture. The interstices are marked by minute transverse ridges. In some specimens 5 spiral lines can be distinguished on the 2nd whorl. Spire short, consisting of 4 whorls.

---

* O. Wilckens, "Die Lamellibranchiaten, Gastropoden, &c., der oberen Kreide Südpatagoniens," Ber. naturf. Gesellsch. Freiburg, Band 15, p. 20, 1907.

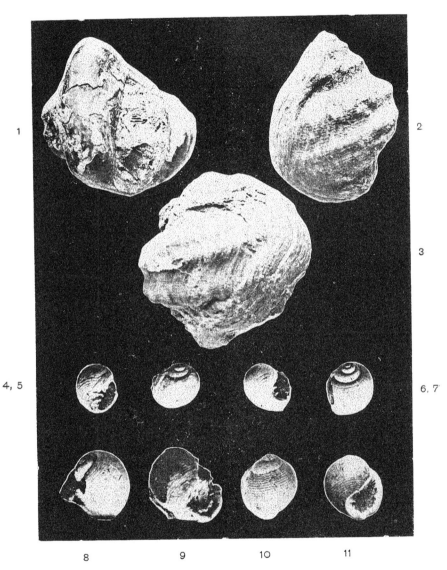

Figs. 1–3.—*Pugnellus australis* n. sp.   Natural size.
Figs. 4–7.—*Avellana paucistriata* n. sp.   × 2½.
Figs. 8, 9.—*Avellana curta* n. sp.   × 2½.
Figs. 10, 11.—*Avellana tertiaria* n. sp.   × 2½.

Each whorl distinctly convex and decreasing rapidly in size. Spire only half the size of the aperture. Suture rather deep. Aperture ovoid, narrowing behind, nearly straight except for an indentation near the base. A wide callosity borders the whole of the outer lip. Inner lip callous. Columella with 2 prominent folds near the base. In one specimen some denticulations can be distinguished on the inner border of the outer lip.

Height, 6 mm. ; width, 5 mm.

Six paratypes in the Otago University Museum, Dunedin.

Locality.—Wangaloa, South Otago.

Specimens sent to the National Museum, Melbourne, with the request to determine whether the mollusc was *Avellana* or *Eriptycha* were returned with the statement that they should be classified with *Eriptycha*. Mr. H. Suter has, however, kindly examined the specimens, and he agrees with me that they belong to *Avellana*.

The genus *Avellana* does not appear to have been collected outside of European countries previously. Thus Cossmann records it from the Albian and Cenomanian of France, Belgium, and England. It does not appear to have been collected from horizons higher than this. The closely allied *Eriptycha*, which is typical of the Upper Cretaceous, occurs somewhat widely in the highest Cretaceous of the circum-Pacific region—viz., California, Chile, and Patagonia ; and Mr. F. Chapman kindly informs me that an undescribed species has been found in the Miocene of Victoria.

Avellana curta n. sp.   Plate XI, figs. 8, 9.

Shell small and globular. Sculpture of body-whorl consisting of about 27 spiral ribs of approximately equal size, flat on the top. Interstices rather narrow, and minute narrow transverse ribs cross the interstices. Sculpture of 2nd whorl 7 spiral ribs. Spire very short, consisting of 4 whorls, each of which is very slightly convex. Suture not strongly marked. Aperture oval. Callosity of the outer lip broad, but apparently without denticulations on its inner surface. Inner lip callous, but the form of the columella is not distinct.

Height, 9 mm. ; width, 7½ mm.

Three paratypes in the Otago University Museum, Dunedin.

Locality.—Wangaloa, South Otago.

This species is very similar to *A. tertiaria* from Hampden, but is distinguished from it by its larger size, shorter spire, more irregular ornamentation, and by the absence of denticulation of the outer lip.

Avellana tertiaria n. sp.   Plate XI, figs. 10, 11.

Shell small, nearly globular. Sculpture 29 flat spiral ribs on the body-whorl. They are of approximately equal size, though somewhat narrower near the base. Interstices narrower than the ribs, and crossed by numerous minute transverse ridges. Spire short, consisting of 4 whorls about ¼ the length of the aperture. Ten small spiral ribs in the 2nd whorl. Outer lip with a callosity 1 mm. wide, marked with 7 striations parallel to its margin. Outer lip marked internally with 13 denticulations, three of which are much larger than the others. Inner lip callous, but the callosity extends only slightly from the aperture. Columella nearly straight, with 2 narrow folds near the base.

Height, 7 mm. ; width, 6 mm.

Holotype, Otago University Museum.

Locality.—Onekakara Beach, near Hampden.

Art. XI.—*Preliminary List of* Mollusca *from Dredgings taken off the Northern Coasts of New Zealand*

By Miss M. K. Mestayer.

Communicated by Dr. J. Allan Thomson.

[*Read before the Wellington Philosophical Society, 27th October, 1915.*]

Plate XII.

Rather more than a year ago my father and I received from Captain Bollons samples of some dredgings he had taken—(A), 15′ S. of the Big King, Three Kings Group, 98 fathoms; (B) 6′ E. of the North Cape, 73 fathoms; (C) 6½′ E., 5′ N. of the North Cape.

When we came to clean and examine the material, that from the Big King proved to be very rich in *Foraminifera* and in the smaller and minute *Mollusca*, there being none of the large forms in it. The other two—(B) and (C)—were not nearly so rich in *Mollusca*, but very rich in *Foraminifera*. So far as I know at present, there is only one mollusc new to New Zealand in either of them, but I have not yet had time to thoroughly work out the whole of the material.

On the other hand, the Big King dredging has yielded far more material, in which are several interesting forms. Two of the genera (*Discohelix* Dunker and *Styliola* Lesueur) are new to the New Zealand fauna. Another one (*Columbarium* von Martens) has only once before been recorded as occurring in New Zealand, a single specimen having been found by the "Terra Nova" Expedition in 1911. Then there are four other species which have not previously been recorded as occurring in New Zealand—viz., *Atlanta lesueuri, Limacina inflata, Brookula* sp. ? and *Liotella incerta*, notes on which will be found at the end of the list.

About the middle of this year Dr. J. Allan Thomson suggested that a list of the species in these dredgings would be of interest. Unfortunately, only a small number of the species are in this list, as owing to lack of time the majority of the bivalves have not been examined (it will be noted that only four or five are included), and there are also a considerable number of species that are very difficult to name, as they do not seem to be described and figured by Mr. Suter in his valuable "Manual of the New Zealand Mollusca," 1914.

Following Dr. Thomson's advice, the list itself is arranged in alphabetical order for easier reference, but the notes are in systematic order. The last note refers to a new species of *Typhis*, of which I have found two specimens—one, chosen as holotype, from a dredging off the Poor Knights Islands, and the other from a dredging off the Hen and Chickens Group, both dredgings taken by Captain Bollons.

In the table which follows, the asterisk before a species shows there is a note upon it. A = 15′ S. of Big King, 98 fathoms; B = 6′ E. of the North Cape, 73 fathoms; C = 6½′ E., 5′ N. of the North Cape, 75 fathoms.

—	A.	B.	C.
Anomia furcata Suter	×		
Atlanta lesueuri (d'Orbigny). Plate XII, fig. 7	×	×	
Brookula sp. ? Plate XII, fig. 4	×		
Cadulus spretus Tate and May	×	×	×
Calyptraea scutum Lesson	×		
Cavolina inflexa (Lesueur)	×	×	
—— *telemus* (Linné)	×	×	
—— *trispinosa* (Lesueur)	×		
Cocculina compressa Suter		×	×
—— *tasmanica* (Pilsbry)	×		
—— *clypidellaeformis* Suter		×	×
Columbarium suteri E. A. Smith. Plate XII, fig. 8	×		
Cominella nassoides (Reeve)	×		
Cylichnella pygmaea (A. Adams)			×
Daphnella amphipsila Suter	×		
—— *aculeata* Webster	×		
Dentalium nanum Hutton	×		
—— *huttoni* T. W. Kirk	×		
—— *arenarium* Suter	×		
—— *ecostatum* T. W. Kirk	×		
Discohelix hedleyi n. sp. Plate XII, fig. 6	×		
—— meridionalis Hedley			×
Divaricella cumingi (Adams and Angas)	×		
Drillia laevis (Hutton)	×		
Epitonium zelebori (Dunker)	×		
Lima lima (Linné)	×		
Limacina inflata d'Orbigny. Plate XII, fig. 1	×	×	×
Liotella incerta (Tenison-Woods). Plate XII, fig. 5	×		
Liotia polypleura Hedley	×	×	
—— rotula Suter	×	×	×
Lissospira corulum (Hutton)	×		
—— micra (Tenison-Woods)	×		
Malletia australis (Quoy and Gaimard)	×		
Marginella albescens Hutton	×		
—— *pygmaea* Sowerby	×		
—— *hebescens* Murdoch and Suter	×		
—— *plicatula* Suter		×	
Monilea semireticulata (Suter)	×		
Natica zelandica Quoy and Gaimard	×		
Neojanacus perplexus Suter	×		
Philine umbilicata Murdoch and Suter	×	×	
Philobrya costata (Bernard)	×		
Protocardia pulchella (Gray)	×		
Pyramidella pulchra (Brazier)	×		
Rissoa suteri Hedley	×		
—— exserta Suter	×		
—— fumata Suter	×		
Scissurella regia n. sp. Plate XII, fig. 3	×	×	
Siphonalia nodosa (Martyn)	×		
Styliola subulata Quoy. Plate XII, fig. 2	×		
Tornatina charlottae Suter	×		
—— *cookiana* Suter	×		
—— *biplicata* Suter	×		
—— *murdochi* Suter		×	×
—— *tenuilirata* Suter	×		
—— *decapitata* Suter	×		×
Turbonilla zelandica (Hutton)	×		
Turris augusta (Murdoch and Suter)	×		
Turritella difficilis Suter	×		
Typhis pauperis n. sp. Plate XII, fig. 9	Poor Knights.		
—— n. sp. Plate XII, fig. 9a	Hen and Chickens.		

Limacina inflata d'Orbingy.   Plate XII, fig. 1, 1a.

"Structural and Systematic Conchology," Tryon, vol. 2, p. 94, pl. xlii, fig. 22.

This species is new to our New Zealand fauna; it is rather plentiful in the Big King dredging, and scarcer in (B) and (C).

Styliola subulata Quoy.   Plate XII, fig. 2.

"Structural and Systematic Conchology," Tryon, vol. 2, p. 91, pl. xlii, fig. 6.

This genus has not apparently previously been recorded as occurring in New Zealand.   It is certainly rather rare, as I do not remember seeing it in any other dredging.

Scissurella regia n. sp.   Plate XII, fig. 3, 3a, 3b, 3c.

Shell small turbinate.   Whorls 3, slightly angled, upper surface lightly convex, under-surface more convex, last whorl large.   Protoconch of 2 whorls, minute, smooth, dull.   Sculpture: 2 fine sharp raised spiral keels, near together on the angle of the whorls.   Fine flexuous riblets on both the surfaces, but on the under-surface they are crossed by fine spiral lirae, which are more distinct round the umbilicus.   Umbilicus small, open.   Aperture obscurely triangular.   Anal slit between the keels on the outer lip, long and more contracted at the edge of the lip than at its upper end.   Operculum unknown.   Colour white.

Material.—The holotype in my collection.   Paratypes: Two perfect, five imperfect, in the Dominion Museum; one perfect in the Australian Museum, Sydney; one perfect in Mr. Suter's collection; six perfect and thirteen imperfect ones in my collection.

Remarks.—Mr. C. Hedley, of Sydney, considers this is a good species, not hitherto described; he says, "It is more finely sculptured than one I described from this coast, *Scissurella australis*" (Mem. Austral. Mus., vol. 4, 1903, p. 329, fig. 63).   Later on he suggested that it might be *S. mantelli* Woodward (*vide* "Manual New Zealand Mollusca," 1914, p. 88, pl. vi, fig. 10), but it does not at all resemble that species, being more depressed and quite differently sculptured.   It slightly resembles *S. crispata*, but is quite distinct from it, while it is very much more nearly allied to *S. australis* Hedley, the general shape of the shells, especially the spires, being very similar.   It seems to be rather a rare form, occurring only in quantity in the Big King dredging, with a single specimen in (B).

Brookula sp. ?   Plate XII, fig. 4.

Mr. Hedley classed this specimen as belonging to this genus, but owing to lack of time it has not yet been specifically identified.   (See Trans. N.Z. Inst., vol. 47, p. 444, for generic name).

Monilea semireticulata (Suter).

Cf. Trans. N.Z. Inst., vol. 47, p. 439.

Liotia polypleura Hedley, and **Liotia rotula** Suter.

Trans. N.Z. Inst., vol. 47, p. 442.

Liotella incerta (Tenison-Woods).   Plate XII, fig. 5.

Proc. Roy. Soc. Tasmania, 1876, fig. 5, p. 148.   For *Liotella* see Trans. N.Z. Inst., vol. 47, p. 442.

Lissospira corulum (Hutton).

Trans. N.Z. Inst., vol. 47, p. 443.

Lissospira micra (Tenison-Woods).

Trans. N.Z. Inst., vol. 47, p. 444.

Cocculina compressa Suter.

Man. N.Z. Mollusca, p. 174, pl. 34, figs. 14, 14A, 1914.
The specimens from (B) are rather variable, being mostly wider and shorter than the type, but one small one agrees fairly well with it. I am almost inclined to think them worthy of being ranked as a variety. It may be possible to settle this point next year, as I have specimens from other localities which could be compared with each other and with the type, which is in my collection.   The single example from (C) is true to type, though considerably smaller.

Rissoa suteri Hedley.

Trans. N.Z. Inst., vol. 47, p. 449.

Rissoa fumata Suter.

Trans. N.Z. Inst., vol. 47, p. 450.

Rissoa exserta Suter.

Trans. N.Z. Inst., vol. 47, p. 453.

Omalaxis Deshayes.

Trans. N.Z. Inst., vol. 47, p. 461.

Discohelix Dunker.

There seems to be some confusion in the use of the generic names *Omalaxis* Deshayes and *Discohelix* Dunker.   In this paper I follow Mr. C. Hedley's use of *Discohelix*.

Discohelix meridionalis Hedley.

One broken specimen from (C), which unfortunately broke still more while under examination, but as it is the first record of it from New Zealand I sha'l preserve it till a better specimen is obtained.   It is undoubtedly very rare.   (Mem. Austral. Mus., vol. 4, 1903, p. 351, fig. 74).

Discohelix hedleyi n. sp.   Plate XII, fig. 6, 6a, 6b.

Shell small, solid, rotate, both upper and under surface slightly concave; sides slightly concave on each side of the central keel.   Colour white.   Whorls almost square in section, with a keel projecting at the upper and lower corners, and another midway between them.   Sutures only lightly marked. Sculpture : Fine flexuous growth-lines. Aperture full size of the whorl. Animal unknown. Operculum unknown.

Material.—The holotype in my collection. One paratype in the Dominion Museum; one paratype in the Australian Museum.

Remarks.—The holotype is a younger but more nearly perfect specimen than the paratype in the Dominion Museum; unfortunately, the other paratype is even more imperfect. There are two Recent species known from the American coast, in deep water—*Discohelix nobilis* Verrill and *D. lamellifera* Dall—which were only mentioned in the Trans. Wagner Free Inst. Sci. Philadelphia, 1890, vol. 3, page 331, but I could not find them figured. The fossil *D. retifera* Dall is slightly similar in general shape, but has no keels. There seems to be also a Mediterranean species, but I could not find a figure of it.

Atlanta lesueuri (d'Orbigny). Plate XII, fig. 7.

Described in " The Zoology of the Bonite " (pl. xx, fig. 8); also recorded in the " Challenger " Rep., Zool. (vol. 23, 1888, p. 40). Mr. Hedley kindly identified this species, giving the references, and adding that it was an interesting record for New Zealand. The genus had been previously known to occur in New Zealand, but it had not been possible to identify the species. (Man. N.Z. Mollusca; p. 352, 1914.)

Columbarium suteri E. A. Smith. Plate XII, fig. 8. Fam. *Fusinae.*

A single example of this species was obtained by the " Terra Nova " Expedition, 1910, at Station 134, near the North Cape, New Zealand, and was described by Mr. E. A. Smith in " The British Antarctic (' Terra Nova ') Expedition," 1910 (vol. 2, Moll. No. iv, 87, pl. 1, fig. 30). The type is in the British Museum. As this volume is not readily accessible, I append Mr. E. A. Smith's account of it. The specimen figured on Plate XII, fig. 8, was found in the dredging off the Big King, and is in my collection : it is about the same size as the type.

" Shell slenderly fusiform, with angular coronate whorls, dirty-whitish, with pale-brown spots between the short spines which adorn the middle of the whorls; periostracum pale-straw-coloured, deciduous; the 2 apical whorls large, smooth, obtuse at the top, the rest sloping above the middle, which is prominently carinate, the keel being produced into short spines or acute tubercles, 10 on the last whorl. Below the keel the volutions are contracted to the suture, which is oblique; above the carina, on the last and penultimate whorls, there are 3 fine spiral threads, and below it, on the last whorl, there are 3 rather coarser threads, below which the rest of the slender rostrum is covered with oblique, very much finer threads. The keel has 1 or 2 spiral striae upon it, and the whole surface exhibits fine but distinct striae or lines of growth : aperture somewhat triangular above, produced below into a very slender straight canal; outer lip thin, angled at the keel, faintly or shallowly sinuated above it : columella covered with a thin glossy callus, which extends from the tip of the canal to the outer lip above.

" Length, 17 mm.; diameter, 6 mm. Aperture, with canal, 11 mm.
" Length, 17 mm.; diameter, 6 mm. Aperture, with canal, 11 mm.
" The unique specimen, judging from the protoconch, is merely the young stage of a shell which attains larger dimensions. It consists only

of 6 whorls, but its characters are so striking that I have not hesitated to found a new species upon it.

"In general form it considerably resembles *C. spinicincta* Martens, from east Australia, but it differs considerably in the details of its ornamentation.

"The genus *Columbarium*, which, as far as at present known, consists of a very few species, has not hitherto been recorded from New Zealand. I have associated with this species the name of Mr. Henry Suter, as a mark of appreciation of the immense industry displayed in the production of his 'Manual of the New Zealand Mollusca,' published in 1913. Although · it may be necessary to revise the nomenclature in a considerable number of instances, and occasionally to correct the synonymy, there can be no doubt that this will always remain *a* standard, or even *the* standard, work on New Zealand *Mollusca*. To have produced such a volume, of 1,120 pages, without the advantage of consulting such complete libraries and collections as we have in this country reflects the greatest credit upon the author."

Siphonalia nodosa (Martyn).

Trans N.Z. Inst., vol. 47, p. 464.

Typhis pauperis n. sp.   Plate XII, fig. 9, 9a.

Shell muriciform, small; body-whorl more than half the length of the shell. Apex bluntly acuminate. Protoconch of 2 whorls, the first transparent, the second opaque and tinged with brown, slightly tilted to the right, smooth. Whorls 4, rapidly increasing in size, convex, sharply angled, bearing tubular spines directed slightly backwards and upwards; of these, there are 4 on the last whorl, and 4 smaller, slightly ragged varices terminating in single spines on the angle, which are recurved towards the preceding whorls. Suture well marked. Aperture roundly ovate, lips raised and free. Canal slightly longer than the aperture, closed, and lightly curved to the right. On the · left there are the remains of 3 former canals. Umbilical fissure very narrow. Operculum unknown. Colour creamy white, with a faint purplish tinge on the angle of the shoulder, which is lightly polished. Height, 8 mm.; breadth, 5 mm.; aperture, 2 mm.; canal, almost 2 mm.

Material.—The holotype, from the Poor Knights Islands, 58–60 fathoms; and one paratype from near the Hen and Chickens Islands, Hauraki Gulf, 25–30 fathoms. The paratype, which is a younger shell with a more narrowly ovate aperture, is in the Dominion Museum, Wellington.

Remarks.—The only certain previous record of a Recent species of *Typhis* is Mr. Suter's, of a specifically indeterminable species from the Great Barrier Island, 110 fathoms. There is, however, a fossil species in the Mount Harris beds, *T. hebetatus* Hutton = *T. McCoyi* Tenison-Woods, for which see Suter, "Revision of Tertiary Mollusca of New Zealand" (N.Z. Geol. Surv. Pal. Bull. No. 3, p. 28), which possesses 5 to 6 spines on each varix, and is thus quite distinct from *T. pauperis*. Of foreign Recent species, *T. yatesi* Crosse, from South Australia, apparently comes nearest, but this is a stouter and more subquadrangular species.

In conclusion, I wish to thank Dr. J. A. Thomson for his generous help in the preparation of this paper, and Mr. J. McDonald for preparing the accompanying plate; also Mr. C. Hedley, of Sydney, who kindly identified some of the specimens. At the same time I should like to acknowledge my deep indebtedness to Mr. H. Suter for his great kindness at other times, and for the immense help his book is to New Zealand shell-lovers.

EXPLANATION OF PLATE XII.

(All figures greatly enlarged, except figs. 8, 9, and 9a.)

Fig. 1. *Limacina inflata* d'Orbigny.
Fig. 1a.   ,,   ,,
Fig. 2. *Styliola subulata* Quoy.
Fig. 2a.   ,,   ,, (larger specimen).
Fig. 3. *Scissurella regia* n. sp. (type).
Fig. 3a.   ,,   ,, (paratype). In H. Suter's collection.
Fig. 3b.   ,,   ,,   ,,
Fig. 3c.   ,,   ,, (portion of sculpture on type).
Fig. 4. *Brookula* sp. ?
Fig. 5. *Liotella incerta* (Tenison-Woods).
Fig. 6. *Discohelix hedleyi* n. sp. (type).
Fig. 6a.   ,,   ,, (paratype). In Dominion Museum.
Fig. 6b.   ,,   ,,   ,, In Australian Museum.
Fig. 7. *Atlanta lesueuri* (d'Orbigny).
Fig. 8. *Columbarium suteri* E. A. Smith. Enlarged 1½ diameters.
Fig. 9. *Typhis pauperis* n. sp. (type) Enlarged 2 diameters.
Fig. 9a.   ,,   ,, (paratype). Enlarged 2 diameters

Art. XII.—*List of* Foraminifera *dredged from 15′ South of the Big King at 98 Fathoms Depth.*

By R. L. Mestayer.

Communicated by Dr. J. Allan Thomson.

[*Read before the Wellington Philosophical Society, 27th October, 1915.*]

Biloculina depressa d'Orbigny.
—— *ringens* Lamarck.
—— *elongata* d'Orbigny.
—— *comata* Brady.
—— *sphaera* d'Orbigny.
Miliolina circularis Bornemann.
—— *valvularis* Reuss.
—— *seminulum* Linné.
—— *auberiana* d'Orbigny.
—— *subrotunda* Montagu.
—— *oblonga* Montagu.
—— *insignis* Brady.
—— *bicornis* Walker and Jacob.
—— *ferrussacii* d'Orbigny.
—— *cuvieriana* d'Orbigny.
—— *agglutinans* d'Orbigny.

Miliolina tricarinata d'Orbigny.
—— *rotunda* d'Orbigny.
Spiroloculina tenuisepta Brady.
—— *excavata* d'Orbigny.
Planispirina contraria d'Orbigny.
—— *celata* Costa.
Cornuspira foliacea Philippi.
—— *carinata* Costa.
—— *involvens* Reuss.
Psammosphaera fusca Schulze.
Jaculella acuta Brady.
Hyperammina friabilis Brady.
—— *vagans* Brady.
—— *elongata* Brady var. *laevigata* Wright.
Botellina labyrinthica Brady.

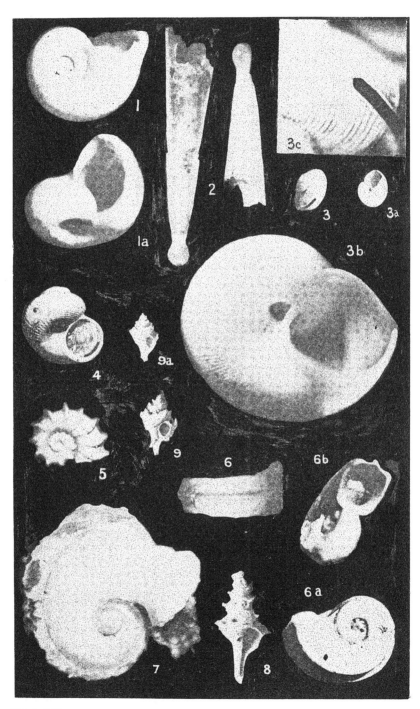

Reophax scorpiurus Montfort.
Haplophragmium agglutinans d'Orbigny.
—— *glomeratum* Brady.
—— *canariense* d'Orbigny.
—— *nanum* Brady.
Haplostiche soldanii Jones and Parker.
Ammodiscus incertus d'Orbigny.
—— *tenuis* Brady.
Textularia rugosa Reuss.
—— *barrettii* Jones and Parker.
—— *conica* d'Orbigny.
—— *turris* d'Orbigny.
—— *concava* Karrer.
—— *agglutinans* d'Orbigny.
—— *trochus* d'Orbigny.
—— *agglutinans* var. *porrecta* Brady.
—— *inconspicua* Brady.
Spiroplecta sagittula Defrance.
Gaudryina rugosa d'Orbigny.
Valvulina fusca Williamson.
Clavulina communis d'Orbigny.
Bulimina ovata d'Orbigny.
—— *pyrula* d'Orbigny.
—— *aculeata* d'Orbigny.
—— *marginata* d'Orbigny.
Bolivina punctata d'Orbigny.
—— *porrecta* Brady.
—— *karreriana* Brady.
Cassidulina subglobosa Brady.
—— *bradyi* Norman.
—— *laevigata* d'Orbigny.
Ehrenbergina serrata Reuss.
Chilostomella ovoidea Reuss.
Lagena hispida Reuss.
—— *sulcata* Walker and Jacob.
—— *orbignyana* Seguenza.
—— *striata* d'Orbigny.
—— *marginata* Walker and Boys.
—— *squamosa* Montagu.
—— *hexagona* Williamson.
—— *distoma* Parker and Jones.
—— *striata* d'Orbigny (apiculate form).
—— *laevis* Montagu.
—— *acuticosta* Reuss.
—— *globosa* Montagu.
Nodosaria filiformis d'Orbigny.
—— *consobrina* var. *emaciata* Reuss.
—— *communis* d'Orbigny.

Nodosaria soluta Reuss.
—— *scalaris* Batsch.
—— *radicula* Linné.
—— (*Glandulina*) *rotundata* Reuss.
Lingulina carinata d'Orbigny.
Marginulina costata Batsch.
—— *glabra* d'Orbigny.
Vaginulina legumen Linné.
Cristellaria latifrons Brady.
—— *tricarinella* Reuss.
—— *gibba* d'Orbigny.
—— *cultrata* Montfort.
—— *rotulata* Lamarck.
—— *orbicularis* d'Orbigny.
—— *tenuis* Bornemann.
—— *crepidula* Fichtel and Moll.
—— *variabilis* Reuss.
—— *acutauricularis* Fichtel and Moll.
Polymorphina gibba d'Orbigny.
—— *gibba* (fistulose form).
—— *problema* d'Orbigny.
—— *compressa* d'Orbigny.
—— *elegantissima* Parker and Jones.
—— *oblonga* d'Orbigny.
—— *angusta* Egger.
Uvigerina asperula Czjzek.
—— *angularis* Williamson.
—— *canariensis* d'Orbigny.
—— *pygmaea* d'Orbigny.
Ramulina globulifera Brady.
Globigerina conglobata d'Orbigny.
—— *bulloides* d'Orbigny.
—— *aequilateralis* Brady.
—— *inflata* d'Orbigny.
—— *dubia* Egger.
—— *bulloides* var. *triloba* d'Orbigny.
—— *sacculifera* Brady.
—— *pachyderma* Ehrenberg.
Orbulina universa d'Orbigny.
Pullenia quinqueloba Reuss.
—— *obliquiloculata* Parker and Jones.
Spirillina vivipara Ehrenberg.
—— *decorata* Brady.
Discorbina globularis d'Orbigny.
—— *parisiensis* d'Orbigny.
—— *parisiensis* (plastogamic form).
—— *rosacea* d'Orbigny.
—— *rosacea* (thin form).
—— *orbicularis* Terquem.

Discorbina patelliformis Brady.
—— *pileolus* d'Orbigny.
—— *vilardeboana* d'Orbigny.
—— *pileolus* (plastogamic form).
Truncatulina praecincta Karrer.
—— *tenuimargo* Brady.
—— *lobatula* Walker and Jacob.
—— *variabilis* d'Orbigny.
Anomalina coronata Parker and
    Jones.
Pulvinulina schreibersii d'Orbigny.
—— *elegans* d'Orbigny.
—— *micheliniana* d'Orbigny.
—— *auricula* Fichtel and Moll.

Pulvinulina crassa d'Orbigny.
Rotalia beccarii Linné.
—— *clathrata* Brady.
—— *soldanii* d'Orbigny.
Polytrema miniaceum Linné.
Nonionina boueana d'Orbigny.
—— *turgida* Williamson.
—— *scapha* Fitchel and Moll.
—— *depressula* Walker and Jacob.
—— *umbilicatula* Montagu.
Polystomella subnodosa Münster.
—— *striato-punctata* Fichtel and
    Moll.
—— *macella* Fichtel and Moll.

In addition to the foregoing recorded species, there are many inter-mediate forms, especially amongst the *Nodosaria, Cristellaria,* and *Lagena,* most, if not all, of which may be referred to one or other of the recorded forms upon fuller examination than they have been able to receive at present.

———

Art. XIII.—*Terminology for Foraminal Development in Terebratuloids* (Brachiopoda).

By S. S. Buckman, F.G.S.

Communicated by Dr. J. Allan Thomson.

The excellent work which is being done by Dr. J. Allan Thomson among the *Brachiopoda* of New Zealand, and his thorough grasp of modern prin-ciples of palaeontology, are abundantly proved in the pamphlets which he has recently published. They are very welcome; they make a genuine and satisfactory advance in knowledge.

There is, however, a slight ambiguity in a certain phrase used by the author, due to a lack of technical terms; and as I have already proposed the necessary terms, which have been in type now for about a year, in a publication on Brachiopods to be issued by the Geological Survey of India, it seems desirable to mention them, so that they can be utilized, because, for various reasons, it may yet be some time before my larger treatise can be published, and it seems advisable to avoid the possible complexity of two sets of terms for the same features.

In his paper "Brachiopod Genera" Dr. Thomson says, "All the known species of *Bouchardia* . . . possess similar and rather unusual beak characters . . . there are sharp beak-ridges uniting in front of the foramen, which is thus behind the apex."* Here is the ambiguity referred to, contained in the words "in front" and "behind"; and perhaps I notice it the more readily because I have myself admittedly stumbled in similar manner on more than one occasion. Now, the ambiguity is this: The above words, "in front" and "behind," are exactly contrary to the

———

* Trans. N.Z. Inst., vol. 47, 1915, p. 397.

use of the terms " dorsal " and " ventral valves " in Brachiopods. As it
happens, these shells are so generally examined and depicted in dorsal view
that one unconsciously comes to think of the dorsal valve as the front.
Really it is the behind valve, and so the beak-ridges of *Bouchardia* unite
behind (dorsally of) the foramen, which is thus in front (ventrally of) the
apex.

The terms which I have proposed are designed to meet this and similar
cases—that is, to express directly the position of the foramen in regard
to the beak-ridges ; for there is development in its position—the pedicle
shifts from the pseudo-area, cuts through the beak-ridges, destroying the
apex, and takes up a position in the ventral umbo. The terms are modi-
cations of Phillips's old generic names *Hypothyris, Epithyris,* bestowed
originally to mark just such differences in the position of the pedicle.
Thus the foramen is—

(1.) *Hypothyrid* when it is in the pseudo-area, and the apex is intact ;
this is a usual condition among *Rhynchonellacea,* but is rare
among *Terebratulacea :*

(2.) *Submesothyrid* when the apex has been absorbed, yet the foramen
lies mainly in the pseudo-area but partly in the ventral umbo ;
this is a frequent condition in *Terebratulina :*

(3.) *Mesothyrid* when the foramen lies about equally each side of the
beak-ridges, a usual condition in Mesozoic *Dallininae :*

(4.) *Permesothyrid* when little of the foramen lies on the pesudo-area
but the main of it is in the ventral umbo, a condition also
found in Mesozoic *Dallininae* and in *Terebratulidae :*

(5.) *Epithyrid* when the foramen lies wholly in the ventral umbo and
the line of the beak-ridges passes dorsally of it ; this is a usual
condition in *Terebratulidae.*

These are five sequent stages in the shifting of the pedicle ventralwards,
of which (1) is the earliest and (5) the latest. The beak-ridges make a
datum-line for observation. The *Dallininae* in general occupy an intermediate position between the bulk of *Rhynchonellidae* and the bulk of
Terebratulidae—at any rate, so far as Mesozoic species are concerned.

There are, however, still other features connected with the foraminal
development which require technical terms. As the pedicle absorbs the
apex, eating through a line of strong beak-ridges in shifting its position,
the ends of the beak-ridges remain like little darts projecting on each side
of the foraminal opening—foramen *telate.* At a later stage these darts
become worn off—foramen *attrite,* well seen in *Magellania.* Still later
the opening is finished off with a deposit of test, a kind of rim—foramen
marginate ; and in further development a lip is projected over the dorsal
umbo—foramen *labiate.* This lip seems to indicate that the pedicle has
more than attained its farthest limit ventralwards, and that it is now
beginning to return on its path to take up a position more dorsalwards.

These modifications of the condition of the pedicle-opening do not keep
step with those other modifications of its position. The consequence is a
series of varied combinations, which may be of considerable utility in the
diagnoses of genera. They also enable the beaks of species to be placed
in developmental position in regard to one another, and, if there is incon-
gruity, suspicion may usefully be aroused as to whether a species is rightly
placed, suggesting investigation of its internal details.

For instance, the beak of *Bouchardia* is, in regard to position of foramen,
just attaining to the epithyrid stage—perhaps something of the permeso-

5*

thyrid stage may be detected, but the condition of the foramen is telate, for the ends of the ridges are not worn off. But in *Rhizothyris*, which Dr. Thomson has, on internal characters, separated from *Bouchardia*, the beak becomes an external tell-tale. According to his figure of *R. rhizoida*,* the foramen is mesothyrid attrite. Therefore, in position the foramen of *Bouchardia* is more advanced than that of *Rhizothyris*, but in condition it is less advanced. These features indicate that these forms are not lineal relatives, but are collaterals. The precedent beak-stages which may be expected in some ancestor of both genera would be mesothyrid telate; but the divergence may have begun earlier.†

The foraminal positions may be diagrammatically illustrated by rules and circles in the following manner, where the rule stands for the line of beak-ridges and the circle for the foramen :—

(5.) Epithyrid

(4.) Permesothyrid

(3.) Mesothyrid

(2.) Submesothyrid

(1.) Hypothyrid

These notes do not exhaust the subject, even for Terebratuloids. A fuller discussion of these and of other development phases is in-type for the memoir in the "Palaeontologia Indica." It is hoped, however, that these preliminary remarks may be of service until that work appears.

ART. XIV.—*High-water Rock-platforms: A Phase of Shore-line Erosion.*

By J. A. BARTRUM, University College, Auckland.

[Read before the Geological Section of the Wellington Philosophical Society, 20th October, 1915.]

Plate XIII.

To most geologists, and certainly to all visitors to the historic Bay of Islands, the "Old Hat" needs no introduction. It has been carved from an emergent knob on a drowned spur, and, as the illustration (Plate XIII) shows, is most aptly named. It was first brought into scientific prominence by Professor J. D. Dana,‡ who visited the district in 1840 as a member of the United States Exploring Expedition of 1838–42.

---

* Trans. N.Z. Inst., vol. 47, 1915, p. 398, fig. 5a.
† According to specimens of *Rhizothyris rhizoida* very kindly sent by Dr. Thomson, the foramen is permesothyrid attrite. This slightly modifies the above statement, but does not invalidate the argument.—2nd March, 1916.
‡ J. D. Dana, Unit. States Explor. Exped. 1838–42, vol. 10, Geology, p. 109, 1849.

[*Winkelmann, photo.*

THE "OLD HAT," RUSSELL, BAY OF ISLANDS.

The rock-platforms are plainly delineated both on the "Old Hat" itself and on the mainland. High-water mark is indicated by a line of driftwood.

This note deals with the origin of the rock-platform constituting the rim of the "Old Hat." Similar platforms are greatly in evidence in many parts of the coast around Auckland and in the more northerly harbours; they are barely covered by mean high tides, and vary in width from a few feet to 30 yards or more. From their seaward margins there is a steep descent for a few feet. Their surface is essentially horizontal but for very minor irregularities, so that they disturb the normal shore profile, which, according to Fenneman, both in building and cutting coasts "is a compound curve, which is concave near the shore, passing through a line of little or no curvature to a convex front."* They are not slightly tilted and uplifted subaqueous shore-terraces, for they are covered at high water.

Dr. von Hochstetter† did not fail to observe these characteristic benches on his visit to Auckland, but neither he nor Professor Dana appears to have realized that they are not normal submarine platforms. Dana's idea of

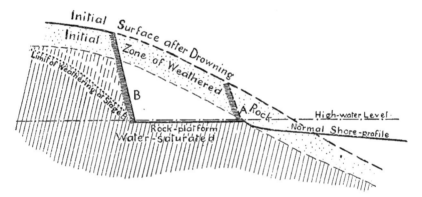

SECTION DIAGRAM ILLUSTRATING THE DEVELOPMENT OF HIGH-TIDE
ROCK-PLATFORMS.

their nature and origin is interesting, particularly as he adduces the "Old Hat" in support of it. He says, "There is, therefore, a level of greatest wear, which is a little above half-tide, and another of no wear, which is just above low tide."‡

It seems essential to the formation of these high-water rock-platforms that wave-attack be not over-vigorous; that the rock in which they originate be moderately resistant to erosion, uniform in texture, and subject to comparatively ready decomposition; and that the coasts have not reached maturity of outline. Given these conditions, they will originate on undestroyed headlands and stacks quite irrespective of the structure of the constituent rocks, for they are particularly well developed at Waiheke Island and at the Bay of Islands, although the shores there are formed by highly disordered sediments.

---

* N. M. Fenneman, "Development of the Profile of Equilibrium of the Subaqueous Shore-terrace," "Journal of Geology," vol. 10, pp. 1–32, 1902.

† F. von Hochstetter, Reise d. o. F. Novara u. d. Erde, Geolog., vol. 1, p. 5, 1864.

‡ J. D. Dana, "Manual of Geology," 3rd ed., p. 677, 1880.

The origin of such platforms is to be explained in great part by the well-known fact that sheets of water protect rocks beneath them from active chemical decomposition.

Surface waters charged with atmospheric gases cannot percolate effectively below sea-level, which must be a level of permanent water-saturation; the rocks below sea-level are thus relatively strong as compared with those above, which have been subjected to subaerial decay, and will resist weak wave-action, which is, however, competent to remove the upper weathered layer.

This fact, apparently, was recognized by Dana, for he says, " The existence of this platform is owing to the protection of the sea from wear and decomposition. Above, the material has disintegrated and been washed away by the action of streamlets and the waves; but beneath the water these effects do not take place."*

But this action alone will not explain all the facts, or so it appears to the writer. He has noted, for example, the occurrence of unweathered rock at a small depth below the surface within a very short distance of well-developed platforms. He would suggest, therefore, that subaerial weathering at the shore-line is progressive, and that, as wave transport removes loosened spoil, fresh impetus is given to weathering, the water-table at the new shore-line is lowered to the level of high water, and the zone of decomposition retrogresses cliffwards. The gist of the hypothesis, then, is that the platforms are due not so much to wave-attack upon a definite zone of weathered rock as to the destruction of the cliff-faces by subaerial erosion and the removal of the resulting waste by weak wave-action. It is evident that the carving must occupy considerable time, and if the wave-action is vigorous the normal shore profile may be established and the sea-cliffs notched at their base in the well-known manner wherever the character of the rocks is favourable.

The high-water platforms of Auckland and North Auckland are invariably surmounted by cliffs of moderate height, a fact that is a necessary corollary of the hypothesis of origin. The districts in question are areas of drowned, maturely sculptured topography, with a climate favouring rapid rock-decay, so that the zone of weathering must have reached a considerable depth before depression gave rise to the present embayed coast-lines. So long as wave-attack is directed wholly against the soft material situated in the zone of weathering of the interrupted earlier cycle, the normal shore profile obviously will result, and not before the rock of this original zone has been removed at the level of wave-attack can the high-tide platforms be initiated. At this stage cliffs with their height equivalent to the depth of weathering will front the sea, and, after this, as they retreat they will increase in height little by little towards the maximum when the slopes of the spurs in which they are cut become reversed. The section diagram illustrates these stages: A represents the sea-cliff at the initiation of the platforms and B is a similar cliff at a much later stage, whilst the dotted line indicates the hypothetical limit of the inwards-advancing zone of decomposition at this stage.

In conclusion, the writer wishes to thank Dr. C. A. Cotton, of Victoria University College, for obtaining references to literature inaccessible in Auckland, and for friendly criticism.

---

*J. D. Dana, Unit. States Explor. Exped. 1838–42, vol. 10, Geology, p. 442, 1849.

ART. XV.—*On the Occurrence of a Striated Erratic Block of Andesite in the Rangitikei Valley, North Island, New Zealand.*

By Professor JAMES PARK, F.G.S., Otago University.

[*Read before the Otago Institute, 7th December, 1915.*]

Plate XIV.

IN 1909 I described* the occurrence of several massive piles of andesitic blocks in the middle and lower portions of the Hautapu Valley. The source of this andesitic material is Mount Ruapehu; and the significance of the occurrence lies in the fact that the Hautapu River does not reach within

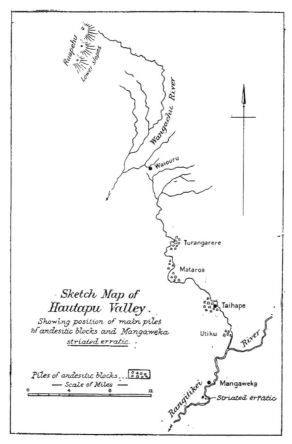

many miles of Ruapehu, but is separated from that mountain by the Wangaehu Valley. The Wangaehu rises on the eastern slopes of Ruapehu, and flows southward through a wide valley fringed on both sides with high terraces.

---

* J. Park, "Glaciation of Hautapu Valley," Trans. N.Z. Inst., vol. 42, p. 577.

There is no evidence that the Hautapu did at any time drain the slopes of Ruapehu; but if it did, then obviously its upper sources have been pirated by the Wangaehu—possibly in the late Pleistocene.

The Hautapu is a tributary of the Rangitikei River. The latter rises in the Kaimanawa Range, composed of Lower Mesozoic greywacke and slaty shale, but in its middle and lower course the river passes over Younger Tertiary marine clays which wrap around the older rocks, and in places rise on the flanks of the range to a height of 4,000 ft. above the sea. The Tertiary formation extends southward to the sea in the form of a great gently sloping plain, now deeply dissected by many rivers and their numerous tributary streams. The slope of this plain coincides approximately with the dip of the strata.

The clays are intercalated at intervals with wedge-shaped beds of impure shelly limestone that attain their greatest thickness in the neighbourhood of the old shore-line.

The Hautapu Stream throughout its whole length lies within the area occupied by the Tertiary clays. Near the sources of the stream the clays are intercalated with a few irregular beds of shelly limestone, seldom more than a few yards thick.

Going southward, following down the Hautapu Valley, the first great pile of andesitic blocks occurs at Turangarere, twenty miles as the crow flies from Ruapehu. Still greater piles occur at Mataroa, thirty miles from Ruapehu, and at Taihape, six miles south of Mataroa. The piles at Mataroa and Taihape are linked up by many isolated blocks, some of great size, perched on the ridges on the north side of the valley.

It is noticeable that the andesitic piles themselves and their constituent blocks become larger and larger as the distance from Ruapehu increases till Taihape is reached.

A small pile of andesitic blocks occurs at Utiku, a few miles south of Taihape, at a point about half a mile from the place where the Hautapu joins the Rangitikei River. This pile lies 1,225 ft. above the sea. Till the present year there was no knowledge of andesitic blocks beyond Utiku, which is thirty-nine miles from Ruapehu.

In February, 1915, I discovered a solitary andesitic erratic in the Rangitikei Valley, at the south end of the Mangaweka Railway tunnel, at a height of 1,070 ft. above the sea. It lies against the foot of a steep ridge of Tertiary clays, 16 ft. above the lower terrace of the Rangitikei, the surface of which is 175 ft. above the bed of the river. (Plate XIV, fig. 1.) This erratic measures 14 ft. long, 6 ft. wide, and 5·5 ft. high, and its weight is approximately 37 tons.

The whole of the underside of the block is smoothed and rounded, and scored with deep grooves and striae, the majority of which run parallel with the longer axis of the mass. At one end the striae extend up the sides for about 2 ft. from the bottom. The diagonal striae are numerous, but, as a rule, not so deep as the longitudinal ones. (Plate XIV, fig. 2.)

This great mass of striated andesite is larger than any block known to occur in the Hautapu Valley. It lies forty-eight miles from its source at Ruapehu, and appears to mark the southern limit reached by the ice-borne blocks.[*]

---

[*] Waterworn boulders of andesite derived from the destruction of the Hautapu moraines are plentiful among the existing river-bed gravels of the Rangitikei River.

[*J. Park, photo.*

FIG. 1.—Striated andesitic erratic near Mangaweka, Rangitikei Valley.

[*J. Park, photo.*

FIG. 2.—Showing striated sides and under-surface of Mangaweka erratic.

In 1909 I expressed the opinion that the piles of andesite in the middle and lower portions of the Hautapu could have been transported and piled in their present places by no known agency but glacier-ice descending from Mount Ruapehu. This mountain is a massive volcano, rising to a height of 9,000 ft. above the sea. The glacier* at present existing on it lies in a basin encircled by peaks ranging from 7,000 ft. to 9,000 ft. high.

Mount Ruapehu is the highest mountain mass in the North Island of New Zealand. It lies in latitude 39° 15' south, and even now possesses a glacier nearly two square miles in extent. The andesitic erratic in the Rangitikei Valley lies in latitude 39° 45' south.

The presence of the morainic piles of andesitic material in the Hautapu Valley and the occurrence of the Mangaweka striated erratic in the Rangitikei Valley show that in the Pleistocene period a valley glacier threaded its way down the Hautapu Valley till it reached the Rangitikei Valley. Both the Rangitikei and Hautapu Rivers run in deep rectangular troughs excavated in the floor of the old glacial valley.

In the Pleistocene, glaciers covered much of the highlands of Tasmania, and, according to Professor J. W. Gregory, F.R.S., descended within 400 ft. of the sea. This view has been confirmed by Professor Edgeworth David, F.R.S., who has stated that the glacier-ice came to within a few hundred feet of sea-level, if not down to sea-level itself.

After all, it does not seem very remarkable that a glacier should have descended to within 1,000 ft. of the sea in the southern end of the North Island during the great Pleistocene glaciation of New Zealand.

ART. XVI.—*The Orientation of the River-valleys of Canterbury.*

By R. SPEIGHT, M.Sc., F.G.S.

[*Read before the Philosophical Institute of Canterbury, 1st September, 1915.*]

THE remarkable orientation of the river-valleys of Canterbury has attracted the attention of numerous writers on physiographic subjects since the main features were first pointed out by Edward Dobson in his report for the year 1865 on "The Possibility of constructing a Road through the Otira Gorge" furnished to the Provincial Council of Canterbury. This report was fully noticed by Haast in his "Geology of Canterbury" (p. 174), and by McKay in his report on Marlborough and Amuri Counties ("Reports on Geological Explorations for 1890–91," p. 26). Reference has also been made to it in Kitson and Thiele's paper on the Upper Waitaki basin (Geographical Journal, vol. 36, 1910, p. 547) and by Gregory in his book on "The Nature and Origin of Fiords," 1913 (p. 366). In view of these references it may not be altogether irrelevant at the present time to re-state the case and to examine in the light of later observations the evidence for and against this hypothesis.

Before dealing with the origin of the valleys it is as well to consider their arrangement in order to understand how far the statement made by Dobson is correct. He says (*loc. cit.*, p. 50), "In addition to the folding

---

* J. Park, "Geology of New Zealand," 1910, p. 203. Fig. 94 shows a portion of the existing Ruapehu Glacier.

above described, the rocks of the central chain have been subject to a variety of upheavals and dislocations which have resulted in the formation of a system of valleys, the direction of which is very remarkable; the principal valleys, from the Taramakau on the north to the Makarora on the south, radiating from a common centre situated about fifty miles to the north of Mount Darwin. It might naturally be imagined that these valleys would form passes through the dividing range, but such is not the case, as, with the exception of the Hurunui Valley, they do not extend through the western portion of the chain, but terminate in glaciers or are bounded by high rugged precipices, as is shown in the sketch of the range at the head of the Waimakariri."

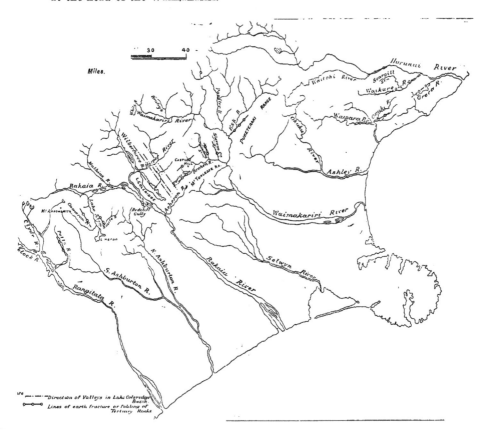

SKETCH-MAP OF CANTERBURY, SHOWING CHIEF LINES OF FAULT OF TERTIARY BEDS.

The fact that the passes through the range do not accommodate themselves as they should renders it extremely doubtful if the existence of a series of great earth-fractures can be postulated to explain the orientation of the valleys. The middle portion of the valleys occupied by the great lakes of Canterbury—viz., Sumner, Coleridge, Tekapo, Pukaki, and Ohau —do correspond with the arrangement indicated by Dobson. Again, though several of the valleys seem to accommodate themselves naturally as far as their main portion is concerned, as we approach the main divide there appears to be a marked departure from the arrangement. This is specially noticeable in the case of the Upper Waimakariri and the Rangitata, and the direction of the line of fracture accounting for the Rakaia is made by Dobson to follow the line of the Mathias River, and not the main Rakaia, from the junction of the former with the latter, although the tributary is of comparatively small size when compared with the main stream, whose upper valley lies right athwart the direction of the suggested fracture in that region. Therefore, if we grant that the dominating influence which determined the course of the main valleys was a system of earth-fractures, some other cause has determined the direction of the headwaters.

Dobson was no doubt conscious of other determining causes, since he has noted very fully the direction of fold and joints of the rocks and their important effect on stream-directions. I think that if full note be taken of the direction of the joints and folds it will explain most of the anomalies in the directions of the passes through the main divide and the departure from the system of earth-fissures—if, indeed, they exist—which is admitted by both Dobson and Haast.

First of all, the general direction of folding in the neighbourhood of Arthur's Pass and farther north is N. 22° E. We have therefore the peculiar circumstance that the direction of folds crosses the range at an angle, and is not parallel with its general trend. This is also the case in the Kaikoura Mountains, as I have been informed by Drs. Allan Thomson and C. A. Cotton. The angle cited by Dobson is the average angle, and it will be found that there are marked divergencies from this value in particular places—for example, in Otira Gorge the line of strike is almost in a line with the direction of the tunnel, being practically due north. In some places the same bed can be seen, owing to slight local variations, following along the top of the inside of the tunnel for several chains—that is, in a direction N. 8° W. It is thus apparent that the line of the Otira and Bealey Rivers follows almost exactly the line of strike of the beds; and, as the dip is in contrary directions on opposite sides of the Otira Valley, so that the valleys of these rivers have been eroded along the axis of an anticline, the formation of the pass between the heads of the valleys is in all probability due to the capture of the eastward-flowing streams originally running through the pass by streams discharging west—a result to be attributed very largely to glacier erosion and the sapping-back of heads of glaciers in the manner first described by Matthes in the report on the "Glacial Sculpture of the Bighorn Mountains" (21st Annual Report U.S. Geol. Surv., 1899–1900, pp. 167–90). This idea has been amplified and extended by W. H. Hobbs in a paper entitled "The Cycle of Mountain Glaciation," published in the Geographical Journal, vol. 35, 1910, p. 146. This piracy has no doubt been hastened by the steeper grade of the rivers and also of the ice on the western side of the main divide.

On considering the directions of the valleys at the headwaters of the Waimakariri the influence of the general direction of folding is quite apparent, especially in the upper waters of the White River, the main southern tributary of the Waimakariri. This is one of the cases noted by Dobson as difficult to explain on the supposition that the valley-directions were determined by a series of earth-fractures, since the strongly defined mass of Mount Greenlaw completely blocks the end of the valley of the main river.

On the other side of the divide between this valley and that of the Rakaia, in the neighbourhood of Browning Pass, the strike of the beds lies right across the main valley leading up to the pass; but when the valley of the Wilberforce is traced up to the bluff which closes its upper end almost completely it turns to the north and follows the line of the strike exactly. The beds here dip north-west, and the scarp slope has formed a line of cliff, composed of loose unstable rock extending for some miles, and quite unscalable from the east except at the pass itself, where the western wall has been broken down to some extent. This line is continued down the valley of the Taipo on the western slope of the range in almost perfect alignment as far as the junction of that river with the Seven-mile Creek, when the river makes a right-angled bend across the strike before it flows through its gorge to join the Taramakau. The parallelism of the upper waters of the Arahura, which are in close proximity to those of the Taipo, is noteworthy in this connection.

In the Whitcombe Pass, between the head of the Rakaia and the Whitcombe River, a tributary of the Hokitika, the direction of the strike of the beds corresponds fairly accurately with the long narrow trench which runs N. 10° E. for a distance of twenty miles through the range, and is occupied in Westland by the Whitcombe River and in Canterbury by the Louper Stream (Bulletin No. 6 (n.s.), N.Z. Geological Survey, 1908; maps). When the direction of the pass is followed south across the Rakaia there is no sign whatsoever of the persistence of the trench. This valley is as marked in the alignment of its valley-walls as that of the Upper Bealey and the Otira, and the agreement in both cases with the folding of the beds seems to indicate that the former of Dobson's causes is the dominating one in determining the directions of the valleys in the neighbourhood of the passes, though Morgan (*loc. cit.*, p. 73) inclines to a fault origin for the Whitcombe Valley.

If we now consider the passes at the head of the Godley, Tasman, Hunter, and Makarora Rivers, they appear to conform more generally to the average strike of the beds in those regions, and this, and not the presence of a series of major earth-fractures, may be the explanation of their orientation.

The actual passes where they cross the main divide seem, therefore, to be controlled in direction by the strike of the folded greywackes. In some cases, however, other general causes must be present which have had a determining effect on the course of the upper streams—*e.g.*, there is such a marked resemblance in the headwaters of the Waimakariri and Rakaia that it can hardly be the result of chance. The chief source of the Waimakariri is the White River, which rises in a glacier on Mount Davie, to the west of Mount Greenlaw, a mass which apparently blocks the valley when looking up-stream from the Bealey Township. The course of the White River Valley is for about three miles in a north-easterly direction, parallel to the strike of the beds, with its south-eastern side a wall of precipitous rocks standing almost vertical in agreement with the high dip; but the river breaks through this barrier in a course cut at right angles past the north-eastern end of Mount Greenlaw, receives a tributary from the north-east, and then flows about south-west, almost parallel to its original direction, past the front of Mount Greenlaw, and then finally turns again almost at right angles to the south-east and follows that general direction past Bealey and the Cass towards the sea.

The case of the Rakaia is analogous. The Lyell Glacier discharges by a river which runs north-east along the strike of the beds behind the spur of Mount Goethe stretching down to Mein's Knob, junctions with the Ramsay, which flows south-west along the strike behind the Butler Range, and the combined stream, forming the main Rakaia, cuts across the beds at right angles, past the northern end of Mein's Knob and south of Jim's Knob, the southern termination of the Butler Range, and thence onward has a general south-easterly trend.

It seems certain that the direction of the folding does exert a determining influence on the upper courses of the river-valleys; but when we examine the middle courses of the streams, before they leave the mountain tract and debouch on to the plains, we find that their orientation has apparently little relation to either. The direction of folding varies so much from point to point that it is almost impossible to arrive at any general direction which is in agreement with the lines of the stream-valleys. In most cases they cut across the strike diagonally, with angles varying from 45 to 90 degrees. The Waimakariri Gorge may be an exception to this. Above the narrow part, near the northern extremity of the Torlesse Range, the flow of the stream is approximately parallel to the strike, as is exemplified by that part of its course along the western flank of the Puketeraki Range, and then it crosses the strike of the beds almost at right angles. This is perhaps an illustration of the features of a young-river gorge to be noted directly; but after a consideration of all the river-valleys and the directions of the fold of the greywackes I can find no agreement of the orientation with even an average direction, and therefore some other cause must in all probability be considered as responsible for the agreement.

When, however, we examine the gorges of the rivers which are of recent origin, we find that in numerous cases their directions either as a whole or in part are directly controlled by both the directions of strike and of jointing. The determination of the latter is at times a matter of some difficulty, since its direction is by no means regular. In general a jointing at right angles to the stratification predominates, but besides this there are additional associated joint planes which intersect the dominating ones at an angle approaching 45 degrees. This is perhaps what we should expect if quadrangular blocks which had already been produced by the normal causes of jointing were subjected to a pressure at right angles to their faces, just as when the strength of a block of cement is tested under a crushing-machine it generally fails on lines which meet the faces of the block at an angle of 45 degrees. The persistence of the cross-joints and the relation of their angle to the dominant ones can probably be explained in this way. This is specially the case where the rock has shown symptoms of failure under earth-pressure; but where it is more resistant, especially where it consists of hard tough greywacke, the prevailing direction of jointing is at right angles to the stratification.

The right-angled bends at times characteristic of young-river gorges probably depend in general on the strike of the beds and the directions of the dominant joints, one reach corresponding to the strike and the next to the joint-direction, and the length of each of these reaches will bear some relation to the importance of the relative effect of the two causes. In general, however, it will be found that the strike is the dominating influence. If they are of equal importance, then the reaches will be approximately the same length. Parallel reaches, such as occur near the heads of the Rakaia and Waimakariri, are due to the river flowing first along the strike, then turning at right angles in direction of the joints, and then continuing the turning in the same sense, and running again along the strike in the reverse direction. As the river passes from the youthful to the mature stage these abrupt right-angled turns will get smoothed out, and the control of rock-structure on the direction may be difficult, if not impossible, to determine ; and specially will this be the case in a country where the direction of the strike, and with it that of the joints, is continually changing, even if we leave out of consideration the disturbing effect of another set of joints crossing at an angle approaching half a right angle. Thus it is that any attempt to correlate our river-directions with the smaller rock structures will lead to doubtful result, although they have really had a dominating effect in the beginning.

Although Dobson states that the directions of the river-valleys was probably due to a series of earth-fractures, he did not submit positive proof of their existence, and I am unaware that any direct evidence has been forthcoming since he wrote his report with regard to the main valleys. In the case of some of the subordinate valleys there is, however, positive evidence of the presence of structural features controlling their direction.

In the valley of the Potts River, a tributary of the Rangitata, there is an outlier of coal-measures whose strike runs north-west parallel to the direction of the valley, and the coal-measures have been faulted down on a line running in this direction. This is altogether different from that of the strike of the greywackes in the locality, which runs approximately north-east. It appears, therefore, that there is an agreement with the line of fault in the direction of the valley. The valley has been enlarged by glaciation till its upper portion has become a typical glacial trough, but it is blocked at the lower end by a bar of greywacke through which the river has cut a gorge 600 ft. deep, where the effects of joints on the direction of short reaches is very apparent. This gorge exhibits for a length of about three miles a series of right-angled zigzags, one set of directions corresponding to the strike of the beds and the other to the cross-joints, the general trend of the gorge being, however, nearly at right angles to the line of the upper glaciated valley. Similarly, in the high country between Lake Heron and the Upper Cameron River there is an analogous occurrence of coal-measures, and, although no large valley runs north-west in association with them, small tributaries of the Cameron, together with well-marked landscape features such as bluffs, which may well be fault-scarps, and a series of peculiar depressions, follow closely the line of dislocation. This direction is parallel to that of the Lake Heron Valley, and remnants of Tertiary beds which are left in it are folded in places on a N.W.–S.E. line.

When we cross the Rakaia and consider the country in the neighbourhood of Lake Coleridge we find a series of five parallel valleys lying with a N.W.–S.E. direction (see map). These are—(1) The main Rakaia, between the gorge and Double Hill, bounded on the west by Mount Hutt and the Palmer Range ; (2) the Lake Coleridge Valley, with its extension up the Wilberforce ; (3) the valley through which the road passes from the southern end of Lake Coleridge to the Harper River, in which lie numerous small lakes, such as Georgina, Eveleen, Selfe, and which is separated from the Lake Coleridge Valley by Kaka Hill and Cotton's Sheep Range ; (4) a valley to the north-east of this, at a much higher level towards the southern end of Lake Coleridge but specially well defined towards the Harper River, in which lies Lake Catherine, the valley being continued across the Harper up the Avoca River ; (5) another valley, of small size, lying right against the western base of Mount Enys and the ridge running south-east from it. The remnants of Tertiary beds which exist at Redcliff Gully, on the Rakaia, at the mouth of the Acheron River, near the Harper River, and at Mount Algidus give little idea of structural peculiarities which may find expression in the shape and position of these river-valleys.

When we cross into the Waimakariri basin we find the same arrangement of parallel valleys in the country between Broken River and the Cass. Taking the most westerly first, there are the following: (1) The valley occupied by Winding Creek and Lakes Pearson and Grasmere; (2) the valley occupied by Sloven's Creek and its continuation following the line of railway over St. Bernard Saddle to Lake Sarah and the Cass; (3) the valley of the Waimakariri.

The directions of the first two of these are certainly determined by structural features—at all events, in their lower portions. In Winding Creek the Tertiary coal-measures form a strip along the floor towards the junction with Broken River, and these have been folded with an axis running along the line of the valley to the north-west. Again, in Sloven's Creek there is a strip of Tertiary sedimentaries and associated volcanics with a north-west strike and a south-west dip which has been tilted and faulted down, and has certainly determined the direction of the valley in its lower reaches. Whether in both these cases the fractures continue north-west past the last exposures of Tertiary beds is quite uncertain; but it is extremely probable that it is so, the soft strata having been removed from the higher parts of the valleys by the increased intensity of erosive agents. There are no Tertiary sedimentary beds in the valley of the main Waimakariri indicating any line of structural weakness which may have determined its direction; but just opposite the point where the Esk River joins it there is proof of a dislocation running up the valley of the tributary in a north-easterly direction, the Tertiary sediments being subjected to folds on that line with a fault running up the western flanks of the Puketeraki Range in that region, tilting the sedimentaries along its line, and determining the direction of the valley either by the channel being formed by the fault-line or by depressing a strip of soft and easily eroded beds. This is the only marked structural line of weakness running north-east, but it is probably continued to the south-west toward the Castle Hill basin, and in the valley of Coleridge Creek, which leads towards Coleridge Pass, in the extreme south-west of the basin, we have a well-marked fault-line where the dislocation can be positively seen in the markedly different levels of beds dipping in the same direction on opposite sides of the valley.

We see, therefore, that, while certain subordinate valleys do show indications of their directions being dependent on deformations of the strata, the main valleys have not furnished any positive evidence of such; but this does not negative the statement that valley-directions are dependent in some way on such a cause. They may be lines of earth-fracture. It will be noted, however, that those cases which I have cited have a general north-westerly trend, whether they are in the Rangitata, the Rakaia, or the Waimakariri basin, and do not correspond with the lines suggested by Dobson. This divergence is most marked in the Waimakariri basin, for, while the map of fractures suggested by that author would make the direction of the upper Waimakariri almost east-and-west, the valleys which are dependent on fractures in that region run N.N.W.–S.S.E.; but it is not impossible that both sets exist, only there is positive evidence of the latter and not of the former. The chief difficulty in detecting fault-lines in the mountain region of Canterbury is that the rocks are of such a monotonous nature, with no well-defined zones of distinctive lithological character or beds which can be

distinguished by their fossil-content. The detection of structural dislocations is therefore a matter of extreme difficulty. Further, they have been
modified by glaciation, and that agent produces landscape features in
valleys which simulate fault-scarps, so that this criterion is very dangerous
to apply. However, the marked agreement in grade. between the main
rivers and their tributaries indicates that any faulting which affected the
main-river valleys and not the tributaries must have dated from a time
long antecedent to the glaciation, or that all discordance in grade has been
rapidly removed. The apparent conformity of the grade of the tributary
with that of the main river is a remarkable feature of the valleys of Canterbury, and differentiates them from those of the West Coast Sounds, hanging
valleys being quite infrequent except towards the heads of the rivers. It
will be apparent, therefore, that this line of investigation does not hold out
at present any great hope of furnishing data by which the problem may be
solved.

It may be pointed out that the fractures which really do exist, with the
exception of the Esk River and Coleridge Creek occurrences, do not follow
the lines as indicated by McKay, and they are divergent from those described
for the Kaikoura region by McKay, Hector, and Cotton, which continue
into north-eastern Canterbury (see map), and from those recorded in the
west of Nelson by Henderson and Morgan.

There is one further suggestion which may be put forward tentatively
to explain the arrangement of the valleys—viz.,·that their directions were
primarily determined by the shape of the land-surface as it emerged from the
sea in late Tertiary times. I have shown elsewhere (Trans. N.Z. Inst.,
vol. 47, 1915, p. 353) that in all probability in late Cretaceous or early
Tertiary times the present mountain region of Canterbury had the form
of a peneplain; that subsequently this surface was depressed and was
covered by a discontinuous veneer of marine sediments; that it was
raised at the close of the Tertiary era and during the Quaternary era
with some amount of folding and faulting. Now, the streams which cut
their beds on this land as it emerged would be consequent on the slope of the
covering beds, and as they were removed by erosion our present drainage
would become a superimposed drainage. The main valley-directions might
then be of great antiquity, while the subordinate ones might be due to
fractures of comparatively recent date. They would, therefore, resemble
what we should get were the central district of the Wellington Province,
between Wanganui and the Ruahine Range, still more elevated than at
present, the mantle of sediments removed, and the base of older rocks
exposed on which they lie unconformably. The present almost parallel
alignment of the rivers would then be perpetuated on the exposed surface,
and would have little relation to the arrangement of the underlying beds.
Even this method of establishment would not negative the existence of
fractures ; but with the complete removal of the covering beds it would be
very difficult indeed, unless the strata exposed were of considerable diversity
in lithological character or of fossil-content, to detect the presence of these
fractures.

ART. XVII.—*Notes from the Canterbury College Mountain Biological Station.*

No. 2.—THE PHYSIOGRAPHY OF THE CASS DISTRICT.*

By R. SPEIGHT, M.Sc., F.G.S.

[*Read before the Philosophical Institute of Canterbury, 1st December, 1915.*]

DR. CHARLES CHILTON, who is in charge of the Cass Biological Station, has asked me to furnish a short description of the most important physiographical features of the locality, with a view to its use by students and others who visit the station. I therefore trust that these notes may be of some service, and that they may perhaps form a basis for reference in the case of future biological work done in the locality.

The station is situated at an elevation of 1,850 ft., in the vicinity of the Cass River. This river rises in the Craigieburn Mountains, and follows a tolerably straight course in a northerly direction, at first through a narrow mountain-valley, afterwards through a gently inclined plain—the floodplain of the river—till it discharges through a short gorge into the Waimakariri River, just below the spot where that river receives from the north the waters of the Hawdon (1,746 ft.). The Cass Valley is bounded on the west by a forbidding rocky ridge running parallel to it, whose most important summits are Mount Misery (5,768 ft.) and Mount Horrible. About a mile from the termination of this ridge it is cut down, and forms a low saddle over which passes the coach-road into the main Waimakariri Valley; this is now known as the Cass Saddle, though in the early days it was called Goldney's Saddle (1,929 ft.). On the east side of the Cass the country is open, but a well-defined line of low hills runs south towards Mount St. Bernard (5,509 ft.), and divides the area into two distinct valleys. In the westerly one lie Lakes Grasmere and Pearson, and in the eastern one lies Lake Sarah. Lake Pearson is the largest of the three; it is about two miles and a half in length, and nearly half a mile wide at its widest part; but it is almost cut into two in the middle, partly by an old moraine, and partly by huge shingle fans, which approach each other from opposite sides of the lake. It is at an elevation of 2,085 ft. above the sea, and has never been known to freeze over. It discharges by Winding Creek towards the Broken River. Grasmere is a much smaller lake, about a mile in length, and nestling close under the ridge which divides the two valleys. It discharges by the Grasmere Stream, which flows through swampy ground close past the station, into the Cass River. Lake Sarah is a small shallow lake, 21 ft. in depth, lying on the opposite side of the ridge from Lake Grasmere, and discharging into the Grasmere Stream. Immediately to the east of the station,

* The places mentioned in this article are indicated in the accompanying locality map.

rising directly from the lake, is the tussock-clad peak called the Sugarloaf (4,459 ft.), along whose northern base flows the Waimakariri.   In this
part of its course this river has an average east-and-west direction, its bed
being a broad flat floor covered with shingle from side to side, over which
the river wanders in anastomosing or braided streams.   It is in places over
a mile in width, and only narrows in this locality where the Hawdon forces
the main stream over to the south bank against the rocky bluff which terminates the ridge to the west of the Cass River.   On the northern side of the

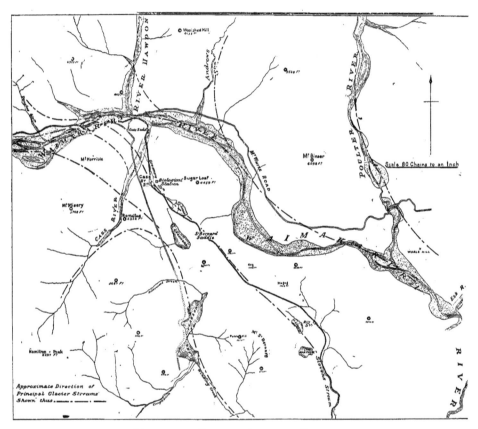

MAP OF THE CASS AND ADJOINING DISTRICTS.

Waimakariri the mountains rise to approximately 6,000 ft., clad with dark-
coloured southern-beech forest on their lower slopes, which passes up into
subalpine scrub and mountain herb-field and fell-field, and finally into bare
rock and loose angular debris with which the summits and adjacent slopes
are usually crowned (see Plate XV, fig. 2).   These mountains and the Craigieburn Range to the south appear to hem in the country near the Cass, and
give it the appearance of being placed in a basin.   The origin of this basin
is no doubt connected in some way with deformation of the earth's crust,
such as faulting and warping; but the details of the landscape are chiefly

due to the modification of such a basin by stream and glacier action. The location of the station affords ample opportunity for the investigation on a minute scale of many of the earth-features produced under such circumstances, and it is to be hoped that these will be undertaken as a corollary to the biological investigation for which the station was primarily established.

The rocks of the neighbourhood consist entirely of folded greywackes, slaty shales, and related sedimentaries of Trias-Jura age, which may be looked on as formed principally from the debris of an ancient granite land lying either to the east or to the west of the present land which constitutes the islands of New Zealand. In places the rocks contain much silica, and become cherty or jasperoid in character, as can well be seen on the hill to the north-west of Lake Sarah, near the Cass Railway-station. The quartzose nature of the rock is so marked at this spot that at one time it was prospected for gold by West Coast miners, and their shafts are still visible on the western end of the hill; no payable gold was encountered. The general strike of the beds as disclosed in the railway and road cuttings is N.E.–S.W., but considerable variations in the direction occur within short distances. The dip of the beds is usually at high angles, the result of folding by powerful earth-pressures operating in a S.E.–N.W. direction, these movements having been attended by faulting, as can be seen from the smoothed and slickensided surfaces exposed at times in the cuttings. A most important feature, which has resulted largely from the same cause as the folding and faulting, is the system of joints which penetrate the rocks and render them especially susceptible to disintegrating agencies, such as frost, under whose influence they break into rectangular fragments and furnish the enormous supply of waste which mantles the mountains in the vicinity, and which gradually moves downward to lower levels, supplies the material for fans, and at times impedes and clogs the normal course of streams.

The rocks are not known to contain fossils, although remains of molluscs have been reported from the road-cuttings on the Waimakariri front; but they have been assigned to the Trias-Jura age, from their continuity with, and lithological similarity to, rocks occurring at Mount Potts, on the Rangitata, at Malvern Hills, and at the Clent Hills, which contain plant fossils of undoubted Lower Jurassic age.

The date at which they were folded was probably Late Jurassic or Early Cretaceous, since rocks of more recent age than this are not affected to the same extent, although they, too, have experienced a moderate degree of deformation, showing that folding movements had not entirely ceased at that date. When they were folded a mountain range was probably formed which was base-levelled subsequently by stream or sea action, and on the surface thus produced Tertiary limestones, sandstones, &c., were laid down, such as can now be seen at Castle Hill, forming in all probability a discontinuous sheet over the area. The region was then raised with its veneer of later sediments, with some amount of differential movement and faulting, which resulted in the formation of such intermontane basins as at Castle Hill and the mid-Waimakariri. The tract was then dissected by stream action, which had advanced to an early mature stage when it was subjected to a somewhat severe glaciation. The agency of ice and the subsequent results of the formation of waste by the action of frost on the jointed sedimentaries have been responsible, either directly or indirectly, for its more distinctive landscape features.

A striking character of the area is the arrangement of parallel valleys which lie in between the Craigieburn Range and the forest-covered mountains to the north of the Waimakariri. These are, starting from the north —(1) The main Waimakariri Valley; (2) the valley in which lies Lake Sarah, and which continues down Sloven's Creek; and (3) the valley in which lie . Lakes Grasmere and Pearson, with an extension down Winding Creek towards Broken River. It is probable that these valleys are of structural origin, and are formed by faults which run in a N.W.–S.E. direction. It is certain that the lower portions of Sloven's Creek and Winding Creek have their directions determined in this way; but it is quite uncertain how far these movements extended to the north, and they may have died out before the heads of the valleys were reached, and therefore in their upper portions the formation of the valleys may be attributable almost entirely to water-action. There does not appear to be any evidence that faulting determined the initial location of the Waimakariri Valley, although it may have done so.

We may take it, then, that before the commencement of the glaciation the drainage directions were as follows : First, there was the main Waimakariri Valley, bounded on the south by the interrupted range of hills of which the Sugarloaf and the ridge stretching behind the present Craigieburn Station are parts. Secondly, there was the valley in which lie the biological station itself, Lake Sarah, the St. Bernard Saddle (where it is partially blocked by moraine), and the upper part of Sloven's Creek. This valley is bounded on the west by the somewhat discontinuous ridges extending from the vicinity of the Cass Railway-station in a southerly direction towards Mount St. Bernard, and continued after a break towards Broken River through the No Man's Land. Thirdly, there is the valley farther west, which contains Lakes Grasmere and Pearson; the valley in which Winding Creek lies may be regarded as an extension of this. In my opinion, then, before the country was glaciated these represented the main drainage-directions. In all probability the ridge extending through Mounts Misery and Horrible, between the heads of the latter two valleys and the Waimakariri Valley towards the Bealey, was in a much more complete condition, and there was no gap through which the present Cass River flows towards the main river; and the Cass River did not exist as such, but probably only as a small stream which rose somewhere near its present site and ran past the Grasmere Station directly into the Grasmere-Pearson Valley. The existing valley of the Cass is, therefore, of more recent formation, and represents the captured headwaters of the streams which discharged into the two westerly of the valleys. The reason for this capture will be given later. The remarkable straight alignment of the sides of the Upper Cass Valley to the west of the Grasmere Station suggests a fault origin for it, but it is more likely due to the erosive action of the ice straightening the walls by the removal of projecting spurs, the marked contrast between a stream-valley with its overlapping spurs and the glaciated portion without them being excellently illustrated by a comparison between the lower levels just above the river and the upper debris-covered slopes. Whatever the reason for the straight alignment of the upper slopes, the actual channel of the river has been determined by water erosion alone. (See Plate XV, fig. 1.)

The changes in the direction of drainage in this locality can be attributed almost entirely to the disturbing effects of glaciation, and it may be useful to summarize at this stage the evidence for the former presence of glaciers.

The following features are indicative of the former presence of ice :—

(1.) *Morainic Accumulations.*—These are well seen on the top of Cass Saddle, in the vicinity of the road; on the slopes of the hill facing the saddle and leading round to the Waimakariri; on St. Bernard Saddle; and at the east end of Lake Sarah, where they are specially well exposed in the railway and road cuttings. There are decided morainic heaps between Lakes Grasmere and Pearson and at the lower end of the latter lake, in this case forming dams behind which the water is ponded, and in the former causing a diversion of drainage. Similar accumulations occur on all the down country in the angle between the Broken and Waimakariri Rivers, especially opposite the mouth of the Esk River. Near St. Bernard Saddle these accumulations have the nature of a boulder-clay, for fine material is mixed with hard tenacious clay, and the angular fragments are frequently striated, giving the deposit the character of a ground moraine.

(2.) *Scratched Surfaces.*—These are well seen on the hard rocks on the faces fronting the Waimakariri; on the top of Cass Saddle, where the surfaces have recently been swept clean of the protective covering of loamy clay; on the northern slopes of the Sugarloaf, facing the Cass River, immediately above the railway. Disrupted striated pavements can be seen in many places where the rocks have been exposed for considerable time to weathering agencies.

(3.) *Roches Moutonnées.*—Almost all the rock ridges in the vicinity of the station exhibit this character. They are elongated in the direction of the movement of the ice-streams, and reduced to elevations with flowing outlines; some, too, have had their cross-sections narrowed by the lateral corrasion of the streams of ice moving along them and attacking their sides. As a result of this they frequently present the appearance of a sugarloaf when viewed end-on and of a long ridge when looked at sideways. Smaller elevations related in character to *roches moutonnées* occur on Cass Saddle, where they form a characteristic dimpled surface. An excellent example of a *roche moutonnée* is the mound (called Romulus on the map) lying close to the Grasmere homestead.

(4.) The cross-sections of the valleys enumerated above are also evidence of the presence of glaciers. Their features are to be seen most perfectly in the main valley. We have the even alignment of their sides, rising at a steady angle from the broad flat floor to about the grass-line, and then sloping back at gentler grade; the absence of overlapping spurs; the truncation or semi-truncation of spur-ends, such as those on the downstream side of the Bealey and Hawdon Rivers and Cass Saddle itself; hanging valleys, such as that in which Andrews Stream lies: all these features, as well as others, indicate clearly that the land suffered a severe glaciation.

We must regard the area, therefore, as covered with a great sheet of ice, formed from snow in the great collecting-basins at the head of the Waimakariri and its tributaries, and moving down the valley to the east, spreading over the undulating country in the triangle between the Waimakariri and Broken Rivers, while through it projected as " nunataks " the summits of the Sugarloaf and Mount St. Bernard.

The volume of this flow may be judged from the size of the cross-section of the main Waimakariri Valley above its junction with the Hawdon and the Cass. Half-way between this point and the Bealey Township the floor is over two miles in width, and this breadth is continued.

with slight diminution up-stream almost to the mouth of the Crow River, about four miles above the incoming of the Bealey River. When one also regards the flare of the valley-walls the actual volume of the basin becomes more impressive. It is no doubt true that this great basin has been enlarged to its present size through the abrasive and excavating power of the ice; but during the ice-flood it would act as a great collecting ground, and materially contribute to the maintenance of the supply of ice, the depth of which, judging from the evidence of ice-action on the valley-walls, must have reached nearly 2,500 ft., even if it did not exceed it, and have covered all the slopes almost to the grass-line.

As this great body moved east it received notable additions from the Hawdon Valley and from Andrews Valley, and at the same time the cross-section of the main valley was much reduced. Hence, driven by the weight of the ice above it and forced south by the tributaries coming in from the north, it crowded over the Mount Misery – Cass Ridge and entered the territory belonging to the valleys lying to the west of the main Waimakariri, lowering the divide continuously by erosion. (See map.) The tendency of such overriding flows of ice to cut distinct channels is well illustrated in the notches in the ridge just mentioned, the upper ones formed at the height of the ice-flood, and the lower one at the Cass Saddle being formed throughout the whole period during which the ice was enabled to surmount that portion of the ridge. Hence it is that the effects of ice are so noticeable in its vicinity, for the erosive action would be very great at that spot. The semi-detached knob which terminates this ridge, and round which the railway goes from the Cass Valley to that of the main Waimakariri, is the remnant of the ridge which has escaped destruction when the overflow cut in behind it. It belongs to one of the "beehive" forms common in glaciated regions where spurs have been semi-truncated. The direction of the movement of ice is shown by that of the striae, and also by the slope of the ridges between Mount Horrible and the Cass River. In the basin which lies immediately to the west of the river these trailing spurs all slope down from the north, even against the flow of the Cass River, and discharge their surplus water south. This peculiarity can be attributed chiefly to the action of ice scouring out channels and leaving ridges pointing in that direction.

The disturbance in the regular and even flow of the ice-stream where it overrode the end of the Cass spur would be very great indeed, and this would be intensified by the stream from the Hawdon coming in at right angles from the north. The effect would closely resemble the eddies or whirlpools seen under similar circumstances in a river. It seems that a great ice eddy occurred between the Cass Saddle and the northern end of the Sugarloaf, and this scoured out a great roundish pothole. A similar landscape form is seen at the junction of the Lake Stream with the Rakaia, formed under exactly similar conditions, but its shape is more circular. In this case, too, a narrow rocky ridge had been left at the outlet of the tributary stream which it has subsequently cut through. The marked similarity of the two cases suggests more than a mere coincidence in shape. It is probable, then, that a rockbound hollow marked the site of the junction of the Cass with the Waimakariri, and when the ice retreated this formed a lake divided from the main stream by a rock bar. (See Plate XV, fig. 1.)

The flooding of the country with ice in the direction of the Broken River is responsible for all the glacial features indicated previously;

almost all the down country in the angle between this river and the Wai-makariri was covered, and at the height of the glaciation the ice reached through the main gorge of the latter river as far as its junction with the Kowhai. Erosion modified the shape of the valleys, and deposits of angular material, such as the moraines at the lower end of Lake Pearson and be-tween that lake and Grasmere, as well as the great mass of angular debris which forms St. Bernard Saddle, marked stages in the recession of the ice-flood.

It must not be assumed that all of this was due to ice which invaded this territory from the Waimakariri. The basin itself must have acted in some measure as a collecting ground, especially the Upper Cass Valley and the hollows of the Craigieburn Mountains ; but the area lies too far to the east of the main divide to allow of any great accumulations of snow, most of which would have been intercepted by the higher range to the west. The Craigieburn Mountains do show, however, the shell-shaped hollows at high levels, some of which certainly held small glaciers of the " corrie " type. On this range they are usually filled with debris at the present time, which is frequently piled in rough ridges across the lower end of the hollow— either a mòrainic accumulation from the glacier period or due to the rolling of loose stones down the frozen snow slopes from the exposed ridges above. In the wetter regions, where plants have a better opportunity for esta-blishing themselves, they form ideal spots for the formation of colonies of alpine plants. The immediate vicinity of the Cass Station is not altogether favourable for examining them, except as to the accumulations of debris, but a short journey to the mountains near the Hawdon enables them to be seen in typical development.

When the ice retreated and left the vicinity of the Cass considerable changes took place in the drainage. These resulted chiefly from two causes : First, the deposit of morainic material caused interference with the normal directions of streams, an interference which might have been slowly overcome had not other causes accentuated it. Transitory lakes, such as Lakes Sarah, Grasmere, and Pearson, probably occupied inequalities in the floors of the valleys, and these would have discharged along the original lines of flow—approximately, at any rate. Now, Lake Pearson does this, but it is slowly disappearing owing to infilling from the sides, and also to its cutting down the dam which blocks it towards Broken River, large areas of drained lake bottom being there visible. Lake Grasmere, how-ever, being more within the sphere of influence of the other cause, has reversed its direction, and has not proved competent to lower the barrier at its southern end. The second important factor has been the over-deepening of the main bed of the Waimakariri in the vicinity of the Cass by the excavating action of the powerful ice-stream which moved down the main valley. The ice here would be much thicker than that in the valley near the middle Cass, and, as power of excavation depends directly on the thickness of the ice, the floor of the main stream would be lowered much below the level of that of the Cass. This over-deepened portion was in all probability occupied by a lake after the ice retreated up-stream towards the Bealey, until the rock bar in the neigh-bourhood of the mouth of the Esk and below it was removed by erosion. When this occurred the effective base-level of all the streams near the Cass would be lowered also, and it is easy to understand, if the ice had not already formed a gap in the rock bar at the mouth of the Cass, why it was cut through, and, if the gap had been formed by the ice or

by subsequent stream erosion, why the headwaters of the streams in the
valleys running south from the Cass were rapidly diverted *to* the main
river, a change which would have been accelerated by the blockage of the
lower parts of these valleys by morainic bars and by shingle fans, as will
be indicated presently. Thus it is that the Cass has captured the stream
issuing from the Craigieburn Mountains near the Grasmere Station, and
also the waters of Lake Grasmere and Lake Sarah, and numerous other
small streams as well. The resemblance to the conditions of the Lake
Stream in the Rakaia Valley is very close indeed in this respect. The
presence of the morainic dam south of Lake Heron, and the overdeepening
of the main Rakaia to the north, has resulted not only in the capture of
small streams in the vicinity of the outlet of the Lake Stream, but also
the waters of Lake Heron itself and its feeders, and, most important of
all, the Cameron River, all of which belong to the Ashburton basin rather
than to that of the Rakaia.* The parallelism between the features of both
localities is indeed most remarkable.

The district furnishes excellent examples of shingle fans in all stages
of development; in fact, I do not know of any locality in New Zealand
where a more representative series can be obtained, the country in the
immediate vicinity of Lake Pearson being especially prolific in the varying
forms of this landscape feature. A common mode of origin, excellently
illustrated by the district, is from the gutters or runnels which sometimes
seam the mountain-side for many hundred feet. These are no doubt
formed at times by a single heavy shower of rain, and their channels are
marked by raised banks and *levées*, and by a cone of detritus at their lower
end on more or less level ground—the deposition of the load of debris being
due in both cases to the check in the velocity of the stream, in the first
instance by friction with the side and the adjoining vegetation, and in
the latter by the lowering of the grade. The deposit at the lower end of
the gutter grows as material is supplied, and the runnel enlarges its basin
by continued erosion of the sides. As the stream increases in volume the
angle of elevation of the fan diminishes and its fringe broadens, and, if
space is allowed, its ultimate size will depend on the volume of water and
the supply of waste. Fans whose diameter exceeds a mile are by no means
uncommon in the Cass region, but they attain greater development in such
unconfined areas as the Canterbury Plains. The stream usually occupies
the highest radius of the fan, contained by *levées* on either side, but sooner
or later it breaks from this unstable position and commences building on
another radius. Construction along successive radii goes on steadily, and
the deposit of one overlaps that of another, so that the surface of the fan
is raised progressively. The establishment and maintenance of the plant
covering has a material influence on its growth, and there is evidently a
close connection between the condition of this covering and the stage of
evolution of the fan, the geological and the ecological factors acting and
reacting on each other. As a rule, when a fan has reached a moribund
condition the detrital matter is completely covered with a closed plant
formation. The biological station is placed on a fan which is practically
dead, and there are good examples of fans in a similar condition cn the
north side of the Waimakariri, where they are covered partly with forest

---

* "The Mount Arrowsmith District: a Study in Physiography and Plant Ecology."
Trans. N.Z. Inst., vol. 43, 1911, p. 341.

and partly with tussock. As a geologist, I should like to know which plant formation was first established on these fans—forest or tussock grassland. The Cass River bed is an active fan on a gentle grade, accommodating itself to the volume of the stream and the supply of waste, both being relatively large. In the valley towards Lake Pearson there are many active and decadent fans, especially on the lower slopes of the Craigieburn Range and Mount St. Bernard. They have spread over the floor of the valley from various points in the sides, dividing it into sections, interfering with the drainage, and thus forming, or helping to form, lakes, such as Grasmere and Sarah, and again filling them up by the lateral extension of the margins of the fan. They are perhaps as important factors in stream-diversion as glaciers, and are the most distinctive agents in the production of landscape features since the ice retreated. I can only suggest, in conclusion, that the position of the biological station affords an admirable opportunity for a detailed study of their true mathematical form and other characters, and a careful investigation into these would furnish results which would be quite worth the time and trouble spent in obtaining them.

BIBLIOGRAPHY.

The following publications may be consulted as regards special aspects of the physiography of the locality :—

1865. Dobson, Edward. " The Possibility of constructing a Road through the Otira Gorge." Report furnished to the Canterbury Provincial Council.

1879. Haast, Julius von. " Geology of the Provinces of Canterbury and Westland." Christchurch.

1907. Speight, R. " Notes on some of the New Zealand Glaciers in the District of Canterbury." Report of Aust. Ass. Adv. Sci.. 1907, p. 285.

1909. Gudex, M. C. " Some Striated Stones from the St. Bernard Saddle, Upper Waimakariri Valley." Trans. N.Z. Inst., vol. 41, p. 33.

1915. Speight, R. "The Intermontane Basins of Canterbury." Trans. N.Z. Inst., vol. 47, p. 336.

1916. Speight, R. "The Orientation of the River-Valleys of Canterbury." Trans. N.Z. Inst., vol. 48, p. 137.

Art. XVIII.—*Notes from the Canterbury College Mountain Biological Station.* .

No. 3.—Some Economic Considerations concerning Montane Tussock Grassland.

By A. H. Cockayne, Biologist, Department of Agriculture.

[*Read before the Philosophical Institute of Canterbury, 1st December, 1915.*]

1. General.

Montane low tussock grassland is not really sharply defined from low-land low tussock grassland, both being very similar associations of the same climatic formation. The term should be applied only to that tussock association in which *Festuca novae-zealandiae* (Hack.) Cockayne is the leading physiognomic plant, while lowland tussock grassland should be restricted to that where the characteristic tussock growth-form is represented by *Poa caespitosa* Hook. f. As implied by the name, lowland tussock grassland is normally developed on land of low elevation, and rarely ascends higher than 1,000 ft. *Poa caespitosa*, however, is not uncommon at higher elevations, but is then mainly confined to moist places along the sides of streams and on the outskirts of forest, and rarely becomes the dominant element of the open grassland. Between true lowland tussock and true montane tussock many intermediate stages can be found, where both *Festuca novae-zealandiae* and *Poa caespitosa* are more or less equal quantity.

Much of the vegetation in the vicinity of the Cass Mountain Biological Station is typical montane low tussock grassland, and, in fact, it is the main climax association, apart from mountain southern-beech forest, of all the area up to an altitude of over 3,000 ft. This is not to say that many other associations are not present; but they nearly all are stages in the ultimate production of montane grassland which appears to be, under natural conditions, the final vegetation of the land-surface, except in sheltered situations, where true forest may be developed. The presence of this uniform climatic formation, characterized by the complete dominance, so far as general appearance is concerned, of the even-sized and almost evenly spaced tussocks of *Festuca novae-zealandiae*, is the special vegetation feature of the valleys and slopes of the mountain-ranges east of the dividing range of the South Island.

2. Area and Distribution.

. Montane tussock grassland is especially characteristic of the South Island,* and comprises an area of over 6,000,000 acres, or, roughly, one-

---

* In the North Island there is a considerable area of the formation here being dealt with situated on the Volcanic Plateau southwards of Mount Ruapehu with *Festuca novae-zealandiae* or a closely related variety or species dominant. The Waimarino Plain, on the contrary, on the west of the volcanoes, although at about the same altitude (3,000 ft., more or less), has *Danthonia Raoulii* Hook f. as its dominant tussock.

seventh of the occupied land of New Zealand. This great plant formation stretches in an almost unbroken line from the Wairau River in Marlborough to the heavily wooded mountains of Southland. It forms a broad belt, interrupted here and there by forest, mainly of the southern-beech — *Nothofagus cliffortioides* (Hook f.) Oerst. — type, between the agricultural land of the lowlands and the subalpine belt. It occupies both valleys and mountain-slopes with its uniform vegetation, and gives a rather monotonous aspect to the landscape, on account of its uniform height and unvarying brownish-yellow hue.

The montane tussock grassland districts of the South Island form one of the few types of vegetation on occupied land that has, in general appearance and in the permanence of its dominant elements, apparently remained comparatively unchanged since the advent of the white man. In certain localities—notably in Central Otago and portions of South Canterbury—this primitive formation has been profoundly modified; but in general the montane tussock grasslands are superficially, and so far as their effect on scenery is concerned, in much the same condition as when first brought under occupation.

3. Utilization.

The montane tussock grasslands have been utilized as sheep-grazing pasture for over half a century. For this purpose they were immediately available in their primitive condition, and they were amongst the first of the areas used for pastoral purposes, ranking in this respect with the original employment of the lowland tussock grasslands. They represent one of the few natural resources of the Dominion that have remained a permanent asset without having to be intentionally modified or specially treated to render them capable of continued production. The same statement would also be true of the primitive lowland tussock associations, but it was early recognized that the land occupied by the latter was in many cases admirably adapted for ordinary farming operations. Thus the replacement of the *Poa caespitosa* association by artificial associations both of a temporary and semi-permanent nature became a general practice. To-day much of the lowland tussock grassland has been converted into arable farms, and even where the plough could not be utilized the extensive surface-sowing of European grasses on the ashes formed by the burning of the tussocks has in many places completely changed the original vegetation. With the montane tussock grassland such has not been the case, and it is only in isolated instances that the original vegetation has been eliminated and replaced by artificial associations.

It may appear strange that in New Zealand, where the general trend of farming operations is in the direction of increasing the carrying-capacity of land, comparatively no effort is made towards any improvement in the utilization of the montane tussock grasslands. For the past fifty years the methods adopted have not varied except in one particular—namely, in a change in the type of sheep that is used. In the early days nearly the whole of the flocks on the upland sheep-stations of the South Island were merinos. This was due largely to the fact that in Australia this was the dominant breed, and that New Zealand was originally mainly stocked from that country. During the past twenty years, however, the merino flocks have suffered a great decline, and their place on the montane tussock-land has been very largely taken

by half-bred and cross-bred sheep. Apart, however, from a change in
the type of sheep carried, no attempt has really been made to improve
the carrying-capacity of the grasslands themselves. Certainly during
recent years a controversy has raged on the advisability of burning
montane tussock grassland, but no finality has been reached regarding
this procedure, although there is a growing tendency to restrict burning
to special times and places.

There are, however, three distinct factors to which must be attributed
the lack of any progression in the utilization of upland tussock grasslands. Briefly expressed, these three factors may be summarized under
the following heads : (1.) The fact that the greater part of the land is
held under a system of short leasehold. (2.) That our knowledge of what
plants could be profitably substituted for the present vegetation is almost
nil, and the methods for the payable establishment of such plants are
quite problematical. (3.) That the individual runs are in general so
large that they furnish ample incomes to their holders without the
adoption of any special methods of soil-utilization.

In general, sheep-stations which consist largely of montane tussock
grassland are situated in mountainous districts, and their higher portions consist of subalpine and alpine associations. The former are
largely used for grazing purposes during the summer; hence the term
"summer country." The lower montane belt is generally called
"winter country," as it is the only ground on which stock can be
carried during the winter months. Thus on many stations the montane
tussock grassland is without stock for considerable periods of the
year, and especially at that period when the grasses are in flower and
seed. During the winter, however, this tussock land, especially that
portion lying towards the sun, is wholly responsible for the carrying
of all stock, as comparatively no provision is ever made for the
production of special winter feed. This is perhaps the most striking difference between the methods adopted on lowland and upland
sheep-runs. At times, in the neighbourhood of the homesteads themselves, a little cultivation is carried on, but this is almost wholly in
the direction of providing chaff for the working-horses of the station.
A few paddocks of European grasses may also be laid down, but the
prevalent custom of using mixtures largely composed of rye-grass (*Lolium
italicum* A. Br.), which dies out in a year or two, has discouraged
any extensive development in this direction. On certain stations,
however, excellent home paddocks of cocksfoot (*Dactylis glomerata* L.)
are established, and it seems reasonable to expect that such a practice would become largely extended were the individual holdings
smaller.

No fattening for market is ever attempted on true montane sheep-
country, but on most stations a certain number of store sheep may be
annually disposed of. In certain seasons, however, the winter mortality
is so high that the normal number of stock carried cannot be maintained
without extensive outside buying.

Reliable figures on the number of sheep carried on the montane
tussock grasslands are not available, but from a careful computation
from the sheep returns I estimate it at about two millions, or, roughly,
one-twelfth of the sheep of the Dominion. As the area under discussion
comprises about 6,000,000 acres, this gives a carrying-capacity of one-
third of a sheep per acre. In point of fact, however, I think this
estimate is on the high side.

4. The Leading Grasses.

The two leading grasses of the montane tussock grassland are the fescue tussock (*Festuca novae-zealandiae*) and the blue tussock (*Poa Colensoi* Hook. f). It is generally conceded that the fescue tussock itself is of little value, but that it affords an indispensable shelter where the more palatable elements of the vegetation are able to develop. The culms of the fescue tussock are from 6 in. to 2 ft. in height. The leaves are hard, harsh, narrow, and involute, persisting in a living condition for a long period, and gradually drying off at the tips; when quite young they are relished by sheep, but in an ordinary tussock they are so protected by the older leaves that it is impossible for sheep to graze on them. This is clearly seen when an area is overstocked; the sheep rapidly lose condition, but there is no palpable alteration in the tussocks, and they are never grazed down. When rabbits are abundant, however, the tussocks are often eaten, so that they look like inverted brushes. It would thus seem that the fescue tussock is of little value as sheep-feed, and as on an average over two-thirds of the vegetation consists of this plant the small carrying-capacity of montane tussock grassland is easily explained. Whether or not the fescue is indispensable for the growth of the other elements of the vegetation is not clear, but it seems feasible to expect that, as the formation is clearly a climatic one, the dominant plant must play an important part in this respect.

The blue tussock (*Poa Colensoi*) is generally looked upon as the most important feeding-element of the montane tussock grassland. It forms much smaller tussocks than does the fescue, and at times fills in most of the intervening spaces. For my part, I think that the feeding-value of the blue tussock has been exaggerated. Even on land recently grazed very little sign of actual feeding-down can be seen, but certainly sheep do much better on land where there is an abundance of this grass than on land where it is scarce. The blue tussock is, however, especially eaten by rabbits, and they graze it almost bare to the ground, totally preventing any seed-production. In the district surrounding the Cass Biological Station rabbits have always been practically non-existent, and it consequently affords a most excellent locality for a complete study of the plants actually eaten regularly by sheep, as these are virtually the only class of stock that has ever been pastured there. In many areas of the montane tussock grassland the blue grass, *Agropyron scabrum* Beauv., is an important constituent, and its flat glaucous leaves are regularly eaten by sheep. In the vicinity of the Cass Station this grass is exceedingly rare, and it has either never been common there or else has been completely eaten out by sheep.

With regard to the other elements of the montane grassland vegetation it is evident that certain of them afford palatable sheep-feed, for I am certain that those plants most abundant—namely, the fescue tussock and blue tussock—are by no means responsible for the sustenance of the stock carried. It, however, has to be remembered that a carrying-capacity of at most one-third of a sheep per acre does not require a large amount of herbage; and were all the vegetation of the montane grassland fed upon, the number of sheep carried would be far and away greater than it is at present. The way that such unlikely food-plants as those belonging to the genera *Aciphylla*, *Discaria* (when burnt down), and *Carmichaelia* are cropped down but left severely alone on land where feed is abundant is a clear indication that palatable feed is really scarce. A really proper investigation of the plants affording *regular* sheep-feed on the montane tussock grassland is badly needed, and for this purpose the fencing-off and feeding-off of small areas is necessary.

5. Changes in the Vegetation.

It is impossible to describe accurately the analytical composition of really primitive montane tussock grassland. The whole area now occupied by this formation has been for sixty years subjected to the modifying influence of grazing, a factor absent in the arrangement and constitution of the primitive vegetation. It is generally assumed that the montane tussock grassland has not altered to any appreciable extent except in so far as the naturalization of exotic species is concerned. It is also generally thought that certain elements of the vegetation not primarily of any importance in the general physiognomy of the formation have become gradually rarer and rarer.

It is, however, fully recognized that over certain areas where through special circumstances the dominant tussock growth-form has been eliminated profound alteration has occurred. This has resulted in the production of a totally different formation, approaching the desert type. Whether or not this new vegetation is a stable permanent one or is merely a transitory type leading to more closed types of associations is not known, but I am inclined to think it is a climax one so long as the present grazing-conditions remain the same. This replacement of montane grassland by a desert or semi-desert type of formation has occurred over wide areas in Central Otago and the Mackenzie country. There the extremely low rainfall, coupled with the various new conditions brought into activity by man's utilization of the land for pastoral purposes, can be held responsible for this remarkable substitution of one formation by another in no way related either taxonomically or ecologically. Over the montane tussock grassland subjected to a rainfall of approximately 30 in. per annum no such radical change has taken place. Nevertheless, with the exception of the rabbit factor (admittedly more important in the so-called " depleted areas ") this grassland has been subjected to the same general pastoral conditions as that where complete replacement has taken place. In general character the montane tussock grassland in areas of moderate rainfall has, apart from the presence of introduced plants, a distinctly primitive appearance, due to the apparently unchanged dominance of the tussock growth-form. It, however, seems impossible to think that profound changes have not taken place apart from the gradual reduction in frequency of occurrence of certain species. At any rate, the wiping-out of certain elements, where this has occurred, must have resulted in their replacement by other plants. If such had not been the case a general opening-up of the formation would have taken place, and steppe would have been produced. The production of steppe from tussock grassland has not, however, taken place in localities of moderate rainfall, except in isolated cases where some specially unfavourable soil or climatic factor, such as wind, has exerted a preponderating influence. Thus, if it is assumed that the relative frequency of certain species has diminished, it is also fair to assume a corresponding increase of other species. It is, however, quite probable that in the depleted areas the grassland degenerated into true steppe before final replacement by desert and semi-desert associations took place. It is generally said that the effect of stock has been to reduce very largely what runholders call " the better and finer grasses,' whatever may be meant by that very general phrase. If a certain proportion of the original vegetation has been more or less eliminated — which there seems no reason to doubt — the remarkable feature of the present-day montane tussock grasslands is that the dominant growth-form has remained permanently the same. *Replacement has been in the direction of an increase in the dominance of the tussock growth-form rather than in any reduction.* This unexpected happening is perhaps one of the most remarkable ecological facts with regard to the influence of stock on a primitive New Zealand association capable of sustaining grazing animals. The lowland tussock grassland, for instance, has in many instances been replaced by a partial sward grass, *Danthonia pilosa.* Fern heath has in many places been turned into meadow land simply by means of stock and the invasion of introduced plants. *Phormium tenax* areas on comparatively dry land have by heavy stocking been replaced by sward grasses. Coastal and rain forest have been destroyed in a similar way. *In these cases the leading physiognomic plants of the primitive associations have been more or less eliminated and replaced by species of a different growth-form, and that not of the tussock type. In the montane tussock grassland areas, with the exceptions already mentioned, the original dominant growth-form has, in spite of grazing, burning, and other introduced factors, become, if anything, intensified rather than reduced.* Again, this tussock growth-form is a climax one, and had attained its dominating position in the vegetation in the complete absence of any grazing animals.

More remarkable still is the fact that many of the areas of southern-beech forest in the montane tussock belt which have been destroyed by man's activity, and not sown intentionally, have developed into typical present-day montane tussock grassland. *It would thus appear, then, as if the tussock growth-form is the only one that is capable of remaining permanently dominant over the montane tussock grasslands.*

In most cases *Festuca novae-zealandiae* is the species showing this great dominance, but in certain instances other plants with a similar growth-form are locally the more important. *Festuca novae-zealandiae* is extremely specialized, and variation in habitat is not followed—as in many other plants, such as *Leptospermum scoparium*, for example—by an alteration in outward form. It seems reasonable to suppose that specialization in form would render a plant capable of occupying only special stations, and that any wide variation in habitat would act unfavourably towards its continued establishment. With the fescue tussock no such restriction to any one special station is apparent. The great area occupied by the montane tussock grassland, although apparently uniform in habitat, is said to be so simply because the dominant physiognomic plants occupying it show no outward response to any change of habitat. That the habitat occupied by the montane tussock grassland varies considerably is shown by the behaviour of introduced plants. Thus in certain places the sweetbrier (*Rosa rubiginosa*) is a tall shrub many feet in height, while in others the individuals of the same race remain as small shrubs not more than 3 ft. high at the outside. In the Cass neighbourhood the great variation in the development of cat's-ear (*Hypochoeris radicata* L.) is most noticeable, and if these variations are looked upon as environmental they indicate considerable differences in the habitat.

Another feature with regard to montane tussock grassland is the small amount of seasonal variation which occurs in the general appearance of the formation. The colour of the tussock is almost invariable during the whole year, and there are no periods of specially rapid growth. This latter, of course, is partly due to the fact that the leaves are long-lived and do not die off each year, so that there does not seem to be any special season for the maturing of the foliage. This almost absolute uniformness of the general appearance of the tussock grassland is one of its most striking features, and sharply marks it off from many other types of grassland in New Zealand.

It has been previously shown that the elements of montane tussock grassland are at most only sparingly grazed upon. It is thus really not peculiar that the dominant physiognomic plants should have retained their position in the formation. In grasses it does not appear unusual that the species least eaten are the most likely to increase. Thus in the pumice soil of the Volcanic Plateau sowings composed of about equal quantities of cocksfoot and fiorin (*Agrostis stolonifera* L.) become at the end of a few years almost pure fiorin pastures. In this case the cocksfoot is kept eaten down, while the fiorin is rejected. In other cases, however, the grass fed on attains the mastery over that not touched, as is seen in artificially induced pastures of one or other of the forms of *Danthonia pilosa* R. Br., where originally the lowland tussock was the more important constituent. In such cases, however, the palatable plant that increases must be better fitted for the conditions than are the species which it displaces.

Before the advent of pastoral occupation, there is no doubt that many areas which are now typical montane tussock grassland were occupied with associations that were gradually turning in course of time into ordinary tussock association. Thus, for instance, *Discaria* thickets were probably more numerous than at present, occupying young consolidated fans and other features of the land-surface that were of comparatively recent origin. Burning has had the effect of more or less eliminating this scrub, and has induced the climax associations of tussock grassland to become developed more rapidly than would have otherwise been the case.

In the neighbourhood of Cass the main feature of the grassland is the very large amount of *Poa Colensoi*. I am inclined to think that this grass was not particularly important in the primitive vegetation, but has increased enormously since the advent of pastoral operations. Its prevalence is especially noticeable on areas formed of material comparatively recently water-transported, such as fans, young river-terraces, and the like. On the hillsides, on the other hand, *Festuca novae-zealandiae* is easily dominant, and the blue tussock (*Poa Colensoi*) is quite a subsidiary element of the vegetation. The last-named grass also appears to be on the increase where burning has been extensively practised. Being smaller than the fescue tussock, it is probably less affected by summer burns than is the taller grass. A notable feature near Cass was its generally uneaten appearance, which was certainly indicative that this grass is not favoured by stock. Of course, it must be recognized that *Poa Colensoi* is an aggregate species, and it is possible that certain forms are more palatable to stock than are others. Another characteristic feature of the Cass tussock-land is the almost complete absence of the true blue grass, *Agropyron scabrum*. In general, I should say that in the upper Waimakariri river-basin the main effect of pastoral operations on the indigenous vegetation has been the more general domination of the tussock growth-form element in the tussock grassland and the reduction in frequency of occurrence of shrubby plants, due to burning, and of certain other shrubby elements, such as the tall species of *Carmichaelia*, due to combined burning and grazing. The scarcity of *Agropyron scabrum* is peculiar, and due either to its having been eaten out or not ever having been an important element of the formation. Over certain areas the injurious effects of summer burns on the fescue tussock are notable, but there is no appearance in general of any opening-up of the grassland and the production of steppe. Unlike what happens on lowland tussock grassland, no invasion with *Danthonia pilosa* has occurred. Certainly a few plants of both that species and *D. semiannularis* are to be seen, but there is no appearance of that replacement of the tussock growth-form by this genus which is so noticeable and characteristic of much lowland tussock grassland in both the main islands of New Zealand.

6. Deterioration in Carrying-capacity.

It is generally accepted that the carrying-capacity of the montane tussock grasslands has seriously deteriorated since they were first brought under occupation. From an examination of the sheep returns covering a considerable period it is evident that deterioration is not a general feature, but is confined to certain special areas. So far as the upper Waimakariri river-basin is concerned, the figures for the four main sheep-stations during the past thirty-five years are as follows :—

Name of Station.	1879.	1889.	1899.	1909.	1914.
Castle Hill	7,500	8,800	8,000*	7,500	7,000
Craigieburn	18,000	21,500	22,500	20,000	17,000
Mount White	18,000	32,000	31,000	38,000	27,500†
Grasmere	8,500	8,500	8,500	7,000	7,500
Totals	52,000	70,800	70,000	72,500	59,000

These figures show that the stock carried thirty-five years ago and to-day is approximately the same, but that during the past six years a considerable decrease has taken place. It will be noticed, however, that the carrying-capacity of the two smaller stations has throughout remained much the same. On all these stations the winter country is montane tussock grassland, but they also possess large areas of summer country.

Thus, so far as the upper Waimakariri area is concerned, it is hardly correct to say that the grazing-land has seriously deteriorated, as it is quite possible that the low figures for 1913 and 1914 may be improved again in a year or two, as has been the case in the past when a diminution in the flocks has been recorded.

For purposes of comparison it is perhaps fairer to take the stock in a whole sheep district for a number of years. As the Mackenzie County sheep district is one that is frequently quoted as showing serious deterioration in the carrying-capacity of its tussock grassland, the

---

* These figures are for 1898, the 1899 ones being just under 5,000, but the reduction is due to excessive mortality.

† 1913 figures. Those for 1914 show a decrease to 24,000, but this is only temporary, as the run is understocked through both sales and winter losses.

following figures are instructive (all the low tussock grassland of this area can be classed as montane) :—

Total Sheep in Mackenzie County.

1884	...	...	...	...	... 422,000
1887	...	...	...	...	... 440,000
1893	...	...	...	...	... .442,000
1899	...	...	...	...	... 394,000
1904	...	...	...	...	... 399,000
1909	...	...	...	...	... 471,000
1914	...	...	...	...	... 495,000

Here again it is seen that thirty years' continuous grazing has not lessened the carrying-capacity of the whole area. Nevertheless there is a very large extent of country in this sheep district where the majority of the tussocks have died out, and considerable stretches of land that were formerly typical montane tussock-land are now virtual desert, with patches or circular low flat cushions of *Raoulia lutescens* as the dominant physiognomic plant. From this it would certainly appear as if the montane tussock-land was not the important grazing association of the area. It is commonly asserted that if only the country could be restored to the condition it was in forty years ago the amount of stock carried would be vastly increased. Yet in the Mackenzie sheep district the destruction of much of the montane tussock has not resulted in diminution in the aggregate numbers of sheep carried. In fact, thirty years ago in that district there were 70,000 fewer sheep than there are to-day. When the figures for Vincent County (Central Otago) are examined it will be seen that serious deterioration in carrying-capacity has occurred. In 1879 this county pastured over 490,000 sheep, while in 1914 the flocks numbered only 330,000, showing a decrease of over 160,000 sheep. In Vincent County almost the whole of the montane tussock grassland association has been destroyed, and its place taken by an almost desert vegetation. To show how seriously the carrying-capacity has been influenced in parts of the Vincent County the case of Galloway Station may be mentioned. In 1879 the number of stock pastured was approximately 70,000, but to-day the number carried on the same area is a trifle over 20,000.

With regard to Vincent County in general, two factors which have not been in operation in the upper Waimakariri river-basin must be recognized. These are a very low rainfall and a superabundance of rabbits from the early " eighties " onwards. The system of sheep-farming in Vincent County and in the upper Waimakariri has been exactly similar, and extensive burning of the tussock grassland a noticeable feature of the management in both localities. In Vincent County extensive and repeated burning of the tussock, coupled with the low rainfall (less than 15 in. per annum at Galloway) and the continued presence of rabbits in large numbers, has virtually destroyed the montane tussock grassland vegetation. On the other hand, the same system of management in the upper Waimakariri has in the presence of a fair rainfall (over 30 in. per annum, and higher in the west) and in the absence of rabbits resulted in comparatively no deterioration in carrying-capacity until after 1909. Since that date the number carried has fallen considerably, but this is due to excessive winter losses, and not to the deterioration of the vegetation. When it is considered that the Waimakariri river-basin has been continually utilized for sheep-grazing for over half a century, and that comparatively nothing effective has been done in the way of intentional sowing of grasses, &c., to improve the feeding-quality, it is remarkable that the carrying-capacity should have been maintained. From this it would appear that in areas of fair rainfall, provided rabbits are not present, the montane tussock grassland will retain its normal comparatively low carrying-capacity for an indefinite period. If, however, any attempt at overstocking is resorted to, the carrying-capacity becomes lessened; but after being stocked for a few years below the average the carrying-capacity again becomes normal.

7. The Rival Theories of Burning.

In the early times of the utilization of the montane tussock grassland it was recognized that the dominant fescue tussock, when in its natural state, afforded but scant feed. Accordingly the practice of burning off the tops of the tussocks each year was resorted to in order to stimulate a fresh young growth that would be readily fed off by sheep. This practice of burning was general from the " sixties " onwards, being carried out at all times of year, including midsummer. After a while it was decided that summer burning had an injurious effect, and this practice was abandoned, but autumn and spring burning still remained popular, and was generally carried out. During the past decade the utility of burning at all has been largely questioned, and at present the montane tussock-land runholders are divided into two distinct schools, one asserting the necessity and the other the fallacy of burning. The arguments adduced by both sides have not been subjected to scientific experimental investigation, so that the truth or otherwise of the premises laid down by both burners and anti-burners has never been accurately determined.

The premises on which the two schools base all their arguments may be summarized as follows :—

(A.) *Arguments for Burning.*

(1.) The tussocks in their natural state afford no feed for stock, the young growth being so protected by the older inedible leaves that it cannot be grazed off.

(2.) The burning results in the production of succulent fresh green herbage readily eaten by sheep.

(3.) Non-burning results in the too rank growth of the tussocks. This causes the leaves of adjacent tussocks to almost completely cover the ground, and increases the difficulty of grazing the areas between the tussocks.

(4.) Burning in early spring and in early autumn does not kill out the tussocks.

(5.) The destruction of tussocks over wide areas can generally be traced to summer burning, a practice now universally rejected.

(B.) *Arguments against Burning.*

(1.) The burning of tussocks results in their gradual weakening, and final death.

(2.) The continued healthy development of the tussock growth-form as the dominant feature of the vegetation is essential for the welfare

6*

of the other elements of the association. Any weakening and killing-out of the tussocks leads to the general destruction of the association, and its replacement by a vegetation of infinitely worse feeding-value.

(3.) .The growth resulting from the burning of the tussocks is capable of yielding feed only for a brief period, as burnt tussocks, even when grazed soon after burning, rapidly reassume their tussocky condition.

(4.) Burning lessens the amount of natural seeding, and this leads to the rapid creation of bare ground.

The evidence on which the anti-burners base their claims is on the wide extent of country, more particularly in Central Otago, where the gradual disappearance of the tussock has been followed by the elimination of the whole of the association and the production of a virtual desert type of vegetation quite useless for grazing purposes. The pro-burners, on the other hand, assert that many stations which do not burn, although they keep the tussock grassland in an apparently unimpaired condition, cannot keep their stock in as good a condition as when burning is carried out. Again, with regard to Central Otago, it is urged that the rabbit factor has played the essential part in the destruction of the vegetation, and this supposition is probably in the main quite correct.

8. The Future of Montane Tussock Grassland.

It is difficult to forecast the future of montane tussock grassland. From present appearances, however, it seems likely that the greater part of the area occupied by this formation is capable of remaining permanently in much the same condition as it is to-day. In other words, it is likely to remain inferior sheep-grazing country, capable of supporting at most one sheep on 3 to 5 acres. On other portions, where both soil and climatic conditions are adverse, a gradual deterioration in carrying-capacity will be experienced, provided special efforts are not made to modify these conditions. Their modification by manuring, cultivation, shelter-tree planting, and other methods is quite possible, but, generally speaking, impracticable from the expense standpoint. The limited experimentation carried out by the Department of Agriculture has all been in the direction of showing the impracticable nature of such work. Other areas, again, where the soil and climatic conditions are good, will probably in time, when great reduction in the size of individual holdings takes place, be converted into better-class grazing-land through cultivation and other farming operations. The production of special crops, more particularly lucerne and also root crops,* may also play an important part in the future. The application of such methods would have a very important bearing on the tussock grassland as a whole. The increased production of feed over certain areas would enable periodical spelling of much of the land without the necessity for reducing the total carrying-capacity.

One point is, however, of great importance, and this is that the present vegetation appears to be specially attuned to the conditions, and this vegetation is of an extremely xerophytic nature. There must be some very powerful factors that have led to the development of such a vegetation, and with these still in operation it seems unreasonable to

* There is a prevalent opinion amongst runholders that sheep grazed on montane tussock grassland refuse to feed on root crops.

expect that ordinary grassland of a more mesophytic type is likely to be produced by artificial means. It is, however, possible that certain grasses that one might term facultative xerophytes, such as certain forms of *Danthonia pilosa*, might become in time important consituents of the montane tussock grassland. When conditions are favourable *Danthonia pilosa* is distinctly a mesophytic grass, with broad flat green leaves; but when conditions are again adverse they dry and roll up, fresh green succulent herbage being again produced on the advent of more congenial conditions. If *Danthonia pilosa* would spread over the montane grassland, and become an important constituent of the vegetation, a great increase in carrying-capacity would be secured.

Spelling the ground during the seeding season has often been advocated, and the increased carrying-capacity over certain areas that has followed understocking for some years seems to indicate that such a procedure would be productive of good results. Against this has to be reckoned the fact that much of the montane tussock grassland has already for many years past been more or less unstocked during the late summer and autumn, when the flocks are mainly located on the higher summer country. Again, the special spelling of definite areas during definite periods would entail a very large expenditure in fencing.

It has been calculated that over one-seventh of the occupied land of New Zealand is composed of montane tussock grassland, from which, roughly, 2½ lb. of wool is produced per acre each year. A certain number of store sheep are also disposed of from this land, but the number is considerably less than is generally thought. *On this huge area the average gross returns are far less than on any other of the occupied lands, and the adoption of any methods which will increase the turn-over would be of far-reaching importance.* In order to determine if any methods can be adopted in this direction, a complete investigation of the vegetation with respect to sheep-grazing is necessary, and should precede any definitely experimental work on the improvement of the carrying-capacity of the land.

The Canterbury College Mountain Biological Station offers exceptional facilities for this all-important research. With the aid of a few hurdles, some sheep, and proper scientific observation much could be accomplished. When it is said that even after sixty years' continuous occupation of the ground for sheep-grazing comparatively no accurate knowledge is available as to what plants are really furnishing the bulk of the feed, the need for a thorough investigation of the vegetation from the economic standpoint becomes apparent. It is safe to say that until such time as this has been done any statements as to the future of the montane tussock grassland are largely a matter of pure speculation.

ART. XIX.—*Notes from the Canterbury College Mountain Biological Station.*

No. 4.—The Principal Plant Associations in the Immediate Vicinity
of the Station.

By L. Cockayne, Ph.D., F.L.S., F.R.S., and C. E. Foweraker, M.A.

[*Read before the Philosophical Institute of Canterbury, 1st December, 1915.*]

Plates XV, XVI.

(A.) GENERAL.

In his general account of the Canterbury College Mountain Biological
Station, Dr. C. Chilton points out—Trans. N.Z. Inst., vol. 47 (1915), p. 333—
that when the boundaries of the botanical reserve are defined and surveyed
a botanical map of the area will be prepared. Such a map would, of course,
be based on the distribution of the florula as determined by that of the
plant associations. A classification and preliminary account of these associations is, then, a necessary prelude to that detailed study which the preparation of a map of the vegetation demands. The associations as defined
below, although the outcome, in part, of the great topographical changes
which have befallen the area during and subsequent to its period of intense
glaciation, are considered from the static and not from the dynamic standpoint, since for purposes of graphic representation they are distinct vegetation entities for the time being. This viewpoint does not in the least
preclude the ultimate necessary study of the plant-covering with regard to
its evolution; it merely offers a convenient basis for present investigations.
No attempt is made at thoroughness in our treatment of the associations,
which is descriptive merely; nevertheless we believe that brief, incomplete
descriptions of undescribed plant communities are better than nothing, and
can be of considerable assistance to those engaged in comparative studies of
New Zealand vegetation. Each association is a collection of species bound
to a definite habitat, the latter being the sum total of the various ecological
factors to which the association is exposed. "Habitat," as thus defined, is
virtually impossible to determine accurately, but it is roughly measured by
climate, by the nature of the soil in its widest sense, by the relation of plant
to plant, and by the animals present. But, even were we able to estimate
the exact results of habitat, no such great advance would be made as might
be supposed, since each species exists under a distinct environment of its
own. Such individual environmental differences may, in fact, easily be
far greater within one association than are the general habitat-influences
of two adjacent distinct associations.

The Canterbury College Mountain Biological Station is situated in the
Eastern South Island Botanical District,* close to its junction with the

_____

* The botanical districts referred to in this paper may be provisionally defined
as follows: (1.) The North-eastern South Island Botanical District includes the north-
eastern portion of the South Island, excepting the wet area in the vicinity of the Marlborough Sounds, and it is bounded on the west by a line marking the average limit of
the western rainfall, and on the south by the River Waiau. (2.) The Eastern South Island
Botanical District is that area extending from the Waiau to the Waitaki Rivers, and
terminating on the west at the line reached by the average western rainfall. (3.) The
Western South Island Botanical District extends from the River Taramakau in the
north to a line not yet determined, lying somewhat to the south of the River Haast,
and on the east it crosses the actual divide and is bounded by a line marking the average
limit of the western rainfall.—L. C.

Western South Island Botanical District, or, in other words, it lies just beyond the influence of the excessive western rainfall. This is most plainly reflected in the physiognomy of the vegetation, where, on the Cass side of the River Waimakariri, the xerophytic grass-tussock form dominates the landscape (see Plate XV, fig. 1); while on the other side of the river, only some two miles distant, stands a vast dark mesophytic forest-mass with clean-cut margin, hardly one tree standing out into the open grassland of the east. Nowhere can a more striking example be seen of the direct effect of climate. The change in physiognomy is instantaneous; there is no transitional phase.

The matter of this paper is the outcome of a few visits to the area made at different times by the authors independently of one another, and especially of a recent brief stay at the station, when the various associations were conjointly examined by the authors, and lists of the species hurriedly taken. Unfortunately, the time available for observations was much too short, while, owing to the season of the year (November), certain species were not in bloom and could not be accurately determined. Therefore, this study is limited to brief descriptions of the associations, and doubtless many species have been omitted. In certain cases " var." is placed after the specific name. This denotes that the species is an aggregate, and that the plant referred to is one of the unnamed varieties of which the aggregate is composed. On the other hand, " var." being absent does not mean definitely that the species is a true entity. It means either that we think the species may be a definite entity or else that we have no proof that it is an aggregate.

The associations dealt with fall under the following heads: (1.) Forest, (2) shrubland, (3) grassland, (4) swamp, (5) rock, (6) river-bed. This classification is obviously faulty, since it contains associations based some on growth-forms and some on habitat; but it is convenient, easy of application, and can be readily used both by those visiting the station and for phytogeographical purposes generally.

Many of the associations are greatly modified at the present time through the action of man, the chief modifying agencies being the grazing of sheep and frequent fires. To a much lesser degree, partly also through the above two factors, various foreign species of plants have gained a firm footing and come into competition with the indigenous element. Much more important, however, is the fact that the degree of frequency of the component species of certain associations has been greatly changed, so that infrequent members of the primitive association may now be of prime importance. But it must be pointed out that, so far as ecological studies at Cass are concerned, the introduced plants are now as much members of the associations as are the indigenous species, and they must be equally considered. This is especially important since a good deal of future research may have a distinct economic bearing. This paper concludes with a list, quite incomplete, of the species belonging to the associations. Also, for the information of those not conversant with the New Zealand flora, a very brief account is given of the growth-form of each species. This, it must be pointed out, refers solely to the plant as it grows at Cass, and has nothing to do with the growth-forms of a species through its entire range, or with its commonest form.

(B.) THE ASSOCIATIONS.

(1.) FOREST.

In the immediate vicinity of the biological station forest is poorly represented. This is not on account of its destruction by fire, as in so many parts of New Zealand, but because in the Eastern Botanical District, in the montane and subalpine belts, the ecological conditions are antagonistic, and forest occurs only in specially well-sheltered situations. Here only the three small pieces of forest situated in the gullies at and near the base of Mount Sugarloaf are dealt with.

The association belongs to the mountain southern-beech (*Nothofagus cliffortioides*) formation, a plant-community of wide distribution in the high mountains of both the North and South Islands. At Cass the association, so far as its undergrowth is concerned, was probably always far from dense, but now, through sheep having sheltered for many years in the forest, undergrowth in no few places is wanting. But the plant-covering of certain spots gives a fair clue to the primitive physiognomy of the forest interior, with its slender tree-trunks and moss-covered floor occupied by numerous *Nothofagus* seedlings, small fronds of *Blechnum penna marina*, and erect or straggling bushes of *Coprosma parviflora, C. microcarpa, Aristotelia fruticosa, Clematis marata* as a ground-plant, and *Rubus subpauperatus.*

The sole tree is the mountain southern-beech, *Nothofagus cliffortioides.* Parasitic on it are the woody hemi-parasites *Elytranthe flavida* and *E. tetrapetala.* The forest-floor is usually dry, and consists on the surface of a brown humus held together by numerous beech-rootlets, while beneath is the usual clay of the locality.

On the margins of the streams the vegetation, as might be supposed, is much richer than elsewhere. The following is a list of the species noted near such streams : *Blechnum penna marina, B. capense* var., *Polystichum vestitum, Lycopodium Selago, Carex dissita* var., *Corysanthes macrantha, Urtica incisa, Cardamine heterophylla* var., *Acaena Sanguisorbae* var. *pusilla, Rubus subpauperatus, R. schmidelioides* var. *coloratus, Aristotelia fruticosa, Hymenanthera dentata* var. *alpina, Epilobium rotundifolium, Galium umbrosum, Coprosma parviflora, Veronica salicifolia* var. *communis, V. Traversii* var., *Erechtites glabrescens.*

(2.) SHRUBLAND.

By "shrubland" is here meant those associations where the shrub-form dominates. It differs greatly in its physiognomy and physiological characters according to the degree of closeness at which the component shrubs occur and their growth-forms. Where the shrubs as a whole do not touch one another, but stand as small clumps or single plants dotted about here and there, the association may be termed "open shrubland." Where the shrubs grow closely, but do not form dense entanglements, the term "thicket" may be used ; but where there is a dense interlacing mass of branches, or the divaricating growth-form plays a notable part, the suggestive term "scrub" may be applied. These names for the various classes of shrubland, as explained farther on, differ in part from those used by L. Cockayne in his previous writings. Obviously, between all the classes intermediate forms must occur.

Shrubland is generally of a more xerophytic character than forest, but its habitats differ so greatly in this regard that some associations are distinctly mesophytic, while others are strongly xerophytic. Of course, it is impossible without physiological, anatomical, cultural, and ecological investigation to declare with confidence that any plant is either a xerophyte or a mesophyte. But, undoubtedly, when it is seen that examples of one and the same true-breeding entity assume different growth-forms under mesophytic and xerophytic conditions, and that these growth-forms ap-

FIG. 1.—Cass Plain with low tussock grassland, greatly modified by grazing. On left, *Discaria toumatou* Raoul. Cragieburn Mountain and gorge of River Cass in background.

FIG. 2.—Prostrate and erect varieties of *Podocarpus nivalis* Hook. growing on stony debris slope at about 3,000 ft. altitude on Mount Sugarloaf. Low tussock grassland in background.

[Face p. 169.

proach nearer and nearer to growth-forms of species limited to the one environment or the other, then it is safe to conclude that the induced apparently epharmonic* forms are truly xerophytic or mesophytic, as the case may be. Certain species from the Cass shrubland exhibit such induced xerophily or mesophily to an astonishing degree, as L. Cockayne showed some time ago.†

A most interesting case of a contrary character came under our notice while studying the Cass shrubland for the purposes of this paper. On coarse stony debris, as will be seen farther on, *Podocarpus nivalis*, the mountain-totara, is extremely abundant, its shoots far-spreading and closely hugging the ground in espalier fashion. This, so far as we had previously noted, was the invariable behaviour of this shrub in such a station. But, according as the plant grows in sun or shade in various parts of New Zealand, so does it assume the prostrate or the more or less erect habit.‡ However, on this debris slope at Cass there are certainly two distinct hereditary races of the species, one the typical *P. nivalis* Hook f., which in the above habitat always remains prostrate, and the other, which grows in apparently exactly the same habitat, is a dense shrub 1·8 m. high or more (see Plate XV, fig. 2). Here, then, a special xerophytic growth-form, the spreading prostrate (espalier), is not, apparently, one whit better suited for its habitat than the more mesophytic, infinitely less wind-resisting, erect shrub-form, which, moreover, from its position should be exposed to a more intense transpiration. The word "apparently" is used advisedly, since the erect shrub may have a longer root-system, and so be under a different environment to the prostrate form.

The Cass shrubland associations are not peculiar to the area, but, with slight floristic modifications, extend throughout the Eastern and North-eastern South Island Botanical Districts, though the latter possesses some shrubland peculiarly its own.§ Excepting the open shrubland, and perhaps the *Leptospermum* thicket, the associations are virtually primitive.

The shrubland associations here dealt with may be designated as follows : (a) *Cassinia* open shrubland; (b) *Discaria* (wild-irishman) thicket; (c) *Leptospermum* (manuka) thicket ; (d) river-terrace and debris scrub.

(a.) Cassinia *Open Shrubland.*

Shrubland where one or other of the closely related species of *Cassinia* dominate is at the present time a common feature in much of lowland and montane New Zealand, especially on coastal areas from which the forest has been removed by man. In primitive New Zealand, associations

---

* The term "epharmonic" is here used as in my former writings—*e.g.*, "Obser-vations concerning Evolution, derived from Ecological Studies in New Zealand" (Trans. N.Z. Inst., vol. 44 (1912), pp. 13–30)—with a somewhat different significance to that of Vesque and Warming (see "Oecology of Plants" (1909), pp. 2 and 369). According to my usage, an epharmonic variation is a change in its form or physiological behaviour *beneficial* to an organism evoked by the operation of some environmental stimulus. Such a change may be called an epharmonic adaptation, as distinguished from such adapta-tions as cannot be traced to any direct action of the environment. To the neo-Darwinian no permanent adaptation according to the above definition would be "epharmonic," whereas to the neo-Lamarckian all would be so considered.—L. C.

† Trans. N.Z. Inst., vol. 32 (1900), p. 123; and *ibid.*, vol. 44 (1912), pp. 15, 17, 20, and 27.

‡ Trans. N.Z. Inst., vol. 44 (1912), p. 17, and for figure see pl. 4, facing p. 21.

§ Especially one on coarse debris where *Senecio Monroi*, elsewhere a rock-plant, is dominant. Another rock species, *Helichrysum microphyllum*, also occurs in a certain form of subalpine scrub in the same botanical district.

of this character must have been comparatively rare. The species of *Cassinia* (*Compositae*) are quick-growing shrubs of ericoid form, tolerant of considerable drought, and bearing in abundance seed of high germinating-power, which can be transported rapidly over considerable distances by the wind.

. The association under consideration contains a variety of the coastal *Cassinia fulvida* as its dominant member. Other shrubby members are *Discaria toumatou, Hymenanthera dentata* var. *alpina, Leptospermum scoparium* var., *Corokia Cotoneaster, Gaultheria depressa, Styphelia Colensoi,* and *Dracophyllum uniflorum* var.

The association occurs on ground formerly occupied by low tussock grassland, where, through burning the tussocks and subsequent grazing, bare ground suitable for invasion by the *Cassinia* has been provided. The shade and shelter furnished by the last-named encourages the establishment of the other shrubs. The erect shrubby members of the association grow in clumps or as single plants, and the interspaces consist of the original tussock association more or less modified. Some of the species of the latter are : *Blechnum penna marina, Poa Colensoi* var., *Festuca novae-zealandiae, Acaena Sanguisorbae* var. *pilosa, Viola Cunninghamii, Pimelea prostrata* var., *Styphelia Fraseri, Celmisia longifolia* var., *Raoulia subsericea, Microseris scapigera,* and *Senecio bellidioides* var. *glabratus.*

Should the *Leptospermum* become dominant, such an association can be rapidly transformed into manuka thicket, or, on the other hand, with increase of *Cassinia*, many of the species mentioned above may be suppressed.

(b.) Discaria *(Wild-irishman)* Thicket.

This is the " *Discaria* shrub steppe " of Cockayne and Laing,* a suitable enough term, but we think it best to abandon the word " steppe " altogether in New Zealand phytogeography, because there is no consensus of opinion amongst ecologists regarding its usage, so that formations† of very different ecological relationships‡ are called " steppe."

The association consists, so far as shrubs are concerned, almost entirely of the spinous, semi-divaricating *Discaria toumatou* (see foreground in Plate XV, fig. 1), of an average height of, say, 1·2 m. In true thicket the shrubs touch or grow into one another, but frequently this association is more or less open. *Clematis marata* may be fairly common as a liane. Other shrubs of the neighbourhood may be present here and there, but they are of no moment.

The association is extremely constant in its distribution throughout the South Island, except west of the Southern Alps, and, when its dark mass is seen at a distance, indicates dry river-terrace and those stony fans where streams issue from their gorges. Eventually grass tussocks and some of their accompanying plants, especially those of most xerophytic structure, become established in the open spaces between the shrubs.

* Trans. N.Z. Inst., vol. 43 (1911), p. 349.
† The term "formation," as used by me, has a much wider signification than "association," and is used to include such associations as are virtually ecologically similar. Thus the whole rain-forest of New Zealand is one formation made up of many associations which differ from one another floristically, but if ecologically, then only to a very limited extent. So, too, low tussock grassland, swamp, and river-bed are ecological units each containing more than one floristic unit. To compare the above phytogeographical conceptions with those of taxonomy, the formation represents the genus and the association the species.—L. C.
‡ See E. A. Rübel, Journ. of Ecology, vol. 2 (1915), pp. 233–35.

(c.) Leptospermum (*Manuka*) Thicket.

This belongs to the formation hitherto called " manuka heath " by L. Cockayne. It seems to us best to abandon the term " heath," partly because it is of loose application, and partly because in many parts of New Zealand the dominant shrub reaches too great a height to permit comparison with the heaths of Europe. At the same time, where climatic and soil conditions combine to produce a low, more or less uniform growth of *L. scoparium,* " heath " is far from being an unsuitable designation.

At Cass fairly wide areas occupied by manuka are not uncommon. How far these are primitive we do not know, but some seem to bear a fairly primitive stamp. The relation of *Leptospermum scoparium* var. to the surrounding vegetation at Cass requires detailed study.

(d.) River-terrace and Debris Scrub.

On the sides of gullies cut by streams through a fan, on the faces of river-terrace throughout the Waimakariri River basin generally, of which the Cass area forms a part, and on the otherwise bare stony debris of the hillside where weathering has led to an accumulation of rather large stones there are dense scrubs of a similar floristic composition, but differing according to habitat in the relative proportions of the important members. These scrubs owe their special physiognomy largely to the presence of the divaricating growth-form in abundance, and to the green- or it may be glaucous-leaved species of *Veronica.* Should *Veronica* dominate—a frequent happening in many parts of montane and subalpine New Zealand—we have a " *Veronica* scrub." These scrubs are usually of considerable density. The shrubs grow intermixed, and lianes may bind them still more tightly together. The divaricating shrubs afford an extremely striking example of convergence, belonging, as they do, to several distinct families—*e.g.*, *Pittosporum divaricatum (Pittosporaceae), Discaria toumatou (Rhamnaceae), Aristotelea fruticosa (Elaeocarpaceae), Hymenanthera dentata* var. *alpina (Violaceae), Corokia Cotoneaster (Cornaceae), Coprosma propinqua* var. and *C. parviflora* var. *(Rubiaceae),* and *Olearia virgata* var. *(Compositae).* If this convergence, which we certainly consider epharmonic, were to be dealt with not merely from one plant association, but from the New Zealand region as a whole, the figures would be not merely eight species belonging to seven families, but fifty-six species belonging to nineteen families and twenty-three genera !

The scrub of shaded gullies is closely related to forest, and with the appearance of *Nothofagus cliffortioides* as a member forest may be considered in process of formation. But the scrub of steep, stony debris, on the other hand, is one of those rather rare communities that is at the same time an initial and climax association on a par with those of shingle-slip, dry rock, and the *Epilobium* association of river-bed.

According as certain species are dominant, so is the facies of the association under consideration more or less altered. But we do not consider such change great enough to warrant the establishment (except in the case mentioned below) of more than one association for Cass. Besides the shrubs already mentioned, the following are generally present : *Discaria toumatou, Gaultheria rupestris* var., *Veronica salicifolia* var. *communis,* and *Veronica Traversii* var. *Olearia avicenniaefolia* var. is occasionally present. If the ground is wet, then *Veronica buxifolia* var. *odora* may become so important as to dominate, in which case certainly the association has changed its character, and may be named *V: buxifolia* scrub.

V. buxifolia, as another variety with erect and but little-branched stems, forms low thicket in wet subalpine western stations, and the two associations must not be confused.

Besides the shrubs there are certain ground-plants in the scrub under consideration. Without giving a special list, the following are characteristic: *Cystopteris novae-zealandiae,* *Blechnum penna marina, Polystichum vestitum, Lycopdium fastigiatum, Acaena Sanguisorbae* var. *pusilla* and var. *pilosa, Epilobium pubens,* and, if the floor is not too dry, *E. chloraefolium.*

The lianes of the scrub are: *Muehlenbeckia complexa* var., *Clematis australis, C. marata, Rubus schmidelioides* var. *coloratus,* and *R. subpauperatus.*

On Mount Sugarloaf there is the interesting open shrub association already referred to growing on a slope of unstable debris, the chief characteristic of which is the dominance of the common prostrate variety of *Podocarpus nivalis*, forming broad patches, orange-brown in colour, which may grow into one another. The above colour is not permanent, but depends upon the direct action of the sun. Here and there are the erect bushy shrubs of another variety of *Podocarpus nivalis,* some of which are 1·8 m. high, and 1·2–1·5 m. through (see Plate XV, fig. 2). Other shrubs present are: *Clematis australis* prostrate on the stones, *Hymenanthera dentata* var. *alpina, Aristotelia fruticosa, Leptospermum scoparium* var., *Gaultheria rupestris* var., *Dracophyllum longifolium* (stunted), *Styphelia Colensoi, S. Fraseri, Pimelea Traversii, Coprosma propinqua* var. Various herbs, &c., occur on the open spaces between the patches of *P. nivalis*—e.g., *Asplenium Richardi, Blechnum penna marina, Lycopodium fastigiatum, Poa sclerophylla* (on the finest debris), *P. Colensoi* var., *Epilobium pubens, Myosotis australis* var., *Celmisia spectabilis, C. Lyallii, Senecio lautus* var. *montanus.* Ecologically the association belongs rather to fell-field than to shrubland, while the finer debris bears true shingle-slip plants.

(3.) Low Tussock Grassland.*

"Tussock grassland" is here used as a substitute for the term "tussock-steppe" hitherto used by L. Cockayne in his ecological publications. The reasons for abandoning the word "steppe" have already been given. All the same, when attention is called to the effect of overstocking and burning in increasing the percentage of bare ground to that clothed by tussock, &c., until, as in Central Otago, desert pure and simple is established, there is certainly an induced steppe association in New Zealand, just as there is induced desert. Such induced steppe, however, we would now term "open low tussock grassland," a term having the merit of defining itself.

The word "low" before "tussock" indicates that the tussocks are either *Poa caespitosa* or *Festuca novae-zealandiae*, and not the much taller tussocks *Danthonia Raoulii* or *D. flavescens*, which, when dominant, one or the other, make "tall tussock grassland," a formation absent in the immediate vicinity of Cass.

Low tussock grassland is far and away the most important association in the vicinity of Cass—as, indeed, it is for the whole of the South Island of New Zealand east of the actual divide. Its presence is clearly in harmony with the drier eastern climate, and when the average rainfall and number

---

* The paper by A. H. Cockayne in this same volume (pp. 154–65) should be read in conjunction with this section of our paper, since it opens up and discusses many matters of phytogeographical interest which we have not touched upon.

of rainy days is known for the Cass Biological Station a fairly accurate measure will be afforded of the maximum toleration of rain by tussock grassland. The description which follows refers chiefly to the association below 3,000 ft. altitude.

The association is at the present time so greatly changed by burning and sheep-grazing that little idea can be formed of its primitive condition, but it seems safe to conclude that the tussocks would generally touch and that *Discaria toumatou* would be quite as abundant as at present, while *Carmichaelia subulata* (so readily eaten by sheep), the now rather rare *Aciphylla squarrosa*, and the grass *Agropyron scabrum*, then growing through the tussocks, would be plentiful. On the other hand, the extremely abundant small tussock, *Poa Colensoi* var., and a number of herbs and semi-woody plants, now common, would be far less in evidence, especially *Scleranthus biflorus* var., *Coprosma Petriei* (two varieties), and *Senecio bellidioides* (two varieties or forms).

Theoretically the dominant plant is the tussock *Festuca novae-zealandiae*, but in reality *Poa Colensoi* var. is far more plentiful in many places, while over considerable areas the *Festuca* is absent (see Plate XV, fig. 1). The following, in addition to species already mentioned as abundant, are common members of the association, the introduced plants being distinguished by an asterisk: *Blechnum penna marina* (where shady), *Lycopodium fastigiatum* (where shady), *Holcus lanatus, *Poa pratensis, Carex breviculmis, Luzula* sp., *Microtis uniflora, Muehlenbeckia axillaris, *Rumex Acetosella, Stellaria gracilenta, Scleranthus biflorus* (two vars.), *Cerastium triviale, Ranunculus multiscapus, Acaena Sanguisorbae* var. *pilosa, Geranium sessiliflorum* var. *glabrum, Viola Cunninghamii, Pimelea prostrata* var., *Epilobium elegans? Hydrocotyle novae-zelandiae* var. *montana, Gaultheria depressa* (where shady), *Styphelia Fraseri* (excessively abundant in many places), *Plantago spathulata, Wahlenbergia albomarginata, Celmisia longifolia* var., *C. spectabilis* (where shady), *Brachycome Sinclairii, Raoulia subsericea, Gnaphalium Traversii, Cassinia fulvida* var. *montana, *Hypochaeris radicata* (many distinct forms apparently quite apart from change in environment). At its higher levels the subalpine element becomes more abundant, *Celmisia spectabilis* being especially noticeable, while species confined to shady stations at lower altitudes here grow in the full sunshine.

Where the association occurs on old flood-plain, and the ground is stony, there is often an abundance of *Carmichaelia uniflora*, and here and there open cushions of *C. Monroi*. In similar situations, forming a distinct subassociation, thanks to its twitch-like underground creeping stems, the grass *Triodia exigua* forms a compact even turf. The pretty blue-flowered *Veronica pimeleoides* var. *minor* is rather common in stony places, and the easily overlooked *Iphigenia novae-zealandiae* is probably fairly common. The above lists by no means exhaust the florula of the association, but they give a fair idea of its present composition for comparative purposes with other parts of the formation in the South Island, and prepare the way for the necessary detailed and experimenal study that should next follow.

Although not all the introduced plants are mentioned, it can be seen that they are far from numerous; in fact, they play comparatively little part as yet in the economy of the association. Nor is it to be expected that they will greatly increase either in number or aggressiveness under the present condition of affairs. The truth seems to be that a balance has been reached and that a plant association unknown in primitive New Zealand is now well established.

(4.) Swamp.

Every transition from lake to tussock grassland is to be seen at Cass; but our data is quite insufficient for even a superficial treatment of the subject, consequently our remarks are confined to swamp. But, even in this regard, the details, on which so much depends, of the relation between the species and the depth of the water cannot be given. By " swamp " we mean an association growing in a habitat where water remains permanently, or at least during the greater part of the year, on the surface of loose, frequently peaty, soil, while it is not too deep to inhibit the presence of plants rooting in the ground. This formation varies considerably in character in different parts of New Zealand, while, except forest, no formation has been so greatly altered by man. The Cass swamps, then, since they are virtually in their primitive condition, stand as most important natural objects, to be jealously guarded and assiduously studied. The piece of swamp close to the biological station supplies a fine example of the gradual change from glacial lake to dry ground by way of swamp.

The swamp belongs to the class " reed-swamp," and to that association where *Typha angustifolia* var. *Muelleri* is dominant—raupo *(Typha)* association.

The vegetation exhibits a fairly well-marked series of girdles which are in harmony with the depth of the water, but the effect is to some extent masked by streams of running water passing through the swamp, and by non-uniformity of depth in places between the centre and margin. Speaking generally, there is a piece of open water in the centre of the swamp too deep as yet for occupation by swamp-plants; next comes a girdle of raupo *(Typha angustifolia* var. *Muelleri)*; next a girdle of Carices and other plants; then the shore girdle, which is subject to periodical submersion; and, finally, boggy ground, usually beyond the reach of flooding. Our notes do not permit exact details of the composition, or even the limits, of these girdles.

The *Carex* girdle has as its dominant species the niggerhead, *Carex secta*, which species in places invades the *Typha* area. Growing with the above *Carex* is more or less *Phormium tenax*. Where the water is shallower the grass-like *Carex Gaudichaudiana* is easily dominant, forming a broad girdle. With it are other plants—*e.g., Epilobium pallidiflorum*—but we have no precise details.

At the margin of the swamp proper there is a considerable assemblage of species, which, though here taken together, do not all grow under the same conditions. The following may be mentioned : *Carex Gaudichaudiana* (stunted), *C. ternaria, Schoenus pauciflorus, Juncus polyanthemos* var., *Luzula campestris* var., *Rumex flexuosus, Montia fontana, Potentilla anserina* var. *anserinoides, Geranium microphyllum, Viola Cunninghamii, Halorrhagis micrantha, Hydrocotyle novae-zelandiae* var. *montana, Oreomyrrhis andicola* var., *Mazus radicans, Plantago triandra, Asperula perpusilla, Celmisia longifolia* var., *Gnaphalium paludosum*.

Schoenus pauciflorus plays an important part in the marginal physiognomy of the swamp, since in no few places it forms almost pure patches. This sedge of the tussock-form is especially characteristic of shallow gullies, where it grows in company with *Sphagnum* and forms a distinct bog association. Common species of such a habitat are : *Blechnum penna marina, Poa caespitosa* var., *Hierochloe redolens, Carex ternaria, C. Gaudichaudiana* var., *Viola Cunninghamii, Hydrocotyle novae-zelandiae* var. *montana, Aciphylla squarrosa* var., *Asperula perpusilla, Wahlenbergia albomarginata, Celmisia longifolia* var., *Olearia virgata* var.

(5.) Rock Associations.

Dry rocks can only support a scanty assemblage of plants consisting of highly specialized, drought-enduring forms, or of species that can epharmonically alter their form. Such, to cite some of the species belonging to the first category found in such a habitat at Cass, are : *Colobanthus acicularis, Pimelea Traversii, Veronica epacridea, V. tetrasticha,* and *Helichrysum Selago* var. But the fissures, hollows, and crevices of rock readily become filled with soil, and so provide a habitat more easily colonized, where many plants belonging to less xerophytic stations can thrive. When such positions are in the shade, or are so situated as to receive an abundant supply of water, there will be a rich florula, but many of the species will have little to do with rock conditions. The following lists include only those species which are either true rock-plants or are commonly found on rocks where soil has accumulated, though also belonging to other associations.

(a.) Montane Rock.

Most of the species belong equally to other associations. The following list does not discriminate between the plants of dry rocks and those of moister rooting-places : *Blechnum capense* var., *Poa Colensoi* var., *P. caespitosa* var., *Festuca novae-zealandiae, Luzula campestris* var., *Colobanthus acicularis, Geum parviflorum, Carmichaelia subulata, Coriaria sarmentosa* var., *Discaria toumatou, Aristotelia fruticosa, Viola Cunninghamii, Hymenanthera dentata* var. *alpina, Pimelea Traversii, Leptospermum scoparium* var., *Epilobium melanocaulon, Angelica montana, Anisotome aromatica* var., *Corokia Cotoneaster, Gaultheria rupestris* var., *Styphelia Fraseri, Dracophyllum uniflorum* var., *Veronica salicifolia* var. *communis, V. Traversii* var., *Coprosma propinqua* var., *Wahlenbergia albomarginata, Olearia avicenniaefolia* var. *Helichrysum bellidioides* var., *H. Selago* var., *Cassinia fulvida* var. *montana.*

(b.) Subalpine Rock.

Here rocks on Mount Sugarloaf are alone considered. The habitat is dry, and frequently wind-swept. The altitude is not sufficient for certain species more or less common on rocks of the neighbouring higher mountains. The following is a list of the species : *Podocarpus nivalis, Aciphylla Colensoi, Anisotome aromatica* var. stunted, *Dracophyllum rosmarinifolium, Styphelia Colensoi, Suttonia nummularia, Veronica epacridea, V. tetrasticha, Celmisia spectabilis, Senecio Bidwillii.* Almost certainly *Colobanthus acicularis* and *Pimelea Traversii* are present, but they are not mentioned in our notes.

(6.) River-bed.

Broad stony river-beds, over which anastomosing streams wander, are a frequent feature of the South Island of New Zealand. They carry a fairly uniform vegetation, which exhibits certain well-defined phases in harmony with topographical and biological changes. The bed proper is that portion of the stony area occupied by the streams, and liable at any time to be flooded and its plant-covering to be eradicated ; it is, in fact, a habitat of extreme instability. Popularly included in the term " river-bed " is the more stable ground, formerly the flood-plain, on either side of the bed proper, and this we likewise include in our treatment of river-bed. Throughout the South Island the vegetation under consideration has a similar life-history.

First of all, there is the extremely open plant-covering of the unstable bed, which, owing to the instability of the substratum, does not pave the

way for future development, but is a distinct beginning and end in itself
—*i.e.*, it is a plant formation closely related to but distinct from the later-
formed river-bed associations. All the same, it is always present, since
when destroyed at one spot it is being renewed elsewhere.

This primary community may be designated the *Epilobium* association,
since one or other species of this genus is the first-comer, thanks to the
wind-borne seeds, their rapid germinating-power, and the quick development of the seedlings. Also, there is present more or less of the mat-
forming *Raoulia tenuicaulis*. The characteristic species of *Epilobium* for
montane river-bed are the erect-branching *E. melanocaulon* and one or
more of the varieties of the mat-forming *E. pedunculare*, a polymorphic
species.

As the bed becomes stable—*i.e.*, as low terraces are built by the river
with their surfaces beyond the reach of the highest floods—another association makes its appearance, and there is now a gradual procession of
events, according to climate and altitude, from an association where mat
and low cushion plants dominate to low tussock grassland, shrubland, or
even forest.

Coming now to the vegetation of the river-bed in close proximity to the
biological station—namely, that of the River Cass—we are face to face
with some apparent discrepancies in the above general remarks. This
river, which rises in the Craigieburn Mountains, is only some eight miles
in length. Its upper portion, flowing through a forest-clad gorge, is of
a more or less torrential character, and does not come within the scope
of this paper. The lower part of the river flows through its ancient flood-
plain, which at the widest part is some two miles across. Here the true
river-bed is remarkably broad in proportion to the water which it carries,
so that it is far more stable than the habitat in general, and in consequence
the primary and secondary plant-communities are not sharply defined,
and there are far more species on that part of the bed, still liable to flood,
than is usually the case in such a habitat. All the same, the actual procession of events is as detailed above, though it is not so evident as on
many river-beds.

The vegetation of the unstable river-bed, now about to be described,
is not one association, but a combination of the true primary *Epilobium*
association and the succeeding *Raoulia* association. The most important
members are *Epilobium melanocaulon*, *Raoulia tenuicaulis*, and *R. australis*
var. On portions of the substratum not swept bare by water for some
time there may be tussocks of *Festuca novae-zealandiae*, and the introduced
Holcus lanatus, *Rumex Acetosella*, and *Cerastium triviale*.

On river-terrace recently formed, but which is of sufficient age to have
enabled plant-colonists to become established, in addition to the species
already mentioned, the following are present : *Muehlenbeckia axillaris*
forming circular wiry interwoven mats, the introduced *Sagina procumbens*,
several varieties of *Acaena inermis* and *A. microphylla* forming rounded
mats, the introduced *Trifolium repens*, rosettes of *Geranium sessiliflorum*
var. *glabrum*, *Discaria toumatou* (here prostrate), *Raoulia Haastii* forming
dense green cushions, and the introduced *Hypochoeris radicata*. On such a
young terrace the vegetation is liable to rapid destruction when attacked
by a stream which has changed its course.

On long-established terrace occupation by plants for a considerable
period has added humus to the soil, so that a more favourable station
for vegetation has gradually developed, thanks to the plants themselves,

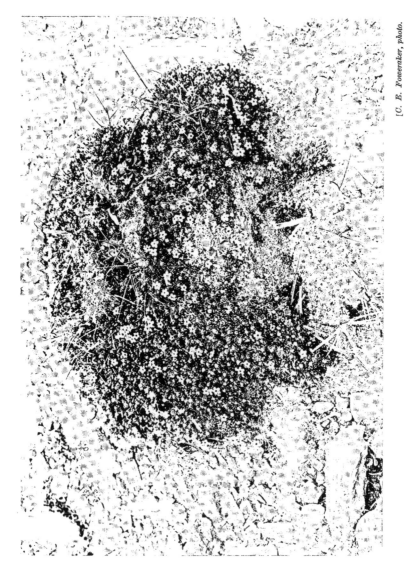

[C. E. Foweraker, photo.

Cushion of *Myosotis uniflora* Hook. f. growing on stony river-bed at Cass, showing the relatively large flower.

especially the species of *Raoulia* and to the moss *Racromitrium lanuginosum.* These Raoulias, according to the species, either collect more or less sand and silt between their branches or the dead parts remain as peat within the cushion. In any case, a seed-bed suitable for the well-being of various species of other plants is provided, which, as they develop, in no few instances kill their host, whose remains, however, add humus to the soil. *Raoulia tenuicaulis* early on is thus suppressed. The following, in addition to most of the species already noted, are important members of this stage of river-bed : *Carmichaelia Monroi,* forming open semi-cushions of short, rigid, flat, leafless stems ; the introduced *Trifolium arvense ; Raoulia lutescens,* forming dense, low, flat, silvery cushions in great profusion ; and the grey, half-dead-looking *Helichrysum depressum,* a low shrub with minute appressed scale-like leaves. The rare herb *Myosotis uniflora* occurs here and there ; it forms small moss-like cushions (see Plate XVI), and produces rather large pale-yellow flowers. As the humus-content of the soil increases, tussocks of *Festuca novae-zealandiae* enter into the association, and by degrees low tussock grassland is established.*

(C.) LIST OF SPECIES BELONGING TO THE ASSOCIATIONS DEALT WITH IN THIS PAPER.

(Unless the contrary be stated, the species are evergreen. Introduced species marked with an asterisk.)

Name of Species, &c.	Growth-form.	Association.
FILICES.		
Cystopteris novae-zealandiae J. B. Armstrong (= *C. fragilis* Cheesem. non Bernh.)	Small tufted fern ..	River-terrace and debris scrub.
Hypolepis millefolium Hook. ..	Summer-green creeping fern	Forest ; river - terrace and debris scrub
Blechnum (*Lomaria*) *capense* Schlcht. var.	Creeping fern ..	Forest ; montane rock.
—— (*L.*) *penna marina* (Poir.) Kuhn (=*Lomaria alpina* Spreng.)	Small creeping fern ..	Forest ; river - terrace and debris scrub ; tus-sock grassland ; bog ; *Cassinia* open shrub-land.
Asplenium Richardi Hook. f. ..	Small tufted fern ..	Coarse debris ; montane rock.
—— *flabellifolium* Cav. ..	Small prostrate fern ..	River-terrace and debris scrub.
Polystichum vestitum Forst. f. (= *Aspidium aculeatum* Swartz var. *vestitum* Hook. f.)	Robust tufted fern ..	Forest ; river - terrace and debris scrub ; montane rock.
Ophioglossum coriaceum A. Cunn. (= *O. lusitanicum* L. in part as defined in Man. N.Z. Flora)	Summer-green fern ..	Tussock grassland.
Botrychium ternatum Sw. var. ..	,, ..	,,

* This is a very brief account of a most interesting series of associations, but a much more detailed description is in course of publication by Foweraker. In the meantime those who desire further particulars may consult L. Cockayne's paper, " On the Peopling by Plants of the Subalpine River-bed of the Rakaia, Southern Alps of New Zealand," (Trans. Bot. Soc. Edin., vol. 24 (1911), p. 104), where the ecological con. ditions of river-bed are discussed, a synopsis of growth-forms furnished, and the ecology of the *Raoulia*-form dealt with.

(C.) LIST OF SPECIES—*continued.*

Name of Species, &c.	Growth-form.	Association.
LYCOPODIACEAE.		
Lycopodium Selago L. ..	Erect branching herb ..	Forest.
—— *scariosum* Forst. f. ..	Prostrate woody plant ..	River-terrace and debris scrub.
—— *fastigiatum* R. Br. ..	Creeping erect woody plant	Forest; tussock grass-land; *Cassinia* open shrubland.
TAXACEAE.		
Podocarpus nivalis Hook. in two vars.	(i) Mat-forming shrub ⎱ (ii) Erect bushy shrub ⎰	Open debris shrubland akin to fell-field.
TYPHACEAE.		
Typha angustifolia L. var. *Muelleri* (Rohrb.) Graebn.	Tall reed - like creeping summer-green herb	Swamp.
POTAMOGETONACEAE.		
Potamogeton Cheesemanii A. Benn.	Water-herb with floating leaves	Swamp in open water.
GRAMINEAE.		
Hierochloe redolens (Forst. f.) R. Br.	Tufted grass	Bog; montane rock; river - terrace and debris scrub.
*Anthoxanthum odoratum L. ..	,, ...	Swamp; tussock grass-land.
Dichelachne crinita (Forst. f.) Hook. f.	Tall tufted grass ..	Tussock grassland; river-bed.
Danthonia semiannularis R. Br. var. *setifolia* Hook. f.	Small tufted grass ..	Tussock grassland.
*Aira caryophyllea L.	Small annual grass ..	Tussock grassland; river-bed.
*Holcus lanatus L.	Tufted grass	Tussock grassland; margin of swamp.
Triodia exigua T. Kirk ..	Turf - forming creeping grass	*Triodia* subassociation of tussock grassland.
*Dactylis glomerata L.	Tussock grass.. ..	Tussock grassland.
Poa caespitosa Forst. f. var. ..	,, ...	Tussock grassland; river-bed; *Discaria* thicket.
—— *Colensoi* Hook. f. var. ..	Small tussock grass ..	Tussock grassland; *Cassinia* open shrub-land; *Discaria* thicket.
—— *sclerophylla* Berggr. ..	Small, rigid, tufted grass	On fine debris.
*—— *pratensis* L.	Turf - forming creeping grass	Tussock grassland.
Festuca novae-zealandiae (Hack.) Cockayne (=*F. ovina* L. var. *novae-zealandiae* Hack.)	Tussock grass.. ..	Tussock grassland; river - bed; *Cassinia* open shrubland.
Agropyron scabrum (R. Br.) Beauv. var.	Tufted grass	Tussock grassland.

(C.) List of Species—*continued.*

Name of Species, &c.	Growth-form.	Association.
Cyperaceae.		
Eleocharis acuta R. Br. ..	Creeping rush-like plant	Swamp.
Schoenus pauciflorus Hook. f. ..	Slender rush-like tussock	Swamp ; bog.
Carex diandra Schrank (= *C. tereti-uscula* Good.)	Creeping grass-like sedge	Swamp.
—— *secta* Boott	Tussock grass-like sedge with stout " trunk "	,,
—— *Gaudichaudiana* Kunth var.	Creeping grass-like sedge	Swamp ; bog.
—— *ternaria* Forst. f. var. ..	,,	,, ,,
—— *dissita* Sol. var. ..		Forest.
—— *breviculmis* R. Br. ..	Small tufted grass - like sedge	Tussock grassland.
Juncaceae.		
Juncus polyanthemos Buchen. var. (= *J. effusus* L. in part as defined in Man. N.Z. Flora)	Rush-like tussock ..	Swamp.
Luzula campestris A. DC. var. ..	Small tufted grass-like plant -	Tussock grassland.
—— sp.	As for *L. campestris* var., but not xerophytic	
Liliaceae.		
Phormium tenax Forst. var. ..	Tall iris-like herb of tussock-form	Swamp.
Iphigenia novae-zelandiae (Hook. f.) Baker	Small summer - green " bulbous " herb .	Tussock grassland.
Orchidaceae.		
Microtis unifolia (Forst. f.) Reichenb. (= *M. porrifolia* R. Br.)	Small summer - green tuberous herb	Tussock grassland ; river-bed.
Pterostylis mutica R. Br. ..	Small tuberous herb ..	Tussock grassland.
Corysanthes macrantha Hook. f.	,, ..	Forest.
Fagaceae.		
Nothofagus cliffortioides (Hook. f.) Oerst. (= *Fagus cliffortioides* Hook f.)	Small canopy-tree ..	Forest.
Urticaceae.		
Urtica incisa Poir.	Low erect herb ..	Forest.
Loranthaceae.		
Elytranthe flavida (Hook. f.) Engler	Shrubby hemi-parasite..	Forest.
—— *tetrapetala* (Forst. f.) Engler	,, ..	,,
Polygonaceae.		
Rumex flexuosus Sol.	Herb with tufted leaves	Margin of swamp.
Muehlenbeckia complexa (A. Cunn.) Meissn. var.	Winding liane ..	River-terrace and debris scrub.
—— *axillaris* Walp.	Mat shrub	River - bed ; tussock grassland.
Rumex Acetosella L.	Creeping herb ..	Tussock grassland ; river-bed
Portulacaceae.		
Montia fontana L.	Branching tufted herb ..	Swamp.
Claytonia australasica Hook. f...	Creeping herb ..	Margin of swamp.

(C.) List of Species—*continued.*

Name of Species, &c.	Growth-form.	Association.
CARYOPHYLLACEAE.		
Scleranthus biflorus (Forst. Hook. f. (two vars.)	Cushion herb	River-bed; tussock grassland; rock.
Colobanthus Billardieri Fenzl var.	Small tufted herb ..	Tussock grassland.
—— *acicularis* Hook. f. ..	Cushion herb	Rock.
Cerastium triviale Link ..	Annual herb	Tussock grassland; river-bed.
Stellaria gracilenta Hook. f. ..	Erect annual ? slender herb	Tussock grassland.
Sagina procumbens L. ..	Small mat herb ..	River-bed.
RANUNCULACEAE.		
Clematis australis T. Kirk ..	Tendril-climbing liane ..	River-terrace and debris scrub; open debris shrubland.
—— *marata* J. B. Armstrong ..	..	River-terrace and debris scrub; *Discaria* thicket; forest.
Ranunculus multiscapus Hook. f. (= *R. lappaceus* Sm. var. *multiscapus* Hook f.)	Small rosette herb ..	Tussock grassland.
—— *depressus* T. Kirk var. ..	Small rosette herb forming patches	Swamp.
CRUCIFERAE.		
Cardamine heterophylla (Forst. f.) O. E. Schulz var. (= *C. hirsuta* L. in part as defined in Man. N.Z. Flora)	Slender erect branching herb	Forest; rock.
PITTOSPORACEAE.		
Pittosporum divaricatum Cockayne (= *P. rigidum* Hook f. in part as defined in Man. N.Z. Flora)	Divaricating shrub ..	River-terrace and debris scrub.
ROSACEAE.		
Rubus australis Forst. f. var. *glaber* Hook. f.	Scrambling liane ..	Forest, outskirts.
—— *schmidelioides* A. Cunn. var. *coloratus* T. Kirk	,, ..	River-terrace and debris scrub.
—— *subpauperatus* Cockayne ..	,, ..	River-terrace and debris scrub; forest.
Geum parviflorum Sm. ..	Rosette herb	Shaded montane rock.
Potentilla anserina L. var. *anserinoides* (Raoul) Cheesem.	Creeping tufted herb ..	Margin of swamp.
Acaena Sanguisorbae Vahl var. *pusilla* Bitter	Mat herb	River-terrace and debris scrub; forest.
—— —— var. *pilosa* T. Kirk ..	,, ...	Tussock grassland; river-terrace and debris scrub; *Cassinia* open shrubland.
—— *hirsutula* Bitter ? (= *A. adscendens* Vahl in part as defined in Man. N.Z. Flora)	,, ...	River-terrace and debris scrub.
—— *microphylla* Hook. f. (one or more vars.)	,, ...	River-bed.
—— *inermis* Hook. f. (several vars.)	,, ...	River-bed; tussock grassland.

(C.) List of Species—*continued.*

Name of Species, &c.	Growth-form.	Association.
Leguminosae.		
Carmichaelia subulata T. Kirk ..	Erect leafless rigid shrub	Tussock grassland; rock; river-bed; river-terrace and debris scrub.
—— *uniflora* T. Kirk.. ..	Mat shrub	Stony dry part of tussock grassland.
—— *Monroi* Hook. f. ..	Open rigid cushion shrub	River-bed; tussock grassland where especially dry.
*Ulex europaeus L.	Spinous dense shrub ..	River-bed.
*Trifolium repens L.	Creeping turf-making herb	Tussock grassland; river-bed.
*—— *arvense* L.	Erect annual herb ..	River-bed.
*—— *dubium* Sibth.	Small prostrate herb ..	Tussock grassland.
Geraniaceae.		
Geranium microphyllum Hook. f.	Prostrate herb ..	Bog; river-terrace and debris scrub.
—— *sessiliflorum* Cav. var. *glabrum* Kunth (possibly there are two vars.)	Rosette herb	Tussock grassland; river-bed; rock.
Oxalidaceae.		
Oxalis corniculata L. var. ..	Mat herb	Tussock grassland.
Coriariaceae.		
Coriaria sarmentosa Forst. f. var. (= *C. ruscifolia* of Man. N.Z. Flora in part)	Creeping summer-green semi-woody plant	Rock; river-terrace and debris scrub.
Rhamnaceae.		
Discaria toumatou Raoul ..	(i.) Semi-divaricating spiny shrub	(i.) *Discaria* thicket; tussock grassland; rock; *Cassina* open shrubland; river-terrace and debris scrub.
	(ii.) Prostrate shrub ..	(ii.) River-bed.
Elaeocarpaceae.		
Aristotelia fruiticosa Hook. f. ..	(i.) Divaricating shrub	(i.) River-terrace and debris scrub, and open debris shrubland.
	(ii.) Bushy shrub ..	(ii.) Forest; river-terrace and debris scrub.
Violaceae.		
Viola Cunninghamii Hook. f. ..	Rosette herb	Bog; tussock grassland; *Cassina* open shrubland.
Hymenanthera dentata R. Br. var. *alpina* T. Kirk. (This may be merely an epharmonic form of *H. crassifolia* Hook. f.)	Open cushion, almost spinous shrub	Rock; river-terrace and debris scrub; scrub; forest, but here of open form; *Cassinia* open shrubland; open debris shrubland.

(C.) List of Species—*continued.*

Name of Species, &c.	Growth-form.	Association.
Thymelaeaceae.		
Pimelea Traversii Hook. f. ..	Small dense shrub ..	Rock ; open d e b r i s shrubland.
—— *prostrata* (Forst. f.) Willd. var. *repens* Cheesem.	Mat shrub	River - bed ; tussock grassland.
—— —— var. erect	Small erect dense shrub	T u s s o c k grassland ; *Cassinia* open shrubland.
Drapetes Dieffenbachii Hook. ..	Turf-making semi-woody plant	Shady tussock grassland.
Myrtaceae.		
Leptospermum scoparium v a r. Forst.	S o m e w h a t fastigiate ericoid shrub	Manuka thicket; *Cassinia* open shrubland ; o p e n debris shrubland.
Onagraceae.		
Epilobium pallidiflorum Sol. ..	Tall erect herb ..	Swamp.
—— *pubens* A. Rich.	Erect herb, woody at base	River-terrace and debris scrub ; rock.
—— *pictum* Petrie	Erect herb	River-terrace and debris scrub.
—— *chloraefolium* Hausskn. ..	,, ...	River-terrace and debris scrub.
—— *rotundifolium* Forst. f. ..	Erect or semi-prostrate herb	Forest.
—— *pedunculare* A.Cunn. (several vars.) (= *E. nummularifolium* R. Cunn. var. *pedunculare* Hook. f.)	Mat herb	River - bed ; tussock grassland.
—— *melanocaulon* Hook. ..	Erect branching herb, woody at base	River-bed ; rock.
—— *microphyllum* A. Rich. ..	Erect branching herb with woody base	River-bed.
—— *elegans* Petrie ? (This may be an undescribed species or a var. of *E. novae-zealandiae* Hausskn.)	Low erect herb ..	Tussock grassland.
Halorrhagaceae.		
Halorrhagis micrantha R. Br. ..	Low prostrate herb ..	Margin of swamp.
Myriophyllum elatinoides Gaud. ?	Partly submerged water-plant	Swamp.
Umbelliferae.		
Hydrocotyle novae-zelandiae DC. var. *montana* T. Kirk	Creeping herb ..	T u s s o c k grassland ; margin of swamp.
Schizeilema Hookeri (Drude) D o m i n (= *Azorella trifoliata* Benth. & Hook.)	Slender creeping herb ..	Forest.
—— (*Az.*) *nitens* (Petrie) Domin	Slender matted herb ..	Margin of swamp.
Oreomyrrhis andicola Endl. var. *Colensoi* (Hook. f.) T. Kirk. (There are probably other vars., or, it may be, all merely dependent on environment)	Low herb	Tussock grassland.
Aciphylla Colensoi Hook. f. ..	Yucca-like spinous rigid herb	Rock.

(C.) List of Species—*continued.*

Name of Species, &c.	Growth-form.	Association.
UMBELLIFERAE—*continued.*		
Aciphylla squarrosa Forst. var.	Yucca-like spinous rigid herb	Tussock grassland; rock.
Anisotome filifolia (Hook. f.) Cockayne and Laing (= *Ligusticum filifolium* Hook. f.)	Grass-like rosette herb ..	River-terrace and debris scrub (probably also grows on stony debris); open forest.
—— *aromatica* Hook. f. var. (= *Ligusticum aromaticum* Hook. f.) (This aggregate species badly needs splitting up into its several very distinct varieties)	Low rosette herb ..	Tussock grassland; rock.
Angelica montana (Forst.) Cockayne (= *A. Gingidium* Hook. f.)	Stout herb with leaves in open rosettes	Rock; side of stream in scrub.
CORNACEAE.		
Corokia Cotoneaster Raoul ..	Divaricating shrub ..	River-terrace and debris scrub; rock.
ERICACEAE.		
Gaultheria depressa Hook. f. (= *G. antipoda* Forst f. var. *depressa* Hook. f.)	Low creeping matted shrub	Shady tussock grassland; *Cassinia* open shrubland.
—— *rupestris* R. Br. var. ..	Small erect or depressed stiff shrub	Rock; river-terrace and debris scrub; open debris shrubland.
EPACRIDACEAE.		
Styphelia (Cyathodes) acerosa Sol. var.	Prostrate ericoid shrub..	Rock.
—— (*Cy.*) *Colensoi* (Hook. f.) Diels	Mat shrub	Rock; tussock grassland; *Cassinia* open shrubland.
—— (*Leucopogon*) *Fraseri* (A. Cunn.) F. Muell.	Low erect shrub with creeping stem	Tussock grassland; rock; river-bed *Cassinia* open shrubland.
Dracophyllum longifolium (Forst.) R. Br. var.	Erect fastigiate shrub ..	Forest-margin; open debris shrubland.
—— *uniflorum* Hook. f. (probably *D. acicularifolium* (Cheesem.) Cockayne is also present)	Erect branching shrub ..	River-terrace and debris scrub; *Cassinia* open shrubland.
—— *rosmarinifolium* (Forst. f.) R. Br.	Prostrate rigid shrub ..	Rock.
Pentachondra pumila (Forst. f.) R. Br.	Small cushion shrub ..	Tussock grassland, in shade.
MYRSINACEAE.		
Suttonia nummularia Hook. f. (= *Myrsine nummularia* Hook f.)	Prostrate trailing shrub	Rock

(C.) LIST OF SPECIES—*continued.*

Name of Species, &c.	Growth-form.	Association.
GENTIANACEAE.		
Gentiana tenuifolia Petrie ..	Erect slender herb ..	River-terrace and debris scrub.
—— sp. (This seems distinct from any species of *Gentiana* with which we are acquainted, but no flowering material is available)	Rosette herb	Tussock grassland.
—— *corymbifera* T. Kirk var. ..	,, ...	Tussock grassland.
BORAGINACEAE.		
Myosotis uniflora Hook. f. ..	Small cushion herb, ..	River-bed.
—— *australis* R. Br. var. (probably quite distinct from the Australian plant)	Erect loose-rosette herb	Tussock grassland; open debris shrubland.
SCROPHULARIACEAE.		
Mazus radicans (Hook. f.) Cheesem.	Creeping mat herb ..	Margin of swamp.
Verbascum Thapsus L. ..	Large - leaved rosette plant	Where scrub has been burned.
Veronica salicifolia Forst. f. var. *communis* Cockayne	Bushy shrub	River-terrace and debris scrub.
—— *Traversii* Hook. f. var. ..	Ball-like shrub ..	River-terrace and debris scrub; *Discaria* thicket; rock.
—— *buxifolia* Benth. var. *odora* T. Kirk	- ,, ..	Scrub of wettish ground.
—— *pimeleoides* Hook. f. var. *minor* Hook. f.	Very low creeping shrub	Tussock grassland.
—— *tetrasticha* Hook. f. ..	Very low shrub, with reduced imbricating leaves	Rock.
—— *epacridea* Hook. f. var.	Prostrate rigid shrub ..	,,
—— *Lyallii* Hook. f. var. ..	Slender mat semi-woody plant	Rock; margin of stream in scrub.
—— *Bidwillii* Hook. f. ..	Ditto	Rock.
PLANTAGINACEAE.		
Plantago spathulata Hook. f. ..	Stout flat rosette herb..	Tussock grassland.
—— *triandra* Berggr. ..	Small flat rosette herb..	Margin of swamp.
RUBIACEAE.		
Coprosma parviflora Hook. f. var.	Divaricating shrub ..	River-terrace and debris scrub; forest; rock.
—— *brunnea* (T. Kirk) Cockayne (= *C. acerosa* A. Cunn. var. *brunnea* T. Kirk)	Rigid prostrate creeping shrub	Rock (should also be on river-bed).
—— *rugosa* Cheesem. ..	Divaricating shrub ..	River-terrace and debris scrub.
—— *propinqua* A. Cunn. var. ..	,, ..	River-terrace and debris scrub; forest.
—— *linariifolia* Hook. f. ..	Bushy shrub	Forest.
—— *cuneata* Hook. f. var. ..	Divaricating shrub ..	Rock.
—— *microcarpa* Hook. f. ? ..	,, ..	Forest.
—— *Petriei* Cheesem. (two vars. with differently coloured drupes)	Mat or cushion shrub ..	River - bed; tussock grassland.

(C.) List of Species—*continued.*

Name of Species, &c.	Growth-form.	Association.
RUBIACEAE—*continued.*		
Galium umbrosum Sol. ..	Straggling slender herb	River-terrace and debris scrub; forest.
Asperula perpusilla Hook. f. ..	Very low slender herb ..	Tussock grassland; margin of swamp.
Nertera Balfouriana Cockayne ..	Small creeping matted herb	Margin of swamp.
CAMPANULACEAE.		
Wahlenbergia albomarginata Hook. (= *W. saxicola* Cheesem. non A. DC.)	Small creeping rosette herb	Tussock grassland; bog; rock.
COMPOSITAE.		
Brachycome Sinclairii Hook. f. ..	Low creeping rosette herb	Tussock grassland.
Olearia avicenniaefolia Hook. f. var.	Tall shrub of tree-composite form	River-terrace and debris scrub; rock.
—— *virgata* Hook. f. var. ..	Divaricating shrub ..	River-terrace and debris scrub; bog.
Celmisia spectabilis Hook. f. ..	Mat or low cushion rosette herb, woody at base	Tussock grassland; rock; open debris shrubland.
—— *Lyallii* Hook. f. ..	Stout erect rosette herb	Open debris shrubland.
—— *longifolia* Cass. var. ..	Rather grass-like erect herb	Tussock grassland; margin of swamp; *Cassinia* open shrubland.
Vittadinia australis A. Rich. ..	Small erect branching herb, woody at base	Tussock grassland.
Gnaphalium Traversii Hook. f.	Low, creeping, woolly mat herb	,,
—— *paludosum* Petrie ..	Slender tufted herb ..	Margin of swamp.
Raoulia australis Hook. f. (two vars.)	Dense silvery semi-woody mat plant	River-bed.
—— *lutescens* (T. Kirk) Beauv. (= *R. australis* Hook. f. var. *lutescens* T. Kirk)	Dense silvery mat or low cushion	,,
—— *tenuicaulis* Hook. f. ..	Dense mat or semi-woody plant	,,
—— *subsericea* Hook. . ..	Dense mat- or turf-making semi-woody plant	Tussock grassland; *Cassinia* open shrubland.
—— *Haastii* Hook. f... ..	Dense green cushion or semi-woody plant	River-bed.
Helichrysum bellidioides (Forst. f.) Willd. var.	Creeping slender mat herb, woody at base	Tussock grassland; bog; rock.
—— *depressum* (Hook. f.) Benth. & Hook. f.	Low, rigid, cupressoid shrub	River-bed.
—— *Selago* (Hook. f.) Benth. & Hook. var.	Low, stout, cupressoid shrub	Rock.
Cassinia fulvida Hook. f. var. *montana* Cockayne var. nov. ined.	Erect bushy ericoid shrub	*Cassinia* open shrubland; tussock grassland.
Cotula squalida Hook. f. ..	Creeping turf-making herb	River-bed; dry tussock grassland.
Erechtites glabrescens T. Kirk ..	Tall erect herb ..	Forest; river-terrace and debris scrub.
*Hypochoeris radicata L. ..	Rosette herb ..	Tussock grassland; river-bed; rock.

(C.) LIST OF SPECIES—*continued.*

Name of Species, &c.	Growth-form.	Association.
COMPOSITAE—*continued.*		
Senecio bellidioides Hook. f. var. *glabratus* T. Kirk. (Many individuals are tomentose on the under surface of the leaf, but we hardly think another variety should be established for this reason)	Low rosette herb ..	Tussock grassland
—— *Lyallii* Hook. f.	Grass-like rosette herb ..	Scrub on bank of streams.
—— *lautus* Forst. f. var. *montanus* Cheesem.	Small rosette herb ..	Open debris shrubland.
—— *Bidwillii* Hook. f. ..	Stout shrub of tree-composite form†	Rock.
Microseris scapigera Sch. Bip. (= *M. Forsteri* Hook. f.)	Low rosette herb ..	Tussock grassland ; *Cassinia* open shrubland.
Taraxacum magellanicum Comm. (= *T. officinale* Cheesem. non Wigg. and *T. glabratum* (Forst. f.) Cockayne)	,, ..	Tussock grassland ; margin of swamp.

† This form is very similar to that of the well-known *Rhododendron ponticum.*

ART. XX.—*Descriptions of New Native Phanerogams.*

By D. PETRIE, M.A., Ph.D.

[*Read before the Auckland Institute, 8th December, 1915.*]

1. Aciphylla trifoliolata sp. nov.

Folia numerosa 20–24 cm. longa lineari-cuneata trifoliolata, foliolis 10–12 cm. longis. Scapi foliis ± duplo longiores ; inflorescentia foeminea oblonga ± 16 cm. longa, bracteis subangustatis ± 4 cm. longis apice trifoliolatis umbellas amplectentibus ac paene celantibus.

Leaves numerous, 20–24 cm. long, linear-cuneate in outline, trifoliolate (rarely unifoliolate or with 2 pairs of pinnae); leaflets 10–12 cm. long, narrow (6–9 mm. in greatest width), rigid and coriaceous, acuminate, pungent-pointed, striate, midrib prominent channelled yellow, edges yellow and finely crenulate, petiole long narrow ; sheaths as long as the petiole or shorter, narrow, membranous, smooth, not striate, expanded towards the base, with 2 short weak linear-subulate lateral spines at the top. Scapes twice as long as the leaves, deeply grooved, ± 6 mm. in diameter ; male inflorescence ± 23 cm. long, narrow lanceolate, bracts numerous with obcuneate rather narrow membranous sheaths ending in a long terminal leaflet (10 cm. long in the lowermost bract), flanked by 2 short weak subulate acicular spines; umbels small, several, distantly seated on a stout peduncle 9 cm. long below but shortening towards the top of the inflorescence, involucral bracts conspicuous, broadly linear-subulate, 3 cm. long or less; female inflorescence $\pm$ 16 cm. long, oblong, bracts broader and shorter than in the males, with broader more rigid and more uniform terminal leaflets; umbels small, as long as the sheaths or less, and almost concealed by them.

Hab.—Rocky spurs on Mount Lyell, western Nelson.

This plant was given me by its discoverer, Mr. William Townson, some years ago. As there seems to be little chance of further specimens being procured at an early date, its description and publication need not be further delayed. It is, I consider, a well-marked species, allied to *A. Lyallii* Hk. f., but by no means closely.

2. Celmisia lanigera sp. nov.

Folia 15–20 cm. longa 4 cm. lata anguste lanceolata ad apicem acutum gradatim attenuata, a marginibus integris $\pm$ recurva, supra lanata vel villosa leviter plicata nervis parallelibus percursa, infra candida.

Scapi complures 25–30 cm. alti moderate graciles lanati, bracteis numerosis elongatis linearibus lanatis.

Capitula $\pm$ 4 cm. in diam., involucri bracteis anguste linearibus dense lanatis.

Achenia linearia costata glaberrima.

Leaves densely tufted, 6–8 in. long, $1\frac{1}{2}$ in. broad a little below the middle, narrow lanceolate, uniformly narrowed to the acute tip, not or scarcely subcordate at the base, little coriaceous, more or less recurved at the entire margins; upper surface woolly or densely villous with creamy-white hairs, longitudinally furrowed but not deeply, veins parallel; below everywhere clothed with appressed white silky tomentum, midrib little conspicuous.

Scapes several, 10–12 in. high, moderately slender, densely cottony or woolly; bracts numerous, woolly, linear, lowermost $3\frac{1}{2}$–4 in. long, diminishing upwards.

Heads $\pm$ $1\frac{1}{2}$ in. across; involucral bracts numerous, thin, narrow linear, woolly on the back and sides; rays numerous, narrow.

Achenes linear, strongly ribbed, glabrous.

Hab.—Garvie Mountains (very common); D. L. Poppelwell! Takitimu Mountains; J. Crosby Smith!

This appears to be a well-marked species, of which I have seen only three or four rather indifferent specimens. In general look it recalls *C. coriacea* Hk. f., but it is readily distinguished from it by the smaller narrower woolly or villous leaves, the slender scapes, and the glabrous achenes. It was, I believe, first collected by Mr. Crosby Smith.

3. Gentiana Astoni sp. nov.

Perennis; caules plures v. complures graciles $\pm$ ramosi glabri 8–30 cm. longi.

Folia in paribus oppositis disposita, linearia vel sursum vix dilatata, acuta vel subacuta flaccida integerrima glabra $1\frac{1}{2}$–2 cm. longa.

Flores pauci vel complures, caulini in foliorum supremorum axillis solitarii, terminales plerumque bini, albi, $1\frac{1}{2}$–2 cm. longi, pedunculati, pedunculis foliis $\frac{1}{3}$ brevioribus valde gracilibus glaberrimis. Calyx obconicus peralte in lobos 5 lineari-subulatos sectus; corolla calycem excedens in lobos 5 oblongo-lanceolatos subacutos apiculatos subalte secta; stamina ac pistillum corolla $\frac{1}{2}$ breviora.

A slender erect or spreading perennial. Root usually slender, but occasionally (in much-branched specimens) as much as 6 mm. in diameter.

Stems several or numerous, slender, sparingly or much branched, glabrous, 8–20 cm. and occasionally even 30 cm. long.

Leaves in opposite pairs, almost connate at the base, linear or somewhat dilated towards the tips, acute or subacute, rather flaccid, entire, glabrous, more or less recurved at the margins, nerveless (midrib obscure), 1½–2 cm. long, 1–2¼ mm. in greatest width.

Flowers few or several; the cauline solitary in the axils of the uppermost leaves, the terminal usually in pairs, white, 1½–2 cm. long, pedunculate; the peduncles ¾ the length of the leaves, glabrous, slender; calyx obconic, cut for ⅔ its length into 5 rather long linear-subulate thin glabrous lobes; corolla white, longer than the calyx, deeply divided into 5 oblong lanceolate subacute apiculate prominently nerved segments; stamens and pistil half as long as the corolla.

Hab.—Limestone ridges, bare or covered with manuka scrub, near the sources of the Ure River, Marlborough; B. C. Aston!

On open ridges, Mr. Aston writes, the stems are erect and stunted; in the scrub they are elongated, and more or less entangled. One specimen rooted in a rock-crevice had a remarkably stout root, giving off a tuft of matted branches that hung down the slope for some 15 in.

4. Myosotis (Exarrhena) eximia sp. nov.

Planta *M. amabili* (Cheesm.) subsimilis; differt caulibus tenuioribus erectis, foliis anguste elliptico-spathulatis acutis vel subacutis pilis candidis brevibus subrigidis sparsis arcte appressis vestitis, floribus majoribus paucioribus late infundibuliformibus, corollae tubo latiore ac calycis lobis breviore.

Perennial, tufted or spreading by slender prostrate more or less rooting branches into patches 2–3 ft. in diameter, everywhere clothed with rather stiff sparse closely appressed white hairs; flowering-stems 1–6 from each branch (commonly 1–3), erect or decumbent at the base, usually simple but sometimes divided at the topmost cauline leaf, slender, leafy for half their length, 5–9 in. high.

Radical leaves numerous, narrow, elliptic-spathulate, acute or subacute, more or less apiculate, ± 2 in. long, nearly ½ in. in greatest width, the blade about as long as the narrow slightly flattened and basally dilated petiole, the lower surface less closely pilose, midrib little conspicuous. Cauline leaves more or less distant or almost overlapping, shorter and narrower, acute, the upper sessile, the lower with progressively longer petioles.

Racemes long-peduncled, not branched, of 10 flowers or fewer; flowers large and showy, white with a yellow eye, ⅝ in. long and equally broad, shortly pedicelled; calyx narrow turbinate, cut for ⅔ its length into narrow linear-subulate lobes; corolla broadly funnel-shaped, cut ⅓ the way down into 5 rounded subacute lobes, the tube wide and shorter than the calyx; stamens inserted just above the narrow rather distant scales, exserted, slightly shorter than the corolla-lobes, the free filaments twice as long as the anthers, their lower half adnate to the corolla-tube. Mature nutlets not seen; in immature state pale and winged.

Hab.—Limestone bluffs and talus slopes of Mount Aorangi (Mangaohane Station), Ruahine Range, 3,900 ft.

This charming species was collected by Mr. Aston at the close of last December. He remarks that in favourable stations it forms continuous patches several feet across, adorned by masses of lovely flowers. It is not closely related to *M. amabilis*, but that is the only northern species with which it might be confounded.

5. Veronica evenosa sp. nov.

Frutex 15–24 dcm. altus glaberrimus, ramis primariis paucis longis late diffusis in ramulos complures graciles ad apicem foliatos subdivisis.

Folia anguste elliptica 1½–3 cm. longa 1 cm. lata, subpatentia imbricata integerrima glabra, acuta vel subacuta, tenuia evenosa sessilia nec petiolata nec punctulata, ramulos basi amplectentia, haud connata, subtus evidenter carinata.

Racemi 2 (raro 3) in axillis foliorum supremorum dispositi foliis duplo longiores, rhachis gracilis sparse pubescens a parte inferiore nuda.

Flores parvi (5 mm. longi) albi; calyx 4-partitus, lobis oblongis tenuibus ad margines scariosis ac ciliatis; corollae tubus calyce duplo longior, limbo in lobos 4 rotundatos secto; antheris ac stylo exsertis; ovario glabro.

Capsula haud visa.

A tall (5–8 ft.) glabrous shrub with stout widely-spreading main branches, towards their tops freely subdivided into slender twigs leafy only at the tips.

Twigs blackish-brown, slender, terete, ringed with the scars of fallen leaves.

Leaves subpatent, imbricating, narrow elliptic, ⅝–1⅓ in. long, ⅜ in. broad at the middle, sessile not connate but completely clasping the younger twigs, glabrous, acute or subacute, thin, veinless, entire, slightly recurved at the edges when dried, dull green above, below yellowish-brown (when dried), midrib grooved above, obviously keeled below, keel reddish-brown slightly excurrent at the tip.

Racemes 2 (rarely 3) in the axils of the uppermost leaves, twice as long as the leaves, dense-flowered; rachis slender, dark brown, sparsely pubescent, the lower third naked; bracts narrow ovate or almost linear, about as long as the short pedicels.

Flowers small, white; calyx 4-partite, the lobes oblong, thin, ciliate along the scarious margins; tube of the corolla rather wide, twice as long as the calyx, the limb cut into 4 broadly rounded lobes a little shorter than the tube; anthers and style exserted; ovary glabrous.

Ripe capsules not seen.

Hab.—Upper edge of forest belt on Mount Holdsworth, Tararua Range, about 2,800 ft.

The present species appears to have been confounded with *V. laevis* (Benth.). In the latter the leaves are shorter, distinctly petiolate, coriaceous and punctulate below; there is always a conspicuous gap between them where they spring from the twigs. It is also a much lower and more compactly growing shrub. I have not seen Buchanan's specimens of *V. laevis* from the Tararuas, but it may be suspected that they belong here.

6. Veronica imbricata sp. nov.

Frutex habitu compactus a vertice rotundatus 40–60 cm. altus ad 40 cm. in diam.

Rami juniores teretes vel obscure tetragoni rigidi erecti sursum ± subdivisi, foliis brevibus imbricantibus dense vestiti.

Folia decussata appressa vel subpatentia, 1½ mm. longa 2½ mm. lata coriacea crassa, a latere inferiore (exteriore) convexa a latere superiore (interiore) concava, basi ad ⅓ connata perobscure carinata, supra subacute rotundata, a marginibus delicate ciliata.

Flores haud visa.

Capsulae in capitulis brevibus densis terminalibus dispositae ca. 1 cm. longis, bracteae.foliis similes sed tenuiores; calyx capsula ± ⅓ brevior in lobos 4 ciliatos coriaceos sulcatos oblongos obtusos apice rotundatos ± inaequales divisus, duobus exterioribus majoribus; capsulae sessiles glabrae obtusae late ellipticae 3 mm. longae.

A round-headed compact erect shrub, 18–24 in. high, and as much as 18 in. across near the top.

Stems moderately stout, freely branched above, older branches terete closely and uniformly ringed by the scars of fallen leaves, younger terete or obscurely tetragonous, stiff, erect, very closely clothed by the short appressed or subpatent imbricating leaves.

Leaves in opposite decussately arranged pairs, 1½ mm. long, 2½ mm. wide, rather thick, coriaceous, convex and shallowly ribbed on the outer (lower) side, concave on the inner (upper) side, connate for about ⅓ their length, subacutely rounded at the tips, very obscurely keeled, delicately ciliate at the edges (except in age), marked by a shallow groove along the base on the outer (lower) side.

Flowers not seen.

Capsules glabrous obtuse, broadly elliptic, 3 mm. long, sessile, forming small compact pale-brown terminal heads about 1 cm. long; bracts similar to the leaves, but thinner; calyx about ⅓ shorter than the capsule, deeply divided into 4 ciliate coriaceous grooved oblong obtusely rounded unequal lobes, the two outer lobes larger.

Hab.—Mount Cleughearn and Mount Burns, Fiord County, at 2,500–4,000 ft.; J. Crosby Smith! Eyre Mountains, Lake County; D. L. Poppelwell!

This plant appears to have been in cultivation for a number of years, but its wild habitat was till recently unknown. Its nearest relative is probably *V. Hectori* Hk. f.

7. Euphrasia Crosby-Smithii sp. nov. ,

Annua ? Caules erecti parce ramosi ± 12 mm. alti graciles teretes sparse glanduloso-pubescentes.

Folia pauca ad 3 mm. longa lanceolato-elliptica in lobos 3–5 subamplos margine revolutos secta.

Calyx 3–4 mm. longus anguste campanulatus delicate glanduloso-pubescens, in lobos 4 breves triangulares apicibus recurvatos scctus; corolla brevis alba basi expansa (tubo calycem vix excedente), labio superiore brevi arcuato late emarginato, inferiore in lobos 3 longiores truncatos secto.

Capsula subacuta apice ciliata calyce brevior; matura haud visa.

Annual ? Stems erect, sparingly branched, ± 12 mm. long, very slender, terete, sparsely glandular-pubescent, usually giving off 2 short branches from the lowermost axils and flowers from the others.

Leaves few, in rather distant opposite pairs, 3 mm. long or less, lanceolate-elliptic in general outline, almost glabrous obtuse or subacute, cut into 3–5 rather large lobes that are recurved at the edges.

Flowers few, on slender sparsely glandular-pubescent peduncles, in fruit about twice as long as the leaves ; calyx 3–4 mm. long, narrow campanulate, more or less glandular-pubescent, cut at the top into 4 short triangular obtuse or subacute lobes with their edges recurved at the tips ; corolla rather short, white, tube expanded at the base scarcely longer than the calyx, upper lips short arched broadly emarginate, lower cut into 3 rather long trunacte lobes.

Capsule subacute ciliate at the top ; fully ripe not seen.

Hab.—Wet alpine meadow on Mount Cleughearn, Fiord County, about 5,000 ft. ; J. Crosby Smith !

Some five or six specimens of this dwarf plant were collected three or four years ago by Mr. Crosby Smith, who kindly communicated them to me. As he expected to revisit the locality where it was found, I put off publishing the species in hopes that a fuller series of specimens might be secured. At the time of this second visit the spot where it was first seen was covered with snow, and no additional material could be collected. In these circumstances it is advisable to publish a description without waiting for additional specimens.

8. Carex filamentosa sp. nov.

Folia 10–20 cm. longa filiformia glabra, supra concava infra rotundata, stolonibus vaginis aphyllis brevibus acutis striatis fulvis vestitis.

Culmi filiformes 4–10 cm. longi teretes, spiculis 3 (raro 4) dense confertis (ima raro subdistante) ovatis paucifloris ± 7 mm. longis ; suprema mascula gracillima breviter pedicellata ; duabus inferioribus foemineis plerumque 2–3 flores masculos apice praebentibus, brevissime pedicellatis ; bracteis filiformibus foliosis valde elongatis, basi in vaginam longam laxam apice truncatam expansis.

Glumae ovatae pallidae subhyalinae haud emarginatae valide carinatae, sensim in mucronem brevem ± scabridum attenuatae.

Utriculi glumas aequantes turgidi elliptici altero latere ± concavi obscure nervati vix stipitati, in rostrum breve sublatum obscure bidentatum contracti. Styli rami 3.

Densely tufted, sending off numerous very slender stolons.

Leaves 10–20 cm. long, excessively narrow, glabrous, concave above, rounded below (midrib little prominent), finely serrate towards the tips, green or with a reddish tinge, young shoots clothed with short acute striate brownish sheathing scales.

Culms filiform, 4–10 cm. long, terete, smooth, more or less striate.

Spikelets 3 (rarely 4), closely crowded or the lowermost somewhat distant, pale ovate ± 7 mm. long ; uppermost male very slender shortly pedicellate, the two lower female usually with 2–3 male flowers at the top very shortly pedicellate ; bracts filiform, leaf-like, long, expanded below into a loose grooved sheath, truncated at the top and with broad scarious borders.

Glumes ovate, pale, membranous, subhyaline, not emarginate, strongly keeled, gradually narrowed into a short more or less scabrid mucro.

Utricles about as long as the glumes, turgid, elliptic, scarcely stipitate, somewhat concave on one side, with 2 prominent lateral nerves and obscure intermediate ones, narrowed into a short rather broad smooth obscurely bidentate beak.

Style branches 3.

Hab.—Table Hill, Stewart Island; W. J. Murdoch! High lands at Port Pegasus; T. Kirk! High lands at head of Paterson Inlet; H. Guthrie-Smith !

Mr. Guthrie-Smith sent me a live plant of this species some years ago. It has grown well, and has supplied some of the material used in preparing this description. Mr. Kirk's specimens were sent me under the MS. name of *C. australis.* This name he afterwards used to designate a small state of *C. litorosa* Bailey, from the coast of Stewart Island. The late Mr. C. B. Clarke referred Mr. Kirk's Port Pegasus plant to *C. uncifolia* Cheesm., but I am unable to acquiesce in this reference.

9. Koeleria Cheesemanii (Hackel) Petrie comb. nov.

A full description of this grass, under the name of *Trisetum Cheesemanii* Hackel, appears in Trans. N.Z. Inst., vol. 35 (1902), p. 381. In his " Manual of the New Zealand Flora " (p. 882) Mr. Cheeseman does not allude to the obvious likeness of this *Trisetum* to *Koeleria*, which Hackel has noted (*loc. cit.*). In an elaborate note to his specific character the latter pointed out that " his species is a very distinct once, like indeed in habit and in the character (indole) of the panicle and spikelets to *Trisetum subspicatum* Beauv., yet by its very shortly bidentate flowering-glumes, that show a very short mucro or awn from between the teeth or from a little below them, it is so divergent not only from *Trisetum subspicatum*, but even from all the genuine Trisetums, that it may preferably be annexed to *Koeleria*. In truth," he adds, " between this genus and *Trisetum* no certain limits are found, and it was for this reason that the celebrated Desvaux was led to reduce *Koeleria* to a section of *Trisetum*." The passage above cited is given in Latin, which I have translated.

It is hard to understand how Hackel could have penned this note, seeing that in his well-known work on " The True Grasses " the genera *Trisetum* and *Koeleria* are maintained, and are not even placed in immediate juxtaposition.

My opinion that *T. Cheesemanii* is a true *Koeleria* is not founded on the statement of Hackel quoted above, but has arisen from the study of the specimens of this grass collected in a number of widely scattered localities in the South Island. In *Trisetum* the awns on the flowering (fertile) glumes are never terminal, nor are they wanting. In *T. Cheesemanii* things are very different. On the same panicle one finds spikelets with short subdorsal awnlets (mucros) on the flowering-glumes which are shortly bidentate at the tip, or the awnlets may be terminal between the short teeth, or the glumes may have acuminate or acute tips with no trace of awnlets. Indeed, two of these conditions may often be found on a single spikelet. In view of these facts, there can, it seems to me, be no question that Hackel's species must be transferred to *Koeleria*. The only alternative is the reduction of *Koeleria* to a section of *Trisetum*, a course which no modern authority on the classification of grasses has followed.

ART. XXI.—*Notes on New Zealand Floristic Botany, including Descriptions of New Species, &c. (No. 1).*

By L. COCKAYNE, Ph.D., F.L.S., F.R.S., Hutton and Hector Memorial Medallist.

[*Read before the Wellington Philosophical Society, 27th October, 1915.*]

1. Acaena novae-zelandiae T. Kirk var. pallida T. Kirk

This variety is dismissed by Cheeseman (Manual, p. 131) with the brief words, " Mr. Kirk distinguishes a var. *pallida*, with paler foliage and the spines often greenish." Kirk himself (" Students' Flora," p. 134) gives more details, stating that it " differs widely from the type in appearance," but he does not emphasize sufficiently the distinctions of this exceedingly striking plant, which some perhaps may prefer to recognize as a valid species.

The plant in question grows abundantly on the dunes at Lyall Bay, Wellington, where it forms broad mats on the sand, a square metre or more in area, or straggles over the bushes of *Coprosma acerosa*. It is almost a true sand-binding plant in behaviour, its stems greatly lengthening when buried. But, even without any sand-advance to speak of, they frequently attain a length of more than 1 m. The older portions of the stem are strongly woody, and extremely stout, often reaching a diameter of more than 10 mm. Even towards the extremity, where, perfectly straight, the stem, unrooted, creeps over the sand with a length of about 50 cm., its diameter is 2–3 mm. The older part of the stem is covered with a fairly thick, cracked, dark-brown bark, tinged with red, while the youngest part is green and strongly hirsute. The leaves are much larger than those of any other variety of *A. novae-zelandiae*, and may measure as much as 11 cm. in length, but about 7·5 cm. is quite a common size. The leaflets are considerably larger than those of *A. novae-zelandiae* in general (2·1 cm. by 10 mm. for the uppermost pair), and differ likewise in their much paler, distinctly glossy yellowish-green, wrinkled upper surface, far paler less glaucous under surface, and thicker substance. The peduncles are extremely stout, stiff, and upright, and frequently measure 15 cm. in length, or twice as much as is common in any other variety of the species. The heads, with the spines, are commonly more than 4 cm. in diameter. The spines are bright pinkish-purple in colour, and never dark purple; but it is extremely rare to see them of a green colour, as described by Kirk and Cheeseman, though they are often quite pale near the base. The plant does not owe its distinctive characters to its special environment, for when growing upon a clay bank it can be recognized at once from neighbouring plants of more widely spread forms of *A. novae-zelandiae*.

2. Acaena Sanguisorbae Vahl. var. viridior Cockayne var. nov.

Folia supra viridia, nunquam nonnihil olivacei-fuscentia; calycis-lobi aculeique pallide virides.

North and South Islands: Probably fairly common. I have already had specimens from Wellington, the neighbourhood of the Marlborough Sounds, Taranaki and Banks Peninsula.

7— Trans.

This variety is easily distinguished from the extremely common var. *pusilla* Bitter by its larger bright-green leaves, which have never brown basal leaflets, and the pale but clear green calyx-segments and spines, whereas the small basal leaflets of the var. *pusilla* are more or less deeply stained with brown, and the spines are pale and stained here and there with light-red. It might be thought that the differences in the colour of the leaves was due to the effect of light, but the two varieties grow side by side with their characters unchanged.

Bitter makes his var. *pusilla* a subspecies of *A. Sanguisorbae* equivalent in rank to his subspecies *novae-zelandiae* (= *A. novae-zelandiae* T. Kirk), but it seems to me simpler to treat the last as an aggregate species and to reduce all Bitter's other New Zealand endemic subspecies of *A. Sanguisorbae* to the rank of varieties, though this step must eventually lead to the establishment of subvarieties.

3. Celmisia Monroi Hook. f.

Celmisia Monroi was first described by Sir Joseph Hooker in the Handbook, p. 133, from specimens collected by Monro on the Upton Downs, Awatere; by Haast near Mount Cook (Hopkins River) and elsewhere in the Alps of Canterbury; and by W. T. L. Travers in some part of the latter. There was no "type" for the species, since it was not only described from a number of individuals, but also the Awatere specimens possessed glabrous achenes, while in those collected by Haast they were hispidulous. The species under consideration has been imperfectly understood by most authors, &c., and, according to Cheeseman, confounded with small forms of *C. coriacea* Hook. f., a not surprising mistake. T. Kirk, in the "Students' Flora," p. 288, enlarged Hooker's conception of the species by including plants from the neighbourhood of Whangarei and the Bay of Islands (Auckland). Cheeseman, in the Manual, p. 313, reduced the compass of the species to that proposed by Hooker, and placed the Whangarei plant under *C. Adamsii* T. Kirk as var. *rugulosa* Cheesem. Recently (Trans. N.Z. Inst., vol. 44 (1912), p. 182) Petrie has removed the Mount Cook plant from the species, describing it as *C. Boweana*, thus limiting *C. Monroi* to the plant of the Awatere.

C. Monroi in its restricted significance, and as examined in the light of fairly abundant material from different stations, does not altogether match the description in the Manual. The leaves are not infrequently much longer (21 in.) than therein described, and broader (1½ in.), though about ¾ in. is perhaps the average breadth. The leaf-sheaths are not always short, but may be of great length—*e.g.*, a lamina 10½ in. long may possess a sheath 9 in. long. The texture of the leaf is frequently thinner than has hitherto been described, and more or less flaccid leaves are not unknown.

Possibly there are two distinct races represented amongst my material. The first possesses leaves varying from linear-lanceolate to broadly lanceolate, and the second, of which I have only one specimen, is a quite different-looking plant with rather short (9 in. long) linear or almost linear rigid leaves, ⅛–¾ in. broad, furnished with a short not densely woolly sheath.

The "race" of *C. Monroi* with the more or less lanceolate leaves is evidently closely related to *C. coriacea*, and some taxonomists may prefer to reduce it to a variety of that aggregate species: but it can be distinguished by its narrower, thinner, and sometimes far shorter leaves, with the lamina never rounded at the base but tapering gradually into the sheath, which may be of considerable length ; the pure-white (not faintly tinged with yellow), rather denser but thinner tomentum of the under-surface of the leaf ; and the distinctly smaller flower-head and shorter rays. The suggested race having the linear rigid leaves could not be confused with *C. coriacea.*

So far as has been observed, *C. Monroi* in its restricted significance is confined to rock, either limestone or greywacke ; nor does it appear to extend beyond the North-eastern Botanical District.* The vertical distri-bution is from sea-level to 600 m. or probably higher.

4. Epilobium rubro-marginatum Cockayne sp. nov.

Herba perennis, pumila, caulibus gracilibus, procumbentibus, plerumque simplicibus, densissime foliosis, ad internodia inferiora radicantibus, basin versus lignosis, pallide viridibus vel apicem versus rubris, obscure bifariam albo-puberulis. Folia parva, brevipetiolata, oblonga, ovato-oblonga vel raro anguste oblonga, cum petiolo circ. 9 mm. longa, 3 mm. lata, sub-disticha, imbricata, glaberrima nisi petioli margine sparsim albo-puberula, subcarnosa, rigida, margine manifeste rubro tincta, remote obscureque dentata, basi vaginata, subtus nervo medio paullo prominente. Capsula erecta, stricta, glaberrima circ. 2·6 mm. longa vel breviora, pedunculis brevissimis.

South Island : Westland and Canterbury — On consolidated stony debris at from 1,000 m. to 1,500 m. and upwards on mountains in the neighbourhood of Arthur's Pass and the Otira Gorge. L. C.

The species is to be recognized by its almost unbranched, far-extending, densely leafy, obscurely bifariously pubescent stems, which are strongly woody near the base, and in the oldest woody throughout, and which root only at the nodes of the lower third ; the small petiolate, stiff, rather fleshy, obscurely dentate, glabrous except on the margin of the petiole, oblong obtuse leaves which are for the most part distichous in arrangement through being turned towards the light, and their distinct red or purplish-red margin ; the small white flowers with acute lanceolate calyx-segments ; and the very short glabrous capsule with extremely short peduncle, which lengthens very little as the fruit ripens.

This species belongs to the series of plants included by Cheeseman in his conception of *E. confertifolium* Hook. f. (Manual, pp. 175–76). But this species was founded by Hooker on Lord Auckland and Campbell Island specimens, which, unlike any of their mainland congeners, have invariably bright-pink flowers. An examination of the subantarctic plant in 1903 convinced me that it should be kept distinct from its allies of New Zealand proper. Cheeseman adopted this view in 1909 (Subant. Islands of N.Z., vol. 2, p. 406), and he thus distinguishes *E. confertifolium* in its restricted sense : " Its distinguishing characters, in the limited sense in which I now understand it, are the creeping and rooting often densely matted stems, the young branches alone rising from the ground ; the densely crowded pale-green and almost fleshy leaves, which are almost sessile, obovate or obovate-oblong, entire or remotely and obscurely denticulate, the lower

---

* This includes the north-eastern portion of the South Island, excepting the wet area in the vicinity of the Marlborough Sounds. It is bounded on the west by a line denoting the average limit reached by the western rainfall, and on the south by the River Waiau, a quite artificial boundary. The area thus defined includes the greater part of Marlborough and the drier portion of Nelson.

ones more or less distichously placed, the upper spirally arranged; the few almost sessile flowers, placed at the tips of the branchlets, and apparently always of a bright-pink colour; and the strict and erect perfectly glabrous capsules, the peduncles of which only slightly elongate after the flowering period."

If the name *E. confertifolium* be confined to the subantarctic plant, then obviously a new name is demanded for those mainland plants referred to that species. But there is already the var. *tasmanicum* of *E. confertifolium* to meet the case. This variety—or, according to Haussknecht, distinct species—was founded by the last-named taxonomist on a plant collected by Gunn in Tasmania, and another plant collected by T. Kirk near Lake Harris, Otago. An excellent figure of his species is supplied by Haussknecht ("Monographie-der Gattung Epilobium" (1884), taf. 20, fig. 84). By the aid of this figure, Haussknecht's description of his species (*l.c.*, p. 296), and an examination of a type specimen in Kirk's herbarium the differences between *E. tasmanicum* and *E. rubro-marginatum* can be ascertained. Thus *E. tasmanicum* is an altogether smaller, more compact plant, with much shorter, more slender, and far less woody stems, which are altogether glabrous, and its leaves are ovate or oblong-ovate and more conspicuously toothed, and the midrib hardly keeled.

Besides the three closely allied plants dealt with above, there are various closely related forms, so that it is difficult to draw strict lines of demarcation. This doubtless arises in part from the fact that, as in the case of many high-mountain plants, both material in sufficient abundance and detailed field observations are lacking, while a full series of specimens is not to be seen in cultivation. Possibly the best course to take would be to once more unite all the species of the *confertifolium* group into an aggregate. If that were done, then there would be vars. *tasmanicum* and *rubro-marginatum*, while a varietal name—*e.g.*, "*vera*," "*subantarcticum*" or "*puniceum*"—would have to be found for the "type." Also, most likely, other varieties would have to be constituted. But, so far as my own knowledge goes, I am not in a position to make further advance.

5. Helichrysum dimorphum Cockayne.

In the description of this species (Trans. N.Z. Inst., vol. 47, (1915), p. 117), the height of the liane is given as 6–8 m. This is a slip of the pen, 6–8 being taken from my note-book and referring to feet. The true height, then, should be 1·8–2·4 m. Of course, any reader who notes that the plant is described as climbing through river-terrace scrub could hardly fail to correct the error for himself, since scrub 8 m. high would be scrub no longer, but forest.

6. Helichrysum Fowerakeri Cockayne sp. nov.

Suffrutex parvus, ramis baso prostratis radicantibus, deinde erectis, junioribus partibus dense albo-lanatis, gracillimis, pauciramosis, ± 8 cm. longis. Folia obovato-spathulata, utrinque albo-tomentosa pilis sericeis, margine integerrima, apice obtusa nonnumquam subacuta, mucro parvo ornata. Capitula terminalia circ. 10 mm. diametro solitaria vel pedunculis corymboso-ramosis; pedunculi circ. 6·2 cm. longi, numerosis bracteis linearis acutis tomentosis praediti; involucri-bractae circ. 5-seriatae lineari-spathulatae 3–6 mm. longae unguibus scariosis parce sericeis et laminis albis patentibus circ. 3 mm. longis. Receptaculum subconicum. Flores numerosi; achenium glabrum.

South Island : Marlborough—In soil on rock, Inland Kaikoura Mountains, at about 1,000 m. altitude ; C. E. Foweraker and L. C.

This plant is related, on the one hand, to *H. bellidioides* Hook. f., from which it differs in its more erect habit, softer leaves tomentose on both surfaces, much smaller flower-heads, and at times branched flowering-stems ; and, on the other hand, to *H. Sinclairii* Hook. f., which it resembles greatly in the leaves and somewhat in habit, but has a very different inflorescence and flower-heads.

Only one plant was observed, so it is possibly extremely rare. Its resemblance to the above two species hints at a hybrid origin.

7. **Lagenophora pinnatifida** Hook. f.

There appear to be two well-marked varieties of this species, which can be distinguished at a glance by differences in the texture and degree of hairiness of the leaf, the extent of indentation of its margin, and the relative size of the flower-heads. The following are diagnoses of these two undescribed varieties, which together make up the aggregate species *Lagenophora pinnatifida* : —

(*a.*) **Lagenophora pinnatifida** Hook. f. var. **hirsutissima** Cockayne var. nov.

Folia pilis albis hirsutis densissime utrinque obtecta, grosse crenato-dentata ; capitula circ. 12 mm. diametro.

North Island : Hawke's Bay—Upper part of Wairoa River ; T. Kirk ! South Island : Fairly common in eastern *Nothofagus* forest, especially of montane belt from Nelson to Otago ; Cheeseman, L. C., and others.

The leaves are most thickly covered with soft white hairs, so that the surface is concealed ; the margin is deeply and coarsely crenate-dentate ; and the heads are about 12 mm. diameter, and borne on straight scapes about 10 cm. long.

The northern plant described below looked so different from that of the south when, in company with Mr. H. Carse, I first saw it in its habitat that I considered it a distinct species. For this reason, looking upon it, in error probably, as the type of *L. pinnatifida* Hook. f., in my "Vegetation of New Zealand" I have given the MS. name *L. sylvestris* to the plant of the South Island. However, it seems best, in view of the close affinity of the two races, to treat them as varieties of an aggregate rather than as distinct species.

(*b.*) **Lagenophora pinnatifida** Hook. f. var. **tenuifolia** Cockayne var. nov.

Planta a varietate hirsuta differt foliis tenuoribus, aliquando subintegerrimis pilis brevioribus haud dense obtectis, capitulis minoribus.

North Island : Auckland—Tauroa, Mongonui County ; H. Carse ! L. C.

This variety is separated from var. *hirsutissima* by its thinner, sometimes almost entire leaves, which are never very deeply cut, their surfaces much less densely covered with shorter hairs, and the smaller flower-heads. In some specimens the leaves are considerably broader than any I have seen of var. *hirsutissima*.

8. **Ourisia calycina** Colenso.

In 1889 Colenso described a species of *Ourisia* collected on " highlands on River Waimakariri, near Bealey, South Island " (which probably means Arthur's Pass), " by a visitor and sent to Napier " (Trans. N.Z. Inst., vol. 21, pp. 97–98). The description clearly shows that the plant in question was identical with one which is extremely common on Arthur's Pass, and in Westland generally from Mount Alexander to, at any rate, the neighbourhood of the Franz Josef Glacier. This plant was included by Cheeseman (Manual, p. 549) in his conception of *O. macrocarpa* Hook. f., while in the herbarium of T. Kirk it is also so designated. But the plant on which *O. macrocarpa* was founded by Hooker was collected by Lyall at Chalky Bay, and, according to the original description in the "Flora Novae-Zelandiae," p. 198, the leaves are cordate at the base. In his description of the same specimens in the "Handbook of the New Zealand Flora," p. 218, Hooker, strange to say, makes no reference to these cordate leaves, so that without hesitation all New Zealand botanists who visited Arthur's Pass referred the large *Ourisia* of that locality to *O. macrocarpa*, and thus it came to be looked upon as typical. In 1912 I visited the Clinton Saddle, and, to my surprise, found that the large *Ourisia* of that locality differed in certain particulars from the Arthur's Pass plant, and answered perfectly to Hooker's description of *O. macrocarpa* in the "Flora Novae-Zelandiae."

The chief differences between the two plants are as follows :—

(1.) Leaf-lamina : *O. macrocarpa*—Frequently rotund with cordate base ; in small specimens broadly ovate with cordate, subcordate, or occasionally rounded but never cuneate base. *O. calycina* (*i.e.*, Arthur's Pass, &c., plant)—Ovate or ovate-oblong, with more or less cuneate base, and never orbicular with cordate base.

(2.) Petiole : *O. macrocarpa*—Rarely more than 4 mm. broad, and frequently less, more than twice as long as the lamina, almost glabrous. *O. calycina*—11 mm. broad or more, hardly as long as the lamina, margined with close white hairs which extend to base of lamina on margin.

(3.) Crenations : *O. macrocarpa*—Large, 5 mm. at base, rounded, continued to leaf-sinus at base of leaf. *O. calycina*—Much smaller, 2 mm. at base, obtuse or almost subacute, wanting on marginal portion of lamina.

(4.) Bracts : *O. macrocarpa*—Small, the largest about 3 cm. long by .1·5 cm. broad, the uppermost in many-leaved whorls. *O. calycina*—Large, frequently 5 cm. long by 2·5 cm. broad, the uppermost in whorls of 2 to 4 leaves.

(5.) Peduncle : *O. macrocarpa*—Comparatively slender, only about 3 mm. diameter. *O. calycina*—Comparatively stout, 5 mm. diameter or more.

(6.) Calyx-lobes : *O. macrocarpa*—6 mm. long, entire, obtuse, thick, veins not evident. *O. calycina*—1·9 cm. long, crenate-dentate above, subacute, thin, 3-veined and reticulating above.

The above clear distinctions show that we are concerned with two distinct plants, which, as they extend virtually unchanged over wide areas, are true-breeding races. Some may prefer to consider them distinct species, while others will group them together, on account of their close relationship, as an aggregate species under the name of *O. macrocarpa*. Possibly, from the phytogeographical standpoint this course should be adopted. Therefore I propose the name *O. macrocarpa* Hook. f. var. *calycina* (Colenso) comb. nov. for the Westland plant which equals *O. calycina* Col., and for the plant of south-west Otago *O. macrocarpa* var. *cordata* var. nov. The latter equals the species *O. macrocarpa* of Hooker in the "Flora Novae-Zelandiae," vol. 1, p. 198, so no further description is needed here. Perhaps a more searching comparison of the flowers of the two varieties may reveal other constant differences, but those cited above are ample to show the distinctness of the two plants.

In the MS. of my " Vegetation of New Zealand " the Arthur's Pass plant is treated as a distinct species, under the name *O. insignis*, I having failed to notice that Colenso had already given it a name.

9. Podocarpus nivalis Hook. var. erectus Cockayne var. nov.

Frutex erectus, ± 1·8 m. altus, 1·2–1·5 m. diametro, haud prostratus, foliis usque ad 2·5 cm. longis.

South Island : Canterbury—Debris-field on Mount Sugarloaf, Cass, in full sunshine, at altitude of 900 m. and upwards. C. E. Foweraker and L. C.

A number of specimens of this shrub were observed growing side by side under exactly the same climatic and edaphic conditions as the ordinary variety of *P. nivalis*, which was everywhere prostrate. For further particulars see paper by Cockayne and Foweraker in this volume (p. 166).

10. Samolus repens (Forst. f.) Pers. var. strictus Cockayne var. nov.

Caules erecti, ± 20 cm. alti, stricti ; folia ± imbricata, linearia vel lineari-lanceolata, acuta.

Kermadec Islands; W. R. B. Oliver. Poor Knights Islands; L. C. Norfolk Island ; R. M. Laing.

This well-marked variety differs from *S. repens* var. *procumbens* R. Knuth, so common on the New Zealand coast generally, in its tall, erect, crowded, straight, unbranched stems, and more or less imbricating much narrower and generally longer leaves. The varietal name was first applied to this plant by me in 1906 (Trans. N.Z. Inst., vol. 38, p. 356), but without any description.

11. Veronica Biggarii Cockayne sp. nov.

Frutex parvus, decumbens, floribundus, ramulis gracilibus, dense foliosis, pauciramosis, strictis, 20 – 24 cm. longis. Folia subdisticha, obovata, obovato-oblonga vel oblonga, subsessilia, glaberrima, coriacea, glauca, saepe purpureo tincta, 1·2–1·8 cm. longa 6–8 mm. lata, margine integerrima, apice rotundata, obtusa vel interdum subacuta. Racemi 6–10, in axillis foliorum dispositi circ. 2·5 cm. infra ramuli apicem, circ. 3 cm. longi, pedunculis subrigidis, 2 cm. longis. Flores parvi, albi. Calycis-lobi ovati, obtusi, minutissime ciliati, corollae-tubo breviores, 1 mm. longi. Corollae-tubus 1·5 mm. longus, fauce glaber ; lobi 4, dorsalis et laterales fere aequantes, obovati, obtusi circ. 3 mm. longi et 2 mm. lati. Capsula ovoidea. compressa, acuta 4 mm. longa, minutissime sparsimque pubescens.

South Island : Otago—On subalpine rocks, Eyre Mountains, at 1,200 m. altitude ; D. L. Poppelwell ! Named in honour of Mr. G. Biggar, of Gore, who has accompanied Mr. Poppelwell on many botanical excursions and rendered material assistance.

This distinct species, which does not seem closely related to any other, is at once recognized by its decumbent or prostrate habit ; spreading, rather stiff, densely leafy branches ; small glaucous, glabrous, almost sessile, oblong, ovate or oblong-ovate leaves with from rounded to subacute apices ; numerous short, rather dense racemes ; small white flowers with short calyx-segments and corolla-tube ; and small slightly pilose capsules. The description is drawn up from a cultivated plant almost past flowering, so some allowance must be made for inaccuracies ; nor can anything be said as to differences between individual plants.

12. Veronica Poppelwellii Cockayne sp. nov.

Frutex parvus ramulis ultimis densis, erectis, cum foliis tetragonis, circ. 3–4 cm. longis, vix 2 mm. diametro, pilosis pilis albis ad foliorum connexum. Folia arcte quadrifariam imbricata, paribus oppositis basi connatis, latissime triangularia, 11-nervosa medio nervo breve carinato, 1 mm. longa, 2·25 mm. lata, crassa, coriacea, apice subacuta vel obtusa, margine pilis albis brevibus ciliata, supra concava, infra convexa. Flores ± 15, in densas spicas circ. 1·2 cm. longas ad apices ramulorum dispositi; bracteolae foliis similes sed 2 mm. longae. Calyx usque ad basin 4-fid; lobi corollae-tubum aequantes, interdum inaequales, obovati, obtusi, 2 mm. longi, coriacei, ciliati. Corollae-tubus vix 2 mm. longus; lobi 4, dorsalis et laterales fere aequantes, obovati, obtusi, 2 mm. longi, lobus anterior angustior. Capsulam maturam non visi.

South Island : Otago—Mount Tennyson, Garvie Mountains; fairly plentiful. D. L. Poppelwell !

Mr. Poppelwell informs me that this species is the *V. Hectori* Hook. f. var. *gracilior* Petrie of Poppelwell's list of Garvie Mountain plants in Trans. N.Z. Inst., vol. 47 (1915), p. 140. It is at once to be recognized by its small stature, short slender tetragonous branchlets, small leaves marked with parallel nerves, dense 15-flowered spikes 1·2 cm. or more long, leaf-like coriaceous, bracteoles, small ciliated obovate obtuse calyx-segments, and short corolla-tube. It does not seeem at all close to any other species of the section to which it belongs.

The species is named in honour of my friend Mr. D. L. Poppelwell, who is doing so much to throw light upon the flora near Lake Wakatipu, and the arrangement of the vegetation.

13. (a.) Veronica salicifolia Forst. f. var. Atkinsonii Cockayne var. nov.

Ab omnibus varietatibus specei discriminanda foliis pallide viridibus, oblongis, ovato-oblongis raro lanceolatis, breve sed manifeste petiolatis, prope apicem ± abrupte breve angustatis sed haud attenuatis, crassis; racemis 'sᴉɹoᴉɹsuǝp cylindricis, obtusis; floribus albis parvis; bracteolis et calycis-lobis brevissimis; corollae-lobis obovatis, apice rotundatis.

North and South Islands : Wellington—Vicinity of Cook Strait. Marlborough—From the Sounds to the mouth of the River Awatere. L. C.

Either a much-branched shrub attaining at times a height of 2 m. or of spreading habit when growing as a rock-plant, and sometimes almost prostrate. The leaves are rather pale green, fairly thick, distinctly petiolate with a short broad concavo-convex pale horny petiole 2–2·5 mm. long, more or less oblong, but occasionally lanceolate, 4–8 cm. long by 1·5–2 cm. or even more broad, narrowed near the apex to a short triangular point with an obtuse or subacute apex but not tapering. The racemes are 2–4 near the ends of the branchlets, dense-flowered, cylindrical, short-peduncled, frequently obtuse, 3–8 cm. long, and the peduncle 8 mm. to 2·5 cm. long. The rhachis is rather stiff, extremely pale green and finely pubescent. The flowers are very numerous, white or very rarely stained with lilac, and sweet-scented; the pedicels are slender, very pale green, minutely pubescent, and almost the length of the calyx; the bracteoles are linear, obtuse, and equalling or rather shorter than the pedicels; the calyx is 4-partite for about three-quarters of its length, about 3 mm. long, very pale green, slightly pubescent, and the segments are ovate or oblong, obtuse or subacute, and possess scarious margins; the corolla-tube is funnel- or sometimes barrel-shaped, 3·5 mm. long, and slightly pubescent at entrance to throat ; the lobes are not wide-spreading, concave, especially the posterior one, which with the lateral are broadly ovate or obovate, 2·5 mm. long, and rounded at the apex, while the anterior lobe is much shorter and narrower.

The above plant has puzzled me greatly for a number of years. Probably the decumbent form of coastal cliffs is the plant collected by Colenso at Cook Strait and referred by Hooker to *V. macroura* Hook. f. This position I took up in 1907 (Trans. N.Z. Inst., vol. 39, p. 361), but considered that it should be separated as a variety. Later on it became obvious that there were no constant differences between the tall shrub, so common in and near Wellington City, and the coastal plant. But both were so different from any form of *V. salicifolia* with which I was acquainted, and also from *V. macroura* proper, that they seemed specifically distinct from either, and so I gave the Wellington plant the MS. name of *V. Atkinsonii*, after my friend Mr. Esmond Atkinson, of the Biological Branch of the New Zealand Department of Agriculture. Later, visiting the Marlborough Sounds, to my surprise there appeared to be no forms in that area of *V. salicifolia*, but *V. Atkinsonii* was abundant. The question at once arose, could this latter be the "type" of *V. salicifolia*, since it can hardly have escaped the notice of the botanists of Cook's second voyage, and from their material *V. salicifolia* was described. But *V. salicifolia*, in one of its well-known forms, would also be collected by the same expedition at Dusky Sound. It seems possible, then, that *V. salicifolia* Forst. f. is a mixture of *V. Atkinsonii*, *V. salicifolia* var. Dusky Sound, and perhaps *V. amabilis* Cheesem. var. *blanda* of the same locality. Most probably, however, judging from the original description, which is as follows, the Dusky Sound *V. salicifolia* is the type : " *V. salicifolia*, racemis lateralibus nutantibus, caule fruticosa, foliis longo-lanceolatis integerrimis." *Nutans* and *longo-lanceolatus* match the Dusky Sound plant, but not that of the Marlborough Sounds. On the contrary, *integerrimus* is not correct, for the leaves of both plants are usually slightly toothed.

Although the Wellington plant is nearly always much as described above, specimens are encountered with longer narrower leaves, with longer more tapering racemes, and with flowers suffused with lilac and less closely placed. Such specimens come much too close to the variety of *V. salicifolia* next described to allow *V. Atkinsonii* to stand as a species, so it is here dealt with as a well-marked variety of the complicated aggregate *V. salicifolia*.

V. salicifolia var. *Atkinsonii* is essentially a rock-plant. This rupestral habit is clearly shown by the rapidity with which rock-cuttings in Wellington City are seized upon by the plant. Such physiological behaviour is a most important character of the race, marking it clearly from var. *paludosa* and var. *communis*, although the latter is also rupestral, but to a much lesser degree.

(*b.*) Veronica salicifolia Forst. f. var. communis Cockayne var. nov.

Folia viridia, lanceolata, apicem versus attenuata acuta vel acuminata, subsessilia, membranacea, ± 7·5–9·5 cm. longi. Racemi graciles, attenuati, ± 12 cm. longi, pedunculis ± 3·5 cm. longis ; flores albi-lilacino tincti ; pedicelli 3 mm. longi ; calycis-lobi lanceolati, acuti ; corollae-lobi ovati, subacuti.

This is the ordinary form of *V. salicifolia* throughout the South Island, where it occurs as a tall or medium-sized much-branched shrub from sea-level to the subalpine belt and from the coast-line to the interior. It probably occurs in the North Island also, but I have no exact information as to its distribution. Its distinguishing characters, subject, however, to considerable individual differences, are: The bright-green but frequently tinged yellowish, willow-like, lanceolate, thin leaves, which towards the apex are narrowed and taper to a fine point; the long slender tapering racemes with peduncles much longer than those of the var. *Atkinsonii;* the lilac-tinged flowers with comparatively long slender pedicels, acute calyx-segments which are wide apart and expose the ovary, and corolla rather larger in all its parts than that of the var. *Atkinsonii,* and with ovate subacute segments.

The form of the West Coast Sounds, if it be a distinct form, has much broader leaves than that of the drier parts of the South Island.

(*c.*) **Veronica salicifolia** Forst. f. var. **paludosa** Cockayne var. nov.

Folia lineari-lanceolata, longe paulatimque apiculata, circ. 6·5–11 cm. longa, 6 mm. to 1·2 cm. lata; racemi cum pedunculis ± 15 cm. longi; rhachis, pedicelli, et bracteoli dense pubescentes; calycis-lobi acuti vel subapiculati.

South Island :. Westland—In lowland swamps.

I have been in the habit of referring this plant to *V. gracillima* Cheesem., but a type specimen of the latter in T. Kirk's herbarium shows the former to be quite distinct. Its quite narrow leaves with long-drawn-out acuminate apex separate it at a glance from the var. *communis.* Probably there are other good distinctions, but my material is too far advanced for an examination of the flower. In general appearance var. *paludosa* greatly resembles the common veronica of the subalpine scrub of Mount Egmont, but this latter, which I propose before long to describe as var. *egmontiana,* has a different capsule, which brings it somewhat near to *V. macrocarpa.*

14. × **Veronica Simmonsii** Cockayne nov. typ. hyb. (*V. salicifolia* Forst. f. var. *Atkinsonii* Cockayne × *V. angustifolia* A. Rich. var.).

Frutex densus, erectus, multiramosus, circ. 1·8 m. altus. Folia lineari-lanceolata, subpetiolata, utrinque glabra nisi petiolo minutissima pubescentia, circ. 6·4 cm. longa, 8 mm. lata, margine integerrima, apice acuta, costa infra paullo carinata. Racemi 2–4, in axillis foliorum superiorum dispositi, circ. 7 cm. longi, 1·8 cm. diametro, rhachibus pedicellisque pubescentibus; pedicelli 2 mm. longi; bracteoli lineares, pubescentes, 1 mm. longi. Flores 4–5 mm. diametro, albi. Calyx parvus 2 mm. longus, profunde 4-partitus lobis oblongis obtusis ciliatis. Corollae-tubi anguste . infundibuliformes, 3 mm. longi; lobi late oblongi, obtusi.

South Island : Marlborough—French Pass and Pelorus Valley; L. C.

It seems to me far safer to treat this plant as a hybrid than as a new species or a variety of *V. angustifolia,* since it was found only when in. company with the latter and *V. salicifolia* var. *Atkinsonii,* but when these are growing separately × *V. Simmonsii* is absent, as I have had frequent opportunity to observe. The plant is named after Mr. G. T. Simmons, ssistant lighthouse-keeper, French Pass.

The hybrid is distinguished at once from *V. angustifolia* by its much broader and longer leaves and dense-flowered racemes, while from *V. salicifolia* var. *Atkinsonii*.it is readily separated by its much narrower leaves and more slender shorter racemes.

Art. XXII.—*Some Hitherto-unrecorded Plant-habitats (X).*

By L. Cockayne, Ph.D., F.L.S., F.R.S., Hutton and Hector Memorial Medallist.

[Read before the Wellington Philosophical Society, 27th October, 1915.]

1. Species from Various Localities.

Acaena saccaticupula Bitter.

South Island: (1.) Nelson—River-bed of the Clarence, near the accommodation-house; C. E. Christensen! (2.) Canterbury—Near Mount Cook; D. Petrie!

Aciphylla Colensoi Hook. f.

South Island: Marlborough—On coarse limestone debris, Woodside Creek, at almost sea-level. L. C.

Aciphylla squarrosa Forst. var.

South Island: Otago—Greatly increasing in amount in the enclosed part of the State plantation at Dusky Hill. L. C.

Angelica geniculata (Forst. f.) Hook. f.

South Island: Canterbury—In remains of forest near Culverden. Miss A. M. Budd!

Anisotome aromatica Hook. f.

South Island: Marlborough—Woodside Creek, at almost sea-level, on face of limestone cliff. L. C.

Aristotelia fruticosa Hook f.

South Island: Canterbury—Banks Peninsula, rare. R. M. Laing.

Arthropodium cirratum (Forst. f.) R. Br.

South Island: Marlborough—Common on rocky headlands, Pelorus Sound. L. C.

Blechnum Banksii (Hook. f.) Mett.

North Island: Taranaki—Coastal cliff near Breakwater, New Plymouth. L. C.

Blechnum nigrum (Col.) Mett.

North Island: Wellington—Montane *Nothofagus* forest, Rimutaka Mountains, east of Whiteman's Valley. L. C.

Celmisia Monroi Hook. f.

South Island: Nelson—On rock-face near accommodation-house, Clarence Valley. C. E. Christensen!

Celmisia petiolata Hook. f.

South Island: Canterbury—Craigieburn Mountains, on Harper Saddle. A. H. Cockayne!

Chordospartium Stevensoni Cheesem.

South Island: Marlborough — Avondale. C. de Vere Teschemaker Shute!

Clematis afoliata Buch.

South Island: Marlborough — (1.) Rocky ground, Avondale; L. C. (2.) On limestone debris, growing over *Coprosma crassifolia*, &c., near mouth of Flaxbourne Stream; L. C. Canterbury—(3.) At base of limestone bluffs between Rotherham and the River Waiau; L. C. (4.) On limestone rocks, Weka Pass hills; L. C.

Clematis marata J. B. Armstg.

South Island: Otago—Dusky Hill. L. C.

Coprosma crassifolia Col.

North Island: Wellington—Near Cape Turakirae. A. H. Cockayne and E. Bruce Levy!

Coprosma foetidissima Forst.

North Island: Wellington—Montane *Nothofagus* forest on Rimutaka Mountains, east of Whiteman's Valley. L. C.

Coprosma tenuicaulis Hook. f.

South Island: (1.) Marlborough—Lowland swamp, Pelorus Valley; L. C. (2.) Westland—In kahikatea forest near Ross, and probably elsewhere; L. C.

Cordyline indivisa (Forst. f.) Steud.

(1.) North Island: Wellington—Subalpine *Nothofagus* forest, Rimutaka Mountains, east of Whiteman's Valley; L. C. (2.) South Island: Nelson—Forest of Motueka hills; F. W. Huffam and L. C.

Corokia Cotoneaster Raoul.

South Island: Nelson—On flattish limestone rocks in full sunshine, flattened to the rock. F. W. Huffam and L. C.

Dactylanthus Taylori Hook. f.

North Island: Taranaki—Forest at "Meeting of the Waters," near New Plymouth, the plants extending thickly over an area of more than one acre. Mrs. L. S. Jennings!

Euphrasia zelandica Wettst.

South Island: Canterbury—Banks Peninsula. R. M. Laing.

Glaucium flavum Crantz.

South Island: Marlborough—Gravel beach at mouth of River Awatere L. C. (This is the first record of this introduced plant for the South Island.)

Hoheria angustifolia Raoul.

South Island: Marlborough—Pelorus Valley. J. Rutland!

Helichrysum Purdiei Petrie.

South Island : Nelson—Bank of River Perceval, Hanmer, in company with *H. glomeratum* and *H. bellidioides*, and nowhere else. C. E. Christensen ! (Only recorded previously from the west side of Dunedin Harbour, between Ravensbourne and Port Chalmers.)

Hymenophyllum ferrugineum Colla.

North Island : Wellington—(1.) On trunks of tree-ferns in kahikatea forest near Levin ; L. C. (2.) On floor of montane *Nothofagus* forest, Rimutaka Mountains, east of Whiteman's Valley ; L. C.

Hymenophyllum Malingii (Hook.) Mett.

(1.) North Island : Wellington—On dead *Libocedrus Bidwillii* in forest on southern slopes of Mount Ruapehu ; L. C. (2.) South Island : Nelson—Subalpine *Libocedrus* forest on Motueka hills ; F. W. Huffam and L. C. (Probably *H. Malingii* occurs wherever there is *Libocedrus Bidwillii*.)

Korthalsella Lindsayi (Oliver) Engler.

North Island : Wellington—Parasitic on *Muehlenbeckia Astoni* near mouth of River Wainuiomata. A. H. Cockayne and E. Bruce Levy !

Linum monogynum Forst. f. var. chathamicum Cockayne.

North Island : Wellington—On rock near Wellington Harbour. Esmond Atkinson ! (This variety, published in Trans. N.Z. Inst., vol. 34, 1902, p. 320, is not cited in Cheeseman's Manual.)

Leptopteris superba (Col.) Presl.

North Island : Wellington—On floor of montane *Nothofagus* forest of Rimutaka Mountains, east of Whiteman's Valley. L. C.

Metrosideros Colensoi Hook. f.

(1.) North Island : Taranaki—Forest near Moumahaki State Farm ; L. C. (2.) South Island : Marlborough—Common in forest near Woodside Creek ; L. C.

Microseris scapigera (Forst. f.) Sch. Bip.

South Island : Marlborough—On coastal limestone rock near mouth of Flaxbourne Stream. L. C.

Muehlenbeckia Astoni Petrie.

South Island : Marlborough—Stony ground near mouth of Flaxbourne Stream. L. C. (Only two plants were noted, but probably others had been destroyed when the ground was " cleared.")

Myosotis Forsteri Lehm. ?

North Island : Wellington—On coastal cliffs between Happy Valley and Cape Terawhiti. L. C.

Myrtus Ralphii Hook. f.

South Island : Marlborough—Kenepuru Sound, in forest. L. C. (In Cheeseman's Manual Marlborough is given as a habitat, but no particular locality is cited.)

Nothofagus Menziesii (Hook. f.) Oerst.

North Island : Wellington—Montane forest of Rimutaka Mountains, east of Whiteman's Valley. L. C.

Nothofagus Solanderi (Hook. f.) Oerst.

South Island : Marlborough—Forest near Woodside Creek. L. C.

Notospartium Carmichaeliae Hook. f.

South Island : Marlborough—Rocky ground, Avondale ; C. de Vere Teschemaker Shute !

Notospartium torulosum T. Kirk.

South Island : Nelson—Hanmer Plains neighbourhood. C. E. Christensen !

Nothopanax anomalum (Hook. f.) Seems.

South Island : Otago—Forest at Dusky Hill. L. C.

Olearia ilicifolia Hook. f.

North Island : Taranaki—Mount Egmont, in subalpine scrub ; local. F. G. Gibson and L. C. (According to Gibson, it occurs (1) here and there near the track from the Dawson Falls House to 'the North Egmont House; (2) at a spot called Holly Flat, near Bell's Falls, on the Pouakai Range; and (3) sparsely near Kahui House. It is quite wanting near the North Egmont House, and in most parts of the subalpine scrub.)

Olearia macrodonta Baker.

North Island : Taranaki—Mount Egmont (very rare) and Pouakai Range, especially at Holly Flat. A. H. Cockayne and L. C.

Orthoceras strictum R. Br.

South Island : Marlborough—Open ground on dry hillside near Kenepuru Sound. L. C.

Pittosporum divaricatum Cockayne.

South Island : Nelson—On almost flat limestone rocks in full sunshine, Motueka hills. F. W. Huffam and L. C.

Ranunculus lobularis (T. Kirk) Cockayne.

South Island : Marlborough—On face of limestone precipice, Woodside Creek, almost at sea-level ; plentiful. L. C.

Ranunculus ternatifolius T. Kirk.

South Island : Otago—Open wet floor of *Nothofagus* forest, Tapanui. L. C.

Senecio laxifolius Buch.

South Island : Nelson—Motueka hills. F. W. Huffam and L. C.

Teucridium parvifolium Hook. f.

South Island : Canterbury—Forest, base of Four Peaks Range. C. E. Foweraker.

Urtica ferox Forst. f.

South Island: (1.) Marlborough—Pelorus Sound; L. C. (2.) Canterbury —Amongst coarse limestone debris, Weka Pass Hills; L. C.

Veronica Cookiana Col. ?

North Island: Taranaki—Sugarloaf, near New Plymouth. W. Waller! (I refer this with some degree of hesitation as above, since I have not been able to compare it with living material of *V. Cookiana.*)

Veronica elliptica Forst. f. var.

North Island: Taranaki—Coast of Mount Egmont peninsula. P. G. Morgan! (This highly interesting discovery extends considerably the hitherto recorded northern limit of distribution of this species. The specimens closely resemble the common form of the Otago fiords, but are not at all like the plant growing at Titahi Bay.)

South Island: Marlborough—Forsyth Island. (I only saw cultivated plants in the garden of Mr. Foote, but that gentleman stated they were collected in the above locality.)

Veronica Menziesii Benth.

South Island: Marlborough—Hill at back of Wilson Bay, Pelorus Sound, according to Mr. Foote.

Veronica rigidula Cheesem.

South Island: Marlborough—Hill at back of Wilson Bay, Pelorus Sound, according to Mr. Foote, in whose garden I noticed the shrub.

Wahlenbergia Matthewsii Cockayne.

South Island: Marlborough—On face of limestone cliff, in rock-crevices, Woodside Creek, almost at sea-level. L. C.

2. Species collected by Captain W. Waller on Motumahanga Rock, Taranaki.

Agropyron scabrum (Lab.) Beauv. var.
Apium prostratum (DC.) Lab. var.
Asplenium bulbiferum Forst. f. var.
—— *flaccidum* Forst. f. var.
—— *lucidum* Forst f.
Blechnum Banksii (Hook. f.) Mett.
Calystegia sepium (L.) R. Br.
—— *Soldanella* (L.) R. Br.
Carex testacea Sol. ?
Crassula Sieberiana (Schultz) Cockayne.
Dactylis glomerata L.; introduced.
Dichondra repens Forst.
Geranium molle L.; introduced.
Halorrhagis erecta (Murr.) Schindl.
Linum monogynum Forst. f.
Lepidium oleraceum Forst. f. var. *acutidentatum* T. Kirk.
Lupinus arboreus Sims.
Luzula campestris DC. var.

Mariscus ustulatus (A. Rich.) C. B. Clarke.
Mesembryanthemum australe Sol. var. *rubrum* Cockayne, ined.
—— —— var. *viride* Cockayne, ined.
Muehlenbeckia complexa Meissn. var. *microphylla* (Col.) Cockayne. var. nov.
  (= *M. microphylla* Colenso in Trans. N.Z. Inst., vol. 20 (1888), p. 204).
—— *australis* (A. Rich.) Meissn.
Phormium tenax Forst.
Phytolacca octandra L.; introduced.
Polypodium diversifolium Willd.
Pteridium esculentum (Forst. f.) Cockayne.
Rubus fruticosus L. var.; introduced.
Salicornia australis Sol.
Sambucus nigra L.
Scirpus nodosus (R. Br.) Rottb.
Senecio lautus Forst. f. var.
Sonchus littoralis (T. Kirk) Cockayne.
Vicia sp. (not sufficient for identification); introduced.

3. List of Species omitted from the " Provisional List of Ferns
   and Flowering - plants of the Port Hills " (Report on
   Scenery Preservation for 1915).

Nearly all the following species were supplied by Mr. R. M. Laing, B.Sc.,
M.A. Plants growing only on the actual coast were purposely omitted
from the list.

Australina pusilla Gaud.
  Rare. R. M. Laing.

Chenopodium triandrum Forst. f.
  Common. R. M. Laing.

Cotula minor Hook. f. ?
  Very rare; only found in one locality, near the summit of the range.
R. M. Laing !

Cotula squalida Hook. f.
  In damp ground, chiefly near the summit of the range. R. M. Laing !

Galium umbrosum Sol.
  Plentiful. R. M. Laing.

Hedycarya arborea Forst.
  Abundant in forest. R. M. Laing.

Hierochloe redolens (Forst. f.) R. Br.
  Near Governor's Bay Road and elsewhere. R. M. Laing.

Lycopodium fastigiatum R. Br.
  Only noted in the patch of forest beyond Kennedy's Bush. R. M.
Laing.

Lycopodium varium R. Br.
　　Slope behind Lyttelton.　R. M. Laing.

Lycopodium scariosum Forst. f.
　　Only noted in the patch of forest beyond Kennedy's Bush.　R. M. Laing.

Mentha Cunninghamii Benth.
　　Plentiful in many places.　R. M. Laing.

Mesembryanthemum australe Sol.
　　Above the track on Mount Pleasant.　R. M. Laing.

Myosotis pygmaea Col.
　　Behind Governor's Bay.

Olearia avicenniaefolia (Raoul) Hook. f.
　　Behind Governor's Bay.　R. M. Laing.

Olearia Forsteri Hook. f.
　　Common on rocks.　L. C.

Podocarpus dacrydioides A. Rich.
　　Kennedy's Bush, but at present time only one tree observed.　R. M. Laing.

Pterostylis graminea Hook. f.
　　Salt's Gully, near Lyttelton.　R. M. Laing.

Pterostylis foliata Hook. f.
　　Heathcote Valley.　Miss Holdsworth.

Ranunculus rivularis Banks & Sol. var. subfluitans Benth.
　　In ponds on the Cashmere Estate.　R. M. Laing.

Scirpus nodosus (R. Br.) Rottb.
　　In the tussock grassland.　R. M. Laing.

Styphelia fasciculata (Forst. f.) Diels.
　　Little Raupaki, &c.　R. M: Laing.

Tetragonia expansa Murr.
　　Above track on Mount Pleasant.　R. M. Laing.

Tetragonia trigyna Banks & Sol.
　　Above track on Mount Pleasant.　R. M. Laing.

Art. XXIII.—*New Species of Plants.*

By T. F. Cheeseman, F.L.S., F.Z.S., Curator of the Auckland Museum.

[*Read before the Auckland Institute, 8th December, 1915.*]

1. Geum divergens Cheesem. n. sp.

Species distincta *G. unifloro* Buch. similis, sed scapis 1–4-floris, bracteis numerosis incisis, floribus flavis.

Herba parvula, 4–10 cm. alta. Rhizoma breve, crassum, procumbens. Folia radicalia, lyrato-pinnata, 2–5 cm. longa ; foliolis lateralibus minutis, terminali maximo orbiculari-cordato irregulariter dentato. Scapi 3·5–5 cm. longi ; bracteis 2–5, lineari-lanceolatis, incisis. Flores 1–5, ratione plantae magni, 1·5–2·5 cm. diam., flavi. Calycis segmenta ovato-lanceolata, acuta vel acuminata. Petala magna, obovata, obtusa. Achenia dense villosa ; stylis glabris, apice uncinatis.

Hab.—South Island : Sheltered places among rocks on the slopes of Mount Captain, Clarence Valley, alt. 5,000 ft., *T. F. C.*

Short, stout, 2–5 in. high. Rhizome creeping, short, thick and woody, clothed with the bases of the old leaves. Leaves all radical, 1–2 in. long including the petiole, lyrate-pinnate ; terminal leaflet very large, ¾–1¼ in. diam., orbicular reniform, indistinctly lobed, coarsely crenate-dentate ; margins densely ciliate with long white hairs ; upper surface sparsely villous, lower almost glabrous ; lateral leaflets 2–4 pairs, minute, lanceolate or ovate-lanceolate, ciliate. Scapes 1½–3 in. high, slender, densely pubescent ; bracts 2–5, the lowest sometimes ¾ in. long, lanceolate, incised or rarely inciso-pinnatifid. Flowers 1–5, large for the size of the plant, ¾–1¼ in. diam., yellow. Calyx-lobes ovate-lanceolate, acute or acuminate, sometimes with small accessory lobes at the base, pubescent. Petals much larger, obovate, obtuse. Achenes villous with long hairs, gradually narrowed into a long glabrous style hooked at the tip.

Although this has something of the habit and appearance of *G. uniflorum*, it is by no means closely allied to that plant. The rhizome is shorter and not so stout ; the leaves are thinner, with a different indumentum ; the scapes are furnished with more numerous and much larger and more deeply incised bracts ; the flowers are more numerous, sometimes as many as 5 to a scape, and are bright yellow, whereas they are solitary and always white in *G. uniflorum.* In size, in the shape of the leaves, and in the colour of the flowers it appears to approach the somewhat vague description given by Buchanan of his *G. alpinum* (a name already occupied). But he describes the flowers of his plant as " minute " and " ⅕ in. diam.," a difference in size so great as to preclude the specific identity of the two plants, whatever " *G. alpinum* " may prove to be.

2. Olearia insignis Hook. f. var. minor Cheesem. n. var.

Differt a typo habitu multo graciliore, foliis minoribus et angustioribus, pedunculis gracilioribus, capitulis minoribus.

Hab.—South Island : Marlborough, between Kaikoura and Blenheim, *H. J. Matthews !*

Much smaller and more slender than the type; branches much less robust and less thickly clothed with tomentum. Leaves 2½–4 in. long including the petiole, 1–1½ in. broad, narrow elliptic-oblong to oblong-spathulate, gradually narrowed at the base into a rather slender petiole 1 in. long, much thinner and less coriaceous than in the type, tomentum on the under-surface much less dense. Peduncles 2 or 3 towards the ends of the branches, 3–5 in. long, slender, angular, sparingly tomentose, furnished above the middle with 2 or 3 linear-spathulate bracts. Heads small, ¾–1¼ in. diam., longer and narrower in proportion than those of the type, almost campanulate in shape; involucral bracts narrower and less densely tomentose.

So far as can be judged from the rather scanty suite of specimens in my possession, this is a well-marked variety, distinguishable from the typical form at a glance by its much smaller size, more slender habit, less copious tomentum, much more slender peduncles, and smaller heads. It appears to keep its characters in cultivation.

3. Celmisia Thomsoni Cheesem. n. sp.

Ab *C. bellidioides* Hook. f. differt caulibus brevioribus et densissime compactis, follis magis coriaceis rosulatis et arcte imbricatis, superne distincte setulosis.

Herba pusilla, caespitosa. Rhizoma breve, ramosum, prostratum; ramis numerosis, brevibus, densissime compactis, 2–5 cm. longis, superne foliosis. Folia numerosa, imbricata, rosulatim disposita, 0·75–1·25 cm. longa, 3·5 mm. lata, anguste obovato-oblonga aut lineari-oblonga, obtusa vel obtuse apiculata, basi attenuata, coriacea, supra copiose et distincte setulosa; marginibus minute denticulatis; petiolis brevibus. Scapi graciles, 4–7 cm. longi, 2–4-bracteati, glabra aut sparse glanduloso-tomentosa; bracteis angustis. Capitula 1–2 cm. diam.; involucri squamae paucae. Achenium sericeum.

Hab. South Island: Eyre Mountains, Central Otago, in rock-crevices on the faces of cliffs, alt. 5,000–6,000 ft., *W. A. Thomson* and *J. Speden !*

Forming compact cushion-like patches 1–3 in. diam., rarely more. Root-stock short, much branched, prostrate; branches usually numerous, densely compacted, leafy above. Leaves numerous, closely imbricate and rosulate, ⅓–½ in. long, ⅛–⅕ in. broad, narrow obovate-oblong or almost linear-oblong, obtuse or bluntly apiculate, narrowed at the base, thickly coriaceous, dull green, glabrous beneath, upper surface setulose with short stiff glandular hairs; margins furnished with minute irregular denticles; petioles short, broad, sheathing at the base. Scapes from near the tips of the branches, 1½–2¾ in. long, very slender, glabrous or more commonly sparsely glandular-tomentose above; bracts 2–4, narrow-linear. Heads ½–¾ in. diam.; involucral bracts few, narrow linear-oblong, acute, green or the outer ones purplish-green. Rays numerous, spreading. Achenes densely silky.

I have much pleasure in dedicating this interesting plant to Mr. W. A. Thomson, of Dunedin, who in company with Mr. J. Speden was the first to collect it. Although allied to *C. bellidioides*, it differs from that species in several important characters. It is much smaller and much more compactly branched, thus assuming a tufted cushion-like habit quite different from the creeping and mat-like appearance of *C. bellidioides.* The leaves are much more closely placed, and are decidedly rosulate, while the upper surface is sprinkled all over with short stiff glandular hairs. The scape is also much more slender than in *C. bellidioides.* When fresh the leaves are of a dull green, quite unlike the bright shining green of *C. bellidioides.* Both Mr. Thomson and Mr. Speden inform me that it usually occurs in crevices on the shaded sides of rocky cliffs; often, in fact, in situations where no direct rain can fall upon it, although drifting snow or fog might reach it. So far as my experience goes, *C. bellidioides* is usually found draping wet cliffs or creeping over wet gravelly slopes.

4. Cotula Willcoxii Cheesem. n. sp.

Species *C. pectinatae* Hook. f. affinis, sed foliis non pectinato-pinnatifidis facillime distinguenda.

Caules 3–15 cm. longi, prostrati, copiose ramosi, radices fibrosas emittentes, sparse sericeo-pilosi. Folia petiolata, cum petiolo 0·5–1·25 cm. longa, glabra vel parce pubescentia, pinnatisecta; lobis 1–2-jugis, lineari-oblongis, subacutis. Pedunculi gracillimi, nudi, sparse pilosi, 2–5 cm. longi, foliis multo longiores. Capitula 0·75–1 cm. diam.; involucri squamae 1–2 seriales, oblongae, obtusae, marginibus scariosis et denticulatis apice purpureis. Flores foeminei 2–3 seriales; flores disci tubuloso-infundibuliformes, 4-dentati. Achenia anguste obovata.

Hab.—South Island: Head of Lake Wakatipu, Otago, near Mount Earnslaw, *W. Willcox!*

Stems 1½–6 in. long, creeping and rooting, copiously branched and usually forming compact patches, sparsely silky-pilose or sometimes nearly glabrous. Leaves alternate, petiolate, with the petiole ⅕–½ in. long, glabrous or faintly pubescent, gland-dotted, pinnatisect; segments usually a single pair with a terminal one, rarely two pairs, flat, linear-oblong, subacute, usually quite entire or very rarely one or both of the lower segments are forked at the tip. Peduncles long, slender, naked, sparingly pilose, ¾–2 in. long, much exceeding the leaves. Heads ¼–⅓ in. diam.; involucral bracts in 1–2 series, oblong, obtuse, purplish at the tips, margins scarious, jagged. Female florets in 2–3 series; florets of the disc narrow funnel-shaped, 4-toothed. Achene narrow-obovoid.

A well-marked species, closely allied to *C. pectinata* in the structure of the flower-heads, but in other respects presenting well-marked points of difference. It is larger, and much more glabrate; the peduncles are longer and more slender; and the leaves are never pectinate-pinnatifid, as is always the case in *C. pectinata.* Usually the leaves have only a single pair of segments with a terminal one, but more rarely there are two pairs of segments, in that case resembling some small forms of *C. pyrethrifolia.* From that plant, however, it is at once distinguished by the bisexual flower-heads, to say nothing of other differences. The only specimens that I have seen are those collected by Mr. Willcox, of Queenstown, and I have much pleasure in associating his name with the plant.

5. Senecio lapidosus Cheesem. n. sp.

Affinis *S. Monroi* Hook. f., sed minore, et capitulis solitariis maxime differt.

Fruticulus humilis, prostratus vel suberectus, densissime ramosus, 15–18 cm. altus, 20–25 cm. diam. Rami numerosi, divaricati et densissime compacti, basi robusti, lignosi, cortice pallido; ramulis apicibus ascendentibus, foliosis, lanato-tomentosis. Folia 2–4 cm. longa, 1–1·5 cm. lata, oblonga vel elliptico-ovata, obtusa, coriacea, supra arachnoideo-tomentosa,

subtus dense et appresse argenteo-tomentosa, marginibus minuti crenatis.
Pedunculi e summis ramulis, solitarii, graciles, 1·5–2·5 cm. longi, dense
argenteo-tomentosi; bracteis multis, linearibus vel lineari-oblongis. Capi-
tulum solitarium, 1–1·5 cm. latum; bracteis involucri 12–15, linearibus,
acutis, tomentosis. Flores radii 12–15, flavi; flores disci numerosi.
Achenia numerosa, longitudinaliter sulcata, hispida.

Hab.—South Island: Faces of rocky cliffs at Hell's Gate, Mason River,
North Canterbury, *H. J. Matthews!* (1909).

A small depressed much-branched prostrate or suberect shrub, forming
dense clumps 6–8 in. high and 8–10 in. diameter. Branches many, closely
compacted, stout and woody at the base, and covered with a pale flaky
bark; branchlets ascending or suberect, leafy, densely woolly tomentose.
Leaves ¾–1½ in. long, ⅓–½ in. broad, narrow oblong or narrow elliptic-ovate,
rarely oblong-spathulate, obtuse, coriaceous, margins minutely crenate,
upper surface thinly covered with cobwebby tomentum, beneath clothed
with dense silvery-white tomentum; petioles somewhat slender, broader at
the base. Peduncles from the tips of the branchlets, stout or slender, 1–2 in.
long, densely silvery tomentose; bracts numerous, linear or linear-oblong.
Heads solitary, turbinate, ⅓–⅔ in. diam.; involucral bracts 12–15, linear,
acute, densely tomentose. Ray-florets 12–15, yellow; disc-florets numerous.
Achenes linear, longitudinally grooved, hispid.

This interesting plant was collected by the late Mr. H. J. Matthews in
1909 in the gorge of the Mason River, North Canterbury, where it was
associated with *Epilobium brevipes, Veronica rupicola,* and other species.
So far, it has not been found elsewhere. It is clearly allied to *S. Monroi,*
but is a much smaller and more compactly branched plant, and the
peduncles are simple with a single terminal head, whereas in *S. Monroi*
the inflorescence is copiously corymbosely branched. It was introduced
into cultivation by Mr. Matthews, and forms an excellent plant for the
rock-garden. I have to thank Mr. A. Bathgate for fresh flowering specimens from his garden in Dunedin, which are rather more slender than the
wild specimens originally forwarded to me by Mr. Matthews.

6. Veronica obtusata Cheesem. n. sp. (*V. macroura* Hook. f. var. *dubia*
Cheesem., Man. N.Z. Flora, 501).

Species *V. macrourae* Hook. f. et *V. divergente* Cheesem. affinis, sed
foliis bracteis et calycis segmentis margine albo-pubescentibus differt.

Frutex parvus, ramosus, 0·75–1·25 m. altus. Ramuli numerosi, patuli
vel decumbentes, juniores teretes, pubescentes sed demum glabrescentes.
Folia patula, sessilia vel brevissime petiolata, 2·5–5 cm. longa, 1·5–2·25 cm.
lata, oblonga vel elliptico-oblonga vel obovato-oblonga, obtusa vel subacuta, subcoriacea, marginibus dense albo-pubescentibus. Racemi ramulorum apicem versus dispositi, 3·5–6 cm. longi, dense multiflori, recti vel
curvati; rhachis cum pedicellis bracteisque dense pubescens. Calycis
segmenta ovato-oblonga, acuta vel obtusa, marginibus conspicue albo-
pubescentibus. Capsula 4 mm. longa, ovata, subacuta.

Hab.—North Island: Sea-cliffs on the coast north of the Manukau
Harbour, *T. F. C.*

A branching shrub 2–3 ft. high; branches spreading or procumbent,
the younger ones terete, at first pubescent, but at length becoming nearly
glabrous. Leaves sessile or very shortly petiolate, 1–2 in. long, ⅗–1 in.
broad, oblong to elliptic-oblong or obovate-oblong, obtuse or subacute,

subcoriaceous, flat, margins edged with a line of short and dense white hairs, midrib above occasionally downy. Racemes longer than the leaves, 1½–2½ in. long, straight or curved, dense-flowered, but not so much so as in *V. macroura ;* rhachis and pedicels densely pubescent. Flowers ⅛ in. diam., pale bluish-white. Calyx 4-partite; segments oblong to ovate-oblong, obtuse or subacute, margins conspicuously ciliate with soft whitish hairs. Corolla-tube exceeding the calyx; limb 4-lobed. Capsule ⅕ in. long, ovate, compressed, not twice as long as the calyx.

In the Manual I treated this as a variety of *V. macroura.* Since then I have obtained a much better series of typical *V. macroura,* mainly collected in the East Cape district by Bishop Williams, and find that the two plants are amply distinct. *V. obtusata* is much smaller, and has a much more diffuse mode of growth ; the leaves are smaller, broader, and more obtuse, and the margins have a dense edging of short white hairs. The racemes are smaller and shorter in proportion ; the flowers are not so dense and slightly larger ; and the calyx-segments have their margins conspicuously ciliate with white hairs. It is probably nearer to my *V. divergens,* but that species has smaller and narrower glabrous leaves, longer racemes, and more glabrous calyx-segments.

With respect to *V. macroura,* I have been informed by Mr. N. E. Brown, of the Kew Herbarium, that the localities of "Whangarei and Cook Strait," given by Hooker in the Handbook, were probably quoted from Colenso's correspondence, as there are no specimens from thence in the Kew Herbarium. Mr. Brown further states that the only locality given on the type-sheet of *V. macroura* is "East Cape," written against a specimen (Colenso, 101) by Sir J. D. Hooker. So far as I am aware, no one has gathered typical *V. macroura* at either Whangarei or Cook Strait of late years.

7· Thelymitra pauciflora R. Br., Prodr. 314 ; Fitzgerald, "Australian Orchids," vol. 1, part 6, t. 2.

Stems slender, wiry, flexuous, 6–12 in. high, rarely more. Leaf much shorter than the stem, narrow-linear, $\frac{1}{10}$–⅕ in. broad, rarely more, thick and fleshy, longitudinally grooved, deeply channelled in front, and thus concave. Flowers 1–6, ⅓ in. long, usually pale-blue or whitish-blue. Sepals and petals narrow ovate or ovate-lanceolate, acute. Column short, stout, the wing continued behind the anther and longer than it, 3-lobed ; the middle lobe much the largest and also the highest, narrower than in *T. longifolia,* thick and swollen, projecting over the anther, deeply emarginate or 2-lobed, brownish-red at the base, bright yellow towards the tip ; lateral lobes smaller at the base, projecting forwards almost horizontally, then suddenly bent upwards and erect, terminated by a dense brush of white cilia. Anther broad, connective produced into a short point.

Hab.—North Island : *Leptospermum*-clad hills in the Auckland district, T. F. C. ; hills near Pukekohe, *W. Townson !* vicinity of Kaitaia (Mongonui County), *H. B. Matthews !* Probably widely distributed in the Auckland Provincial District. Flowers from the middle of October to mid-November.

I have been acquainted with this for many years, and have been accustomed to regard it as a variety of *T. longifolia,* to which species Mr. Bentham reduced Brown's *T. pauciflora.* But the structure of the column is very different from that of *T. longifolia,* and as long as the species of the genus are principally founded on deviations in the form of that organ it is difficult to avoid the belief that the two plants are distinct. The late Mr. R. F. Fitzgerald, in his magnificent work on Australian orchids, unhesitatingly accepted this view; and a comparson of the figures of the two species given by him shows how great the differences are. The receipt of a large parcel of fresh specimens of *T. pauciflora* collected by by Mr. Townson near Pukekohe gave me an opportunity of reviewing the matter, with the result of fully supporting the correctness of Mr. Fitzgerald's opinion. *T. pauciflora* differs from *T. longifolia* in its smaller size and more slender habit; in its narrower and deeply channelled leaf; in the smaller flowers; and especially in the middle lobe of the column-wing, which is deeply emarginate or 2-lobed, whereas it is much broader, more hood-shaped, and barely emarginate in *T. longifolia.* I should perhaps say that in all essential points New Zealand specimens agree with the drawing given by Mr. Fitzgerald of *T. pauciflora.*

Hooker's *T. Colensoi*, which has not been collected of late years, and which was originally referred to *T. pauciflora*, differs from that plant, according to Hooker, "in the very narrow sepals and petals, very short column, and very long erect appendages."

8. Rhopalostylis Cheesemanii Beccari in Herb. Kew., MS.

In August, 1887, I accompanied the expedition sent by the New Zealand Government to annex the Kermadec Islands to the Colony of New Zealand. During this visit I was able to give rather more than a week to the exploration of the flora of Sunday (or Raoul) Island, and to form a fairly complete collection of the phaenogamic plants and vascular cryptogams of the island. Included in this series was a palm which I then supposed to be identical with the Norfolk Island *Kentia Baueri* (now known as *Rhopalostylis Baueri*), although, as no actual comparison of specimens could be made, an element of doubt still remained. A short time ago, however, the veteran botanist Dr. O. Beccari, one of the chief authorities on palms, compared my specimens with the Norfolk Island species, and informs me that the Kermadec plant is specifically distinct. He states that "it is closely allied to *R. Baueri*, but is easily distinguishable by its larger spherical fruits and larger leaves with considerably longer leaflets." I hope that Dr. Beccari will shortly publish a full diagnosis; but in the meantime it is well to publish the fact that the Kermadec plant is a distinct and endemic species, and that *R. Baueri* does not extend beyond the confines of Norfolk Island.

ART. XXIV.—*The Species of the Genus* Pinus *now growing in New Zealand, with some Notes on their Introduction and Growth.*

By T. W. ADAMS.

[*Read before the Philosophical Institute of Canterbury, 7th August, 1907.**]

No genus of trees is more easily recognized than *Pinus*, but it is not so easy to identify the species, especially in the case of trees without cones.

Here, for convenience, the species are divided into the following classes : (1) Those with two leaves in a sheath ; (2) those with three leaves in a sheath ; (3) those with five leaves in a sheath. The above characters are fairly uniform in the different species, and are of considerable service for purposes of identification.

(1.) TREES USUALLY WITH TWO LEAVES IN A SHEATH.

Pinus austriaca Link.

This was probably first introduced in 1866. Potts, in the "New Zealand Country Journal," vol. 3, p. 38, reports a plant 6 in. high growing at Governor's Bay in that year, and in a paper read before the Wellington Philosophical Institute on the 22nd July, 1871, he states that a tree growing in Wellington had reached the height of 8 ft. 9 in. A tree of this species planted at Greendale in 1877 is now 43 ft. high. Mr. Potts reports on *P. austriaca* as one of the best to withstand salt breezes when planted by the sea. It is also a good drought-resister.

Pinus Banksiana Lambert.

This is a distinct and hardy pine. Plants at Greendale raised from seed in 1904 are 11 ft. high, and produce cones.

Pinus bruttia Tenore.

By some botanists this is considered to be only a geographical form of *P. halepensis*, but trees here 12 ft. high, while showing a relationship to that species, are distinct in habit, and the cones are distinctly different.

Pinus contorta Dougl.

This was introduced about 1880 in three forms, and two of these are so distinct that I prefer to follow those botanists who separate them into the two species *P. contorta and P. Murrayana.*

Lemmon, in "West American Cone-bearers," says of *P. Murrayana*, "Until recently confounded with *P. contorta*, but clearly distinct." G. B. Sudworth, United States Government Dendrologist, however, in a quite recent letter to me puts his view of the matter thus : "The stable botanical characteristics of the different forms of this tree, as now constituted under *Pinus contorta*, do not differ sufficiently, in my judgment, to justify a separation into distinct species, notwithstanding the fact that the crown-habit and even the size of the cones of individual trees appear to indicate

---

* Since the paper was read in 1907 it has been brought up to date by the inclusion of later measurements, &c.

specific distinctions. One very remarkable fact is that our Rocky Mountain lodgepole pine (*P. Murrayana*) is a distinctly narrow-crowned tree, while the Pacific Slope form of it is distinctly a broad-crowned tree, there being no other essential differences."

Here in New Zealand we have another very marked difference in the two trees, as under the same conditions the coastal or broad-top tree grows everywhere much faster than the mountain or spire-shaped tree.

Two trees measured here, and fairly representing the best growth of each, give the following result: *P. contorta*, 49 ft., and *P. Murrayana*, 32 ft. In the plantations here there are probably fifty trees of each of the same age, so that the difference in the rate of growth is not accidental.

Pinus densiflora S. & Z.

A tree planted under this name in 1890 is now 18 ft. high. A common forest-tree in Japan, this tree has not proved successful here.

Pinus edulis Engelm.

This New Mexican pine is represented in my collection by several trees, some of which are now 12 ft. high, and bear cones. The seeds are edible.

Pinus halepensis Mill.

This was early introduced into New Zealand, as trees were growing in 1868 both at Governor's Bay, Canterbury, and at the Hutt, Wellington, and were then nearly 4 ft. high. A tree of this species was 36 ft. 6 in. in height in 1885 at Governor's Bay.

Pinus laricio Poir.

This tree is grown successfully in many places in New Zealand. A tree planted at Greendale in 1877 is now 62 ft. high, and of a fine form.

Pinus leucodermis Antoine.

The white bark is the chief feature which distinguishes this pine from *P. austriaca*. By some it is considered only a variety of that pine. A tree raised from seed in 1904, presented to me by Dr. L. Cockayne, F.R.S., is now 6 ft. 6 in. high.

Pinus monophylla Torr. & Frem.

This western American pine was cultivated in Duncan's nursery (Christchurch) many years ago, but I do not know whether there are any living examples in New Zealand at the present time. This should properly not come in this class, with two leaves in a sheath, since it has merely one and not two. In habit it is a low round-headed tree 8 ft. to 45 ft. high. Its usual habitat is very dry stations of a desert character.

Pinus montana Mill.

This was early introduced in one or other of its forms, and large bushes of this species are growing in the Christchurch Botanical Gardens.

Pinus mugho Poir.

Some consider this only a variety of the last species. At Greendale *P. mugho* grows much faster than *P. montana*.

Pinus muricata D. Don.

This was planted in 1870 by Potts in Governor's Bay, and he reports that the tree had reached 35 ft. in height in 1885. Trees planted at Greendale in 1876 are now 51 ft. high. *Pinus muricata* cannot be recommended as a timber-tree, but makes a valuable shelter-tree.

Pinus Parrayana Engelm.

This pine, a native of California, is a short-trunked low tree 15 ft. to 30 ft. high in its natural habitat. Cone-bearing trees are to be seen in both the Ashburton and Tinwald Domains. I am placing this pine here because it has sometimes two leaves in a sheath, although four is a commoner number, and even three is not unknown.

Pinus pinaster Sol.

This was probably the first pine successfully introduced into New Zealand, as very old trees may be seen growing wild in the scrub in North Auckland, and seeds of it were sent Home at a very early date as a new pine, and named by Loddiges *Pinus nova-zelandica*. It is one of the best for seaside planting. The variety *Hamiltoni*, growing here, is about 12 ft. high.

Pinus pinea L.

Trees of this species were growing in Auckland, Wellington, and Canterbury in 1868, and Potts that year reported a tree in Governor's Bay to be 3 ft. 6 in. high, and when measured again in 1885 the tree was 33 ft. high.

Pinus pyreniaca Lap.

Raised from seed in 1904, this is now 12 ft. high, and is a distinct pine.

Pinus resinosa Sol.

Introduced about 1880, this has not generally succeeded, although the late Mr. Threlkeld spoke highly of its success at Flaxton, Canterbury.

Pinus sylvestris L.

Seed of this and several other pines was imported by the General Government and distributed by the Geological Survey Department about 1864. The tree at first gave good promise of success, until attacked by an aphis, when most of the trees became sickly and stunted in growth, only a very few surviving the attack and growing into fair specimen trees.

Pinus Thunbergii Parl.

This is a Japanese tree which gives some promise of success in New Zealand. Trees planted at Greendale in 1890 are now 21 ft. high, and bearing cones.

Pinus virginiana Mill.

Trees of this species have grown well at Greendale, and are bearing cones.

(2.) TREES HAVING USUALLY THREE LEAVES IN EACH SHEATH.

Pinus australis Michx.

This has the longest leaves of all the pines growing at Greendale, being longer than *P. longifolia* Roxburgh. Trees planted in 1889 are only 15 ft. high, but, being almost destitute of side branches, they have a very singular appearance for pine-trees.

Pinus Benthamiana Hartw.

This was introduced as early as 1865. It is generally considered a form of *P. ponderosa*, which was introduced the same year.

Pinus Bungeana Zucc.

Was first planted at Greendale in 1903. It is a native of China. Plants are of a lively green, and of slow growth, but are quite healthy.

Pinus canariensis C. Sm.

Was introduced as early as 1865, and was growing at the Hutt, Wellington, and Governor's Bay, Canterbury. A tree at the latter place measured, in 1885, 43 ft. in height. There is a tree of this species near the Museum, Christchurch, 9 ft. 9 in. in girth and 76 ft. high. During the severe winter of 1889 all the examples of this species at Greendale were damaged by the frost to the extent of losing all the previous summer's growth, but they have since quite recovered. This pine grows much better near the coast than at Greendale.

Pinus Coulteri D. Don.

This has the largest cones of all the pines; the leaves also are long and stout. *P. Coulteri* is generally adapted for conditions here, and trees of the second generation have commenced to bear cones. Trees planted in 1877 are now upwards of 60 ft. high.

Pinus Gerardiana Wall.

This is a healthy but slow-growing tree here. In India its seeds are valued as an article of food. Plants at Greendale are 7 ft. 6 in. high.

Pinus Jeffreyi A. Murr.

By some this is considered only one of the many varieties of *P. ponderosa*, but the habit of the tree in New Zealand makes it distinct enough for all forestry purposes, and it is easily recognized by any ordinary observer. The cones and seeds are larger and of a different shape from those of *P. ponderosa*; the leaves are also of a different shade. The bark, too, is of a different shade, and is divided into smaller checks than that species.

Seed of this tree was early distributed in New Zealand, and trees 50 ft. or 60 ft. high are not uncommon. A tree on Mr. Albert Adams's farm at Sheffield measures 7 ft. 6 in. in girth.

Pinus khasya Royle.

Trees of this species planted at Greendale in 1900 only lived through a few winters, not being sufficiently hardy to withstand the frosts of a severe season.

Pinus longifolia Roxb.

This was most likely introduced with other pines by the General Government about 1864, when seeds were distributed by the Geological Survey Department. Plants of this species are in the earliest lists of trees that I have been able to discover, and were growing at several places in 1866, and mentioned by Messrs. Potts, Pharazyn, Ludlam, and Mason as small plants at that time. While young the trees are beautiful, but when older are rather open and rugged-looking. The largest trees I have noticed of this species are growing in the Wellington Botanical Gardens.

Pinus luchneulis.

This is a recently introduced pine from the mountains of Formosa. It is of doubtful hardiness. Small plants have passed through two winters with but slight injury.

Pinus mitis Michx.

This tree is more variable in its leaves than most pines, having sometimes two leaves and at others three. Both cones and leaves are small, and the growth is slow. Trees planted in 1881 are only 25 ft. high.

Pinus patula Schiede & Deppe.

A tree supposed to be this species, planted in 1877, grew to a large size before it died, about ten years ago. Trees planted later are growing fast, and are extremely ornamental. It is a native of Mexico.

Pinus ponderosa Dougl.

Small plants were mentioned as growing at Governor's Bay in 1866, and these had reached the height of 37 ft. in 1885. It is also mentioned by Mr. Ludlam. At least four fairly distinct pines have been introduced to Canterbury under this name. *Pinus ponderosa* gives promise of being a profitable timber-tree in New Zealand.

Pinus radiata D. Don.

This is better known in New Zealand as *P. insignis*. This pine was planted as early as 1866 at Governor's Bay by Potts, and also by Mr. Gillies the same year. The specimens were then about 1 ft. in height, and when measured in 1885 were 67 ft. high. Trees planted at Greendale in 1873 have reached 128 ft. in height, and are still growing vigorously. No other species of pine—or, indeed, tree of any other genus—yet planted in New Zealand can compare in rate of growth with this pine.

Pinus rigida Mill.

This is a slow-growing pine of a distinct shade of green. It has one peculiar feature for a pine-tree, in that the stumps of trees which have been felled will sprout again. Some years ago, thinning a plantation, trees of this species that had been felled and were lying on the ground showed the above habit, for they were found to have sent forth sprouts along the trunk, as broad-leaved trees occasionally do.

Pinus Sabiniana Dougl.

This grows well in Canterbury, but the trees have a great tendency to divide into several leaders. The cones are large, second only to those of *P. Coulteri*, while the seeds are larger than in that species. It was early introduced, being mentioned in 1866. At Governor's Bay in 1885 it had reached 34 ft. in height. Trees planted at Greendale in 1881 are now 58 ft. high.

Pinus scopulorum Lemmon.

By many authorities this is considered a variety of *P. ponderosa*, but trees growing here more resemble some forms of *P. Jeffreyi*. The rate of growth is much slower than in *P. ponderosa*.

Pinus taeda L.

This grows well in New Zealand. It is said to have been introduced by the late Mr. Rolleston while Superintendent of the Canterbury Province, but I know of no trees planted so early. Trees here are 20 ft. high.

Pinus teocote Cham. & Schlecht.

Small plants of this Mexican pine are growing at Greendale, and will probably prove to be hardy.

Pinus tuberculata Gord.

This is a fast-growing pine in New Zealand, and also remarkable for retaining its cones on the tree unopened for many years, while occasionally cones may be seen almost entirely enveloped in the trunk of the tree. Trees here planted in 1880 have still all their cones on them, not a seed having been released, and cones twenty years old when broken open give forth seed that readily germinates. Introduced about 1869.

Pinus yunnanensis Franchet.

This species was discovered in western China by the Abbé Delavay, and first described in 1899. Through the kindness of Professor C. Sargent, of the Arnold Arboretum, seeds were sent to me in 1909, and a goodly number of plants were raised. These have been planted in many places in New Zealand, and everywhere are making good growth. Trees planted at Broadwood, Hokianga, are now bearing cones: This new pine is one of the most beautiful, if not the most beautiful, yet introduced, and will be very much admired for its rapid growth and fine form when better known.

(3.) TREES WHOSE LEAVES ARE USUALLY FIVE IN A SHEATH.

Pinus aristata Engelm.

This is one of the American alpine pines; it is of very distinct appearance while young. Trees here are 4 ft. high, and growing slowly but quite successfully.

Pinus Armandi Franchet.

This comes from western China. Seeds were received from Professor C. Sargent in 1909, and plants raised from them are now 3 ft. 6 in. high. It appears to be related to the Korean pine (*P. koraiensis* S. & Z.) in leaf and cones, two of which I received from Kew, though distinct, reminding one of that species.

Pinus cembra L.

This European mountain-pine grows slowly here. At Greendale the Siberian form grows much better than the Swiss form.

Pinus excelsa Wall.

This was introduced by Potts in 1868. Trees measured in 1885 were 27 ft. high. At Greendale *P. excelsa* grows faster than *P. strobus* or *P. Lambertiana*, and is the most promising of the pines having five leaves in a sheath.

Pinus flexilis James.

This was introduced about 1903 by the Hon. R. H. Rhodes. Two trees of this species given to me at the time are now each 8 ft. high.

Pinus koraiensis S. & Z.

Small plants of this pine introduced in 1906 are growing here, and appear quite hardy.

Pinus Lambertiana Dougl.

This comes from western America, and grows there to a larger size than any pine in any other part of the world. Some of the oldest and largest trees at Greendale have died, probably through drought. Trees planted here in 1881 are now 23 ft. high, but give no promise of becoming the giants they are in California.

Pinus Mastersiana Hayata.

Small trees of this rare species have been raised from seed sent from Formosa two years ago.

Pinus Montezumae Lamb.

For plants of this species I am indebted to Mr. Shaw, of the Arnold Arboretum, who spent two years in Mexico studying the pines of that country, and sent me seeds of this and other pines in 1911. Trees growing here are very distinct in appearance, and give promise of successful growth.

Pinus monticola Dougl.

Trees of this species growing at Greendale are 8 ft. high, and appear to be capable of growing successfully under local conditions.

Pinus peuce Grisebach.

This Macedonia pine was raised from seed sent to me by Dr. Augustine Henry, who in his turn received the seed from King Ferdinand of Bulgaria in 1908. The young plants look healthy and promising.

Pinus rudis Endl.

This is considered by Mr. Shaw a variety of *P. Montezumae*. Plants raised here from seeds sent by Mr. Shaw can be distinguished, while young, from that species by a different shade of green; otherwise in length of leaf and habit they are the same.

Pinus strobus L.

This was early introduced, as it was growing at Governor's Bay, Canterbury, in 1866, and was reported in 1868 from the Hutt, Wellington. *P. strobus* is considered a valuable forest-tree in the eastern States of America, but does not seem suitable for general planting here, although trees at Greendale have reached a height of 48 ft.

Pinus Torreyana Parry.

This is quite unlike any other five-leaved pine, the leaves being long and much stouter, reminding one of some forms of *P. Jeffreyi* or *P. Coulteri*, but the tree has a more open head. The species was introduced here in 1870, and has grown rapidly. The largest tree in the Christchurch Botanic Gardens is 90 ft. high, and there is a large one also in the public gardens, Timaru. These heights are much greater than the floras give, but the tree is unlike any other in its appearance, so that there can be no doubt as to its identity.

In addition to the pines cited above, plants have been raised here of pines under the following numbers—1370, 1378, 1390, and 1396—from seeds collected by Mr. E. H. Wilson in western China, and kindly sent to me by Professor Sargent, of the Arnold Arboretum.

Articles on the introduction and growth of pines will be found in the "Transactions of the New Zealand Institute," as follows:—Vol. 1 (1869): Here Mr. Ludlam mentions seventeen species as growing in his pinetum; no heights are given. Vol. 4 (1872), p. 368, in a paper read before the Wellington Institute, Mr. Pharazyn gives the height of seven species of pines growing in 1871. In vol. 12 (1880), p. 357, there is a list of trees given that were planted in 1866, and their height; the list contains eight species. In vol. 29 (1897) a list of twenty-six species growing at Taita, Hutt, is given, with their height in 1896.

In the "New Zealand Country Journal," vol. 3, p. 37, will be found a list of the pines growing at Ohinitahi, Governor's Bay. The list contains twenty-one species of pines, the date of planting, their rate of growth, and much other valuable information. In discussing some peculiarities of *Pinus tuberculata*, Potts incidentally mentions that his plants were raised from seeds imported by the New Zealand Government and distributed through the Geological Survey Department.

In an appendix to the Journals of the House of Representatives, 1877, an account is given of a number of plantations in New Zealand, mostly of eucalypts, but twelve species of pines are mentioned as having been observed growing in them.

ART. XXV.—*On an Exhibit of Acorns and Leaves of Oaks grown by the Author at Greendale, Canterbury, New Zealand.*

BY T. W. ADAMS.

[*Read before the Philosophical Institute of Canterbury, 7th July, 1915.*]

THE collection contained the acorns of thirty species, and the leaves of sixty-three species or varieties, all of which are grown by the author. The species were as follows:—

Quercus acuta Thunb.

The leaves are more suggestive of a laurel than an oak. It is quite hardy at Greendale. Planted, 1893. Height, 5 ft. A native of Japan.

Quercus aegilops L. (the Valonia Oak).

The peculiar cups are said to yield a greater percentage of tanning-matter than any other known plant. According to Mr. Maiden, this oak was introduced to Australia by Mr. George Cunnack, tanner, Castlemaine, Victoria. Under the name of "Valonia" considerable quantities are imported into New Zealand from Greece. Only small trees are known to me.

Quercus agrifolia Née (the Coast Live Oak).

This is a common tree on the low country of California, and is growing well in Canterbury. Trees at Greendale have borne acorns. At best, only a small tree. Planted, 1892. Height, 11 ft.

Quercus alba L. (the White Oak).

Approaches the nearest of any of the American oaks to the common English oak in the value of its timber, but the tree in cultivation has not been a success either here or in England. The autumn foliage is very fine, and distinct in colour. Planted, 1892. Height, 10 ft.

Quercus ambigua Kit.

This is by some botanists considered a hybrid oak. Plants here do not answer to Louden's description of this species or variety, so my specimens may be wrongly named.

Quercus aquatica Walt. (the Water Oak).

A tree at Greendale on ordinary soil, without the aid of excess of water, is making fair progress. Almost evergreen, with leaves of an unusual shape. A native of eastern North America.

Quercus Banisteri Michx. (the Bear Oak).

This is a remarkable little tree, with distinct foliage. It has flowered but not yet borne acorns at Greendale. Planted, 1893. Height, 11 ft.

Quercus bicolor Willd. (the Swamp White Oak).

This is said to be a large and valuable tree in America. Here the trees grow moderately well, with large leaves, downy on the underside. Planted, 1892. Height, 16 ft.

Quercus castaneaefolia C. A. Mey.

A native of Asia Minor. It is quite a promising tree to grow in New Zealand. Here it grows better than any of the chestnut-leaved trees from America. Planted, 1900. Height, 13 ft.

Quercus Cerris L.

The Turkey oak was early introduced, and there are large trees in Canterbury. In England the tree is said to grow faster than the native oak. It is a valuable tree, but here of slower growth than the common oak. Planted, 1880. Height, 46 ft.

Quercus chrysolepis Liebm. (the Maul Oak).

This is one of the most successful in growth here of the west American oaks. Trees growing at Greendale have for several years borne acorns. Although not a large tree, in California the wood is considered of superior quality.

Quercus cinerea Michx.

The upland willow oak was early introduced to New Zealand, and trees have borne acorns for many years in Canterbury.

Quercus coccinea Wangenh. (the Scarlet Oak).

This is a valuable tree, furnishing the " quercitrin," so highly valued as a yellow dye and a tanning-material. In recent years largely planted in New Zealand for its fine autumn foliage.

Quercus cuspidata Thunb.

This is a very desirable evergreen tree from Japan. It has flowered here several years, but not yet borne fruit. In Japan the acorns are considered good to eat. Planted, 1893.

Quercus dentata Thunb.

This has probably the finest foliage of any oak. Although not so large as those of *Q. macrocarpa*, the leaves resist the wind much better, and are retained on the tree until the spring growth commences.

Quercus dilatata Lindl.

A tree of this species in the Botanical Gardens, Christchurch, is now 57 ft. high and 6 ft. in girth. Trees raised at Greendale from acorns received from Darjeeling in 1904 are 10 ft. to 12 ft. high, and have successfully withstood the winters.

Quercus dumosa Nutt. (the Scrub Oak).

This has very small holly-like leaves, which are probably the smallest leaves of any oak. Trees only 8 ft. high have borne acorns at Greendale. A native of California.

Quercus esculus L. (the Italian Oak).

This is said to be a handsome tree ; only small plants are growing here. The " Index Kewensis " considers it synonymous with *Q. Cerris*.

Quercus falcata Michx.

An American tree, there called the " Spanish oak." The wood is not much valued, but the bark is reputed to be of great value for tanning. The tree is growing well at Greendale.

Quercus Garrayana Dougl. (the Oregon Oak).

This is a native of California. It is making very slow progress here.

Quercus glauca Thunb.

A beautiful evergreen tree from Japan, not hardy in England. It was damaged by frost at Greendale a few years ago.

Quercus Ilex L. (the Evergreen Oak).

This evergreen oak was early introduced, and is growing well throughout New Zealand; there are large trees at Riccarton, the Three Kings College, Auckland, and many other places.

8—Trans.

Quercus imbricaria Michx.

The shingle oak of the Americans is a deciduous tree with laurel-like leaves. It grows well at Greendale; it is a most desirable tree. Planted, 1901. Height, 19 ft. 6 in.

Quercus incana Roxb.

This is a very distinct and beautiful oak, but not perfectly hardy here. In severe winters the late growth is destroyed by frost. Trees introduced in 1905 are now 7 ft. high. It is a native of Himalayan India.

Quercus infectoria Oliv.

The gall-nut oak of the Mediterranean region. Hundreds of tons of these galls are imported into England annually for use in manufactures. There is much that is interesting about these galls on the oaks of Asia and Europe. Why should the punctures of the different species of insects each produce a different-shaped gall? Who can tell?

Quercus Kellogii Newb. (the California Black Oak).

This is the Californian representative of the American red oaks, and is growing well in Canterbury. Planted, 1896. Height, 22 ft.

Quercus laevigata Blume.

A distinct oak from Japan with the habit of a shrub. Plants here have produced flowers, but no acorns. Planted, 1895.

Quercus lamellosa Sm.

This evergreen oak from the Himalayas bears large acorns in very singular-shaped cups. Only small plants are growing here.

Quercus Libani Oliv.

The Lebanon oak has distinct and pretty leaves; it is quite hardy here, and trees have several times borne acorns. Planted, 1905. Height, 11 ft.

Quercus lineata Blume.

Small trees growing here have large evergreen leaves of a beautiful shape. A native of the Himalayas and Java.

Quercus lobata Née (the Valley Oak).

This is decribed as the most beautiful oak in California. Several trees here are making fair progress, and are now from 8 ft. to 10 ft. high.

Quercus Lucombeana Sw.

This is supposed to be a hybrid oak raised in England. Two small trees raised here from imported acorns are uniform in appearance, and are growing well.

Quercus macrocarpa Michx. (the Burr Oak).

This has the largest leaves of any oak growing at Greendale; the acorns are also large, and enclosed in mossy cups. Planted, 1903. Height, 13 ft. A native of eastern North America.

Quercus macrophylla Née.

This has fine leaves, but not so large as those of *Q. macrocarpa*. The tree, too, does not grow at all well here. Planted, 1903. Height, 7 ft. The "Index Kewensis" considers the species synonymous with *Q. magnoliae-folia* Née. It is a native of Mexico.

Quercus Merbeckii Durhamel.

This, a native of Algeria, is a fine oak. It grows well in New Zealand, and is almost evergreen. Large trees are growing in the Hutt Valley, Wellington.

Quercus mexicana Humb. & Bonpl.

A beautiful evergreen tree, which grows rapidly in Canterbury.

Quercus Michauxii Nutt. (the Basket Oak).

Considered by some to be only a variety of *Q. bicolor*, but it is held distinct in the "Index Kewensis." Small trees at Greendale are quite distinct.

Quercus Mühlenbergii Englm. (the Yellow Chestnut Oak).

One of the American chestnut-leaved oaks, which make a fine autumn display. The trees grow satisfactorily here. The "Index Kewensis" considers it synonymous with *Q. prinus* L.

Quercus nigra L. (Black Jack).

A small tree, almost evergreen, with leaves of a distinct and unusual shape.

Quercus obtusiloba Michx. (the Post Oak).

A small plant of this oak promises well. A native of eastern North America.

Quercus palustrus Du Roi (the Pin Oak).

The wood of this oak is not considered of much value, but the autumn foliage is very fine. It has been planted in considerable numbers in New Zealand. A native of eastern North America.

Quercus paniculata and Q. pulverulenta Hort. ex C. Koch.

These, which are growing here from imported acorns, are probably only varieties of *Q. robur*.

Quercus phellos L. (the Willow Oak).

This is one of the most distinct of all the oaks, with narrow willow-like leaves and twiggy branches. A tree here is healthy, and promises to make a good tree. Planted, 1900. Height, 10 ft. It is a native of eastern North America.

Quercus phillyraeoides Gray.

A native of Japan. It is a very distinct and elegant shrub; when not in fruit most difficult to recognize as an oak.

8*

Quercus prinus L. (the Chestnut Oak).

This requires some shelter ; it then grows well in New Zealand. In America this is a valuable tree. Planted, 1895. Height, 16 ft.

Quercus purpurea Lodd. var. elegans.

This tree has very large leaves of the shape of *Q. robur*, and in autumn makes a grand display. A hybrid oak. Planted, 1904. Height, 13 ft.

Quercus robur L. var. pedunculata (the Common Oak).

This grows well in New Zealand, and already some fine specimens may be seen. A tree planted here in 1869 is now 59 ft. high, with a spread of branches equalling the height, and the bole at 4 ft. from the ground measures 76 in. in girth.

Quercus rubra L. (the Red Oak).

This grows freely in New Zealand, and is being planted in considerable numbers on account of its fine autumn tints. Planted, 1895. Height, 21 ft. It is a native of eastern North America.

Quercus serrata Thunb.

. This was at first introduced as *Q. dentata*, an oak it very little resembles. *Q. serrata* grows well in New Zealand. Planted, 1895. Height, 16 ft.

Quercus sideroxyla Humb. & Bonpl.

This is a native of Mexico, and is a distinct oak both in leaf and fruit. A small tree is growing here.

Quercus suber L. (the Cork Oak).

This is growing successfully in New Zealand, particularly near the sea, and was early introduced. Planted at Greendale, 1895. Height, 12 ft.

Quercus tinctoria Bartr. (the Black Oak).

One of the black oaks of America. A small tree growing here has large leaves, which fade to a dark red.

Quercus toza Gillet.

An oak growing here under this name is probably *Q. Merbeckii* Durhamel. Planted, 1895. Height, 24 ft.

Quercus variabilis Blume.

This possesses leaves of the same shape as *Q. serrata*, but distinguished from it by the dense down on the underside of the leaf. Planted, 1910.

Quercus vibrayana Franchet.

This grows naturally on the mountains of Formosa, and is said to be there a valuable tree. Small plants of this oak are growing at Greendale.

Quercus virens Ait. (the Live Oak).

A hardy evergreen oak from America, with very dark acorns. It grows well in New Zealand. Planted, 1894. Height, 22 ft.

Quercus Wislizeni A. DC.

A native of the mountains of California. Trees here are bearing acorns of a long narrow shape. It is quite hardy. Planted, 1900. Height, 17 ft.

Art. XXVI.—*The Norfolk Island Species of* Pteris.*

By R. M. Laing, M.A., B.Sc.

[Read before the Philosophical Institute of Canterbury, 7th July, 1915.]

Since drawing up the list of ferns for my paper on the flora of Norfolk Island (Trans. N.Z. Inst., vol. 47, p. 1) I have received a large number of additional specimens from my father, Mr. W. Laing, resident on the island. These necessitate a reconsideration of the species of the genus *Pteris*. This revision need not include the common *Pteridium esculentum* (Forst. f.) Cockayne and *P. comans*, as to whose occurrence on the island all are agreed. The following table shows roughly how Endlicher's species have been viewed in recent literature :—

Maiden.	Endlicher.	Christensen.
? *P. tremula*	= *P. Baueriana*..	= Doubtful species.
? *P. incisa*	= *P. Brunoniana*	= *P. incisa.*
? *P. quadriaurita* ..	= *P. Zahlbruckneriana* ..	= *P. comans.*
P. tremula var. *Kingiana* ..	= *P. Kingiana* ..	= *P. tremula.*
? *P. quadriaurita* ..	= *P. Trattinickiana* ..	= Doubtful species.

Now, it is probably impossible without direct reference to the type specimens at Vienna to determine with certainty the identity of Endlicher's plants; but I think that it can be done more accurately than has yet been done, if fairly full material is available.

The first important point to consider is the question of the venation, whether forking or anastomosing; and an examination of Endlicher's descriptions gives the following results: Veins forked — *P. Baueriana, P. Kingiana, P. Trattinickiana;* veins anastomosing — *P. Brunoniana, P. comans, P. Zahlbruckneriana.* It is true that no mention is made in the case of *P. Baueriana* as to whether the veins are free or anastomose; but as the plant is said by Endlicher to be very near *P. tremula*, in which the veins are free, we may consider both the same in this respect.

A. Species with Forked Venation.

(*a.*) Pteris tremula R. Br. = *Pteris Baueriana* Dies., Endl., No. 37.

Let us consider the forms with forked venation first. I have three of these which agree well with Endlicher's descriptions of *P. Baueriana, P. Kingiana,* and *P. Trattinickiana.* I have little doubt that they are distinct species. Endlicher's diagnosis scarcely serves to separate *P. Baueriana* from *P. Kingiana*, but his detailed description enables this to be done with a fair amount of accuracy, and there seems to be but little doubt that *P. Baueriana* Dies. = *P. tremula* R. Br. In this I follow Maiden, though Endlicher himself admits that *P. Baueriana* is " *Pteridi tremulae* R. Br. proxime affinis." Hooker and Baker do not mention this species.

A comparison of specimens from Norfolk Island of this plant with the New Zealand *Pteris tremula* shows that the two forms are almost indistinguishable.

---

* Specimens of these and other Norfolk Island plants collected by me will be found in the Canterbury Museum.

(*b*.) Pteris **Kingiana** Endl., No. 40.

" *Pteris Kingiana*, frondis coriacea 3-partitae ramis pinnatis pinnis pinnatifidis glaberrimis, laciniis lineari-subfalcatis acutiusculis [integerrimis."

There is not very much in this diagnosis to separate the [plant from *P. Baueriana*. *P. Baueriana* is said to be bipinnate; but both forms

Fig. 1.—*Pteris Kingiana* Endl.
A fertile pinna (⅔ natural size), with an enlarged fertile pinnule, *a*.

I have are bipinnate only at the base, and tripartite generally; and the distinction in this case seems to me to be one that is unimportant, depending a good deal upon the luxuriance of growth. However, the two species, if I understand them rightly, are easily separated by the shape of the frond and of the pinnules, and by the consistency of the frond, colour of the stipes, &c. They are quite distinct in appearance, and not likely to be confused. These points are mostly noted in the detailed description of Endlicher, and I have little hesitation in stating that the species *P. Kingiana* will have to be revived. The following is the fuller description of the species as given by Endlicher : "Filix subbipedalis, stipite pennae columbinae crassitie rubro-fusco, glaberrimo, nitido, antice profunde sulcato. Frons coriacea glaberrima, 3-partita. Rami pinnati ; pinnis pinnatifidis ; laterales in specimine observato 9-pollices longi, erecto-patentes, utrinque pinnis 3–4 onusti ; pinnae approximatim alternae, sub-sexpollicares, inferior supremi et infimi paris dimidio fere brevior v. plane nana. Rami medii pinnae alternatim collaterales, utrinque 5–6 erecto-patentes, omnes usque ad costam pinnatifidae. Laciniae coriaceae lineari-subfalcatae acutiusculae integerrimae glaberrimae suboppositae, rami medii pollicares, 2½ lineam latae, lateralium 7-lineas longae, latitudine 3-lineari, obtusiusculae, omnes basi deorsum dilatatae sinu acuto disjunctae ; terminalis elongata. Nervus laciniarum prominulus, venulas alternas, prope basim late 2-furcas exserens. Indusia membranacea, ½ lineam lata, paullo supra basim laciniae axorta, infra apicem desinentia."

This is readily distinguished from *P. tremula* by its much smaller size, much more broadly deltoid frond, much more coriaceous texture, smaller number of (3–4) pinnae on each side. The pinnules also are quite distinct from those of *P. tremula*, being much more falcate in outline, less rounded at the tips, and increasing in breadth towards the base. Hooker and Baker (Syn. Fil., p. 161) refer to *P. Kingiana* as a variety (B) of *P. tremula* "with the ultimate segments larger, sometimes 1½ in. long, nearly ¼ in. broad, without being toothed" ; and state it "was originally published from Norfolk Island, but some New Zealand specimens agree with it."

I do not understand this except on the assumption that the form they describe is a form of *P. tremula*, as it might well be, and not of *P. Kingiana*. I have seen nothing from New Zealand at all matching *P. Kingiana*.

(c.) Pteris biaurita L. var. quadriaurita = ? *Pteris Trattinickiana* Endl., No. 42.

This appears to me to be a form coming very close to the preceding. Maiden considers it as possibly the same as the widespread subtropical *P. quadriaurita*, which it undoubtedly closely resembles. Endlicher's diagnosis is as follows : "*P. Trattinickiana*, frondis membranacea 3-partitae ramis pinnatis, pinnis pinnatifidis, laciniis oblongo-linearibus obtusis discretis, argute serrulatis, venulis furcatis, soris interruptis."

In *P. Kingiana* the non-fertile portions of the frond are either entire or only slightly serrulate at the tips ; in this form they are regularly serrulate throughout. In *P. Kingiana* the pinnae are exceptionally more than 4 on each side ; in this they are usually 5 or 6. The pinnules in *P. Kingiana* are subfalcate ; here they are oblong. The frond is also much more membranous than in *P. Kingiana*. It will thus be seen that it is intermediate between *P. Kingiana* and *P. tremula*, but in general outline and character of frond and pinnules it approaches much more nearly to the former than to the latter.

A note is necessary regarding the indusia. They are thus described by Endlicher : "Indusia membranacea, interrupta, saepius unilateralia." In this form and the preceding, very frequently towards the apex of the pinna the sorus is formed only on one side of the pinnule, and is much

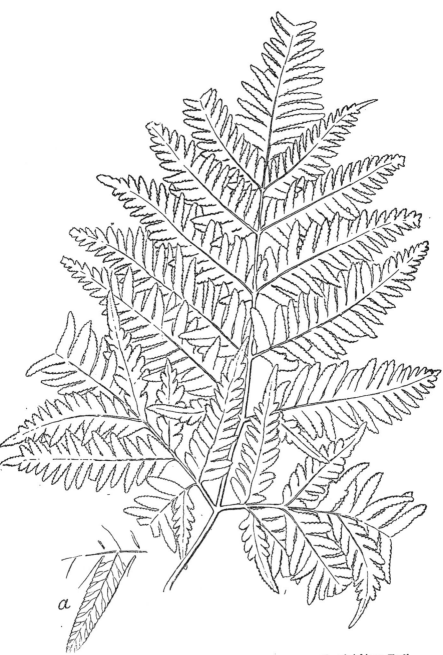

FIG. 2.—*Pteris biaurita* L. var. *quadriaurita* Retz. = *? Pteris Trattinickiana* Endl.
A portion of a pinna (½ natural size), with a magnified sterile pinnule, *a*.

abbreviated. I have seen no specimens, however, in which it could be said to be interrupted; and this makes the identification of my plant with *P. Trattinickiana* somewhat doubtful.

I have seen only one authentic specimen of *P. quadriaurita*, and that a South African one. A comparison shows that the form from Cape Colony has longer, narrower, and more acuminate pinnules and much more interrupted sori than the one under consideration. I am usually averse

Fig. 3.—*Pteris comans* Forst. f. = ? *Pteris Zahlbruckneriana* Endl.
A pinna of the Norfolk Island form (⅓ natural size), with a magnified fertile pinnule, *a*.

to identifying two species which differ somewhat and come from different areas, but think that as *P. quadriaurita* is the name of a well-known and widely distributed form it had better be retained here. My specimens show no fertile fronds.

B. SPECIES WITH NETTED VENATION.

(*d*.) Pteris comans Forst. f. = ? *Pteris Zahlbruckneriana* Endl., No. 41.

Excluding *Pteris comans*, I have only two forms from Norfolk Island coming under this head. One of these agrees well with the description of

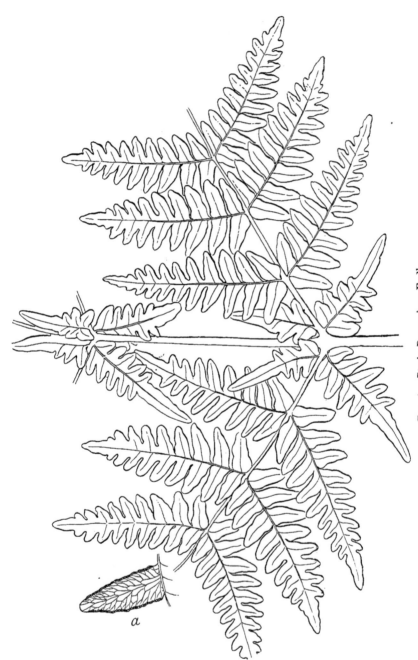

Fig. 4.—*Pteris Brunoniana* Endl.

A portion of the frond (⅓ natural size), with a magnified fertile pinnule, *a,*

P. Brunoniana; and the other is *P.* (*Histiopteris*) *incisa,* which was apparently not collected by Bauer, but was first recorded by Müller from specimens collected by Carne. Thus I have no form corresponding to *P. Zahlbruckneriana* Endl. Maiden considers this as perhaps the same as *P. quadriaurita* and *P. Trattinickiana;* but as these are both forked-veined this cannot be so. Christensen regards it as a synonym of *P. comans,* and there is certainly little in Endlicher's description to separate it from *P. comans.* Hooker and Baker do not record it. I have no opinion of my own to offer on the matter, and think that it may well be left as a doubtful synonym of *P. comans.* *P. comans* of the island is a much more luxuriant form than that of New Zealand, and much more membranous in texture, but scarcely otherwise different.

(e.) Pteris Brunoniana Endl., No. 38.

I quote Endlicher's description of *P. Brunoniana:* "*Pteris Brunoniana* frondibus 3-pinnatis membranaceis glaberrimis subtus glaucescentibus; foliolis alternis oppositisque, pinnulis oppositis sessilibus lanceolatis obtusi-usculis, infimis basi repandis, superioribus confluentibus, terminali elongato, venulis anastomosantibus. *Pteridi Vespertilionis* affinis. Pinnas video pedales sesquipedalesque, utrinque foliolis sessilibus subdenis duodenisve oppositis v. in altero specimine alternis, patentissimis 4–6 pollicaribus infimis sterilibus summisque brevioribus onustas. Pinnulae sessiles oppositae, patientiusculae, 10–14 lin. longae, semi pollicem latae, basi inter se coadunatae dilatatae, sub-repandae, superiores breviores confluentes, terminalis elongata angustata, 2-pollicaris, latitudine 3-lineari, basi obsolete repanda, apice integerrima acutiuscula, omnes utrinque glaberrimae, supra obscure virides, subtus glaucescentes. Venulae anastomosantes, sori continui v. rarius interrupti; indusia angustissima, demum patientiuscula. Rhachis communis partialesque glaberrimae."

It seems to me that *P. Brunoniana* is quite distinct from *P. incisa.* with which Christensen regards it as synonymous, and Maiden doubtfully so. The species *P. Brunoniana* Endl. will therefore have to be re-established.

The following show some of the points of distinction between the Norfolk Island forms of these two species:—

Pteris incisa.	*Pteris Brunoniana.*
1. Pinnules rounded, lanceolate to oblong or ovate.	1. Pinnules deltoid.
2. Sori much interrupted.	2. Sori usually continuous from base of sinus to a short distance from the apex.
3. Sinus (in Norfolk Island forms) often nearly closed, adjacent segments sometimes overlapping.	3. Sinus open and spreading.
4. Indusium poorly developed.	4. Indusium well developed.

Undoubtedly on Norfolk Island the two species are distinct, though it is possible that elsewhere there may be intermediates between *P. incisa* and *P. Brunoniana.* I have left this fern under the genus *Histiopteris,* though further examination may show that it should be included under *Pteris.*

(*f.*) Histiopteris incisa (Thbg.) J. Sm.

A well-marked form of this fern is fairly common on Norfolk Island. It is much more luxuriant and stouter than the New Zealand plant, but otherwise scarcely different.

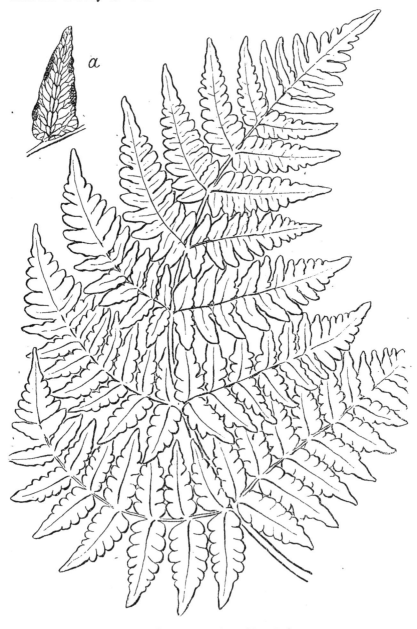

FIG. 5.—*Histiopteris incisa* (Thbg.) J. Sm.
A pinna (½ natural size), with a magnified fertile pinnule, *a*.

CONCLUSION.

The results of this paper may thus be summed up in a list showing the chief forms of *Pteris* on the island :—

(1.) *Pteridium esculentum* (Forst. f.) Cockayne = *Pteris esculenta* Forst. f.

(2.) *Pteris tremula* R. Br. = *P. Baueriana* Dies.

(3.) *Pteris Kingiana* Endl. (*vide* fig. 1).

(4.) *Pteris biaurita* L. var. *quadriaurita* Retz. = ? *P. Trattinickiana* Endl. (*vide* fig. 2).

(5.) *Pteris comans* Forst. f. = ? *P. Zahlbruckneriana* Endl. (*vide* fig. 3).

(6.) *Pteris Brunoniana* Endl. (*vide* fig. 4).

(7.) *Histiopteris* (*Pteris*) *incisa* (Thbg.) J. Sm. (*vide* fig. 5).

ART. XXVII.—*Some Further Additions to the Flora of the Mongonui County.**

By H. CARSE.

Communicated by T. F. Cheeseman, F.L.S., F.Z.S.

[*Read before the Auckland Institute, 8th December, 1915.*]

1. Hypericum gramineum Forst.

On rather bare slope of moorland between Kaitaia and Awanui. Rare in the north.

2. Melicytus micranthus Hook. f.

Both var. *longiusculus* Cheesem. and var. *microphyllus* Cheesem. occur freely.

3. Geranium pilosum Forst.

Plentiful throughout the district.

4. Leptospermum scoparium Forst. (two forms, as under).

A very curious " sport " of this plant occurs in a maritime morass near Ahipara. It is a low, densely branched shrub under 1 ft. in height. The branches are little more than 1 mm. in diameter; many less than that. The leaves are 2–4 mm. in length. I have not seen flowers or fruit.

A form of *Leptospermum scoparium* is not uncommon in which, in place of 5 petals and a 5-valved capsule, as is usual, many of the flowers have 6–10 petals and the same number of valves. I have not, however, seen any with more than 5 sepals.

5. Crantzia lineata Nutt.

A form of this plant with unusually large leaves occurs in fresh water in several places in the North Cape Peninsula. Leaves 20–35 cm. long, 5 mm. wide. Usually this plant occurs in damp sand. The present form is found in fairly deep running water.

6. Coprosma spathulata A. Cunn. (form with red drupes).

In the " Manual of the New Zealand Flora " the drupe of this species is described as " black, very rarely red." In woods in this district, where the

---

* This is a continuation of my former papers : see Trans. N.Z. Inst., vol. 47 (1915), pp. 76–93 ; vol. 45 (1913), pp. 276–77 ; and vol. 43 (1911), pp. 194–224.

species is very plentiful, not only is the drupe almost as often red as black, but it is frequently of a deep-orange colour. I have watched these plants for a good many years, and am convinced that the red and orange drupes are permanent, not changing to black when dead-ripe.

7. Coprosma tenuicaulis Hook. f. var. major Cheesem.

This variety, which I originally discovered in the Manukau County, occurs sparingly in damp lowland woods.

8. Coprosma rhamnoides A. Cunn.

Both varieties *vera* Cheesem. and *divaricata* Cheesem. are abundant. Intermediate forms also occur which are difficult of discrimination.

9· Coprosma sp. ined.

Mr. H. B. Matthews has in his garden a *Coprosma* which differs from any yet described. It was found along with *C. Baueri* Endl. growing on a rock on the west coast. It forms a dense spreading shrub, closely appressed to the ground. Male plant only seen.

10. Dracophyllum latifolium A. Cunn. var. Matthewsii Carse var. nov.

Frutex habitu *D. latifolia*, sed in omnibus partibus minor. Rami graciliores, non verticillati. Folia similes sed minora, 10–30 cm. longa, 12–25 mm. lata. Panicula 10–28 cm. longa, semper pendula. Flores minores, minus densae, rubro-purpureae vel nigro-purpureae. Capsula 2 mm. diam.

Hab.—Dry ridges in hilly forests in the county.

This neinei, which was discovered by Mr. H. B. Matthews, whose name I have much pleasure in associating with it, bears a general resemblance to *D. latifolium*, but is in all respects smaller. It occurs as a shrub or small tree, rarely more than 15 ft. in height. The branches, which in the type are usually whorled and more or less dichotomously divided, issue irregularly from the trunk; they are much more slender, and, in comparison, longer. At the extremities where the tufts of leaves occur they are only $\frac{1}{8}$–$\frac{1}{4}$ in. in diameter.

The leaves are 4–12 in. long and $\frac{1}{2}$–1 in. wide at the broadest part.

The panicles are 4–12 in. long, always drooping. The rhachis and its branches are yellow-green in colour, as also are the sepals. The petals are purplish-red or almost black; the anthers white, tinged with pink.

These marked differences, as well as the fact that its flowering period is from September to October, distinguish this plant from the type.

11. Dracophyllum Sinclairii Cheesem.

On ancient landslide, Tauroa, H. B. Matthews.

This is a somewhat rare plant in this district. What may prove to be a variety of it occurs in a kauri forest near Peria. It is a shrub or small tree, 12–30 ft. high, with much narrower leaves.

12. Dracophyllum Urvilleanum A. Rich. var. filifolium Cheesem.

On open ridges in forests and on moorlands, Kaiaka, Kaitaia, Peria, Tauroa.

13. Chenopodium triandrum Forst.

On rocks near the sea, not common.

14. Atriplex Billardieri Hook. f.

On sandy shores, Rangaunu Harbour ; rare.

15. Muehlenbeckia complexa Meissn. var. grandifolia Carse var. nov.

Frutex ramosissimus scandens, habitu *M. australis* Meissn. Caudex 12–25 mm. diam. Rami teretes, ultimati pubescentes. Folia 25–75 mm. longa, oblongo-ovata, acuminata, ad basim cordata vel truncata, coriacea, glaucosa infra. Flores in spicas 50–75 mm. longas. Perianthus in fructu non succulens.

At first sight this well-marked plant might be mistaken for *M. australis*, but an examination of the ultimate branches shows that they are terete and densely pubescent, which at once settles the point. The leaves, too, are much more coriaceous than those of *M. australis*. In short, save in size and habit, it bears no resemblance to that plant.

This variety is confined to damp alluvial situations, and though, no doubt, connected by intermediates with the numerous and varied forms which this species assumes, is worthy of varietal distinction.

16. Phyllocladus glaucus Carr.

A small grove of this very handsome taxad occurs near Peria, and scattered trees are to be found within a few miles.

17. Earina sp. nov.

This new orchid will be described shortly by Mr. Cheeseman. It occurs in several places in the district, but is by no means common, and elsewhere has been found sporadically as far south as Wellington.

18. Pterostylis barbata Lindl.

In open moorland, Peria. H. B. Matthews! Very rare. Previously reported from Kaitaia.

19. Pterostylis trullifolia Hook f. var. gracilis Cheesem.

Not uncommon in open woods and moorlands.

20. Calochilus paludosus R. Br.

Moorland, Kaimaumau. H. B. Matthews!
This orchid, which is very rare here, was also reported from Kaitaia.

21. Astelia sp.

A small *Astelia* with leaves 3–12 in. long, $\frac{1}{4}$–$\frac{1}{2}$ in. wide, is found in flower occasionally. The scape is 2–5 in. long, flowers solitary in the angles of the bracts. As I have not yet seen ripe fruit of this, I defer describing it for the present. I have gathered it also in the Whangarei district.

22. Juncus tenuis Willd. var. secundus Engelm.

A troublesome weed, much more plentiful than the type.

23. Juncus plebeius R. Br.

Has been known from Kaitaia for many years; also occurs near Ahipara.

24. Mariscus ustulatus (A. Rich.) C. B. Clarke forma grandispiculosus Kükenth. in litt.

This is a form in which the spikelets are much longer than usual, and they are so set as to give the spike a fan-like appearance.

25. " Scirpus Carsei " Kükenth. sp. nov. (In litt. Nov., 1913.)

" Proxima *S. pauciflorus* Lightf. utriculis plano-convexis marginibus non incrassatis evidenter rostratis distincte."

Growing on muddy margin of channel leading to Lake Tangonge, Kaitaia. H. B. Matthews !

This is a small leafy *Scirpus*, a few inches high. It got, almost by accident, among a lot of sedges I was sending to Oberpfarrer Kükenthal some years ago, and was described from one small specimen. My own opinion, in which I have the support of Mr. Cheeseman's opinion, is that it is merely a depauperated state of *Scirpus lenticularis* Poir., which occurs plentifully in the adjacent lake.

26. Schoenus Carsei Cheeseman.

In morass, Wharekia, Rangaunu Harbour. Not previously reported from north of Whangarei district, where I originally discovered the plant.

27. Cladium glomeratum R. Br. forma major Kükenth. form. nov.

" Foliis crassioribus, inflorescentia longiore, habitu elatiore." (Kükenthal in litt., 1913.)

A much larger plant than the usual form. Leaves more numerous, 5–7 ft. long, compressed below ; stems 4–6 ft. high, much less numerous than in the type ; panicle 6–16 in. long. Nutlets rather large.

In wet places, Kaitaia and Tauroa ; not common.

28. Cladium Huttonii T. Kirk.

In swamps, usually near the sea ; plentiful.

29. Cladium junceum R. Br. var. elatior Carse var. nov.

Culmus multo longior, 1–1·5 m., panicula ramosa 5–10 cm. longa.

A very tall slender form with culms 3 ft. to 4 ft. 6 in. long. The panicles are more branched than usually, and are 2–4 in. long.

Not uncommon in shaded woods, Tauroa.

30. Lepidospermum filiforme Labill.

I first noted this handsome sedge in New Zealand from the Peria Gum Hills. It also occurs sparingly near Kaitaia, and plentifully on moorlands near the Rangaunu Harbour.

31. Uncinia uncinata (L. f.) Kükenth. var. laxior var. nov.

A more robust plant than the type. Culms far overtopping the leaves, 2–3 ft. long. Spikes slender, 6 in. long. Lax throughout.

Near to Kükenthal's *U. pedicellata*, but a much larger plant in all points.

From the North Cape to Ahipara ; rare.

32. Uncinia riparia R. Br. var. affinis Kükenth.

In woods, Tauroa and Kaiaka ; not common.

One form of this, identified by Kükenthal, appears to me almost identical with type specimens from the South Island ; this grows in woods near the sea. The Kaiaka form, also identified by Kükenthal, is barely, if at all, distinguishable from the very plentiful var. *Banksii* C. B. Clarke; perhaps the leaves are slightly broader.

33. Carex testacea Sol.

In a small wood at Tauroa I gathered a form of this with culms 7–11 ft.!! in length—much longer than I have ever seen before.

34. Carex lucida Boott.

A form elongating in fruit to more than 8 ft. is not uncommon on the coast near Ahipara.

35. Carex comans Berggr. forma subsessilis Kükenth. form. nov.

" Spiculis inferioribus breviter-pedunculatis." (Kükenthal in litt., 1913.) Ahipara ; not common.

36. Carex pumila Thunb. var. macrocarpa Carse var. nov.

C. typo similis sed in omnibus partibus major. Foliae 5–8 dm. longae. Culmus 3–4 dm. altus. Spiculae 25–35 mm. longae. Utriculus longior et latior.

A much larger plant than the type, with culms 8–15 in. long; leaves equalling typical form in breadth, but much longer, 20–30 in. The spikelets are 1–1½ in. long, and broader in proportion. The utricles are slightly longer and a good deal broader.

In damp hollows, among sand-dunes, Tauroa (Reef Point).

37. Danthonia pilosa R. Br. var. racemosa Buch.

On hilly slopes ; common.

38. Poa anceps Forst. var. gracilis Cheesem.

On creek-banks, Ahipara.

39. Agropyron multiflorum T. Kirk var. longisetum Hack.

On the coast ; common.

40. Asperella gracilis T. Kirk.

Near Kaitaia ; rare. H. B. Matthews !

41. Hymenophyllum Cheesemanii Bak.

In hilly forests ; not uncommon.

42. Trichomanes reniforme Forst.

I have lately found a curious form of this fern, with the frond distinctly lobed ; in some cases it is clearly bifurcated.

43. Pellaea falcata Fée.

Near Kaitaia ; rare. H. B. Matthews !

44. Asplenium lucidum Forst. var. Lyallii Hook. f.

On rocky slope, Kaiaka ; rare.

Naturalized Plants.

Ranunculus arvensis Linn.

Kaiaka ; not common.

Ranunculus Flammula Linn.

In wet kahikatea forest, Kaitaia.

This rare plant was discovered by Mr. H. B. Matthews. Previously recorded only from the Waiharakeke Stream, Piako (Cheeseman, T. F., "Contributions to a Fuller Knowledge of the Flora of New Zealand: No. 1," Trans. N.Z. Inst., vol. 39, p. 450, 1907).

Berberis sp.

Spread from a hedge in Kaiaka.

Fumaria muralis Sond.

A troublesome weed in gardens.

Brassica nigra Koch.

Fields and waste places ; common.

Lepidium ruderale Linn.

Kaitaia, Pukepoto ; not common.

Polygala virgata Thunb.

Mangatete, Pukepoto ; rare.

Polygala myrtifolia Linn.

Sand-dunes, west coast ; rare.

Tillaea trichotoma Walp.

Near Kaitaia. H. B. Matthews!

Lythrum Groefferi Tenore.

Not common.

Carum petroselinum Benth. and Hook. f.

Otukai ; rare.

Valerianella olitoria Pollich.

Not common.

*Aster subulata Michx.

Spreading rapidly in all soils and situations.

Helenium quadridentatum Labill.

Chiefly on coast : common.

Carduus pycnocephalus Linn.

Otukai ; rare.

---

* Now first recorded as occurring in New Zealand.

·*Erechtites valerianaefolia DC.

Otukai ; not common.·

*Erechtites Atkinsoniae F. v. M.

Spreading : a troublesome weed.

Anagallis arvensis Linn. var. coerulea Lamk.

Plentiful.

Solanum auriculatum Ait.

Near Kaitaia. H. B. Matthews !

Verbascum Thapsus Linn.

Victoria Valley ; spreading.

Chenopodium urbicum Linn.

In waste places ; not common.

Tradescantia fluminensis Vell.

Creek-banks and lowland woods ; spreading rapidly.

Panicum Linderheimeri.

Near Kaitaia ; rare. H. B. Matthews!

Polypogon monspeliensis Desf.

Common in wet places.

Holcus mollis Linn.

Not uncommon.

Glyceria aquatica Wahlb.

Sown many years ago in Fairburn. Has spread in swamps. I have
seen no ripe seed, but the plant spreads by its creeping rhizome.

Poa nemoralis Linn.

Not uncommon ; in shaded spots.

Festuca ovina Linn.

Not uncommon.

Lolium perenne Linn.

As a rule, this grass is not very enduring, being subject to "rust," a
fungoid disease which weakens the growth or even kills the plant ; but near
Kaitaia is a large area of river-flat, sown with this grass more than sixty
years ago, where it has held its place without deterioration.

Lolium italicum A. Br.

Plentiful.

---

* Now first recorded as occurring in New Zealand.

Art. XXVIII.—*Notes on the Plant-covering of Pukeokaoka, Stewart Island.*

By D. L. Poppelwell.

[*Read before the Otago Institute, 7th September, 1915.*]

Pukeokaoka is one of the mutton-bird islands situated in Foveaux Strait. It lies about ten miles from Half-moon Bay, and is passed on the trip from Bluff to Stewart Island. It is roughly triangular in shape, and its greatest diameter is only about half a mile. It lies between Motunui and Herekopere, and is less than half a mile from the former and about one mile from the latter.

On the 1st January, 1915, I spent some hours on the island, accompanied by my son, noting its vegetation. As the plant-covering differs considerably from that of Herekopere, a short description of it may prove of interest.

The name Pukeokaoka was, according to Mr. J. Bragg, given to this scrap of land by the Maoris because of the abundance on the island of the tree-nettle *Urtica ferox.* This plant is called " ongaonga " in the North, but the name is hardened to " okaoka " in the South. " Pukeokaoka " should therefore be translated " nettle hill," a reasonable enough designation for this island. The sides of the island are rocky, and consist of steep cliffs in several places. These are covered with vegetation. On the east side there is a rough accumulation of large boulders, and it was on these that we landed.

Approaching the shore, the physiognomy shows a dull grey-green colour, with a smooth surface, although here and there can be seen a green patch. This dull colour is brought about by the abundance of *Olearia angustifolia* which fringes the water, and must in its season be a mass of white from its beautiful blossoms. The greener patches consist of *Veronica elliptica,* which is also fairly abundant on the cliff-side. Where the cliffs are steep, or a slip has occurred, great curtains of *Tetragonia trigyna* mantle the surface, the reddish stems, and, where exposed to strong light, the red leaves, contrasting strongly with the surrounding plants.

Senecio rotundifolius is common in this association, and on a closer view *Stilbocarpa Lyallii* peeps through in patches, its large leaves giving a striking characteristic to the physiognomic appearance.

The usual shore fern *Asplenium obtusatum* is also common among the undergrowth, and *Apium prostratum* is not infrequent. On the steeper cliffs *Mesembryanthemum australe* is common, its reddish stems and pink flowers contrasting beautifully with its dark-green succulent leaves.

On a more open part of the cliff-side the association consisted of *Poa Astoni, Linum monogynum, Gnaphalium luteo-album,* and *Sonchus littoralis,* all growing very rankly, chiefly on account of the nesting habits of the various petrels which frequent the island. The petrel-burrows serve to drain and aerate the soil, while their droppings enrich it very much.

As the hill was ascended *Senecio lautus* became common, and a few plants of *Scirpus nodosus* (?) were noted, together with a mixed association consisting of *Coprosma lucida, Rapanea Urvillei, Pittosporum Colensoi, Erechtites prenanthoides, Sonchus littoralis, Muehlenbeckia australis, Hierochloe redolens, Griselinia littoralis,* and *Polypodium diversifolium.* At first these plants were very stunted, showing the effect of the strong wind, but as the top of the hill was reached they became taller, and a forest association commenced.

Among the scrub for considerable areas the floor-covering consisted of *Polypodium diversifolium* as a pure association. On the edge of the forest the scrub was about 12 ft. tall, and consisted chiefly of *Rapanea Urvillei, Coprosma lucida, C. areolata,* together with the ferns *Asplenium lucidum* and *Polystichum vestitum,* and the climbing-plants *Rhipogonum scandens* and *Rubus australis.* The floor of the forest was covered for considerable areas with *Polystichum vestitum, Asplenium lucidum,* and *Stilbocarpa Lyallii,* and occasional patches of *Blechnum durum* from 8 ft. to 10 ft. in diameter were seen.

The forest consisted of *Metrosideros lucida, Schefflera digitata, Pittosporum Colensoi, Aristotelia racemosa, Fuchsia excorticata, Melicytus lanceolatus, Nothopanax Edgerleyi, Griselinia littoralis, Carpodetus serratus,* and *Veronica elliptica.* The undergrowth was 3 ft. or 4 ft. high, and consisted of *Asplenium bulbiferum, A. flaccidum, A. scleroprium, A. lucidum, Polystichum vestitum,* and *Polypodium diversifolium,* while *Rhipogonum scandens* was common.

The interior of this forest presents a tangled mass of *Rhipogonum,* through which can be seen the crooked trunks of the forest-trees, grey with lichens and mosses. *Metrosideros lucida* grows about 30 ft. high, and *Coprosma areolata* here forms a straight tree about 20 ft. high, with a trunk 6 in. in diameter. *Urtica ferox* was also growing rankly in this forest. It was about 6 ft. tall, and had leaves 9 in. or 10 in. long, including the petiole of 2 in. to 3 in. The only tree-fern was *Dicksonia squarrosa.* Epiphytes were not common, but consisted principally of the ferns *Asplenium flaccidum* and *Polypodium diversifolium.* *A. lucidum* was noted growing on *Dicksonia squarrosa* and also on the sloping trunks of *Metrosideros lucida.*

Near a cliff-edge I noted *Hypolepis tenuifolia, Calystegia tuguriorum, Stellaria media, Apium prostratum, Carex trifida,* and *Poa imbecilla.* On this steep side *Urtica ferox* was also very plentiful, associated with *Muehlenbeckia australis.*

At the northern end of the island there is a steep cliff exposed less to the prevailing south-west wind and subject to more direct sunlight. The principal vegetation here consisted of *Veronica elliptica* (common), *Apium prostratum* (in patches), *Mesembryanthemum australe* (plentiful), with quantities of *Tetragonia trigyna, Muehlenbeckia australis, Linum monogynum,* and *Poa Astoni.* Several strong-growing patches of *Hierochloe redolens* in full bloom were also noted, with *Sonchus littoralis, Dichelachne crinita, Halorrhagis erecta,* and *Lepidium oleraceum* var. *acutidentatum.* Throughout these plants the burrows of the petrels were very plentiful.

SUMMARY.

The top of this island is almost flat, and is perhaps less than 200 ft. high. The peat seems much drier than on Herekopere, and is apparently deeper. There does not, on the whole, appear to be nearly so much evidence of bird traffic, except on the steep sides. The difference in the vegetation is quite marked, inasmuch as there is here a " forest " association which is entirely absent from Herekopere. (See my "Notes of a Botanical Visit to Herekopere Island," Trans. N.Z. Inst., vol. 47, pp. 142–44, 1915.)

The absence of *Poa foliosa* and *Senecio Stewartiae* from Pukeokaoka is also strange, as these plants are a marked feature of Herekopere, only about a mile distant. Possibly they grow here also, and were overlooked by me; but, if so, they are certainly rare. The only way I can account for the absence of these plants is by suggesting that the close forest formation of Pukeokaoka has prevented them getting a hold. *Poa foliosa* seems to be present on all the exposed parts of the mutton-bird islands where the scrub is unable to exist, but does not grow where the light has not full access. The restricted habitat of these plants is of more than passing interest.

As the plants noted are all mentioned in the text, I have not appended a list.

ART. XXIX.—*Notes on the Plant-covering of the Breaksea Islands, Stewart Island.*

By D. L. POPPELWELL.

[*Read before the Otago Institute, 7th September, 1915.*]

THE Breaksea Islands consist of a group of six small islands lying on the east of Stewart Island, between Port Adventure and Lord's River. Their latitude is 47° 6' S. and longitude 168° 15' E. As the eastern coast of Stewart Island trends away towards the south-west from this latitude, these islands receive no protection from the southerly weather, which strikes and breaks on their southern sides with full force; hence the name of the group. On the 7th January, 1915, by the kindness of Mr. Henry Hansen, I had the good fortune to be able to visit the principal islands of the group, and take some notes of their plant-covering.

Comparatively little has been recorded of the botany of these outlying islands; therefore a short description of their plant associations may be of interest.

RUKAWAHA-KURA, OR JOSS'S ISLAND.

This is the largest of the group, and lies only about 500 m. from the mainland of Stewart Island. It was the first one visited by me, and some time was spent in examining its vegetation. The island is almost circular in shape. It is not more than about 50 m. high, and is about 500 m. in diameter in its widest part. There is a good landing on the north-east side.

The plant formations are best considered under three heads—namely, (1) rocks and cliffs; (2) forest; (3) heath.

(1.) *Rocks and Cliffs.*

Most of the coast-line of the island consists of steep rocky faces, covered from above high-water level with a close association of almost pure *Olearia angustifolia* where there is sufficient soil to give it a hold, but in other places with *Poa Astoni*, which covers the cliff-faces with a grey-green drapery. Here and there, where the wind strikes less directly, great green patches of *Stilbocarpa Lyallii*, with their large leaves overlapping one another, can be seen, giving quite a tropical appearance to the vegetation. On exposed rocks *Crassula moschata* is plentiful, its reddish stems contrasting strongly with the greyish-white of the granitic rock. At the rear of the frontal rocks the shore ferns *Asplenium obtusatum* and *Blechnum durum* are abundant, the former in more or less isolated clumps and the latter in large closely matted patches. On the cliff-faces *Olearia Colensoi* is plentiful, especially behind the fringe of *O. angustifolia;* while *Veronica elliptica,* although a little past full flowering, still gave by its bloom a whitish tinge to the cliff-side, especially where the soil is deepest. Occasionally *Dracophyllum longifolium* pushes its brown head through the other plants, especially where the stunted shrubs bear testimony to the power of the wind. In parts of the island the cliff vegetation also contained *Mesembryanthemum australe* and *Tetragonia trigyna,* both plants giving a warm colour to the association, the former by its bright-pink flowers and the latter by its reddish leaves when exposed to extreme light. At a short distance away, however, the most dominant feature of the sloping cliff-covering is *Olearia angustifolia,* which is the principal plant in exposed situations. It protects its close relation *O. Colensoi* from the devastating effect of the salt-laden wind. The steep rocky faces are, on the other hand, covered with *Poa Astoni*, while here and there in the crevices *Veronica elliptica* ekes out a precarious existence. On the south side of the island the principal plant near the water's edge is *Olearia angustifolia*, but as we get farther in from the shore on the steep face *O. Colensoi* is found, with an occasional *Senecio rotundifolius* and *Dracophyllum longifolium*. At an elevation of about 40 m. above sea-level there is an open heath on this side, which is dealt with under its appropriate heading.

(2.) *Forest.*

When approaching the islands the physiognomy of the forest nearest the coast presents a greenish-grey appearance. The roof is close, and its wind-shorn appearance bears testimony to the severe gales which characterize this region. This fringe is composed almost wholly of *Olearia angustifolia* and *O. Colensoi*. A little farther back isolated patches of bright green attest the presence of *Griselinia littoralis*, while on the summit of the island dark patches of *Metrosideros lucida* are visible. Upon a closer acquaintance it is found that the soil consists of a deep layer of peat, fairly dry, and honeycombed in all directions with the burrows of mutton-birds (*Puffinus griseus*). There can be little doubt of the effect these birds produce on the vegetation. Not only do their tracks ramify in all directions, showing the physical effect of their traffic, but their manure is everywhere in evidence, while their burrows must greatly assist in the quick drainage of the surface, and also help to aerate the soil and make it suitable for plant-growth. The forest proper is chiefly *Metrosideros lucida*, which is not more than from 8 m. to 10 m. high, and consists of excessively gnarled trunks, usually prostrate for part of their length, and then turning their close twiggy tops upwards. Growing among these trees, and frequently epiphytic upon them, one finds *Griselinia littoralis*, *Pittosporum Colensoi*, *Nothopanax Colensoi*, *Coprosma areolata*, *Fuchsia excorticata*, *Rapanea Urvillei*, and *Olearia Colensoi*, with occasional plants of *Aristotelia racemosa*, *Olearia arborescens*, *Coprosma foetidissima*, and *Nothopanax Edgerleyi*. Here and there the tree-fern *Dicksonia squarrosa* grows in clumps, but is nowhere plentiful.

The most marked feature of the undergrowth is the strong growth of *Stilbocarpa Lyallii*, which practically covers the whole forest-floor. It is very luxuriant, and in several places is over 1·5 m. high. The leaves, from actual measurement, attain occasionally the great width of 63 cm. This plant forms in most parts great continuous growths, acres in extent, and in some places it is difficult to get through. Its great leaves, spreading horizontally, exclude the light, and prevent other growth on the peat, which is frequently so pierced by the mutton-bird burrows as to sink under one's steps. Where sufficient open space allows, strong clumps of the ferns *Asplenium lucidum*, *A. obtusatum*, and *A. scleroprium* will be found. The last-named fern grows particularly strongly in several places where it is undermined by burrows. In these places the ferns often have a distinct caudex, and stand nearly 1·5 m. high.

Patches of *Polystichum vestitum* are also occasionally seen, together with patches of *Histiopteris incisa*. The tree-trunks are covered with *Polypodium diversifolium* and *Asplenium flaccidum*. In several instances *Olearia Colensoi* had a trunk nearly 75 cm. in diameter near the base, but quickly branched into several strong-growing limbs which ultimately reached a height of nearly 10 m. Near the edge of the cliff—and, in fact, in most places where the strong light can reach the soil—clumps of *Poa foliosa* are common, and in open places this becomes a thick mantle, under which the mutton-birds burrow for nesting purposes. I also noted *Rubus australis*, but it is not very common. Here and there *Dracophyllum longifolium* grew in the forest, but was not plentiful in this association. Along some of the open tracks I also noted the following small plants : *Carex lucida, Plantago Raoulii, Cardamine heterophylla, Blechnum durum, B. capense;* and the orchids *Pterostylis australis, Thelymitra uniflora, T. longifolia, Caladenia bifolia, Microtis unifolia,* and *Prasophyllum Colensoi.* In damp places *Carex trifida* is not uncommon, and I noted one or two specimens of *Carex lucida.*

(3.) *Heath.*

On the south side of this island, where the full force of the prevailing wind and sea strikes the coast, there is a considerable area of open heath. This is on one of the highest parts of the island, and presents a somewhat bleak appearance. The general aspect is as if fire had run over the ground, the dead branches of the low scrub being bleached and white. I do not think, however, that fire has ever touched the island, the dry appearance being a characteristic result of the stormy conditions. The soil is peaty and in parts fairly damp, but could hardly be called boggy. The plant association of this part is very similar to that found in exposed places on some of the mountain peaty heaths in Otago where the water-content is not high. Low stunted *Leptospermum scoparium* forms the principal shrubby vegetation in the most exposed parts, the shrubs being fairly open. *Dracophyllum longifolium* is the next most conspicuous plant among the shrubby vegetation. These plants are dotted about everywhere, but do not attain more than about 1 m. in height, except in the hollow and sheltered places. Here and there *Styphelia acerosa* and *Olearia arborescens* are common, while between these plants *Oreobolus pectinatus* cushions are common. Red patches of *Drosera spathulata* are plentiful. I also noted one plant of *Dacrydium biforme.* In the drier places *Lycopodium volubile* was noted, together with patches of *Gentiana saxosa.* Stunted *Phormium Cookianum* is also present in isolated bushes. *Oreostylidium subulatum* is not uncommon, while *Anisotome intermedia* (?) is also seen, together with *Thelymitra uniflora, Pentachondra pumila,* and *Leptocarpus simplex.* I also noted one patch of *Nertera depressa.* The southern coast of the island is steep, and is protected by the usual fringe of *Olearia angustifolia, O. Colensoi, Senecio rotundifolius,* and *Dracophyllum longifolium,* while the shore ferns *Asplenium obtusatum* and *Blechnum durum* are common. There were no signs of bird-burrows on the heath, no doubt on account of the wetness it must experience in bad weather.

(4.) *Introduced Plants.*

Near the mutton-birders' huts, where a small clearing has been made, I noted the following introduced plants, all of them growing luxuriantly : *Dactylis glomerata, Holcus lanatus, Rumex obtusifolius, R. Acetosella, Stellaria media, Sonchus asper, Poa pratensis,* and *Trifolium repens.* I also noted *Cotula coronopifolia* near the door of one of the huts, but, although this is an indigenous species, it had all the appearance of having been introduced. The above species have no doubt been accidentally introduced to these islands on the annual visits of the Maoris when mutton-birding. None of them have spread beyond the clearing.

BREAKSEA ISLAND, OR WHAREOTEPUAITAHA.

This island stands farthest out of the group, and consequently is in a much more exposed position. From a little distance it presents a remarkably grey uniform appearance, the tops of the trees being as close as possible and seeming as though trimmed. Upon landing on the north-west side the same shore association is met with as is mentioned above ; but when the low forest is entered it is at once seen that much less undergrowth is visible, except along the north side, where it is somewhat sheltered. There a pure association of *Stilbocarpa Lyallii* is visible as far as the eye can reach among the weird and gnarled trunks of *Olearia angustifolia* and *O. Colensoi*, which practically make up the shrubby vegetation of this island. Here and there near the shore *Veronica elliptica* pushes out its green head. *Poa Astoni* and *Crassula moschata* grow on the rocks. In a damp place *Carex trifida* was growing with a patch or two of *Asplenium obtusatum* and *Blechnum durum*. Towards the weather side of the island practically no undergrowth is found. Very few young plants are seen. Some young *Olearia Colensoi* plants, however, caught my attention on account of the great size of the leaves, which by actual measurement were 28·5 cm. long by 14 cm. wide. The above length included the petiole, which measured 2·5 cm.

Here and there among the scrub are some odd plants of *Poa foliosa*. In open places *Tetragonia trigyna* is common. In parts of this island, and especially on the south side, the interior of the scrub presented the most extraordinary sight I have ever seen. The trunks of *Olearia angustifolia* were about 90 cm. in diameter at the base, gnarled and bent by the storms in an extraordinary manner. The bark was polished white, and in places was furrowed to a depth of 5 cm., presenting as weird a sight as possible. The ultimate branches turned upwards and closely roofed in the whole plant. Where exposed to the wind, in many cases it was stunted to a height of less than 1 m. On the north side of the island the trunks were fairly straight and the shrubs much higher.

On this side also I noted several specimens of *O. Traillii*, a plant that has, so far as is at present known, a distribution almost as restricted as any in our flora. The specimens seen by me in their general appearance differed little from *O. angustifolia* and *O. Colensoi*, among which they grew, but upon closer examination at once showed the difference in foliage and flower. From the situation in which this species grows, and from its intermediate characteristics, I am inclined to think that *O. Traillii* may yet prove to be a hybrid between the other two olearias above named. Of course, careful experiments in artificial pollination only can prove whether this theory is correct, and I merely mention the matter for the purpose of drawing attention to the intermediate characteristics of this species.

(1.) *Rocks and Cliffs.*

On this island there is a special cliff association on the southern side. Here the cliffs are sheer down for about 50 m., and the worn and weather-beaten rocks are very noticeable. The Messrs. Hansen Brothers, who have sailed in these waters for many years, inform me that during southerly weather the waves strike these cliffs and splash right over the top. They were under the impression that there was a " blowhole " in the island, but upon examination of the spot with myself it was discovered to be only a sheer-down cliff which in one part slightly overhung.

Some idea of the force of the waves in this place may be gathered from the fact that bits of sawn timber were seen by us right on the top of the cliff, having been apparently carried there by the sea. Along the top of this cliff—and, indeed, for the greater part of the southern side of the island—practically no scrubs grow. The association is a low-growing one, chiefly consisting of *Poa Astoni*, which covers the ground almost as a sward. Among this grow numerous large, succulent plants of *Myosotis albida*, much of it in flower, but still with many buds. Round cushions of *Gentiana saxosa* from 10 cm. to 15 cm. in diameter are common. *Apium prostratum* is also scattered about, while *Selliera radicans* and *Crassula moschata* cling to the rocky surface in parts. Here also I noticed one patch of *Anisotome intermedia* (?) about 2·5 m. in diameter, and a common sedge, probably *Scirpus aucklandicus*, is also dotted about. In places low bushes of *Olearia angustifolia* also grow, and in their shelter *Blechnum durum* and *Asplenium lucidum* are sparingly found. On the brow of the cliff hang screens of *Mesembryanthemum australe* in full bloom, and *Tetragonia trigyna* in flower also sprawls over the surface. Here and there *Carex trifida* grows strongly, with patches of *Anisotome intermedia* (?). On the level top, where the peat is deeper, there is a considerable patch of *Poa foliosa*, dotted throughout with isolated plants of *Lepidium oleraceum* var. *acutidentatum*. *Senecio lautus* and *Sonchus littoralis* were also noted. The above grass is in strong tussock form, about 80 cm. high, with a stem-like base of more or less decayed leaves, and the peat between the tussocks is everywhere full of mutton-bird burrows.

(2.) *Introduced Plants.*

Near the mutton-birders' huts a few introduced grasses and weeds grow, also some *Phormium tenax*, the latter probably planted by the Maoris on account of its useful qualities.

The Remaining Islands.

The other islands of the group, except one, are quite small, being only about 30 m. to 40 m. in diameter. I did not land on them, but passed close along shore in a boat, thus securing a good view of their general plant formations. The small islands have their summits and more level parts covered with *Olearia angustifolia*, mixed here and there with *O. Colensoi*.

The cliff-sides are covered with *Poa Astoni*, with here and there small patches of *P. foliosa* where the exposure is too much for the scrub formation. *Veronica elliptica* clings sparingly to the rocks, and in the sheltered places *Stilbocarpa Lyallii* can be seen. The largest of these other islands, known as King's Island, or Kaihuka, has, so far as I might judge, much the same covering as Breaksea Island, the associations of which are, I think, fairly typical of the exposed situations throughout the whole group.

Conclusions.

The facts of most importance gathered on my visit are those connected with the distribution of the plants in their relation to soil and climate. *Olearia angustifolia* seems to be *par excellence* the plant of the exposed seashore on all these islands. *Stilbocarpa* owes its position to the wind-still atmosphere of the interior of the forest and the rich soil of the bird-manured areas. The wind-resisting powers of *Poa foliosa* are well known, and there can be little doubt that both it and *Stilbocarpa* confer great benefit on the birds for nesting purposes, and in turn receive assistance from the increased nutrition of the bird-droppings.

Poa foliosa seems to require free access of light, as it is not found under the scrub except on exposed points where the scrub is tolerably open. In these situations occasional tufts of this grass will be found, increasing in number in proportion to the opening among the scrub. This is especially so where the depth of the peat gives special facilities for the bird-burrows. *Poa foliosa* does not, however, seem to me to make a stable formation, as young plants of both *Olearia angustifolia* and *Veronica elliptica* are found amongst it. These plants must gradually take its place by overtopping it, and thus shutting out the light.

As regards *Olearia Traillii*, I can form no very convincing theory for its being restricted to such a narrow habitat. It is nowhere plentiful, nor are there any very specialized conditions in the association in which it is found. My previous remarks as to the possibility of its being a hybrid constitute the only explanation I can suggest of its presence. Its habit of growth is similar to that of both *O. Colensoi* and *O. angustifolia*. Its leaves are intermediate in form between these species. Its flowers are racemed like *O. Colensoi*, but, unlike the latter, which has practically discoid heads, *O. Traillii* has short ray-florets, thus placing it in an intermediate position between *O. Colensoi* and *O. angustifolia*, which has comparatively long ray-florets. As previously mentioned, however, this matter can only be settled by actual experiment.

The number of indigenous species noted on these islands is sixty-nine, belonging to fifty-four genera and twenty-nine families. Appended is a list of them.

LIST OF INDIGENOUS PLANTS NOTED.

PTERIDOPHYTA.

Cyatheaceae.
Dicksonia squarrosa (Forst. f.) Sw.

Polypodiaceae.
Polystichum vestitum (Forst. f.) Presl.
Asplenium obtusatum Forst. f.
—— *scleroprium* Homb. & Jacq.
—— *lucidum* Forst. f.

Asplenium flaccidum Forst. f.
Blechnum durum (Moore) C. Chr.
—— *capense* (L.) Schlecht.
Histiopteris incisa (Thbg.) J. Sm.
Polypodium diversifolium Willd.

Lycopodiaceae.
Lycopodium ramulosum T. Kirk.

SPERMOPHYTA.

Taxaceae.
Dacrydium biforme (Hook.) Pilger.

Gramineae.
Hierochloe redolens (Forst. f.) R. Br.
Poa foliosa Hook. f.
—— *Astoni* Petrie.

Cyperaceae.
Scirpus aucklandicus (Hook. f.) Boeck.
Carpha alpina R. Br.
Gahnia procera Forst.
Oreobolus pectinatus Hook. f.
Carex lucida Boott.
—— *trifida* Cav.

Restionaceae.
Leptocarpus simplex A. Rich.

Liliaceae.
Phormium tenax Forst.
—— *Cookianum* Le Jolis.

Orchidaceae.
Thelymitra longifolia Forst.
—— *uniflora* Hook. f.
Microtis unifolia (Forst. f.) Rchb.
Prasophyllum Colensoi Hook. f.
Pterostylis Banksii R. Br.
—— *australis* Hook. f.
Caladenia bifolia Hook. f.

SPERMOPHYTA—*continued.*

Aizoaceae.
Mesembryanthemum australe Sol.
Tetragonia trigyna Banks & Sol.

Cruciferae.
Cardamine heterophylla (Forst. f.)
O. E. Schulz. var.
Lepidium oleraceum Forst. f. var.
acutidentatum T. Kirk.

Droseraceae.
Drosera spathulata Labill.

Crassulaceae.
Crassula moschata Forst. f.

Pittosporaceae.
Pittosporum Colensoi Hook. f. var.

Rosaceae.
Rubus australis Forst. f.

Myrtaceae.
Leptospermum scoparium Forst. var.
Metrosideros lucida (Forst. f.) A. Rich.

Onagraceae.
Fuchsia excorticata L. f.

Araliaceae.
Stilbocarpa Lyallii J. B. Armstrong.
Nothopanax Edgerleyi (Hook. f.) Seem.

Umbelliferae.
Apium prostratum Labill.
Anisotome intermedia Hook. f. (?).

Cornaceae.
Griselinia littoralis Raoul.

Epacridaceae.
Pentachondra pumila (Forst. f.) R. Br.
Styphelia acerosa Sol.
Dracophyllum longifolium (Forst. f.)
R. Br.

Myrsinaceae.
Rapanea Urvillei (A. DC.) Mez.

Gentianaceae.
Gentiana saxosa Forst. f.

Boraginaceae.
Myosotis albida (T. Kirk) Cheesem.

Scrophulariaceae.
Veronica elliptica Forst. f. var.

Plantaginaceae.
Plantago Raoulii Decne.

Rubiaceae.
Coprosma lucida Forst. f.
—— *areolata* Cheesm.
—— *foetidissima* Forst.
Nertera depressa Banks & Sol.

Goodeniaceae.
Selliera radicans Cav.

Stylidiaceae.
Oreostylidium subulatum (Hook. f.)
Berggr.

Compositae.
Olearia angustifolia Hook. f.
—— *Traillii* T. Kirk.
—— *Colensoi* Hook. f.
—— *arborescens* (Forst. f.) Cockayne
and Laing.
Senecio lautus Forst. f.
—— *rotundifolius* Hook. f.
Sonchus littoralis (Kirk) Cockayne.
Cotula coronopifolia L.

ART. XXX.—*Studies in the New Zealand Species of the Genus* Lycopodium : *Part I.*

By the Rev. J. E. HOLLOWAY, M.Sc.

[*Read before the Philosophical Institute of Canterbury, 3rd November, 1915.*]

Plates XVII, XVIII.

IN these studies I hope to put together the results of my observations on the life-history of the New Zealand species of *Lycopodium.* In several of the papers which have been published on the subject of the genus *Lycopodium* the paucity of information available for the purpose of a comparative study has been commented upon. The present study seeks to bring into the arena of *Lycopodium* problems additional data with regard to the occurrence and habit of the mature plant, the occurrence and structure of the prothallus, the nature of the dependence of the young plant upon the prothallus, and the vascular anatomy of both the " seedling " and the full-grown plant.

The chief modern writers on the subject are Treub, Goebel, Bruchmann (6), Lang (14), and Thomas (16) on the prothallus; Jones (13) on the stem-anatomy; and Sykes (15) on the morphology of the sporangium-bearing organs. Bower (4) has thoroughly analysed and co-ordinated all known *Lycopodium* facts in connection with his well-known theory of the origin of the sporophyte : he concludes that the genus is to be read as a progression from a simple *Selago*-like form. Lang (14) has also given a comparative analysis of the facts, and, following Treub, has arrived at conclusions in which the species which belong to the *L. cernuum* cycle of affinity are viewed as the most primitive in the genus, the whole genus being read as a reduction series. Thomas and Sykes support this second interpretation, but Goebel doubts the primitive nature of the protocorm, on which, of course, this view is largely based. Lady Isabel Browne (5) has lately proposed a third interpretation, according to which the protocorm is to be regarded as a reduced stem. Bruchmann's inter-pretation of the various types of *Lycopodium* prothallus would separate so widely the different sections of the genus from one another as practically to deny that they are interrelated at all. New Zealand contributions on the subject, in addition to Thomas's preliminary note on the prothallus of *Phylloglossum* (16), have been Miss Edgerley's account (8) of three species of prothallus, and my own papers on *Lycopodium* stem-anatomy (11) and on the protocorm (12). Unfortunately, I have not had the advantage of consulting, for the purpose of this present study, either Treub's or Goebel's original publications. My knowledge of the work of the former has been derived from various references to and figures from it in Bower's " Origin of a Land Flora " and in other standard books of reference, and also from Treub's own preliminary note in the " Annals of Botany " (17), and Bower's " Review of *L. Phlegmaria* " (2).

After this paper had been put into the printer's hands I discovered the prothallus and young plants of the endemic *L. ramulosum.* Through the courtesy of the Editors I have been able to add an account of these, with several figures.

I desire to express my gratitude to Dr. L. Cockayne, F.R.S., and to Dr. C. Chilton, C.M.Z.S., Professor of Biology in Canterbury College, Christchurch, for the encouragement and advice they have given me during the last two or three years, and to the former also for many suggestions with respect to this paper; also to Mr. C. E. Foweraker, M.A., of the Biological Laboratory, Canterbury College, who has given me great assistance in the preparation of objects for the paraffin bath.

I. Occurrence and Habit.

L. Selago Linn.

This species is common in Europe, and is well known from the investigations of H. Bruchmann (6) and C. E. Jones (13) and others, so that it will need but slight notice in this paper. It occurs commonly throughout the South Island of New Zealand in damp places as a member of fell-field, herb-field, and sometimes subalpine *Nothofagus* forest, and elsewhere on the high mountains.

L. Billardieri Spring.

This species is found in the forest throughout New Zealand. Though typically an epiphyte, growing pendulous from tree-fern trunks and the upper branches of forest-trees, it also not uncommonly grows on the ground. In North Auckland it is met with on peaty soil in groves of *Leptospermum*. On the volcanic islet of Rangitoto, Auckland Harbour, it grows on patches of humus among the blocks of scoria. In Southland and Stewart Island it occurs frequently on the damp forest-floor on patches of humus at the base of large trees. In one instance I found a single "seedling" plant growing on a bush-road clay-cutting. The individual plant consists of a main stem, which in its short underground region is sparingly branched. This underground portion is covered with scale leaves, and bears a number of roots at its lower end. These roots are much branched, and are covered in their terminal portions with a mat of rhizoids. At or near the surface of the soil the stem branches dichotomously several times to form a tuft of aerial stems. When the plant is epiphytic these aerial stems hang in tresses from 1 ft. to 4 ft. in length (see Cockayne, L., fig. 6, in " Report on a Botanical Survey of the Waipoua Kauri Forest," 1908); when terrestrial the plant is more or less upright, and in some instances it is then hardly to be separated from *L. varium*. Several plants generally grow together, their subterranean portions thickly interpenetrating the patch of humus. The appearance of the plant, with its dichotomously branched aerial stems and numerous long terminal tetragonal fertile spikes, is well illustrated by Pritzel's figure of *L. Phlegmaria* in Engler and Prantl (9).

L. varium R. Br.

This plant is stated by Cheeseman (7) to be probably only an extreme form of *L. Billardieri*. It is sometimes epiphytic, but it occurs in those parts of New Zealand subject to a semi-subantarctic climate, and on Stewart Island and the Subantarctic Islands of New Zealand more commonly as an erect, stout, rigid terrestrial plant. On Stewart Island I found it growing on the forest-floor in large patches, some of which were as much as 12 ft. across.

L. Drummondii Spring.

There is only one locality, so far as is known, where this species occurs—viz., the *Sphagnum* bog at the east end of the small Lake Tongonge, at Kaitaia, North Auckland. I desire to thank Messrs. H. Carse and R. H. Matthews, of that neighbourhood, for kindly conducting me to the spot in January, 1914. In a recent letter the former states that the lake and bog are to be drained, so that probably this species, so rare in New Zealand, will disappear. The main stem of the plant is never more than 6 in. in length, and is generally branched several times. It creeps above ground, but is tightly bound down to the mossy surface by the adventitious roots, one of which is borne at the junction of each branch with the main stem. The cones are borne singly on erect peduncles, and stand from 2 in. to 4 in. in height. In several instances a fertile region was seen to be divided into two lengths by the interposition of a short sterile region, and in other cases the old cone of the previous year was observed to have grown on to form the new one.

L. laterale R. Br.

I have studied this species more especially on the clay "gum-land" in boggy localities around Kaitaia and on the Auckland Isthmus. In the latter locality it occurs at the margins of the small bogs which occupy the numerous hollows among the clay hills, growing amongst certain *Cyperaceae* and *Gleichenia dicarpa* var. *hecistophylla*. Around Kaitaia and on the Peria Gumfields it grows extremely abundantly on the open damp hillsides. The surface soil of these gum-lands consists of a peaty humus, which for the greater part of the year holds much water, but which during the summer months is generally more or less dry. The adult plant consists of an irregular and much-branched colourless rhizome, which ramifies through the soil in all directions. The shorter branches emerge at the surface to form the erect aerial shoots. These latter, when growing amongst thick scrubby vegetation, are extremely slender, and attain a height of from 2 ft. to 3 ft. On open ground they are short and stout, and often reddish in colour. The cones are short, and are normally lateral and sessile. In some cases, however, individual cones are borne on short leaf-covered peduncles, and they must then be regarded as terminal. Pritzel's description of this species in Engler and Prantl is rather misleading. He there states that the *Cernua* section comprises forms without a widely creeping main axis, mostly like a little tree. It is to be noted that it is only the aerial branches which are tree-like in the two New Zealand species *L. laterale* and *L. cernuum*. The main body of the plant in *L. laterale* is subterranean and widely ramifying.

L. cernuum Linn.

This species is well known from the writings of Treub and of Jones (13). It grows very abundantly throughout the northern part of the North Island of New Zealand on clay moorlands as described for the preceding species. It thrives especially in North Auckland amongst scrub vegetation of the *Gleichenia-Leptospermum* association, individual plants often attaining to a length of 12–15 ft., and the upright branches to a height of 1–4 ft. It is also extremely common on the Volcanic Plateau, in the neighbourhood of hot water and near fumaroles. The main stem of the adult plant is above ground, and has a serpentine habit of growth. It extends in a succession of loops and nodes, at each node the stem being fastened to the ground by a group of adventitious roots, which arise in the first place immediately behind the stem-apex on its ventral side, while from the loops arise the branches of limited and unlimited growth. The branches are borne laterally right and left on the main stem, but the erect tree-like fertile branches take their origin from its dorsal side. Here, again, Pritzel's description is misleading: only certain of the branches—namely, the fertile ones—are tree-like; the main body of the plant is widely creeping and branched. The figure which he gives in Engler and Prantl illustrating this species is that of an erect fertile branch only; the group of roots at the base should not be so figured. There are numerous short cones borne at the ends of the branchlets in the fertile regions.

L. densum Labill.

This species grows abundantly throughout the Auckland Province on clay land amongst light open " scrub " vegetation, and also in the more open parts of kauri forest near its outskirts. (For general habit see Cockayne, L., *loc. cit.*, fig. 16.) The main rhizome, which is from 4 ft. to 10 ft. in length, is stout, and there are also subterranean branches both of limited and also unlimited growth, borne laterally on the main stem. The branches of limited growth emerge from the ground as rigid, erect, much-branched, tree-like, aerial shoots, generally from 1 ft. to 3 ft. in height. I have often observed that when the plants are growing amongst tangled scrub the aerial shoots may be as much as 8 ft. or 9 ft. in height, and in some cases remain totally unbranched. Stout adventitious roots arise ventrally from the main rhizome and branches, and are borne singly, generally immediately behind a point of branching. There are three distinct varieties of this species, corresponding to differences in the general size of the leaves with which the aerial branches are covered. The numerous short cylindrical cones are solitary and terminal on the ultimate branchlets. In the particular variety which is characterized by the acicular form of its leaves the branchlets on which the cones are borne are more or less modified as peduncles.

L. volubile Forst.

This species is common throughout New Zealand, excepting in the driest districts, growing freely amongst *Leptospermum* and other heath-like vegetation. It has a scrambling habit, or at times is a winding liane, spreading over the ground or over low-growing bushes. Individual plants are often so much as 12–15 ft. in length. There is a main axis of growth, on which are borne laterally, in the plane of the surface on which the plant is growing, branches of limited and also of unlimited growth. The leaves are dimorphous, as has been noticed and commented upon by Boodle (1), the larger laterally borne, sickle-shaped, and distichously spreading; the smaller linear, scale-like, and borne dorsally and ventrally. The distichous character is for the most part confined to the smaller branchlets on the branches of limited growth. An account of the development of heterophylly in this species and in *L. scariosum* has been given in a former paper by the writer (11, p. 366). Here it need only be mentioned, as Goebel has already pointed out, that certain leaves on the main shoots are hook-like,

and so probably climbing-organs (10, p. 346).   Adventitious roots
are borne ventrally at intervals along the main axis.   In places
where the plants are scrambling over low-growing bushes these roots
may attain a length of 3–5 ft. before reaching the ground.   In the
late winter and early spring a very characteristic feature is the thick
envelope of mucilage which covers from 3 in. to 12 in. of the growing
root-tip before it has reached the ground.   The fertile spikes are thin and
cylindrical, and from 1 in. to 4 in. in length, and occur in large terminal
much-branched panicles.   They are figured in Engler and Prantl (9).
The fertile branches are to be found for the most part in those regions
of the plant which are elevated on some low-growing vegetation, and they
are thus generally pendulous.   Although the panicles of spikes in their
normal form are very distinct from the ordinary vegetative branches,
yet close observation shows that all stages of transition may occur.
Isolated sporangia are sometimes to be found on sterile branches; in
other cases fertile and sterile branchlets are indiscriminately mixed;
while in others, again, sterile tracts may appear in the spikes themselves.

L. ramulosum T. Kirk.

I have gathered this species from the peaty flats at the head of the
Rakiahua Arm of Paterson Inlet, Stewart Island, and from bogs in
the neighbourhood of Hokitika and Kumara, Westland.   It occurs
abundantly, covering the ground with mats of interlacing plants.   It is
common in such a habitat all over Stewart Island, ascending to the
summits of the mountains; and is found also in bogs, both lowland and
subalpine, throughout Westland and north-west Nelson.   The individual
plant is very short—from 2 in. to 9 in. in total length—and is irregularly branched both above and below ground.   The study of this species
as growing in Stewart Island shows that the subterranean portions are
of two distinct kinds.   Those stems nearer the surface are whitish in
appearance, and are thickly covered with scale leaves; from these the
aerial branches arise.   Those portions which penetrate the peaty humus
more deeply are brownish in colour, and are more or less naked of scale
leaves, and are the ones which more frequently bear the adventitious
roots.   The two kinds of stem arise from one another without transitions.   The aerial branches are procumbent or ascending, and are much
branched.   The short solitary cones are borne terminally on erect
branches, but occasionally they may occupy a lateral position.

L. fastigiatum R. Br.

This species is common on open mountainous country throughout the
South Island, especially on tussock grassland.   It also occurs in subalpine
Nothofagus cliffortioides forest, where it assumes a more mesophytic habit.
In Southland and Stewart Island it also descends to low levels.   There is
a subterranean creeping main axis, which is usually from 1 ft. to 3 ft. in
length, but which may be as much as 5 ft. long.   Branches of limited and
also of unlimited growth arise from the main axis, the former emerging
from the surface to form the erect greenish or reddish tree-like aerial
branches.   These latter are from 6 in. to 12 in. in height, and are densely
branched, but when growing amongst thick tussock-grass are much more
slender in habit, and are taller.   When growing on sour peaty soil the
branches may be flattened to the ground.   The cones are from 1 in. to
2 in. in length, and are borne, usually singly but sometimes two or
more together, at the end of the branchlets on distinct peduncles.

L. scariosum Forst.

This species, like the last, occurs fairly commonly in open situations at fairly high elevations, especially throughout the South Island. In such a situation it is always creeping on the surface of the ground. Individual plants may have an extreme length of 5–6 ft., but are generally shorter. The stout and rigid main stem bears branches both of unlimited and of limited growth. The former are closely adpressed to the surface of the ground. and bear the adventitious roots. The branches of limited growth are heterophyllous, and are markedly flattened in the plane of the ground. The development of the heterophylly is different in this species from that in *L. volubile*, and has been fully described else-where (11, p. 366). The cones are from 1 in. to 2 in. in length, and are borne singly at the ends of branchlets on peduncles, in the same manner as has been described for *L. fastigiatum,* and as is so well known in the European species *L. clavatum*. When growing amongst thick fern vegetation—as, for example, on the tailing-heaps in the neighbourhood of the old alluvial gold-mining claims in Westland—it is noticeable that this species may show an almost entire absence of the usual dorsiventral appearance of its branches. The lateral branches are erect, and, except in the older parts of the branches, the leaves are scattered and tend to be acicular in form, while in the ultimate branchlets they are reduced to mere scales. The general habit of the plants in these cases is almost scrambling, and the long, rigidly erect, naked, and closely crowded ultimate branchlets present a very characteristic and forest-like appearance. The tips of certain of these branchlets become fertile and develop as cones, while isolated fertile regions may also be found occasionally on other branchlets.

Summary.

Of the foregoing eleven species of *Lycopodium* native in New Zealand, one is generally an epiphyte (*L. Billardieri*); another is sometimes epiphytic, but more frequently terrestrial (*L. varium*); five occur in more or less wet habitats (*L. cernuum*, *L. laterale*, *L. Drummondii*, *L. ramulosum*, and *L. Selago*); and four in dry and at times fully exposed localities (*L. volubile*, *L. scariosum*, *L. fastigiatum*, and *L. densum*). All of these species except *L. Selago* and *L. cernuum* are confined to the countries and islands of the South Pacific Ocean, so that possibly they are not well known either to European or to American botanists. A somewhat detailed description of their occurrence and habit has therefore been given, but more exact ecological studies are demanded before a true estimate of their life-requirements can be gained. Some of these species are amongst the largest of modern Lycopodiums, as, for example, *L. volubile*, *L. scariosum*, *L. densum*, and *L. cernuum;* and also, as will be described below, the vascular cylinder of the stem in these species is greatly developed. Two show dimorphism in their leaf-structure (*L. volubile* and *L. scariosum*), but other characters show that these two species are not closely related. In several of the species the character of the fertile regions is variable, and also the habit of the whole plant in one or two cases varies, perhaps epharmonically, under differing conditions. Not a few of the New Zealand species occur most abundantly in localities formerly occupied by forest. For example, *L. densum* and *L. laterale* find their most luxuriant development on the clay gum-lands of the Auckland Province, areas occupied at no greatly distant bygone period by the kauri-tree forests; *L. Billardieri* and *L. volubile* also occur freely in these localities; *L. cernuum* is also abundant in the same locality, but it shows still greater luxuriance in the neighbourhood of hot-water streams. *L. scariosum* and *L. fastigiatum* frequently occur in mountain areas of both Islands, on hillsides from which the forest has retreated owing to climatic changes or to the hand of man, but the latter species is also a common plant of subalpine southern-beech forests. The two epiphytic species very readily adopt a terrestrial habit both on the forest-floor and also in more open situations in which the conditions are practically epiphytic. In Westland *L. volubile* and *L. scariosum* are very common on the heaps of tailings around abandoned alluvial gold-mining claims, and the endemic *L. ramulosum* is abundant in the man-induced *Sphagnum* bogs in the same localities.

II. OCCURRENCE OF PROTHALLI AND YOUNG PLANTS.

L. Billardieri.

Prothalli and young plants associated together were found in two different localities, and young plants alone in two others. On one occasion a large number of young plants were found growing at the top of a nikau palm in the Waikumete Bush, Auckland, in a mass of humus through which the stems and roots of the adult plants were ramifying. One or two old prothalli attached to well-grown young plants were obtained in this case. On another occasion twelve prothalli were obtained in the month of January in the forest at Pipiriki, Wanganui River, from a mass of humus in the fork of a tree. Here, also, old plants were present. Of these prothalli, seven were without young plants attached, and five bore young plants in different stages of development. The prothallus of this species grows completely buried in the humus. The prothalli are easily seen, by reason of their whitish appearance, in the midst of the dark humus when the latter is dissected in water, but may be mistaken for a mass of young root-tips.*

L. laterale.

Young plants and prothalli of this species were discovered by me in three different localities on the Auckland Isthmus during the summers of 1905 and 1914. In two cases they were growing on patches of peaty humus which had been overturned by gum-diggers, and occurred close to the edge of boggy ground in the immediate vicinity of adult plants. In the third case they were growing fairly abundantly on a patch of damp soil, which was sparsely covered with short moss, in the midst of a clump of mature plants. Altogether eight prothalli were found, as also were many young plants in all stages of development. Diligent search was made in several localities in the Mongonui† County, North Auckland, on the open hillsides where this species grows luxuriantly, but in no case were the young plants seen. I adopted the plan of cutting out small turves of the soil in which young plants could be seen, and reserving them for examination under a dissecting microscope. The process of examination proved exceedingly tedious, for the peaty humus was closely inter-

---

* I have found also the young plants of *L. Billardieri* var. *gracile* growing on the trunk of a tree-fern near Lake Kanieri, Westland. The distinguishing characters of this variety are its small, slender form, almost unaltered character of the fertile leaves, and the fact that the fertile regions are not confined to the ends of the branches.

† This is the official spelling of the name of the county according to the Counties Act of 1908, although the correct spelling is Mangonui.

9*

mixed with a small moss, and had to be dissected in water very small portions at a time. The prothalli are almost colourless, and are very small and delicate, and hence not easy to distinguish. They were never found attached to young plants which bore more than two protophylls. The youngest plantlets are also very minute, but are more easily seen amongst the moss and soil by reason of their vivid green colour. Many of the localities where *L. laterale* grows become dry during the summer, and probably in some summers the young plants all die; but during spring and early summer they doubtless are growing in large numbers in favourable spots. A close examination of disturbed soil—*e.g.*, gum-diggers' holes, hoof-prints, &c.—along the edge of the bogs around Waikumete and Henderson, on the Auckland Isthmus, at that season of the year, and in all probability generally, results in the discovery of colonies of the young plants of this interesting species.

L. cernuum.

On many occasions, and in many different parts of the Auckland Isthmus, as also in the Mongonui County, I have noticed young plants of this species growing in the vicinity of older plants. The spores would seem to germinate very freely. But, as in the case of *L. laterale*, a dry summer would bring about the destruction of most of the plantlets. A damp clay bank or a shaded roadside cutting in the neighbourhood of adult plants is the best place to search for the young plants and pro-thalli—*i.e.* they are more in evidence under artificial than under natural conditions. Such a clay bank, if damp, shaded, and old enough for a thin covering of moss and slime fungus to have appeared on it, invariably contains the young plants of this species, generally in great abundance. The same method of search for the prothallus was adopted as in the case of the last species. The very young plants and prothalli of *L. cernuum* are difficult to clean owing to the intimate penetration of the clay and slime by their numerous rhizoids. *L. cernuum* and *L. laterale* both grow commonly from the Auckland Isthmus northwards, but their habitats do not overlap. The former species keeps to the higher parts of the undulating clay gum-lands, while the latter is to be found in the hollows and on the damp lower parts of the hillside. I have never found the prothalli and young plants of both species growing together. However, even if this had been so, it would not have been very difficult to dis-tinguish between them, for in spite of the similarity in appearance both of their prothalli and of the very young plantlets there are some characteristic differences.

L. densum.

Frequent and long search for the young plants of this species in various parts of the Auckland Province, where it grows abundantly, has never met with much success. On one occasion, in the neighbourhood of the Bay of Islands, I found two young plants, one of which still showed a large " foot." The plants were growing in a grove of *Leptospermum* on a patch of thick damp moss. At the time when they were found it was impossible to spend more than a few minutes in searching, but I have no doubt that a closer examination of the ground would have resulted in the discovery of still other young plants. The rarity of occurrence of prothallial plants of this species may be due simply to the fact that its usual habitat is a dry open one, where damp shaded spots, favourable to the development of the prothallus, infrequently occur.

L. volubile.

I have been able to find young plants and prothalli of this common species in several different localities in both the North and South Islands, in some cases in abundance. Several young plants and one very large prothallus were dug out of a clay bank at Kaiaka, North Auckland. Between sixty and seventy prothalli were unearthed from an old clay cutting at the edge of the forest near Henderson. Some of these were very young, the smallest being about 1 mm. in size. In several localities in Nelson also I have found both young plants and prothalli, the habitat being in some cases a disturbed damp clay soil in the open, and in others a damp spot on the forest-floor. Old plants on several of these occasions were not to be seen in the vicinity of the young plants. Probably they had died out, or had been destroyed when the land was cleared. In not a few other parts of New Zealand I have seen young prothallial plants of this species both in southern-beech forest where the soil consists of decaying vegetable material and in shaded clay situations on open hillsides. *L. volubile* would seem to propagate itself quite commonly by the dispersal and germination of its spores. The prothalli are buried at a depth of $\frac{1}{2}$–3 cm. Rarely they occurred at the surface, and then the upper part of the prothallus is a bright green. In an old gold-mining claim near Hokitika I found several prothalli both of this species and of *L. scariosum*. In both cases they were of an abnormally large size, and the young plants attached were as much as 6 in. in length and several times branched, and bore from one to three large adventitious roots.

L. ramulosum.

During the month of April, 1916, I found the prothalli and young plants of this species growing in an old shingle-pit on the side of the main road between Waimea and Kumara, Westland. In the near vicinity four species of *Lycopodium* occurred — viz., *L. volubile*, *L. scariosum*, *L. fastigiatum*, and *L. ramulosum*. This particular shingle-pit was in a damp but warm situation. There was an abundance of young plants of *L. ramulosum* of all stages of growth, and up to the present I have been able to dissect out from the mossy turves which I gathered twenty-four prothalli. The moss was a short brown variety, and when dissected was seen to be full of slime fungus and of several varieties of microscopic algae, with which the rhizoids of the prothalli were closely intermatted.

L. fastigiatum.

In one locality on a high exposed ridge near Mount Oxford, Canterbury, I have discovered sixty-one prothalli of this species during three years. Prothalli and young plants of both *L. fastigiatum* and *L. scariosum* were growing on the leeward side of the highest point on this ridge in patches of moss and *Helichrysum filicaule* on clayey soil amongst rocks and boulders. The area in which the young plants occurred was very limited in extent, nor were they to be seen elsewhere along the ridge. The adult plants of these two species grew abundantly on Mount Oxford, which lies a mile or two to the windward of this ridge, but nowhere have I found them on the ridge itself. The great majority of the prothalli of *L. fastigiatum* were obtained from turves of clayey soil which came from the lower sides of embedded rocks and boulders.

In one instance I found a little group of fourteen prothalli actually touching one another in the humus underlying the moss, two of these prothalli being very small, not more than ½ mm. in size, and two others being old and exhausted. The prothalli were always buried to the depth of 1–5 cm., and occurred either in the humus underlying the covering of moss and *Helichrysum* or deeper still in the clayey substratum. I visited this spot frequently during three years, and on the last occasion found many very small and young prothalli. It was noticeable that the young plants invariably died after they had attained a length of 2 in. or 3 in. The patches in which the young plants and prothalli occurred did not die during the heat of the summer. All the ridges and hills in this neighbourhood were once covered with southern-beech forest and were still covered with the stumps of the trees, and it was ascertained that this particular ridge was burned off about fifteen or sixteen years before. Judging from the facts here stated, I am inclined to think that the spores were not blown from Mount Oxford, but had been shed from some mature plants which grew on the spot before the forest was removed. When the forest was burned the soil naturally would be very much disturbed and overturned, and favourable conditions would be set up for the germination of the spores and development of the prothalli. This would indicate that a long period of time is necessary for the development of sexually produced plants, a conclusion which is in accord with the observations of Bruchmann (6) in the case of the European species *L. clavatum*, *L. complanatum*, and *L. annotinum*. In several lowland localities near Hokitika, Westland, I have found young plants of *L. fastigiatum*.

L. scariosum.

I have collected about sixty prothalli of this species from two different localities on the Dun Mountains, Nelson, and from the Mount Oxford ridge mentioned above, and from a roadside clay cutting at Lake Kanieri, Westland, and have observed the young plants in various other localities. This species also would seem to propagate itself freely from spores when favourable conditions are present. The prothalli are always subterranean, and are more deeply buried than are those of *L. volubile* or *L. fastigiatum*, in some cases lying at a depth of 8–10 cm. Several times I have found the young plants of *L. volubile* and *L. scariosum* growing together, and in one instance dug the prothalli also of both species from the same patch of soil. Both the prothalli and also the youngest plants of these two species are easily to be distinguished from each other, the former because they belong to different types, and the latter because in the case of *L. scariosum* they are much stouter and coarser in appearance than in the other species, and also develop their characteristic form of heterophylly much earlier.

Summary.

The chief mode of propagation of the Lycopodiums is no doubt the vegetative one. Compared with other Pteridophytes, even with the Ferns, this character is most extensively developed in the Lycopodiums. One finds large areas covered by some one species or other, as, for example, by *L. volubile*, *L. densum*, *L. laterale*, *L. fastigiatum*, or *L. ramulosum*, where in all probability the original individuals were the only ones which had been produced sexually. In this connection we notice that the form of most of the terrestrial species is well adapted to this end. It is long drawn out with branches of unlimited growth, in many cases the stems and branches being subterranean, while at frequent intervals large adventitious roots are borne.   It is in the epiphytic species that this character of extensive vegetative propagation is noticeably absent, and there, of course, the opportunity is wanting owing to the confined area in which the plants grow.   The usual habitats of most of the terrestrial species are not suitable for the germination of the spores.   Such forms as *L. densum, L. fastigiatum, L. volubile,** and *L. scariosum* luxuriate in open situations which are dry and unpromising. *L. cernuum, L. laterale, L. ramulosum,* and *L. Drummondii* also grow most abnndantly in peaty sour land which is alternately waterlogged or dried up, or even permanently wet.   From my own observations, extended over a good number of years and in many parts of New Zealand, I have found it to be an almost invariable rule that young plants and prothalli are not to be met with in localities in which the adult plants are abundant.   It is only in special localities, such as a damp shaded clay bank or roadside cutting, or some other patch of recently disturbed soil in the neighbourhood of adult plants, that the young plants occur.   But it must be added that when the favourable conditions are present prothalli occnr often in great abundance.   In the case of several of the above-mentioned New Zealand species, prothalli and young plants were found in several localities in different years, while in certain species, as, for example, in *L. cernuum, L. fastigiatum, L. volubile,* and *L. scariosum,* they were discovered in large numbers.   Bruchmann (6, p. 5) records that when once he had discovered the right kind of locality he was able to collect over five hundred prothalli of each of the species *L. clavatum* and *L. annotinum.*   The epiphytic species grow under conditions which are normally more favourable to the development of the prothallus—namely, in shaded, damp, well-drained patches of humus.   Treub (17) was able to find the prothallus of four different epiphytic species.   Neither Miss Edgerley nor the present writer experienced great difficulty in discovering the prothallus of *L. Billardieri.*   It is certain that in the case of some of the species—for example, *L. fastigiatum, L. scariosum,* and *L. volubile*—a long period of time, with consistently favourable conditions, is required before the prothalli are fully grown and the young plants are established.   It has been shown above that there is reason to believe that a period of fifteen years may elapse before the spores of *L. fastigiatum* have germinated and the prothalli have developed.   This is very similar to the conclusion reached by Bruchmann (6, p. 10) in those instances in which data to work upon were forthcoming.   It would seem, however, that the prothalli of such species as *L. cernuum, L. ramulosum,* and *L. laterale,* which are exceeding minute and delicate, require only a single season for their full development.   These are the species which occur in a typically damp habitat.   However, when the prothalli, and even also the young plants of terrestrial species, have succeeded in developing, a change of conditions, as, for example, a more than usually dry summer, may result in the destruction of all the young plants and of the prothalli also.   Thus we conclude that although the spores of the *Lycopodium* species germinate freely under suitable conditions, yet the long period of the growth of their prothalli and young plants, and the uncertain conditions under which they live, have brought it about that the Lycopodiums have to depend mainly upon the vegetative mode of propagation.   Probably we may also conclude that in the case of the terrestrial species propagation

---

* In parts of Westland, which possesses a wet climate, there are large breadths of this species where road cuttings have been made.

from the spore is now of importance only in the initial establishment of
a certain species in new localities, as, for example, at the edge of the
retreating forest, and that the subsequent peopling of these localities by
such species, and their continued existence and spread, is due to vegetative propagation alone.*

III. External Features of the Prothalli.

L. Billardieri.

The prothallus of this species has recently been described in detail
by Miss Edgerley (8, p. 104). External examination of my own specimens
showed that it conforms more or less closely to the type of *L. Phlegmaria.*
Miss Edgerley states that the internal structure also, except in one point,
closely resembles that of the latter species. In each prothallus there is
a central mass of tissue, in some cases compact and bulky, in others
more extended and thin, from which long club-shaped processes arise
(figs. 1–11, on p. 277). The prothallus is dingy white in colour, the ends

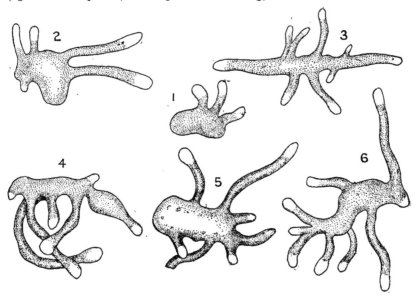

Figs. 1–6.—*Lycopodium Billardieri.* Complete prothalli. 1, 2, 3, and 5, × 8;
4 and 6, × 4.

of the processes being more translucent in appearance. Rhizoids arise
over the whole of the prothallus except at the extreme ends of the processes.
Besides the rhizoids, numerous short and thin processes, each of which
is several cells in length but only one in thickness, arise from the
surface of the prothallus. These will be probably the paraphyses described by Treub in *L. Phlegmaria* and by Miss Edgerley in this species.
The youngest prothallus found by me (fig. 1) was 1 mm. in total length,
and the largest from 12 mm. to 15 mm. (figs. 4–6).

* Where the land-surface was deeply covered by volcanic ash during the eruption of
Tarawera in 1886, *L. cernuum* was growing in abundance some years ago near certain
streams of hot water. Colonies of this description must have originated from spores,
but, unfortunately, no exact data are available as to the first appearance of young plants.

L. laterale.

This prothallus (text figs. 13–16, and Plate XVII, fig. 3) corresponds, with certain secondary differences, to the type of *L. cernuum.* All the prothalli found were colourless, except for a very faint tinge of green in the lobes. They are about 1 mm. in length. There is a lower rounded portion ("primary tubercle") darker in appearance than the rest of the prothallus, and which in the solid stains more deeply with haematoxylin. Examination of serial sections showed that the cells of this region were occupied by a fungus. This lower part of the prothallus bears numerous long rhizoids. The middle region of the prothallus ("shaft") in two cases was very short, giving the prothallus a stout and solid and opaque appearance (fig. 14). In the other prothalli it was slightly longer, these appearing more cylindrical and drawn out. The leafy expansions on the crown of the prothallus are filamentous and less lobe-like than in *L. cernuum.* A characteristic feature of all the prothalli of *L. laterale*

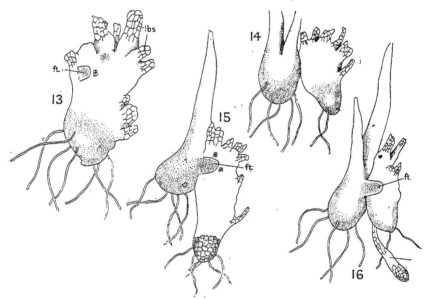

Figs. 13–16.—*Lycopodium laterale.* Complete prothalli, with young plants attached. 13, × 20; 14, × 10; 15 and 16, × 14.

examined was the presence of a filamentous outgrowth or a group of outgrowths on the shaft, either lower down towards the tubercle or higher up beneath the crown. None of the prothalli of *L. cernuum* seen by me showed any such leafy expansions on the shaft. Two prothalli, one of which is shown in fig. 16, possessed a long process attached to the primary tubercle. The end of this process was club-shaped, and consisted of numerous cells, but the remainder was thinner, and showed two or three rows of longish cells. Goebel (10, p. 194) quotes from Treub that in *L. salakense,* and occasionally also in *L. cernuum,* several branches may arise from the primary tubercle. Perhaps the process found in the two prothalli of *L. laterale* is of the nature of such a branch. It is interesting, however, to compare this with the long-drawn-out prothallus of *L. ramulosum*, described later. In the case of those prothalli which bore a young plant the foot could be plainly discerned through the semitransparent prothallial tissues. The prothallus of this species is very short-lived, and decays away after the young plant has developed two, or at the most three, protophylls.

L. cernuum.

Treub has fully described this prothallus. There are three regions.—viz., the lower primary tubercle, the intermediate shaft, and the crown of lobes (figs. 17–21). In some cases the tubercle is pointed, in

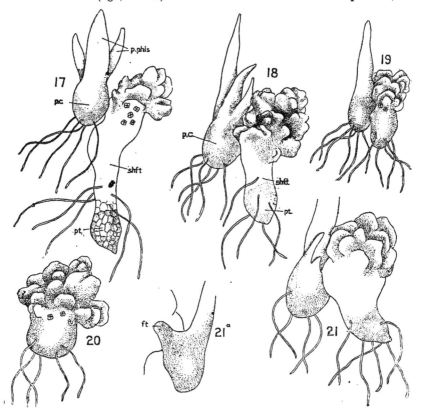

Figs. 17–21.—*Lycopodium cernuum.* Cnmplete prothali, with and without young plants. attached. 17, 19, and 2 , × 15; 18 and 24, × 18.
Fig. 21a.—*Lycopodium cernuum.* Young plant, showing foot. × 18.

others rounded, at the lower extremity. I examined a large number of specimens, and noted that the chief variation in them was in the length of the shaft. This is dependent upon the depth of the germinating spore below the surface of the ground, for the crown of lobes is always at the surface. Fig. 17 shows one prothallus with an abnormally long shaft; on the other hand, in fig. 20 it is seen that the shaft is almost absent. The crown in every case is very much lobed, the lobes of some prothalli being slightly greenish, and of others quite colourless. Rhizoids spring out of both the upper part of the tubercle and the lower portion of the shaft. The tubercle is always thickly infested with a fungus, as also are the rhizoids. The necks of old archegonia were seen in surface view on the upper portion of the shaft immediately under the lobes. As in *L. laterale*, the foot of a young plant attached to a prothallus could be plainly seen through the tissues of the latter. The total length of the prothallus of this species varies between 1 mm. and 2 mm. No instances of branching from the tubercle were observed, such as has been described in the last species.

L. densum.

As already stated, no prothalli of this species were found, but judging from the fact that a large foot was seen on a young plant, which was over 4 in. in height, and that this foot was about an inch below the surface of the ground, it is concluded that the prothallus is large, firm, and long-lived, and more or less deeply buried.

L. volubile.

A description of the prothalli of this species and of *L. scariosum* has been given by Miss Edgerley (8, pp. 95–99) from material supplied some years ago by the present writer to the Botanical Laboratory of the Auckland University College. The following account incorporates the chief points which I noted in the external examination of these prothalli and of others of the same species subsequently discovered. The smallest prothallus of *L. volubile* was about 1 mm. in height (fig. 22). There was a long tapering cylindrical projection below, and an upper more bulky

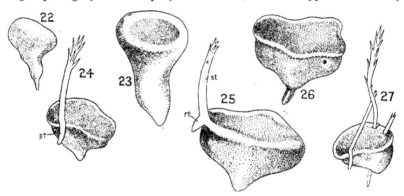

FIGS. 22, 23.—*Lycopodium volubile.* Young prothalli. 22, × 12; 23, × 15.
FIGS. 24–27.—*Lycopodium volubile.* Older prothalli. 24, × 6; 25, × 12; 26, × 8; 27, × 6.

portion. The prothallus was opaque and whitish in colour, except for its upper surface, which was more translucent in appearance. Fig. 23 is that of a slightly older specimen, in which a saucer-like depression surrounded by a thick rim is developing on the upper surface. In this prothallus the upper portion was above ground, and the translucent rim and depressed surface were green with chlorophyll. Still older prothalli are shown in figs. 24–28. In most of these it will be seen that the first-formed tapering portion is still present, in some cases as a blunt and in others as a pointed projection on the lower surface. The greater number of the prothalli showed a simple saucer-like depression on the

upper surface, surrounded by a folded rim, elongated in one direction.
The rim was absent at the two ends of the depression, which, owing to
the continued growth of the upper region of the prothallus, had in some
cases become curved outwards and downwards. This agrees in all main
points with the descriptions which Lang (14) and Bruchmann (6) have
given of the prothallus of *L. clavatum,* and Bruchmann also in the case
of the prothallus of *L. annotinum.* The oldest and largest specimens.

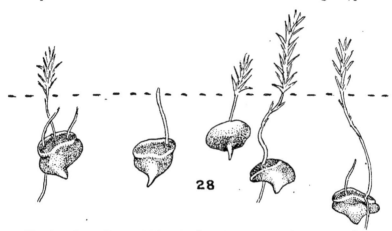

FIG. 28.—*Lycopodium volubile.* Prothalli in position relative to surface
of ground. × 4.

sometimes showed one or more secondary folds on the upper surface, and
rounded protuberances, which had the effect of almost filling up the
original depression (fig. 29). The prothallus shown in fig. 26 was growing at the surface of the ground, and its upper portion was coloured a
dark green. Rhizoids are present on well-grown prothalli, especially
on the region above the first-formed tapering portion immediately below
the rim; but many of the prothalli seemed to be quite destitute of them,.

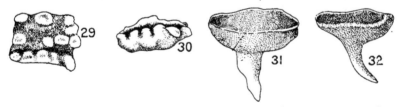

FIGS. 29, 30.—*Lycopodium volubile.* A very large prothallus; top and side
views respectively. × 2½.
FIGS. 31, 32.—*Lycopodium volubile.* Exhausted prothalli. × 8.

and were easily brushed clean of any adhering particles of soil or pieces.
of moss. In not a few cases two or three young plants were seen on
the same prothallus. Figs. 31 and 32 show two exhausted prothalli,
in which the long tapering first-formed erect-growing portion is very
obvious, the upper part of the prothallus being flattish and slightly
concave, and the rim sharp. The depleted prothallus shown in fig. 32,
when viewed from below, showed between the tapering portion and
the upper flat region a narrow ridge which corresponded in direction to·

the elongation of the upper surface. This would seem to indicate that the horizontal extension of the upper region of the prothallus follows a more or less definite axis of growth. The largest prothallus found—that shown in figs. 29 and 30—was 6 mm. in length, the average size of a mature specimen being from 3 mm. to 4 mm.

L. ramulosum.

The figures illustrating the prothallus of this species are numbered from 32A to 32G inclusive. The youngest prothallus found was about 1 mm. in height, and is shown in fig. 32A. It will be seen that in general form it corresponds with the young prothallus of *L. inundatum* figured from de Bary by Goebel (10, p. 191, fig. 3). The lower rounded region was infested by a fungus, which extended somewhat up one side of the shaft. At its upper end the prothallus was extended into two

FIG. 32A.—*Lycopodium ramulosum.* Young prothallus. × 20.
FIGS. 32B–32D.—*Lycopodium ramulosum.* Prothalli of long-drawn-out form, showing fungal regions, &c. × 12.
FIGS. 32E, 32F.—*Lycopodium ramulosum.* Prothalli of stout, massive form, attached to young plants. × 12.
FIG. 32G.—*Lycopodium ramulosum.* The lower extremity of prothallus shown in fig. 32F. × 35.
FIG. 32H.—*Lycopodium ramulosum.* Protocormous rhizome, showing stem-axis, first root, and two vegetative bulbils. × 4.

processes, which had a slightly brownish appearance, as if withered. At the base of one of these processes was a darkly staining cell, possibly an antheridium. Below the upper end of the prothallus, on the under side of a short lateral process, three epidermal cells were seen to be infested by the fungus, and a rhizoid was borne on one of the adjacent cells. Rhizoids were also present on the lower end of the prothallus. Three young prothalli of this age were found. All the prothalli found were, in general, of the *L. cernuum* type, but they presented some remarkable variations of it. In two or three instances old pro-thalli attached to young plants were very similar in appearance to those of *L. cernuum* shown in figs.. 18 and 21. The other pro-thalli, however, were either long-drawn-out, being from 2½ mm. to 4 mm. in length, or short and comparatively massive. Eighteen prothalli of the former kind were found, two of which bore very young plants. Three of these are shown in figs. 32B, 32C, and 32D. Of the latter kind I have discovered six, all of which were attached to young plants. In figs. 32E and 32F are shown two of them. The prothallus shown in fig. 32B possessed a lower rounded extremity, which bore rhizoids, and whose cells were infested with fungus. It was evident, however, that this portion of the prothallus, which may be spoken of as the "primary tubercle," since it corresponds to the region in the prothallus of *L. cernuum* which has been given that name by Treub, is not the first-formed part of the prothallus, for its cells were continued below into an empty-celled process which had been broken off short. The shaft is very much longer than in any of the prothalli of *L. cernuum* or *L. laterale.* Half-way up the shaft was a filamentous projection which bore sexual organs. On the opposite side of the shaft to this process two epidermal cells were infested by the fungus. At its upper extremity the prothallus was expanded into a bulky mass of tissue. The cells of the underneath region of this massive tissue showed the presence of fungus, and a group of rhizoids sprung from the same region. Imme-diately below the upper surface of the prothallus the necks of archegonia could be seen. Fig. 32C is that of a prothallus in which there were no fewer than five separate fungal regions along one side of the shaft. It was noticeable that each one of these regions was swollen and rounded, as if the presence of the fungus had served to stimulate the growth of the cell-tissue. A group of rhizoids was borne on each one of these fungal regions. Filamentous outgrowths with archegonia at their base occurred in two places on the main shaft, opposite to the fungal zones. This prothallus was 4 mm. in length. I did not observe whether it was growing upright or horizontal. All of the extended prothalli were dissected out of short moss and slime fungus which was free from soil and decayed humus, and possibly the abnormal length of the shaft is to be put into connection with the depth at which the spores germinated below the surface of the moss, while the presence of several fungal regions bearing rhizoids is due to the absence of humus in the layer in which the prothalli were growing, and to the consequent dependence of the latter upon the intracellular fungus for much of the required food. In fig. 32D is shown another prothallus, whose total length was about 4½ mm. It bore two groups of filamentous processes, at different places on the shaft. There were several fungal regions, all of which bore rhizoids, and were slightly swollen, though to a less extent than in the prothallus shown in fig. 32C. The four lower groups of fungus-contain-ing cells were situated on the same side of the shaft, and were so close together that they almost formed one continuous zone. It will be noticed that in this particular prothallus the lowest fungal region is scarcely tubercular in form, and that it is continued below into a tapering empty-celled filament, which was probably the region of the prothallus first formed from the spore. The two uppermost fungal regions are on the side of the shaft opposite to the others. The long shaft is expanded above into a somewhat bulky crown, which bears filamentous and lobe-like foliar expansions. A diminutive young plant which had developed a single protophyll was attached to the crown of this prothallus, the uppermost fungal region being situated immediately below it. In figs. 32E and 32F are shown two solid massive prothalli. The latter of these was $2\frac{1}{2}$ mm. in total height, and about 2 mm. in diameter in its upper region. These prothalli (and the other four of the same nature which were found) seem to correspond to the upper bulky region of the long-drawn-out prothallus of this species, described above. It would appear, however, that they owe their compact form not to the decaying-away of the shaft, but to its almost complete suppression. That this is so is evident from fig. 32F, which shows the first-formed region of the prothallus in a remarkably intact condition. Fig. 32G is a much-enlarged view of the same. These short stout prothalli were in three instances found attached to young plants which bore four or five proto-phylls. Half-decayed prothalli were observed attached to young plants of still greater size and age. All the prothalli of *L. ramulosum* were green in their upper region and in the upper parts of the shaft. After studying the prothallus of this species I carefully examined again serial sections of several prothalli of *L. cernuum*, and in the case of one of them found that the primary tubercle was continued below into such a filamentous process as that shown at the base of the prothallus in fig. 32B. I am inclined also to regard the club-shaped process described on the primary tubercle of two of the prothalli of *L. laterale*, and illustrated in fig. 16, as corresponding to the first-formed region of the prothallus of *L. ramulosum.* It would seem thus that in these species, in some instances at least, and perhaps also as a rule, the spore on germinating gives rise to a delicate filament of cells which at some point or other soon becomes infected with the fungal element and then swells to form the so-called " primary tubercle." One prothallus found was shaped like the letter Y, it having branched into two more or less equal branches about half-way up the main shaft. At the point of branching the shaft was swollen and showed a fungal area, there being also the usual " primary tubercle " at the base of the shaft.

The prothallus of *L. ramulosum* presents some important features which serve to emphasize the great variability of the *Lycopodium* prothallus, and which suggest links between the different prothallial types much in the same. way that the variations in form of the prothallus of *L. Selago* have been interpreted by Lang (14, pp. 303–5). The form with the long shaft and scattered fungal areas indicates how the long-drawn-out prothallus of *L. Selago*, and so also the epiphytic type of prothallus, could have arisen from an ancestral *L. cernuum*-like type; and, on the other hand, the short massive prothallus of *L. ramulosum*, with its longer life and greater capability of the horizontal extension of its upper region, is intermediate in form between the *L. cernuum* type and the short variety of the *L. Selago* prothallus, and is suggestive of the subterranean *L. clavatum* and *L. complanatum* types.

L. fastigiatum.

The general form of this prothallus in the mature state would seem to be very like that of *L. volubile*, but there is an important develop-mental distinction to be noted. Figs. 33–40 illustrate specimens in which the manner of development of the adult form can be traced. The youngest prothalli found were about $\frac{1}{2}$ mm. in height. They showed a rounded tubercular opaque body surmounted by a smaller and more translucent region. Fig. 33 is that of a young prothallus 1 mm. in height showing these two regions. Very early in the development the

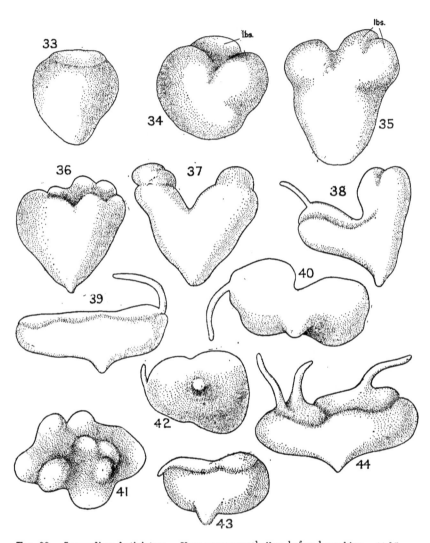

Fig. 33.—*Lycopodium fastigiatum.* Very young prothallus, before branching. × 15.

Figs. 34–38.—*Lycopodium fastigiatum.* Young prothalli, showing stages in branching. 34 and 35, × 15 ; 36 and 37, × 12 ; 38, × 9.

Figs. 39, 40.—*Lycopodium fastigiatum.* Older prothallus; side and underneath views respectively. × 10.

Fig. 41.—*Lycopodium fastigiatum.* Full-grown prothallus, showing lobing on upper surface. × 5.

Figs. 42, 43.—*Lycopodium fastigiatum.* Prothallus; underneath and side views respectively. × 7.

Fig. 44. — *Lycopodium fastigiatum.* Large prothallus, with three young plants attached. × 5.

upper region becomes lobed (fig. 34), and these lobes soon show themselves to be two main branches (figs. 35, 37, 38), each of which also develops secondary lobes. A large number of young prothalli of this age were collected, and they were all seen to show more or less clearly the Y form. In older specimens there is an extended and slightly concave flattened upper surface, the concavity being more cr less obscured by irregular lobing and folding (figs. 41 and 44). The original two main branches can always be distinguished in an underneath view of the prothallus (figs. 40 and 42). The view from above also in some cases shows this, and shows as well, in spite of the complexity of the folding of the upper surfaces, that each main branch has divided into two branches, so that the upper surface is somewhat quadrangular in form (fig. 41). The first branching of the prothallus always takes place very early, so that in the older prothalli the peg-like projection is never as long and tapering in this species as it is in *L. volubile*. In the young prothalli the rhizoids arise especially from the upper ends of the two branches. The lobes are always clearer in appearance than the rest of the prothallus, which is in colour dingy white. None of the prothalli showed any trace of chlorophyll, being always more or less deeply buried.

L. scariosum.

Miss Edgerley (8, pp. 100–2) states that the prothallus of this species resembles that of *L. clavatum*. I would unhesitatingly, however, say that it approaches more closely to the type of *L. complanatum* as described by Bruchmann (6). The material examined by Miss Edgerley had been collected by the present writer, and had been killed and fixed in 80 per cent. alcohol, and there is no doubt that it had shrunk considerably by the time that Miss Edgerley studied it. Hence she describes the upper surface of the prothallus as "concave . . . with a ridge running round the margin." In figs. 45–53 in the present paper are depicted fresh prothalli of this species, and it will be see that they correspond very closely in general form to those of *L. complanatum* figured by Bruchmann on plate V and described by him on pp. 57–59 of his work. There is, however, one main difference—namely, that old prothalli of the New Zealand species grow to a great size and become very irregular in form, so that a pseudo-branching is sometimes to be observed in them. The prothalli show a conical region below, in some cases blunt and rounded, in others more tapering, while above they are more massive. In the prothallus of this species there is no definite horizontal extension of the upper portion or lobing, such as there is in that of *L. volubile*. The upper region is surmounted by a bulky convex mass of tissue, which is semitranslucent in appearance, and is separated from the main body of the prothallus by a well-marked neck. In partially depleted prothalli, and in prothalli which have shrunk in alcohol, this upper tissue becomes slightly concave, but in fresh material it is seen to be invariably rounded and massive, and is not surrounded by a rim, but bulges out over the main body of the prothallus. In such an old specimen as figured in fig. 53 this upper tissue constitutes the major bulk of the prothallus. A longitudinal section of a medium-sized prothallus, such as any of those shown in figs. 45–48, corresponds almost exactly with that figured by Bruchmann (6, pl. V, fig. 25). My observations are, however, in agreement with Miss Edgerley's statement that the various layers of

·fungus-infested tissue in the prothallus of *L. scariosum* occupy a characteristically small proportion of the total bulk of the prothallus. Several old large prothalli showed a right and left portion separated by

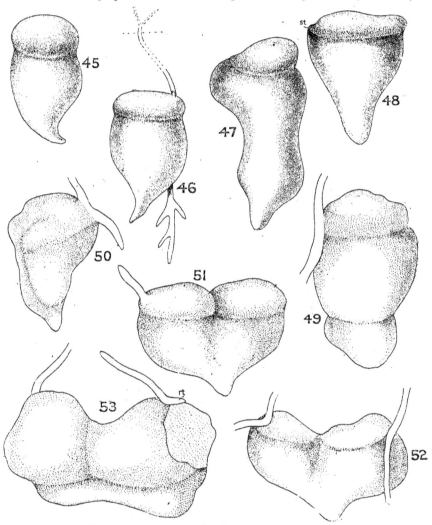

Fig. 45.—*Lycopodium scariosum.* Young prothallus. × 10.
Figs. 46–48.—*Lycopodium scariosum.* Prothalli. 46, × 6; 47, × 6; 48, × 8.
Figs. 49–52.—*Lycopodium scariosum.* Large irregularly grown prothalli. 49, × 8; 50, 51, and 52, × 5.
Fig. 53.—*Lycopodium scariosum.* Very large irregularly grown prothallus. × 6.

a cleft which extended through the upper translucent tissues well down into the body of the prothallus (figs. 51–53). One of Miss Edgerley's figures also illustrates the same feature. This is probably to be regarded as a pseudo-branching resulting from the continued and irregular growth in size of the prothallus, it being characteristic of the prothallus of this species that it goes on growing till it has quite lost its original form. The specimen shown in fig. 47 was $8\frac{1}{2}$ mm. in height. The old prothallus in fig. 53 was $5\frac{1}{2}$ mm. in height and 5 mm. in width, and no less than 13 mm. (including the broken-off piece) in length. Fig. 49 illustrates an irregularly grown bulky prothallus, where the main lower portion has become divided into two by a horizontal constriction.

Summary.

In summing up this section I would lay stress first of all upon the fact that the study of the prothalli of the New Zealand species described above introduces no new type differing in any great degree from the five main types enumerated by such writers as Bruchmann, Lang, Pritzel, Bower, and others. This is noteworthy, for it might have been expected that, since the discovery of the prothalli of only eleven out of the large number of modern species of *Lycopodium* had resulted in five distinct types being recognized, a considerable increase in our knowledge of the Lycopod gametophyte would reveal the existence of still further types. The prothallus of *L. ramulosum* certainly presents remarkable features, but these are to be regarded probably only as variations from the *L. cernuum* type. Thus the results of the present study lend additional weight to the theory that in the sexual generation of the genus *Lycopodium* we are to trace the influence of two main factors—viz., epiphytism on the one hand, and a subterranean mode of life on the other—the development of the various types from a relatively primitive, chlorophyllous, and surface-living form having proceeded along these two main lines.

A second point to be noted is that although the New Zealand species of *Lycopodium* prothalli conform to the types so well known from the study of European and tropical species, yet in every case there are interesting modifications to be observed. This emphasizes the fact that the *Lycopodium* prothallus is a most variable one. But, further still, the nature of these various modifications may possibly be regarded as having significance in indicating how the epiphytic and the subterranean types have evolved from an original surface-living form, or even as supplying actual links connecting the different types together along two such lines of evolution. Lang and Bower both seem to consider the possibility of the genetic relationship of the different subterranean forms of prothallus, although they also suggest that they may be independent developments from a common ancestor. The most noteworthy variations from the ordinary types which occur in the New Zealand species of prothalli are as follows: The shaft of the prothallus of *L. cernuum* may be relatively long drawn out, suggesting the long cylindrical branches of the prothallus of *L. Selago* and of the epiphytic prothalli. The leafy expansions of the prothallus of *L. laterale* are distinctly filamentous and more branch-like than in the case of *L. cernuum*, and, moreover, these filamentous processes arise normally from other parts of the shaft as well as from its terminal region. The prothallus of *L. ramulosum* may be either long drawn out, with a somewhat massive crown, or short and bulky, with the shaft almost entirely suppressed and the crown greatly developed. Lang (14, p. 306) has also emphasized the variability of characters in the *L. cernuum* type which is known from the study of *L. inundatum* and *L. salakense*, such as the presence or absence of the crown of lobes and the manifold branching from the primary tubercle, features which seem to indicate a relationship between the *L. cernuum* and the *L. Selago* types. It will be remembered that Thomas (16, p. 287) draws attention to the variation in the length of the shaft in the prothallus of *Phylloglossum*. Miss Edgerley (8, p. 107) in her description of the prothallus of *L. Billardieri* has noted that it is simpler in organization than that of *L. Phlegmaria*, in that the elongated central cells in the branches do not show the presence of pits on the walls, and that the paraphyses are only three cells in length and are never branched. The prothalli of *L. fastigiatum* and of *L. volubile* are in their mature form very much alike, being flattish and saucer-like; but, as has been described above, this form has been arrived at differently in the two cases. The branching of the prothallus of *L. fastigiatum* would seem to be reminiscent of the branching which is found in the *L. Selago* type; and also in the epiphytic and the *L. cernuum* types.

The close relationship between the form of the *Lycopodium* prothallus and its habitat, as illustrated in the species dealt with in this paper, must also be briefly noticed. The prothalli of *L. cernuum*, *L. ramulosum*, and *L. laterale* are delicate, minute, and short-lived, and are surface-growing. The variation in length of the shaft of their prothalli is probably to be put in connection with the varying depth at which the spores germinate below the surface of the ground. These three species occur in wet habitats where the establishment of the young plant on the surface of the ground can readily take place. The epiphytic species of prothallus has an extensively branched and ramifying form. Its typical habitat is a loose matted humus. The greatest variety is to be found amongst those terrestrial species whose prothalli are subterranean. Some of these subterranean prothalli, such as those of *L. volubile* and *L. fastigiatum*, occur chiefly in the loose humus in the top layer of soil immediately beneath the overlying moss and vegetable growth. These are the prothalli whose form is flat and saucer-like owing to the superposition of a horizontal mode of growth on the original vertical radial habit. The prothallus of *L. scariosum*, on the other hand, occurs as a rule much deeper and in more compact soil, and here the upright manner of growth is preserved and the saucer-like depression on the upper surface is absent. That view which would look upon the *Lycopodium* prothallus as a very plastic one would seem to be more in accordance with the foregoing observations than that other view which would regard the different types of prothallus as having been genetically distinct from a very remote period.

IV. The Morphology of the Young Plant.

L. Billardieri.

The young plant takes its origin from the central mass of tissue of the prothallus. It first appears as a simple cylindrical stem (figs. 7 and 9), which it quite destitute of leaves for a length of 1–6 cm., according to the depth at which the prothallus is buried in the humus. The first-formed one or two leaves are generally scale-like, but the succeeding ones are large and of the mature form (figs. 10–12). A longitudinal section of the prothallus and young plant shown in fig. 7 revealed the fact that the latter possessed a fairly large firm foot, and that there was a distinct epithelial layer of cells where the latter came in contact with the prothallial tissues (fig. 8). It also showed the "first root" (throughout this paper this term is used to signify the first functional root) making its appearance as a protuberance at the base of the stem.

The vascular strand of the stem and first root did not enter the foot. It was observed that the foot still persists on young plants which show as many as six to eight full-sized leaves. There was no trace of any swelling on the young plant which could be interpreted as the rudiment of a protocorm. The first root is developed relatively late. In the plants

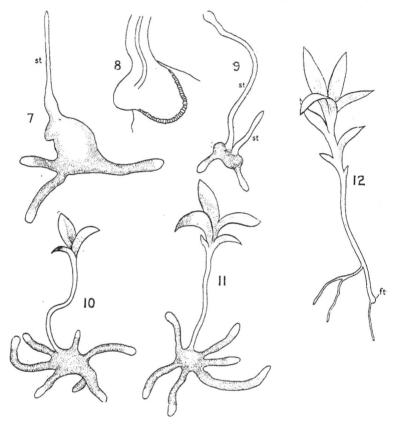

FIGS. 7–11.—*Lycopodium Billardieri.* Complete prothalli, with young plants attached. 7; × 8; 9–11, × 4,

FIG. 12.—*Lycopodium Billardieri.* Young plant with foot and first roots. × 4.

shown in figs. 10 and 11 it had not appeared, although several full-sized leaves had been formed. All roots in the young plant subsequent to the first root are adventitious, and make their appearance on the stem above the foot (fig. 12).

L. laterale.

In two previous papers (11, pp. 357–61, and 12) the writer has given an account of the young plant of this species. The chief points in its development will here be given in reference to several micro-photographs and drawings which are included in this paper. The prothallus of this species has already been described as belonging to the *L. cernuum* type. During the early stages in its development the young plant is very similar to that of *L. cernuum.* It consists of a basal tuberous protocorm surmounted by one or two protophylls, and is connected with its parent prothallus by a foot which can clearly be distinguished through the prothallial tissues (figs. 14–16). The protocorm and protophylls are a vivid green in colour, and stomata occur on the latter. The protocorm bears numerous long rhizoids. From this stage onwards the young plant of *L. laterale* differs in its development from what normally takes place

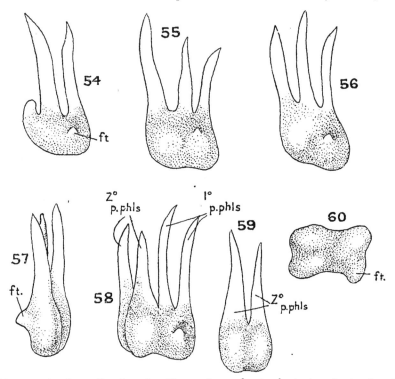

FIGS. 54–56.—*Lycopodium laterale.* Young plants, showing beginning of extension of protocorm. × 16.

FIG. 57.—*Lycopodium laterale.* Young plant shown in fig. 56; side view. × 16.

FIG. 58.—*Lycopodium laterale.* Young plant with four protophylls. × 16.

FIG. 59.—*Lycopodium laterale.* Young plant shown in fig. 58; the latest-formed protophylls in end view. × 16.

FIG. 60.—*Lycopodium laterale.* Young plant shown in fig. 58; underneath view. × 16.

in the case of *L. cernuum.* The third protophyll arises in a lateral position (fig. 54), and when it is full-grown its base shows as a swelling clearly to be distinguished from the original protocormous tuber (figs. 55 and 56). In fig. 57 the young plant shown in fig. 56 is depicted in a sideways position, in which the two distinct swellings are clearly seen. The fourth protophyll arises alongside the third, and forms a pair with it (fig. 58). Here again it is to be observed that the swollen bases of these two protophylls are distinct from one another. In fig. 59 are shown in end view the two latest-formed members of the plant illustrated in fig. 58. Fig. 60 is an under view of the same plant, showing the foot on the first-formed protocorm proper, and the bases of the two latest-formed protophylls. A transverse section of such a plant through the thickest region of the plant-body reveals clearly the original tuber and the later extension of it, but also shows that internally the swollen bases of the third and fourth protophylls are

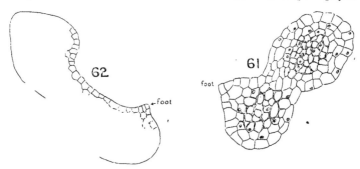

FIGS. 61, 62.—*Lycopodium laterale.* Transverse section of two such young plants as shown in fig. 58. 61, × 60 ; 62, × 35.

completely fused, although externally they can be distinguished (figs. 61 and 62). In no case was the prothallus found still attached to a young plant which bore more than the two original protophylls. The plant-body continues to grow sideways owing to the lateral development of new protophylls (figs. 63–65). In the majority of the young developing plants that were examined it was observed that the first-formed protocorm proper could clearly be distinguished from the later-formed protocormous extension, there being a well-marked constriction between

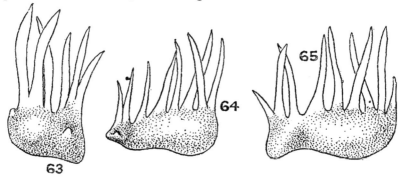

FIGS. 63–65.—*Lycopodium laterale.* Young plants, showing further extension of the protocorm. × 12.

the two parts (figs. 64 and 65). In the cleaning process the two portions in not a few cases broke away from each other. The protocormous rhizome continues to elongate laterally, owing to the further development of protophylls, till there are as many as eight to twelve of the latter. The protophylls arise in pairs, and in many cases they are arranged in two more or less distinct rows along the dorsal side of the rhizome. This, of course, indicates that their development, and so also that of the whole plant-body, has taken place very regularly. The rhizome is covered on its ventral surface with a mat of rhizoids. In the earlier stages of its development it is bright green in colour and semi-translucent, but later it becomes yellowish and opaque and very firm. The

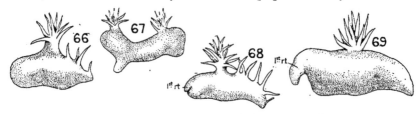

FIG. 66.—*Lycopodium laterale.* Young plant with fully developed protocorm and a very young stem-axis. × 4.
FIG. 67.—*Lycopodium laterale.* Young plant with a branched protocormous rhizome and two stem-axes. × 4.
FIGS. 68, 69.—*Lycopodium laterale.* Young plant with young stem-axis and first root just showing. 68, × 4; 69, × 5.

total length of the fully developed rhizome is from 3 mm. to 5 mm., and its thickness from 1 mm. to 2 mm. In one instance it was observed to have forked into two equal branches, and on each of these a young stem-axis was developing (fig. 67).

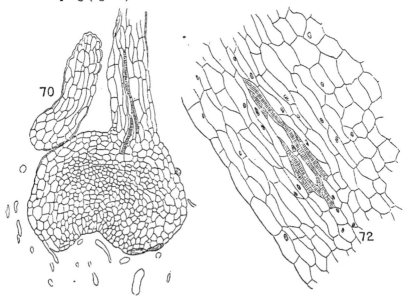

FIG. 70.—*Lycopodium laterale.* Transverse section of fully grown protocormous rhizome before development of vascular strand, showing two protophylls above and groove on ventral side. × 30.
FIG. 72.—*Lycopodium laterale.* Portion of rhizome shown in fig. 71, showing relation of vascular strand to the tissues of the protocormous rhizome. × 225.

The rhizome consists of parenchymatous tissues throughout, the cells of the central region being smaller and more compact, while those nearer the surface are larger and sometimes show air-spaces at their angles (fig. 70). The centrally placed cells stain much darker with haematoxylin than those nearer the surface, owing to the presence in them of fairly abundant cell-contents. Each protophyll possesses a simple vascular strand, which penetrates into the upper region of the rhizome,

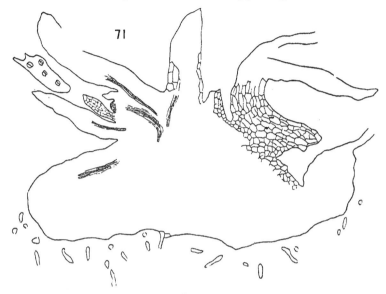

FIG. 71.—*Lycopodium laterale.* Longitudinal section of fully grown protocormous rhizome, showing young stem-apex and course of vascular strand. × 25.

and there ends blindly (fig. 70). A transverse section of a fully grown rhizome, such as is shown in fig. 70, shows a well-marked groove running ventrally along the length of the rhizome. This is probably a conse-

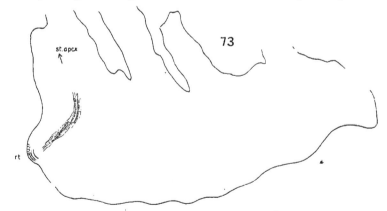

FIG. 73.—*Lycopodium laterale.* Longitudinal section of fully grown protocormous rhizome, showing initiation of first root. × 25.

quence of the fact that the body of the rhizome is formed by the swollen bases of the two rows of its protophylls. The stem-axis is initiated eventually at some point on the dorsal surface, and is marked by the aggregation of several protophylls. This takes place either close to the growing end of the rhizome or at some point farther away from it, although the position of the young stem is always nearer to the growing end than to the first-formed protocorm proper (figs. 66–69, 71, 73, 75–78). Almost immediately vascular tissues are initiated from the stem-apex, and extend down into the upper region of the rhizome, receiving on the way strands from the neighbouring protophylls (fig. 71). In the rhizome these vascular tissues bend round at an angle which is more or less sharp according to whether the stem-apex is farther away from the growing point of the rhizome (fig. 71) or close to it (fig. 73), and take a course through the tissues of the rhizome, near to its dorsal surface, towards its growing end. At a later stage this vascular strand is surrounded by a slight zone of sclerenchyma (Plate XVII, fig. 1). The arrangement of the

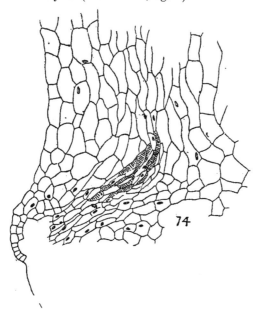

Fig. 74.—*Lycopodium laterale.* Portion of rhizome shown in fig. 73, showing relation of vascular strand to the apex of root. × 225.

elements in the strand is described in the next section of this paper. Fig. 72 is a highly magnified drawing of a portion of fig. 71, and shows that the vascular strand of the stem and first root as it passes through the body of the rhizome is developed from the actual tissues of the rhizome. The same fact is also apparent from fig. 74, which is a magnified drawing of a portion of fig. 73, and from Plate XVII, figs. 1 and 2. At the same time that the vascular strand from the stem-apex is taking form along the body of the rhizome in the direction of its growing end, the latter at a point on the surface towards the dorsal side begins to grow outwards and downwards to form a finger-like protuberance (figs. 68, 69, 73, 74). Into this protuberance the vascular strand passes. This is the first root. The extension in length of the rhizome is brought to a close by its initiation. All subsequently formed roots in the young plant emerge adventitiously from the stem, and do not pass through the

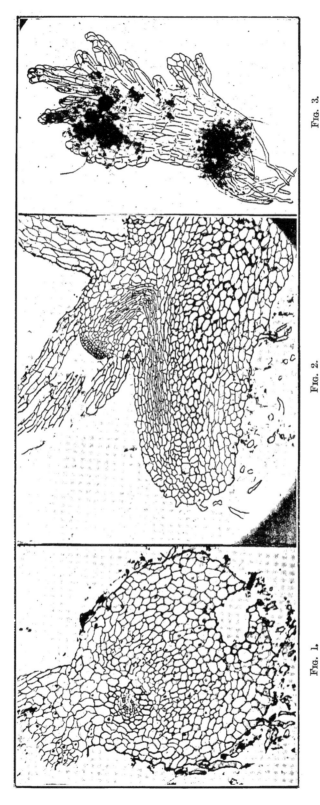

Fig. 1.

Fig. 2.

Fig. 3.

Fig. 1.—*Lycopodium laterale.* Transverse section of protocormous rhizome, showing relation of vascular strand to tissues of rhizome. × 40. (N.B.—In this figure and in the other plate figures which accompany this paper the cells have all been carefully outlined on the photograph with indian ink, in order to ensure clear reproduction.)

Fig. 2.—*Lycopodium laterale.* Longitudinal section of portion of protocormous rhizome, showing course of vascular tissue from stem-apex to first root. × 35.

Fig. 3.—*Lycopodium laterale.* Complete prothallus. × 30.

tissues of the protocormous rhizome. The leaves on the young stem-axis are in nowise different from the protophylls on the protocorm proper or on the rhizome; in fact, the ordinary vegetative leaves of the adult plant also have much the same form. The protocormous rhizome of this species is a persistent organ, owing to its large size and firmness. It may be recognized at the base of the stem of young plants which are 2 in. or even more in height.

L. cernuum.

The young plant of this species has been fully described by Treub, so that in this paper mention will be made only of such features as seem

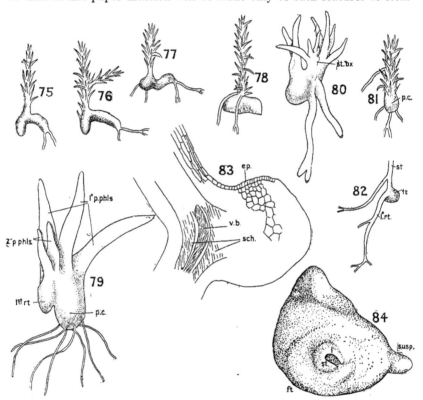

FIGS. 75–78.—*Lycopodium laterale.* Young plants with developing stem-axis and persisting protocorm. × 3.

FIG. 79.—*Lycopodium cernuum.* Young plant, showing initiation of first root. × 12.

FIGS. 80, 81.—*Lycopodium cernuum.* Young plants with developing stem-axis and adventitious roots. 80, × 6; 81, × 3.

FIG. 82.—*Lycopodium densum.* Lower region of stem of young plant, showing foot and first roots. × 4.

FIG. 83.—*Lycopodium densum.* The foot shown in fig. 82. × 40.

FIG. 84.—*Lycopodium fastigiatum.* Model of embryo of young plant, showing relative positions of foot and suspensor to the rudiments of the root, stem, and first leaf. × 105.

to be noteworthy for the purpose of a comparison with the young plant of the last species. The general form of the protocorm and first protophylls will be seen in figs. 17–19, 21, and 21A. In *L. cernuum,* as a rule,

the protophylls are all developed from the top of the protocorm, so
that there is no noticeable lateral extension of the latter (figs. 79–81).
In one or two instances, however, out of a great number of young plants
examined I observed that there was a certain amount of lateral growth,
the latest-formed protophylls occupying a position not on top of the
protocorm, but at the side of it farthest from the foot.   One young
plant showed seven protophylls in all on such an extended protocorm
before a stem-axis had been initiated.   Plate XVIII, fig. 1, is a photograph of a section of a young plant whose protocorm has thus grown
sideways.   In this particular plant the stem-axis had already been
initiated on the end of the protocorm farthest from the first-formed
portion, but it is not included in the section.   The first root had
also appeared, and is shown in the photograph.   There is a distinct differentiation to be observed in the body of this protocorm
between the protocorm proper and the laterally extended portion.   As
a rule, however, the protophylls are aggregated on the summit of the
protocorm, and after a certain number have been developed—generally
four or five—the stem-apex appears amongst them at the base of several
rapidly growing new protophylls.   At the same time that the stem-apex
is initiated a finger-like protuberance appears in that region of the
protocorm which lies immediately below the position of the stem, and
grows outwards and downwards (fig. 79).   A longitudinal section of the
young plant shown in fig. 79 revealed the fact that vascular tissues
were already developing from the stem-apex, and were declining into
the root-like protuberance at a slight angle.   A photograph of this is
given in Plate XVIII, fig. 2.   The section shown in this photograph
is not exactly median either for the stem-apex or for the root-apex,
but it shows the course of the vascular strand, and also the position
of the first root on the protocorm.   The root-apex was protected
by a cap of cells which had originated from periclinal divisions in
the outer layer of cells of the apex.   In the growing plant new leaves
are rapidly developed from the stem-apex, and a stem-axis is formed
(fig. 80).   Subsequently formed roots emerge either at the base of the
stem through the tissues of the protocorm (fig. 80) or higher up the
stem (fig. 81).   The protocorm can be distinguished at the base of
young stems which are from 1 cm. to 3 cm. in height.   As in *L. laterale*,
the prothallus does not persist after the first two or three protophylls
have developed.

L. densum.

As has been stated in a preceding section of this paper, I have not
been able to discover more than a single young plant of this species.
This was a plant about $4\frac{1}{2}$ in. in height, erect, and branched above.
One other developing plant was also discovered in the same patch of
damp moss, but this was older, and had already assumed the creeping
habit.   These young plants occurred in the immediate vicinity of adult
plants of *L. densum*, and there were no other species of *Lycopodium* to be
seen in the neighbourhood.   On the first-named " seedling " plant a large
firm foot was present at a depth of about 1 in. below the surface of
the ground (fig. 82).   The first root was borne at the base of the stem,
and an adventitious root on the stem itself a short distance above the
foot.   Longitudinal sections through this foot failed to reveal any
vascular tissue passing into it from the main stele.   The main body
of the foot consisted of large-sized parenchymatous cells, and there was

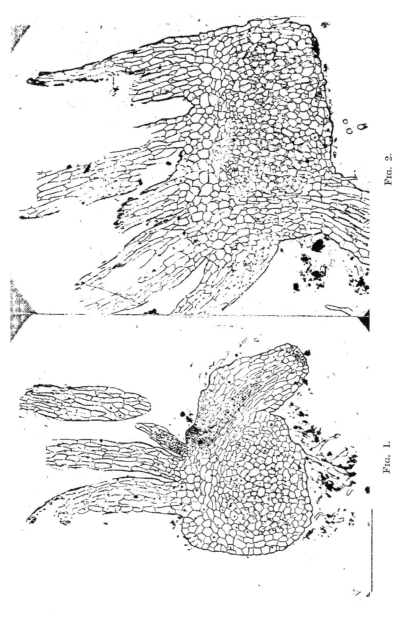

Fig. 1.

Fig. 2.

Fig. 1.—*Lycopodium cernuum.* Longitudinal section of protocorm, showing course of vascular strand from stem-apex to first root. × 40.
Fig. 2.—*Lycopodium cernuum.* Longitudinal section of protocorm which had developed first root and stem-apex, showing unusually large size of protocorm. × 45.

a distinct epithelial layer of small regularly arranged cells (fig. 83). There was also a central core of lignified cells which was continuous with the sclerenchyma zone of the stem. Although the prothallus of this species was not found, it is evident from the presence of the foot and its depth below the surface of the ground that the prothallus would belong to one of the subterranean types, as in the case of the three succeeding species, and that the young plant is dependent upon its parent prothallus for a considerable time.

L. ramulosum.

The young plant of this species corresponds very closely in its structure and manner of development to that of *L. laterale*. There is a protocorm which attains to a comparatively large size before the stem-axis is originated, and, as in the case of *L. laterale*, this protocorm may even branch. One peculiar feature, however, in the young plant of the present species must be referred to. In the young plant shown in fig. 32H it will be seen that, contrary to what normally takes place, the growing end of the protocormous rhizome has continued its growth after the stem-axis and first root have developed. In fact, the end of this rhizome seems to have branched, the two developing branches appearing as bulbous outgrowths, each surmounted by a single protophyll. I am inclined to regard these bulbous branches as vegetative bulbils which would be capable of independent existence. In another instance seven young plants were found bunched together. They had to be extricated from one another. Two of them were exceedingly small, showing respectively one and two protophylls each; three others were of slightly larger size, having three or more protophylls; while the two remaining were much larger and apparently older, with somewhat irregularly grown rhizomes. Each of these plantlets or portions was developing, its growth being localized in some particular spot, which was a vivid green, while the rest of the plantlet was browner in colour. From the appearance of these plantlets and from the manner of their occurrence I concluded that they were vegetative outgrowths from, or portions of, one original and irregularly grown protocormous rhizome. In several other instances I have found two or more young plantlets in the closest proximity to an older and brown-coloured rhizome. These plantlets almost invariably consisted of a brownish basal portion surmounted by a few protophylls, one or two of which were semi-decayed, while at some point or other on the plantlet there was a bluntly rounded vividly green area which was obviously the growing region. In none of these instances was a prothallus or the remains of a prothallus to be seen anywhere near the young plants, although some of the latter were exceedingly small. Further investigation of this point is necessary, but it would appear to be probable that the protocormous rhizome of *L. ramulosum* under certain circumstances gives rise to vegetative bulbils which develop into young plants. ·

L. volubile, L. fastigiatum, L. scariosum.

In these species the prothallus is subterranean, and is large, firm, and long-lived, so that it supports the young plant till the latter has attained to a considerable size. In fact, two or even three plantlets may arise on the one prothallus in the case of each of these three species, although eventually only one continues its development. It will suffice to state that, as in the last species, and also in *L. clavatum* as described by Lang (14), the foot in the embryo plant of these three species is of a large size, and consists of a uniform parenchymatous tissue bounded where it is in contact with the tissues of the prothallus by a distinct epithelium. In view of Miss Wigglesworth's statement (18) that in *L. complanatum* a short strand of vascular tissues passes from the main stele into the foot, I carefully examined transverse sections of the foot in several plantlets of different stages of growth of the present three species. In the case of *L. volubile* my results bear out very closely Miss Wigglesworth's statements. In the smaller plantlets of this species a strand of small thin-walled cells with abundant cell-contents penetrates well into the centre of the foot. It is in connection with the vascular tissues of the main stele. In older plantlets a few tracheides make their appearance in this strand. In one case, in the sections nearest to the main stele, these tracheides were in two groups separated by a single group of thin-walled cells, while towards the centre of the foot they gradually disappeared till there was only one left, much in the same manner as described by Miss Wigglesworth in the case of *L. complanatum*. In these larger plantlets the sclerenchyma in the foot closed in the vascular tissues together. In the young plants of *L. scariosum* and *L. fastigiatum* that I examined there is very little development of vascular tissues from the main stele into the foot. In the former species the foot is vary large, but only in the sections nearest to the main stele was any small-celled tissue seen to lead off from the stele into the foot, and no tracheides were present. In *L. fastigiatum* the foot-strand was slightly more developed, but not to the same extent as in *L. volubile.* In all three species the epithelial cells of the foot remain intact even on the largest plantlets, but the outer walls of the cells become strongly thickened. The prothalli of these species belong to one or other of the large sub-terranean types, and they continue to grow in size in many cases long after the young plant has begun to develop. The development of vascular tissue in the foot of the young plant varies in extent in different individuals of the same species, and possibly this is dependent simply upon the size to which the parent prothalli grow. There is no indication of a swelling comparable in any way with the protocorm of *L. cernuum.* The young plants take their origin from the upper region of the prothallus, in the case of *L. volubile* and *L. fastigiatum* generally at one end or the other of the main groove or depression which is present upon the upper surface of the prothallus, and in the case of *L. scariosum* at some point or other on the margin of the upper bulging region. In each case the first root shows as a peg-like outgrowth at the base of the young stem. The stems are cylindrical, colour-less, and naked (except for a few minute scale-like leaves) for a greater or lesser length according to the depth of the prothallus below the surface of the ground. In *L. scariosum*, whose prothalli are the most deeply buried, the stems may be as much as 2–3 in. in length before the ordinary assimilating leaves are formed. Young plants of each of these species are figured on the prothalli which have been referred to in Section III of this paper. The relative positions of the foot, stem, first leaf, and first-root rudiments in the young plant of *L. fastigiatum* are shown in fig. 84. This is a model of a developing embryo of this species viewed from above. It is tilted slightly in order to show the full sweep of the large foot and the projecting first root. In this case the apex of the stem and first leaf had just emerged from the tissues of the prothallus, but the root would not be apparent externally.

Summary.

The conclusions arrived at by Lang (14), Bower (4), and others with regard to the significance of the mode of dependence of the *Lycopodium* young plant upon its parent prothallus may here be briefly stated. Lang, following Treub, would regard not only the *L. cernuum* type of prothallus, but also the *L. cernuum* type of young plant, as being primitive for the genus. He says (14, p. 302), "The form of the young plant of *L. cernuum*, &c., is not to be regarded as recent and adaptive, but as possessing an important phylogenetic bearing." With regard to the epiphytic species and *L. Selago*, he suggests that the protocorm stage has there been lost owing to the subterranean habit, and quotes Treub's statement of the existence of a rudimentary protocorm in *L. Phlegmaria*. In the subterranean types also of *L. clavatum* and *L. complanatum* the protocorm has been lost, and the large development of the foot is to be regarded as an adaptation in accordance with the large size of the prothallus and the lengthy dependence of the young plant upon it. Bower looks upon the *L. Selago* type of embryogeny, where the only extraordinary feature is the varying length of the hypocotyl, as being the most simple and primitive within the genus *Lycopodium*. In the types of *L. clavatum* and *L. complanatum* he would see an adaptation from the *L. Selago* type in accordance with the subterranean saprophytic specialization of the prothallus. He inclines to deny (3, footnote on p. 248) that the swelling in the embryo of *L. Phlegmaria* stated by Treub to be a rudimentary protocorm is to be interpreted as such. The protocorm of the *L. cernuum* cycle of affinity he would regard merely as a specialization, classing it rather as a parenchymatous swelling such as is the foot, and doubting any general application of the protocorm theory in the whole genus. The protophylls are, in accordance with his view, to be judged simply as turgid outgrowths from the protocorm. Goebel, in the first edition of his "Organography" (10, pp. 231–33), does not favour Treub's theory of the protocorm, and lays emphasis on the fact that protocorms are found in certain epiphytic orchids. One other writer's views must also be briefly stated in order to bring forward the main lines of discussion along which the consideration of the *Lycopodium* embryogeny is being directed. Lady Isabel Browne (5, p. 223) has suggested that the protocorm is to be regarded as a modified form of stem due to reduction. She lays emphasis, on the one hand, on the great development of the stem in the oldest fossil Lycopods known to us, and, on the other hand, she concludes that "since vascular tissue penetrates for some distance in the protocorm of *L. laterale*, this organ cannot, at least in that species, be dismissed as a mere parenchymatous swelling."

No new facts of great importance emerge from the study of the morphology of the young plants of *L. Billardieri*, or of *L. volubile*, *L. scariosum*, *L. densum*, and *L. fastigiatum*. The foot of the young plant of *L. Billardieri* seems to be larger than that figured by Treub for *L. Phlegmaria*. In the young embryo plant of *L. fastigiatum* there is only one leaf-rudiment encircling the apex of the stem, and not a pair as figured by Bruchmann for *L. clavatum* and *L. annotinum*. In *L. volubile*, and to a less extent in *L. scariosum* and *L. fastigiatum*, a slight strand of vascular tissue is given off from the main stele into the foot.

But the study of the protocorm in L. laterale *and* L. ramulosum *brings to light facts of considerable interest.* First of all, it must be emphasized that, as in the other species in which a protocorm has been found, it is here associated with the *L. cernuum* type of prothallus. In these two species it would seem to be much more than a mere "temporary substitute for a root-system delayed in its development" (Bower), for it constitutes the plant-body for a whole season, attaining to a considerable size, and even branching. Moreover, the fact that vascular tissues develop within the main body of the protocormous rhizome suggests that in origin it is not a mere parenchymatous swelling. On these grounds it might be argued that the facts brought forward in this paper lend weight to the theory which regards the protocorm as a very ancient organ, possessing great phylogenetic importance. On the other hand, there are certain considerations which suggest that too much stress must not be laid upon the comparatively large size of the protocorm in *L. laterale* and *L. ramulosum.* In the first place, the manner of development of the protocormous rhizome in these species lends credence to the idea that it is merely a physiological specialization suited to carry the young plant over the dry season. The original protocormous tuber, surmounted by its two protophylls, corresponds closely with that in *L. cernuum*, &c. The rhizomatous extension of the protocorm would seem to be, however, an added feature, to be interpreted apart from the original tuber. In its development it is markedly distinct from the latter, being separated from it by a constriction, and, moreover, it is initiated in *L. laterale*, and in certain cases in *L. ramulosum*, subsequently to the decaying-away of the prothallus, these two facts suggesting that a certain interval elapses before it begins to develop. The manner of growth of the rhizome strongly suggests that it is merely a specialized swelling, for it is the swollen bases of each new pair of protophylls which add to its length, and even the fully grown rhizome bears witness to the manner of its development in the arrangement of the protophylls in two more or less obvious rows on its dorsal surface and in the median groove running the length of the rhizome on the ventral side. Also, in *L. cernuum* a somewhat similar lateral extension of the protocorm has been observed to take place, although to a less important extent, and this extension would seem to be a swelling distinct from the original tuber. There is a strong suggestion that the Lycopod protocorm is more plastic than an ancient and highly primitive organ would be expected to be, and that stress must not be laid from a phylogentic point of view upon the fact of its normally large development in *L. laterale* and *L. ramulosum.*

In the young plant of *L. cernuum* the vascular strand from the developing stem-apex takes a course through a corner of the protocorm. The short region of the latter which lies between the stem and the first-root apices may possibly be regarded as the rudiment of the stem-axis, retarded in its development and pushed out of its place through the intercalation of a tuberous stage. It is to be noted that the stem-apex and the first root always originate close together, and on the side of the protocorm farthest away from the prothallus. In *L. laterale*, and in *L. ramulosum* also, the stem and first root arise at the growing end of the protocorm, sometimes in close juxtaposition, though generally farther apart than in the case of *L. cernuum.* Here, too, it is possible to look upon that region of the protocorm through which the vascular strand passes as the rudiment of the stem-axis, very much postponed in development, and varying in size, owing to the intercalation in the embryogeny of the abnormally large rhizomatous swelling.

On the whole, the present writer inclines to the opinion that *the large size and other abnormal features of the protocorm of* L. laterale *and* L. ramulosum *are to be regarded simply as a special adaptation, and would conclude that this lends weight to the theory that the Lycopod protocorm in general may best be interpreted in this way.* The fact that the protocormous species are representative of three distinct sections of the subgenus *Rhopalostachya*, in which there is a considerable variation in the character both of the gametophyte and of the mature sporophyte generation, and is found also in the allied genus *Phylloglossum*, is significant, as indicating a considerable degree of antiquity for the protocorm within the genus *Lycopodium*.

V. GROSS ANATOMY OF STEM AND BRANCHES.

L. Selago.

The vascular cylinder of the mature stem of this species in cross-section is stellate in appearance (fig. 85). The xylem rows and protoxylem groups are stout, the latter being extended around the periphery of the cylinder in thick masses. At the base of the adult stem there are five to seven such protoxylem groups. The configuration of the vascular cylinder is variable: the centre may be occupied by xylem, or, as in the figure given, by an isolated group of phloem. The cortex is differentiated into three different zones; the innermost of these consists of cells whose corners are lignified, the outermost is developed as a broad zone of sclerenchymatous thick-walled tissue, and the middle cortex consists of a loose tissue of large thin-walled parenchymatous cells, through which several roots take their course. The vascular strand of the root generally consists of a large crescentic group of xylem embracing a single group of phloem; but in some roots there are two groups of xylem separated by a band of phloem; while in others, again, besides the main xylem groups there are a number of small isolated groups of protoxylem, in these cases the phloem being also in more or less isolated patches. A transverse section through the upper part of the stem shows the xylem in six stout rows radiating from a common centre, the protoxylem being in thick groups at their extremities. In this part of the stem there is no development of sclerenchymatous tissue in the cortex.

L. Billardieri.

In all the prothallial plants of this species shown in figs. 7–12 the vascular strand of the stem consisted of a single small crescentic group of protoxylem, enclosing between the horns of the crescent a single group of protophloem. In the plant shown in fig. 11, which was about $1\frac{1}{2}$ in. in height, and possessed five leaves, the layer of cells immediately surrounding the pericycle had become slightly sclerenchymatous. The plant in fig. 12 showed two roots, each surrounded by a zone of well-developed sclerenchyma two cells in width, traversing the cortical tissues of the stem. The strand in these roots consisted of a single crescentic group of protoxylem embracing a single group of protophloem. By the time the young plant has developed a dozen leaves the stem vascular cylinder shows two groups of protoxylem, which have been formed by the two horns of the crescent separating and the phloem extending between them. At this stage also the outermost layer of cells of the cortex has become slightly sclerenchymatous, as well as the layer immediately surrounding the pericycle. As the plant develops, the central cylinder becomes triarch by the splitting of one of the groups of protoxylem into two, the three groups of protoxylem alternating with three groups of phloem, the centre also being occupied by phloem.

Further development in complexity takes place by the splitting of the groups of protoxylem. In the young plants of this species, although the groups of protoxylem and the bands of phloem are compact and definite in form, yet from the very first a constant tendency towards rearrangement of the elements is apparent. The single crescentic group of protoxylem may split up into two or even three groups of single elements separated by single groups of phloem, and then join together again. And in the typical two-group stage the groups may be either compact and small or broad and extended around the periphery. The vascular cylinder in the mature stem (fig. 86) may be best described as stellate, and corresponds more or less closely with that described by Jones (13, p. 23) for *L. squarrosum*, the chief difference being that in the New Zealand species the groups of protoxylem are much larger and the outer ends of the xylem rows stouter than in the other species. The phloem is in bands or in isolated islands, as also is the xylem. The configuration of the cylinder has a tendency to alter owing to cross-connections taking place, between the xylem bands and groups. The phloem consists of sieve tubes surrounded by a well-differentiated phloem parenchyma with abundant cell-contents. The bands of xylem vessels are not accompanied by any small-celled xylem parenchyma. The innermost and the outermost zones of the cortex consist of cells with more or less thickened walls, but it is only a layer or two of cells at both the extreme outer and inner edges which stain at all noticeably with safranin, and these only at the cell-corners. There is no marked rearrangement of the tissues of the vascular cylinder preparatory to a dichotomy. In one plant an exact trichotomy of the vascular cylinder was observed in the lower part of the stem. In the ultimate branchlets the number of protoxylem groups is reduced to four or three, the xylem being arranged sometimes radially and sometimes in isolated groups. In these branchlets the leaves, which throughout the plant in this species are comparatively large, are arranged in four to six orthostichies. In the fertile regions the sporophylls are always in four orthostichies, and the number of protoxylem groups is normally three.

L. varium.

In this species the stem is thick, but it consists almost entirely of thin-walled parenchymatous tissue. A narrow zone, five or six cells in width, immediately surrounding the vascular cylinder, is slightly sclerenchymatous. The central cylinder itself (fig. 87) is much smaller than in the last species. In the older parts of the stem there are seven or eight massive groups of protoxylem, and these are joined across by metaxylem or left isolated in a varying manner. The configuration of the stele thus cannot be definitely described. The groups of protoxylem not uncommonly join together as in the figure, and thus become greatly extended around the periphery of the cylinder. The cells in the centre of the phloem bands are empty, but are no larger than those of the phloem parenchyma which borders them on both sides. The latter have abundant cell-contents, as also do the cells in the pericyclic zone.

L. Drummondii.

The main stem here consists of a very loose parenchymatous cortex, the narrow innermost zone of which is sclerenchymatous, and a medium-sized vascular cylinder (fig. 88). The metaxylem elements in the latter are arranged in not very compact bands, and sometimes single isolated xylem vessels occur. The protoxylem is very much extended peripherally in narrow bands, several of which may join and so form a thin unbroken band extending a considerable distance around the cylinder. The phloem tissue is homogeneous, there being no differentiation into sieve tubes and phloem parenchyma. There is a pericyclic zone of cells lying between the vascular tissues and the sclerenchyma of the cortex, the innermost layer of this pericycle being composed of small phloem-like cells staining darkly with haematoxylin. The sporophylls are arranged in eight orthostichies in alternate whorls of four, but there is no correspondence whatever between the leaf-system and the configuration of the vascular cylinder in those parts. Both in the cone and in its pedicel

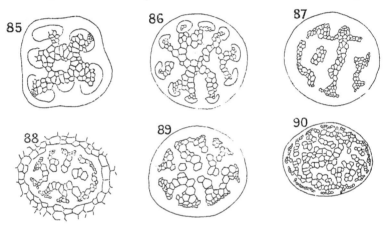

FIGS. 85–96. — *Lycopodium*, New Zealand species. Transverse sections of vascular cylinder of main stem. Semi-diagrammatic. Xylem and protoxylem elements indicated throughout. The circle in each case represents the inner limit of the cortex.

FIG. 85.—*Lycopodium Selago.* Transverse section of vascular cylinder of lower region of stem. × 70.

FIG. 86.—*Lycopodium Billardieri.* Transverse section of vascular cylinder of lower region of stem. × 40.

FIG. 87.—*Lycopodium varium.* Transverse section of vascular cylinder of main stem. × 60.

FIG. 88.—*Lycopodium Drummondii.* Transverse section of vascular cylinder of main rhizome. × 60.

FIG. 89.—*Lycopodium laterale.* Transverse section of vascular cylinder of stout rhizome. × 40.

FIG. 90.—*Lycopodium cernuum.* Transverse section of vascular cylinder of main stem. × 20.

the characters of the vascular tissues of the main stem are reproduced, although on a smaller scale. It is to be noted that in this species a cone with its pedicel forms a not inconsiderable portion of the bulk of the whole plant, so that the vascular cylinder in the fertile regions is not greatly reduced in size.

L. laterale.

Fig. 89 depicts the vascular cylinder of a strongly growing main rhizome, such as is typical when this species occurs in an open situation. The inner cortex, comprising about one-half of the entire cortical tissues,

10*

is slightly sclerenchymatous, but the outer cortex right up to the epidermis is thin-walled and spongy. In between the vascular tissues
and the zone of sclerenchyma there is a pericyclic layer about three cells
wide. The metaxylem elements are arranged in rows and groups, and
the phloem in like manner. The latter is not differentiated into sieve
tubes and phoem parenchyma, but is homogeneous. The protoxylem is
in broad groups, very much extended around the periphery of the
cylinder. In the sterile ultimate branchlets and in the cones the protoxylem is in three or four groups, which may be in connection at the
centre or isolated : they are more or less extended peripherally. In
the cones the sporophylls are arranged in six orthostichies in alternate
whorls of three. There is no constant relation between the leaf-arrangement in those parts and the configuration of the vascular tissues. In a
previous publication (11, p. 362) I stated that there was such a correspondence to be traced, but this statement must now be withdrawn, for I have
observed that, while the arrangement of sporophylls does not vary, the
vascular cylinder of the cone may be either quadrarch or triarch, according to the number of branchings that have taken place. A transverse
section immediately behind the apex of the main rhizome shows a full-
sized central cylinder, at the periphery of which there are tangentially
extended protoxylem groups, while the centre of the cylinder consists
of phloem and unthickened metaxylem elements. The differentiation of
the metaxylem takes place in a regular manner from the protoxylem
inwards. A transverse section of the protocormous rhizome of this
species on which the young stem-axis and first root have appeared
shows that the vascular strand as it passes along the protocorm consists
of two broad groups of protoxylem separated by a single elongated
group of protophloem. In the developing stem the vascular cylinder
increases in complexity through the separation of the two original protoxylem groups into three and more, the phloem extending between them.

L. cernuum.

In the very young plant the vascular strand, which leads from the
stem-apex through the protocorm into the first root, consists at first of
a single small crescentic group of protoxylem which includes a small
group of protoxylem between its two horns. Lower down in this first
root the protoxylem separates into two groups and the group of protophloem extends between them, while higher up in the young stems the
same takes place. A further stage in the young developing stem is
reached by the two groups of protoxylem broadening out, and several
metaxylem elements being formed between them, and at length joining
the two groups together, thus dividing the phloem into two groups.
The further growth in complexity of the vascular cylinder takes place
by the protoxylem separating into three and more groups, the phloem
extending between them. It is to be noted that from the very earliest
stage the vascular elements do not preserve constant relative positions,
but tend to separate easily from one another and then join together in
a somewhat different arrangement. The vascular cylinder of the mature
stem of this species has been described by Jones (13, p. 25), and the figure
he gives corresponds very closely with that given in the present paper
(fig. 90). Jones likens the vascular cylinder of *L. cernuum* to that of
Gleichenia, and Boodle (1) describes that of the allied species *L. salakense* in the same way. The outer cortex of *L. cernuum* is sclerenchymatous, but the main inner bulk of the cortical tissues consists of thin-

walled parenchyma right up to the vascular cylinder. The metaxylem elements in the central cylinder are more irregularly disposed even than in *L. laterale* and *L. Drummondii.* Possibly it is more noticeable in the present species simply because of the much larger size of its cylinder. The protoxylem is extended peripherally, this being especially marked in the smaller stems and branches. In the main rows and groups of phloem the centrally placed elements are large, and have the appearance of sieve tubes, while those which surround them are smaller. There is no rearrangement of the vascular tissues preparatory to branching, but the main cylinder simply constricts into two more or less equal parts. As noted above in the last species, behind the apex of a main stem or large branch the differentiation of the metaxylem proceeds regularly from the protoxylem inwards; in these regions also the larger-sized phloem elements are seen to be empty, whilst the smaller phloem elements which surround them have abundant darkly staining contents. There is no definite radial arrangement of the vascular tissues in the ultimate branchlets or the cones.

L. densum.

The single " seedling " plant found, which was branched in its upper region, showed a radial arrangement in its vascular cylinder, there being six protoxylem groups connected in the centre by metaxylem and alternating around the periphery with six groups of phloem. Both the first root and the first adventitious root showed a triarch structure, the phloem in both cases extending between the groups of xylem and occupying the centre of the cylinder. The main rhizome of this species is stout and firm, and its vascular cylinder is among the largest in modern Lycopodiums (fig. 91). The cortical tissues are throughout more or less sclerenchymatous, and increasingly so towards the centre. The middle region of the cortex is stored with starch. The xylem and phloem are arranged in alternate bands or plates, which lie parallel to one another in the plane of the ground. On the ventral side of the cylinder, however, this parallel arrangement is disturbed by the giving-off of adventitious roots. The division of the cylinder at a branching takes place right and left of a line at right angles to the plates of tissue. The number of groups of protoxylem is large, in the case figured there being seventeen, while at the base of some of the larger aerial branches the number may be as great as twenty-one. The xylem is differentiated into vessels and xylem parenchyma, and the phloem into large sieve tubes and phloem parenchyma. There is a pericylic zone of cells three or more in width. The aerial branches arise right and left of the rhizome, and then immediately turn upwards and grow erect. At their base the parallel structure is generally to be found, but this passes into the radial form higher up. In one strongly growing aerial stem, at a height of 4 ft. from the ground and below the first branching, it was observed that there were as many as twenty-one protoxylem groups, and that the configuration of the cylinder was markedly radial. Immediately behind the growing apex of the main rhizome the central cylinder is of the full size, and the differentiation of the metaxylem takes place from the protoxylem inwards. In the ultimate branchlets the number of orthostichies of leaves varies from six to eight, and the protoxylem is in three or four massive groups. A common condition is where the leaves are in six orthostichies in alternate whorls of three, and there are three groups of protoxylem. However, it cannot be said that there is any constant relation between the leaf-trace system and the number of protoxylems, the latter in each ultimate branchlet depending upon the order of the branch. In the cones both the number of ortho-stichies of leaves and the number of protoxylem groups is variable : the vascular cylinder may be either pentarch or quadrarch. In the stouter regions of the adventitious roots there are from six to nine groups of xylem arranged round the periphery of the central cylinder, which are either joined up with each other or isolated. The protoxylem elements are very few in number. The innermost zone of the cortex is sclerenchy-matous, as in the rhizome. The smallest roots show a single crescentic group of protoxylem, with a single group of protophloem. Larger roots show two and three groups of xylem, the phloem occupying the centre of the cylinder and extending between each of the xylem groups. The triarch condition is the common one in all medium-sized roots. It was noted that lateral rootlets are frequently borne in pairs, and that they arise by the trichotomous branching of the root-apex and vascular cylinder. A transverse section of a root at such a point shows that the vascular elements of each rootlet are derived from two adjacent groups of protoxylem and the intermediate group of phloem.

L. volubile.

The vascular strand in the stems of the youngest prothallial plants of this species examined always showed two small groups of protoxylem separated by a group of protophloem. The inner cells of the cortex from an early stage are slightly sclerenchymatous. The two groups of proto-xylem in the developing stem join across, and the cylinder thereafter passes through a triarch and then a quadrarch radial stage through the splitting of the protoxylem groups. I have previously published an account of the development of the parallel arrangement of the plates of vascular tissue in the adult stem both of this species and of *L. scariosum* (11, pp. 362–64), in which the conclusion is reached that this arrangement is initiated by and persists from the branching of the stem in the plane of the ground. Also, the different mode of the development of hetero-phylly in the lateral branchlets of these two species has there been given. In *L. volubile* the first indications of heterophylly in a lateral branchlet do not make their appearance until the plant is 4 in. or 5 in. in height (fig. 97). In slightly older plantlets all stages in the development of the heterophylly may be traced in the various branchlets on the same plant (fig. 98). In figs. 99*a* and 99*b* is shown a portion of the distichous region of a branch of a mature plant from the dorsal and the ventral sides respectively. The vascular cylinder of the main stem shows a more or less parallel arrangement of the xylem and phloem plates, this being disturbed, however, on the ventral side owing to the giving-off of the adventitious roots (fig. 92). There are from ten to sixteen proto-xylem groups in the main stems. Each xylem plate is differentiated into vessels and adjacent xylem parenchyma, and each phloem plate into a row of very large sieve tubes bounded on either side by small-celled phloem parenchyma with abundant cell-contents. The whole of the cortex consists of thick-walled sclerenchyma. As in the other species which show the parallel disposition of the vascular tissues — viz., *L. densum, L. scariosum,* and *L. fastigiatum*—branching of the vascular cylinder always takes place in such a way that the division of the plates of tissue is in a line at right angles to the plane of their arrangement. In the large adventitious roots the configuration of the central cylinder

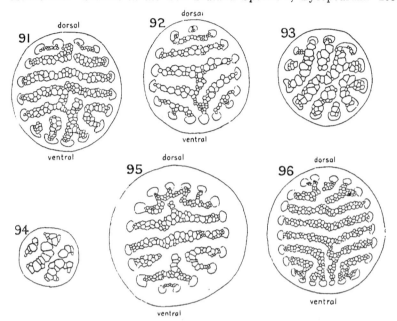

FIG. 91.—*Lycopodium densum.* Transverse section of vascular cylinder of main rhizome·
  × 25.
FIG. 92.—*Lycopodium volubile.* Transverse section of vascular cylinder of main stem.
  × 28.
FIG. 93.—*Lycopodium volubile.* Transverse section of vascular cylinder of large adventitious root. × 40.
FIG. 94.—*Lycopodium ramulosum.* Transverse section of vascular cylinder of scaly
  rhizome. × 40.
FIG. 95.—*Lycopodium fastigiatum.* Transverse section of vascular cylinder of main
  rhizome. × 30.
FIG. 96.—*Lycopodium scariosum.* Transverse section of vascular cylinder of main
  rhizome. × 25.

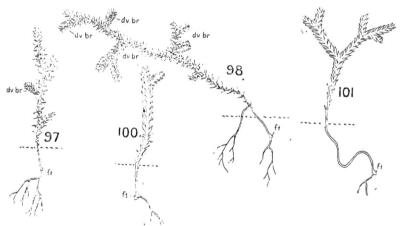

FIG. 97.—*Lycopodium volubile.* Young erect plant, showing beginning of heterophylly.
  Natural size.
FIG. 98.—*Lycopodium volubile.* Young plant with plagiotropic habit, showing different
  stages in development of heterophylly. Natural size.
FIGS. 100, 101.—*Lycopodium scariosum.* Young erect plants, showing early and sudden
  development of hetrophylly. Natural size.

is stellate (fig. 93), and there are about a dozen groups of protoxylem. As in the main stem, the whole of the cortex here also in sclerenchymatous. Branching of the adventitious roots and also of the lateral rootlets is by dichotomy, and not infrequently by a more or less exact trichotomy. As noted above for the last species, the smallest rootlets

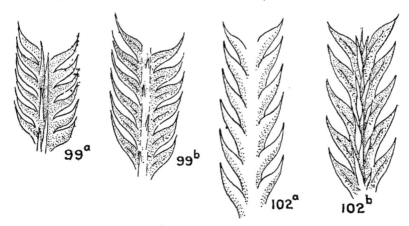

Figs. 99a, 99b.—*Lycopodium volubile.* Portion of distichous region of ultimate branch of mature plant, from dorsal and ventral sides respectively. × 4.
Figs. 102a, 102b.—*Lycopodium scariosum.* Portion of distichous region of ultimate branch of mature plant, from dorsal and ventral sides respectively. × 4.

show a single small crescentic group of protoxylem, while others show two or three xylem groups. In the ultimate branchlets there is no correspondence between the distichous habit and the number of protoxylem groups; the vascular cylinder in these regions of the plant is generally pentarch.

L. ramulosum.

In the stem of the young plant the xylem elements are more definitely coherent than in the case of *L. cernuum,* but the disposition of the xylem groups varies greatly. Both the cortex and the vascular cylinder of the more deeply buried scaleless stems of this species are larger than those of the scaly less deeply buried stems from which the aerial branches arise. In both the middle region of the cortex is sclerenchymatous, and in the more deeply growing stems the inner cortical zone is stored with starch. In fig. 94 is shown the vascular cylinder of the smaller scaly rhizome. In the larger rhizomes the number of protoxylem groups is generally eight or nine. The elements of the xylem are arranged in rows and groups, and do not show any tendency to separate from one another. There is no differentiation either of the phloem into sieve tubes and phloem parenchyma or of the xylem into vessels and xylem parenchyma. In the cones the sporophylls are arranged either in four orthostichies in alternate whorls of two or in six orthostichies in alternate whorls of three. There are four groups of protoxylem in the central cylinder of the cones, the groups being so broad and extended as to form the four sides of an almost complete square. The metaxylem elements, however, vary greatly in their arrangement.

L. fastigiatum.

In the stems of the youngest prothallial plants examined there were two groups of protoxylem connected across by one or two elements of metaxylem, and a group of protophloem on either side of this median plate. In slightly older plantlets the number of groups of protoxylem is three or four, and they are joined together at the centre by the metaxylem elements. It frequently happens, however, that one or two of these protoxylem groups become isolated, and then two groups of the phloem become joined to form a band across the centre of the cylinder. In the full-grown rhizome the cortex is markedly differentiated into zones. There is an outer narrow zone of very slightly sclerenchymatous cells, which is separated from the inner main cortex by a narrow layer of very large thin-walled parenchymatous cells. The inner main cortical zone is strongly sclerenchymatous towards the centre. The vascular cylinder is large, there being from fifteen to eighteen groups of protoxylem (fig. 95). The xylem and phloem are disposed in parallel plates. This parallel arrangement, as in the other plagiotropic species which show it, is disturbed on the ventral side of the cylinder, probably in relation to the giving-off of adventitious roots. The xylem bands are differentiated into vessels and xylem parenchyma, and the phloem bands into sieve tubes and phloem parenchyma, the latter having abundant cell-contents. The pericyclic zone in this species is very broad, and is differentiated into an outer layer, three or four cells in width, whose cells are empty, and an inner layer, five or six cells in width, staining brown with safranin, whose cells show abundant contents. The arrangement of leaves on the cones is constantly in eight orthostichies in alternate whorls of four. The vascular cylinder of the cones also seems to have a very constant configuration, consisting of five groups of protoxylem joined together in a stellate manner, the centre of the cylinder being occupied commonly by an island of phloem.

L. scariosum.

In the young prothallial plants of this species the vascular cylinder is much stouter than in either of the last two species. The very smallest and youngest stems sectioned showed never less than three protoxylem groups, while both four and five groups are commonly found in the stems of young prothallial plants. The groups of phloem also are large, and the centre of the cylinder generally shows several large-sized elements of undifferentiated metaxylem. A series of transverse sections of a prothallus and young plant presented an opportunity for observing the changing configuration of the vascular strand as it passes from the stem into the first root in the neighbourhood of the foot. In this instance the young stem above showed five protoxylem groups. These groups in the region immediately above the foot joined up with one another around the periphery of the cylinder and separated again, regrouping themselves in a most irregular manner. One striking configuration was that of a somewhat irregular but continuous semicircle of protoxylem elements, on the periphery of the cylinder opposite to it there being a single compact group, and in the centre three isolated elements of protoxylem among the undifferentiated metaxylem elements. In the sections median through the foot the protoxylem groups were observed to be arranged in a triarch manner, but below it they adopted the form of a Greek omega, the root evidently preparing to fork. The characteristic heterophylly of this species makes its appearance very early, even in plants which are still attached to their prothalli. In figs. 100 and 101 are shown two such young plants. It will be seen that, in marked contradistinction to what is found in the young plants of *L. volubile*, there are no gradual stages in the development of the heterophylly to be traced in these young plants, but the transition is sudden. In figs. 102*a* and 102*b* are shown the dorsal and ventral sides respectively of an ultimate branchlet in the mature plant of *L. scariosum*. The vascular cylinder of the mature rhizome is very large. The number of protoxylem groups varies from eighteen to twenty-seven, in the case figured (fig. 96) there being twenty-three. The xylem and phloem are markedly disposed in parallel plates. The xylem bands show an absence of xylem parenchyma, and the xylem vessels are much smaller than in most of the other species described. In each phloem band the row of large sieve tubes is bounded on either side by phloem parenchyma, with abundant cell-contents. The entire cortex is more or less sclerenchymatous. In the ultimate branchlets, where the heterophylly is most marked, the configuration of the vascular cylinder is radial and quadrarch. In the cones the leaves are in eight orthostichies in alternate whorls of four, and the vascular cylinder is constantly hexarch.

Summary.

From the comparative study of the gross anatomy of the stem, roots, and branches of the eleven New Zealand species of *Lycopodium* the following leading facts stand out : In the most simple form, in the " seedling " stem and in the first root and all young rootlets, there is a single crescentic group of protoxylem embracing a single group of protophloem. In slightly older plants the vascular cylinder both of the stem and of the first adventitious root is either diarch or triarch, as it is also in most rootlets in the adult plant, and the number of protoxylem groups increases by splitting, so that a quardrarch or pentarch radial stage is attained. This is also the normal condition of the vascular cylinder in the ultimate branchlets and in the cones of the adult plant, except in *L. cernuum* and certain other allied species, in which the arrangement of the vascular elements is indefinite and mixed throughout the plant. In the ultimate branchlets the radial condition has resulted from the reduction in size of the mature vascular cylinder by sucessive branchings. The leaf-arrangement on the ultimate branchlets exercises no influence upon the configuration of the vascular cylinder in those parts, the number of protoxylem groups in the latter being dependent simply upon the order of the branch. In the main stems and branches and in the large adventitious roots the configuration of the central cylinder is stellate. The bands and groups of xylem and of phloem in the central region of the cylinder are more or less cross-connected, so that the actual centre may be occupied by a band or group either of phloem or of xylem, but at the periphery the disposition of both xylem and phloem is more or less ray-like. This form of vascular cylinder is found in species which belong to Pritzel's *Selago* and *Phlegmaria* sections of the genus. There are certain important modifications of this arrangement which are to be noted. In the first place, in certain species belonging to the sections *Inundata* and *Cernua*—viz., *L. cernuum*, *L. laterale*, and *L. Drummondii*—the protoxylem groups are very much extended peripherally, so that the vascular cylinder is more or less enclosed in a thin shell of protoxylem. This has also been described by Boodle (1) in the case of *L. salakense*. Moreover, the disposition of the xylem and phloem groups in these species is very mixed, and the individual elements cohere loosely together to form the rows and groups. This is most marked in the two species *L. cernuum* and *L. salakense*. It has been noted that in *L. cernuum* and *L. laterale* the vascular strand in the very young plant is indefinite in configuration, owing to the fact that it is largely made up of the leaf-traces from the neighbouring protophylls and the first-formed leaves. Moreover, in the adult stem of *L. cernuum* branching takes place not only right and left of the stem, but also from its dorsal side, while in the rhizome of *L. laterale* there is no regular plane of branching at all.

Another important variation from the typical stellate or radial vascular cylinder is the arrangement of the xylem and the phloem in more or less parallel plates. This is found in certain plagiotropic species which Pritzel has grouped in the *Clavata* and *Cernua* sections. It will be shown below that there are reasons for removing *L. densum* and *L. volubile* (both of which show the parallel character) from the vicinity of *L. cernuum*. Beside the two species just mentioned, *L. scariosum* and *L.·fastigiatum* both show the parallel arrangement of their vascular tissues, and Jones (13) has enumerated various other plagiotropic species which do the same, all of which Pritzel has grouped in his section *Clavata*. This parallel character is due probably to the branching being restricted to the plane of the ground. In these plagiotropic species the growth in length of the main axis is unlimited and the growing apices are broad, so that the vascular cylinder at each branching is of the full size. Thus the parallel disposition of the vascular tissues, when once initiated, naturally persists.

The adventitious roots bear a marked resemblance to the rhizome in the plagiotropic species not only in vascular structure, but also in other characters. There is the same differentiation of the cortical tissues in both. The more deeply growing almost naked rhizome of *L. ramulosum* is interesting as suggesting an intermediate form between the typical scaly rhizome and the leafless adventitious root. The long aerial adventitious roots of *L. volubile* also might be termed either stems or roots.

The New Zealand species of *Lycopodium* vary markedly from one another in the differentiation both of their vascular and of their cortical tissues. The xylem elements may be either all of one kind, as also the phloem, or there may be a marked differentiation into vessels and accompanying parenchyma. Generally speaking, it is in the plagiotropic species that this differentiation is found in the greatest degree.

The view is here taken that the stellate or radially banded type of vascular structure, such as is found in the *Selago* and *Phlegmaria* sections of the genus, as also in the young plants and ultimate branches and the roots of most of the species, is the primitive one for the genus. The mixed *L. cernuum* type stands by itself, and has resulted perhaps partly from the dominance of the leaf-tree system over the cauline vascular tissues in the very young plant, and partly from the fact that along with the plagiotropic habit and the consequent fact of the presence of the vascular cylinder in its full size and complexity in the growing regions of the plant there is an irregular branching of the cylinder (branching taking place from all four sides of the growing apex), and hence a continual tendency to an indiscriminating mixing of its tissues. In the type which is characterized by the parallel arrangement of the vascular tissues we may recognize a modification of the primitive stellate or radial form of structure, and one which is not so far removed from it as is the mixed type. Here, along with the plagiotropic habit, branching has become restricted to one plane, so that whereas in the young orthotropic stem the vascular tissues are arranged radially, in the mature stem a directive tendency is continually present at the growing apices towards a disposition of the bands of tissue in the plane of branching.

Jones (13, pp. 31, 32) would connect the epiphytic type of vascular cylinder with that of *L. cernuum*, but it seems to me from the study of the New Zealand species that the former is rather of the same nature as that of *L. Selago*. Pritzel's description in Engler and Prantl (9) of *L. cernuum* and *L. laterale* as erect-growing tree-like forms without widely creeping main axis is obviously incorrect, and his inclusion of *L. densum* and also of *L. volubile* in the section *Cernua* is open to objection. In her criticism of my former account of the anatomy of six New Zealand species of *Lycopodium*, Lady Isabel Browne has (5, pp. 224–25) combated the suggestion that the mixed type of *L. cernuum* is to be regarded as relatively primitive. In this present paper it will be seen that I have accepted this criticism, and now speak of the *L. cernuum* type as a much modified one.

General Conclusions.

It remains now to bring together the main facts and conclusions noted in the different sections of this paper, and to attempt to estimate their value with respect to a natural classification of the genus *Lycopodium*, and also to put them into relation with the different theories which have been advanced regarding the interrelationships of the various species.

The prothalli of seven New Zealand species have been described— viz., *L. Billardieri*, *L. cernuum*, *L. laterale*, *L. volubile*, *L. ramulosum*, *L. fastigiatum*, and *L. scariosum*—while certain facts concerning the probable nature of that of another species—viz., *L. densum*—have been given. These prothalli have been found to belong to one or another of the types so well known from the descriptions of Treub, Bruchmann, and others: that of *L. Billardieri* to the *L. Phlegmaria* type; those of *L. laterale* and *L. ramulosum* to the *L. cernuum* type; those of *L. volubile* and *L. fastigiatum* to the *L. clavatum* type; and that of *L. scariosum* to the *L. complanatum* type; while we may assume that the prothallus of *L. densum* will be found probably to belong to one or other of the two latter types. At the same time it has been found that interesting modifications are present in all of these New Zealand species of prothalli, such as may have significance in a biological or even in a genetic sense in supplying connecting-links between the various types. In *L. volubile*, *L. densum*, *L. fastigiatum*, and *L. scariosum* it was found that a large foot is a characteristic feature of the embryogeny, and in *L. laterale* and *L. ramulosum* an exceptionally large and long-lived protocorm has been described. The conclusion was reached from the study of its development that the large size of this latter organ in these two species is merely a physiological adaptation, and that this suggests that a similar interpretation is to be applied in the cases of the other species in which the protocorm is found. It was seen that a stellate or radial configuration of the vascular cylinder is characteristic in *L. Selago*, *L. Billardieri*, and *L. varium;* a mixed type in *L. cernuum,*

L. laterale, and *L. Drummondii;* and a parallel type in *L. volubile, L. densum, L. fastigiatum*, and *L. scariosum*. From the comparative study of the main stems, the roots, and the branches of all of these species, and of the development of the vascular strand in the young " seedling " plants of eight of them, it has been concluded that the radial or stellate type is probably to be regarded as most closely representing the primitive one for the genus, and that the mixed and the parallel types are modified forms of it.

In the main, the particular subdivision of the genus into sections adopted by Pritzel is in accordance with the facts known concerning both the sexual and the asexual generations of the various species. The *Selago* and *Phlegmaria* sections comprise species which are orthotropic when growing terrestrially, and pendulous when epiphytic. The prothalli of six of the species—viz., *L. Selago, L. Hippuris, L. Phlegmaria, L. Billardieri, L. nummularifolium*, and *L. carinatum*—are known: they belong to one or other of the two types, *Selago* and *Phlegmaria*, which both Bower and Lang would consider to be more or less closely related. The form of the embryo plant is simple, being uncomplicated by the presence of a large foot or a protocorm. The vascular cylinder in eleven species has been investigated by Jones (13), and also in *L. Selago, L. Billardieri*, and *L. varium* in the present paper, and has been seen to be constantly of the radial-stellate type. In the *Inundata* and *Cernua* sections occur those species whose prothalli are of the *L. cernuum* type, and whose young plants pass through the protocorm stage. This is known in five species—viz., *L. inundatum, L. cernuum, L. laterale, L. ramulosum*, and *L. salakense*. The structure of the vascular cylinder is mixed with extended protoxylem groups, this being especially characteristic of *L. cernuum* and *L. salakense,* and also of *L. laterale* and *L. Drummondii;* and Jones has noted the broad protoxylem groups in *L. inundatum*. Two species, *L. volubile* and *L. densum*, included by Pritzel in the *Cernua* section possess characters which possibly would justify their removal from this section. They both show the parallel structure of the vascular tissues, and both possess a large subterranean prothallus. The species in the *Clavata* section would seem to be closely connected with regard to their main characters. The prothallus has been found in five species—viz., *L. annotinum, L. clavatum, L. fastigiatum, L. complanatum*, and *L. scariosum*—and belongs to one or other of the two large subterranean types, *L. clavatum* and *L. complanatum*. The large foot is characteristic of the embryo plant of these species. The vascular cylinder has been investigated by Jones in five species and by the present writer in *L. fastigiatum* and *L. scariosum*, and shows constantly the parallel structure.

The question arises, are these main sections of the genus which correspond so closely with the chief characters of both the gametophyte and the sporophyte generation to be regarded as more or less nearly related to one another, or as widely separated? The latter is the conclusion arrived at by Bruchmann (6, pp. 108–9) from his study of the prothalli of four European species. Lang (14, pp. 305–6), on the other hand, believes that the different prothallial types are " all more or less profound modifications of a type not unlike that of *L. cernuum*. The two forms of prothallus found in *L. Selago* give the clue " to the specialization of the subterranean saprophytic types on the one hand and of the epiphytic type on the other. With this view Bower (4) is in close agree-

ment.   Various modifications of and variations from the main types have
been enumerated by Lang, and others also, in this paper, which appear
to supply connecting-links between them.   The present writer would
emphasize especially the variations in the prothallus of *L. ramulosum*
in this connection.   In respect of the character of the embryo types
also, Bower and Lang both contemplate a certain degree of affinity
between the different parts of the genus, with this difference in their
views : that the latter would look upon the *L. cernuum* type of embryo,
with its protocorm. as primitive or the genus, the protocorm having
been lost from the life-history of the species which belong to the other
sections, whereas the former regards the *L. Selago* type of embryo as
being the most primitive, there being derived from it the type which
shows the large foot, while the *L. cernuum* type takes a middle position.
In the present paper the conclusion has been arrived at from the study of
the protocorm of *L. laterale* and *L. ramulosum*, and also of *L. cernuum*,
that this organ, as Bower suggests, is a physiological specialization rather
than a highly primitive feature.   The comparative study of the vascular
cylinder in the genus *Lycopodium* also suggests that the different sections
as dealt with above are more or less closely related.   Jones (13, pp. 27–28)
believes that a simple radial, root-like structure of the vascular cylinder is
to be regarded as the primitive one for the genus, and my own view,
derived from the study of the above eleven New Zealand species, is the
same.   The growth in size of this type of cylinder would either leave the
configuration strictly radial or (on account of the constant changes in the
disposition of the individual vascular elements and of the groups and
rows in which they are arranged, a feature which is a characteristic one
right through the genus) would cause it to adopt a stellate form in which
cross-connections are present.   Now, this is the type characteristic of
the *Selago* and *Phlegmaria* sections.   From this the mixed type of the
Cernua and *Inundata* sections on the one hand, and the parallel type
of the *Clavata* section on the other hand, would be derived, as has
been described in Section V of this paper.   Lastly, Miss Sykes (15)
has shown that Pritzel's sections of the genus can be arranged in a
series in respect of the morphology of the sporangium-bearing organs.
From her study, however, she would conclude that the *Cernua* section
comprises the most primitive members of the genus, and that those of the
Selago section have been derived from them by reduction, while the type
of *L. inundatum* is " an interesting intermediate step."

Thus a belief in the interrelationship of the different sections of
the genus *Lycopodium* is more generally held than the opposite one—
that they have been widely separated from an ancient period.   The
view which the present study seems to suggest is that on a general sum
of characters the *Selago* section must be held to comprise the most
primitive members of the genus, and that the *Phlegmaria* and *Clavata*
sections have been independently derived from it, the former being less
modified than the latter.   The exact relation of the *Cernua* and *Inundata* sections to the *Selago* section is difficult to gauge.   The consensus of
opinion seems to be that the *L. cernuum* type of prothallus is to be
regarded as primitive for the genus, but the plagiotropic habit, the mixed
type of vascular structure, and the protocorm, on the other hand, would
all seem to be highly specialized characters.   Hence the *Cernua* and
Inundata sections may perhaps best be placed in a group by themselves,
as having been derived from ancestors common to themselves and to the
Selago section, but independently of the latter.

Literature Consulted.

1. Boodle, L. A. "On the Structure of the Stem in Two Species of *Lycopodium.*" Ann. of Bot., vol. xiv, pp. 315–17, 1900.
2. Bower, F. O. "Lycopods." (Review of "Etudes sur les Lycopodiacées," par M. Treub, Part II, Annales du Jard. du Bot. de Buit., vol. v.) *Nature,* vol. 34, pp. 145–146, 1886.
3. —— "Imperfect Sporangia in certain Pteridophytes: Are they vestigial?" Ann. of Bot., vol. xv, pp. 225–67, 1901.
4. —— "The Origin of a Land Flora." 1908.
5. Browne, Lady Isabel. "Review: A Colonial Contribution to our Knowledge of the Genus *Lycopodium.*" "New Phytologist," vol. xii, pp. 222–25, 1913.
6. Bruchmann, H. "Uber die Prothallien und die Keimpflanzen mehrerer europäischer Lycopodien." Gotha, 1898.
7. Cheeseman, T. F. "Manual of the New Zealand Flora." 1906.
8. Edgerley, K. V. "The Prothalli of Three New Zealand Lycopods." Trans. N.Z. Inst., vol. 47, pp. 94–111, 1915.
9. Engler and Prantl. "Pflanzenfamilien," I Teil, iv Abteilung, *Lycopodiaceae* (Pritzel). 1900.
10. Goebel, K. "Organography," Pt. ii. 1st ed. Eng. Transl., 1905.
11. Holloway, J. E. "A Comparative Study of the Anatomy of Six New Zealand Species of *Lycopodium.*" Trans. N.Z. Inst., vol. 42, pp. 356–70, 1910.
12. —— "Preliminary Note on the Protocorm of *Lycopodium laterale.*" Trans. N.Z. Inst., vol. 47, pp. 73–75, 1915.
13. Jones, C. E. "The Morphology and Anatomy of the Stem of the Genus *Lycopodium.*" Trans. Linn. Soc., series ii, Bot., vol. vii, pp. 15–35, 1905.
14. Lang, W. H. "The Prothallus of *L. clavatum.*" Ann. of Bot., vol. xiii, pp. 279–317, 1899.
15. Sykes, M. G. "Notes on the Morphology of the Sporangium-bearing Organs of the *Lycopodiaceae.*" "New Phytologist," vol. vii, pp. 41–60, 1908.
16. Thomas, A. P. W. "Preliminary Account of the Prothallium of *Phylloglossum.*" Proc. Roy. Soc. London, vol. 69, pp. 285–91, 1901–2.
17. Treub, M. "Some Words on the Life-history of *Lycopodium.*" Ann. of Bot., vol. i, pp. 119–23, 1887.
18. Wigglesworth, G. "The Young Sporophytes of *Lycopodium complanatum* and *L. clavatum.*" Ann. of Bot., vol. xxi, pp. 211–34, 1907.

ART. XXXI.—*The Vegetation of the Tarawera Mountains, New Zealand.*

By B. C. ASTON, F.I.C., F.C.S.

[*Read before the Wellington Philosophical Society, 27th October, 1915.*]

Plates XIX–XXII.

INTRODUCTION.

ON the 10th June, 1886, New-Zealanders were startled from their quiet lives by a volcanic eruption, the sounds of which reached from three to four hundred miles north and south from the seat of the outburst, while the ashes fell over an area variously estimated at from four thousand to six thousand square miles. The greatest depth of the matter ejected, measured at the lip of the great vent, was 170 ft., the top of the range being increased by this amount. It has been estimated that from two-fifths to one cubic mile of material was thrown out. The eruption took place along a flat-topped range, 3,600 ft. altitude, composed of rhyolitic lava-flows, known as Tarawera Mountains, and comprising the peaks Wahanga (northernmost), Ruawahia, and Tarawera (southernmost). This range is situated on the eastern side of Lake Tarawera, in the thermal district of the North Island, where are situated Lakes Rotorua and Taupo, well known as tourist resorts.

A gigantic rent opened along the axis of the range, running roughly north-east and south-west, commencing at the north end and extending to Tarawera, thence in a line more westerly to Lake Rotomahana, the waters of which are thought to have caused further explosions. This rent finally ended at Lake Okaro in the south, taking but three or four hours to form. Along this immense chasm in the earth, nearly nine miles long, 900 ft. deep at its greatest and 300 ft. at its least depth, and a mile and a half to an eighth of a mile wide, were no fewer than seventeen points of eruption. The fissure is not continuous, but is bridged in several places by the original surface remaining in position. This titanic feature of the North Island thermal district is at the highest points certainly the most impressive of the sights, and probably the least visited by tourists.

The north-western face of the Tarawera Range at present rises from the shores of Lake Tarawera (which is 1,032 ft. above sea-level) to a height of 2,738 ft. above it, the summit of the range being 3,770 ft. above the sea. S. Percy Smith (2, 4, and 23), who had the advantage of thoroughly exploring the mountain both before and after the eruption, describes the contour of the mountain as being unlike any other in the district, with the plateau-like summit sloping down to, say, 400 ft.; the steep "mural crown" below descending for, say, another 400 ft.; the sharply inclined talus at an angle of 30° extending 1,000 ft. below the rock-faces forming the "mural crown"; and the beautifully easy slopes, lowest of all, grading down to the lake-margin as a pumice beach or ending in low cliffs at the water's edge. This is a fair average of the aspect it presents on every side, except towards the east, where a range of less elevation joins it and spoils its symmetry. The forests which once clothed the slopes of this mountain, according to the above authority, were of considerable extent, especially over the south-eastern flanks—that is, on the opposite side of the range to those described in this paper. The eruption has utterly destroyed these forests. Instances of a totara (*Podocarpus totara*) and rimu (*Dacrydium cupressinum*) standing on the edge of the chasm are given (2, p. 52).

A. P. W. Thomas (11) has written perhaps the most comprehensive report of the eruption, although he had not the advantage of having visited the mountain prior to that event. He states that forests composed of large trees grew upon the sides of the Tarawera Mountain, and these, of course, were wholly wrecked by the eruption. Those on the site of the chasm disappeared altogether. Fortunately, T. Kirk had ascended the mountain in 1872, and his published papers (1 and 27) make it plain that where favourable to plant-life the surface of the mountain was clothed with vegetation, though on the highest parts it was scanty and stunted. A dwarf shrubby vegetation was found on the very summit of Tarawera, in sheltered places affording cover for a luxuriant growth of mosses and lichens, in which the orchids *Caladenia bifolia*, *Thelymitra longifolia*, and *Orthoceras Solandri* occurred. Elsewhere on the summit were *Metrosideros hypericifolia*, *Corokia buddleoides*, *Coprosma lucida*, *Olearia furfuracea*, *Raoulia tenuicaulis*, *Dracophyllum strictum*, *D. Urvilleanum*, *Polypodium serpens*, *Tmesipteris tannensis ;* while near the summit were noticed *Lycopodium volubile* (2,800 ft.), *Astelia trinervia* (3,300 ft.), *Cyathodes empetrifolia*, and *Gaultheria oppositifolia* (3,200 ft.).

On the margins of Tarawera Lake Kirk noticed abundant trees of *Metrosideros tomentosa* of large size, and also *Astelia Cunninghamii*, *Scirpus maritimus*, *Ranunculus acaulis*, and *Chenopodium glaucum* var. *ambiguum*. At the entrance to the gorge separating Ruawahia from Wahanga were observed large terrestrial specimens of *Metrosideros robusta*, and elsewhere he saw *Panax Colensoi*, the most prominent shrub, forming handsome dwarf bushes sheltering *Hymenophyllum bivalve*, *H. multifidum*, and other ferns ; *Danthonia* sp., *Deyeuxia quadriseta*, and *Pittosporum tenuifolium* were also observed. He states that the total number of species collected above 3,000 ft. did not exceed seventy, and considered that the vegetation of the mountain comprised a remarkably limited number of species.*

It would be too much to assume that all vegetation was entirely killed out at the base of the mountain by the eruption of 1886. The dead stumps of large *Metrosideros tomentosa* trees may still be found standing at the lakeside, and many species of the family *Myrtaceae* are so tenacious of life that stumps with several feet of mud or scoria round them might have sprouted and produced seed. Moreover, seed might have become uncovered by the rain cutting gulches in the soft mud and sand, and have found the absence of competition from other plants a factor favourable to growth.

Thus, on account of the volcanic origin of Tarawera Mountain and the isolated position of the north-west face, the foot of which, save for one narrow isthmus, is washed by the waters of Lakes Rotomahana and Tarawera, and on account of the eruption of 1886, this area presents peculiar facility for the study of the spread of species on new ground, since only an infinitesmal fraction of the plants could have survived the 1886 eruption.†

THE NORTH-WEST FACE.

In the spring of 1913 the author twice visited Tarawera, on the first occasion (14th September) ascending to the summit of the range by the

---

* Kirk instances other sea-littoral plants, such as *Convolvulus Soldanella*, *Juncus maritimus* var. *australiensis*, *Leptocarpus simplex*, *Carex pumila*, *Zoysia pungens*, and *Bromus arenarius*, which are found in this thermal district, as supporting the theory of the submarine origin of the lowlands of the central portion of the North Island; but when one considers that wild water-fowl, such as shags, black swans, and wild duck, travel frequently between coast and lakes, and that shags nest in the *Metrosideros tomentosa* trees on the coast, a simpler explanation of the presence of these sea-littoral plants in fresh-water littoral situations far inland becomes apparent.

† A lithographed reproduction of a photo of Tarawera seen from the south-west before the eruption, in A. P. W. Thomas's report (*loc. cit.*), shows dense forest to about one-third of the height above Lake Tarawera.

broad valley shown in Plate XIX, fig. 1, lying to the north of the landing
for Rotomahana Lake, and on the second visit landing on Kuaehape Beach,
and, after examining the *Metrosideros* forest, proceeding to a point about
1,700 ft. above sea immediately above the beach.

In the midsummer of 1915–16 the author again twice visited Tarawera,
on the first occasion ascending the range by the route taken in 1913, traversing the summits of Tarawera and Ruawahia, and returning from the sand
crater by the scoria slope separating Ruawahia from Wahanga, and on the
second occasion visiting the area called for convenience the northern face,
which is separated from the area called the north-west face by a comparatively unbroken scoria slope which was traversed on the previous visit,
descending from the sand crater to the lake.*

At Kuaehape Beach within a chain of the lake there are two beach-levels,
the lower consisting of white pumice, barren of plant-life, but mixed with
driftwood, and the upper beach, a few feet higher, composed of red pumice
and sand with patches of *Raoulia australis* dotted about, while growing
at the edge of the beach are bushes of *Veronica salicifolia* var. and *Coriaria
ruscifolia.* Scattered about are young plants of a naturalized *Erigeron.*
Shrubs of *Myrsine Urvillei, Cyathodes acerosa,* and *Muehlenbeckia complexa*
are near at hand, and the herb *Hydrocotyle asiatica* also occurs. A little
to the right of the beach are lava cliffs about 250 ft. high, which are covered
with a growth of young *Metrosideros tomentosa* forest, fringed at its margin
with *Coriaria ruscifolia* and *Veronica salicifolia* var., and containing also
the following : *Knightia excelsa, Coprosma lucida, Leptospermum ericoides,
Panax arboreum, Pittosporum tenuifolium, P. Colensoi, Geniostoma ligustrifolium, Leucopogon fasciculatum, Gaultheria antipoda* var., *Cyathodes acerosa,
Halorrhagis erecta, Poa anceps.* Where the forest reaches the shore there
are many dead stumps of *Metrosideros tomentosa* which had been killed by
the eruption. Plate XX, fig. 2, gives a very good idea of the vigorous
young growth on the lake-side.

In the broad valleys of the lower slopes, where the soil is better, the
dominant plant is *Coriaria ruscifolia,* forming pure shrubberies, 15 ft. to
20 ft. high, the plants having numerous trunks, 6 in. to 8 in. in diameter,
springing from the ground. On the gravels of temporary watercourses
patches of *Raoulia australis* are attempting to form a covering. In the more
exposed situations, where the soil is poorer and not so moist and the altitude greater, the *Coriaria* is replaced wholly or in part by *Leptospermum
scoparium* (see Plate XXI, fig. 1).

Ascending a deep gully above Kuaehape Beach, where surface waters
had cut the beds of ash into a vertical-walled ravine, the following were
noticed between lake-level (1,040 ft.) and 1,500 ft. above sea-level : Shrubs
or young trees—*Metrosideros tomentosa, Weinmannia racemosa, Pittosporum
tenuifolium, Leptospermum scoparium, Veronica salicifolia* var., *Olearia
furfuracea, Fuchsia excorticata, Melicytus ramiflorus, Griselinia littoralis,
Leucopogon fasciculatus, L. Fraseri, Gaultheria oppositifolia* (in large masses
6 ft. across on the walls of the gorge), *Pimelea laevigata, Solanum aviculare,
Coprosma robusta, Melicytus ramiflorus, Muehlenbeckia axillaris;* lianes—
Rubus australis, Muehlenbeckia australis; herbs—*Anagallis arvensis* (naturalized), *Epilobium rotundifolium, Dianella intermedia, Gahnia pauciflora,
Cladium Vauthiera, Acaena Sanguisorbae* var.; ferns—*Pteris esculenta, P.*

---

* The author is much indebted to Judge Brown, Mr. Tai Mitchell (Government
Surveyor), and Mr. L. D. Foster for their company and assistance in one or more of
these journeys, and to Mr. Warbrick (in charge of the tourist traffic on the lake) for
assistance in landing at different points.

[B. C. Aston, photo.

Fig. 1.—Valley of ascent to summit, looking west. Foreground, scoria slopes; middle distance, scattered *Coriaria* bushes; background, almost pure *Coriaria* association on mud-covered hills deeply furrowed by surface waters; distance, Lake Tarawera.

[G. D. Valentine, photo.

Fig. 2.—Tarawera Mountain, looking north-east, from site of Te Ariki Village (destroyed), showing Green Lake crater and mud-covered hills. Taken shortly after the eruption.

[B. C. Aston, photo.

FIG. 1.—Looking south-east across deepest part of Tarawera chasm, showing on opposite
wall the lavas which compose the scoria slopes—the lighter, rhyolitic; the
darker, andesitic.

[B. C. Aston, photo.

FIG. 2.—Taken from boat on Lake Tarawera, looking south-east, near Kauehape Beach.
Fringing lake are dead stumps of *Metrosideros tomentosa*. On the beach is a
fringing shrubbery of *Coriaria* bushes merging into a tangled mass of shrubs

[B. C. Aston, photo.

Fig. 1.—At 1,700 ft. above sea. looking south-east. Foreground of light and dark scoria, dotted with patches of *Pimelea laevigata* near walking-stick stuck in ground. Scattered shrubbery of *Leptospermum ericoides*, *Coriaria ruscifolia*, and *Veronica salicifolia* var. extending up the easy scoria slopes to the steeply inclined unstable slopes below the "mural crown" of stable lavas. On the highest point of Tarawera may be seen a cap of red scoria. In the distance is the gully separating the range into Ruawahia (left) and Tarawera proper (right).

[B. C. Aston, photo.

Fig. 2.—At 1,500 ft. above sea, the broad valley separating the portion called "north-west face" from that called "northern face," and more northerly to the view in fig. 1 above. Foreground, *Muehlenbeckia axillaris* and *Pimelea* patches with *Leptospermum scoparium* var. growing out of them, and scattered plants of *Oenothera;* middle distance, scattered *Coriaria* and *Leptospermum* bushes, with background on either side of forest showing dead stumps of *Metrosideros robusta;* distance, unstable slopes of steeply inclined scoria becoming stabilized by *Raoulia* and other patch-plants. On the extreme left of the picture, skirting the young forest, is the track to the sand crater.

[B. C. Aston, photo.

FIG. 1.—Taken near view shown in Plate XXI, fig. 2. Foreground, white and black scoria mixed. Slopes becoming stable by growth of *Raoulia* patches and *Coriaria* association developing, showing individuals growing from *Raoulia* patches.

[B. C. Aston, photo.

FIG. 2.—At 1,700 ft. above sea. looking north. Young forest on northern face, showing dead stumps of trees killed by the eruption.

tremula, Lomaria capensis, Asplenium flaccidum, A. adiantoides, A. lucidum, Polypodium Billardieri, P. pennigerum, P. serpens, Hemitelia Smithii, Cyathea dealbata, Pellaea rotundifolia.

Now succeed scoria flats which support a sparse growth, 2 ft. to 6 ft. high, of *Coriaria ruscifolia, Pteris esculenta, Veronica salicifolia* var., *Olearia furfuracea, Coprosma robusta, Weinmannia racemosa, Leptospermum scoparium*, while *Pimelea laevigata* in patches a foot or more in diameter are closely appressed to the pumice-gravel.

Finally, at 1,550 ft., open stony slopes are reached where the rise is so gentle as to be hardly perceptible. Here the growth of shrubs is still scantier. The *Pteris* disappears, and the vegetation is *Leptospermum ericoides, Coriaria ruscifolia, Veronica salicifolia* var., *Pimelea* patches, and numerous young plants of the naturalized *Oenothera odorata*, evidently seedlings from last year's old plants. From this point to the "mural crown," which can be seen in Plate XXI, fig. 1, the vegetation was not inspected, the journey to the summit being made by a more southerly route, where, by avoiding the "mural crown," and by travelling for a while on the edge of the crater, the top of the range is more easily approached (see Plate XIX, fig. 2). On this route, in addition to many of the species already mentioned, the following were seen up to 1,800 ft.: Shrub—*Aristotelia racemosa;* giant reed — *Arundo conspicua;* herbs — *Epilobium nummularifolium, Raoulia australis, Gnaphalium luteo-album, Wahlenbergia gracilis, Thelymitra* sp., *Danthonia semiannularis* var., *Raoulia glabra, Carex* sp., *Erechtites scaberula*, and naturalized plants of *Hypochoeris radicata, Sonchus oleraceus, Trifolium repens, T. pratense;* ferns—*Pteris esculenta.*

At 1,800 ft. there were noted a few stunted shrubs of *Weinmannia racemosa, Gaultheria oppositifolia, G. antipoda, Leucopogon fasciculatus, L. Fraseri, Dracophyllum subulatum, Muehlenbeckia axillaris, Cyathodes empetrifolia, Metrosideros robusta, Griselinia littoralis, Dodonaea viscosa*, and *Lycopodium densum;* the herbs *Raoulia australis* and *Drosera auriculata;* and the ferns *Lomaria penna marina, L. capense, Polypodium Billardieri.*

On compacted scoria slopes at 3,000 ft. patches of *Raoulia tenuicaulis* (in flower), *R. australis, Danthonia semiannularis* var., and *Hypochoeris* seedlings were the higher plants, while occasionally in a damper cavity than usual a fern would be found, and patches of moss.

On the summit of the range there is a cap of red scoria, on which the only growth at present is patches of silvery *Raoulia australis.*

On account of the Tarawera Mountain-range having been built up by successive outflows of lava, ashes, and mud, and also on account of its isolated position, its flora might be expected to show species which are specially adapted to spreading easily by means of wind and water and birds; and this we find to be the case. Practically the whole of the plants found on the mountain are those the seed of which is thus spread. The species having succulent edible fruits — *e.g., Coriaria* — and therefore which are spread by means of birds, in the number of individuals hold first place on the lower slopes; while those which have light seeds, or seeds furnished with special structures enabling them to float in the wind, and capable of travelling long distances in the air, are a good second, and on the higher slopes are dominant — *e.g., Raoulia* and *Leptospermum.* And this is true of the plants observed by Kirk in 1872, before the eruption in 1886, as well as of those recorded by me in 1913, twenty-seven years afterwards.

The following are the principal forest-trees of the Rotorua district, which should, of course, be growing on the lower western slopes of Tarawera;

but those which are not probably lack a means of transport for their heavy seeds :—

Beilschmiedia tawa Benth. & Hook. f. This tree constitutes 75 per cent. of the millable timber in some parts of the Rotorua district. The berry is 1 in. long, and solitary-seeded. Its distribution is effected probably entirely by rare large birds (pigeons and kaka parrots).

Kirk records the following as " chief trees of Ngongotaha (a forest-clad mountain, 2,554 ft., seven miles west of Rotorua) and the adjacent hills " (1, p. 327) :—

Dacrydium cupressinum Sol. Nut ovoid, about $\frac{1}{8}$ in. long.
Metrosideros robusta A. Cunn.
Beilschmiedia tawa Benth. & Hook. f.
Knightia excelsa R. Br. Pubescent follicles $1\frac{1}{2}$ in. long, tapering into the persistent style, ultimately splitting into two boat-shaped valves containing 3 or 4 winged seeds.
Litsaea calicaris Benth. & Hook. f. Solitary-seeded berry, $\frac{3}{4}$ in. long.
Laurelia novae-zelandiae A. Cunn. Achenes hairy, narrowed into long plumose styles. (This tree is common in the forests on the west side of Lake Tarawera.)

Abundant in the forest at the north end of Lake Rotorua, according to Kirk, are the following trees :—

Elaeocarpus dentatus Vahl. Drupe about $\frac{1}{2}$ in. long, oblong, ovoid, stone rugose, 1-celled, 1-seeded.
Metrosideros robusta A. Cunn. Capsule coriaceous, 3-celled, 3-valved, or irregularly dehiscent; seeds densely packed, numerous, linear.
Podocarpus spicatus R. Br. Drupe $\frac{1}{3}$ in. in diameter.
Podocarpus ferrugineus D. Don. Drupe $\frac{3}{4}$ in. long.
Knightia excelsa R. Br.
Litsaea calicaris Benth. & Hook.

Much less frequent, he says, are—

Weinmannia racemosa Linn. f. Capsule $\frac{1}{5}$ in. long, 2- to 3-valved, seeds hairy, minute, and numerous.
Fusanus Cunninghamii Benth. & Hook. f. Drupe $\frac{1}{3}-\frac{1}{2}$ in. long.
Ixerba brexioides A. Cunn. Capsule $\frac{3}{4}$ in. diameter; seeds large, oblong, compressed.
Carmichaelia sp., probably *C. juncea* Col. Common in many parts of the Rotorua district. Leguminous seeds.

In addition to those given by Kirk might be mentioned from my own observations—

Persoonia toru A. Cunn. Common on Karamea (Rainbow Mountain). Drupe $\frac{1}{4}-\frac{1}{3}$ in. long; 1- or 2-celled, with single seed in each cell.

The forest near Te Wairoa, the Maori village buried at the time of the eruption, now contains—

Laurelia novae-zelandiae A. Cunn. This species, which exhibits a decided preference for swampy land, may require soil-conditions which do not occur on Tarawera.

At the Te Ngae forest, about ten miles farther away, I noticed—

Podocarpus dacrydioides A. Rich. Fruit a black ovoid nut, about $\frac{1}{8}$ in. long, seated on a red fleshy receptacle.

Carpodetus serratus Forst. Fruit globose, size of small pea, almost fleshy, indehiscent 3–5-celled, seeds numerous, pendulous.

Clematis indivisa Willd. Achenes with a plumose tail often more than 2 in. long.

List of Species found on Tarawera Mountain, North-western Face, in September, 1913, and the Summer of 1915–16.

[A letter "B" prefixed to the name of the species denotes the probability of its being spread by birds, and "W" by wind. "?" before a name denotes that the method of disposal is doubtful, but possibly wild animals (rabbits or hares), or floated across by water, or carried by water-fowl.]

B. *Melicytus ramiflorus* Forst. Small berry, $\frac{1}{5}$ in.

B. *Pittosporum tenuifolium* Banks & Sol. Capsule woody and seeds sticky.

B. —— *Colensoi* Hook. f. Capsule woody and seeds sticky.

B. *Aristotelia racemosa* Hook. f. Berry size of a pea.

W. *Dodonaea viscosa* Jacq. Above 1,400 ft. Membranous compressed capsule, very broadly 2–3-winged ; wings membranous.

B. *Coriaria ruscifolia* Linn. Above 1,400 ft. Crustaceous achenes invested by juicy petals.

B. *Rubus australis* Forst. Many succulent 1-seeded drupes crowded upon a dry receptacle.

? *Acaena Sanguisorbae* Vahl. var. Achenes attached to fruiting calyx provided with 4-barbed bristles.

? *Weinmannia racemosa* Linn. f. Above 1,400 ft. Capsule containing small hairy seeds.

W. *Drosera auriculata* Backh. Above 1,400 ft. Seeds minute.

W. *Halorrhagis erecta* Schindl. Seed small, dry, 2–4-seeded, nut $\frac{1}{10}$ in. long with 4 ribs dilated into wings.

Gunnera monoica Raoul. Small fleshy drupe.

W. *Leptospermum scoparium* Forst. Above 1,400 ft. Woody or coriaceous capsule, containing numerous linear seeds.

W. —— *ericoides* A. Rich. Above 1,400 ft. Linear seeds.

W. *Metrosideros florida* Smith. Above 1,400 ft. ⎫ Coriaceous or woody capsule ; seeds numerous, linear.
W. —— *robusta* A. Cunn. Above 1,400 ft. ⎬
W. —— *tomentosa* A. Rich. ⎭

W. *Epilobium junceum* Sol. ⎫
W. —— *rotundifolium* Forst. ⎪
W. —— *pubens* A. Rich. ⎪ Capsule 4-angled, seeds numerous, furnished with a tuft of long hair at the summit.
W. —— *nummularifolium* R. Cunn. ⎬
W. —— *melanocaulon* Hook. ⎪
W. —— *microphyllum* A. Rich. ⎪
W. —— *glabellum* Forst. ⎭

B. *Fuchsia excorticata* Linn. f. Fleshy, many-seeded berry.

? *Hydrocotyle asiatica* Linn. 2 dry, indehiscent, cohering carpels.

B. *Panax arboreum* Forst. 2–4-celled succulent exocarp.

B. *Griselinia littoralis* Raoul. Above 1,400 ft. Small 1-seeded berry.

B. *Coprosma lucida*, Forst. 2-seeded fleshy drupe.

B. —— *robusta* Raoul. Above 1,400 ft. 2-seeded fleshy drupe.

W. *Celmisia longifolia* Cass. var. Composite linear achene.

W. *Olearia furfuracea* Hook. f. Above 1,400 ft. Composite ; achenes small, with pappus hairs.

W. *Gnaphalium japonicum* Thunb. ⎫ Achene minutely papillose.
W. —— *luteo-album* Linn. ⎭

W. *Vittadinia australis* A. Rich. Linear pubescent achene; pappus copious.
W. *Raoulia australis* Hook. f. var. *lutescens* T. Kirk. Achenes with numerous extremely slender pappus hairs.
W. —— *tenuicaulis* Hook. f. Above 1,400 ft. Achenes with copious pappus hairs.
W. —— *glabra* Hook. f. Puberulous achenes.
W. *Erechtites scaberula* Hook. f. Achenes with many series of copious soft slender pappus hairs.
W. *Senecio lautus* Forst. Achenes linear, pappus copious.
W. *Sonchus oleraceus* Linn. Achenes with many series of copious soft slender pappus hairs.
? *Wahlenbergia gracilis* A. DC. Capsules 2–5-celled; seeds numerous, small, compressed.
B. *Gaultheria antipoda* Forst. var. Above 1,400 ft. Capsule included in large and succulent calyx and lobes; seeds minute.
W. —— *oppositifolia* Hook. f. Above 1,400 ft. Capsule dry; seeds minute.
B. *Cyathodes acerosa* R. Br. A baccate succulent drupe.
B. —— *empetrifolia* Hook. f. Above 1,400 ft. 3–5-celled small drupe.
B. *Leucopogon fasciculatus* A. Rich. Above 1,400 ft. Small baccate drupe.
B. —— *Fraseri* A. Cunn. Above 1,400 ft. Small baccate drupe.
W. *Dracophyllum subulatum* Hook. f. Above 1,400 ft. 5-celled capsule with numerous seeds. Rare.
B. *Myrsine Urvillei* A. DC. Fruit small, globose, drupaceous, dry or fleshy.
? *Geniostoma ligustrifolium* A. Cunn. Capsule splitting into 2 boat-shaped valves; seeds numerous.
B. *Solanum aviculare* Forst. Large many-seeded berry.
W. *Veronica salicifolia* Forst. var. Above 1,400 ft. Capsule.
B. *Muehlenbeckia australis* Meissn. ⎫ Small nut enclosed in a succulent
B. —— *complexa* Meissn. ⎬ perianth.
B. —— *axillaris* Walp. Above 1,400 ft. ⎭
? *Knightia excelsa* R. Br. Coriaceous 1-celled, 4-seeded follicles, 1½ in. long, tapering into a persistent style, ultimately splitting into 2 boat-shaped valves; seeds winged at the top.
B. *Pimelea laevigata* Gaertn. Above 1,400 ft. Fruit usually baccate.
W. *Thelymitra longifolia* Forst. Orchid; seeds very minute.
W. *Microtis porrifolia* R. Br. Orchid; seeds very minute.
B. *Dianella intermedia* Endl. Berry.
? *Juncus* sp. Capsule, small-seeded.
? *Typha angustifolia* Linn. var. Aquatic.
? *Gahnia pauciflora* T. Kirk. Hard and bony nut.
? —— *Gaudichaudiana* Steud. Above 1,400 ft. Nut small.
? *Carex* sp. Nut.
W. *Deyeuxia filiformis* Petrie. Caryopsis.
W. *Danthonia semiannularis* R. Br. var. Above 1,400 ft. Caryopsis.
W. *Dichelachne crinita.* Caryopsis.
W. *Arundo conspicua* Forst. Caryopsis.
W. *Poa anceps* Forst. Caryopsis.
W. *Cyathea dealbata* Swartz. Spores.
W. *Hemitelia Smithii* Hook. Spores.
W. *Pellaea rotundifolia* Hook. Spores.
W. *Pteris aquilina* Linn. var. *esculenta.* Above 1,400 ft. Spores.
W. —— *tremula* R. Br. Spores.

W. *Lomaria lanceolata* Spreng.   Above 1,400 ft.   Spores.
W. —— *penna marina* Trev.   Above 1,400 ft.   Spores.
W. —— *capensis* Willd.   Above 1,400 ft.   Spores.
W. *Asplenium adiantoides* C. Chr.   Spores.
W. —— *lucidum* Forst.   Spores.
W. —— *flaccidum* Forst.   Spores.
W. *Polypodium pennigerum* Forst.   Spores.
W. —— *serpens* Forst.   Spores.
W. —— *Billardieri* R. Br.   Spores.

Naturalized Plants.

? *Trifolium repens* Linn.   Possibly introduced by rabbits, hares, horses, or pigs.
W. *Erigeron canadensis* Linn.
W. *Hypòchaeris radicata* Linn.
W. *Anagallis arvensis* Linn.
W. *Oenothera odorata* Jacq.
? *Erythraea Centaurium* Pers.
? *Rumex Acetosella* Linn.

If the above list be analysed it will be seen that of ninety-one species observed on the isolated north-western face, twenty-four (or 26 per cent.) may be called bird-distributed, fifty-three (or 58 per cent.) wind-distributed, and only fourteen (or 15 per cent.) are difficult to account for.

The absence of certain species may be noted. Perhaps, of forest-trees, the absence of all species of *Nothofagus* would be dismissed lightly, as few patches of *Nothofagus* forest are known in the Rotorua area ; but on looking more carefully into the matter it is certainly singular that the highest mountain in that neighbourhood should not have been peopled by *Nothofagus*, were it not that its seeds are not obviously spread by birds or wind. Indeed, a knowledge of how *Nothofagus* seed is distributed may perhaps help to solve many points in the distribution of the genus at present a puzzle to ecologists. The genus occurs at Waimarino and on the volcanic mountain Ruapehu (four species), on the lower slopes of the Kaimanawa Mountains (three species), on Mount Hikurangi, on Te Aroha Mountain (volcanic), in the Mangorewa and Omanawa Gorges, and on the Matai Road near Te Puke.

Buchanan (29) in 1866 noted that the forest on Mount Egmont (an isolated volcanic mountain more distant from any *Nothofagus* forest on non-volcanic mountains than is Ruapehu, Tarawera, or Ngongotaha, and rising from sea-level to 8,200 ft.) is chiefly peculiar through the absence of *Nothofagus*, and his observations, which related to a limited portion of the mountain, have been confirmed for the remainder of the area by subsequent observers.* On the other hand, Ruapehu (9,175 ft.) and Te Aroha (3,126 ft.) are not many miles distant in an air-line from *Nothofagus*-clad geologically ancient mountains, and therefore, judging from analogy, *Nothofagus* should some day appear on Tarawera.

Dracophyllum is a genus which occurs very sparingly on this portion of Tarawera, and *Ixerba*, *Dacrydium*, *Podocarpus*, *Melicope*, *Carmichaelia*, *Carpodetus*, *Phyllocladus*, *Fusanus*, and *Parsonsia*, all plentiful in the

---

* The author ascended Mount Egmont to the summit by the Inglewood track on the 14th February, 1901, and again on the 20th March, 1910.

Rotorua district, have not been observed on any portion of the north-west or northern faces. They are large-seeded plants, although some have fleshy fruits, and are no doubt distributed by large birds. One of the two naturalized rosaceous species may possibly have been spread by wild horses, which are said to fatten on the fruits.

The Northern Face.

From the sand-crater which separates the northernmost peak (Wahanga) from the middle peak (Ruawahia) after a few hundred feet of steep scoria descends a beautifully easy slope of scoria down to the lake-edge. This slope is comparatively bare of continuous scrub growth until near the 40 ft. terrace above the lake, when a shrubbery of *Leptospermum*, &c., is encountered. This slope, where it descends without a break, forms a natural boundary separating what may be called the north-western area from the northern area. The area north of this boundary has been visited on two occasions. Numerous dry vertically walled ravines about 40 ft. deep run roughly parallel to each other, making progress across country slow. Landing near the eucalypts at the north end of the lake, several aquatic and semi-aquatic plants were noticed. In small stagnant pools was the submerged *Nitella* (two species), *Potamogeton Cheesemanii*, and *Myriophyllum* sp.; while on the margins were *Glossostigma elatinoides* (with pinkish flowers) and *Typha angustifolia* var.

Progress on to the 40 ft. terrace (the lake rose 40 ft. at the time of the eruption and afterwards subsided; this may be the beach formed at that time) was made easy by the tracks of wild horses. Several naturalized plants were noticed—*e.g.*, *Rumex crispus*, *Oenothera* (in beautiful yellow flower), *Anagallis arvensis*, *Rumex Acetosella*, *Carduus pycnocephalus*, *Sonchus arvensis*, *Erigeron canadensis*, *Verbascum Blattaria*, *Bartsia viscosa*, and, what is lamentable, *Rosa rubiginosa* and *Rubus fruticosus*. There is no doubt if these latter two get thoroughly established on the mountain they will be a fearful curse in days to come. The *Oenothera* seems to be able to thrive on the bare scoria without the aid of any humus, and is rapidly travelling up the mountain. A most interesting fact is that the great *Coriaria* association is here dying out. The tree form of *Coriaria ruscifolia*, 15 ft. high, with many large trunks springing from one root, as described above, are now dead. This accords with an observation of Mr. C. Way, of the Wairoa accommodation-house, that, although the *Coriaria* is the first shrub to take possession of volcanic ejectamenta, it is comparatively short-lived. Here, where it is obviously dying out from old age, it is giving place to vigorous thickets of *Aristotelia racemosa* and *Fuchsia excorticata*. *Aristotelia* is entirely dominant in places, and is 20–25 ft. high; but associated with it are *Solanum aviculare*, *Melicytus ramiflorus*, *Cyathodes acerosa*, *Muehlenbeckia complexa*, *Asplenium flaccidum*, and *A. adiantoides*. Some 50–60 ft. above the lake the open scoria slope is reached, which leads to the sand-crater arete of Tarawera. Here a most remarkable change is going on, which is unfolded to one as the gradual ascent is made. The gentle slope is covered in places with uneven patches of *Raoulia australis* and *R. glabra*. These are the dominant patch plants. Others are *Pinelea laevigata*, *Muehlenbeckia axillaris*, and even *Leucopogon Fraseri*. *Olearia furfuracea*, the dominant composite shrub of the mountain, in full flower is a fine sight. Occasionally, where the water-supply is favourable, lichens and moss may perform their usual function of transforming the barren rock into fertile soil, but the *Raoulia* must be accounted the great humus-maker of this mountain.

As it languishes in vigour, owing to age, from it grow other plants, the chief woody ones being *Coriaria* and *Leptospermum*, and sometimes *Pittosporum*, but also herbaceous plants such as *Trifolium* and *Rumex Acetosella*. Four stages may thus be predicted for the repeopling of the plant-covering of this open area (excluding the ravines, which are able to jump the first and possibly the second stages): First, the patch plants; secondly, the *Coriaria;* thirdly, the *Aristotelia*, with possibly *Fuchsia* and *Melicytus;* fourthly, forest.

The scoria slope is now left at about 150 ft. above the lake, where the track towards the spur is found overgrown with *Aristotelia*, *Rubus australis*, *Coprosma lucida*, *C. robusta*, *Veronica salicifolia* var., and *Coriaria*.

A densely wooded ridge on the north of the scoria slope is investigated. It consists of *Weinmannia racemosa*, which is perhaps the dominant growth, but the composition is very varied. *Knightia excelsa* stands out above all the other young growth in cylindrical or slightly conical tops. A patch of tree-ferns is noticed. Giant *Fuchsia* and *Weinmannia*, with *Melicytus ramiflorus*, all about 30 ft. high, with a bare floor, form a subassociation of their own, supplanting the dying *Coriaria*, nearly as high. Higher on the spur, 500 ft. above the lake, the growth is more varied, being a thick mass of *Panax arboreum*, *Olearia Cunninghamii*, *Brachyglottis repanda*, *Lomaria capensis*, *Veronica salicifolia* var., young *Weinmannia*, *Gaultheria oppositifolia*, *G. antipoda* var., *Carex Gaudichaudiana*, and *Cyathodes acerosa*. At 675 ft. above the lake numerous dead stumps of *Metrosideros* occur. These may either be *M. tomentosa* or *M. robusta*, young plants of both being common in flower on the open slopes at this altitude and higher. The ravines, four or five of which were crossed diagonally in returning to the scoria slope, yielded some valuable information as to the composition of the forest before the eruption. One dead *Beilschmiedia tawa* with a broken trunk 15–20 ft. high was 3 ft. in diameter, and springing from its side was a young tree about 18 in. in diameter 4 ft. from the ground, where it forked into a tree 35–40 ft. high. Close to this was a dead giant *Litsaea calicaris* stump 4 ft. in diameter at the surface, and 9 ft. from the ground, where broken, 3 ft. in diameter; and attached to it was a young living shoot. Other dead *Litsaea* stumps showed that there had been a forest containing many of these, and young thickets of saplings showed that it was being re-established. *Melicytus* was also found springing from old dead stumps. Only two Beilschmiedias, both with young live growth springing from the dead stump, were, however, seen. Other plants not hitherto met with on the mountain are *Rhipogonum scandens*, *Schefflera digitata*, *Hemitelia Smithii*, *Hedycarya arborea*, *Asplenium bulbiferum*, *Lomaria lanceolata*, *Gaultheria rupestris*, *Astelia Cunninghamii*, *Elaeocarpus dentatus*, *Cyathea dealbata*, *Lycopodium volubile*. Also present were *Fuchsia excorticata*, *Asplenium lucidum*, *Polypodium Billardieri*, *P. pennigerum*, *Myrsine Urvillei*, *Dianella intermedia*, *Metrosideros robusta*, *Gunnera monoica*, *Vittadinia australis*.

On the northern face of the mountain the following additional species were noted: *Elaeocarpus dentatus* Vahl., *Olearia Cunninghamii* Hook. f., *Gaultheria rupestris* R. Br., *Hedycarya arborea* Forst., *Beilschmiedia tawa* Benth. & Hook., *Litsaea calicaris* Benth. & Hook., *Cyathea dealbata* Swartz, *Hemitelia Smithii* Hook., *Lomaria lanceolata* Spreng., *Asplenium flabellifolium* Cav.; together with the following naturalized species—*Rubus fruticosus* Linn., *Rosa rubiginosa* Linn., *Sonchus arvensis* Linn., *Carduus pycnocephalus* Linn., *Verbascum Blattaria* Linn., *Bartsia viscosa* Linn., *Rumex crispus* Linn.

BIBLIOGRAPHY.

1. "Notes on the Flora of the Lakes District of the North Island." T. Kirk. Trans. N.Z. Inst., vol. 5, 1873, p. 322.
2. "The Eruption of Tarawera." S. Percy Smith. Wellington, 1886.
3. "Phenomena connected with Tarawera Eruption as observed at Gisborne." W. L. Williams. Trans. N.Z. Inst., vol. 19, 1887, p. 380
4. "Tarawera Eruption." J. A. Pond and S. Percy Smith. Trans. N.Z. Inst., vol. 19, 1887, p. 342.
5. "Tarawera Eruption as observed at Opotiki." E. P. Dumerque. Trans. N.Z. Inst., vol. 19, 1887, p. 382.
6. "Tarawera Eruption as seen from Taheke, Lake Rotoiti." W. G. Mair. Trans. N.Z. Inst., vol. 19, 1887, p. 372.
7. "Observations on Tarawera Eruption." J. Hector. Trans. N.Z. Inst., vol. 19, 1887, p. 461.
8. "Tarawera Eruption: Criticism of Explanations of its Causes." J. Hardcastle. Trans. N.Z. Inst., vol. 20, 1888, p. 277.
9. "The Geology of New Zealand." James Park. 1910, p. 166.
10. "Geology of New Zealand." P. Marshall. 1912, p. 106.
11. "Report on the Eruption of Tarawera and Rotomahana." A. P. W. Thomas. Wellington, 1888.
12. "The Great Tarawera Volcanic Rift, New Zealand." J. M. Bell. Geog. Journ., vol. 27, 1906, p. 369.
13. "A Visit to Mount Tarawera." H. M. Cadall. Scott. Geog. Mag., xiii, 1897, p. 246.
14. "A Visit to the New Zealand Volcanic Zone." Trans. Edin. Geol. Soc., v, 1897, p. 183.
15. "The Recent Volcanic Eruption in New Zealand." A. Geikie. *Nature*, vol. 34, 1886, p. 320.
16. "Preliminary Report on the Recent Volcanic Eruptions, with Appendix." W. Skey, Wellington, 1886; also, *Nature*, vol. 34, 1886, p. 389.
17. "On the Recent Volcanic Eruptions at Tarawera: Preliminary Report." J. Hector. 1886–87.
18. "The Volcanic Eruption in New Zealand." H. S. Johnston-Lavis. Geol. Mag., 1886, iii, p. 523.
19. "Volcanic Ash from New Zealand." J. Joly. *Nature*, vol. 34, 1886, p. 595. Note: J. W. Judd, Brit. Assn., 1886, p. 644.
20. "The Volcanic Eruption in New Zealand." J. H. Kerry-Nicholls. Journ. Soc. Arts, 1887, p. 174.
21. "The Volcanic Eruption of Tarawera." T. W. Leys. Auckland, 1886.
22. "History of Volcanic Action in New Zealand." P. Marshall. Trans. N.Z. Inst., vol. 39, 1907, p. 542.
23. "On the Tarawera Eruption and After." James Park, Geog. Journ., vol. 35, 1910; S. Percy Smith, Proc. Roy. Geog. Soc. Lond., 1886, p. 783.
24. "Tarawera." *Nature*, vol. 34, pp. 275 and 301.
25. "Volcanic Eruption in New Zealand." Amer. Journ. (iii), vol. 22, 1886, p. 162.
26. "Volcanic Eruption in New Zealand." Proc. Roy. Geog. Soc. (n.s.), 8, 1886, p. 783.
27. "Official Report on the Flora of the Lake District." T. Kirk. *New Zealand Gazette*, No. 43, 4th September, 1872.
28. "New Zealand Plants and their Story." L. Cockayne, 1910, p. 54.
29. "Botanical Notes on the Kaikoura Mountains and Mount Egmont." John Buchanan. N.Z. Geological Survey Report, 1866–67.

ART. XXXII.—*Observations on the Lianes of the Ancient Forest of the Canterbury Plains of New Zealand.*

By J. W. BIRD, M.A.

[*Read before the Philosophical Institute of Canterbury, 3rd November, 1915.*]

Plates XXIII–XXVI.

I. INTRODUCTION.

LIANES form, in the New Zealand forest, an ecological group of prime physiognomic importance, as pointed out by L. Cockayne (1908, p. 24). Further, within quite limited areas the various life-forms of this class of plants occur side by side, so that it is comparatively easy to study and compare their forms, adaptations, behaviour, and life-histories. Up to the present time such studies have been for the most part neglected; nor is this to be wondered at, since in a newly settled country the earliest botanical investigations must of necessity be floristical, while even in the Old World ecology is yet in its infancy. The work of L. Cockayne stands. out as a notable exception to the above statement, for in a series of writings, commencing in 1898 and extending to 1915, he has given a good many details as to the methods of climbing and life-histories of lianes, and the evolution of the climbing habit; but he has not published any comprehensive account of the group. Nor in this paper is it possible for the writer to attempt. anything really comprehensive—a matter that would have required several. years' experience in the field in all parts of the botanical region. On the contrary, the study, for which no completeness is claimed, since New Zealand contains no less than forty-seven lianes, which belong to sixteen families. and twenty-two genera, is confined to one definite forest area, Riccarton Bush, the last remnant of the ancient forest of the Canterbury Plains.*

Nor is the study of the lianes in this area complete. A complete account must of necessity include stem-anatomy. This in itself is of such importance that it should be made the subject of a separate paper, for thus only can it receive its deserved attention. There are experiments to be made dealing with water-conduction, mechanical principles, &c., and for this. considerable time and observation are necessary. Moreover, the forest investigated is of such importance that it is rightly preserved closely; and the amount of material required for a complete stem account cannot be obtained without considerable destruction to the forest.

Riccarton Bush is a portion of an ancient kahikatea (*Podocarpus dacrydioides*) forest. The members of this species do not here grow very closely together, but they attain a great size, their trunks, often 1·6 m. in diameter, rising straight up to a height of 30–40 m., with their lowest branches 20 m. from the ground. On the surface of the ground their roots spread widely, many of them twisting up and forming knees about 50 cm. in height, and others, through the displacement of the soil by which they were formerly partially covered, forming a kind of reticulate platform round the base of the tree (Plate XXIV, fig. 2). Growing to the same level in the forest, but present in small numbers, are two other species of *Podocarpus—P. spicatus* (matai) and *P. totara* (totara)—and two species of *Elaeocarpus— E. dentatus* and *E. Hookerianus.*

* For an account of this association see Armstrong, J. F., 1869, and Cockayne, L., 1914.

The smaller trees and the shrubs are represented by many more species which grow chiefly in the more exterior portions of the forest, where the light-conditions are more favourable. The shrub or low tree *Paratrophis microphylla* is especially abundant.

On the forest-floor grow certain herbs and ferns, the diffuse light causing elongated thin shoots and broad membranous leaves, while seedlings of the most diverse inhabitants of the forest are here found in considerable numbers, forming a living floor of vegetation.

From their tall supports hang the rope-like stems of the lianes, often 12 cm. in diameter. These stems, which belong to the various species of *Rubus* and *Muehlenbeckia*, act as supports for still more lianes, the whole often forming impenetrable tangled masses. Where the support has died, or the liane slipped down, other stems form great coiling masses, extending along the ground for several metres. The various twining-lianes, tendril-bearers, and many scramblers climb on the smaller trees and the shrubs, while the small shrubs and the herbs are excellent supports for the growing juvenile lianes.

The number of species of vascular plants in the forest is sixty-eight, consisting of—Trees and shrubs, 35 ; lianes, 12 ; parasites, 2 ; herbaceous plants, 12 ; ferns, 7.

The lianes dealt with in this paper may be classified as follows : (1.) Scramblers : *Rosaceae—Rubus australis* Forst. f. var. *glaber* Hook. f., *R. schmidelioides* A. Cunn., *R. cissoides* A. Cunn.,* *R. subpauperatus* Cockayne ; *Onagraceae—Fuchsia Colensoi* Hook. f. (2.) Root-climbers: *Myrtaceae—Metrosideros hypericifolia*. (3.) Twining-plants : *Polygonaceae—Muehlenbeckia australis* (A. Rich.) Meissn., *M. complexa* (A. Cunn.) Meissn. ; *Apocynaceae—Parsonsia heterophylla* A. Cunn., *P. capsularis* (Forst. f.) R. Br. var. *rosea* (Raoul) Cockayne. (4.) Tendril-climbers : *Ranunculaceae—Clematis indivisa* Willd. ; *Passifloraceae—Tetrapathaea australis* Raoul.

Before concluding this introduction, I wish to express here my indebtedness to those who gave me assistance in the preparation of this paper. I especially wish to thank Mrs. Deans, through whose permission to make investigations in Riccarton Bush the work was able to be carried out; Dr. C. Chilton, C.M.Z.S., who supervised the preparation of the paper and the laboratory-work; and Dr. L. Cockayne, F.R.S., who suggested the outlines of the paper and gave valuable assistance in connection with the field-work.

II. AUTECOLOGY OF THE LIANES.

1. Fuchsia Colensoi.

A. LIFE-FORM.

Although fairly abundant in the forest, the distribution of this species indicates that the conditions required for its growth are an abundance of moisture, but at the same time a well-drained soil. The species is most abundant on the sides of and in the vicinity of the drains which run through the forest, and, further, at the edge of the swampy portion of the forest. In this swamp itself the species is entirely absent. The individual plants, upon leaving the ground, give off numerous shoots, which may scramble up among the branches of an overhanging shrub, or trail along the forest-

* Dr. L. Cockayne informs me that there is some doubt as to what *R. cissoides* A. Cunn. really is, but the plant here dealt with is *R. cissoides* as defined by Cheeseman, 1906, p. 125.

floor for a distance of 5 m. The stems often attain a thickness of 3 cm. in diameter, and are covered with a papery bark. At intervals along these stems shoots arise, which may reach a support, or, failing this, they bend over till they touch the ground, where they continue their growth. As these shoots spread out in all directions, their branches are a considerable distance apart, and thus it is almost impossible that none of them reach a support. Roots arise in large numbers from these trailing-stems, and thus the food-supply available for the shoots is largely increased. The value of these roots may be seen from the fact that stems with roots attached which had been cut through by the writer continued to live, and the roots gained enough food-material from the soil to maintain in vigorous growth the shoots which were given off near the roots.

Shoots arising from the adult stem in the shade are very thin (5 mm. in diameter) and elongated (internodes are 7–8 cm. long). The stems which straggle upwards through the supporting branches are mostly un-branched, but maintain their upright position by twisting and turning through the network of supporting branches, and finally may attain a height of 4 m. The leaf-petioles are very thin and fragile, and thus, though they project from the stem at right angles, cannot be of great assistance in climbing. But where the plants grow in more strongly illuminated positions the stems branch freely, and, as the branches project at right angles, they are of prime importance in obtaining and maintaining new positions of support. On the top of supporting shrubs, branches may stand erect for fully 50 cm., but most of the branches lie horizontally on the support and form a dense covering. Others, again, hang down from the edge of the support, and thus the leaves of the plant are borne in all positions suitable for assimilation.

In exposed situations the plants form dense, low bushes, about a metre high, and usually 2–3 m. across. The primary stem gives off, near its base, numerous branches, which spread out and branch further, the branches interweaving and forming masses of the divaricating life-form. From the tops of these masses stems may rise up, and by their mutual support gain a height of a metre above the main mass of branches. And thus, should they come into contact with the branches of any overhanging tree or shrub, the liane-branches may push their way into the support, and there continue their scrambling growth. Branches from these exposed plants at the edge of the forest may trail along the ground, and, rooting freely, give off shoots in the shade, where supports are abundant. And so efficient are these roots that sometimes the attenuated shoot may give off near these roots new shoots, which attain a thickness fully double that of the parent shoot.

Fuchsia Colensoi is one of the few New Zealand indigenous plants which are deciduous. Plants in exposed situations lose their leaves early in winter, but in the shade they are devoid of leaves for only a few weeks.

B. LEAF.

(i.) *Leaf-form.*

Leaves alternate, petiolate, thin, membranous; upper surface pale green, dull; lower surface greyish-green, shiny; veins purple; 2–4 cm. long; orbicular or orbicular-ovate; cordate or rounded at base; minutely and remotely serrate. Petioles usually slightly longer than blade, translucent, upper surface slightly grooved.

Sun leaves are slightly smaller and less membranous, and have shorter petioles.

(ii.) Leaf-anatomy.

(*a*.) *Shade Leaf.*—Epidermis: Cells flattened, thin-walled, outer walls very convex ; lower epidermal cells smaller ; cells beneath midrib thick-walled. Stomata: Lower surface only, level with epidermal cells. Chlorenchyma : Palisade—1 layer cells, slightly elongated, and rather loosely packed Spongy—5–6 layers ; upper cells nearly spherical, lower cells more or less elongated, parallel to leaf-surface ; intercellular spaces large. Chlorophyll abundant both in palisade and in spongy tissues. Vascular bundles surrounded by poorly developed bundle-sheath. Calcium oxalate in raphides.

(*b*.) *Sun Leaf.*—Epidermis : Walls slightly thickened. Chlorenchyma : Palisade—Cells more elongated and more closely packed than in shade leaves. Spongy—Cells more regular ; intercellular spaces smaller.

2. THE SPECIES OF RUBUS.

There can be no doubt that the species of *Rubus* in New Zealand are poorly defined. Cheeseman did not recognize *R. subpauperatus* in the Manual, but from the following descriptions it will be seen that there are sufficient grounds for the establishment of Cockayne's species (1909A, p. 42). The different species show many points of similarity in their life-form, but there are some characteristics which, though common to all, are more noticeable in particular species.*

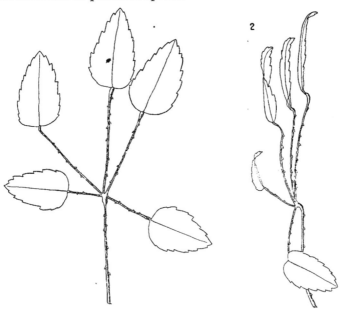

FIG. 1.—Rubus australis. Leaf ; top view.
FIG. 2.—Rubus australis. Leaf ; side view.

The leaves are alternate and palmately compound (see figs. 1 and 2), with 3–5 leaflets, which vary in size. Margins serrate, with incisions to

---

* In the descriptions below, therefore, the special points of each species are more fully dealt with. Further, the descriptions cannot be rigidly applied to all the members of a species, on account of the diversity of form of the individual plants ; and, in fact, it is probable that hybrids are of common occurrence.

varying depths. The distal end of subpetioles is bent up about 0·5 cm. from the lamina, and considerably thickened. Further, the lamina is at right angles to this thickened portion, and so lies parallel to the proximal portion of the subpetiole. Prickles may be present on the under surface of petiole, subpetiole, and midrib.

(*a*.) Rubus australis var. glaber.

A. LIFE-FORM.

This species is fairly abundant in the forest, and is found chiefly in the drier situations. Since these parts of the forest are the habitat of plants of varying habits, the liane is presented with abundant support, usually climbing up among the interlacing branches to a height of 12 m. or more. Very often the falling of trees forms large gaps in the forest vegetation, which provide excellent situations for plants which thrive best in bright sunlight. And it is on the plants surrounding such gaps that *R. australis* is found growing at its best. Its thick stem, often 4 cm. in diameter, covered with a hard, brown, scaly bark, gives off at right angles numerous lateral shoots, which branch further, and, finally, leafy masses borne on these branches form an impenetrable wall 3–4 m. high.

In the more shaded portions of the forest the stems give rise to very few lateral branches, but stretch up like ropes into the branches overhead, where their foliage is spread out in the light. In such cases, where the expanse of foliage is on the branches of a kahikatea, whose lowest branches, as stated above, are usually 20 m. from the ground, it is impossible to conceive how the liane could reach the support directly. However, the frequent presence of decaying stumps of trees near these taxads suggests that the lianes commenced their growth by scrambling up these trees, and from their top branches reached those of the neighbouring pines. Such a process as this is actually seen occurring on other trees which form supports for the liane. From almost any part of the liane-stem there arise adventitious shoots similar to those of *R. schmidelioides*, and if these shoots happen to arise within 2–3 m. of the branches of a neighbouring tree the shoots very often bridge the intervening space; then by the grasping action of prickles, which are borne on the stems, the new supports are firmly held. The stems of *R. australis* which have never reached a support or which have fallen away from such, and lie along the ground, show great tendency to adventitious rooting wherever in contact with moist soil. From these trailing-stems adventitious shoots often arise, which may reach a support, or which may bend over and, continuing their growth along the ground, give off still more roots. Now, the writer has proved by experiment that when these roots are well developed they are themselves capable of supplying sufficient nutritive material and water for the further growth of any adventitious shoot which arises near them. Thus an injury to the primary root or stem need not result in the death of the whole plant. On the contrary, this rooting is a means by which the plant can increase by vegetative reproduction, one such plant noted representing at least seven potential individuals.

B. LEAF.

(i.) *Leaf-form.*

Leaves 3–5-foliate; leaflets coriaceous, glabrous, 8–12 cm. long, ovate-oblong or ovate-lanceolate; base rounded or truncate, apex acute, margin serrate. Petiole 6–8 cm. long, deeply grooved on upper surface along its entire length. Prickles in small numbers on petiole and subpetiole, absent from midrib. The distal end of the petiole is often swollen and bent, and from it the leaflets project at varying angles so that they face the light.

Both seedlings and adventitious shoots bear leaves which show all transitions from simple to compound. The first leaf is simple, and is often succeeded by a simple leaf with a lobe near its base. Later, leaves with 2 lobes may appear, and may be succeeded by a leaf with a lateral leaflet in place of one of these lobes. From this stage to the adult 3–5-foliate leaf there are numerous transitional forms. Injury to the growing-point by frost or other means may lead to the development of lateral shoots, which bear leaves in an order similar to that on the primary stem, and, on lateral shoots from these, leaves may again develop in a corresponding order.

(ii.) *Leaf-anatomy.*

(*a.*) *Shade Leaf.*—Epidermis : Cells slightly flattened ; well-developed cuticle ; stomata under surface only, level with epidermal cells. Hypoderm : One layer isodiametrical cells, continuous on upper surface ; on lower surface confined to the regions near the vascular bundles. Chlorenchyma : Palisade — 2 layers, cells rather elongated. Spongy — 5–7 layers ; cells mostly elongated in direction of palisade, and so no sharp boundary between the tissues ; intercellular spaces fairly large. Vascular bundles : Surrounded by sheath of collenchyma, which connects with the upper and the lower hypoderm ; a small band of stereome is present at the base of the larger bundles. Leaf-margin : Very thick cuticle ; hypoderm 2 layers ; chlorenchymatous cells elongated and arranged radially. Calcium oxalate : Aggregate crystals.

(*b.*) *Sun Leaf.*—Epidermis : Upper epidermis, cells with small lumina and much-thickened walls ; cuticle very thick ; lower epidermis, walls much thickened, but to a less extent than those of upper epidermis. Chlorenchyma : Palisade—Cells very elongated and closely packed ; tissue occupies almost one-half of leaf-thickness. Spongy—Cells fairly regular ; intercellular spaces smaller than in shade leaf. Vascular bundle : Stereome at base of each bundle well developed, and composed of cells with walls greatly thickened. Leaf-margin : Hypoderm forms a mass of 3–4 layers of cells with thick walls.

(*b.*) Rubus schmidelioides.

A. LIFE-FORM.

In the shade of the forest this liane often grows to a height of 8 m. or more before it reaches the sunlight. It may be growing upright beside its support, or, if the support is a tree with smooth bark and few lateral branches in the shade, the liane-stems may be coiled in a tangled mass at the foot of the support, indicating that the liane has probably fallen down through the increased weight of its own body. Between the liane's root and the support there is often a distance of 4–5 m., and in this intervening space the stem lies close along the ground, perhaps covered with debris. Also, adventitious roots may be given off which anchor the stem, and doubtless are of further use with regard to the food-supply.

The stem, growing to 4 cm. in diameter, is dark brown in colour, and smooth ; the nodes are 7–9 cm. apart, and usually slightly swollen. Prickles are practically absent. From almost any portion adventitious shoots may arise, which are characterized by very vigorous growth. Although very slender, the stems may grow erect without support to a height of 1 m. or more, further growth resulting in the shoot bending over and finally touching the ground. Here they continue their rapid growth, and bear leaves, which stand upright with their expanded laminae facing the incident rays of light. As adventitious shoots frequently arise high up on the plant, they are important in reaching the new positions of support, either on the same tree or on adjacent trees. Lateral branches normally arise where the stem is growing in the sun. They are at first usually at right angles to the main axis, but upon reaching the branches of the support grow upwards and give rise to more shoots. At other times, failing to come into contact with a support, they hang down, forming with other branches a divaricating reticulation, upon which masses of leaves are borne. The divisions of the leaves assist further in the formation of a " leaf-mosaic," whose efficiency may be judged by the manner in which the supporting tree is often hidden from sight.

In the more exposed portions of the forest the species scrambles over low-lying shrubs, forming a dense mass, with interlacing branches, which have a tendency to droop, and thus often hide the support.

B. LEAF.

(i.) *Leaf-form.*

Leaves 3–5-foliate ; leaflets 4–5·5 cm. long, orbicular-ovate or orbicular-oblong, coriaceous, acute, rounded or cordate at base, irregularly toothed, usually pubescent beneath. Petioles 4–6·5 cm. long ; subpetioles vary in length, terminal 4–5·5 cm. long, basal 0·5–1 cm. long. Prickles numerous on petioles and less numerous on subpetioles, usually absent from midrib.

Shade leaves : Adult, subcoriaceous ; juvenile, membranous, and often brightly pigmented, being reddish-brown in colour.

(ii.) *Leaf-anatomy.*

(*a.*) *Sun Leaf.*—Epidermis : Both upper and lower epidermis have a well-developed cuticle, the upper being especially thick ; stomata on lower surface only, and level with the epidermal cells. Hypodermis : A single layer of large cells with collenchymatous walls inside both upper and lower epidermis. Chlorenchyma : Palisade tissue—2–3 layers of elongated cells, rather loosely packed. Pneumatic tissue—6–8 layers of irregular cells, more or less rounded ; intercellular spaces very large. Vascular bundles : Collateral, surrounded by a well-developed sheath of thick-walled parenchyma, the cells below the vascular bundle having their walls especially thick. Leaf-margin is strengthened by 2–3 layers of collenchyma, which extend for a few cells' length along both surfaces. Crystals of calcium oxalate are present in the hypoderm and the chlorenchyma in the form of aggregate crystals.

(*b.*) *Shade Leaves.*—Shade leaves have most characteristics in common with exposed leaves. The variations are—(1) Many cells of the hypoderm have chloroplasts ; (2) cuticle is not so well developed ; (3) less strengthening of the leaf-margin ; (4) *pigment* may be present. The cells of the upper hypodermis and the first row of palisade cells often contain abundant anthocyan, which is scattered to a less extent throughout the rest of the chlorenchyma.

11—Trans.

(c.) Rubus cissoides.

A. LIFE-FORM.

This species occurs in the forest in most situations, and though not growing to a great size, nevertheless exhibits greatly varying habits of growth. On the smaller trees and on the shrubs its growth resembles that of other species of *Rubus*, but, like *R. subpauperatus*, it abounds on the outskirts of the forest, where it forms straggling masses with leaves reduced to midribs.* In numerous cases certain shoots scramble into a tree, in whose shade the leaf-blades are well developed. Other branches of the same plant stretch along the ground and form the wiry masses characteristic of the plant in these situations, while branches between these two portions bear leaves which exhibit all gradations in leaf-reduction (Plate XXIII). The prickles, unlike those in *R. subpauperatus*, are absent from the stem, and are practically absent on the leaf-lamina, although as many as three may sometimes be present. The stems are covered with a greyish-brown bark, and grow to a thickness of 5 cm. in diameter. Leaves

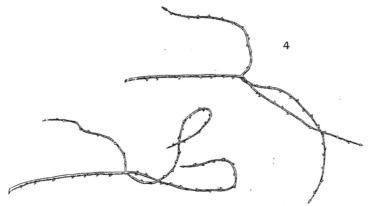

Figs. 3, 4.—*Rubus cissoides.* Climbing-leaf.

with smaller laminae possess more prickles, while in the final stages of reduction the midrib may bear as many as 25. The prickles form a feature by which this species can be readily distinguished from *R. subpauperatus*, for in *R. cissoides* they are yellowish, but in the latter red.

Especially characteristic of *R. cissoides* is the production of adventitious shoots. These are green, thus being fitted for photosynthesis, and the internodes are relatively long — *e.g.*, 15 cm. or more — and in some cases 1·7 cm. in diameter, and stand erect for 2·5–3 m. The leaves of such shoots are almost reduced to greatly elongated midribs. A typical shade leaf has a terminal leaflet 5–7 cm. in length, and rather shorter lateral leaflets ; but the leaves of adventitious shoots may have their terminal midrib 21 cm. long and their lateral midribs 13 cm. As these leaves project from the stem at right angles, and bear large numbers of prickles, it can be seen how important they are in aiding the plant to reach new supports.

Very often the liane is found straggling along the forest-floor for nearly 10 m. In places where the undergrowth is scanty *Rubus cissoides* forms

---

* This is the so-called var. *pauperatus* (J. B. Armst.) T. Kirk.

Rubus cissoides. Leaves, showing stages in leaf-reduction.

FIG. 1.—*Tetrapathaea australis* growing on support.

FIG. 2.—Interior of the forest, showing *Podocarpus dacrydioides* with reti
culating roots.

FIG. 2.—*Muehlenbeckia australis.* Twining-stem round former support.

FIG. 1.—*Rubus subpauperatus* stretching up beside a supporting white-pine.

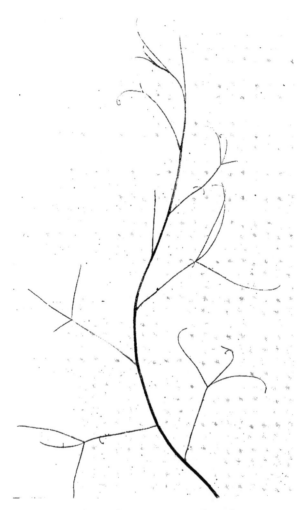

Rubus cissoides. Showing curving of climbing-leaves.

extensive flattened masses, but where there is more abundant vegetation it may push its way among the tangled branches and increase the density of these masses.

It has been stated above that in exposed situations the leaves are reduced to midribs. However, the reduction in this species cannot always be a reaction to environmental factors, for in very damp shady stations certain individuals occur which have their leaves similarly reduced, while adjacent plants show no reduction. Possibly there may be two varieties— one in which the leaves show reduction in the open only in response to some environmental stimulus, and another in which leaf-reduction is a fixed hereditary character.*

B. LEAF.

(ii.) *Leaf-anatomy.*

(*a.*) *Shade Leaf.*—Leaf 3–5-foliate; laminae 4–7 cm. long, subcoriaceous, ovate-oblong or ovate-lanceolate, base truncate; margin deeply serrate. Yellow prickles on under surface of petiole; few or absent on subpetioles and midribs.

(*b.*) *Sun Leaf.*—Leaf reduced almost to midribs; terminal 9–12 cm., lateral 7–9 cm. Petiole 6–7 cm. Prickles in large numbers on both petiole and midribs.

(i.) *Leaf-form.*

(*a.*) *Shade Leaf.*—Epidermis: Cells almost isodiametrical; well-developed cuticle; stomata lower epidermis only. Hypoderm: One layer of thick-walled collenchyma, extending a short distance from leaf-margins. Chlorenchyma: Palisade — 2 layers cells; upper rather closely packed, second layer looser. Spongy — 4–5 layers irregular cells; intercellular spaces large. Calcium oxalate in aggregate crystals. Vascular bundles surrounded by collenchymatous sheath; stereome scattered irregularly round the bundles; on under surface of midribs a strengthening mass of 4–5 layers of collenchyma.

(*b.*) *Sun Leaf.*—Epidermis: Cuticle very thick; stomata lower epidermis only. Hypoderm: One continuous layer beneath epidermis; cells large, isodiametrical; beneath lower epidermis it forms patches of cells beneath the vascular bundles, and extending a short distance on each side. Chlorenchyma: Tissue separated by vascular bundles and their sheaths, which extend from the upper to the lower epidermis. Palisade — Cells elongated and very closely packed. Spongy—Not sharply separated from palisade; the cells are elongated in the same direction as the palisade, and intercellular spaces are small. Mechanical tissue is more strongly developed than in shade leaves. Leaf-margin protected by 3–4 layers of collenchyma; chlorenchyma at the margin composed of palisade cells with small ends facing the leaf-surface.

(*c.*) *Assimilatory Midrib.*—Epidermis: Thick walls; outer walls convex; stomata distributed round whole surface. Hypodermis: One layer isodiametrical cells. Chlorenchyma: A continuous ring 2–3 cells thick; on upper surface cells elongated, forming a palisade tissue; on under surface irregular cells and intercellular spaces which connect with the stomata through breaks in the hypoderm. Vascular system: A ring of bundles lies close to the chlorenchyma, the dorsal bundle being greatly enlarged. Medulla composed of large spherical cells with thin walls.

---

* For a fuller discussion of this important topic see Cockayne, L., 1901, pp. 291–95.

11*

(*d.*) Rubus subpauperatus.

A. LIFE-FORM.

Members of this species are scattered throughout most parts of the forest, and attain a size at least equal to that of any other of the lianes. Their foliage may be expanded upon the branches of the highest trees, and between the foliage and the ground, the stems stretching up, as in *R. australis*, beside the bare stems of supporting taxads, attain considerable thickness; sometimes they are as much as 16 cm. in diameter, and are thickly covered with rough, dark-brown, scaly bark (Plate XXV, fig. 1). Even from such stems as these, adventitious shoots arise, which in the more open situations do not grow to a great length, but give off numerous leafy branches, so that the adult stem is often covered for a considerable portion of its entire height by these hanging leafy festoons. Among the low-lying trees and shrubs the liane, as in other species of *Rubus*, scrambles over the support and among the branches. In the more shaded stations the internodes are longer than those of more exposed plants, and from stems trailing along the ground numerous adventitious roots arise. Beyond the outermost fringe of forest-trees *R. subpauperatus* forms dense masses 1–2 m. in height. The plants here very quickly cover any low-growing shrub with their wiry interlacing branches, and spread over extensive areas. This is due to the rapid elongation of their shoots, which may bridge the spaces between adjacent shrubs, or which may trail along the ground, and thus reach supports many metres distant. Both leaves and stems are thickly beset with prickles, which by their grasping action are especially suitable for this straggling growth over low supports. Moreover, the prickles enable the plants to use supports of greatly varying nature, such as branching shrubs and trees, tree-trunks with rough bark, and even vertical rock-faces.* The growth of plants in the open results in a shortening of the internodes and excessive reduction of leaf-surface. This reduction appears to be correlated with increased transpiration, for when a plant is exposed to southerly winds the greatest reduction is on its south side, although this is shaded from the bright sun. Again, when the prevailing winds are from the north it is on the north side of the plant that the leaves are most reduced.

B. LEAF.

(i.) *Leaf-form.*

(*a.*) *Shade Leaf.*—Leaf 3–5-foliate; terminal leaflet, lamina 6–8 cm. long; lateral leaflets slightly shorter; when 5-foliate the 2 basal leaflets very small (2–3 cm.). Leaflets subcoriaceous, glabrous, linear-lanceolate or linear, acute, base truncate or obtuse, margin serrate. Petiole 7–9 cm.; terminal subpetiole 3–4 cm., but other subpetioles at most 5 mm. Prickles numerous on back of petioles, in smaller numbers on subpetioles and midribs.

(*b.*) *Sun Leaf.*—Plan of structure similar to that of shade leaf, but much smaller. Lamina at most 5 cm. long, and very narrow; but there is no sharp distinction between the two types of leaves, for a single plant may bear leaves which show all gradations in leaf-reduction.

---

* This does not refer to Riccarton Bush.

(ii.) *Leaf-anatomy*

(*a*.) *Shade Leaf.*—Epidermis : Cells slightly elongated in axis parallel to the leaf-surface ; thick cuticle ; lateral walls thinner and irregular. Hypoderm : A single layer of collenchymatous cells on both leaf-surfaces in the regions of the vascular bundles ; cells are large and isodiametrical. Chlorenchyma : Palisade — In regions between two adjacent hypoderm layers the tissue touches the epidermis and is 3 layers in thickness ; beneath the hypoderm it is composed of 2 layers. Cells are long and closely packed. Tissue occupies nearly one-half of leaf-thickness. Spongy—5–6 layers thick ; cells are elongated and abutting by their ends, forming large intercellular spaces ; chloroplasts are comparatively numerous. Vascular bundles surrounded by thick-walled parenchymatous sheath. Mechanical tissue : Each bundle-sheath is connected with the hypoderm by collenchyma, the whole forming an I-shaped girder, which is further strengthened by stereome at the base of each bundle ; each half of a leaf-blade contains about 5 of these girders ; at the base of the midrib the mechanical tissue is very abundant. Leaf-margin is strengthened by thickening of the walls of the epidermis and hypoderm ; chlorenchyma cells are elongated and radially arranged. Calcium oxalate in aggregate crystals.

(*b*.) *Sun Leaf.*—Epidermis : Cuticle very thick. Hypodermis forms an uninterrupted ring beneath the epidermis. Chlorenchyma : Palisade—Cells more elongated, and more closely packed. Spongy—Cells more spherical, and intercellular spaces thus lessened.

The Leaf of the Species of Rubus as a Climbing-organ.

The value of the leaf as a climbing-organ arises from four characteristics : (*a*.) The leaves extend from the shoot at right angles, and are therefore in the position most suitable to overcome vertical strains, so that the shoot may maintain its upright position and thus grow to a greater height. This is shown by the great height of many adventitious shoots. In many cases in which the leaves can find a support, and so prevent the shoot from slipping down, shoots as slender as 1·7 cm. in diameter may grow upright for 7 m. (*b*.) Lateral strains are overcome by means of the prickles, which, on account of their hooked shape, form efficient anchors. The effect of the strain is minimized by the position of the prickles—a position which causes the strain to be distributed along the whole petiole. It is evident that the prickles also are an important factor in preventing the slipping-down of an upright shoot. (*c*.) Assisting in overcoming such strains is the downward curvature of the apex of the lamina. The power of resistance of this is perhaps very small in comparison with that of a prickle, but it has the advantage that it acts where there is little resistance by prickles, the prickles being more commonly absent from the lamina. (*d*.) The strains may also be slightly overcome by the top of a leaf coming into contact with a support. This is due to the hooked arrangement which results from the irregular growth of the petioles at the base of each lamina.

The *Rubus* leaves of most importance in climbing are those of *R. cissoides* var. *pauperatus*. These are practically devoid of laminae, and the midribs, armed with sharp recurved prickles, grow to a great length—often 24 cm. These usually project from the petiole at right angles, or may be variously curved and twisted, and so are highly efficient in grasping supports (figs. 3 and 4). In a few cases studied this tendency of the midribs to curve was so great that they were practically encircling supporting twigs to which they had become attached. *It may, therefore, be said that in this species the leaf has changed its function, and has become a distinct climbing-organ.*

Not only are the leaf-prickles of importance to the plants in climbing, but the stem-prickles, which persist for a great length of time, function to some extent in this regard. *They are practically absent from the stems of R.* cissoides, *and it is interesting to note this absence going hand-in-hand with the great development on the leaf-midribs of that species.* In *R. schmidelioides* a few prickles are present. In *R. australis* they are more numerous, and in *R. subpauperatus* they are very abundant. On stems of the last-named species 1 cm. in diameter, many of the prickles can individually withstand a strain of 900 grams. The average for the species is about 650 grams, and in the other species this same strain is the maximum which they can withstand.

3. Metrosideros hypericifolia.

A. LIFE-FORM.

This species is by no means common in the Riccarton Bush, there being only about six individuals. These are confined to the damp shaded portions of the interior, where they grow on the trunks of large trees to a height of 10 m. The stems are 3–4 cm. in diameter, and at about 3 m. from the ground they usually give off numerous branches, whose growth is often more or less erect, but which may be inclined to varying degrees. These branch again, so that the whole plant forms a mat-like growth round the support. Leaves borne on these branches are arranged in two lateral rows; though small, they are in large numbers, forming a dense mass of foliage. When brightly illuminated the branches hug the support, against which the leaves are closely pressed; but in shaded places growth is more vigorous, and branches may project from the support for 50 cm., and by their interweaving form dense masses. Usually the climbing-roots at the basal portions of the liane are dead, and the stem swings freely. Branches which arise near the ground often grow out over semi-exposed surface roots, and finally on to the surrounding forest-floor, where they become covered with debris and bear roots which penetrate the soil for a depth of 10–12 cm. From these stems leafy branches may arise erect for a few centimetres, or, as is more common, they lie along the ground-surface.

These surface-growing shoots are mostly found in contact with the roots of the kahikatea, which extend along the surface of the ground for many metres. The main shoots of the liane keep in contact with the roots, while lateral shoots branching off usually grow over the forest-floor, giving off absorbing-roots. The stems which thus grow along the taxad-roots are often broken by animals and other means, but the severed portions continue their growth, thus showing the efficiency of the absorbing-roots.* The anchoring-roots which attach the liane to the support are usually diageotropic, but it is evident that contact with the support may overcome the influence of gravity. The factors influencing the development of roots

---

* Cockayne, L. (1909A, p. 14), describes how in the forest of Stewart Island there is often a stout creeping stem beneath the loose peaty soil many yards in length, from which climbing shoots may be given off, "the plants of adjacent trees in this manner being at times merely branches of one plant."

seem to be shade and moisture. Root-hairs are usually absent from the anchoring-roots, but in a moist chamber they arise in large numbers. If the roots fail to come in contact with a support they become attenuated, and remain unbranched, but in other cases they branch freely and grow to a length of 3–4 cm. In more exposed situations the plants thus growing along taxad-roots do not elongate very much, but tend to become more bushy. The leafy branches are very numerous, and by their interweaving support each other, so that the plants form small leafy clumps up to 40 cm. in height.

B. LEAF.

(i.) *Leaf-form.*

Leaves distichous, glabrous, subcoriaceous, sessile or subsessile, 0·8–1·5 cm. long, ovate-oblong or ovate-lanceolate ; acute, apiculate, or obtuse.

(ii.) *Leaf-anatomy.*

(*a.*) *Shade Leaf.*—Epidermis : Cells in transverse section of leaf isodiametrical; outer walls slightly thickened. Stomata : Lower surface only ; slightly raised above level of epidermal cells. Chlorenchyma : Palisade—Usually 3 layers cells, elongated, but not closely packed. Spongy—Cells roundish or irregular ; intercellular spaces very large ; tissue comprises two-thirds of leaf-thickness. Vascular bundle : Surrounded by a sheath of thick-walled stereome, 1–2 layers thick. Calcium oxalate : In chlorenchyma in aggregate crystals. Secretory cavities occur beneath the epidermis, and are lined by a distinct epithelium ; they are described by Solereder (1908, p. 353) as schizogenous in origin ; more abundant on upper surface. Leaf-margin : Epidermal cells are here larger, and have thicker walls.

(*b.*) *Sun Leaf.*—Epidermis : Cell-walls thicker. Stomata level, with epidermal cells ; ratio of palisade to spongy tissue is greater than in shade leaves. Palisade cells are more closely packed, and intercellular spaces are smaller.

(*c.*) *Climbing-organ.*—(i.) Climbing-roots : The vessels of the central cylinder have small lumina and very thick walls. Endodermis forms a distinct ring of large, thin-walled cells. Cortical cells in young roots contain chloroplasts ; but in older roots the walls of the outer cells become lignified, and the cells lose their contents, forming a strong peripheral band of mechanical tissue. Root-hairs are absent, but can be induced by placing the roots in a moist chamber.

(ii.) Absorbing-roots : The vessels have slightly larger lumina, and the cortical cells continue their functions, this resulting in the absence of peripheral mechanical tissue. This distribution of mechanical tissue is in both types of root advantageous for resisting the particular strains to which the roots are exposed. Thus the anchoring-roots must resist vertical strains, due to the weight of the plant, and lateral strains, due to the action of the wind and to growth in circumference of the support. In the absence of peripheral mechanical tissue the roots would be subject to injurious torsion ; and, further, the tissue prevents injury to the living cells through crushing or attrition ; but in the absorbing-roots the axile arrangement of mechanical tissue is the more advantageous, for they have to withstand pulls from various directions, and by the axile arrangement of mechanical tissue the stress is most evenly distributed.

4. MUEHLENBECKIA AUSTRALIS.

A. LIFE-FORM.

This is one of the most widespread inhabitants of the forest, growing both in the shade and in the bright sun at the forest-edge; in the damp undrained area and in the neighbourhood of drains, and in the driest exposed situations. In the forest-interior it is found twining round the thin trunks of young trees and shrubs. Its direction of twining is not uniform, as in *Parsonsia*, but different individuals twine some right and some left. Further, a single plant may commence its growth as a sinistrorse twiner, and after a few turns become displaced by a projecting branch of the support, beyond which it continues its twining in a dextrorse manner. The highest position of the liane is on the branches of the tallest trees, 30–40 m. from the ground; but in no case does the liane twine round their trunks. On the contrary, the liane-stems usually hang freely from the supporting branches at a distance of 1–5 m. from the large trunk, or perhaps quite near the latter, either growing fairly erect from the root or with their stems coiling first along the ground for a distance of fully 12 m. From these coiling stems adventitious roots pass off, which sometimes grow near the surface and attain a length of 1 m. The hanging stems almost invariably are greatly twisted, suggesting that they commenced their growth on smaller supporting trees, which they finally strangled; indeed, the climbing-stem frequently encircles a portion of some stem which has been strangled (Plate XXV, fig. 2). The condition of these decaying remains shows how intense has been the "struggle" between liane and support. Both stems may be greatly compressed, and the tissues of the support between the liane-coils may be growing out and nearly surrounding the liane. But it is a significant fact that in no place has a tree been found which indicates that it has ever been tightly surrounded by one of these *Muehlenbeckia* stems.

On the top of the supporting tree the liane-shoots branch freely, the branches, interweaving and twining round each other, forming an efficient platform for the display of foliage. Many shoots project beyond the support, and, continuing their growth, gradually hang down. The growing apex, however, is apogeotropic, and upon coming into contact with the hanging portion commences to twine upward round this same shoot.

In the more exposed portions of the forest, where the supporting trees have numerous branches, the *Muehlenbeckia* shoots hang down in large numbers, and on their lateral branches bear large masses of foliage. The species is very abundant at the edge of the forest, where it is found growing over low shrubs. Here the liane forms large straggling masses, whose slender stems do not grow so long as in the forest, but bear numerous branches which, unable to grow erect in the absence of a support, interweave freely, and thus form cushion-like masses. In these exposed situations the species is almost deciduous. A few plants lose all their leaves; others, which are more sheltered, are semi-deciduous; finally, in the shelter of the forest, the plants are evergreen.

A most noticeable feature of the species is the production of adventitious shoots. These arise in autumn from almost any part of the adult, being found on the leafy stems at the top of the support, and on the adult stems trailing along the forest-floor. Although at most 0·7 cm. in diameter, they can grow erect for 1–5 m., after which they bend over until they touch the ground; but if the terminal shoot be injured a lateral shoot may arise from any part of the main one, and thus finally may reach a height of 2 m. or more. At other times they project from the liane almost horizontally, and in many places finally rest on plants which are 2–3 m. distant. They thus illustrate the suitability of the term "searcher shoot," which is applied by Goebel to such organs (1905, pp. 453–54), for, in obtaining new supports for the liane, they are of immense importance. During the first metre of their growth they exhibit no nutation or torsion, but after attaining this length the shoot, even in the absence of a support, slowly commences sinistrorse spiral growth. If it comes into contact with a support the position of the latter determines the direction of the twining—*i.e.*, it may be sinistrorse or dextrorse.

A. LEAF.

(i.) Leaf-form.

Leaves alternate, petiolate, 4·5–6·5 cm. long; ovate, apiculate, cordate or truncate at base; membranous or subcoriaceous; glabrous, entire, margin undulate. Petioles 1–1·5 cm. long, bent in varying manner, so that the lamina usually lies facing the light. Dorsal groove along entire length. Stipules membranous, closely pressed to stem, deciduous. Shade leaves differ but slightly from sun leaves; they are slightly larger and more membranous, while the petioles are a little longer.

The above description applies to an ordinary adult leaf, but on an adult two other distinct types of leaves occur—viz., (*a*) reniform, (*b*) trilobed with acuminate apex—which arise in a definite order. The first leaf to appear on a lateral shoot is small and reniform. Either at the next node or two or three nodes distant there is borne a leaf which approaches more or less closely to the 3-lobed acuminate type, this being succeeded by the usual adult type. The same sequence is apparent in developing seedlings, and there the three types, which are very distinct, are connected by intermediate forms. The series, however, is sometimes broken by the omission of any stage *except the first.* Succeeding the reniform leaf is one in which the indentation at the leaf-tip is slightly reduced, and two small incisions are present, one on each side of the tip. These incisions in later forms become more and more marked, and at the same time the leaf-tip becomes at first flat, and then produced into an acuminate apex. At this stage the leaf belongs to the 3-lobed type. In leaves succeeding this the incisions become smaller, and disappear in the final type, which has an entire undulate margin.

(ii.) Leaf-anatomy.

(*a*.) *Shade Leaf* (Juvenile).—Epidermis: Upper epidermis consists of large ovoid cells with thin walls; lower epidermis, smaller cells; stomata, lower surface only, slightly raised. Chlorenchyma: Palisade is poorly developed; cells in uppermost layer of mesophyll are almost isodiametrical, and are loosely packed; a collecting layer beneath this connects with a spongy tissue of about 3 layers of spherical cells loosely arranged; these cells differ from the palisade in their tendency to elongate parallel with the leaf-surface, and in their containing few chloroplasts. Leaf-margin: Outer walls of epidermal cells very convex and thickened; chlorenchmatous cells almost spherical. Vascular bundle poorly developed; no distinct sheath. Calcium oxalate: A few aggregate crystals.

(*b*.) *Sun Leaf* (Adult).—Epidermis: Striated cuticle; lateral walls also thickened. Stomata level with epidermal surface. Hypoderm: Above and

below midrib ; one row of spherical thick-walled cells.   Chlorenchyma :
Palisade—3 layers of elongated cells arranged in vertical rows ; tissue
comprises one-half of leaf-thickness. Spongy cells nearly spherical ; intercellular spaces small. Leaf-margin : Epidermal walls thicker than in
shade ; chlorenchyma cells more closely packed, and showing tendency
to elongate as in palisade tissue. Vascular bundle : One layered parenchymatous sheath. Calcium oxalate : Crystals in upper spongy tissue.

5. MUEHLENBECKIA COMPLEXA.

A. LIFE-FORM.

The members of this species are confined to the outer portions of the
forest, where the plants receive a large amount of light ; they are much
branched and of comparatively low growth. The species is most abundant
in the drier, well-drained soil. The climbing-stems — at most 2 cm. in
diameter—are covered with a rough blackish bark and trail along the forest-
floor for many metres, rooting freely at the nodes. When they reach any
thin stem they twine round it, and, as in *M. australis*, the twining is either
sinistrorse or dextrorse, the former being the more common, and, likewise,
any stem may change its direction of twining. The twisting of adult stems,
as well as the position of the lianes in relation to their supports, show how
former supports have been strangled ; but these occurrences are not so
well marked as in *M. australis*. This is due partly to the small growth of
M. complexa, which nowhere rises higher than 9 m., and partly to the nature
of the supporting trees and shrubs. These, as above mentioned, are much
branched, and often form dense masses of the divaricating life-form. In
reaching to the tops of these plants the liane *shows a tendency to scramble rather
than to twine*, interlacing branches being more favourable for scramblers than
for twiners ; but, although twining by *M. complexa* is by no means common
in such places, the twining, when it does occur, prevents the liane from
slipping from the support. In the light, at the edge of the support the stems
branch freely, and the lateral branches bear large numbers of leaves, which,
with the stems, form compact masses. These stems, although very slender,
arise in large numbers, and their tendency to twine round each other is
marked. By mutual support they may thus project for nearly 1 m., and
by so doing come into contact with new supports, which soon become
covered by the dense masses of liane-stems. These are greatest on the
tops of the supports, where they grow so close that no light can penetrate
the dense mass. Such masses are especially characteristic of the plants
of the forest-margin. Here the only supports are shrubs, which vary in
height up to 3 m. On these the lianes form masses which by their weight
bend the shrub down to the ground. They continue their branching, and
the intertwining branches soon obscure the support from view, the whole
mass then resembling a rounded cushion, often from 1 m. to 2 m. high, and
from to 2 m. to 4 m. across.

B. LEAF.*

(i.) *Leaf-form.*

Leaves alternate, petiolate, varying in shape and size, 0·5–1·5 cm. long,
orbicular or obovate, rounded or obtuse at tip, borne singly or in pairs

---

* Like all the descriptions in this paper, this refers to the Riccarton Bush plant
alone. Were the " species " being dealt with for the whole of its area of distribution,
the differences in leaf-form would be much greater.

at ends of minute arrested branches, base cordate or truncate, slightly coriaceous, glabrous, margin entire. Petioles 3–5 mm., rounded at base, towards the lamina grooved dorsally, puberulous. Stipules membranous, closely pressed to stem, deciduous.

(ii.) *Leaf-anatomy.*

(*a.*) *Shade Leaf.*—Epidermis : Cells of upper epidermis large ; walls slightly thickened, outer walls convex ; lower epidermal cells smaller and outer walls straighter ; stomata on lower surface only. Chlorenchyma : Palisade and spongy tissues not sharply differentiated ; two upmost layers of cells isodiametrical and loosely packed ; beneath these, 3–4 layers of cells loosely packed and with large intercellular spaces. Leaf-margin : Cells larger ; walls more convex and more thickened. Bundle-sheath poorly developed. Calcium oxalate in aggregate crystals.

(*b.*) *Sun Leaf.*—Epidermis : Cell-walls thicker ; anthocyan often in both upper and lower epidermis. Chlorenchyma : Palisade — 2 layers cells slightly elongated, and more closely packed. Spongy—Intercellular spaces smaller.

6. PARSONSIA HETEROPHYLLA.

A. LIFE-FORM.

Although not so abundant as *Muehlenbeckia australis*, this species is quite as widely distributed. It is present in its largest numbers on trees and shrubs at the forest-margin. The stems, covered with a rough greyish bark and slightly swollen at the nodes, are at most 3–4 cm. in diameter. They reach to a height of fully 20 m., and before doing so may trail along the forest-floor for 12 m. or more, adventitious roots often arising from the trailing portions. The stems of plants in the forest-interior hang from the supports in a manner similar to those of *Muehlenbeckia australis*, and, like these, they are often twisted and encircle portions of some former support, which has been strangled. The stems differ in behaviour from those of *M. australis* in that the twining of *Parsonsia* is always sinistrorse. They are thus of less efficiency in gaining support, but they possess a marked superiority in that they can twine round much thicker supports, the climbing-stems often being found coiled round tree-trunks up to 25 cm. in diameter. The stems show a marked tendency to twine round each other, in many cases lateral shoots inter-twining with the primary. By this mutual support they have a better chance of reaching higher positions. In bright light the stems bear leafy branches, which form a dense covering on the support. The stems may stand erect, but more commonly they lie horizontally on the top of the support. Others project from the latter, and gradually bend down until they hang almost vertically. The stems then commence to twine back round themselves, and should they reach any one of the many shoots which arise from these hanging stems—shoots which at first stand erect and then stretch outwards—the primary shoots may twine round one of these, and the two together thus reach some new support. In such positions on a support the tendency of stems to intertwine is very marked, and " ropes " of as many as 8 stems often project from the support for fully 2 m. These " ropes " are the chief means by which distant supports are reached, for, although adventitious shoots arise from various parts of adult plants, they are not numerous, and, unlike those of *Muehlenbeckia australis*, never attain a great length, while their growth is very slow.

B. LEAF.

(i.) *Leaf-form.*

Each adult plant may be said to have a typical leaf-form, but this is not constant. Leaves are alternate, petiolate, coriaceous, upper surface dark green and slightly glossy, under surface pale green or yellowish-green and dull, 4–7·5 cm. long, ovate, oblong-ovate, or ovate-lanceolate, acute, margin entire; petiole 1–2 cm. long. Besides any one of these forms, any individual may bear at the base of lateral shoots leaves which exhibit many juvenile leaf-forms, the most common being orbicular-ovate and lanceolate or linear with 2–6 rounded lobes on each side.

The leaves of seedlings present a remarkable diversity in shape, and at first no connection between the forms is apparent. But in a careful study L. Cockayne demonstrated that there is a definite process of development and change: "This complexity arises from the fact that there are two distinct types of leaf—a primary short, broad leaf, and a secondary long and narrow one. Between these two there are all kinds of intermediates, and, moreover, 'reversion shoots' freely occur, thus bringing primary leaves quite out of their proper place in the sequence [see fig. 5]. The leaves which succeed the cotyledons are certainly variable in size and shape, but they are always of what may be called the short, broad type. Sometimes they are quite small and almost circular, at other times various varieties of oblong predominate. The next phase of development is an increase in length and narrowing of the base of the lamina, so that in the most extreme cases a well-marked spathulate leaf is the result. Then the circular leaf-apex of this latter is lost, and the second leaf-form, a long and narrow leaf, comes into being. This second stage persists for some considerable time—*i.e.*, there is a prolonged juvenile form—but sooner or later, when by the twining of the ever-lengthening stem round its support the bright light is gained, the adult and third form appears, the leaves large and broad, and of a more or less oblong character" (Cockayne, 1908, pp. 486–87).

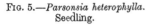

Fɪɢ. 5.—*Parsonsia heterophylla.* Seedling.

The direction in which seedling-leaves face exhibits to a marked extent the arrangement of leaves so as to receive a maximum amount of light.

(ii.) *Leaf-anatomy.*

(*a.*) *Shade Leaf* (Juvenile).—Epidermis: Cells in transverse section iso-diametrical, outer walls slightly convex and a little thicker than lateral walls; in upper epidermis cells larger than in lower; stomata on under surface only, level with epidermal surface. A few unicellular hairs are scattered on both leaf-surfaces. Chlorenchyma: Palisade—1 layer cells cuneate, with base towards upper epidermis. Chloroplasts most numerous towards this base. Tissue occupies about one-sixth of leaf-thickness. Spongy—Beneath palisade 1 layer of fairly regular collecting-cells, with numerous chloroplasts; remainder of spongy tissue is composed of irregular cells, 5–6 cells thick, whose connections leave large intercellular spaces; chloroplasts few in number. Vascular bundle surrounded by well-developed sheath of spherical parenchymatous cells. Leaf-margin: Epidermal cells with thick cuticle; chlorenchymatous cells more or less spherical. Antho-cyan often present in palisade layer. Calcium oxalate: A few aggregate crystals scattered in upper spongy tissue.

(*b.*) *Sun Leaf* (Adult).—Epidermis: Thick cuticle; lateral walls also thickened. Hypodermis: Present beneath upper epidermis in regions of vascular bundles; cells more or less spherical and thick-walled. Chloren-chyma: Palisade—Beneath hypoderm 2 layers, in other places 3 layers; cells elongated and very closely packed; chloroplasts most abundant along lateral walls; each cell contains a large oil-drop; tissue comprises about one-third of leaf-thickness. Spongy—Cells more spherical than in juvenile; intercellular spaces smaller; cells with few chloroplasts. Vascular bundle: Parenchymatous sheath thick-walled. Leaf-margin: Epidermal cells greatly thickened; beneath epidermis a mass of hypoderm composed of thick-walled cells; for a short distance from the margin the hypoderm ex-tends in 2 layers; chlorenchymatous cells spherical or slightly elongated; intercellular spaces very small. Calcium oxalate: Crystals more numerous than in juvenile. Anthocyan: Often abundant in top palisade layer and lowest spongy layer.

7. PARSONSIA CAPSULARIS *var.* ROSEA.

A. LIFE-FORM.

This species is similar in its distribution to *P. heterophylla,* but it is present in much smaller numbers. The life-form of the two species is similar, except that *P. capsularis* var. *rosea* does not grow to such a height, being found chiefly on the dense shrubs. The young stems, brownish in colour, are very elastic, and their tendency to intertwine is even greater than in the stems of *P. heterophylla.* Groups of 4–8 stems form " ropes," and, further, a number of these "ropes" may come into contact and together form a single large "rope," which thus is fairly rigid and can project from a support for a considerable distance.

B. LEAF.

(i.) *Leaf-form.*

(*a.*) Adult.—Leaves vary in size and shape from narrow-linear, 3–10 cm. long and 0·2–0·3 cm. broad, to oblong or oblong-lanceolate, 3–7 cm. long and 1–1·5 cm. broad; obtuse or subacute, coriaceous, margins usually entire or occasionally slightly lobed.

(*b.*) Juvenile.—Numerous seedlings were grown by the writer, nearly all of which were similar in their leaf-development. Seed-leaves are 1–1·5 cm. long, oblong or oblong-lanceolate ; they are succeeded by a pair of leaves 2–3 cm. long and 0·5–0·8 cm. broad at the middle, either tapering to a point both at base and at apex or broadening out at the apex and thus being almost spathulate. The next 3 or 4 pairs of leaves which arose are narrow-linear, 4–5 cm. long and 0·2–0·3 cm. broad. In 2 or 3 plants the leaves preceding this linear stage passed through stages similar to those of *P. heterophylla.*

In connection with this heterophylly, L. Cockayne writes, "At this [linear] stage, in one very distinct form of *P. capsularis,* further development always stops, and the adult leaf in this case is indentical with that of the second stage of *P. heterophylla*—*i.e.,* the adult of one species is merely a fixed juvenile form of the other " (1908, p. 487).

All the leaves of the seedlings grown were variegated and this variegation is of common occurrence in wild plants. Along the centre of the leaf is a white strip, while at the margin the leaf is brown or reddish-brown. The upper surface of the seed-leaves is uniformly brown, and in all the lower surface is pale green.

(ii.) *Leaf-anatomy*

(*a.*) *Shade Leaf.*—Epidermis : Cells slightly flattened ; outer walls thickened ; cells of lower epidermis smaller than upper ; thin cuticle ; stomata on lower surface only. Hypoderm : Well developed beneath mid-rib ; at leaf-margin there is a single layer which extends for a short distance along both surfaces. Chlorenchyma : Palisade—3 layers cells elongated, but not very closely packed ; tissue comprises one-fourth of leaf-thickness ; each cell with a large oil-drop. Spongy—Cells irregular, and intercellular spaces large ; cells bordering on lower epidermis more regular and more closely packed. Vascular bundles have a well-developed sheath of spherical cells with thickened walls. Anthocyan : Abundant in upmost palisade layer and lowest spongy layer. Calcium oxalate in aggregate crystals. Leaf-margin : Strengthened by thicker-walled epidermis and by the layer of hypoderm.

(*b.*) *Sun Leaf.*—Epidermis : Cells with thicker walls ; cuticle thicker. Chlorenchyma : Palisade—Occupies fully one-third of leaf-thickness ; cells more elongated and more closely packed. Spongy—Intercellular spaces smaller. Bundle-sheath much strengthened.

8. CLEMATIS INDIVISA.

A. LIFE-FORM.

Growing in the interior of the forest, the species attains its greatest size in places where the smaller trees form its support. The position of the plants seems to indicate that normally the upward growth is upon shrubs, from the top of which the liane continues its growth into the branches of overhanging trees. It may thus reach a height of 10 m. or more, and then the frequent branching of stems expanded along the top of the supporting tree results in an effective display of leaves. The ascending climbing-stem, 2–3 cm. in diameter, and covered with a buff-coloured wrinkled bark, may be close to the trunk of the support, or it may be among the leafy branches. The latter is by far the more common arrangement, being associated with the best conditions of growth—*i.e.*, nearness to the underlying shrubs, from which transition is easy, and abundance of supporting branches for the climbing-organs. Branches of the liane may bridge the space between adjacent trees, and the new support be utilized for further display of foliage. Should this intervening space be small, it may be bridged by a single branch ; but in many cases two or more projecting branches may intertwine and be held in position by reciprocal grasping of their petioles. By the strength thus attained these branches may reach a support at a distance of 1·5 m.

Where low shrubs form the only support for the *Clematis*, the liane grows horizontally along the tops of these shrubs. The action of the petioles prevents the displacement of the stem by the wind, and ensures the maintenance of the leaves in the favourable positions they have taken up. The primary arrangement of leaves is decussate, and in cases where the stem stands erect this position is maintained ; but in stems growing along the top of a shrub, and on stems which trail along the ground, there occurs torsion of the stem, as a result of which the leaves lie in two rows. In whatever position a leaf arises, the petiole twists so that the laminae of the leaflets lie at right angles to the incident rays of light.

From stems which are lying on the ground adventitious roots arise where there is an abundance of moisture. These roots grow chiefly near the surface of the ground, and often attain a considerable length.

B. LEAF.

(i.) *Leaf-form.*

Leaves decussate, petiolate ; on adult plants 3-foliate, coriaceous, glossy, upper surface dark green, under surface paler green ; leaflets 4–8 cm. long,

FIG. 6.—*Clematis indivisa.* Variations in leaf-form.

ovate-oblong or ovate-cordate ; margin entire or lobed. On *juvenile plants* the earliest leaves are simple, usually ovate-oblong or ovate-lanceolate, 3–5 cm. long ; in rare cases, linear-lanceolate, 7–10 cm. long ; subcoriaceous.

Succeeding leaves show all transitions between simple and 3-foliate, and in certain plants later leaves are still more divided, many being biternate; but this form in Riccarton Bush is always followed by the 3-foliate form of the adult (fig. 6).

(ii.) *Leaf-anatomy.*

(*a.*) *Shade Leaf.*—Epidermis : Cells regular and slightly flattened ; outer walls firm, with thin cuticle ; stomata lower surface only, level with epidermal surface. Hypoderm : Above midrib 1 layer ; below midrib forms a supporting mass, 4–5 cells in thickness. Chlorenchyma : Palisade—3 layers isodiametrical cells rather closely packed ; arm-palisade cells, mentioned by Solereder (1908, p. 15), are numerous in the top layer and are present in smaller numbers in the other layers ; wall-infolding is confined to the upper portion of each cell ; chloroplasts irregularly scattered ; tissue comprises about one-half of leaf-thickness. Spongy—Cells are more irregular than palisades, but no sharp distinction between the tissues ; intercellular spaces large ; chloroplasts rather numerous. Leaf-margin : Epidermal cells enlarged ; outer walls very convex and much thickened. Vascular bundle : Well-developed parenchymatous sheath. Calcium oxalate : In aggregate crystals. Anthocyan : Sometimes in lower epidermis.

(*b.*) *Sun Leaf.* — Epidermis : Cell - walls much thickened, and well-developed cuticle, thicker on upper epidermis. Chlorenchyma : Palisade—Cells more closely packed. Spongy—Intercellular spaces smaller and cells more spherical. Leaf-margin : Strengthened by 2–3 layers of thick-walled hypoderm cells. Vascular bundle : Sheath more developed.

(*c.*) *Climbing-organ.*—(i.) Form and Behaviour.—The pairs of leaves at first project beyond the growing apex of the stem. The petioles elongate in this position until they attain a length of 3 cm., the leaflets remaining small, being at most 1 cm. in length. The leaves then bend downwards and outwards, the petioles at the same time increasing to 4 cm. in length. When in the horizontal position, the leaf-blades increase in size, and by curvature and torsion of the subpetioles the blades become placed at right angles to the light. If in its downward curvature a leaf-petiole comes into contact with some object the stimulus causes the petiole to coil round the object. The under surface of the petiole is most commonly in contact with the support, but curvature caused by the upper surface of the petiole being sensitive to contact is by no means uncommon. Whether the greater response of the under surface is due to its being more sensitive to contact, or whether it is due to this surface being in a position where it will most often be exposed to stimulation, has not been investigated. But it is obvious that as this results in the leaves being bent downwards it is most suitable in preventing the liane-shoots from slipping down from any support which they have reached. The subpetioles are quite as sensitive, and it seems that they are equally sensitive on all sides. Like the petioles, they are stimulated to curvature by contact with foreign objects, or with parts of the same plant—stem, leaf, or other petioles. After their contact they become much strengthened, and the portions which are in contact with an object attain a thickness considerably greater than that of the rest of the petioles. The small size of the leaf-blades during the growth of the petioles and subpetioles is of great importance in gaining supports, for a large leaf-blade would tend to become entangled in twigs, &c., and by its resistance prohibit a petiole from efficiently coiling round the support.

(ii.) Leaf-petiole: Anatomy (see fig. 7). — Before the petiole attains a length of 1·5 cm. it is devoid of stereome. Four vascular bundles are at this stage well developed, and the rigidity of the petiole is further increased by abundant collenchyma. Stereome first appears between the bundles in a narrow ring, which gradually increases in thickness, and which, extending outwards, joins more masses of stereome which meanwhile have

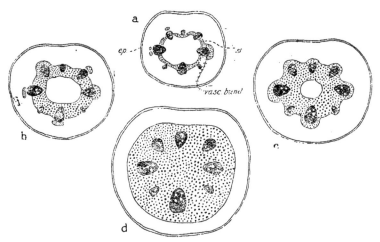

FIG. 7.—*Clematis indivisa.* Diagrams of tendril, showing development of stereome. *ep.*, epidermis; *st.*, stereome; *vasc. bund.*, vascular bundle.

developed outside the phloem. At the same time the stereome develops more and more towards the centre of the petiole, until finally all the ground - tissue except the 5 or 6 rows adjoining the epidermis becomes lignified.

These outer cells contain chlorophyll, while in a few cells anthocyan may be present. The epidermis has a thick cuticle. Stomata are sunk below the surface.

9. TETRAPATHAEA AUSTRALIS.

A. LIFE-FORM.

Although the species is found chiefly at the edge of the forest, many plants grow in the inner portions and even reach the summit of the tallest trees. The stem of the liane, growing up to 12 cm. in diameter, usually ascends close to the trunk of its support, but it sometimes trails for 10 m. along the ground before gaining the support. It is evident that a tree-trunk is a poor support for a tendril-climber, but the presence of decaying stems of species of *Rubus* at the bases of trees indicates a means by which *Tetrapathaea* may have reached the lofty branches. The liane produces practically no branches in the shade, but on the exterior of the supporting tree it bends over, and ultimately gives off numerous branches, the foliage of which is often so thick as to obscure from view the supporting tree (Plate XXIV, fig. 1). Continuing their growth, these branches hang down, and the weight of the masses is frequently sufficient to break the branches of the support. The long rope-like stems may then be swung to and fro by the wind and become tangled in the branches of an adjacent tree. Recommencing its upward growth on this new support, the liane may ultimately reach its greatest height. In places where shrubs form the only available support, the liane may bridge the spaces between the shrubs, and continue its growth along the tops of the shrubs for a distance of 20 m. or more. As a rule, the liane forms no coiled masses at the base of a support, and it seems probable that this is due to the efficiency of the tendril as a climbing-organ. And that the tendril is not an ephemeral structure is shown by its persistence on stems which have attained a thickness of 6 cm.

Adventitious shoots are conspicuous on adult shoots in the shade. They may arise singly or in groups of 2 or 3, and, though very slight, can rise erect for about 60 cm. The first leaves are very small, being at most 4 cm. long and 2–3 cm. broad, but the size of succeeding leaves gradually increases. However, on the distal portion of the stem the leaves for a long time remain quite small, thus facilitating free movement of the tendrils. The tendrils arise early on the shoots, often being present on the second node, and thus these adventitious shoots are of great value to the plant in gaining new supports.

B. LEAF.

(i.) Leaf-form.

Leaves alternate, petiolate, 5–9 cm. long, oblong-lanceolate or ovate-lanceolate, acuminate, coriaceous; upper surface very dark green and glossy, lower surface lighter green, margin entire. Petiole 0·5–1 cm., dorsal groove, often twisted. Shade leaves are larger than sun leaves (sometimes 12–13 cm. long), more membranous, and with longer petioles (1·5–2 cm.).

(ii.) Leaf-anatomy.

(*a.*) *Shade Leaf.*—Epidermis: Slight cuticle; lateral walls also thick: stomata, lower surface only, slightly elevated. Chlorenchyma: Choroplasts large, spherical. Palisade — 3 layers; cells almost as broad as long; not very tightly packed; tissue occupies slightly less than half the cell-thickness. Spongy—5–7 layers; majority of cells more or less oblong, with long side parallel to leaf-surface. Leaf-margin: Towards the margin the cells of palisade and spongy tissues become more and more alike, and form a homogeneous tissue which near the margin itself is composed of oblong cells resembling those of the spongy tissue, but with smaller intercellular spaces, and containing very few chloroplasts. Vascular bundle: Surrounded by a parenchymatous sheath. Calcium oxalate: Present in aggregate crystals.

(*b.*) *Sun Leaf.* — Epidermis: Cuticle thicker; stomata not raised. Chlorenchyma: Palisade—Cells more elongated and more closely packed; tissue occupies fully half the leaf-thickness. Spongy—Intercellular spaces smaller. Leaf-margin: Hypoderm 3-4 layers; cells of chlorenchyma spherical.

(*c.*) *Climbing-organ.*—(i.) Form and Behaviour: The tendrils arise singly in the axils of leaves, and it is probable that, like those of other members of the *Passifloraceae*, they are modified inflorescence branches. Lateral shoots of the plant arise from buds situated in the axils of the tendrils. The tendrils at first project beyond the growing apex of the stem, and then bend outwards and downwards, at the same time increasing in size until finally they attain a length of 9–11 cm. If in this downward curvature a tendril comes into contact with some suitable support it encircles the latter, and, being thus held firmly in position, the rest of the tendril contracts into a double spiral (fig. 8). But in the absence of a support a tendril coils into a continuous spiral, and after remaining for a time in this posit:on 'it again straightens, and later shrivels up. In tendrils which have obtained a hold on some support the part which is actually in contact with the latter becomes greatly thickened, often attaining a

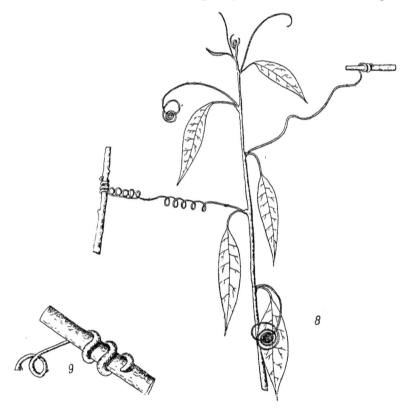

FIG. 8.—*Tetrapathaea australis.* Young shoot, with tendrils.
FIG. 9.—*Tetrapathaea australis.* Tendril, showing thickening.

diameter fully twice that of the rest of the tendril (fig. 9). By the development of mechanical tissue these tendrils persist for a great length of time, being common on stems which have grown to 6 cm. in diameter. At this age they are still extremely tough, being able to withstand a tension of 3 kg. These old tendrils are usually almost straight, and thus it is evident that, other things being equal, a younger tendril which is spirally coiled will be able to withstand a still greater strain.

As tendrils are present upon a plant in large numbers, and as they are widely distributed, the plant, in growing upwards to new positions of support, is admirably adapted for withstanding all strains—vertical strains due to the weight of the plant, and lateral strains due to the action of the wind. Further, the whole plant is thus adapted to maintain its position upon the loftiest supports in the forest.

(ii.) Tendril-anatomy: The structure of the tendril shows how in its stages of development it is admirably suited for performing its functions (fig. 10). In its earliest stages the best position is an extended one, which will increase its chances of coming into contact with a support, and the arrangement of tissues is such as to favour the maintenance of this position. The collenchyma forms a band near the periphery of the tendril, and the

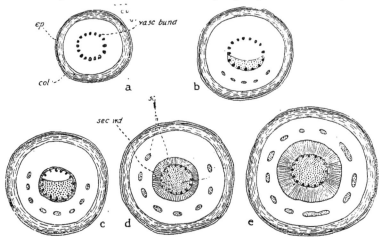

Fig. 10.—*Tetrapathaea australis.* Diagrams of tendril, showing development of mechanical tissue. *col.*, collenchyma; *ep.*, epidermis; *sec. wd.*, secondary wood; *st.*, stereome; *vasc. bund.*, vascular bundle.

vascular bundles are distributed in a ring at a considerable distance from the centre. Then at a later stage the development of stereome is equally suitable. On the future convex side, where growth and expansion of the tissues is essential, stereome is at first absent; but on the concave side it develops rapidly, and thus by its resistance to tension it prevents the tendril being unrolled from the support. Finally, the injury to structures from external pressure is prevented by the ring of collenchyma, which is a constant character of the tendril-structure..

III. METHODS OF GAINING SUPPORT.

1. FUCHSIA COLENSOI.

Seedlings are numerous in the forest throughout well-drained but moist positions.

The primary shoot does not usually grow more than 35 cm. high without support, for lateral shoots are early given off and tend to drag the plant down to the ground. However, it often comes into contact with some support against which it rests. Should the support be very low, the *Fuchsia* branches, which project at right angles, soon spread over it. Shoots stand up from the top of the support, and, supporting themselves, can reach a height of fully 50 cm.; other shoots project from the sides of the support, and thus a new support is often reached.

Again, if the seedling-shoot rests against the branches of some shrub, the shoot pushes its way up between the branches to the top of the shrub. The leaves, being very flexible, are of little assistance in climbing, but the sinuous growth through divaricating branches seems to afford sufficient means of keeping the plant erect, and the branches which are given off at the top of the shrub finally secure a firm anchorage.

Should a seedling not reach a support, the thin stems trail along the ground for 1–2 m., and in favourable conditions root freely. From these trailing-stems lateral shoots rise up, and thus often reach a support a considerable distance from the position where the plant emerged from the ground ; in fact, it sometimes happens that a single plant reaches more than one support by means of these spreading stems. On the other hand, no support at all may be gained, and the branching stems then form a prostrate mass on the forest-floor, or else they interweave and by their mutual support the plant takes the form of a cushion-like mass, which is often as much as a metre high. Shoots projecting from the sides or from the top of such a mass may then often reach some adjacent shrub or tree, and through this rise to a greater height.

2. The Species of Rubus.

The seedlings of the various species of *Rubus* show great similarity in their early behaviour. The primary shoot rises erect for 60 cm., and when the growing-point is injured a lateral shoot arises which continues the upward growth. It will be seen that this is the greatest height to which the primary shoot of any of the lianes here dealt with can stand unsupported, this being due to the relative thickness of the *Rubus* stem, seedlings 50 cm. in height having a stem 5 cm. in diameter. By resting against any near object the seedlings can reach greater heights, and in doing so the leaves are of extreme importance. Prickles are well developed at a very early stage, and, as leaves project from the stem often for 10 cm., a fairly wide area is thus in reach of the prickles. Any support is immediately held by a touching prickle, and by this means the seedlings are well suited to rapidly reach high supports. Any two shoots of a seedling rising up together support each other by the grasping of the prickles, and thus can reach to a height of nearly a metre. But when there are no supports near a seedling, the shoots trail along the ground, and form low, flat, straggling growths, which are very characteristic of certain portions of the forest. They are often 3–4 m. across, and so at the circumference of such a growth there is usually some plant which is suitable for the support of *Rubus* shoots.

3. Metrosideros hypericifolia.

Young plants of this species have been observed during the course of this present investigation only in moist situations in the immediate vicinity of roots of forest-trees, which roots are present in large numbers. The *Metrosideros* plants show a marked tendency to scramble up the sides of and over these roots, applying themselves closely to the bark, and putting forth adventitious roots which fix the plants firmly. By continued growth the distal shoots extend over the tree-roots, such as those shown in Plate XXIV, fig. 2, and some of these shoots reach the base of the main tree-trunk. Here they may branch copiously, applying themselves closely to the bark, in some instances encircling the entire tree. Their climbing-roots are given off, and by this means, and by continued growth, a gradual ascent on the support is made. The growth is slow, and in the limited time at the writer's disposal a careful investigation of the species has been impossible, nor is Riccarton Bush suitable for such a study. The greatest height of observed specimens is 10 m.

4. MUEHLENBECKIA AUSTRALIS.

Seedlings of this species are common throughout all parts of the forest, and in many places form more than 50 per cent. of the forest-floor vegetation. They are able to grow erect for 40 cm., and if the growing-point is injured a lateral shoot arises and continues the upward growth and twines round any suitable object with which it comes into contact. The twining is usually sinistrorse, but the common occurrence of dextrorse twining seems to indicate that the direction is at least partly due to the manner of contact with the support. The stems cannot twine round any support which is more than 5 cm. in diameter.

After the seedling reaches the top of the support it continues to grow erect for a short distance, and then bends down, until finally the growing apex, which continues to point upwards, comes into contact with the more proximal hanging portion of the same shoot. It then commences to twine round this, but the behaviour does not seem to aid in reaching a new support. But lateral shoots may rise from the hanging shoot, and these, by projecting out, as they do, for 40 cm., often reach an adjacent support.

In the largest number of seedlings two or more often come early into contact, and by supporting each other can reach a height of 70 cm. or more, and thus have a better opportunity of coming into contact with a support. If no support is reached the seedlings sink to the ground; but yet, through their elongation, they may ultimately reach some support. From the prostrate stems lateral shoots stretch up, and by their ascent in a position different from that of the primary shoot a support is often reached. Many of these lateral shoots arising in autumn have vigorous growth, and unaided can reach a height of 65 cm. In stations with little undergrowth the branching stems form a reticulation on the forest-floor in the same manner as those of *M. complexa;* but in *M. australis* these reticulations are seldom more than 1 m. across. They are, however, far denser, and often form compact masses with an effective display of foliage.

5. MUEHLENBECKIA COMPLEXA.

Seedlings of this species are most numerous in the vicinity of adult plants—*i.e.*, in the outskirts of the forest and near the forest-margin.

In the forest they grow erect for 30 cm., and upon touching a support they commence to twine. In the absence of a support they bend over to the ground, and trail along the forest-floor. These trailing-stems root freely at the nodes, and often attain a length of 5 m. From any of the nodes lateral shoots may rise up, and should these reach no support they bend over to the ground, where they continue their growth. By a repetition of branching the plants thus soon form a loose reticulation on the forest-floor, and, unless the floor here is very bare, shoots from the network of stems will sooner or later reach a support. Indeed, it more often happens that a single liane-seedling reaches numerous supporting young plants. Again, it is common for any two shoots to come into contact; these then twine round each other, and by their mutual support they grow to a greater height than can a single shoot, and thus have greater opportunity of reaching a support.

At the edge of the forest there are two sets of conditions to which developing seedlings are exposed—(1) In places where the belt of introduced trees* at the forest-edge is broken the ground is partially covered by rank grasses, chief of which is *Dactylis glomerata ;* (2) in the vicinity of the trees the grasses are absent, but large numbers of seedlings of oak, &c., are present, and the ground is covered with a layer of dead leaves.

In the former open situations the developing liane-seedlings upon .emerging from the ground grow erect for about 30 cm., and bend over. In so doing they usually come into contact with the flowering-stalks of the grasses, which stand up for fully a metre. Round these the liane-stems twine, and after reaching the top of these supports they continue to grow up for a short distance until they are unable to support their own weight. They then cause the supporting grass-stalks to bend, and thus they are able to reach any adjacent shrub, which then forms a new support for the liane. Should the liane-seedling in such places reach no support it trails along the ground, and by the lateral shoots which arise along this stem the liane usually reaches at least one support at some distance from its original position.

But it is under the second set of conditions that the tendency for the surface trailing-shoots to elongate is most marked. The shoots become more or less covered by dead oak-leaves, and although under this covering the shoots do not branch so much as do the trailing-shoots in the forest, yet they elongate to a greater extent, and thus the area in which shoots may rise up from the plant and reach supports is greater than in the forest. By means of the rooting at the nodes the shoots are not dependent for their food-supply upon the primary root, and, as a result of this, injury to the main stem does not cause the death of lateral shoots. The oak-seedlings which surround any liane-seedling here are, on account of their small size, of little importance as supports for the lianes. Therefore the only plants which reach the adult stage are those whose growth has brought them close to the shrubs at the forest-edge, and the support obtained from these shrubs enables the lianes to reach a height from which climbing-shoots can soon reach a permanent support.

6. Species of Parsonsia.

A young seedling at first normally grows erect, and independently can reach a height of 45 cm. In the absence of a support, further growth results in the apex gradually bending over, and, with increasing growth, bending still further, until it finally touches the ground. Then in the dense undergrowth it sooner or later reaches a support up which it twines, the twining always being sinistrorse. In seedlings whose growing-point becomes injured, a lateral shoot arises and grows erect, ultimately bending over in a manner similar to that of a primary axis. However, at an early stage a seedling is likely to come into contact with some support ; especially is this the case in the interior of the forest, where in places of dense growth seedlings are often present in countless numbers. Any two seedlings may thus early come into contact, and, as the tendency to twine is soon manifested, their twining round each other gives support, which enables them to reach a height of 75 cm., after which their apices bend over in the manner

---

* European trees have been planted in many places on the outskirts of the Riccarton Bush for the purpose of sheltering it from the wind.

of a single seedling, but, of course, moving through a larger circle, and thus giving a better chance of touching a support. This tendency to twine round one another is not confined to the movements of any two individuals only, for very often large numbers are seen thus giving mutual support— in some cases as many as 15. These then form a rope-like mass, whose resistance to bending is very effective; and the resistance is further aided by the positions of the roots of the individuals, they being arranged round the ascending shoots in a circle, whose radius may be as much as 10 cm. By such means the seedlings often attain a height of fully 1 m.—a height at which the vegetation available as support is usually very abundant. There is no doubt that the leaf is of importance as an aid to climbing. The most common leaf-form of the seedling is the linear form. The leaves are borne in pairs, and project at right angles from the stem to a distance of 6 cm., and by torsion of the stem they point in all directions. There is thus a probability that any one of these leaves may touch a support, and by resistance to bending allow the seedling to reach a greater height. From any portion of a shoot which is bending over towards the ground, lateral shoots may arise for fully 40 cm., and, as the primary shoot at the place of origin of these lateral shoots may be anything up to 20 cm. from the ground, a comparatively high position is reached.

7. CLEMATIS INDIVISA.

The seedlings of this species are not widespread in the forest, being confined chiefly to a small part near one of the few adult female plants. There they are well sheltered, and are surrounded by numerous supports.

Without any support they can stand erect for 40 cm. The leaves, which are decussate, are at the lowest nodes simple, and project from the stem for 5-6 cm. They are thus of assistance in lessening vertical strains on the seedling. The third pair of leaves usually have petioles which upon continued contact with an object bend slightly. They can thus over-come any slight lateral strain, and, as any two consecutive pairs of leaves will operate at right angles to each other, they assist in maintaining the seedling in an erect position and enabling it to reach a greater height. The petioles of leaves succeeding the third pair are more sensitive, and encircle any small branch which they touch; in doing this they become much thickened and strengthened. When a pair of leaves arise from a stem they at first project beyond the growing-point. The leaf-blades for a long time remain small, but the petioles elongate to a length of 2-3 cm. The leaves then gradually bend down until they project straight out from the stem, and if in this downward movement they touch any branch they entwine it. As the leaf-blades at this stage are still small, their size does not retard the action of the petioles by their becoming arrested in the encircling movement, but they later expand, as do also the blades of leaves which have reached no support in the downward curvature. Above the anchorage secured by any of the petioles the free end portion of the seed-ling can rise erect for fully 30 cm., and so, on account of the numerous supports which usually surround the seedlings, the latter easily make their way into the sunlight.

However, should a seedling reach no support, and thus bend down, any lateral shoot from the primary axis will rise up in a position perhaps 20 cm. from the former position of seedling, and so reach a support.

In many cases seedlings are found which, in addition to their petiole-action, ascend a support by twining, the twining being in all cases sinistrorse. The petioles entirely surround the support, one of each pair circling to the right and the other to the left, and they thus firmly grasp the support.

8. TETRAPATHAEA AUSTRALIS.

The seedlings are found chiefly in the exterior portions of the forest, where the vegetation is dense, and thus they are surrounded by numerous supports of varying nature. Without any support the seedling can stand erect for 40 cm., but a greater height can be reached near any support by means of the leaves. These are borne alternately, and project straight out from the stem for 6–8 cm. ; and, being fairly rigid, they support the seedling, and maintain it in its erect position. But most assistance is gained from the tendrils which arise while the seedling is still standing erect, usually appearing first from about the 18th node. They at first project beyond the growing-point, with tip bent outwards, this bending, no doubt, assisting the tendrils to retain their hold upon any object with which they come into contact. These tendrils do not remain in this projecting position, but gradually bend down until they lie along the stem of the seedling, with the apex pointing towards the ground. And in this downward bending they may come into contact with a support. The support is firmly encircled, and by spiral contraction of the tendril the seedling is pulled towards the support. Having thus obtained an anchorage, the seedling may rise up from this point to a still greater height, and in so doing may obtain further support. Should the young seedling reach no support, it bends over to the ground, and grows for a while horizontally. In this way it may reach a support; and, further, any lateral shoots springing from the primary shoot, and rising up as they do for fully 30 cm., increase the chances of the seedling coming into contact with some shrub or other support. From the lowest branches of the shrubs the growth of the liane to the top of the shrub is easy, and thence the liane has little difficulty in reaching higher adjacent plants, and thus rising to the sunlight.

IV. THE EVOLUTION OF LIANES.

The question of the evolution of lianes cannot be gone into in much detail, for this would entail a close acquaintance with all the New Zealand lianes and with various classes of forest. However, the lianes present in the forest investigated exhibit an interesting series from non-specialized to highly specialized forms.

1. SCRAMBLERS.

Characteristics common to all the classes of lianes are the great elongation of stem and the absence of lateral branches. Now, it must be noted that in positions in the forest-interior where the liane-stems, attenuated and unbranched, elongate rapidly, this elongation is characteristic also of other plants, certain species of *Coprosma*—bushy shrubs as distinct from lianes as possible—being most noticeable in this regard.* From such observations

---

* Subsequent to the writing of this, Dr. L. Cockayne has drawn the writer's attention to the fact that in certain parts of the forest near Wellington stems of *Coprosma robusta,* normally a bushy shrub, are so greatly elongated that without careful attention they are easily mistaken for true liane-stems.

it is generally agreed that the chief factors determining stem-elongation are moisture and diminished light, and thus the variation in the life-form of *Fuchsia Colensoi* can be readily understood. *This species, which forms a compact, much-branched shrub in the open, in the forest-shade has stems which are greatly elongated and which have few lateral branches.* These stems can rapidly push their way into vegetation above them, and by leaning against supports attain a height of fully 4 m.

That this scrambling habit was at first the effect of environmental stimuli—moisture and shade—cannot, in the face of the behaviour of other plants in similar conditions, be doubted ; and it is possible that races were evolved in which the scrambling habit became hereditary. On this point L. Cockayne (1912, p. 21) says, " It is possible that there may be climbing and non-climbing races. This is the more likely as the ' species ' is considered variable, and large forms are said to ' almost pass into *F. excorticata* ' (Cheeseman, 1906, p. 187), which is a small tree or shrub, but never a liane." Such a view receives support from the very great elongation of stems of *F. Colensoi* in positions in the forest where supports are abundant —positions where the stems often lie along the ground for fully 5 m.

When species of *Rubus* and other plants possessing prickles commenced growth in the forest-interior they would, on account of their prickles, be better suited for reaching the light than would plants without such appendages.* They would thus be able to propagate their kind more quickly, and, with their larger number of descendants, races in which the scrambling habit was hereditary might soon arise. And in comparing the relative abundance of *Rubus* plants in the forest, and the luxuriant growth of the plants there, with those in more open situations, it seems probable that natural selection would by itself result in the climbing habit being retained. In fact, it seems that in the species of *Rubus* found in the Riccarton Bush the scrambling habit is in all cases hereditary, for those plants which grow in the open do not form shrubs, but low straggling masses, which resemble somewhat the exposed forms of more specialized lianes. In the forest-scramblers the importance of leaves in climbing has been pointed out above, and so *it is interesting to note how in one of the species of* Rubus, *R. cissoides, the leaf has changed its rôle, or, rather, has acquired a new rôle, being now a special climbing-organ.*

2. Root-climbers.

In this class of climbers the small number of representatives studied by the writer makes it difficult, using only the facts derived from the Riccarton Bush, to conceive means by which the climbing habit has been gradually adopted, but the variations in the life-form of *Metrosideros hypericifolia* may have some bearing on the question. In this species it is evident that there is no sharp distinction between climbing-roots and absorbing-roots. Both are at first similar in structure ; and it seems probable that the nature of the environment determines whether the roots shall elongate and act as absorbing-organs, or whether they shall remain comparatively small and fasten the liane to the support.

It will be noticed from accounts of the life-form of the lianes that in all the species *the important fact stands out that there is a tendency to adventitious*

---

* Before the scrambling growth was commenced, however, the prickles would possibly be of no use to the plants possessing them.

rooting by creeping stems in moist positions. In *M. hypericifolia* the roots arising from stems lying on the forest-floor are very numerous. Now, the surface roots of the forest-trees are often semi-exposed, and covered to a greater or less extent by debris and bryophytes, so that moisture-conditions on these roots may not be very different from those in the surrounding soil. If so, we should expect that if it is moisture which acts as a stimulus to the formation of the adventitious roots, then the rooting of *Metrosideros* stems lying over these surface roots would be a natural occurrence. And such is actually the case. It is probable that certain physiological races would arise which required a smaller amount of moisture to stimulate adventitious rooting, and so roots might be given off when stems were in contact with the moist bases of tree-trunks. By secretions the roots might easily be attached to the trunks, and thus the plants could raise themselves from the forest-floor. Greater heights would be reached by the plants whose tendency to rooting was greatest under the drier conditions which would be met at the increasing distance from the forest-floor.* And even now, as has been stated above, *M. hypericifolia* is found only in the most shaded positions of the Riccarton Bush, and these plants do not attain a very great height. But other species seem to thrive under drier conditions. For example, the writer has observed in the Botanical Gardens at Wellington a plant of *M. scandens.* This was growing in the open, and formed a much-branched dense shrub about 50 cm. high. On one side, however, a branch has come into contact with the stem of a shrub, and by means of its climbing-roots it has ascended this stem to a height of 1 m.

3. TWINING-PLANTS.

When we turn to twining we are dealing with a phenomenon which is of frequent occurrence in plants, and it is found not only in those which can at once be classed as true lianes, but also in others in which the twining is perhaps only slightly marked. It is these latter which are of importance in trying to trace the development of twining.

The fern *Pteridium esculentum* (Forst. f.) Cockayne, which usually grows in exposed stations, becomes in the shade a scrambler with elongated stems, which sometimes show a slight tendency to twine. " So, too, with the scrambling liane *Lycopodium volubile* Forst. f., which, gaining a thin support, winds freely, the winding being in this case an hereditary characteristic " (Cockayne, 1912, p. 21).

An interesting case, not hitherto reported, is that of *Carmichaelia subulata* T. Kirk, which in the open is a rigid erect xerophytic leafless shrub 0·5–1 m. high; but it has been found by the writer in the shade as a prostrate plant, with its stems hanging down over the edge of a rock and showing a marked tendency to twine round one another. From this plant a shoot rose upright for about 50 cm. Towards the distal end it was twisted and coiled into a spiral, giving the appearance of a " searcher shoot " of *Muehlenbeckia australis* which has commenced its spiral growth. Unfortunately, no opportunity arose of visiting this remarkable plant later to note the further growth of the shoot. The discovery of this plant seems to be of considerable importance. Climbing is not unknown in the genus *Carmichaelia:* it is hereditary in *C. gracilis,* a rare scrambling liane growing

---

* It must be remembered that the tree-trunks are kept comparatively moist by the dripping of water during atmospheric precipitations.

in damp stations in the South Island, and hereditary also in *C. exsul*, of Lord Howe Island, the only species of the genus outside New Zealand. So the presence of a scrambling form showing a tendency to twine in such a rigid species as *C. subulata* undoubtedly has close bearing on the question of liane-evolution.

Another interesting case is that of *Mueklenbeckia complexa*, already noted, where, it may be remembered, the liane, having gained the upper-most twigs of a dense shrub, shows a tendency to scramble rather than to twine, interlacing branches, as pointed out, being more favourable for scramblers than they are for twiners.

In connection with such cases as the above we may note Darwin's remarks : " As in many widely separated families of plants single species and single genera possess the power of revolving, and have thus become twiners, they must have independently acquired it, and cannot have inherited it from a common progenitor. Hence I was led to predict that some slight tendency to a movement of this kind would be found to be far from uncommon with plants which did not climb, and that this had afforded the basis for natural selection to work on and improve " (1878, p. 197).

In any view of the development of twining the case of *Antirrhinum majus* is of interest. In a race of this well-known garden-plant a form has been found which gives evidence of the inception of twining, the form having, too, the characteristic anatomical features of twiners. It appears to be a mutant, and it comes true to seed.

The question next arises as to whether the tendency to twine has been lost in certain plants. Darwin stated that this has happened in many tendril-bearers, and in his observations on *Clematis* showed that the power of twining was in this genus poor (1865, pp. 26–34). This point is brought forward here on account of the behaviour of *Clematis indivisa*.

In the section of this paper dealing with the methods of obtaining support it has been stated that the developing seedlings of *Clematis indivisa* twine round a thin support with which they come into contact; but the seedlings, being young, not one of those observed has made more than 5 spirals round the support. However, the writer has found an older plant which has attained a height of 2 m., the plant having made 14 spirals and still twining regularly. Adult plants also are occasionally found which exhibit a slight twining round their supports; but the twining in these cases is so slight that it may possibly be due solely to irregular growth in the ascent of the liane.* And we may notice here *Calystegia Soldanella*, a common inhabitant of the sandy shore, where it forms a low compact mass (Cockayne, L., 1910, p. 67, fig. 27). This species has been seen by the writer at Day's Bay, Wellington, growing in a sandy position where a number of low shrubs have become established. Here numerous shoots of the *Calystegia* rise up among the branches of the shrubs, and twine freely and regularly. Just as in *Clematis*, it is probable that we have here a species whose ancestors were twining-plants, and the capacity to twine, through inheritance from these, lies latent in the plants which now occupy the exposed places of the sea-shore.

---

* Since this was written Dr. L. Cockayne informs me that he has noted distinct twining in a large example of *C. indivisa* in the forest on the southern slopes of Mount Ruapehu.

4. TENDRIL-CLIMBERS.

The evolution of tendril-climbers has been referred to by Darwin, who states, " With respect to the sensitiveness of the foot-stalks of the leaves and flowers, and of tendrils, nearly the same remarks are applicable as in the case of the revolving movements of twining-plants. As a vast number of species, belonging to widely distinct groups, are endowed with this kind of sensitiveness, it ought to be found in a nascent condition in many plants which have not become climbers " (1878, p. 197).

In support of this statement is a series of plants explained by Müller (Journ. Linn. Soc., vol. ix, p. 344)—plants which represent stages from those which climb by obtaining support from their branches stretched out at right angles, such as *Chiococca*, to those whose branches form true tendrils, as with *Strychnos*.

Transitions from leaf to tendril are also common, and so it is interesting to note the behaviour of many leaves of *Rubus cissoides*. As stated on p. 326, the leaves of this species are often reduced to midribs, which, with their strong recurved hooks, form distinct climbing-organs. Many of the midribs are much curved at the distal end (Plate XXIII), and in a few cases the midribs had actually encircled a twig with which they had come into contact. It may therefore rightly be asked whether such behaviour denotes the inception of tendril-formation. The plants possessing the peculiarity have a marked advantage in gaining supports, so we should expect that natural selection will preserve these plants, and that the " rudimentary tendrils " will by this means be gradually perfected.

V. GENERAL CONCLUSIONS.

The descriptions given above of the life-forms of the lianes dealt with show to a certain extent how luxuriant is their growth. Moreover, this luxuriant growth is common throughout all the New Zealand forest, to which the lianes are often said to give an appearance similar to that of tropical forests. Undoubtedly there is a Malayan element in the New Zealand forest, so that at first thought it might appear that the lianes were of tropical origin. But this is by no means the case. It is not the great heat alone of a tropical rain-forest that is the primary cause of the liane habit, but rather it is this combined with the excessive moisture of the atmosphere. In New Zealand the moisture is also great, and the climate is equable without extreme heat and cold. This condition of affairs is probably responsible for the high development of lianes in New Zealand, and this view is supported by the remarkable fact that there are here a large number of climbing *Myrtaceae*, a family without lianes in the rest of the world.*

The individual lianes studied exhibit many characters in common, of which we may first note the tendency to form adventitious roots. The formation of such roots is, of course, common with many plants, being most

* Cheeseman (1914, pl. 50) offers as an objection to this theory, for which I am not primarily responsible, the fact that Polynesian and Malayan species of *Metrosideros* have not become lianes, although the climate they are exposed to is "even more humid and equable." I possess no exact details as to the life-forms or the exact environment, on which all depends, of each of the nine (mostly New Caledonian) Malayan and Polynesian species; but, unless some of these are shrubs, the lianoid form would not be expected. On the other hand, in New Zealand *M. scandens* (and probably *M. florida*), *M. diffusa*, and *M. albiflora* possess both shrub and lianoid forms.

noticeable in plants with creeping or underground stems. It is considered probable that moisture is the chief factor determining their production, and this view is supported by the positions of the roots in the lianes observed. They are of great value in obtaining food for the lianes, and by their efficiency in this respect they enable the plants to reproduce vegetatively to a marked extent. The vigorously growing shoots which may arise near these roots receive a sufficient supply of food from them, and thus can soon reach the adult stage, so that in a short time any portion of the forest may contain at least a dozen potential individuals all derived from a single plant.

A point of equal interest is the striking heterophylly of many of the species. These plants possess a juvenile leaf differing in form to varying extent from the adult leaf, and between the two forms there are all transitional stages. Much attention has been directed to the phenomenon by L. Cockayne, whose views in part seek to explain the different leaf-forms by reference to past changes in the environment of the plants. Thus, after dealing with *Parsonsia*, he writes, " This, taken in conjunction with the fact that about 200 species of New Zealand plants—*i.e.*, some 12 per cent. of the spermophytes—belonging to most diverse genera and natural orders, exhibit heterophylly of a more or less striking character in their life-histories, seems to distinctly point to there being some reason in New Zealand itself for this special phenomenon ; and this reason, it seems to me, must be sought for in the manifold changes which the geological history of the New Zealand Archipelago has brought about " (1908, p. 488). And with regard to the leaf-form of *Parsonsia* itself, " It seems clear that the possibilities of both juvenile and adult are latent in the one plant, but each requires its necessary stimulus to set it free in its entirety. If the stimulus is not sufficient, then one or the other form may persist, or there may be a combination of characters, as in the transitional forms " (1912, p. 24).

In reviewing the features of the leaves of the lianes it is seen that there are many characters common to them all. Also the shade leaves present many differences from the sun leaves.

In shade leaves the characters are as follow : (1.) The leaf-blades are relatively expanded and membranous. In *Rubus cissoides*, whose leaves are reduced to midribs, these midribs are much longer than are those in the open. (2.) Petioles are elongated. (3.) Cuticle is not well developed. (4.) Lateral walls of epidermal cells are more wavy on the under leaf-surface. (5.) Palisade cells are not closely packed. (6.) Spongy tissue is well developed, and has irregular cells, between which are large intercellular spaces. (7.) Stomata are on lower surface only.

In sun leaves, on the other hand : (1.) Leaf-blades are smaller and more coriaceous. (2.) Petioles are shorter. (3.) Lateral walls of epidermal cells are straighter than in shade leaves. (4.) There is in most cases a thick cuticle. (5.) Palisade tissue is well developed, with the cells elongated and very closely packed. (6.) Spongy tissue comprises a smaller portion of the thickness of the leaf, and has small intercellular spaces. (7.) Stomata are on lower surface only, but more numerous than in shade leaves.

The above features appear to present the following advantages :—

The expansion of the leaf in the shade is advantageous in that a larger number of light-rays will fall upon the assimilating surface ; and the utilization of light is further aided by the broad palisade cells, which give the chloroplasts a more superficial position than they occupy when the cells are narrow and elongated.

The development of large intercellular spaces is more connected with transpiration. Cowles (1911, p. 554) states that it is highly probable that the feature is caused by the small transpiration which characterizes such damp shaded regions. But at the same time it cannot be doubted that the spaces are of great advantage to the plant. They necessarily result in an increase in leaf-size, and at the same time, although the transpiration-rate is low, the intercellular spaces ensure an efficient aeration of the photosynthetic tissues. This view of the development of intercellular spaces explains their absence—or, rather, reduction—in the exposed sun leaves, which it is obvious must also have tissues aerated. The aeration is here aided by the greater transpiration-current, due to the dryness and warmth of the surrounding atmosphere; for, though transpiration may result in partial closure of the stomata, Cowles states that at the same time the increase of transpiration may cause an increase in the number of stomata. This view is in accord with the increase noted in all the liane sun leaves. It is further remarked by Cowles, " Stomatal structures and activities cannot stop transpiration; at best there is only retardation " (p. 567). Now, if this be correct, then the larger number of stomata on the leaves in the sunlight will, by increasing the transpiration-rate, be of advantage to the liane in aiding the conduction of water, with its contained salts, through the vessels of the stem. And it is obvious that this conduction in so small a stem to so great a height is for the liane a matter of supreme importance.

The great development of cuticle in the open has also been interpreted as due to increased transpiration. While the loss in external transpiration due to cuticularization may be fully balanced by the gain caused by the increased number of stomata, the cuticle serves an important rôle in strengthening the leaf, and therefore ensuring protection from wind and storms.

The cells of the epidermis can be roughly classed into four types: Type 1, walls of cells very wavy (*e.g., Fuchsia Colensoi*, shade leaf); type 2, walls wavy, but to a far less extent than in type 1 (*e.g., Rubus cissoides*, shade leaf); type 3, walls straighter than in type 2, but not so straight as in type 4 (*e.g., Clematis indivisa*, shade leaf); type 4, walls straight, and perhaps slightly rounded at the corners (*e.g., Clematis indivisa*, sun leaf).

The importance of variation in the regularity of the lateral walls of epidermal cells is not so evident; but since they are more wavy in the lower epidermis than in the upper, and more regular in sun leaves than in shade leaves, it is probable that transpiration is a factor determining their regularity.

Finally, we must notice how the structure of the stem must be related to all questions dealing with water-conduction and transpiration. In lianes especially do the stems have to be particularly adapted for rapid water-carriage, and without knowledge of these adaptations no comprehensive conclusions can be obtained.

APPENDIX.

1. Table of Leaf-characters.

	Number of Stomata per Square Millimetre.		Percentage of Number in Shade Leaf of Sun Leaf.	Type of Epidermal Cell, Lower Surface.	
	Shade Leaf.	Sun Leaf.		Shade Leaf.	Sun Leaf.
1. *Fuchsia Colensoi* ..	64	80	80	1	2
2. *Rubus australis* ..	240	296	81	2	3
3. —— *schmidelioides* ..	216	288	75	2	3
4. —— *cissoides* ..	230	294	78	2	3
5. —— *subpauperatus* ..	256	300	85	2	3
6. *Metrosideros hypercifolia* ..	220	264	83	1	2
7. *Muehlenbeckia australis* ..	64	88	73	1	2
8. —— *complexa* ..	96	136	71	1	2
9. *Parsonsia heterophylla* ..	225	280	80	2	3
10. —— *capsularis* var. *rosea* ..	256	320	80	2	3
11. *Clematis indivisa* ..	96	128	75	3	4
12. *Tetrapathaea australis* ..	160	224	71	2	3

2. Bibliography.

(*a.*) *Special.*

Armstrong, J. F.   1870.   " On the Vegetation in the Neighbourhood of Christchurch, including Riccarton Bush, Dry Bush, &c." Trans. N.Z. Inst., vol. 2, p. 118.

Cheeseman, T. F.   1906.   " Manual of the New Zealand Flora."

—— 1914.   " Illustrations of the New Zealand Flora."

Cockayne, L.   1899.   " An Inquiry into the Seedling Forms of New Zealand Phanerogams and their Development." Trans. N.Z. Inst., vol. 31, p. 354.

—— 1900.   " An Inquiry into the Seedling Forms of New Zealand Phanerogams and their Development." Trans. N.Z. Inst., vol. 32, p. 83.

—— 1901.   " An Inquiry into the Seedling Forms of New Zealand Phanerogams and their Development." Trans. N.Z. Inst., vol. 33, p. 265.

—— 1905.   " Notes on the Vegetation of the Open Bay Islands." Trans. N.Z. Inst., vol. 37, p. 368.

—— 1907.   " Report on a Botanical Survey of Kapiti Island." Government Printer, Wellington.

—— 1908.   " A Preliminary Note on Heterophylly in *Parsonsia.*" Australasian Association for Advancement of Science, vol. 11, p. 486.

—— 1908a.   " Report on a Botanical Survey of the Waipoua Kauri Forest." Government Printer, Wellington.

—— 1909.   " Report on a Botanical Survey of Stewart Island." Government Printer, Wellington.

—— 1910.   " On a Non-flowering New Zealand Species of *Rubus.*" Trans. N.Z. Inst., vol. 42, p. 325.

—— 1910a.   " New Zealand Plants and their Story."

Cockayne, L. 1912. "Observations concerning Evolution, derived from Ecological Studies in New Zealand." Trans. N.Z. Inst., vol. 44, p. 1.

—— 1912a. "Some Examples of Precocious Blooming in Heteroblastic Species of New Zealand Plants." Australasian Association for the Advancement of Science, vol. 13, p. 217.

—— 1914. "The Ancient Forest of the Canterbury Plains." Journal of the Canterbury A. and P. Association, p. 19.

Haast, J. von. 1870. "Introductory Remarks on the Distribution of Plants in the Province of Canterbury." Trans. N.Z. Inst., vol. 2, p. 118.

Kirk, T. 1872. "A Comparison of the Indigenous Floras of the British Islands and New Zealand." Trans. N.Z. Inst., vol. 4, p. 247.

—— 1893. "On Heterostyled Trimorphic Flowers in the Fuchsias of New Zealand, with Notes on the Distinctive Characters of the Species." Trans. N.Z. Inst., vol. 25, p. 261.

Laing, R. M., and Blackwell, E. W. 1906. "Plants of New Zealand."

(b.) Other Works consulted.

Coulter, Cowles, and Barnes. 1911. "A Text-book of Botany," vol. 2, "Ecology."

Darwin, C. 1865. "On the Movements and Habits of Climbing Plants." Journ. Linn. Soc., vol. 9, p. 1.

—— 1878. "The Origin of Species." 6th edition.

Goebel, K. 1905. "Organography of Plants," Part II. English translation by I. Bayley Balfour.

Kerner, A. 1894. "The Natural History of Plants." Translated by F. W. Oliver.

Schenck, H. 1892. "Beitrage zur Biologie und Anatomie der Lianen."

Schimper, A. F. W. 1903. "Plant Geography." Translated by P. Groom and I. Bayley Balfour.

Solereder, H. 1908. "Systematic Anatomy of the Dicotyledons." Translated by L. A. Boodle and F. E. Fritsch.

Pfeffer, W. 1906. "The Physiology of Plants," vol. 3. Translated by A. J. Ewart.

Warming, E. 1909. "Oecology of Plants." Translated by P. Groom and I. Bayley Balfour.

Art. XXXIII.—*A New Species of* Orchestia

By Charles Chilton, M.A., D.Sc., LL.D., M.B., C.M., F.L.S., C.M.Z.S.,
Professor of Biology, Canterbury College, New Zealand.

[*Read before the Philosophical Institute of Canterbury, 3rd November, 1915.*]

Of the numerous species of *Orchestia* that are found on the coasts of New
Zealand one of the commonest is *Orchestia chiliensis* Milne-Edwards, usually
found under stones, seaweed, &c., on rocky shores, but not on sandy beaches.
In general characters it resembles *O. mediterranea* A. Costa, of Europe, and
it agrees well with the short description given by Stebbing in " Das Tierreich
Amphipoda," p. 537. The species can usually be recognized in the fully
developed male by the stout peduncle of the lower antenna and by the
somewhat widened meral and carpal joints of the fourth and fifth peraeopoda ; the females and immature males are, however, much more difficult
to distinguish from those of allied species.

In August, 1915, I received from Mr. T. B. Smith, of the Stephen
Island Lighthouse, to whom I am indebted for many interesting *Crustacea*,
a large number of specimens of an *Orchestia* which I at first thought to
be *O. chiliensis* M.-E., one undoubted male of which was, indeed, present.
Among them, however, a few of the largest males had the meral and
carpal joints of the last two peraeopoda widened into large flat plates,
thus differing markedly from the form usually met with in *O. chiliensis*.
It is, of course, possible that this is only an extreme development of the
tendency shown to a less extent in the ordinary specimens of *O. chiliensis*,
for it is well known that in several *Amphipoda* the structures specially
modified in the male may in certain individuals be developed to an extent
that makes them look quite different from the ordinary form. I have
described an example of this in the case of *Cerapus flindersi* Stebbing,*
and other examples could be quoted. In the present case, however, until
the relationship between the Stephen Island form and *O. chiliensis* is
better known it will be safer to consider the former to be a distinct species,
and I am therefore describing it under the name *Orchestia miranda* sp. nov.

Orchestia miranda sp. nov.   Figs. 1 to 6.

Specific Diagnosis.—In general resembling *O. chiliensis* M.-E., the male
differing from the female in the stouter lower antenna, in the gnathopoda,
and especially in the enlarged joints of the last two pairs of peraeopoda.
In the fourth peraeopod the merus is of normal width proximally but
widens distally to fully twice this width, thus forming a triangular plate;
the carpus is greatly dilated into a large oblong plate with rounded corners,
rather wider than the greatest width of the merus ; the propod is not
dilated, but of normal width. The fifth peraeopod is modified in a similar

---

* Rec. Australian Museum, vol. 2, p. 1, 1892.

way, but the carpus is much more dilated, especially toward the posterior margin, its greatest width being half as great again as that of the merus.

Length of body of largest specimens, about 20 mm.

Hab.—Stephen Island, Cook Strait, New Zealand, on rocky shores.

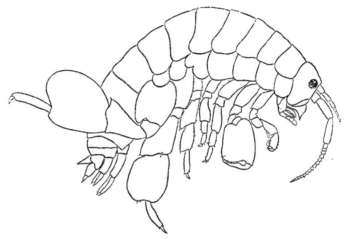

FIG. 1.—*Orchestia miranda*, male; side view.

This brief diagnosis may be supplemented by the following more detailed description of a fully developed *male* :—

Body rather compressed. First side plate smaller than the second, by which it is overlapped, fifth as deep as the fourth. Third pleon segment

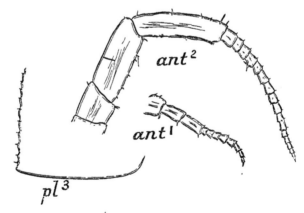

FIG. 2.—*Orchestia miranda*, male. *ant¹*, first antenna ; *ant²*, second antenna ; *pl³*, inferior and posterior margins of third pleon segment.

(fig. 2, *pl³*) with postero-inferior angle quadrate, its posterior margin with small low serrations each with a minute seta. Eyes black, round or slightly oval, the distance between them about equal to their greatest width.

First antenna (fig. 2, *ant¹*) reaching to the end of penultimate joint of the lower ; second and third joints subequal and a little longer than the

12*

first ; flagellum as long as peduncle, and containing about 8 joints. Second antenna (fig. 2, *ant*²) fully one-third the length of the body ; penultimate joint of peduncle rather shorter than the ultimate, both rather broad ; flagellum stout, subequal in length to the peduncle, of about 20 joints.

Mouth parts apparently not presenting any distinctive features.

First gnathopod (fig. 3, *gn*¹) with side plate subtriangular, somewhat produced downwards anteriorly, its lower margin with a few stout setae, its inner surface with an irregular row of more slender setae extending from the insertion of the basal joint to the infero-anterior angle ; merus with a small rounded pellucid process, carpus much longer than the propod, the pellucid area on each marked off from the rest of the joint by a row of stout setae, palm transverse, finger not extending beyond the true palm.

Second gnathopod (fig. 3, *gn*²) with side plate produced at about the middle of the posterior margin into a subacute point, basal joint not much expanded, merus and carpus very short, propod very large, widening distally, anterior and posterior margins without setae, palm only slightly

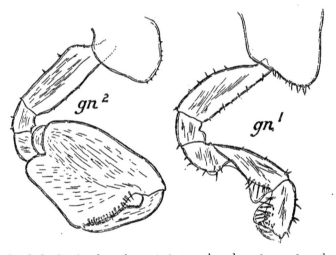

Fig. 3.—*Orchestia miranda*, male. *gn*¹, first gnathopod ; *gn*², second gnathopod.

oblique, straight or a little convex, provided with a double row of short stout setae, finger strong with an enlargement at about its proximal third, end curved and fitting into a short groove at the end of the palm.

First and second peraeopoda alike and presenting no special features, the side plates subrectangular with rounded corners, posterior margin in each produced into a subacute process, setae on the various joints few and short, two or three stout ones on anterior border of propod at base of finger. Third peraeopod (fig. 4, *prp*³) subequal in length to the two preceding, side plate with anterior lobe as deep as the fourth, basis broadly expanded, its posterior margin convex and serrate, merus and carpus somewhat widened, about twice as wide as the propod. Fourth peraeopod (fig. 4, *prp*⁴) with basis similar to that of the third, ischium normal, merus widening greatly toward distal end, which is rather oblique and nearly as wide as the joint is long, carpus forming a large rectangular plate with rounded corners, nearly as broad as long, propod and finger normal, not expanded.

Fifth peraeopod (fig. 4, prp^5) longer than the fourth, and similarly expanded but with carpus fully as wide as long, its posterior margin being greatly produced and very convex.

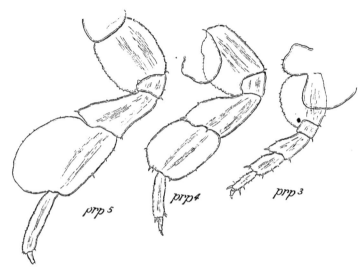

FIG. 4.—*Orchestia miranda*, male. prp^3, third peraeopod; prp^4, fourth peraeopod; prp^5, fifth peraeopod.

First and second uropoda (fig. 5, urp^1 and urp^2) normal, the peduncle and both rami in each bearing short stout spines. Third uropod (fig. 5, urp^3) with peduncle laterally compressed, its depth near the base nearly equal to the length, a few stout setae at distal end of upper margin, ramus

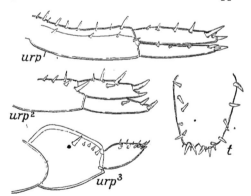

FIG. 5.—*Orchestia miranda*, male. urp^1, first uropod; urp^2, second uropod; urp^3, third uropod; t, telson.

much shorter than peduncle, with a double row of small stout setae along upper margin. Telson (fig. 5, t) longer than broad, narrowing slightly distally, posterior margin with small triangular notch, lateral and posterior margins supplied with short stout setae.

The *ovigerous female** differs from the male in having the second antenna more slender, in the absence of dilatation of the joints of the peraeopoda, and in the gnathopoda. In the first gnathopod (fig. 6, gn^1) the basal joint is longer than any of the others, and is of equal width throughout, being about three times as long as broad, the carpus is considerably longer than the propod and widens slightly towards the distal end, the propod is oblong, rather narrower than the carpus, the palm is transverse, finger slender, not reaching beyond the palm; all the joints provided with a few stout spinules arranged in the usual manner. In the second gnathopod (fig. 6, gn^2) the basis is much widened, its hind margin straight, anterior sinuous and fringed with fine spinules, the greatest width at the proximal third being about one-half the length; in the remaining joints, in the branchia, and in the large incubatory plate the appendage presents the characters common to the genus. None of the peraeopoda have any joints

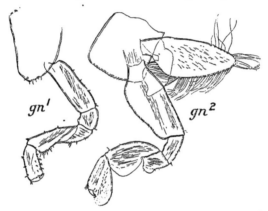

FIG. 6.—*Orchestia miranda*, female. gn^1, first gnathopod; gn^2, second gnathopod.

expanded; in the fifth the basis has the posterior margin convex and serrate and not produced downwards beyond the extremity of the joint; the following joints are slender and of the usual form.

The *immature males* resemble the females except in the gnathopoda, which appear to acquire the special character of the male early, though they are for a time much smaller than in older individuals . In quite young males, though the gnathopoda have already acquired the characteristic form, the antenna and the peraeopoda show no sign of dilatation, but are quite like those of the females. Among my specimens are numerous transitional stages from this up to the widely dilated peraeopoda of the adult males as already described; some of these would be difficult to distinguish from *O. chiliensis*.

The species now described, *O. miranda*, appears to belong to that section of the genus containing *O. mediterranea*, *O. gammarellus*, *O. chiliensis*, &c.,

---

* I am assuming that the numerous females sent with the equally numerous males belong to *O. miranda*, for, with one exception, all the recognizable males belong to that species, and I am unable to divide the females into two groups. The one exception is, however, a fully developed male of *O. chiliensis* M.-E., so that possibly some or all of the females may belong to that species. In the meantime I am unable to give any characters that would differentiate between the females of these two species.

in which the male tends to differ from the female in the dilatation of the meral and carpal joints of the fifth peraeopod, but in it, as to a less extent in *O. chiliensis* also, the same tendency is seen in the fourth peraeopod as well. *O. miranda* differs from *O. chiliensis* in the much greater expansion of the joints of the last two peraeopoda, and also in the second gnathopod, which has the palm less oblique and without the large obtuse tooth near the finger-hinge characteristic of *O. chiliensis*. It must be remembered, however, that all the examples of *O. miranda* at present known come from a single restricted locality; if it is found to be more widespread an examination of specimens from other localities will probably show that the distinctions drawn above between it and *O. chiliensis* will not invariably hold.

ART. XXXIV.—*Some Australian and New Zealand* Gammaridae.

By CHARLES CHILTON, M.A., D.Sc., LL.D., M.B., C.M., F.L.S., C.M.Z.S.,
Professor of Biology, Canterbury College, New Zealand.

[*Read before the Philosophical Institute of Canterbury, 3rd November, 1915.*]

THE following paper deals with a few species of *Gammaridae* found in Australian and New Zealand seas. Nearly all of them are widely distributed, and show considerable local variation. Hence the delimitation of the species and of their varieties is difficult, and will call for much more investigation than can be devoted to the subject at the present time.

Three of the species are now recorded from New Zealand for the first time.

Melita festiva (Chilton). Figs. 1 and 2.

Moera festiva Chilton, 1884, p. 1037, pl. 46, fig. 2; Stebbing, 1910A, p. 642. *Moera rubromaculata* Haswell (part), 1885, p. 105. *Ceradocus rubromaculatus* Della Valle (part), 1893, p. 720; Stebbing (part), 1906, pp. 430 and 732.

Specific Diagnosis.

Male.—Peraeon smooth. Pleon with fourth segment produced dorsally into a single tooth, fifth segment into 2 small teeth with 1 or 2 setae. Third pleon segment with postero-lateral angle produced, acute, lower margin bearing 2 setae anteriorly and being indistinctly serrate posteriorly.

First gnathopod small, merus bearing posteriorly short furry setae as well as some long hairs; carpus slightly longer than propod, bearing a distinct row of long setae and some furry setae near antero-distal angle and many long setae arranged in short transverse rows on the posterior margin and on the inner surface; propod somewhat narrowed at the base; palm short, nearly transverse, hardly defined, tufts of long setae on posterior margin and along the palm, smaller tufts on the anterior margin and at base of finger; finger curved, acute, fitting closely on to palm when closed. Second gnathopod having merus produced distally into a sharp tooth with a small tuft of setae near the apex ; carpus short, triangular, cup-like, its posterior margin bearing many transverse rows of long setae ; propod very large, longer than the whole of rest of limb, oblong, margins parallel, posterior margin with about 10 small tufts of long setae, similar tufts or short transverse rows sometimes present on anterior margin and inner surface ; palm nearly transverse, well defined, and usually with 2 irregular teeth, the larger flat-topped ; finger arising from near the centre of distal end of propod, very short and stout, not longer than palm, its end roundly truncate.

In other respects showing the characters usual in *Melita*.

Female.—First gnathopod similar to that of male. Second gnathopod much smaller than in the male, the carpus longer than in male and fully half as long as the propod, propod oblong but with palm oblique and not very clearly defined, irregularly toothed ; finger of normal shape, long and acute.

Colour.—Pale brown, with tints of green.

Locality.—Sydney Harbour, New South Wales; and Auckland Harbour, New Zealand.

Remarks.—In 1884 I (1884, p. 1037) described under the name *Moera festiva* an Amphipod of which I had collected a few specimens in Sydney

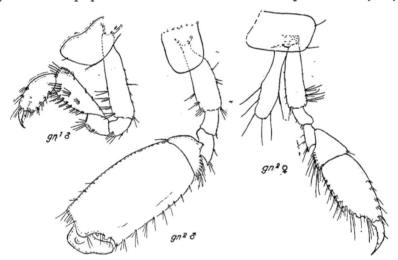

Fig. 1.—*Melita festiva.* *gn*[1] ♂ , first gnathopod of male ; *gn*[2] ♂ , second gnathopod of male ; *gn*[2] ♀ , second gnathopod of female.

Harbour. The species was distinguished especially by the very short and peculiar truncate finger of the second gnathopods of the male. The antenna and gnathopoda of both male and female were described in some detail, and I stated that in all my specimens the " terminal pleopoda "—*i.e.*, the third uropods—had been lost, and they were therefore probably of large size, but that in their absence it was impossible to decide whether the species should be placed under *Maera* or *Melita*. In the next year Professor Haswell (1885, p. 105), in revising the Australian *Amphipoda*, united *Moera spinosa* Haswell and *M. ramsayi* Haswell with *Moera rubromaculata*

Stimpson, which he described as a species with varying forms of the second gnathopods. He added, " *Moera festiva* Chilton also belongs to this very variable species."   In *M. rubromaculata* the third uropoda are large, with both rami equally developed, so that in this character the species agrees with the generic characters of *Maera* and differs from *Melita*.   The dactyl of the second gnathopod of the male of *Moera festiva* differed so much from those of the other forms referred by Haswell to *M. rubromaculata* that I felt very doubtful of the correctness of referring *M. festiva* also to this species, but at that time I had no further specimens or other means of definitely settling the question.

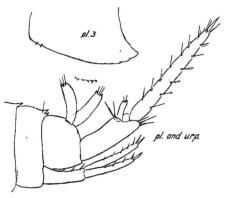

FIG. 2.—*Melita festiva*.   *pl. 3*, lower portion of third segment of pleon; *pl. and urp*, terminal portion of pleon, with uropoda.

In 1893 Della Valle (1893, p. 720) placed *Moera. rubromaculata* Stimpson under the genus *Ceradocus* A. Costa, and so did Stebbing in 1899. Accordingly in "Das Tierreich Amphipoda," Stebbing (1906, p. 430) gave the species as *Ceradocus rubromaculatus* (Stimpson), and followed Haswell in considering *Moera ramsayi* Haswell, *M. spinosa* Haswell, and *M. festiva* Chilton as synonyms.   More recently, however (1910a, p. 643), he says, " The position of all three should rather be regarded as still doubtful. *M. festiva* shows agreement with Haswell's *M. hamigera*."

Meanwhile Mr. Walker (1904, p. 276) had described from Ceylon an Amphipod which he named *Elasmopus dubius*, and had drawn attention to the resemblance between the second gnathopod of his species and that of *M. festiva*, saying, " This species is certainly very near *Moera festiva* Chilton."   As, however, the description of that species differed somewhat from his specimens, he thought it better to consider them as distinct, adding, " It is unfortunate that in both cases the third uropods, so important in this family, should be wanting."   The loss of these appendages in *Elasmopus dubius* raises some doubt as to the correctness of the genus to which the species was assigned by Walker, as in my experience the third uropods in *Elasmopus* are usually not much longer than the first and second, and are not so readily lost as in species of *Maera* and *Melita*.

Recently I have been going over and naming an extensive collection of *Crustacea* gathered by Mr. W. R. B. Oliver, and in one small tube of specimens collected under stones on Rangitoto Reef, Auckland Harbour, I found three species of Gammarids—viz., *Ceradocus rubromaculatus* (Stimpson), *Maera viridis* Haswell, and another that is without doubt the same as *M. festiva* Chilton. Of this species there were only two specimens, a male and a female, but fortunately the male still bears one of the third uropods, and this has the inner branch quite short and the outer one long, thus showing the form typical of the genus *Melita ;* in other points also the species agrees well with the characters of *Melita,* and must be placed in that genus. It is therefore quite different from *Ceradocus rubromaculatus* (Stimpson). The peculiar structure of the second gnathopod is quite consistent with *Melita,* for in that genus the second gnathopod in the male is usually large, and differs greatly from that of the female, assuming in several species very peculiar and even bizarre forms, as, for example, in *Melita fresnelii* Aud.

My Auckland specimens of *M. festiva* differ slightly from the Sydney specimens which have the appendages, and especially the second gnathopod, more setose ; the resemblance in other respects, however, is so great that I have no hesitation in considering them as the same species. In another Amphipod, *Elasmopus subcarinatus* (Haswell), I have noted that some of the Australian specimens are more setose than the New Zealand ones.

Maera hamigera Haswell, with which, as Stebbing has pointed out, *M. festiva* shows some agreement, has the rami of the third uropods long and equal, and is a true *Maera,* and therefore quite distinct. Moreover, although the second gnathopods show considerable resemblance to those of *M. festiva,* they differ in being unequal, and in the larger one the finger, though short, has not the peculiar truncate end that it has in *M. festiva. Moera hamigera* has been recorded by Walker (1909, p. 335) from Suez and from Khor Dongola ; and both he and Stebbing, who examined specimens from the " Thetis " Expedition (1910A, p. 600), have added to and amended the original description given by Haswell.

In the absence of any knowledge of its third uropods it is impossible to come to any conclusion as to the position of *Elasmopus dubius* Walker. The second gnathopods appear to show as much resemblance to the larger one of *M. hamigera* as to those of *M. festiva,* but in the absence of any note to the contrary it is to be presumed that those of the right and left sides are equal.

Maera viridis Haswell. Figs. 3 and 4.

Moera viridis Haswell, 1879, p. 333, pl. 21, fig. 1. *Moera incerta* Chilton, 1883, p. 83, pl. 3, fig. 3. *Elasmopus viridis* Stebbing, 1906, p. 445, and 1910A, p. 643 ; Chevreux, 1908, p. 482 ; Chilton, 1912, p. 131.

The species *Maera viridis* was described in 1879 by Professor W. A. Haswell from specimens collected at Clark Island, Port Jackson. He added to his description by pointing out the differences between this species and *M. truncatipes* (Spinola), *M. quadrimanus* Dana, and *M. ramsayi* Haswell.

In 1883 I described from specimens obtained at Lyttelton Harbour a species *Maera incerta,* pointing out that it closely resembled *M. viridis* Haswell and the other species mentioned by Haswell, but differed from all in the form of the second pair of gnathopods, in this respect closely resembling *M. blanchardi* Spence Bate.

Both the species *M. viridis* and *M. incerta* were included with others under *M. truncatipes* by Della Valle in 1893.

In 1899 Mr. Stebbing transferred *M. viridis* to the genus *Elasmopus*, and this view was also taken in his " Das Tierreich Amphipoda," published in 1906, where *M. incerta* is ranked as a synonym of *Elasmopus viridis*.

In 1908 Chevreux recorded the species under the name *Elasmopus viridis* from several localities in the Gambier and Tuámotu Archipelagoes, and in 1912 I accepted Mr. Stébbing's identification of *M. incerta* with *M. viridis* and retained the species under the genus *Elasmopus*, pointing out that in it the second gnathopod was almost the same in the female as in the male.

I have recently been examining the New Zealand species of *Elasmopus*, and have come to the conclusion that *M. viridis* Haswell is too closely related to *M. inaequipes* (A. Costa) to be placed in a different genus, and that it is best left under *Maera*.

M. inaequipes, the name now used for *M. truncatipes* (Spinola), is recorded in "Das Tierreich Amphipoda " from the Mediterranean and from the North Atlantic (Azores); but in 1904 Mr. Walker had already recorded it under the name *M. scissimana* (Costa) from the west of Ceylon, and in subsequent papers he recorded it from the Maldive Archipelago and other localities in the Indian Ocean. It also occurs in Australia and New Zealand (see below). In describing the Ceylon specimens Mr. Walker says that the species (*M. inaequipes*) " forms a connecting-link between the genera *Maera* and *Elasmopus*. The fore part, including the third peraeo-pods, is typical *Maera*, while the massive and very spinous fourth and fifth peraeopods (a character that is much more marked in Ceylon than in Mediterranean specimens), and the comparatively short rami of the third uropods, resemble *Elasmopus*. Another peculiarity of the species is that the size and shape of the hand of the second gnathopods is much the same in males and females." In describing the Maldive specimens he mentions that in them the third to fifth peraeopods are less robust and more like the Mediterranean than the Ceylon forms.

It will be seen from what has been already said that the close relation-ship of *M. viridis* to *M. inaequipes* has been pointed out more than once. In describing *M. incerta* I stated that it seemed to come nearest to *M. blanchardi* Spence Bate; but *M. incerta* is now considered the same as *M. viridis* Haswell, and *M. blanchardi* is ranked in "Das Tierreich Amphipoda " as a synonym of *M. inaequipes*, being presumably the female.

The resemblance between the two species *M. inaequipes* (Costa) and *M. viridis* Haswell is emphasized by one or two special points. Thus in 1904 Walker pointed out that in *M. scissimana* (Costa)—*i.e.*, *M. inaequipes* —the second gnathopods in the female are of much the same size and shape as in the male. In my MS. notes I had previously recorded the same peculiarity in *M. viridis*, though this was not published till 1912. In the genus *Elasmopus*, to which *M. viridis* is assigned by Stebbing, there is usually a marked difference between the sexes in the second gnathopods, and there is considerable difference in some of the other species of *Maera*. Both species, again, have a slight depression or emargination towards the distal end of the anterior border of the carpus in the first gnathopod.

There is similarity also as regards colour. Thus in "Das Tierreich Amphipoda " Stebbing gives the colour of *M. inaequipes* as " dorsally green bronzed with a little red, gnathopods 1 and 2 tinged with green, other appendages pellucid pinkish." Haswell gives the colour of *M. viridis* as " light green "; the New Zealand specimens are also a light green, occasionally tinged with pink on the appendages.

From the discussion given above it will be evident that *M. inaequipes* and ₰M. *viridis* present many points of resemblance,₰and that they cannot be placed in different genera. I prefer to keep them both in *Maera*, though

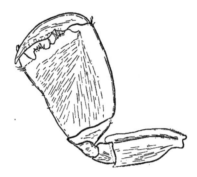

Fig. 3.—*Maera viridis.* Second gnathopod of male, from a Lord Howe Island specimen.

in the short third uropoda and in the widened joints of the last pair of peraeopoda *M. viridis* certainly approaches to *Elasmopus*, and naturally in the present state of our knowledge the distinctions between the genera are somewhat artificial.

In the New Zealand specimens of *M. viridis* that I have examined the palm of the second gnathopod is transverse or a little projecting, and it is usually straight or even, though sometimes showing slight indications

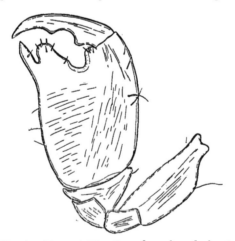

Fig. 4.—*Maera viridis.* Second gnathopod of male, from a Port Jackson specimen.

of a central notch and of a narrower depression next to the defining tooth. In a specimen from Lord Howe Island that I think must belong to the same species the defining tooth (fig. 3) is much longer and better marked and the palm more uneven, the median notch, however, being shallow and divided by a small projection. Again, in a specimen from Port Jackson sent to me many years ago as *M. viridis* by Professor Haswell, the palm (fig. 4) is much more projecting, and has the central notch much deeper and wider, extending almost to the base of the finger. In the Lord Howe Island and Port·Jackson specimens the propod is nearly free from setae, while in the New Zealand specimens there are short tufts along both anterior and posterior margins; there are also equally marked differences in the structure of the finger.

Maera quadrimana Dana, described from specimens collected at the Fiji Islands, is evidently very closely allied.to *M. viridis*, and may prove to be identical with it. Both species were united with *M. inaequipes* by Della Valle in 1893.

Maera rathbunae Pearse (1908, p. 29) as further described by Kunkel (1910, p. 46) is also very near to *M. viridis,* and may prove indentical with it.

Maera inaequipes (A. Costa). Figs. 5 and 6.

Amphithoe inaequipes A. Costa in Hope's Catal. Crost. Ital., 1851, p. 45, *Maera quadrimana* G. M. Thomson, 1882, p. 235, pl. 17, fig. 4a (part). *Maera truncatipes* Della Valle, 1893, p. 725, pl. 1, fig. 2, and pl. 22, figs. 26–40 (part); Miers, 1884, p. 569. *Maera hirondellei* Chevreux, 1900, p. 84, pl. 11, fig. 1. *Maera scissimana* Walker, 1904, p. 273, pl. 5, fig. 32. *Maera inaequipes* Stebbing, 1906, p. 435, and 1910A, p. 599; Kunkel, 1910, p. 44.

Specific Diagnosis.

Dorsal surface of body smooth. First side plate with front corner produced, acute. Third pleon segment with posterior margin smooth, inferior margin smooth or obscurely serrate posteriorly, angle acute, slightly produced. Eyes round. First antennae with flagellum about the same length as peduncle, accessory flagellum about half as long. Second antennae with flagellum subequal to ultimate joint of peduncle. First gnathopod with depression in anterior border of carpus, inner surface of carpus with numerous oblique rows of setae arranged as in fig. 5, gn^1.

Second gnathopod large in both sexes, palm somewhat·oblique, defined by an acute tooth, convex and serrulate or irregularly toothed in the female, more toothed and usually with a deep central notch in the male. Third uropod with rami subequal, considerably longer than the first and second. Telson deeply cleft, each lobe bidentate, the inner tooth being longer than the outer.

Of this species I have specimens from the following New Zealand localities: Paterson Inlet, Stewart Island (these are some of the specimens referred to *M. quadrimana* by G. M. Thomson in 1882); Chatham Islands, Miss S. D. Shand.

Distribution.—The species has long been known under various names from the North Atlantic and the Mediterranean, and has been recorded from the Indian Ocean by Walker, and more recently from Australia by Stebbing.

I feel fairly confident that the New Zealand specimens rightly belong to this widespread species. It appears to be very closely allied to *M. viridis* Haswell, and, as in that species, it has the characteristic depression on the anterior margin of the carpus of the first gnathopod.   The two species were united by Della Valle, and in each there appear to be so many varieties that it may be difficult to draw a distinction between them in all cases.   The more typical examples of *M. inaequipes,* however, appear

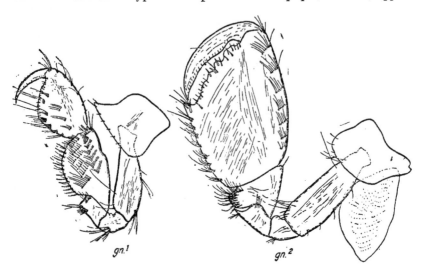

Fig. 5.—*Maera inaequipes,* female—a specimen from Chatham Islands.   *gn¹,* first gnathopod (inner surface) ; *gn²,* second gnathopod (inner surface).

to be distinguished by the more oblique palm of the second gnathopod, and by having the third uropods considerably longer than the preceding pairs.   Stebbing (1906, p. 436, and 1910a, p. 599) describes the palm of the second gnathopod as " almost transverse," and this usually appears to

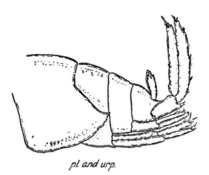

pl. and urp.

Fig. 6.—*Maera inaequipes,* female —a specimen from Chatham Islands. *pl. and urp.,* terminal portion of body.

be so in the males, but in the female it is distinctly oblique, though not greatly so, and apparently this is also the case with some forms of the male, as, *e.g.,* in *M. hirondellei* Chevreux.   The third uropods, again, appear fairly long.   Stebbing describes the rami as " not very long," but they seem always to reach well beyond those of the preceding pairs, and they are considerably elongated in some of the New Zealand specimens; in one of the specimens from the Chatham Islands they are much shorter on both sides, hardly reaching beyond the second and third, but I think this condition is abnormal, and due to the regeneration of appendages that had previously been lost. Walker has pointed out that the fourth and fifth peraeopoda are more massive and spinous in the Ceylon than in the Mediterranean specimens, and has also drawn attention to the fact that the species forms a connecting-link between the genera *Maera* and *Elasmopus*, mentioning among the characters that resemble the latter genus the comparatively short rami of the third uropods. In the New Zealand specimens, however, they are quite long enough to justify the inclusion of the species in *Maera*, and the species appears to differ also from *Elasmopus* in having the second gnathopods of approximately the same size in both sexes, and in having the setae on the inner surface of the carpus of the first gnathopod arranged on a different pattern.

Maera hirondellei Chevreux differs from the more typical forms in having the accessory appendage shorter, and in the oblique palm of the second gnathopod of the male; but these differences are perhaps only of varietal importance. The general resemblance in other characters is very close.

Miers mentions (1884, p. 569) that in the British Museum there is a specimen from the Korean Seas which cannot, he thinks, be distinguished from *M. truncatipes* (Spinola)—*i.e.*, from *M. inaequipes* (A. Costa). *M. diversimana* Miers appears also to be closely allied to the present species and to *M. viridis* Haswell, but apparently differs in having the segments of the pleon dorsally toothed, and in having the right and left second gnathopods unequal in size, while in *M. inaequipes* they are, according to Stebbing, as a rule, " quite alike in size and sculpture." The name of the species, however, indicates that this is not always so, and probably in this as in other species with large gnathopods the appendage on one side may differ from that on the other in some individuals, while in others they are alike.

Maera mastersii (Haswell).

Megamoera mastersii Haswell, 1879, p. 265, pl. 11, fig. 1. *Moera quadrimana* G. M. Thomson (part), 1882, p. 235, pl. 17, fig. 4b (not *M. quadrimana* Dana). *Megamoera thomsoni* Miers, 1884, p. 318, pl. 34, fig. B. *Maera mastersii* Stebbing, 1906, p. 439; 1910A, p. 642; 1910B, p. 457: Chilton, 1911, p. 564.

Specific Diagnosis.

Body smooth, none of the segments being dorsally produced into teeth. First side plate produced anteriorly into a rounded lobe. Third pleon segment with lower portion of posterior margin serrate, inferior margin with 2 setae but not serrate. Eyes narrowly reniform. First antenna with accessory flagellum short, usually of not more than 4 joints, primary flagellum as long as peduncle, fairly stout, setose. First gnathopod with carpus as long as propod, its inner surface with tufts and comb-like rows of setae, propod slightly narrower than carpus, palm smooth, somewhat oblique, not defined. Second gnathopod considerably larger, merus produced into sharp tooth, carpus about half the length of propod, which is oblong with margins parallel and provided with many tufts or transverse rows of setae on both margins, palm oblique, defined by a small tooth, irregularly denticulate, the teeth varying in number and being acute or rounded at end ; right and left second gnathopods often slightly unequal in size, those of the male apparently larger and with more distinctly toothed palm than in the female. Third uropods much longer than the first and second, rami equal, broadly lanceolate. Telson cleft to base, lobes bluntly conical, produced into an acute tooth on each side the terminal notch, which bears 2 or 3 spinules.

Colour. — Usually light yellowish - brown ; " dirty green " (G. M. Thomson).

Localities. — Moeraki (Dunedin Museum collection) ; Stewart Island (H. B. Kirk) ; Paterson Inlet, Stewart Island (G. M. Thomson—recorded as *Maera quadrimana* Dana) ; Chatham Islands (H. B. Kirk) ; off Cape Maria van Diemen, dredged in 50 fathoms (C. Chilton).

Distribution. — Australia, Kermadec Islands, New Zealand, Chatham Islands.

The species was originally described from Port Jackson by Haswell, and afterwards recorded under the name *Megamoera thomsoni* by Miers from various localities in the north of Australia. These two species were united by Haswell (1885, p. 105). I have a specimen from St. Vincent Gulf, South Australia, sent to me by Mr. S. W. Fulton, and in 1911 I recorded it from Kermadec Islands, the specimens having been collected by Mr. W. R. B. Oliver.

In 1912 I examined a few specimens from Cape Colony collected by the " Scotia " Expedition, and with some hesitation referred them to *Maera mastersii*, mentioning the points in which they differed from the description given by Stebbing in " Das Tierreich Amphipoda." Mr. Barnard has since kindly sent me further specimens of the same species from Cape Town, and has pointed out that they differ in some specific characters from *M. mastersii*. I find that this is so, and that they belong to *Maera bruzelii* Stebbing, a species which I had overlooked, as it was accidentally omitted from " Das Tierreich Amphipoda." Mr. Barnard considers the Cape Town specimens slightly different from the description of *Maera bruzelii*, and looks upon them as a separate variety or a closely allied species, and places them in the genus *Elasmopus*, to which, as I pointed out in 1912, they show considerable resemblance.

Stebbing (1910B, p. 457) gives four points of difference between *M. bruzelii* and *M. mastersii*. Two of these hold for my specimens—viz., the accessory appendage is only 4-jointed in *M. mastersii* but about 8-jointed in *M. bruzelii*, and the third uropoda reach considerably beyond the others in *M. mastersii* but only a little beyond them in *M. bruzelii*. The other two distinctions do not hold ; thus the first side plate is produced forward in *M. mastersii* about as much as in *M. bruzelii*, and the palm of the first gnathopod is not quadridentate but smooth. The teeth on the palm of the *second* gnathopod are sometimes 4 in number.

Maera mastersii was first recorded from New Zealand by Mr. G. M. Thomson, under the name *M. quadrimana* Dana, in 1882. Mr. Thomson obtained several specimens with the dredge in Paterson Inlet, Stewart Island, and another from between tide-marks in the same locality. He pointed out that the specimens differed in certain respects from Dana's species, and that they differed among themselves in the structure of the second gnathopods. I find that Mr. Thomson's specimens belong to two species, the shore specimen and some of the dredged specimens being *M. mastersii*, the second gnathopod of the shore specimen being figured by Mr. Thomson in pl. 17, fig. 46 ; the other dredged specimens belong to *M. inaequipes* (A. Costa).

The second gnathopoda of the female in *M. mastersii* are of moderate size, and probably not markedly different from those of the male, but the specimens at my disposal are not sufficient for the satisfactory working-out of the sexual differences.

Ceradocus rubromaculatus (Stimpson)

Ceradocus rubromaculatus Stebbing, 1910A, p. 598 (with synonymy) ; Chevreux, 1908, p. 479. *Moera spinosa* Chilton, 1883, p. 81, pl. 2, fig. 3a.

This appears to be a common species on the coasts of Australia, and it is occasionally found in New Zealand.

I have specimens from Auckland (H Suter and W. R. B. Oliver); Rangitoto Reef, Auckland (W. R. B. Oliver) ; Dunedin (G. M. Thomson) ; Chatham Islands (H. B. Kirk) ; and Akaroa (C. Chilton).

The species can be distinguished from others likely to be confused with it by the serrations on the segments of the pleon. The first gnathopod has the inner surface of the carpus covered with short tufts and comb-like rows of setae arranged on the same general plan as in *Maera inaequipes*, but there is no excavation on the anterior margin. The second gnathopoda have the propod large in both sexes, but larger in the male, in which the palm bears 2 or more flat-topped teeth ; but there appears to be considerable variation in the armature of the palm. In the specimens I have examined the palm is quite oblique, but M. Chevreux has described a form from the Gambier Archipelago in which the gnathopoda of the two sides are unequal, the larger having the palm transverse with 2 narrow excavations, the smaller being not unlike that of the female. I had pointed out the same inequality in the gnathopoda of the male in the specimens from Auckland recorded in 1883 under the name *Moera spinosa*. In M. Chevreux's specimens the basal joint of the last pair of peraeopoda is rather narrow, and is produced downwards into an acute narrow tooth. The same tendency is noticeable in Australian and New Zealand forms, but in them the joint is wider and the downward prolongation not so narrow.

REFERENCES.

Chevreux, E. 1900. "Amphipoda" in Résultats des Campagnes Scientifiques par Albert 1er de Monaco, fasc. 16.
—— 1908. Mém. Soc. Zool. de France, vol. 20, pp. 470–527.
Chilton, C. 1884. Proc. Linn. Soc. N.S.W., vol. 9, pp. 1035–44, pl. 46 and 47.
—— 1883. Trans. N.Z. Inst., vol. 15, pp. 69–86, pl. 1–3.
—— 1911. Trans. N.Z. Inst., vol. 43, pp. 546–73.
—— 1912. Trans. N.Z. Inst., vol. 44, pp. 128–35.
Della Valle, A. 1893. "Gammarini" in Fauna und Flora des Golfes von Neapel, Monographie 20.
Haswell, W. A. 1879. Proc. Linn. Soc. N.S.W., vol. 4, pp. 319–50, pl. 18–24.
—— 1885. Proc. Linn. Soc. N.S.W., vol. 10, pp. 95–114, pl. 10–18.

Kunkel, B. W. 1910. '' The Amphipoda of Bermuda.'' Trans. Connecticut Academy of Arts and Sciences, vol. 16, pp. 1–116.

Miers, E. J. 1884. '' Crustacea '' in Report Voyage H.M.S. '' Alert,'' 1881–82. London, 1884.

Pearse, A. S. 1908. Proc. U.S. Nat. Mus., vol. 34, pp. 27–32.

Stebbing, T. R. R. 1906. '' Amphipoda Gammaridea '' in Das Tierreich, Lieferung 21.

—— 1910A. Australian Museum, Memoir 4, pp. 567–658.

—— 1910B. '' Annals South African Museum,'' vol. 6, pp. 281–593.

Thomson, G. M. 1882. Trans. N.Z. Inst., vol. 14, pp. 230–38.

Walker, A. O. 1904. '' Amphipoda '' in Pearl Oyster Fisheries. Suppl. 17, pp. 229–300, pl. 1–8.

ART. XXXV.—*Notes on the Occurrence of the Genus* Trachipterus *in New Zealand.*

By H. Hamilton.

[*Read before the Wellington Philosophical Society, 22nd September, 1915.*]

The object of this paper is to record the occurrence of five specimens of *Trachipterus* from the New Zealand coast, to bring together what is already known about their distribution in the Australasian region, and to offer comparisons with Trachipterids found in the Northern Hemisphere.

Early in 1914 a large specimen of *Trachipterus*, or deal-fish, was found on the beach at Waikanae, near Wellington, and forwarded to the Dominion Museum by Mr. Watt, a local fisherman. In March, 1915, a smaller specimen was donated to the Museum by Mr. Foster, of the Wellington Meat Export Company, who obtained it from the Chatham Islands. Both specimens were considerably damaged, as is usually the case, but, being of such rare occurrence, were carefully preserved for future reference. Professor Benham, Curator of the Otago University Museum, and Mr. R. Speight, Curator of the Canterbury Museum, have kindly placed at my disposal three specimens not previously recorded, thereby allowing a survey of all known New Zealand occurrences. I am much indebted to these gentlemen for their co-operation, and also extend my thanks to Mr. E. R. Waite, of the Adelaide Museum, for his sound advice on the arrangement of the subject-matter.

Previous writers on the Trachipterids have laid stress on the fact that all original observations relating to their appearance and distribution should be recorded to help to solve the problem of their life-history and economy, for only by recording apparently simple facts and examining in detail long series of variable species can a definite conclusion be arrived at.

The Trachipterids of the Mediterranean, once regarded as four species, have been proved by Emery(2), after examination of twenty-three specimens in all stages of growth, to belong to one species only. He showed that the nominal species—viz., *T taenia* Bloch & Schn., *T. filicauda* Costa,

T. spinolae Cuv. & Val., and *T. iris* Walb.—were merely successive stage-growths of a very variable species, and must all be considered juvenile stages of the adult *T. taenia*, a species now more correctly known as *T. trachypterus.*

No similar work has yet been done on the Australasian and Pacific forms, but no doubt when sufficient material has accumulated and local libraries offer better facilities for reference our species of this genus will be thoroughly revised.

In a paper on the Southern Pacific forms of the Trachipterids, Ogilby(1) endeavoured to correlate many of the described New Zealand and Australian forms. He opposed the contention that similar species could inhabit widely disconnected areas of ocean, and for this reason did not compare the Australasian Trachipterids with northern forms, but qualified his remarks with the observation that, as far as our present knowledge extends, the conditions which regulate animal-life at great depths below the surface of the ocean are everywhere more or less identical as far as temperature is concerned. This being so, there should be no obstacle to the cosmopolitan distribution of similar forms. He admits that certain Australasian Trachipterids may be comparable to the Valparaiso species, *T. altivelis* Kner.

Since Ogilby wrote in 1897, American and Japanese authors have described several species from the Japanese and Californian coasts, and these have a distinct similarity to those found in the Australasian region.

A list of the fifteen occurrences of *Trachipterus* in the Australasian and New Zealand regions, as far as can be compiled from all sources, and arranged in chronological order, follows:—

1873. *T. altivelis* Kner? Recorded by Hutton(6). Dried specimen in the Auckland Museum. (Since lost.)

1876. *T. altivelis* Kner. Identified by Hutton. Specimen in the Otago University Museum, Dunedin.

1881. *T. arawatae* Clarke(8). Type specimen in the Dominion Museum, Wellington.

1881. *T. jacksonensis* Ramsay(3). Type specimen in the Australian Museum, Sydney.

1882. *T. altivelis* Kner. Recorded by Johnston(7). Specimen in the Hobart Museum.

1886. *T. taenia* Bloch & Schn. Recorded by McCoy(5). Three specimens in the National Museum, Melbourne.

1897. *T. jacksonensis polystictus* Ogilby(2). Type specimen in the Technological Museum, Sydney.

1903. *T. taenia* Bloch & Schn. Specimen in the Otago University Museum, Dunedin, from Purakanui.

1905. *T. taenia* Bloch & Schn. Specimen in the Canterbury Museum, Christchurch.

1908. *T. jacksonensis* Ramsay. Caught at Nelson in November, 1908. Only a drawing preserved.

1911. *T. taenia* Bloch & Schn. Specimen in the Otago University Museum, Dunedin, obtained from Port Chalmers.

1914. *T. jacksonensis* Ramsay. Specimen from Waikanae, now in the Dominion Museum.

1915. *T. taenia* Bloch & Schn. Specimen from Chatham Islands, now in the Dominion Museum.

From the Northern Pacific area the following species have been recorded :—

1881. *T. altivelis* = *rex-salmonorum* Jordan and Gilbert(9). Specimen in the United States National Museum.

1894. *T. rex-salmonorum* Jordan and Gilbert(9). Type specimen in the Museum of Leland Stanford, Jr., University.

1901. *T. ishikawae* Jordan and Snyder(4). Type in the Imperial Museum, Tokyo.

1901. *T. ijimae* Jordan and Snyder(4). Type in the Imperial Museum, Tokyo.

1908. *T. misakiensis* Tanaka(11). Four specimens, including the type, in the Zoological Institute Museum, Tokyo.

The following species have been recorded from the west-coast areas of South America :—

1859. *T. altivelis* Kner. Type specimen in Vienna Museum.

1874. *T. weychardti* Philippi. (Arch. f. Nat., xl, 1874, p. 118, pl. iii; described from a photograph.)

These lists are given at length to show the total number of specimens known from the Pacific, and to assist in making comparisons with New Zealand forms. Another specimen, from Station 207, near the Philippine Islands, is recorded by Günther in the " Challenger " Reports, xxii, p. 72, as being similar to *T. repandus* Costa (Faun. Napol., tab. ix). Perhaps this may be included in the Pacific regional forms.

Detailed descriptions of the several specimens are appended.

Trachipterus jacksonensis (Ramsay). Fig. 1.

1881. *Regalecus jacksoniensis* Ramsay, Proc. Linn. Soc. N.S.W., v, p. 631, pl. xx. 1886. *Trachypterus jacksoniensis* Ogilby, Cat. Fishes N.S.W., p. 43. 1901. *Trachypterus ishikawae* Jordan and Snyder, Jour. Coll. Sci. Tokyo, xv, 1901, p. 310, pl. xvii.

B. 6; D. VI–174; C. ?; P. II; V. 2.

Body long and slender, tapering backwards, and not constricted behind the vent. Greatest depth of body, half-way between snout and vent, and slightly greater than the length of head. Height of body at vent is contained 4 times in the distance from the snout. Abdominal profile is conspicuously tubercled. Vent is before the middle of the body. Head, 9 in length. Preorbital wide and rugose. Teeth, 8 in upper jaw, 8 in lower jaw, and 5 in the vomer. Eye large and round, 3¼ in head.

Fins.—The first spines of the dorsal slender and not detached, beginning just behind the posterior margin of the eye. The complete dorsal composed of 174 rays, the longest opposite the vent and about ⅓ the height of the body. All rays smooth. Membrane uniting the rays not attached to the body, and composed of 2 distinct layers. The ventral fin is worn off, but shows the insertion of two spines in a triangular space opposite the base of the pectoral. The pectoral fin is composed of 11 rays, the first being the longest, about 2½ in. The tail is damaged. The lateral line is armed with bony scutes, having short stout spines towards the posterior end. Anterior surface of head and snout an intense black; the rest of the body a mottled grey. Darkened oblique bands, formed by bony scutes, run from the interspaces of the dorsal rays half-way to the lateral line.

The specimen, a female, measures approximately 6 ft. in length, and was found on the sandy beach at Waikanae, near Wellington.

A comparison of this description with Ramsay's original notes indicates that the species are identical. *T. ishikawae* Jordan and Snyder is synonymous with this species.

Ogilby(2), in summarizing the occurrence of *Trachipterus* in the Australasian region, includes under the name *jacksoniensis* specimens recorded by McCoy(3) as *T. taenia* and those of Hutton and Johnston as *T. altivelis.* He considers *T. jacksonensis* to be the adult form of *T. taenia* and *T. altivelis,* the dark markings being merely indicative of immaturity, yet he gives subspecific rank to a specimen he describes, differing, as he states, from *T. jacksonensis* in having the head and body dappled.

Unless the adult forms of this family lose the granulations on the fin rays, and also the spiny tubercles at the base of the spines, there can be no identity of *T. taenia, T. altivelis,* or *T. polystictus* with a form like *T. jacksonensis,* which has no spinules or granulations on the spine rays. As far as at present known, no radical change takes place on the surface of the fin rays with increasing age.

I have not seen either of Hutton's specimens of *T. altivelis,* but it is extremely probable that he based his identification on the description in Günther's catalogue, and would have noticed whether the fin rays were granular or not. As *T. altivelis* or some closely allied species seems to be the most common New Zealand form, it may be assumed that the granulations were present in Hutton's specimen.

The specimen under consideration appears to be the second recorded occurrence of *T. jacksonensis* in the Australasian region.

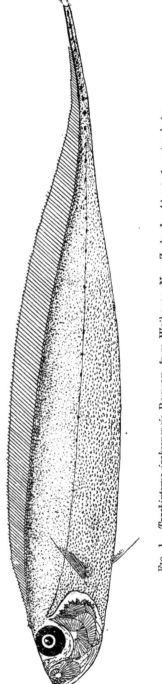

Fig. 1.—*Trachipterus jacksonensis* Ramsay, from Waikanae, New Zealand. About ⅒ natural size.

Trachipterus jacksonensis Ramsay? (Juvenile). Fig. 2.

In November, 1908, a ribbon-fish, preserved in formalin, was for-warded to the Dominion Museum from Nelson, South Island. I was present when it arrived, and remember the extraordinary development of the dorsal and ventral fin rays. Unfortunately, I can find no trace of the specimen, but from a careful drawing made at the time I have reason to think that it was a young example of *T. jacksonensis*. The specimen was

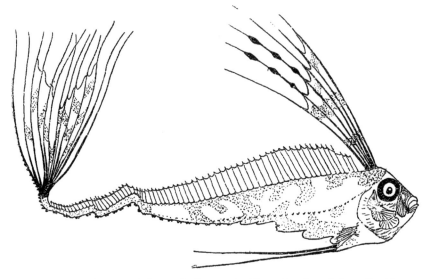

Fig. 2.—*Trachipterus jacksonensis* Ramsay? From drawing of juvenile fish taken at Nelson in 1908. About ⅔ natural size.

13 in. long, and exhibits distinctive characters, as shown in the accompany-ing illustrations, especially in the crenulations of the abdominal profile. As it also bears a strong resemblance to Jordan and Snyder's figure of *T. ijimae* (now stated to be the young of *T. ishikawae*), and as I consider *T. ishikawae* to be identical with *T. jacksonensis*, it may be that this is a young specimen of *T. jacksonensis*.

Trachipterus trachypterus (Gmelin). Fig. 3.

1873. *Trachypterus altivelis* Hutton, Trans. N.Z. Inst., vol. 5, p. 264.
1876. *T. altivelis* Hutton, Trans. N.Z. Inst., vol. 8, p. 214.
1882. *T. altivelis* Johnston, Proc. Roy. Soc. Tasmania, p. 123.
1886. *T. taenia* McCoy, Prod. Zool. Victoria, dec. 13, pl. cxxii.

B. 6; D. VI–175; C. 10; P. 11; V. 8.

Body long and tapering, slightly constricted behind the vent; greatest depth of body just behind the head, contained 5¾ times in the total length (excluding the caudal); abdominal profile studded with a double series of small tubercles; vent situated a little before the middle of the body. Head short and the muzzle truncated; jaws protractile; eye situated near the upper profile, and a little behind the middle; all the bones of the head thin, and radiatingly ridged. Eight teeth in the upper jaw, 8 in the lower jaw, and 3 in the vomer. First 6 rays of the dorsal detached, spinous, and fragile; probably elevated; separated from the continuous dorsal by membrane; hinder portion of the dorsal composed of 175 rays, attaining their greatest height opposite the vent, and being contained 2½ times in depth of body; all rays spinous, and having a spiny tubercle at the base; membrane connecting the rays not attached to the body. Ventral fins composed of 6 to 8 rays, the first being the longest and strongest, with spines on anterior surfaces. The pectoral fins are damaged, but show basal portion of 11 rays, the first being the longest and strongest; all rays roughened. Caudal fin composed of 10 rays, directed obliquely upwards and backwards; all rays granulated and connected by membrane; signs of a lower rudimentary caudal. Lateral line composed of a row of conical spines, directed forwards and more strongly developed towards the posterior portion.

Colour generally bright silvery; transverse darkened lines correspond with neural spines; 3 large round spots above lateral line, and a fourth near the abdominal edge, a little behind the first on the back. Nuchal crest and top of head darkened.

Measurements.—Diameter of eye, 26 mm.; depth of body at vent, 107 mm.; greater length of dorsal rays, 45 mm.; height of body at insertion of ventral fin, 115 mm.; proportion of height to length (excluding caudal fin), 1–6·38; longest caudal rays, 140 mm.

Locality.—Chatham Islands.

Professor W. B. Benham, Curator of Otago University Museum, has kindly placed at my disposal two specimens from the neighbourhood of Dunedin Besides being able to examine the specimens themselves, I have the benefit also of his original notes made when the fishes were fresh. I beg to thank Professor Benham for his kind assistance in

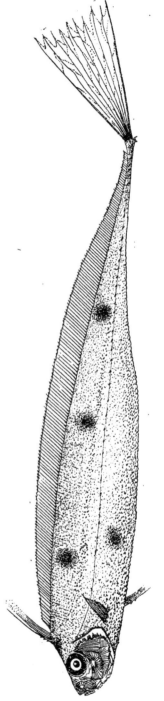

FIG. 3.—*Trachipterus trachypterus* (Gmelin), from Chatham Island. About ¼ natural size.

allowing me to use his notes and offer descriptions of his specimens. These specimens are distinctly similar to the one described from Chatham Islands. The first was taken by Mr. Ewart at Purakanui, near Dunedin, on the 11th November, 1903, and was in good condition, except for having the caudal fin and part of the tail missing.

The following description is based on notes by Professor Benham and myself (fig. 4):—

B. 6; D. VI–168; C.—; P. II; V.—.

Body long and tapering, constricted behind the vent; greatest depth at a point just posterior to the pectoral fin, contained 5 times in total length, excluding the caudal; abdominal profile studded with tubercles; vent situated 228 mm. from tip of snout; height of body at vent, 73 mm.; width of tail, 7 mm. Head short and truncated. Eye a little behind the middle of the head; diameter of the eye, 17 mm. I was unwilling to examine the dentition for fear of damaging the specimen. First 5 rays of dorsal elevated and detached; spinous and granulated; in length about ⅜ height of body. Remainder of dorsal composed of about 166 rays, all granulated and having a spiny tubercle at the base; longest rays opposite the vent, being about ½ height at vent; membrane connecting

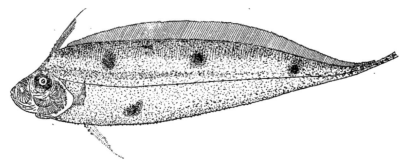

Fig. 4.—*Trachipterus trachypterus* (Gmelin), from Purakanui, New Zealand,
About ⅔ natural size.

rays apparently connected to the body. The ventral is worn off. The pectoral is composed of 11 short, granulated rays, connected by membrane, inserted horizontally. Tail absent. The lateral line rather above the middle of the body—viz., 40 mm. below base of dorsal fin and 45 mm. above ventral margin of body at level of first black spot; armed with spiny tubercles directed forwards, but comparatively small and weak as compared with other specimens; more strongly developed towards posterior portion of tail. Colour silvery, with 3 black spots on dorsal surface and 1 near ventral margin. Spot A: Its middle is 65 mm. from anterior end of base of predorsal fin; spot B is 100 mm. from centre of A; spot C is 100 mm. from centre of B—it is less well defined than A or B. Top of head and crest, black; dorsal fin, red. Eye, silver with a pink iridescence.

The second specimen was caught at Port Chalmers in November, 1911, and presented to the Otago University Museum. It much resembles that described from Purakanui, but differs in having a fourth black spot on the tail in addition to three on the sides of the body.

B. 6 ; D. VI-168 ; C. 9 ; P. — ; V. 4. (Fig. 5.)

Body long and tapering, constricted behind the vent ; greatest depth at position of pelvic fin is equal to one-fifth of length, excluding the caudal ; depth at vent is greater than length of head ; abdominal profile is tuber-culate ; vent situated about the middle of the body. Mouth inclined upwards ; gape nearly vertical ; jaws very protractile; teeth, 6 in upper jaw and 4 in the lower. The first dorsal spine nearly equal to the length of the profile ; it is serrated, granular, and elevated. The insertion of 5 spines can be counted, and indications point to there being a sixth. Apparently the spines are connected by membrane with the remaining dorsal. Continuous dorsal composed of 168 rays, the longest being opposite the vent and $1\frac{7}{8}$ in. in height ; all rays connected by membrane, not attached to the body. The ventral and pectoral fins are not represented, as the abdominal cavity has been ruptured. The caudal fin is well preserved, and shows two distinct lobes ; the upper lobe carries 9 rays, directed upwards and backwards. The outer rays are longest and strongest ; all rays are granulated, and connected by membrane ; anterior ray, 5 in. long ; posterior ray, $6\frac{1}{8}$ in. long ; intervening rays less in height. Below the caudal fin is a well-defined lobe with 2 rudimentray rays, $\frac{1}{4}$ in. long, directed

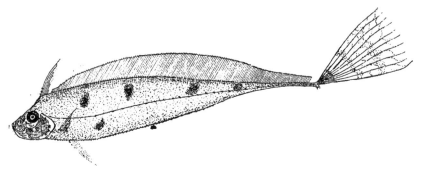

Fig. 5.—*Trachipterus trachypterus* (Gmelin), from Port Chalmers, New Zealand. About $\frac{3}{10}$ natural size.

backwards and downwards ; the posterior surface of this lobe terminates in a strong spine directed forwards. Lateral line armed with bony scutes, directed forwards, and stronger towards the posterior ehd. Both generally bright silver ; 4 black spots above the lateral line, 2 on the body and 2 on the tail ; a fifth spot on the side of the abdomen, below the lateral line and a little behind the 1st dorsal ray. Top of head and crest, deep black ; fins, red.

The fifth specimen I have the opportunity of describing was kindly lent by Mr. R. Speight, Curator of the Canterbury Museum. It is labelled, " *T. altivelis*, New Zealand," and bears the registered number P. 441-0. The fish was in a remarkably good state of preservation, the only damage being the loss of the anterior dorsal fin rays (see fig. 6).

B. 6 ; D. VI-165 ; C. $\frac{8}{5}$; P. 11 ; V. 7.

Body long and tapering, constricted behind the vent ; greatest depth of body at the insertion of the ventral fin, contained about $4\frac{1}{4}$ times in length, excluding the caudal ; depth at vent equal to the length of head ; vent situated 158 mm. from tip of snout ; abdominal profile studded with tubercles. Head 43 mm. in length, short and truncated; mouth inclined upwards; gape nearly vertical; eye situated in posterior third of head; 13 mm. in diameter. Six teeth in the upper jaw and 4 in the lower.

Fins.—Anterior portion of dorsal fin composed of 6 elevated, spiked, and granulated rays, united by membrane; remainder of dorsal composed of 165 rays, attaining their greatest length opposite the vent, being there 35 mm. long; all rays granular, and having a spiny tubercle at the base; membrane uniting the rays not attached to the body; no signs of articulations in the rays. The caudal fin is well developed, composed of 8 rays, united by membrane, the longest being 99 mm. long; the two outer rays are strong and granulated; whole fin directed upwards and backwards. On the inferior portion of the caudal lobe there is a well-defined rudimentary adipose caudal fin, as is illustrated in fig. 6. Besides having the hair-like

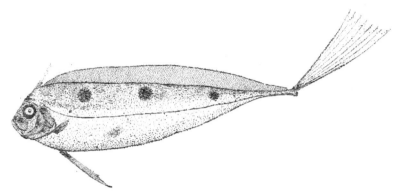

Fig. 6.—*Trachipterus trachypterus* (Gmelin), in Canterbury Museum.
About $\frac{3}{10}$ natural size.

appendages found in many specimens, there are 3 distinct membrane-connected spines, curved upwards and directed backwards, situated below these appendages. The ventral fins are well preserved. Each fin composed of 7 rays, apparently branched and fissile; the anterior rays being armed with 2 spines directed forwards; all rays roughened and granular; longest ray equals the depth of the body at the pectoral. The pectoral fin is small and delicate, composed of 11 rays, inserted horizontally just below the apex of the outer gill-cavity; the 1st ray strongest and curved; length of pectoral about equals diameter of the eye. The lateral line armed as usual, but showing a spike, directed forwards at the confluence of the lateral lines on the caudal lobe.

Three black spots on the upper surface, the first opposite the 25th dorsal, the second opposite the 60th, and the third opposite the 94th dorsal; a fourth spot on the abdominal surface about opposite to the 46th dorsal. All spots and markings on the head of brownish colour, due to the action of preservatives. Rest of body bright silvery.

COMPARISONS OF NEW ZEALAND TRACHIPTERIDS.

The descriptions of the last three specimens from the New Zealand coast show that they are closely allied, if not identical. That from Pura-kanui agrees almost exactly with the Chatham Island individual, not only in regard to the position and size of the spots, but in the relative proportions of the body. The Purakanui specimen is deeper, the body being contained 5 times in the total length, while the depth of the Chatham Island specimen is contained $5\frac{3}{4}$ times therein (excluding the caudal fin in both cases). This may be accounted for by the fact that the smaller specimen is young, while the Chatham Island fish is probably the adult form. In other respects —viz., shape of head and position of eye, nature of fin rays and spines, position and shape of spots, texture of integument, and other characters —the fishes are identical.

The specimen from Port Chalmers has a fifth spot not present in the other examples.

In discussing the variations in *T. taenia*, Emery writes, " The distribution of spots is not constant, and is therefore fallacious for specific purposes. . . . In one example I have seen four spots in the dorsal series."

The three specimens under consideration illustrate successive stage-growths of the same species, the Chatham Island specimen the mature form, the Portobello example being more developed than the juvenile Purakanui fish.

Professor Benham has drawn my attention to the similarity of Clarke's drawing of *T. arawatae* to Emery's illustration of *T. spinolae* Cuv. & Val. The main difference lies in the fact that *T. spinolae* has 3 spots that are not represented in *T. arawatae*.

The outstanding features of this last-named species, the type of which I have examined, are the rudimentary dorsal and anal fins. The ventral rays are extremely long, and, with other characters, indicate that the specimen is young.

McCoy's smaller figured specimen of *T. taenia* shows an apparently rudimentary anal fin. The larger example, admitted to be the same species, shows only traces of such a fin. Clarke made an error in separating the rudimentary dorsal from the continuous dorsal, for the specimen shows no signs of a break.

Throughout the genus *Trachipterus* there are indications pointing to the degeneration and disuse of the anal and ventral fins, and the ultra-development of the dorsal appendages. Even the caudal fin is assuming an almost vertical position, and usurping the functions of a dorsal fin. It is only in the juvenile forms that any signs of the existence of ventral and anal appendages can be recognized. The cause for the degeneration of these fins must remain obscure until we know more about the life-history of these extraordinary fishes.

In discussing his specimens of *T. taenia*, McCoy(5) observes that " the young are deeper and shorter in proportion than the old, and consequently specific differences founded on the greater number of times the length of the head or the depth of the body are contained in the whole length are not to be trusted for specific characters when the length of the specimens is different." It may be noted that McCoy's young specimen of *T. taenia* has no black spots, as is also the case in *T. arawatae*.

Emery(1) demonstrated that in *T. taenia* the fin rays commence to grow when the young is about 6 mm. long, and continue to lengthen till the fish is about four times that size, after which period a shortening of the rays takes place.

Jordan and Snyder(12) now think that *T. ijimae* is a young specimen of *T. ishikawae.* On consulting the original descriptions and the two figures, one would pronounce the specimens to be distinct, for no two fishes ever looked more unlike. These ichthyologists must have additional evidence in favour of the view expressed, although they do not state on what grounds it is based. *T. ijimae* has an extraordinary development of the pre-dorsal fin rays, while *T. ishikawae* has no pre-dorsal appendages whatsoever. The conditions show what remarkable and incomprehensible changes take place in the life-history of these fishes. No excuse is, therefore, needed for advancing the opinion that *T. arawatae* Clarke is the young of *T. taenia.*

COMPARISONS WITH AUSTRALIAN TRACHIPTERIDS.

Excluding *T. jacksonensis*, which is obviously a distinct species, the only Australian records I can find are *T. taenia* by McCoy, Victoria ; *T. altivelis* by Johnston, Tasmania ; and *T. polystictus* by Ogilby, New South Wales.

Johnston(7) only briefly describes the specimen from Tasmania, and Ogilby(2) states that it should be associated with those from Victoria.

In recording the occurrence of *T. taenia*, McCoy(5) gives descriptions and excellent figures, and there can be no doubt that the Victorian examples are co-specific with those found round the New Zealand coast.

With *T. polystictus* Ogilby it is more difficult to deal. Ogilby published a description, but gave no figure. His specimen differs in the main from *T. taenia* McCoy in having no spiny granules on the outer rays of the caudal fin and a markedly different coloration, but in many other characters the specimens agree. Why Ogilby should regard his specimen as being closely allied to *T. jacksonensis* is not clear, as it is most certainly more nearly related to *T. taenia* McCoy.

COMPARISONS WITH OTHER REGIONAL FORMS.

T. rex-salmonorum, from the Californian coast, as described by Jordan and Gilbert(9), differs from the local species in the position and size of the black spots. In nearly all other respects there is close agreement. As previously stated, Emery showed that the position and number of spots cannot əq relied on for specific purposes. Perhaps further investigations will show identity between the Californian and New Zealand species, for the two regions have much in common zoologically.

With *T. altivelis*, described by Kner(15) from Valparaiso, one local specimen shows close similarity. Jordan and Gilbert originally identified a Californian fish as *T. altivelis*, but in 1894 named it *rex-salmonorum*. The reasons for so doing were based on the angle of the nuchal crest, the height of the dorsal and ventral fins, the texture of the skin, and the size and position of the black spots—all characters, however, that depend on the state of development, and, as previously mentioned, change greatly during the growth of the fish.

NOMENCLATURE OF AUSTRALASIAN TRACHIPTERIDS.

Until more evidence has been collected regarding the life-history of the Southern Pacific forms of these fishes all efforts at correct nomenclature must be regarded as tentative only.

In consideration of the wide distribution of these deep-sea fishes, previous writers have compared their specimens with those found chiefly in the Mediterranean seas. It has been shown that these Mediterranean species—*T. iris, T. spinolae,* and *T. taenia*—should be correctly identified with Gmelin's *Cepola trachyptera.* Since *Cepola* of Gmelin(13) had to give way to *Trachipterus* of Gouan(14), the correct name for the type of the genus should be *Trachipterus trachypterus.*

After comparing the descriptions of the fishes from Chatham Island, Purakanui, Port Chalmers, Victoria, and Tasmania, it is concluded that they are all referable to *T. taenia* Bloch & Schn., the conclusion arrived at by McCoy(5) when describing the Victorian example. As previously stated, *T. taenia* Bloch & Schn. must be looked on as a synonym of Gmelin's *Trachypterus trachyptera,* and must now be known as *Trachipterus trachypterus* (Gmelin).

Ogilby's specimen from Newcastle, to which he gave the subspecific name of *polystictus,* is evidently very closely related to the fish described by McCoy from the Victorian coast. If entitled to the rank of a subspecies it must certainly be referred to *Trachipterus trachypterus,* and would therefore be known as *Trachipterus trachypterus* var. *polystictus* Ogilby.

A table of comparative measurements of four specimens of *T. trachypterus* is appended. (All measurements are in millimetres.)

	Chatham Islands (fig. 3).	Purakanui (fig. 4).	Port Chalmers (fig. 5).	Canterbury Museum (fig. 6).
Total length, excluding caudal fin	730	375	461	274
Depth at pectoral fin	115	92	90 ?	62
Length of head	94	73	59	43
Depth at vent	107	73	73	42
Tip of snout to vent	371	228	230	158
Dorsal fin rays opposite vent	45	28	45	35
Caudal fin rays	140	?	151	99
Pectoral fin rays	?	20	?	14
Diameter of eye	26	17	16	13
Ventral fin rays	?	?	?	61
Proportion of height to length	6·38	4·01	5·1	4·42

LIST OF AUTHORS QUOTED.

(1.) Emery, C. "Contribuzioni all' ittiologia: Le metamorphosi del *Trachypterus taenia.*" Atti R. Accad. Lincei, Roma, Mem., 3, 1879, pp. 390–97.

(2.) Ogilby, J. D. "Remarks on the South Pacific Species of *Trachypterus* Gouan." Proc. Linn. Soc. N.S.W., xxii, 1897, pp. 646–59.

(3.) Ramsay, E. P. "On a New Species of *Regalaecus* from Port Jackson." Proc. Linn. Soc. N.S.W., v, 1881, p. 631, pl. xx.

(4.) Jordan, D. S., and Snyder, G. O. "Descriptions of Nine New Species of Fishes contained in Museums of Japan." Journ. Coll. Sci. Tokyo, xv, 1901, p. 310.

(5.) McCoy, F. "*Trachypterus toenia* (Bloch): the Southern Ribbon-fish." Prod. Zool. Vict., dec. 13, ii, 1886, pl. 122.

(6.) Hutton, F. W. "Contributions to the Ichthyology of New Zealand." Trans. N.Z. Inst., vol. 5, 1873, p. 264; vol. 8, 1876, p. 214; vol. 22, 1890, p. 281.

(7.) Johnston, R. " On *Trachypterus altivelis* from Spring Bay, Tasmania." Proc. Roy. Soc. Tas., 1882, p. 123.

(8.) Clarke, F. " On a New Species of *Trachypterus : T. arawatae*." Trans. N.Z. Inst., vol. 13, 1881, p. 195.

(9.) Jordan, D. S., and Gilbert, C. H. " Description of a New Species of Ribbon-fish, *Trachypterus rex-salmonorum*, from San Francisco." Proc. Calif. Acad. Sci. (2), iv, 1894, p. 144.

(10.) Snyder, J. O. " Description of *Trachypterus seleniris*, a New Species of Ribbon-fish from Monterey Bay, California." Proc. Acad. Nat. Sci. Phila., vol. 60, 1908, p. 319.

(11.) Tanaka, S. "Notes on some Japanese Fishes." Journ. Coll. Sci. Tokyo, vol. 23, 1908, art. 7, p. 52.

(12.) Jordan, D. S., and Snyder, J. O. " On a Collection of Fishes made by Mr. Alan Owston in the Deep Waters of Japan." Smith's. Miscell. Coll., vol. 45, 1904, p. 240, pl. lxiii.

(13.) Gmelin, J. F. " Linnaeus Systema Naturae." Ed. 13, cura J. F. Gmelin, Lips., 1789, p. 1187.

(14.) Gouan, A. Hist. Piscium. Argent., 1770, p. 104.

(15.) Kner, R. " Ueber *Trachypterus altivelis* und *Chaetodon truncatus* n. sp." Sitzungsler d. math.-wiss. Cl. d. K. Akad. d. Wiss., Wien, Bd. xxxiv, 1859, pp. 437–45.

List of Papers referred to and pertaining to the Genus Trachipterus.

Macleay, W. Proc. Linn. Soc. N.S.W., vi, 1882, p. 55, and ix, 1884, p. 120. (List of names.)

Jordan, D. S., and Gilbert, C. H. Proc. U.S. Nat. Mus., 1881, p. 52.

Ogilby, J. D. Catal. Fishes N.S.W., 1886, p. 43.

Lucas, A. H. S. Proc. Roy. Soc. Vict. (2), ii, 1890, p. 32.

Hutton, F. W. Trans. N.Z. Inst., vol. 22, 1890, p. 281. (List of names.)

Gill, T. Mem. Acad. Nat. Sci. Wash., vi, 1894, p. 120.

Jordan, D. S., and Gilbert, C. H. Proc. Calif. Acad. Sci. (2), iv, 1894, p. 144.

Waite, E. R. Mem. N.S.W. Nat. Club, ii, 1904, p. 54.

Poey, F. Mem. Cuba, ii, 1861, p. 420.

Günther, A. C Cat. Fish, iii. 1861, p. 303.

ART. XXXVI.—*On the Much-abbreviated Development of a Sand-star* (Ophionereis schayeri ?).—*Preliminary Note.*

By H. B. KIRK, M.A., Professor of Biology, Victoria College, Wellington.

[*Read before the Wellington Philosophical Society, 27th October, 1915.*]

Plates XXVII, XXVIII.

I OBSERVED at Island Bay in August of this year several clusters of small round eggs attached to the underside of stones. Some of these were brought in on the 24th August, and observed from time to time to determine, if possible, to what animal they belonged. Only slight attention was paid to them, but they were probably not left unobserved for more than two days together. To my great surprise, there began to emerge from the eggs of one group small Echinoderms having the general appearance of Asteroids with a perfectly formed disc that had the barest suggestion of arms. Each had the primary tube-feet developed beyond the extremity of the radial groove. The tube-feet had somewhat club-shaped extremities, without suckers, but with a number of stiff, bristle-like processes. Each of the five points of the disc consisted of a single, grooved terminal plate like that of *Ophionereis schayeri* Müller and Troschel. There could be no doubt that the young animals were Ophiuroids. *O. schayeri* is very common in the neighbourhood. I have often examined the bursae, and have never found them to contain embryos.

I carefully observed the remaining capsules, and obtained others from the same locality ; but the time for the very early stages had gone by. From these later observations, and from the earlier but less thorough observations, I am able to give the following preliminary account of this very extraordinary instance of Echinoderm development.

The eggs are spherical, 0·5 mm. in diameter, each with a perfectly transparent, thin, but extremely tough chitinous envelope. They are deposited in irregular clusters of from 10 to 100 or·more. The embryo does not occupy the whole of the space within the envelope, but is surrounded by colourless, apparently mucilaginous, matter. The envelope was so tough that I could not tear it away satisfactorily, the embryo getting crushed in the process. The embryo is of a buff or pinkish-brown colour, and is so opaque that no internal structure is to be observed. The pigment is very refractory to all the ordinary solvents. For knowledge of internal structure sectioning is likely to be the only process, and for this I shall have to obtain fresh material next year.

The earliest embryos observed were apparently late gastrulae, in which the blastopore had closed or become indistinguishable. No movement was at any time observed until the disc and tube-feet were formed. I do not think that any external cilia were ever developed, or that there was any stage that could be recognized as corresponding at all closely to a typical larval stage of any Echinoderm. My present view is that development is absolutely direct, but I cannot yet speak with certainty.

The first movement observed was that of the tube-feet. At the time these appear the disc is concave on the ventral surface, the tube-feet showing as rounded structures where they project beyond the margin of the cup. As the disc flattens somewhat, pressing the egg-envelope with its incipient rays, the envelope is drawn flatter dorsally and ventrally. Against the ventral portion of the envelope the tube-feet are worked with a stamping or pushing movement, much as an infant works with its hands at the mother's breast. Presently the envelope gives way, and the young animal struggles out. The disc soon completes its ventral flattening. Locomotion is effected by the animal raising itself upon its almost-rigid tube-feet and toddling, as it were, occasionally falling forward. The movements suggest those of a fat puppy making its first real attempt to walk. From under the grooved terminal plate of each arm these extends a blunt, rigid tentacle, capable of being retracted.

As there is yet no marine laboratory here, and I wished to keep these young animals alive as long as possible, I placed a number in a specially contrived glass cage, which I sank in a rock-pool, concealing it by means of stones. When I visited the pool three weeks later I found that the stones had been lifted out and the cage taken away.

In the laboratory I was able to keep some alive and under observation for thirty days. The mortality was, however, great, and only two specimens reached the stage at which a second pair of tube-feet developed.

For the photo-micrographs that illustrate this note I am greatly indebted to the kindness of Mr. P. G. Harwood, of Auckland.

ART. XXXVII.—*On the Gonoducts of the Porcupine-fish* (Dicotylichthys jaculiferus *Cuvier*).

By H. B. KIRK, M.A., Professor of Biology, Victoria College, Wellington.

[*Read before the Wellington Philosophical Society, 22nd September, 1915.*]

IN the Teleost fishes generally the gonoducts, when distinct ducts are developed, either open together to the exterior, or both open into the base of the ureters or of the combined urinary duct. I have been unable to find a recorded case in which the two gonoducts discharge separately to the exterior, the condition that exists in *Dicotylichtys jaculiferus* so far as the male is concerned.

The mesonephridia are large, and in each the ureter arises from the posterior end, toward the ventral aspect. The two ureters converge as they run backwards, and about 4 cm. before they unite they become closely bound together by connective tissue (removed in the preparation figured). Their union forms the urinary bladder, which discharges by a median opening into a shallow urino-genital sinus situated just behind the anus.

The testes are large, and their ducts are very short indeed. Each has its own opening, difficult to distinguish, into the urino-genital sinus. The two openings are slightly in front (ventrad) of the urinary opening, and

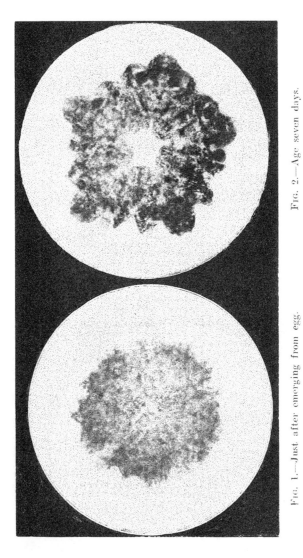

Fig. 1.—Just after emerging from egg.

Fig. 2.—Age seven days.

Ophionereis schayeri.

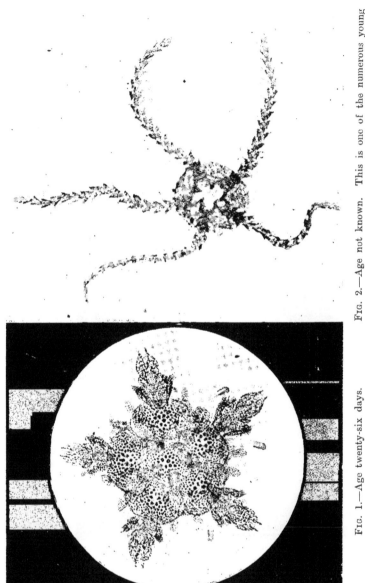

FIG. 2.—Age not known. This is one of the numerous young specimens found in rock-pools in October.

FIG. 1.—Age twenty-six days.

OPHIONEREIS SCHAYERI.

are situated one on either side of it. By opening the testes a probe can be passed along the duct, and I have verified the observation by drawing a thread through each duct.

In the female the ovisacs have very short oviducts of considerable width, which open separately at the very mouth of the urinary opening.

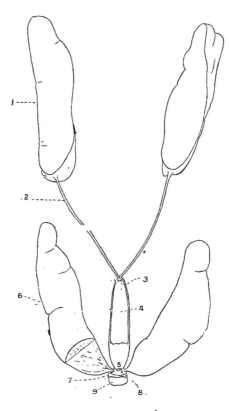

Dicotylichthys jaculiferus : Male urinary and genital organs dissected from dorsal aspect 1 left kidney; 2, left ureter; 3, opening of left ureter into urinary bladder; 4, urinary bladder, part of dorsal wall removed; 5, opening of urinary bladder into urino-genital sinus; 6, left testis, part of posterior end removed; 7, opening of left sperm-duct into urino-genital sinus; 8, opening of right sperm-duct into urino-genital sinus; 9, anus.

For several of the specimens of *Dicotylichthys* examined I am indebted to Professor Prince, of Ontario, and Mr. T. Anderton, of the Portobello Marine Fish-hatchery, the specimens having been obtained during Professor Prince's investigations on the New Zealand coast.

— — — —

ART. XXXVIII.—*Notes on New Zealand* Polychaeta *(II).*

By Professor W. B. BENHAM, D.Sc., F.R.S., Hutton Memorial Medallist.

[*Read before the Otago Institute, 7th September, 1915.*]

Fam. EUNICIDAE.

Eunice pycnobranchiata McIntosh, "Challenger" Reports, xii, 1885, p. 294.

> *Eunice antennata* Ehlers, Neuseeland. Annelid., ii, 1907, p. 12 (*nec E. antennata* Savigny).

AMONGST the Eunicids in the "Endeavour" collection from the Australian seas which was submitted to me for description were a good number of specimens of this species, and in comparing them with the species from our coasts it became evident that they are identical with the worm which Ehlers has recorded under the title "*E. antennata* Sav." from specimens sent to him by me, of which I retain duplicates. But this identification does not accord with Crossland's* investigation into the true *E. antennata* from the Red Sea, for the latter worm has golden acicular chaetae, and the gills meet almost across the back where fully developed; whereas in the New Zealand worms these chaetae are black, and the gills are small, as they are in McIntosh's species, with which it agrees in other respects. For a fuller discussion of the matter see my account† of the "Endeavour" *Polychaeta,* pp. 216 and 224.

When writing that report I had forgotten the fact that Ehlers had identified his *E. antennata* with Quatrefages' *E. gaimardi.* But of this I feel sceptical, for when I was engaged in working out the New Zealand Annelids, some twelve to fourteen years ago, I tabulated the characters given by Quatrefages to his two species from New Zealand—namely, *E. gaimardi* and *E. australis*—and compared these with our two common species of *Eunice.* I came to the conclusion at the time—which I see no reason now to alter—that it is impossible from the data given to identify either of our two common species with either of these two descriptions.

The only difference which may be regarded as of importance referred to by Quatrefages is the character of the jaws. In *E. gaimardi* the upper jaw—*i.e.,* forceps, or "*Zangen*" of Ehlers—is described as "*gracilis.*" The large dental plate (his upper jaw) has 6 teeth, and the denticula—*i.e.,* "*Sageplatte*"—are undulations rather than teeth. On the other hand, he states that the upper jaw of *E. australis* is "robust," the dental plate has 10 teeth, and the denticula are dentate.

Ehlers (p. 31) says of *E. australis* that "der linke Zahn hat 5, der rechte 6, die unpaare Sageplatte, 10," &c., and describes the forceps as slender—"die Zangen schlanke"—which can scarcely be a translation of Quatrefages' words, "*maxillae superae robustae.*" Ehlers' "slender forceps" would equally apply to those of *E. pycnobranchiata.*

---

* Proc. Zool. Soc., i, 1904, p. 316.

† Biolog. Results of Fishing Experiments of F.I.S. "Endeavour," 1909–14, vol. iii. Commonwealth of Australia : Fisheries Department, 1915.

It seems to be mere guesswork to go further than to acknowledge that Quatrefages probably had before him our two common species; but to decide which of his two names apply to our two species, without a re-examination of the types, seems impossible. For instance, in *E. pycnobranchiata* the gills extend practically throughout the length of the body, while in the other species, which Ehlers identifies as *E. australis*, they are limited to some 20 to 30 segments. But Quatrefages says nothing as to the extent of the gilled region ; he merely states that the gill commences on the 6th segment in the one and on the 7th in the other.

I fail to understand how Ehlers has managed to sift the two species from the brief diagnoses given. I am not aware whether any zoologist has re-examined Quatrefages' species in the Paris Museum, or whether Ehlers himself has had access to them. But, so far as the records go, it seems to me that in the meantime it would be better to adopt McIntosh's specific name for this species, as he gave a good account of it, accompanied by figures.

As to the worm called by Ehlers *E. australis*, I must defer any remarks to some future article.

Localities.—Foveaux Strait, 17 fathoms, on the oyster-bed ; Tasman Bay ; Pegasus Bay ; Timaru, 10–20 fathoms ; Massacre Bay.

Distribution.—Bass Strait ; Tasmanian waters ; South Australia ; New South Wales ; in addition to New Zealand.

Fam. APHRODITIDAE.

Physalidonotus thomsoni sp. nov. Figs. 1–5.

The genus was founded by Ehlers in 1904* for a large Polynoid which is fairly common on our shores, and described many years ago under the name of *Aphrodita squamosa* by Quaterfages, and later by T. W. Kirk as *Lepidonotus giganteus*.† Till recently the genus was represented only by this species, but Moore had described two worms under the generic name *Lepidonotus* from the coast of Japan which undoubtedly belong to Ehlers' genus ; and the " Endeavour " collection contained four new species. The present species I name after Mr. George M. Thomson, who has done so much good work in natural history and for zoology in New Zealand, especially by the establishment of the Portobello Fish-hatchery. It serves also to recall the fact that his son Malcolm worked out the anatomy of *P. squamosus*.‡

The new species was found some years ago by the late Mr. A. Hamilton in Dunedin Harbour, though under what circumstances—whether on shore or in a dredge—I do not know. For a long time I regarded it as the young of the common species, than which it is much smaller ; but closer examination recently shows that it is quite distinct from it.

P. thomsoni is short and relatively broad, measuring 18 mm. in length by 10 mm. over the elytra and 12 mm. over the ventral chaetae. These are of the usual rich golden-brown colour.

The elytra are nearly white, with pale-brown star-like tubercles with 8–10 rays. The tops are flat or feebly convex. These tubercles are sparsely scattered over the exposed surface, more numerous and rather

---

* Ehlers, Neuseeland. Annelid., p. 9.
† For a fuller history see my report in " Endeavour "⸮*Polychaeta*, p. 185.
‡ Proc. Zool. Soc., 1900, p. 974.

13*

larger in the region of the areola, with an irregular row of rather smaller ones near the posterior and external margins, and between these two rows are a few intermediate in size (figs. 1 and 2). Seen under the microscope, the concealed area, which appears smooth to the naked eye, is found to be covered with rounded tubercles, constricted at their bases, and terminating in 2–3 points (fig. 3). The exposed surface also between

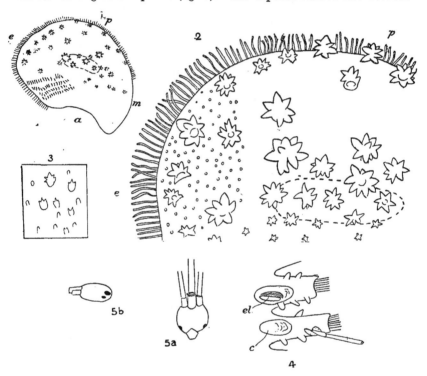

Fig. 1.—An elytron, enlarged, drawn under a dissecting-lens, freehand. *a*, the anterior margin; *e*, external; *m*, the mesial, or internal; *p*, posterior margin. Between the letters *a* and *e* are seen the rows of cylindrical hair-like papillae on surface.

Fig. 2.—A portion of the posterior and external margins (camera, × 10). *e*, the external margin. The outline of the "areola," or area of attachment, is indicated by the dotted lines. The whole surface of the elytron is covered with small tubercles, which are shown only towards the external margin.

Fig. 3.—A portion of the anterior surface of an elytron (× 17), showing the gradual development of the spinose tubercles from the simpler conical ones. This seems to be characteristic of the genus.

Fig. 4.—Two consecutive parapodia (enlarged), showing the arrangement of the papulae. *c*, cushion at the cirriferous segment; *el*, elytrophore.

Fig. 5.—The prostomium (× 4) from above and from the side, merely to show the position of the eyes

the large star-shaped tubercles is covered with numerous small rounded transparent smooth papillae. Further, the anterior region of the exposed surface a little anterior to the excavation of the margin is densely covered with long hair-like papillae resembling those of the marginal fringe.

The parapodial papulae, or gills, are few in number, and rather difficult to detect owing to the poor state of preservation of the worm, for the cuticle readily separates from the underlying body-wall; but so far as I can make out from examination of several feet the arrangement is as follows (fig. 4): On the cirriferous feet there is 1 short rounded papula about midway along the anterior face; in one case I noted a second smaller papula near the base of the foot. On the posterior face there are 2 papulae, one close to the cushion, the second close to the base of the cirrus. On the elytriferous feet the anterior face carries 2 and the posterior 3 papulae, of which latter one springs from below the cushion, the second about midway along the foot, the third on the upper surface just behind the notopodial lobe.

The chaetae present no special features; they are quite like those of other species, except that one or two of the most ventral chaeta are smaller than the rest (which is no unusual thing), and present the same constriction below the bearded region that I have described and figured for *P. pauci-branchiatus*.* (In the figure of this chaeta the constriction is exaggerated; it is to be remarked that it is less noticeable in glycerine mounts than in Canada balsam; and perhaps I have in my account laid too much stress on this feature.)

In the present species the long fringes of the " beard " are broken or worn away, as is the case in most of the specimens of *P. squamosa*. It is perhaps due to the fact that these worms normally live in rather deep water, and those that we find on the shore have been washed up, and so damaged.

The prostomium is about as broad as its length. Both pairs of eyes are very far back, and quite lateral in position (fig. 5). Only the posterior eye is visible from above, and only the upper edge of this. When the prostomium is viewed from the side, the two eyes are seen to be close together; the hinder and upper eye is larger than the anterior lower eye, as in *P. paucibranchiatus*.

The median tentacle is about 3 times the length of the prostomium, and the laterals about $2\frac{1}{2}$ times.

Locality.—Otago Harbour.

Remarks.—In the structure of the head this species bears considerable resemblance to *P. paucibranchiatus*, as also in the general arrangement of the elytral tubercles. But in that species the supra-areolar tubercles are much more conspicuous, owing to their larger size and very definite linear arrangement; and the latter is true of the marginal tubercles.

The rays are narrower, more regular in size, and more sharply pointed. The upper surface of the tubercles when seen from above or in side view is studded with small rounded prominences.

The new species differs entirely from the ordinary *P. squamosus* in the form and arrangement of these tubercles, which in that species are long and subcylindrical, and especially numerous on the external region.

The gills, however, the general form of the body, and the chaetae are different.

As to the papulae, we are ignorant as to how far these are good specific characters—how far they may vary at different ages of one and the same individual; but so far as my studies have gone they seem to be fairly constant. Of the elytral tubercles it is known in other Polynoids that there may be a great range of variability, and it may turn out that this New Zealand worm is identical with *P. paucibranchiatus*.

---

* Benham, "Endeavour" *Polychaeta*, p. 196.

Fam. AMPHINOMIDAE.

Chloeia inermis Quatrefages, Hist. Nat. des Annelées, 1865, vol. i, p. 389.

Since the publication of this comprehensive work on the Annelids there has been no further record of the occurrence of this worm. Nevertheless, I have received several specimens from time to time, and wrote an account of it some years ago, which has not been published. It was, however, by an oversight, not included amongst the Polychaetes which I sent to Ehlers.

So far as our knowledge went, it was confined to New Zealand waters, in which it is evidently by no means uncommon. But amongst the " Endeavour " worms I find a specimen from the South Cape of Tasmania.* A brief account of the species may be given, though it is unnecessary to describe it in detail, for McIntosh has given an excellent account, with figures, of a typical species, *C. flava* Pallas, in the " Challenger " Report on the Polychaetes, p. 8, pl. iii.

The genus may readily be recognized by its general form. Its body is spindle-shaped, blunter anteriorly than posteriorly. The belly is very convex, and curves upwards to meet the narrow and flat back. It is fringed on either side by two series of long, glassy, brittle, white or lemon-coloured chaetae, which are directed outwards and backwards, and the upper bundles partially upwards also. Along the inner, or dorsal, side of the upper bundles is the series of pinnate gills, which commence in the 5th segment, although on the 4th there may be a small and simple gill.

The body-colour is yellowish-brown or pale buff, after long preservation, with a white narrow band along the mid-dorsal surface. This is bordered on each side by a narrow yellow line, and extends along the whole length of the body. In one specimen the buff colour of the back gives way to a pale-violet tint on the hinder segments.

The caruncle, typical of the family, is attached to the first two segments, but its free pointed end overhangs the next two ; it is pale yellow in colour.

As in some other species, the dorsal cirri, as well as the prostomial and peristomial tentacles, are dark-maroon-coloured or violet, even after years of preservation in alcohol. The ventral cirri are white.

The chaetae of this species are exceptional in structure, in that they are without the serrations usual in the genus, and without the fork near the tip. It was, no doubt, from this simplicity in structure that Quatrefages named the species " *inermis* "—the bristles are unarmed with outgrowths.

The majority of the chaetae in the dorsal bundles, both of the mid-body and of the anterior segments, are perfectly smooth, without any trace of serration or of forking (fig. 7) ; but one or two, which are longer and finer than the rest, exhibit a minute step-like trace of a subapical spur.

The ventral chaetae are much thinner than the dorsals, and are of three sizes — (*a*) the stoutest, few in number, are perfectly smooth ; (*b*) the majority, about half the thickness of the dorsals, have a minute obsolescent spur (figs. 8, 9) ; and (*c*) extremely fine ones, with a similar spur.

I guarded myself against overlooking this small spur in the dorsals, as I recognized, of course, that so small a feature might, if it lay above or below the main stem, be invisible under a low power ; but I was unable, even by focussing carefully with a high power, to observe any sign of its presence in the majority of the chaetae.

---

* Benham, *loc. cit.*, p. 206.

It occurred to me that possibly in the young condition some evidence of the typical serrations and fork might exist; but the examination of the smallest, and therefore youngest, of the worms (one which measures 14 mm.) shows no trace of any serration. But in the mid-body most of the dorsals do present an obsolescent spur, resembling that of the ventral chaetae of the adult (fig. 10); but it is situated rather farther from the apex. In a few this was totally absent; in a few others—two or three in the bundle —a definite fork is present (fig. 11).

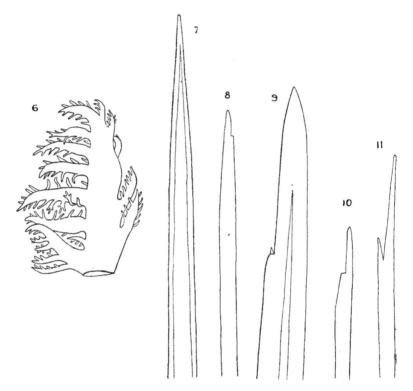

FIG. 6.—A gill (camera outline, × 10).
FIG. 7.—One of the chaetae from a dorsal bundle of an adult worm (× 45).
FIG. 8.—One of the chaetae from a ventral bundle of an adult worm (× 45).
FIG. 9.—The tip of a ventral chaeta (× 180).
FIG. 10.—A dorsal chaeta from the mid-body of a very young specimen, with a "step" below the apex (× 45).
FIG. 11.—Another dorsal chaeta from the same bundle, showing the bifurcation more usually present in the genus, but in this species only occasionally present, and only in the young stages (× 45).

The origin of the shorter limb of the fork is farther removed from the apex than is the step-like trace of spur in the other chaetae, suggesting that the tips of the latter are worn away, reducing the tip of the main axis as well as the shorter limb of the fork, for these forked chaeta appear to be newly formed young bristles; but in the adult I see no indication in the interior of the chaetae of any cavity leading into the short spur whereas in the young forked chaeta the shorter limb is hollow ; so possibly the suggested explanation is not the true one. In the anterior segments of this young worm I find no forks.

The gill is " bipinnate," consisting of a relatively stout axis, bearing about 6 pinnae along each side, each pinna carrying a double series of slender pinnules (fig. 6). The number of the pinnae and the size of the gill decrease at each end of the animal.

The dimensions of the worm vary from 14 mm. by 4 mm. with 24 segments in the specimen (*e*) to 57 mm. by 12 mm. with 34 segments in specimen (*h*). The more usual size, however, of the above individuals is 40–45 mm. by 10 mm. with 30–33 segments.

I have examined specimens from the following localities : (*a*) D'Urville Island, Cook Strait ; coll. Captain F. W. Hutton. (*b*) Wellington (2 individuals) ; coll. Captain F. W. Hutton. (c) Lyall Bay, Wellington ; coll. Miss Mestayer. (*d*) Orepuki, Southland ; coll. G. E. Howes. (*e*) Off Otago Heads, 100 fathoms ; coll. W. B. Benham. (*f*) Stewart Island ; coll. E. Waite. (*g*) Stewart Island ; coll. W. Traill. (*h*) Chatham Island ; coll. Miss S. Shand.

Remarks.—It seems to me probable that Baird's species, *C. spectabilis,* is identical with this. Baird[*] himself notes that it resembles both the above and Grube's *C. egena.* Quatrefages suggested that his species was identical with Grube's *C. egena.*[†] It would be better to drop this latter name entirely. It was found, according to Quatrefages, in a bottle without any indication of its locality. The diagnosis as quoted by him seems to me insufficient to settle its identity, the only important feature being the simplicity of the chaetae.

C. pinnata Moore,[‡] from the south coast of California, has also non-serrate chaetae, with an obsolescent spur ; but a few of the chaetae show minute traces of serrations: The worm is altogether smaller, but in many respects seems related to the present species.

Fam. NEREIDAE.

Cheilonereis peristomialis Benham, " Endeavour " *Polychaeta.*[§]

I obtained the first specimens of this peculiar Nereid in 1899, from trawlings from the s.s. " Plucky," and I wrote an account of it at the time, which has not been published. The worms were found in the upper whorls of a large Gastropod (? *Neptunea*) inhabited by a hermit crab. Since that date, however, a closely allied species has been described from the Pacific coast of North America by Harrington and others[||], under the name *Nereis cyclurus.* The New Zealand species turned up in the " Endeavour " collection.

The striking feature of the new genus is the great development of the peristomium, the ventral and lateral portions of which are much pleated,

---

[*] Baird, Journ. Linn. Soc., x, 1868, p. 234.
[†] Grube, Beschreib. Neu. od. wenig. bekannt : Ann., p. 91, 1855.
[‡] Proc. Acad. Nat. Sci. Philadelphia, 1911, p. 239.
[§] The second part of my account of the " Endeavour " worms has not yet (April, 1916) been published by the Commonwealth Fisheries Department.
[||] Harrington, Trans. N.Y. Acad. Sci., vol. xvi, 1898, p. 214 ; H. P. Johnston, Proc. Boston Soc. Nat. Hist., vol. xxix, 1901, p. 400 ; Moore, Proc. Acad. Nat. Sci. Philadelphia, 1908, p. 343, and in same, 1911, p. 246.

and form, when fully expanded, a large hood or collar, which nearly reaches to the tip of the palps, and hides the base of the everted pharynx.

·My friend Mr. T. D. Adams, whom I consulted as to a suitable prefix to *Nereis* which would indicate this peculiarity, suggested the Greek word *cheilos*, a lip. This great lip is not the only feature which marks it out from other species of *Nereis*. It is accompanied by peculiarities in the form of the parapodia and in the possession of rather exceptional chaetae in the ventral bundle of the posterior feet, as has been recently pointed out by L. N. G. Ramsay,* which I had already noted in my MS. account. It seems to me that these features warrant the creation of a new generic name to mark it off from the various other genera into which the old genus *Nereis, sensu latu,* has. been divided.

In my account of the " Endeavour " specimen I have described the species fully, and have indicated the similarity to and differences from the *Cheilonereis cyclurus* of the eastern shores of the Pacific. At the time I wrote the account of the " Endeavour " specimens I had not seen Ramsay's paper. He, like myself, would unite *N. shishidoi* Izuka† with *N. cyclurus.*

It will suffice here to note the general coloration of the living worm. The ground-tint is a light chocolate-brown, with a pinkish tint, due no doubt to the blood-vessels in the body-wall ; but each segment is traversed near its anterior margin by a narrow cross-bar of white, which extends outwards on to the upper surface of the foot. The head, its appendages, the cirri, and the lobes of the parapodia are brown. But in the mature epitokous female, filled with eggs, the colour is very different. It is slaty-blue, owing to the blue eggs, which fill the cavity of the body and of the parapodia, and so distend the body-wall that its brown pigmentation is obscured by the blue eggs seen through it. In alcohol this blue colour of the eggs changes to brown, while in formalin it turns yellow.

The size of the worm when alive is about 8 in.—*i.e.,* 200 mm.—which shrinks to 175 mm. when preserved. Its breadth in this state is greatest at the 8th segment, where it measures 10 mm., or, including the feet, 17 mm. From this point backwards it decreases in diameter.

The body is flat; the parapodia are relatively large and high, and are remarkable for the great size of the lamelliform expansions not only of the various lobes, but also of the whole upper surface of the foot, so that the dorsal cirrus is carried upwards and outwards in a notch in a lamella which is higher than the rest of the foot, and which increases relatively towards the hinder end.

Fam. STERNASPIDAE.

:Sternaspis scutata Ranzani.

    S. thalassemoides Otto ; ? *S. princeps* Selenka.

Hitherto the only specimens of *Sternaspis*‡ which have been recorded from the sea around New Zealand are the two individuals described by Selenka under the title *S. princeps,* from Station 179 of the " Challenger " Expedition, which is situated due east of East Cape ; they were obtained from a depth of 700 fathoms. To this species I allude later on.

---

* Proc. Zool. Soc., 1914, p. 237.
  † Izuka, Journ. Coll. Sci. Imp. Univ. Tokyo, vol. xxx, 1912, p. 177.
  ‡ A figure of this peculiar Annelid may be seen in the Cambridge Natural History Museum.

During the present year I received from Dr. Chilton two specimens of *Sternaspis scutata* which were obtained during the cruise of the G.S.S. " Hinemoa," off the Akitio River, on the east coast of the North Island, in from 20 to 36 fathoms ; and some years ago Mr. Suter was good enough to give me several specimens which he had obtained off Akaroa in 6 fathoms of water.* These are, no doubt, the same species, though, as will be seen, they differ a good deal in size; and that they belong to the common species from the European and American waters there can be as little doubt. The Akitio specimens are the larger : one measures 15 mm. in length, with a breadth near the hinder end (" abdominal breadth ") of 7 mm.; the breadth near the anterior end, in the region of the rows of chaetae, is 5 mm. The worms are a good deal contracted, so that these measurements are below those of the living individual.

The characteristic posterior ventral " shield " is of very dark colour —in one, a vandyke brown ; in the other, of a deep purple-brown. It measures 5·5 mm. from side to side, with a length of 3·25 mm. at the side, while the median line is 2·5 mm. This shield is fringed externally and posteriorly by 15 or 16 bundles of long bristles ; it is difficult to make out whether the former or latter number is correct, for at the hinder corners the bundles are so close together that under a lens it is difficult to distinguish them. The anterior rows of strong chaetae contain 11 in each row.

The specimens from Akaroa are much smaller : the largest is only 10 mm. in length by 4 mm. across the abdomen. The shield is brick-red, is 4 mm. from side to side, and 2·25 mm. in length.

The anterior rows contain 9 or 10 chaetae, the ventral ones being slenderer and paler, indicating that they are young. There are 17 bundles of bristles at the margin of the shield in one individual, and 15 in a second, of the same dimensions.

Both in the Akitio and Akaroa specimen, as in the Naples specimen, the skin of the body is rough, being covered with groups of minute sand-grains, which are visible only when a piece of the skin is mounted and examined under a microscope. The fact that these grains are in groups seems to indicate the presence of glands in the skin, to the secretion from which the grains have adhered.

I have, fortunately, some specimens of the European species, obtained some years ago from the biological station at Naples (under the name *S. thalassemoides*, which by most authorities is now replaced by Ranzani's earlier name), so that I was able to make a comparison of the external features between them and our New Zealand specimens. They vary in size from 13 mm. to 21 mm. in length, with an abdominal breadth of 4·5 mm. to 9 mm., and anterior diameter of 4 mm. and 6 mm. respectively. The dimensions of the shield vary in proportion.

I wished to ascertain whether there are any points of specific difference between this and the New Zealand specimens, but can find none. For instance, not only does the size of the shield vary, as one would expect, with the size of the animal—that is, with age—but the number of chaetae in the anterior rows around the margin of the shield vary likewise. Thus in the smaller specimens the rows of chaetae contain 9, in the larger

---

* Mr. Suter wrote me that he had sent some of these to Professor Ehlers, of Got tingen, who has, however, not published anything about them.

I count 10 and 11 ; the posterior bundles in the smaller are 17 pairs, in the larger 19 pairs. The skin, too, exhibits the same groups of sand-grains.

From time to time in various parts of the world, even in the European seas, specimens of *Sternaspis* of larger or smaller size, or with different-coloured shield, &c., have been made into distinct species, but sooner or later, as more careful examination has been made, and as our knowledge of variation and the factors in geographical distribution have progressed, many of these have been absorbed into the type species. So it has been on the American coast, both east and west. It is, indeed, becoming doubtful what are the specific characters of the genus. Even the discovery by Sluiter of a specimen provided with a bifid proboscis (*S. spinosa*) has led some authors, such as Selenka, to suggest that this feature is present in all species, but that owing to its fragility and sensitiveness it drops off when the animal is preserved, or even when it dies.

So far, then, as externals go, it appears that size of body and shield, the colour of the latter, the number of chaetae in the anterior rows and around the margin of the shield, are mere matters of age. For that reason I refer these specimens from New Zealand waters to the type species of the genus.

What about *S. princeps* Selenka ? The account* is very brief, but he prefaces it with the words (on p. 5), " It does not seem to be beyond question whether this form . . . can be regarded as really the representative of a new species." Certainly his few lines describing it (on p. 6) do not carry conviction as to its specific separation from *S. scutata*.

Only two individuals were obtained, and no measurements are given ; but his figure (of the larger) is said to be three times the natural size, which makes the worm, therefore, 30 mm. in length by 12 mm. in abdominal breadth. As it was " imperfectly preserved," it may be that these dimensions are greater than in life. His account, short as it is, is vague in one or two points. His first sentence—" Along the middle of the ventral surface there runs a shallow furrow "—really applies, as the context shows, not to the body of the worm, but to the shield. Now, this furrow is always present ; it is a line of division between the two halves of the shield. Further, he notes the existence of " an oblique ridge," separating the shield into an anterior larger and a posterior smaller triangular area. This, also, is present in the Naples specimens as in our own. It may be remarked in passing that this feature is not shown in his figure (pl. i, fig. 1). There are " about 40 bundles " of bristles around the margin of the shield—that is, about 20 on each side. In the larger Naples specimens I find at least 19 bundles. So that this is no specific character.

There is only one other statement : " The whole body is studded with fine scattered chitinous setae, each having at its base a number of smaller chitinous pieces grouped together into wart-like protuberances." If this is really the case, it would be diagnostic of the species. Unfortunately, I have neither Vejdovsky's† nor Rietsch's‡ memoirs available here, so that I am unaware whether this histological feature has been described ; but my examination of the skin of the Naples specimens does not support it. Has Selenka confused the sand-grains under a hand-lens ?

---

* Selenka, "Challenger" Report, vol. xiii, 1885, *Gephyrea*.
† Vejdovsky, Denksch. d. Wien. Akad. Math. Naturw. cl., vol. xliii, 1882.
‡ Rietsch, Ann. Sci. Nat., 6th ser., Zool., vol. xiii, 1882.

In the light of our knowledge of the structure of the worm it would be surprising to find such " chitinous setae " springing haphazard from the skin. At that date (1885) *Sternaspis* was included among the *Gephyrea*, and in the Sipunculids there are tufts of " chitinous setae " scattered over the skin : it would not be anything unusual for them to occur. But we now recognize that *Sternaspis* is a Chaetopod, and their occurrence can scarcely be accepted from a mere inspection, as one may gather was the case with this worm.

Weighing all the facts, I think it would not be unreasonable to suggest that *S. princeps* is nothing but a large specimen of *S. scutata*.

It may be useful to summarize in the following tabular form the facts recorded in this paper (measurements in millimetres) :—

Specimen.	Size.	Shield.	Anterior Chaetea.	Shield Chaetae.	Colour of Shield.
Akaroa ..	10 × 4	4 × 2·25	9–10	15–17 pairs	Brick-red.
Akitio ..	15 × 7·5	5·5 × 3·25	11	15–16 pairs	Dark brown. Purple-brown.
Naples ..	13 × 4·5 to 21 × 13	4 × 2·5 to 7 × 3·25	9–11	17–19 pairs	Pale brown to dark brown.
S. princeps ..	30 × 12	..	..	20 pairs ..	..

ART. XXXIX.—*Notes on the Marine Crayfish of New Zealand.*

By GILBERT ARCHEY, M.A., Assistant Curator, Canterbury Museum.

[*Read before the Philosophical Institute of Canterbury, 3rd November, 1915.*]

Plate XXXIX.

THESE notes are intended to bring together the various scattered references to the marine crayfishes of New Zealand, and thus to have definitely recorded in the " Transactions of the New Zealand Institute " the correct names and complete descriptions of these forms. Descriptions of the larval stages, so far as they are at present known, have also been included.

There are only two species of New Zealand marine crayfish, both belonging to the same genus. They were first assigned to the genus *Palinurus*, to which the English crayfish belongs, but T. Jeffrey Parker (1883, p. 190) pointed out that the genus, as then understood, could be divided into three subgenera, which he named *Jasus*, *Palinurus*, and *Panulirus*, the New Zealand species belonging to the first-named, which was distinguished chiefly by the absence of the stridulating organ. The full text of Parker's paper was published in the following year (Parker, 1884, p. 304). Parker subsequently claimed priority for the name *Jasus* as a generic name over *Palinosytus*, described by Spence Bate (1888, p. 85), and quoted by Stebbing (1893, p. 197), and so the generic name *Jasus* now stands for the New Zealand crayfishes.

Of the two species of *Jasus* known from New Zealand, the first is the common crayfish *Jasus lalandii* (M.-Edw.) sold in the shops, and the other, *J. hügelii* (Heller), is the Sydney crayfish, which is only met with occasionally in New Zealand seas, and then only on the northern coasts.

The common New Zealand crayfish was described by Hutton (1875, p. 279) under the name of *Palinurus edwardsii;* he distinguished it from *P. lalandii* Milne-Edwards, previously known from the Cape of Good Hope, " by its much smaller size, the shape of the beak, its having no spine on the penultimate joint of the anterior legs, and in having a second small spine at the distal extremity of the third joint of the last four pairs of legs." Miers (1876, p. 74) stated that he had hitherto believed *P. lalandii* to be common in New Zealand, but he found that all specimens so named, from New Zealand, in the British Museum agreed with Hutton's *P. edwardsii,* and so· he was doubtful whether *P. lalandii* were really to be found in New Zealand. He therefore placed it in the catalogue with the habitat " New Zealand " queried. *P. edwardsii* was included in the " Catalogue of the Australian Crustacea " by Haswell (1882, p. 171), who added, " Found also at St. Paul and in New Zealand."

In order to· settle the question of the identity of *P. lalandii* and *P. edwardsii,* Parker (1887, p. 150) obtained specimens of the undoubted *P. lalandii,* which should now be known as *Jasus lalandii,* from Cape Town, and compared them with Hutton's specimens in the Otago Museum. He saw that the characters relied upon by Hutton to distinguish the two species were individual variations common to specimens of both *Jasus lalandii* and *J. edwardsii,* and therefore could not be used to separate the two species ; but he found in the sculpturing of the segments of the abdomen characters which would distinguish his Cape Town specimens from Hutton's New Zealand forms. At the same time he pointed out that the differences were slight, and that the examination of specimens from other localities might show that the species would have to be·united.

In the following year Dr. F. McCoy (1887, preface) stated that the Melbourne crayfish, which he referred to *Palinurus lalandii,* was the same as the crayfish from the Cape of Good Hope, and that it was found at New Zealand and at the Island of St. Paul ; later (1888, p. 222, note) he referred to Parker's paper (1887), remarking that he also had obtained specimens from Cape Town to examine, and had independently come to the same conclusion as Parker as to the characters given by Hutton, and that his observations further showed that the sculpturing of the abdomen could not be used to distinguish the species. McCoy did not formally unite the two species, however. This has since been done by Ortmann (1891, p. 16), who has united *J. frontalis* (M.-Edw.) from Chile, *J. paulensis* (Heller) from St. Paul, and *J. edwardsii* (Hutton) with *J. lalandii* (M.-Edw.), and the name as understood by Ortmann has been quoted by Stebbing (1902, p. 38) ; the New Zealand marine crayfish should therefore now be·known as *Jasus lalandii* (Milne-Edwards).

In 1880, T. W. Kirk (1880, p. 313) described under the name *Palinurus tumidus* a very large crayfish obtained from Waingaroa, in the north of Auckland. He distinguished it from *Palinurus hügelii,* described and figured by Heller (1868, p. 96, pl. vii), by its larger size, the beak, supraorbital and antennal spines being turned upwards, and by the telson being less triangular, and rounded instead of scarped.

Haswell (1882, p. 172) pointed out that these differences do not exist, except in regard to the telson, where he supposed they were due to the wearing, or other mutilation, of this part in Heller's specimen. Haswell's opinion was later confirmed by McCoy (1888, p. 222), whose description and plate of *Palinurus (Jasus) hügelii* from Sydney agreed with T. W. Kirk's description of his *Palinurus tumidus.*

I have examined specimens of *Jasus hügelii* from Auckland, and also one specimen from Sunday Island, in the Kermadec Group, this being the one mentioned by Dr. Chilton (1911A, p. 549). I have concluded that *J. tumidus* (Kirk) is the same as *J. hügelii* (Heller), which is thus the second species of New Zealand marine crayfish.

Genus Jasus T. Jeffrey Parker, 1883.

1883. *Jasus* Parker, *Nature*, vol. 29, p. 190. 1884. *Jasus* Parker, Trans. N.Z. Inst., vol. 16, p. 304. 1888. *Palinosytus* Bate, "Challenger" Reports, *Crustacea Macrura*, vol. 24, p. ix. 1888. *Palinostus* Bate, "Challenger" Reports, *Crustacea Macrura*, vol. 24, pp. ix, 85. 1891. *Jasus* Ortmann, Zool. Jahrb., vol. 6, pp. 14, 16. 1893. (*Palinosytus*) Stebbing, "A History of the Crustacea," p. 196. 1893. *Jasus* Stebbing, *ibid.*, p. 197. 1897. *Jasus* Ortmann, "American Journal of Science," vol. 4, p. 291. 1900. *Jasus* Stebbing, "Marine Investigations in South Africa: Crustacea," pt. i, p. 30. 1902. *Jasus* Stebbing, *ibid.*, pt. ii, p. 38.

The following is the generic diagnosis as given by Parker: "Stridulating organ absent; rostrum well developed, clasped by paired pedate processes of the epimeral plates; procephalic processes present; coxocerites imperfectly fused; antennulary flagella short."

The New Zealand species may be distinguished by the following key :—

1. Rostrum much smaller than the supra-orbital spines; tubercles on carapace each surrounded by a ring of small hairs .. .. .. *J. lalandii.*
2. Rostrum at least as large as supra-orbital spines; no hairs on carapace *J. hügelii.*

Jasus lalandii (Milne-Edwards), 1837.   Plate XXIX, fig. 3.

1837. *Palinurus lalandii* Milne-Edwards, Hist. Nat. Crust., vol. 2, p. 293. 1843. *P. lalandii* Krauss, Südafrik Crust., p. 53. 1868. *P. lalandii* Heller, Reise der "Novara," Crustacea, p. 96. 1875. *P. edwardsii* Hutton, Trans. N.Z. Inst., vol. 7, p. 279. 1876. *P. edwardsii?* Miers, Cat. Stalk- and Sessile-eyed Crust. of New Zealand, p. 74. 1882. *P. edwardsii* Haswell, Cat. Australian Stalk- and Sessile-eyed Crust., p. 171. 1884. *Jasus edwardsii*, Parker, Trans. N.Z. Inst., vol. 16, p. 304. 1887. *Palinurus lalandii* McCoy, "Prodromus of the Zoology of Victoria," vol. ii, dec. xv, p. 193, pl. 149 and 150. 1887. *P. edwardsii*, Parker, Trans. N.Z. Inst., vol. 19, p. 194, pl. x, figs. 1, 2, 5, 6, 7, 8, 12. 1888. *Palinostus lalandii*, Bate, "Challenger" Reports, *Crustacea Macrura*, vol. xxiv, p. 86. 1888. *Palinurus lalandii* McCoy, "Prodromus of the Zoology of Victoria," vol. ii, dec. xvi, p. 222, note. 1891. *Jasus lalandii* Ortmann, Zool. Jahrb., vol. 6, p. 16. 1902. *J. lalandii* Stebbing, "Marine Investigations in South Africa, Crustacea," pt. ii, p. 38. 1911. *J. edwardsii* Chilton, "Records of the Canterbury Museum," vol. i, p. 303.

Diagnosis.

Rostrum much smaller than the supra-orbital spines, depressed, and rapidly narrowing anteriorly to the apex, which curves upwards; *clasping processes* from the epimeral plates curving round the rostrum, and meeting or nearly meeting above it.

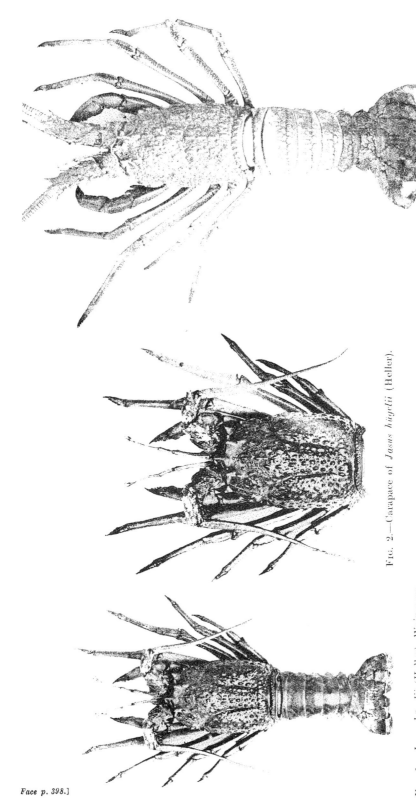

Fig. 1.—*Jasus hügelii* (Heller), Waingaroa.
Auckland.

Fig. 2.—Carapace of *Jasus hügelii* (Heller).

Fig. 3.—*Jasus lalandii* (Milne-Edwards).
Banks Peninsula.

Carapace armed with spines, and large, oval, depressed tubercles, each surrounded by rows of short hairs.

Abdomen sculptured with flat tubercles separated by rows of short hairs; *abdominal pleura* with a single large, backwardly curved spine below, and toothed posteriorly.

Antennules longer than the peduncle of the *antennae; antennae* longer than the body.

First pair of peraeopods with a large spine on the under-surface of the ischium and the merus; *last four pairs* with 1 or 2 spines on the upper surface of the distal end of the merus.

Colour : Carapace dark brownish-purple; abdomen the same, marbled with yellow; legs and caudal appendages reddish-orange more or less marked with purple.

Length, 12 in. to 18 in.

Hab.—Chile, Tristan da Cunha, St. Paul Island, Cape of Good Hope, southern shores of New Zealand.

Jasus hügelii (Heller), 1861. Plate XXIX, figs. 1 and 2.

1861. *Palinurus hügelii* Heller, Sitzungsb. der Weiner Akad. der Wissenschaften, Bd. 45, S. 393. 1868. *P. hügelii* Heller, Reise der " Novara," Crust., p. 96, pl. 8. 1879. *P. tumidus* Kirk, Trans. N.Z. Inst., vol. 12, p. 314, pl. xi. 1882. *P. hügelii* Haswell, Cat. Australian Stalk- and Sessile-eyed Crust., p. 172. 1884. *Jasus hügelii* Parker, Trans. N.Z. Inst., vol. 16, p. 304. 1888. *Palinurus hügelii* McCoy, " Prodromus of the Zoology of Victoria," vol. ii, dec, xvi, p. 221, pl. 159. 1888. *Palinostus hügelii* Bate, " Challenger " Reports, *Crustacea Macrura,* vol. xxiv, p. 85. 1911. *Jasus hügelii* Chilton, Trans. N.Z. Inst., vol. 43, p. 549.

Diagnosis.

Rostrum as large as, or larger than, supra-orbital spines; *clasping processes* from the epimeral plates small, attached only to the under-surface of the rostrum.

, *Carapace* armed with forwardly directed sharp spines only, which are not separated by rows of hairs; there are no flattened tubercles on the carapace.

Abdomen not sculptured, but provided with scattered small blunt spines, without hairs; *abdominal pleura* as in *J. lalandii*, except that the large spine is not so markedly curved backward.

Antennules longer than the peduncle of the antenna; *antennae* shorter than the body.

First pair of peraeopods with 3 small spines on the under-surface of the ischium, and with a large spine on the under-surface of the merus; *last four pairs* with 1 spine on the distal end of the upper surface of the merus.

Colour : Upper surface dark olive-brown; sides of carapace and abdominal pluera chestnut, with a few small scattered cream spots; antennae and legs reddish-brown.

Length : Commonly 12 in.; large specimens up to 2 ft.

Hab.—Indian Ocean, New South Wales, Port Phillip (rare), New Zealand, Kermadec Islands.

Larval Stages.

There are four well-marked stages known in the life-history of *Jasus lalandii*, which might briefly be noted as — (*a*) the *naupliosoma,* a form swimming rapidly by means of the antennae; (*b*) the *phyllosoma,* which swims more slowly, and uses the exopods of the thoracic legs; (*c*) the "*natant*" stage, or *puerulus,* which either walks about the bottom, or swims by means of the abdominal pleopods, which are specially hooked together for the purpose; and (*d*) the *adult,* which makes use of the telson and uropods occasionally for swimming.*

I have had for examination specimens of the three larval stages of *J. lalandii,* and the following notes will be in the form of brief descriptions of them. The first two stages—the naupliosoma and the phyllosoma— are described from specimens reared by Mr. T. Anderton, of the Portobello Marine Fish-hatchery, Dunedin ; and the pueruli were collected at Stewart Island by Mr. Walter Traill. I have also a specimen of the puerulus of *J. hügelii,* collected at Cuvier Island, near Auckland, by Mr. P. W. Grenfell. All the specimens were in the possession of Dr. Chilton, who very kindly forwarded it to me to describe in this paper.

(a.) The Naupliosoma (fig. 1).

The naupliosoma larva has only recently been described by Gilchrist (1913, p. 225)†, who observed it hatching from specimens of *Jasus lalandii* kept in tanks. This stage lasted only a very short time (four to six hours), and would, he said, be readily overlooked, especially if hatching takes place in the night. In the letter accompanying the specimens I have Mr. Anderton wrote that the stage with setae-bearing antennae could not be found after an hour or more.

The naupliosoma is a much more advanced stage than the nauplius, though it resembles it in the possession of large biramose and setose antennae, and a median eye-spot. In it the mouth parts are all well developed, except the 1st maxillipedes, which are rudimentary ; there is a distinct thorax, bearing 3 pairs of biramose walking-legs, and a short, indefinitely segmented abdomen is also present.

The *cephalic region* is rounded, and is nearly as deep from above downwards as it is long and broad ; it is rendered opaque by the presence of yolk-granules.

The *antennules,* of normal length, project between the eyes. The *antennae* are large, consisting of a fairly long basal portion, bearing a protuberance posteriorly at the point of origin ; they divide without visible segmentation into a longer exopod, bearing 7 setose processes, and a shorter endopod, with only 2 processes ; 2 spines project into the two processes which arise at the tip of both endopod and exopod.

---

* In the proceedings of the Linnean Society's meeting of the 2nd March, 1916, there is an abstract of a paper by Dr. Gilchrist on "Larval and Post-larval Stages of *Jasus lalandii.*" Dr. Gilchrist is now of the opinion that the term *naupliosoma,* which he applied to the first stage was rather inappropriate, "since it tends to obscure the reasonable presumption that the *nauplius* stage has 'been passed long before in the development of the embryo.'" The interesting and important information given in Dr. Gilchrist's paper will no doubt be soon available in complete form in the publications of the Linnean Society.

† Since the above was written my attention has been called to a description of this stage by G. M. Thomson (Trans. N.Z. Inst., vol. 39, 1907, p. 484, pl. 20). In the same place the above author has also described the phyllosoma, and gave a drawing of this second stage.

The *mandibles* (fig. 2) appear to have 2 segments, and are short and broad; they are provided at the inner end with a pad of spines and a 3-toothed process.

The *1st maxillae* are biramose, but unsegmented, and each ramus bears a couple of deeply set spines. The *2nd maxillae* are separated from the other mouth parts and from each other; they are long, almost straight processes, with the convex inner margin bearing a spine, and with 3 deeply set spines at the end.

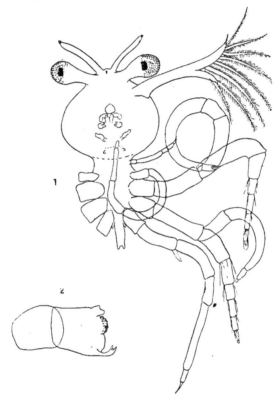

FIG. 1.—Naupliosoma larva of *Jasus lalandii* (M.-Edw.).   × 26.
FIG. 2.—Mandible of naupliosoma larva.

The *1st maxillipedes* are rudimentary, but there is a distinct segment in this region. The *2nd* and *3rd maxillipedes* are similar in shape, but the 3rd are much the larger; they have 5 segments, with spines at the end of the last.

The *1st and 2nd peraeopods* are broader and stronger than the 3rd maxillipedes, and do not, as yet, bear setae, the swimming being performed, as before mentioned, by the setae on the antennae. The *3rd peraeopod* differs from the others by having only a rudimentary exopod; the *4th p raeopo1* is rudmentary, and the *5th* is not developed at all. In the figure the *peraeopods* are shown extended, whereas in life they are carried closely folded under the thorax.

(b.) The Phyllosoma (fig. 3).

This stage has been known for some time, and need not be fully described here, so only the chief points wherein it differs from the naupliosoma will be mentioned. The *antennae* have become segmented, but have lost their setae, although they are still biramose, and have spines at the tips of each branch. The mouth parts are the same as in the preceding

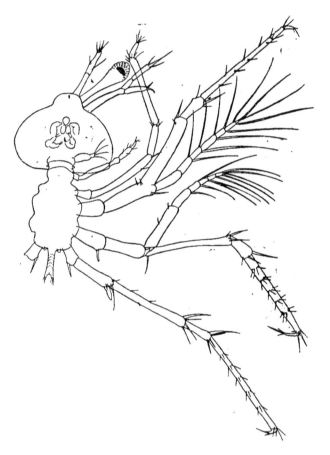

Fig. 3.—Phyllosoma larva of *Jasus lalandii* (M.-Edw.).   × 23.

stage, except that the spines now project freely beyond the ends. The *first three pairs of peraeopods* are now completely unfolded, and the exopods of the first two pairs bear long swimming-setae. The exopods of the 3rd pair are rudimentary, as in the naupliosoma. The *thorax* and *abdomen* are longer, and 6 segments can be seen indistinctly in the latter. There is still a median eye. I have specimens of the phyllosoma sixteen days old, but there is no sensible difference between them and the ones just described. Mr. Anderton stated in his letter that he had not detected a moult up to this stage.

(c.) The Puerulus.

The puerulus is, at first sight, simply a small adult, and, indeed, some of these larval forms have been described as species of a genus *Puerulus* Ortmann (1897, p. 290, footnote). Dr. Calman (1909, p. 441) is of the opinion that this genus is a perfectly valid one, the species of which are really quite distinct from the puerulus larva. The characters which can be recognized as larval characters are the presence of vestiges of exopods on the thoracic limbs, the absence of the cervical groove, the separation of the 3rd maxillipedes at their base, the carapace having few spines and having lateral ridges giving it a prismatic instead of a cylindrical form. More important is the presence, at the end of the appendix interna of the pleopods, of small coupling-hooks, which enable the pleopods to be joined together in pairs, and thus to make efficient swimming-organs.

The specimens which I have are undoubtedly pueruli, those from Stewart Island being 30 mm. in length, and that from Cuvier Island 20 mm. There are certain differences between the Cuvier Island specimens and those from Stewart Island, such as leave no doubt in my mind that the Stewart Island forms are pueruli of *Jasus lalandii*, while the Cuvier Island puerulus is the larva of *J. hügelii*. For instance, the rostra show exactly the differences which have been set down above in the key to the species, that of the Cuvier Island puerulus being as large as the supra-orbital spines, and projecting straight forward, while the larvae from Stewart Island have a small rostrum, bent down towards the antennules and slightly turned up at the tip. Certain details of the spinulation of the telson mark further differences, the large spines in the Stewart Island larvae being placed exactly as in the adult *J. lalandii*, while in the Cuvier Island larvae there is a distinctly different pattern, which, though not arranged exactly as in the adult *J. hügelii*, certainly tends towards that arrangement.

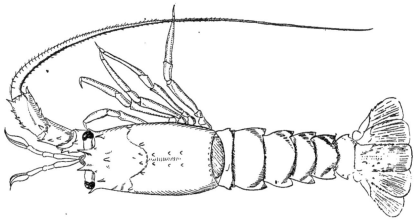

Fig. 4.—Puerulus of *Jasus lalandii* (M.-Edw.), Stewart Island. × 3.

Puerulus of Jasus lalandii (M.-Edw.) (fig. 4).

Carapace with lateral ridges giving it a somewhat prismatic form, wider posteriorly than anteriorly; spines very few, but, where present, distinct. *Cervical groove* very indistinct; *supra-orbital spines* large, projecting upwards and outwards, with a very small spine immediately at their bases; *post-orbital spines* smaller than supra-orbital, and each with a still smaller spine behind.

The *rostrum* is small, and is bent down towards the antennular peduncle, which it does not reach, however, and is slightly upturned at the tip. The *clasping processes are not visible.*

The *abdominal pleura* end in a point projecting downwards and backwards, and have a smaller tooth projecting from the posterior border.

The pattern of the *spines on the telson* (fig. 5) is as follows: Behind the anterior border of the telson there is a submedian pair of large spines,

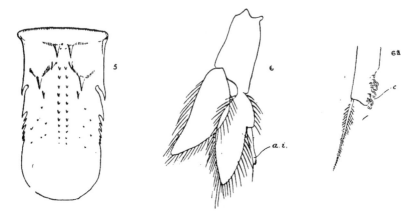

Fig. 5.—Telson of puerulus of *Jasus lalandii*, showing the arrangement of the spines. × 9.
Fig. 6.—Pleopod of puerulus of *Jasus lalandii*. *a.i*, appendix interna.
Fig. 6a.—Appendix interna of pleopod of puerulus of *Jasus lalandii* (M.-Edw.), showing coupling-hooks (*c*).

behind which is a double median row of small spines running back to near the posterior end. Behind the first pair, and wider set, is a second pair, these being followed by a third pair. The edges of the telson and uropods are deeply serrated.

The *pleopods* (figs. 6, 6a) have an appendix interna armed at the tip with small, strongly curved coupling-hooks.

.The *3rd maxillipedes* are sparated at the base; the *antennal peduncle* is longer than the peduncle of the *antennules.* There is no trace of thoracic exopods.

The *colour* in alcohol is creamy white.

Length, 30 mm.

Locality.—Stewart Island.

Puerulus of *Jasus hügelii* (Heller) (fig. 7).

This larva is smaller than the preceding, being only 20 mm. long The *carapace* is of the same shape as that of the Stewart Island pureuli, except that it has stronger lateral angles, and has only supra-orbital and post-orbital spines; the fairly large *rostrum* projects straight to the front.

The *abdominal pleura* end in a strong, backwardly and downwardly projecting spine, and have one tooth on the posterior border.

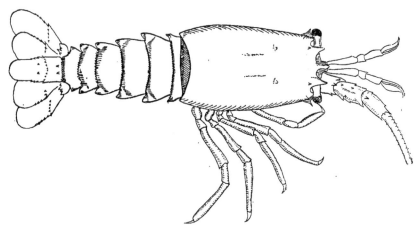

Fig. 7.—Puerulus of *Jasus hügelii* (Heller), Cuvier Island. × ⅔.

The *spines on the telson* (fig. 8) have the following arrangement: There are 2 strong submedian spines behind the anterior border, behind them is a row of 8 spines stretching across the telson, with the inner teeth slightly in advance of the outer their outer neighbours, giving the line a very broad arrow-head shape. There is no sign of a double narrow median row, as in the puerulus of *J. lalandii*. Other larval characters, such as the coupling-hooks on the pleopods, the separation of the mandibles at the base, and the short antennular peduncle, are as in the puerulus of *J. lalandii*. There are no thoracic exopods visible.

Length, 20 mm.
Locality.—Stewart Island.

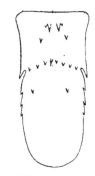

Fig. 8.—Telson of puerulus of *Jasus hügelii* (Heller), showing arrangement of the spines. × 12.

BIBLIOGRAPHY.

1861. Heller, C. Sitzungsb. der Weiner Akad. der Wissenschaften, Bd. 45, S. 393.

1868. Heller, C. Reise der " Novara," Crustacea, pp. 96–99, pl. vii.

1875. Hutton, F. W. " Descriptions of Two New Species of Crustacea from New Zealand." Trans. N.Z. Inst., vol. 7, pp. 279–80.

1876. Miers, E. J. " Catalogue of the Stalk- and Sessile-eyed Crustacea of New Zealand," p. 74.

1880. Kirk, T. W. "Description of a New Species of *Palinurus*." Trans. N.Z. Inst., vol. 12, pp. 313–14, pl. xi.

1882. Haswell, W. A. "Catalogue of the Australian Stalk- and Sessile-eyed Crustacea," pp. 171–72.

1883. Parker, T. J. "On the Structure of the Head in '*Palinurus*,' with Special Reference to the Classification of the Genus." *Nature*, vol. 29, pp. 189–90.

1884. Parker, T. J. "On the Structure of the Head in '*Palinurus*,' with Special Reference to the Classification of the Genus." Trans. N.Z. Inst., vol. 16, pp. 297–307, pl. xxv.

1887. Parker, T. J. "Remarks on *Palinurus lalandii* M.-Edw. and *P. edwardsii* Hutton." Trans. N.Z. Inst., vol. 19, pp. 150–55, pl. x, figs. 1–14.

1887. McCoy, F. "Prodromus of the Zoology of Victoria," dec. xv, vol. 2, preface, p. 193, pl. 149, 150.

1888. McCoy, F. "Prodromus of the Zoology of Victoria," dec. xvi, preface, pp. 221–23, pl. 159.

1888. Bate, C. Spence. "Challenger" Reports, Crustacea Macrura, vol. xxiv, pp. 85–88, pl. xi, figs. 1, 1Q; pl. xiA; pl. xii, figs. 1, 1H, 1I.

1891. Ortmann, A. E. Zool. Jahrb., vol. 6, pp. 14, 16.

1893. Stebbing, T. R. R. "A History of the Crustacea," pp. 196–97.

1897. Ortmann, A. E. "On a New Species of the Palinurid-genus *Linuparis* found in the Upper Cretaceous of Dakota." Amer. Journ. Sci. (4), vol. 4, p. 290, footnote.

1902. Stebbing, T. R. R. "Marine Investigations in South Africa, Crustacea," pt. ii, pp. 38–40, pl. vii.

1909. Calman, W. T. "The Genus *Puerulus* Ortmann, and the Post-larval Development of the Spiny Lobsters (Palinuridae)." Ann. Mag. Nat. Hist. (8), vol. 3, p. 441.

1911. Chilton, C. "Scientific Results of the New Zealand Government Trawling Expedition, 1907: Crustacea." "Records of the Canterbury Museum," vol. 1, p. 303.

1911A. Chilton, C. "The Crustacea of the Kermadec Islands." Trans. N.Z. Inst., vol. 43, pp. 544–73.

1913. Gilchrist, J. D. F. "A Free-swimming Nauplioid Stage in *Palinurus*." Journal of the Linnean Soc., vol. 32, pp. 225–31.

ART. XL. — *Contributions to the Entomology of New Zealand : No. 8—*
Parectopa citharoda *Meyr.* (*Order* Lepidoptera).

By MORRIS N. WATT, F.E.S.

[*Read before the Wanganui Philosophical Society, 1st November, 1915.*]

Parectopa citharoda Meyr.

Parectopa citharoda Meyr., Trans. N.Z. Inst., vol. 48, p. 418.

Ovum.

Class.—Flat.

Shape.—Oval, wafer-like; base flat; upper portion rounded. The shell appears to extend beyond the egg proper, forming a wide flat transparent rim.

Dimensions. — Total width, 0·40 – 0·50 mm.; width, excluding rim, 0·28 – 0·35 mm; total length, 0·59 – 0·75 mm.; length, excluding rim, 0·41 – 0·47 mm.; height, about 0·10 mm.

Sculpture.—Devoid of sculpturing, except for very minute white spots (elevations) scattered over the surface of the shell in regular hexagonal formation.

Micropyle.—Not distinguished.

Shell.—Transparent, thin and flexible, but fairly strong; shiny, almost glossy; covered with minute wrinkles.

Colour.—White; the growth of the embryo can be clearly followed.

Note.—The ovae are well cemented to the leaf-surface; are large for the size of the moth. Period of incubation, about fourteen days. (Described 14th February, 1915.)

Egg-laying.

The eggs are laid indiscriminately on either side of the leaf of the food plant; this follows from the fact that the leaves naturally grow in a vertical position, and there is little, if any, difference between the two surfaces. The eggs are also laid with practically no fixed position between the margin of the leaf and the midrib. There is a distinction, however, in that they are laid invariably on young tender leaves, towards the upper (as opposed to the basal) end. The young seed-pods are also favoured by the parent moth for the deposition of her ovae.

Mine.

The mine is not difficult to detect, being most conspicuous. It has no definitely fixed position, but often follows the midrib or the margin of the leaf for some distance. It is to be found on either side of the leaf, for the same reason as the deposition of the ovum; for the same reason, also, the larva, when full grown, leaves the leaf from whichever side is most convenient to it. The gallery is long and narrow, in many cases first starting from the site of the egg in a small semicircle and then continues more or less straight till some obstruction is reached, as, for instance, the hard midrib or the margin of the leaf. The obstruction may be followed for some distance, and in this way the mine may become much twisted and contorted. As the larva grows, the mine naturally becomes gradually wider till the last three or four days of the larval existence in the leaf; during this latter period the gallery is much widened, and becomes a large irregular blotch or chamber. The total length of a mine

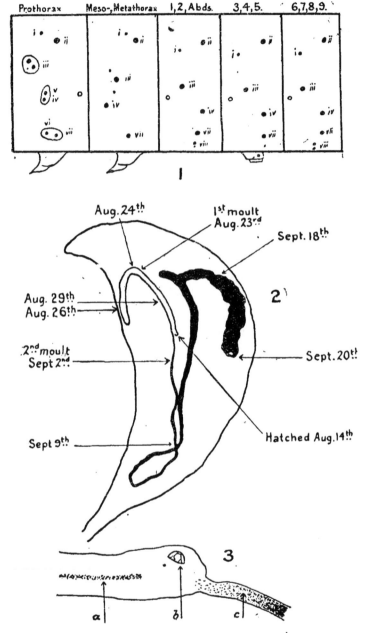

Fig. 1.—Larval tubercles, 3rd stadium.
Fig. 2.—The mine; natural size.
Fig. 3.—Portion of mine at second moult. *a*, frass granules occupying centre of
mine; *b*, cast-off head-piece of larva; *c*, beginning of mine, 3rd stadium.

· seldom falls short of 7 in. or 8 in. The deposition of the frass in the mine is an important point for identification. In this case it at first occupies the central portion of the mine, forming a narrow slightly waved line of more or less detached granules, leaving a pale space ·on either side of it. In some cases this line of frass is continuous without indication of a granular consistency, and looks as if it had been deposited in a soft condition and had run together into a homogeneous thread; at times it is vermiform in character, according to the manner in which the larva was feeding while advancing. This arrangement is suddenly stopped by a break or clear space in the mine, after which the frass granules become more prolific. This continues for some distance, when there is another change—the mine goes deeper into the leaf and becomes choked throughout its whole length with the frass granules. These changes in the character of the frass denote the larval moults. At first the mine is white and in marked contrast to the rest of the leaf, but as moisture gets under the cuticle it becomes much less conspicuous; however, the leaf soon becomes brown along the track of the mine, and so it becomes even more conspicuous than before. The latter portion of the mine is first dark green, but as the frass dries and the cuticle of the leaf withers it becomes dark brown.

The Larva.

1st Stadium.—The prothorax is greatly enlarged; the remaining segments are well rounded, and rapidly diminish in size towards the anus, giving the grub a very attenuated appearance. Colour light green, head light brown. There appear to be none of the primary setae present; to discover these a ¼ in. objective was used without effect. The head is triangular equilateral, flattened, somewhat retractile.

2nd Stadium.—Head large; the cheeks are margined with black; the clypeus has also the black margin, is broad against the labrum, and slightly widens till about two-thirds of its length, when it narrows considerably till it meets the angle formed by the cheeks on the top of the head; mandibles prominent; the cheeks are pale green, with a darker area extending their whole length against the clypeus; labrum dark brown. Prothorax very large, meso- and meta-thorax smaller than the prothorax, the latter smaller than the former, all three flattened dorsally and ventrally. Abdominal segments well rounded, of almost equal size, slightly diminishing towards the anal end, somewhat flattened dorsally and ventrally. Colour light green, darker in centre. Primary setae still absent.

3rd Stadium.—Colour greenish-yellow, a faint white spiracular line. Length just before emerging from leaf (fully extended), about 6·20 mm.; breadth, about 0·90 mm. Body covered with a fine pile. Spiracles circular, minute, very inconspicuous. Prolegs on abdominal segments 3, 4, and 5; armed with single row of 7 brown, well-formed hooks. Primary tubercles bearing short, simple, and in most cases single transparent setae. Tubercle i is minute, above ii in the thoracic segments, but some way in front of and below ii in the abdominal segments. ii is well developed. iii is coalesced with two setae on the prothorax, where it is prespiracular; in the meso- and meta-thorax is beneath but in front of i, and has a minute secondary seta above and in front of it, is post-spiracular in the abdominal segments. iv in the prothorax is prespiracular, and is coalesced with a second seta (probably v), is beneath and in front of iii in meso- and meta-thorax, is post-spiracular and behind iii; almost directly beneath ii, in the abdominal segments. v is absent in the meso-, meta-thorax, and abdominal segments. vi appears to be present in the prothorax

in conjunction with vii, but is absent in the other segments.   vii is·
beneath iv in the abdominal segments.   (**Fig. 1.**)

Larval Habits.

During the first two stadiums the larva burrows close against the outer
cuticle of the leaf, but after the second moult descends at once into the
parenchyma, gradually eating all that portion of the leaf between the two
outer cuticles.   It is easy to see where the moults have taken place, not
only from the appearance of the frass, as shown in a former paragraph,
but also from the fact that at such places the mine is slightly enlarged and
the next portion of the mine is directed at a slight angle from the old one,
as if the larva had tried to avoid its cast skin, the head-piece of which can
be readily distinguished in the clear portion of the mine.   After the second
moult the mine is at first narrower than before, and is difficult to follow
on account of its descending into the denser portion of the leaf (see fig. 3).
The larva appears to undergo only two moults within the mine.

The first two stadiums are of equal duration, being about ten days,
though this period may be shortened or lengthened according to meteorological conditions.   The 3rd stadium is longer, lasting, on an average, a
fortnight.   When full-grown the larva leaves the leaf and descends to the
ground to pupate.

The Cocoon.`

The cocoons are exceedingly pretty little structures, and their construction is most fascinating to watch under low powers of the microscope.
The usual length is 7 mm., and the width varies according to the situation
of the cocoon ; when constructed on a flat surface it may be as much as
5 mm.   The shape is oval and flat.   The construction is exceedingly delicate, the cocoon proper having the appearance of a fine white shiny skin.
The term " cocoon proper " is used because when completed the whole
exposed surface is thickly covered with numbers of minute white floccy
globules.   These look like small bubbles and when seen through the
microscope each one has the appearance of being a collection of minute
transparent cells, the whole forming a somewhat lengthened sphere.   These
globules are exceedingly delicate, and many appear to be lightly attached
to the cocoon, though numbers can be removed by blowing upon them.
The object of this elaborate superstructure is not known.

The globules are ejected from the anal aperture of the larva during
the construction of the cocoon proper.   The larva takes no notice of the
newly ejected globule till it comes across it while weaving the silken
canopy ; on finding the globule it roughly tears an opening in the structure
it has been taking so much trouble in making, and forces it out through
the opening so made, not taking the trouble to mend this ugly rent with
any degree of care.   Between seventy and eighty globules are ejected in
this way from the cocoon before the work is completed, and these give to
it a most curious and beautiful appearance.   Watching two larvae constructing their cocoons at the same time, I was astonished to find that the
globules were ejected by each at regular intervals of about twenty minutes.
Having observed the time of one ejection, I was able to observe subsequent
ejections at the minute of the operation throughout the two days during
which the cocoons were being constructed.   The cocoons are to be found
in numbers upon the ground in the dead leaves under the food plant :
the smallest and most curved and twisted leaves are the most favoured ;
even crannies in the lumps of dirt are taken advantage of, besides blades
of grass, and sometimes crannies in the bark of the tree itself.

The Pupa.

Colour at first almost transparent crystal white, later assuming a light yellow. The eyes are the first to change colour, changing to a reddish-brown ; a couple of days or so before emerging the pupa becomes black, excepting those portions that are white in the imago, and the markings on the wings are very plain. The pupa is roughly cylindrical in shape, slightly flattened dorsally and ventrally ; the greatest width is opposite the end of the 1st legs ; from here the sides converge slightly towards the head, which is broad and well rounded, and somewhat more so towards the last abdominal segments. In profile the dorsal outline is practically straight, head well rounded ; between the head and the mesothorax there is a deep incision, at the bottom of which is situated the prothorax.

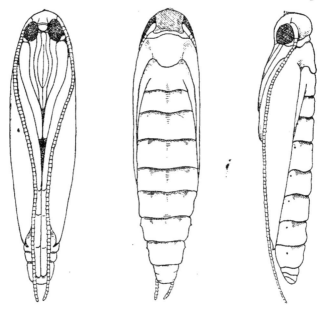

Figs. 4–6.—The pupa (ventral, dorsal, and lateral views).

Ventrally there is a depression opposite the bottom of the eyes, and from here the maxillae and 1st legs and other ventral appendages form a rounded prominence, interrupted by the termination of the 1st legs, from whence the outline gradually follows the antennae towards the anal segments.

Such is the rough outline of the pupa. On top of the head, situated anteriorly, is a small dark prominence used in pushing open the cocoon on the day of emergence. The eyes are fairly large and prominent. The mandibles are present, and occupy two-thirds of the outer edge of the labrum between the eyes ; between these, and occupying the remaining margin of the labrum, are the labial palpi ; these are well developed, and, though at first slightly constricted, widen out considerably at about half their length, and then gradually taper off. On either side of these, their base resting against the mandibles and a small portion of the eye, are the maxillae ; these are very narrow, and are prolonged beyond the labial palpi a short distance. On either side, beneath and somewhat behind the eyes, are the maxillary palpi, forming a sort of eye-collar. Of this Tutt says, " The palpus on leaving the maxilla passes backwards in the angle between the head and the prothorax, until it is situated deeply beneath the antenna, then it turns forwards to the antenna, and only reaches the surface by emerging from beneath the antenna, and, turning inwards, forms the ' eye-collar,' which contains only its terminal joints, the others being concealed deeply."

The 1st legs occupy all the space between the maxilla and the antenna, and even follow the antenna for a short distance ; they are fairly broad throughout, and end a little above half the length of the pupa ; the terminal joints occupy almost the whole length of the maxillae. The 2nd legs extend along the antennae to about two-thirds of the total length of the pupa, and are broadest opposite the termination of the maxillae. The 3rd legs make their appearance beneath the 2nd legs, and, at first constricted, are fairly wide opposite the end of the wings, and terminate just above the end of the anal segments.

The antennae are narrow, and, commencing above the eyes, extend some distance beyond the terminal segments, more in some pupae than in others ; their ends are free ; the joints are most conspicuous.

The forewings are broadest opposite the 1st abdominal segment, and rapidly narrow to a narrow, pointed, incurved tip just below the 6th abdominal segment. The hindwings occupy a very narrow strip, which is lost in the 2nd abdominal segment.

The dorsal head-piece is large, prominent, well rounded posteriorly. The prothorax is extremely narrow, and forms a thin sunken strip behind the head-piece. The mesothorax is large and prominent, slightly extended behind. The metathorax is medium-sized. The abdominal segments are of about equal length, and gradually decrease in width towards the anal end. There are indications of a dorsal suture in the metathorax and upper abdominal segments. The free incisions are between segments 5–6, 6–7, 7–8. Segments 8, 9, and 10 appear to be soldered together. Spiracles are minute and raised. There are indications of neuration on the wings.

Duration of pupal stage from specimens obtained in August, twenty-nine days ; in December, eleven days.

Chief Measurements.

Measurement at	Length from Front.	Transverse Diameter.	Anterior-posterior Diameter.
	Mm.	Mm.	Mm.
Top of eye	0·16	0·49	0·55
End of labrum	0·35	0·69	0·52
„ labial palpi	1·18	0·97	0·83
„ maxillae	1·63	0·97	0·83
„ 1st legs	1·94	0·97	0·76
„ 2nd legs	2·69	0·83	0·72
„ 3rd legs	4·00	0·28	0·31
„ last abdominal segment ..	4·07	..	..
„ antennae	4·34	..	..

The above measurements are, of course, subject to slight variation.

Imago.

See page 418 of this volume.

Dehiscence.

When the imago is matured, and ready for emergence, the pupa forces its anterior segments out of the cocoon. The pupa-case bursts down the centre of the ventral surface; the head-piece, to which the eye-cases and maxillae and palpi are attached, is thrown off, but does not get lost owing to its being held by the tips of the maxillae.

Food Plants.

The Australian broad- and narrow-leaved wattles (*Acacia pycnantha*, *Acacia saligna*). Indigenous food plants unknown.

Distribution.

Wanganui (M. N. W.), July to March.

ART. XLI.—*Description of a New Species of* Melanchra *from Mount Egmont.*

By MORRIS N. WATT, F.E.S.

[Read before the Wanganui Philosophical Society, 1st November, 1915.]

Melanchra olivea n. sp.

EXPANSE of wings slightly under 1¾ in. *The antennae of the male are deeply serrated, with the serrations finely ciliated.* The forewings are very rich brown, strongly tinged with claret-colour; there is a short blackish-brown basal streak, very broad at the base; the first line is slightly curved oblique, extending from ¼ of costa to ⅓ of dorsum; the claviform is large; *the orbicular is large, oval, oblique, not closed towards the costa; the reniform very large, ear-shaped, also open towards the costa, and inwardly edged with whitish towards the termen;* both reniform and orbicular are sharply outlined in very dark rich brown; there is also a darker brownish cloud between them; the second line is rich brown finely waved, indistinct except towards the dorsum; the subterminal line is rusty-brown, obscurely edged with whitish ochreous; the terminal area is obscurely clouded with blackish; the hindwings are dark brown, paler towards the base.

The type specimen was captured in the early part of January, 1915, and was the only specimen obtained during that trip.* The food plant of this species is not yet known. It may be worth noting here that very few of the large-bodied moths were seen till about 9 p.m., after which time they would be flying in numbers. This happened every night and with great punctuality. Curious, too, the light-bodied moths, which were not attracted by light, though plentiful before 9 p.m., would then become very scarce.

---

* Since writing the above I have obtained a good series of this moth from Mount Egmont during a short trip in January, 1916. These were all caught at light on the veranda of the Mountain House (altitude, 3,000 ft.). During this trip the female was caught. It is a beautiful moth, with the antennae simple, expanse of wings greater, colour richer, and markings more distinct than in the male. I have lately seen the collection of Mr. George Howes, of Dunedin, and was surprised to find a series of a moth which appears to be the southern form of *M. olivea*. The striking points are: Palpi densely covered with grey hair; *forewings uniformly bluish-grey* with markings generally red-brown; outer margin of reniform strongly bent inwards. It is interesting to note that two specimens of this latter form were caught on Egmont during the January, 1916, trip. I had considered these to belong to a new species till I saw the intermediate varieties in Howes's collection.

ART. XLII.—*Descriptions of New Zealand* Lepidoptera.

By E. MEYRICK, B.A., F.R.S.

Communicated by G. V. Hudson, F.E.S.

I AM again enabled, by the kindness of my correspondents Messrs. G. V. Hudson and A. Philpott, to describe some interesting new forms of *Lepidoptera,* including two new genera of remarkable peculiarity.

TORTRICIDAE.

Tortrix melanosperma n. sp.

♂. 21 mm. Head, palpi, and thorax light grey, palpi moderate, ascending. Antennal ciliations 1. Abdomen pale ochreous-grey, anal tuft ochreous-whitish. Forewings elongate, costa moderately arched, without fold, apex tolerably pointed, termen nearly straight, rather strongly oblique, pale grey, with scattered black scales tending to form rows ; costa rather broadly suffused with whitish ; a conspicuous black dot in disc at ¾ : cilia whitish-grey. Hindwings light grey : cilia whitish-grey.

Arthur's Pass, 3,000 ft., in December (Hudson) ; one specimen. Allied to *indigestana.*

Harmologa trisulca n. sp.

♂. 29–31 mm. Head, palpi, and thorax with ferruginous hairs, beneath these dark fuscous ; palpi 3. Antennae without ciliations. Abdomen dark fuscous, segments margined with grey-whitish scales, anal tuft grey-whitish. Forewings elongate, moderate, posteriorily dilated, costa gently arched, with narrow fold on basal sixth, apex obtuse, termen straight, little oblique, rounded beneath ; deep ochreous-yellow ; a broad pale ochreous-yellowish median streak from base to ⅘, suffused beneath and posteriorly, sharply defined above, costal area above this wholly deep red-brown, costal edge suffused with dark leaden-grey, dorsal third of wing suffused with ferruginous : cilia light leaden-grey, basal third with yellowish scales, tips whitish. Hindwings dark grey, towards apex and termen blackish-tinged : cilia grey-whitish, basal third grey.

♀. 33 m. Head and palpi pale ochreous, palpi 4. Thorax with pale-ochreous hairs, blackish-scaled beneath. Abdomen whitish-ochreous. Forewings elongate, moderate, slightly dilated, costa gently arched, apex obtuse, termen faintly sinuate, little oblique ; pale ochreous-yellowish ; two dark-fuscous dots in disc before and beyond middle, just below upper margin of cell : cilia whitish, basal third pale-yellowish. Hindwings pale whitish-yellowish sprinkled with grey, dorsal ⅔ suffused with grey : cilia yellow-whitish.

Arthur's Pass, 3,500 ft., in December (Hudson) ; four specimens. The ♂ is much like *siraea,* but a larger and finer insect, more brightly coloured, the costal area of forewings very much darker than dorsal, the head and thorax without grey suffusion, whilst in *siraea* the costal and dorsal areas are nearly the same in colour, the costal edge distinctly whitish, and the median streak forms a distinct projection along fold ; the ♀, however, are entirely different, *siraea* having grey-whitish forewings and white hindwings, whilst this species is much more like *aenea.* A ♂ sent by Mr. Hudson from the Hunter Mountains is true *siraea.*

Harmologa lychnophanes n. sp.

♂. 28 mm. Head dark fuscous, with some ferruginous hairs. Palpi 3, dark fuscous mixed with brownish. Antennae minutely ciliated. Thorax dark fuscous, with ferruginous hairs. Abdomen dark fuscous, segments margined with whitish-yellowish scales, anal tuft pale ochreous-yellowish mixed with fuscous. Forewings elongate, moderate, posteriorly somewhat dilated, costa gently arched, with moderate fold extending from base nearly to ¼, apex rounded-obtuse, termen slightly rounded, hardly oblique; ochreous-yellowish, somewhat paler in middle longitudinally, dorsal area deeper, with a ferruginous-brown dorsal streak from ¼ to near tornus; costal third bright deep ferruginous, along costa rather broadly and irregularly deep leaden-grey: cilia pale ochreous with grey median shade, towards base ferruginous-yellow. Hindwings ochreous-yellowish sprinkled with grey, becoming yellowish-ferruginous towards apex and termen, base somewhat suffused irregularly with light grey: cilia pale-yellowish, basal third ferruginous-yellow.

Mount Arthur, 4,500 ft., in January (Hudson); one specimen. Formerly identified by me incorrectly as *siraea*, of which my series was taken by myself in the same locality (broadly speaking); I now see that it must be regarded as quite distinct. There is evidently a not inconsiderable group of allied species, and other mountains should be searched for them. As the sexes are always very dissimilar, both should be obtained from the same locality if possible.

GELECHIADAE.

Phthorimaea plemochoa n. sp.

♂ 11–12 mm., ♀ 8–9 mm. Head pale bronzy-grey, side tufts whitish. Palpi grey, sometimes tinged with ochreous. Thorax pale bronzy-grey. Abdomen grey. Forewings in ♂ elongate-lanceolate, in ♀ rather abbreviated and more acutely pointed; lighter or darker bronzy-ochreous, more or less tinged or suffused with grey; a broad streak of whitish suffusion along costa, becoming subcostal for more or less distance beyond middle; plical and second discal stigmata sometimes dark fuscous, sometimes obsolete: cilia grey, more or less mixed with whitish. Hindwings bluish-grey: cilia grey.

Otira River, in December (Hudson); seven specimens. Distinct and interesting.

OECOPHORIDAE.

Borkhausenia xanthomicta n. sp.

♂♀. 13–16 mm. Head, palpi, thorax, and abdomen dark fuscous. Antennal ciliations of ♂ 1. Forewings elongate, widest before middle, costa gently arched, apex pointed, termen extremely obliquely rounded; dark fuscous, sometimes obscurely whitish-sprinkled; markings light yellow-ochreous, more or less tinged with ferruginous in disc; a thick oblique streak from near base in middle to above dorsum at ⅖; oblique narrow more or less incomplete fasciae before and beyond middle, usually not reaching costa; an inwardly oblique transverse spot from costa at ⅘: cilia yellowish, becoming ferruginous-yellow towards base, on costa, and tornus fuscous mixed with ferruginous-yellow towards base. Hindwings dark grey: cilia grey, basal third darker.

Wellington and Wadestown, in November (Hudson); six specimens. Intermediate between *chrysogramma* and *siderodeta*.

Borkhausénia thalerodes n. sp.

♂. 17 mm. Head, thorax, and abdomen blackish, apex of patagiā pale ferruginous-yellowish. Antennal ciliations 2½. Forewings elongate, costa gently arched, apex obtuse-pointed, termen faintly sinuate, rather strongly oblique; deep ferruginous, somewhat mixed with grey; a suffused light-yellowish streak along basal third of dorsum; a whitish-yellowish inwardly oblique transverse spot from dorsum beyond middle, reaching ⅖ across wing: cilia ferruginous, with two grey shades. Hindwings blackish-grey: cilia dark grey, with blackish-grey subbasal shade.

Arthur's Pass, 3,000 ft., in December (Hudson); one specimen. A distinct species, allied to *monodonta*, but with different antennal ciliations.

Izatha huttonii Butl.

This is a good species, and not a faded form of *peroneanella* as I had supposed. I had not possessed a specimen, but have now received one from Mr. Hudson, who states that he has a good series, and that it is constant and distinct; it is white, with scattered scriptiform brown (not black) markings. Taken at Wellington.

Locheutis vagata n. sp.

♂. 14–15 mm. Head, palpi, and thorax bronzy-fuscous. Antennal ciliations nearly 4. Abdomen dark grey. Forewings elongate, rather narrow, costa gently arched, apex obtuse, termen very obliquely rounded; rather glossy fuscous-grey, suffusedly mixed with coppery-bronze; plical and second discal stigmata obscurely darker: cilia bronzy-grey. Hindwings dark grey: cilia grey, with darker basal shade.

Tararua Ranges, 4,000 ft., in November (Hudson); two specimens. An inconspicuous species, and the examples are not in very good condition. The genus *Locheutis* Meyr. is new to New Zealand; it is allied to *Eulechria*, from which it is distinguished by the absence of pecten on basal joint of antennae. There are three Tasmanian species, and one in Ceylon.

Trachypepla chloratma n. sp.

♂. 18 mm. Head ochreous-yellowish, face fuscous. Palpi pale-yellowish, second joint dark fuscous except apex, base of terminal joint dark fuscous. Antennal ciliations 1. Thorax dark fuscous somewhat mixed with pale-yellowish. Abdomen grey. Forewings elongate, rather narrow, posteriorly slightly dilated, costa gently arched, apex tolerably pointed, termen rounded, rather strongly oblique; rather dark fuscous; a suffused ochreous-yellow streak along fold throughout, dilated into a blotch towards extremity, interrupted by a small raised dark-fuscous spot representing plical stigma; posterior ⅗ of wing irregularly mixed or sprinkled with ochreous-yellowish, with an undefined cloudy fascia of ochreous-yellowish suffusion from ⅔ of costa to tornus: cilia fuscous, sprinkled with pale-yellowish. Hindwings grey: cilia light grey.

Table Hill (Invercargill), in December (Philpott); one specimen. May be placed next *lichenodes*.

GLYPHIPTERYGIDAE.

Philpottia n. g.

Head with appressed scales; ocelli absent; tongue absent. Antennae ½, in ♂ moderately ciliated, joints about 26 in number, rather elongate, with slightly expanded whorls of scales, basal half of stalk thickened with rough scales, more strongly towards base, basal joint short, stout, thickened with dense scales projecting on anterior edge. Labial palpi extremely small, rudimentary. Maxillary palpi absent. Posterior tibiae with appressed scales. Forewings with 1*b* furcate, 2 from towards angle, 3 from angle, 4 and 5 approximated at base, 7 to termen, 8 and 9 approximated at base, 11 from middle. Hindwings somewhat under 1, elongate-ovate, cilia $\frac{2}{3}$; 3 and 4 somewhat approximated towards base, 5 tolerably parallel, 6 and 7 somewhat approximated towards base, transverse vein rather strongly oblique.

A very remarkable genus. On the neural and antennal characters I can only regard it as belonging to this family, but it differs widely from all the other New Zealand genera in the minute labial palpi, which are only perceived with difficulty. It has, however, many points of resemblance to the Australian genus *Cebysa*, in which the labial palpi are very short, and there seems to be a true relationship. Mr. Philpott sent me at first a good ♀; on perceiving its singular interest, being doubtful whether the palpi and tongue might not have been broken off, I wrote at once asking him to examine his own specimens on this point, and he very kindly forwarded to me his only ♂ specimen (unfortunately damaged), and also informed me that he possessed two other ♀, in which the structure of the mouth parts was quite as in the one originally sent. I have had pleasure in naming this curious genus after the captor. Assuming the relationship to *Cebysa*, that genus is itself no less singular and isolated, so that the general affinity still remains to be elucidated by the discovery of allied forms.

Philpottia iridoxa n. sp.

♂♀. 14–15 mm. Head and thorax purple-coppery-metallic. Antennae deep purple. Abdomen dark fuscous. Forewings elongate, posteriorly slightly dilated, more so in ♂, costa gently arched, apex obtuse, termen obliquely rounded; coppery-purple, with strong peacock-blue gloss; markings ochreous-whitish; slender transverse fasciae at $\frac{1}{4}$ and middle, triangularly dilated on dorsum, more strongly in ♂, first not reaching costa; a triangular or wedge-shaped spot on costa at $\frac{2}{3}$, one more elongate on costa towards apex, and a narrow posteriorly oblique mark from just before tornus: cilia bronzy-grey, basal third coppery-blue-purple. Hindwings in ♂ dark grey, in ♀ grey: cilia grey.

"The specimens were all taken on the same date, 29th December, 1914, on Mount Burns, Hunter Mountains, at an elevation of about 3,250 ft. The locality was a sheltered slope with a dense carpet of native grasses. A grove of birch (*Nothofagus*) was near at hand, and one of the moths was beaten from one of the trees; the others were disturbed from the herbage. With regard to your suggestion that the larva might be a lichen-feeder, it may be noticed that these mountain birches are generally covered with lichens of various species, the forest on the Hunter Mountains being particularly noticeable in that respect. The day was sunny and calm." (Philpott.)

The larva of *Cebysa* feeds in a portable case of silk and refuse on lichens on rock-faces.

Simaethis zomeuta Meyr.

A pair from Arthur's Pass, 3,000 ft. (Hudson), must be referred to this, but I now doubt whether *zomeuta* is anything more than a large mountain form of *combinatana*. The ♂ sent (in fine condition) is larger than any of my examples of *combinatana*, and there is a ferruginous spot beneath the white dot on lower angle of cell connecting it with the following line, and the costal edge towards middle is suffused with ferruginous.

Glyphipteryx brachydelta n. sp.

♂. 8 mm. Head and thorax rather dark bronzy-fuscous. Palpi with four whorls of blackish white-tipped scales. Abdomen dark fuscous. Forewings elongate, rather narrow, costa slightly arched, apex obtuse-pointed, termen slightly sinuate, very oblique; rather dark bronzy-fuscous, becoming lighter bronze posteriorly; an oblique-triangular white spot on dorsum beyond middle, apex shortly acute and slightly bent over posteriorly; five silvery-whitish strigulae from costa, margined with dark fuscous, first in middle, first three little oblique, reaching half across wing, each followed by an obscure whitish costal dot, last two short, direct; two similar strigae from before and beyond tornus, first erect, almost united by a dot with apex of second costal, second shorter, outwardly oblique; a whitish dot on termen beneath apex; apex suffused with black: cilia whitish, within a dark-fuscous line whitish-bronzy-fuscous, indented with white on subapical dot. Hindwings dark fuscous; cilia fuscous.

Karori, Wellington, in March (Hudson); three specimens. Easily distinguished from all others by the small size and broad-triangular white dorsal spot; next *leptosema*.

GRACILARIADAE.

Parectopa citharoda n. sp.

♀. 10 mm. Head probably white (injured). Palpi white. Thorax white, patagia dark fuscous. Abdomen dark grey, sides obliquely striped with white, ventral surface white. Forewings very narrow, moderately pointed; dark bronzy-fuscous, towards apex lighter and more bronzy; five slender white blackish-edged streaks from costa, first three very oblique, first from ¼, reaching half across wing, second from middle, reaching more than half across wing, its apex closely followed by a short fine dash, third shorter, fourth fine, direct, reaching termen, dilated on costa, fifth just before apex, fine, inwardly oblique, cutting through a small round blackish spot; a white dorsal streak from base to middle, terminated by an oblique projecting streak reaching nearly half across wing; a white triangular spot on dorsum beneath apex of second costal streak; a short outwardly oblique white streak from tornus: cilia greyish, with white bars on costal markings, and dark-fuscous median and apical lines above apex separated with whitish. Hindwings dark slaty-grey: cilia fuscous.

Wanganui and Wellington, in November (Hudson); one specimen.

PLUTELLIDAE.

Circoxena n. g.

Head smooth, rounded; ocelli distinct; tongue developed. Antennae ¾, basal joint very long, slender, thickened towards apex, with slight pecten. Labial palpi very long, slender, recurved, terminal joint longer than second, acute. Maxillary palpi very short, filiform, porrected. Posterior tibiae with rough projecting hair-scales above. Forewings with 1*b* apparently simple, 2 from ⅘, 3 from angle, 3–5 somewhat approximated at base, 7 and 8 long-stalked, 7 to costa, 9 and 10 rather approximated to 8 at base, 11 from middle. Hindwings under 1, lanceolate, cilia 1½; 3 and 4 connate, 5 and 6 stalked, 7 parallel.

A singular form, structurally nearest to *Acrolepia*, but quite peculiar in appearance, and very interesting.

Circoxena ditrocha n. sp.

♀. 11 mm. Head dark shining bronze. Palpi and antennae white lined with black. Thorax dark bronzy-fuscous, external edge of patagia white. Abdomen dark grey. Forewings elongate-lanceolate, fuscous, suffusedly streaked longitudinally with blackish; a fine white longitudinal line just beneath costa on basal fourth, costal edge black; dorsum white towards base; a light yellow-ochreous patch occupying basal third of wing from near costa to fold, marked with three fine whitish longitudinal lines diverging from base; two large fine whitish rings in disc before middle and about ⅔; a small black apical spot preceded by some whitish suffusion: cilia on costa dark grey, basal half barred with ochreous-whitish, on termen ochreous-whitish tinged with grey towards tips and basal area barred with grey. Hindwings grey, becoming dark grey towards apex: cilia whitish-grey.

Wainuiomata, in December (Hudson); one specimen.

NEPTICULIDAE.

Nepticula cypracma n. sp.

♀. 7 mm. Head whitish-yellowish, collar grey-whitish. Antennal eyecaps whitish. Thorax dark grey. Abdomen grey. Forewings lanceolate; prismatic grey-whitish, irregularly sprinkled with dark grey; basal fourth dark purple-grey; a deep coppery-bronze apical spot mixed with blackish: cilia violet-grey sprinkled with black. Hindwings and cilia grey.

Karori, Wellington, in November (Hudson); one specimen.

LYONETIADAE.

Erechthias macrozyga n. sp.

♂. 14 mm. Head white, crown mixed with dark fuscous. Thorax white marked with dark fuscous (injured). Abdomen grey. Forewings narrow-lanceolate; dark bronzy-fuscous sprinkled with blackish; an irregular-edged white streak along dorsum and termen from base nearly to apex; a white mark from costa just before apex: cilia fuscous mixed with white, with a dark-fuscous median line round apex. Hindwings purple-fuscous, darker towards apex: cilia fuscous.

Tisbury, in February (Philpott); one specimen.

TINEIDAE.

Tinea accusatrix n. sp.

♂. 8–10 mm. Head white, face grey. Palpi whitish, lined with blackish. Thorax white mixed with fuscous. Abdomen light-greyish. Forewings elongate, narrow, costa gently arched, apex tolerably pointed, termen faintly sinuate, extremely oblique; dark fuscous; a short fine white median longitudinal line from base; oblique white streaks from costa at ¼ and middle reaching more than half across wing, and shorter oblique marks from dorsum opposite; seven white wedge-shaped marks from costa on posterior half, anteriorly somewhat oblique, posteriorly direct, one from tornus and a dot on termen beneath apex, space between these with violet and bronzy reflections; a round deep black spot at apex: cilia whitish, basal third within a blackish line bronze, indented with whitish beneath apex, above apex with a black apical line which projects at apex as a fine straight point, in subapical indentation with two fine dark-fuscous bars. Hindwings light grey, with bronzy and purple reflections: cilia whitish-grey.

Kaitoke (Hudson); two specimens. A curious development of the *margaritis* group, distinct from all others by the singular apical caudate projection of cilia. The type of markings and of cilia seems modelled on certain forms of *Glyphipteryx*.

14*

ART. XLIII.—*Descriptions of New Species of* Lepidoptera.

By Alfred Philpott.

Communicated by Dr. W. B. Benham, F.R.S.

[*Read before the Otago Institute, 7th September, 1915.*]

CARADRINIDAE.

Aletia cuneata n. sp.

♂♀. 38–39 mm. Head, palpi, and thorax grey tinged with ochreous. Antennae in ♂ serrate, ciliated to apex, 1. Abdomen ochreous mixed with fuscous. Forewings, costa almost straight, apex subacute, termen oblique, evenly rounded; ochreous-grey mixed with fuscous; lines often very obscure; subbasal whitish, margined with fuscous on both sides, deeply indented beneath costa; first line outwardly oblique, strongly but irregularly dentate, whitish-ochreous, posteriorly black-margined; orbicular annular, ochreous; a suffused fuscous, irregularly dentate median shade; reniform filled with dark fuscous, pale-margined; second line indicated by a series of black-margined crescentic ochreous marks, excurved from costa to below middle, thence incurved to dorsum; veins more or less clearly marked with alternate black and white or ochreous dots; an obscure ochreous subterminal line, dilated above tornus: cilia ochreous, densely mixed with greyish-fuscous. Hindwings fuscous-brown, becoming ochreous on basal half: cilia ochreous with an obscure darker basal line.

Near *A. griseipennis* Feld., but smaller and more ochreous-tinged; the dark reniform is a good distinguishing character.

Ben Lomond, Macetown. Not uncommon from 2,000 ft. to 3,000 ft. in January and February.

HYDRIOMENIDAE.

Chloroclystis erratica n. sp.

♂♀. 19–23 mm. Head and palpi fuscous mixed with greyish. Palpi, 2¼. Antennae in ♂ fasciculate-ciliate, ciliations 3½. Thorax grey-fuscous with broad central stripe reddish-brown. Abdomen reddish-brown sprinkled with fuscous and grey, sides fuscous. Forewings elongate-triangular, termen obliquely bowed, greyish-fuscous; a curved grey fascia near base, often obscure; median band limited anteriorly by curved grey fascia slightly indented below middle and margined more or less with blackish on both sides; preceding this a parallel, suffused, reddish-brown fascia; 3 or 4 thin waved grey fasciae in posterior portion of median band; posterior margin of median band from ⅔ costa to ⅘ dorsum, outwardly oblique to near middle of termen, thence abruptly bent inwards for about half the breadth of band and from vein 2 almost right-angled to dorsum; a rather broad fascia of reddish-brown parallel to median band, paler anteriorly and sometimes traversed by a thin darker line; a thin serrate grey subterminal line, close to termen on lower half; a black line round termen: cilia fuscous-grey, obscurely barred with black. Hindwings, termen unevenly rounded; fuscous-grey; numerous lighter and darker fasciae from dorsum and a broader reddish-brown shade before termen; a strong black terminal line: cilia as in forewings, but paler.

In the form of the median band this species comes nearest to *C. lichenodes* Purd., but is at once differentiated by the structural differences in the antennae. It is probably more closely related to *C. magnimaculata* Philp., but the darker ground-colour easily distinguishes it from that species.

Mr. C. C. Fenwick took a ♀ on Bold Peak, Humboldt Range, in December, 1913, and in the same month of 1914 I found the species very plentiful on the Hunter Mountains at an elevation of 3,250 ft. It was associated chiefly with *Veronica buxifolia*, and it is remarkable that in a series of about thirty specimens there was no example of the ♀. Type, ♂ in coll. A. Philpott; ♀ in coll. C. C. Fenwick.

Chloroclystis rivalis n. sp.

♂♀. 17–20 mm. Head, palpi, and thorax fuscous mixed with red and white scales. Palpi 2. Antennae in ♂ fasciculate-ciliate, ciliations 3½. Abdomen fuscous, densely sprinkled with reddish and grey. Forewings triangular, termen obliquely rounded; reddish-fuscous with some slight ochreous admixture; median band not clearly defined, anterior margin indicated by a pair of pale curved fasciae; several similar fasciae within band; posterior edge of band broadly and bluntly projecting at middle, margined on upper half by a bluish-white fascia followed by a thin dark fascia, which is in turn followed by a rather broad ochreous fascia, these fasciae becoming almost obsolete on lower half; apical area more strongly reddish; subterminal line serrate, interrupted, whitish or greenish; a black terminal line: cilia fuscous, mixed with grey and obscurely barred with black on basal half. Hindwings, termen unevenly rounded; fuscous mixed with grey and some reddish scales; numerous alternate light and dark fasciae obscurely indicated: cilia as in forewings.

Nearest to *C. sandycias* Meyr., but easily distinguished by the much darker colour of both fore and hind wings.

A single ♂ taken by Mr. C. C. Fenwick on Bold Peak in December, 1912. I took several of both sexes on the Hunter Mountains (3,250 ft.) in December, 1914, and Mr. C. E. Clarke obtained a ♂ at about the same date at the Routeburn.

TORTRICIDAE.

Pyrgotis consentiens n. sp.

♂♀. 12–15 mm. Head, palpi, and thorax dark purplish-red. Abdomen fuscous. Forewings oblong, costa gently arched, apex rounded, termen subsinuate; purplish-red; sometimes a white fascia from ¼ costa to before middle of dorsum, narrowest towards costa, sometimes upper portion obsolete: cilia reddish-ochreous. Hindwings fuscous: cilia fuscous-grey with a darker basal line, reddish-ochreous round apex.

Abundantly distinct form other forms of the genus. The examples having the white fascia are not common, occurring in about the proportion of 1 to 12.

Table Hill, Stewart Island. One specimen at 2,000 ft. in December. Hunter Mountains in December. Common amongst *Veronica* and *Cassinia* scrub at 3,000 ft. to 3,500 ft.

Eurythecta varia n. sp.

♂. 9–10 mm. Head, palpi, and thorax brownish-ochreous. Antennae dark brown, annulated with ochreous. Abdomen dark fuscous. Forewings with vein 6 present, to termen, costa almost straight, apex subacute, termen rounded, strongly oblique; dull to bright ochreous, mixed with dark fuscous; a white or pale-ochreous streak from base to ⅓, much dilated posteriorly, margined beneath by a dark-fuscous blotch, and with a similar blotch above extending beyond ; a white, posteriorly oblique, cuneate striga from dorsum beyond middle, its apex almost reaching costa, followed by a suffused dark-fuscous blotch ; 3 dark-fuscous spots on apical third of costa, sometimes margined with white, often obsolete ; usually some orange scales round tornus : cilia grey, mixed with ochreous and orange at base. Hindwings with vein. 4 absent, greyish-fuscous : cilia greyish-fuscous with a darker basal line.

♀. 10–12 mm. White markings of ♂ obsolete, and forewings almost wholly suffused with greyish-fuscous.

Probably with most affinity to *E. potamias* Meyr., from which it is easily separated by the white markings.

Discovered by Mr. C. C. Fenwick on the Kaikoura Range (Marlborough) in the month of December. Mr. Fenwick informs me that the species was common in marshy spots, and that when disturbed from the herbage it almost invariably alighted on the water.

Tortrix fastigata n. sp.

♂♀. 21–24 mm. Head, palpi, and thorax pale ochreous mixed with brown. Antennae ochreous annulated with brown. Abdomen ochreous. Forewings in ♂ elongate-triangular, in ♀ oblong, costa hardly arched in ♂, in ♀ strongly arched at base and indented slightly before middle, apex rounded, termen sinuate, hardly oblique ; pale yellow, irrorated with purplish-brown ; markings in ♂ very obscure, median fascia brownish, outwardly oblique beneath costa, bent inwardly to disc, thence obliquely to before tornus ; a triangular brownish blotch on costa before apex ; a slightly curved linear mark beneath this at middle ; in ♀ a basal striga with a blunt projection outwardly at middle is obscurely indicated, and the triangular costal patch is connected with the curved median mark beneath it : cilia yellow, mixed with brown near apex. Hindwings and cilia pale whitish-yellow sprinkled with purplish-brown.

Near *T. acrocausta* Meyr., but the ♂ is much longer- and narrower-winged and the markings of the ♀ are dissimilar. Probably *fastigata* is attached to open country, while *acrocausta* is confined to the bush.

Longwood Range, a ♀ at 3,000 ft. in December. Hunter Mountains, one of each sex at 3,500 ft. in January.

OECOPHORIDAE.

Borkhausenia hastata n. sp.

♂. 19 mm. Head and thorax whitish - ochreous. Palpi whitish-ochreous, mixed with purplish-brown beneath. Antennae whitish-ochreous, annulated with purplish-brown, ciliations 1. Abdomen ochreous-brown, tuft ochreous. Forewings elongate, costa moderately arched, apex produced, termen strongly oblique ; whitish-ochreous ; markings purplish-brown ; an obscure streak from base along fold to ½ ; a dot above dorsum at base and one in disc at ⅔ ; a rather suffused series of spots along termen and some scales on costa near apex : cilia whitish-ochreous with some brownish scales near base. Hindwings and cilia shining white, ochreous tinged.

Not far from *B. chloradelpha* Meyr. in coloration, but wholly different in form of wing, in which it is more nearly approached by the much smaller *B. maranta* Meyr.

Seaward Moss, in October. A single specimen.

GLYPHIPTERYGIDAE.

Glyphipteryx plagigera n. sp.

♂♀. 9½–11 mm. Head ochreous-white. Thorax fuscous-brown. Palpi moderately tufted beneath, ochreous-white with 4 obscure fuscous rings. Abdomen blackish-grey, obscurely annulated with white. Posterior legs black, tibiae and tarsi annulated with white. Forewings elongate, moderately dilated posteriorly, costa moderately arched, apex obtuse, termen oblique ; dark greyish-fuscous, purplish tinted and more or less sprinkled with white posteriorly ; a narrow, outwardly oblique, white streak from costa at ¼, reaching to near middle of wing ; 5 similar streaks between this and apex, the last two less oblique ; a broad white blotch on dorsum near base and a similar one before middle, sometimes uniting at apex ; an obscure white streak from tornus, sometimes uniting with third costal streak : cilia fuscous-grey with median line and apical hook darker. Hindwings broadly lanceolate ; fuscous-grey : cilia fuscous-grey.

Distinguished from *G. leptosema* Meyr. and *G. iochaera* Meyr., which have somewhat similar markings, by the tufted palpi. *G. oxymachaera* Meyr. has tufted palpi, but differs from the present species by the absence of the dorsal blotches and the presence of a submedian stripe.

Bluff, in November. Three specimens.

ART. XLIV.—*Notes on some* Coccidae *in the Canterbury Museum, together with a Description of a New Species.*

By G. Brittin.

[*Read before the Philosophical Institute of Canterbury, 3rd November, 1915.*]

Since my last paper on New Zealand *Coccidae* was read before this Institute I have had an opportunity of making an examination of the Coccid slides deposited in the Canterbury Museum by the late Mr. W. M. Maskell, and the following are some notes on the different species.

Eriococcus multispinus Mask., Trans. N.Z. Inst., vol. 11 (1879), p. 217 ; vol. 23 (1891), p. 21 ; vol. 24 (1892), p. 31.

In dealing with this species, Mr. Maskell has undoubtedly mixed up *E. multispinus* with *E. pallidus*. In his original diagnosis of *E. multispinus* (vol. 11, p. 217) he states, "The insect is seen to have several rows of large conical spines. . . . The antennae have 6 joints, the 3rd being the longest, the 4th and 5th equal to each other and nearly round." Again, in giving the generic and group distinctions among the *Acanthococcidae* and *Dactylopinae* (vol. 24), "Passing now to the genus *Eriococcus*, figs. 11, 12, and 14 of Plate IV show that there are three antennal forms, and figs. 16 and 17 that there are two forms of marginal spines in the genus. Fig. 14 (*E. raithbyi*) differs from fig. 11 (*E. multispinus*) only in having 7 joints: in both the joints are subequal. But in fig. 12 (*E. pallidus*) the 3rd joint is much longer than any of the others. As regards the spines, those of *E. pallidus* and its allies are shown in fig. 16 to be much larger and more slender than those of *E. multispinus* and its allies, fig. 17." In treating of the variations of *E. pallidus* (vol. 23, p. 21), he says, "This species exhibits several variations in the arrangement of the dorsal spines, and slightly in the size and colour of the sac. . . . I leave all these as variations of one species, chiefly on account of the antennae, which I find similar in all, with 6 joints, of which 5 are subequal; but the 3rd joint is longer, usually equal to any two of the others. This character, together with the slenderness of the spines, distinguishes *E. pallidus* from *E. multispinus*, irrespective of variations in the sac, which are not, indeed, important."

There can be no doubt after reading these different descriptions that Maskell's first description of *E. multispinus* undoubtedly referred to what he afterwards called *E. pallidus*, the antennae of which have 6 joints, the 3rd being the longest, and the spines long and slender. On the other hand, the antennae of *E. multispinus* consist of 6 subequal joints, and the spines are, in comparison, short and broad.

Now, this alteration in the diagnosis would not have been of so much consequence but for the fact that Mr. Maskell made the same mistake when forwarding specimens to others who were working on the *Coccidae*, and the slide in the Museum shows that such a mistake has evidently been made, and it needs only a slight glance at this specimen to see that the 3rd joint of the antennae is much longer than any of the others and that the spines are long and slender, both of which facts point to its being *E. pallidus*. Unfortunately, there are no slides in the Museum labelled "*E. pallidus*," so that the two species cannot be compared.

Some time ago I forwarded to Mr. E. E. Green, F.E.S., one of the leading English authorities on the *Coccidae*, some specimens, under the name of *E. multispinus*, and received an answer saying that it differed in several important characteristics from Maskell's *multispinus*. On my still pointing out several characteristics mentioned by Maskell in Trans. N.Z. Inst., vol. 23, I received the following, which I here quote : "I see that you are right about the 6-jointed antennae of *E. multispinus*, but on comparing your specimen with typical examples (received from Maskell himself) I am still of opinion that your insect is distinct. The character and arrangement of the spines is very different in the two insects. If the difference should be considered to be insufficient to warrant the erection of a new species, your insect should at least be distinguished by a varietal name." I have lately forwarded Mr. Green specimens of both species, and at the same time pointed out the mistake made by Mr. Maskell when describing his species.

Dactylopius poae Mask., Trans. N.Z. Inst., vol. 11 (1879), p. 220 ; vol. 23 (1891), p. 23.

This is another species that will want investigating. The normal generic characters of this genus are as follows : Adult female with antennae of 8 joints, the last joint almost invariably longer than the penultimate ; mentum biarticulate ; legs persistent ; anal lobes small or rudimentary. In the genus *Ripersia* the antennae consist of 6 joints, rarely of 5 or 7. Maskell's diagnosis of *D. poae*, given in Trans., N.Z. Inst., vol. 11, p. 220, is practically useless as a means of identification. In his book on "New Zealand Scale Insects " he simply mentions that the antennae consist of 8 joints, and are very short. The slide in the Museum labelled "*D. poae*" contains only the anterior portion of the insect, with the antennae, rostrum, and anterior pair of legs. I made a very careful examination of the antennae, and found that it consisted of only 6 joints, the 3rd and apical joints being longest and about equal to each other. This material difference would, of course, at once place the insect in the genus *Ripersia*, and, if all the other characteristics are the same, would agree with my description of *Ripersia globatus*, published in Trans. N.Z. Inst., vol. 47, p. 155, which I think will eventually become a synonym of Maskell's species under the name of *Ripersia poae*.

Eriochiton spinosus Mask.

> *Ctenochiton spinosus* Mask., Trans. N.Z. Inst., vol. 11 (1879), p. 212. *Eriochiton spinosus* Mask., Trans. N.Z. Inst., vol. 19 (1887), p. 47. *Lecanium armatus* Brittin, Trans. N.Z. Inst., vol. 47 (1915), p. 152.

This species is undoubtedly the same as the one reported by me under the name of *Lecanium armatus*. Maskell's diagnosis and diagrams of *E. spinosus* are very imperfect, and it was little wonder that I was unable to recognize it from his description. In vol. 11, p. 212, when describing it under the name of *C. spinosus*, he states that the abdominal lobes are as usual. This is not correct, and, in fact, the lobes appear to be very unusual for the genus *Ctenochiton*. Again, in *E. spinosus* the anal ring is situated between the lower half of the abdominal lobes. Since my last paper was read before this society I have been able to find specimens covered with a thin cottony test, similar to that of an *Eriococcus*. These specimens will, I think, turn out to be Maskell's *E. hispidus* (Trans. N.Z. Inst., vol. 19, p. 47). They appear to agree very well with his description, but the difference is so slight that they may be eventually placed as a variety of *E. spinosus* rather than as a distinct species. Up to the present time I have been unable to find any signs of a test covering my original species. Maskell's slide in the Museum is a typical example of the species.

Mytilaspis drimydis Mask., Trans. N.Z. Inst., vol. 11 (1879), p. 196.

I have carefully examined the slide of this species deposited in the Museum, and can come to no other conclusion than that it is the second instar of one of the *Fiorinia*, and most probably *F. stricta*. The extremity of the pgyidium, with its broad flat squames, small narrow lobes, and the presence of few dorsal spinnerets, together with the absence of grouped circumgenital glands, all point to its being the early stage of the second instar of some species of the *Fiorinia*. I should mention that Maskell's *F. stricta* and *F. asteliae* have been both placed in the genus *Leucaspis*.*

Subfam. DIASPINAE.
Genus ODONASPIS?

Odonaspis ? leptocarpi sp. nov.

Puparia of the females situated underneath the ligules of the plant on which they live; they are always found packed closely together, and consequently it is very hard to distinguish the separate puparia. Ventral scale complete, white, and remains firmly attached to the plant; dorsal portion white, elongate. Exuviae yellow, and appear to be situated rather to one side at the anterior extremity. Dorsal portion generally attached to ligule, and separates from the ventral scale on detaching the ligule from the stem, thus leaving the insect uncovered.

---

* E. E. Green, F.E.S.: "Some Remarks on the Coccid Genus *Leucaspis*, with Descriptions of Two New Species." Trans. Entom. Soc., London, February, 1914, pt. iii, iv.

Adult female elongate-ovate, being equally rounded at both extremities ; colour pink, turning bright green on maceration in potash. Segmentation very distinct. Body covered with minute wrinkles. Eyes in living insects very prominent, and appear as large dark granulated spots. Rudimentary antennae with two or three rather long hairs. Rostrum rather more chitinized than is usually found in the *Diaspinae;* mentum appearing as a round ring. Spiracles widely dilated at outer extremity and tapering inward to a small round orifice. Parastigmatic glands absent. Pygidium broader than long, slightly chitinized, with 5 groups of circumgential glands ; anterior group 6–8 glands ; anterior laterals 16–18 ; posterior laterals 18–24 ; anterior group widely separated from the rest ; lateral groups almost joining. Anal orifice situate midway between anterior lateral groups. Margin of pygidium with a crenulate appearance, without lobes or large marginal tubular spinnerets. Dorsal tubular spinnerets in series ; the first six series are immediately beneath the grouped glands, and each consists of from 6 to 8 spinnerets, extending directly upwards towards the grouped glands ; the next series on each side extends upwards along the outer side of the circumgenital glands to a level with the anterior group. Each segment above has numerous tubular spinnerets at the outer margin, and on each articulation there is a series of rather larger spinnerets extending directly in towards the body ; there are also a few spinnerets on the lower half of the cephalic segment. Above the antennae there are a few short spiny hairs.

Length, about 1·08 mm ; width, 0·54 mm.

Hab.—On *Leptocarpus* sp., at present only from New Brighton.

Note.—The genus *Odonaspis* has not hitherto been recorded as occurring in New Zealand.

Art. XLV.—*New Light on the Period of the Extinction of the Moa (according to Maori Record).*

By T. W. Downes.

[*Read before the Wanganui Philosophical Society, 26th January, 1916.*]

It would be hard to find anything connected with the natural history of New Zealand that has attracted more attention than the extinct moa. Over seventy papers, containing several hundred pages of matter, have been published in the "Transactions of the New Zealand Institute," and of this matter a large portion deals with the period of extinction ; yet the date is by no means settled, and I have no doubt that any additional light that will assist in arriving at a solution of this much-discussed question will be welcomed by those who are interested in the subject.

Arguments relating to this discussion have been carefully studied and summarized by F. W. Hutton in a fine paper entitled "The Moas of New Zealand" (Trans. N.Z. Inst., vol. 24, p. 93), and the writer, after careful deliberation, comes to the conclusion that in the North Island the moa was exterminated by the Maori not very long after their arrival in New Zealand—that is, not less than four or five hundred years ago—and that they existed for about one hundred years later in the South Island. .

Among the gentlemen who held a contrary view—namely, that the moa was exterminated in quite recent years—I would mention J. W. Hamilton, J. and W. Murison, J. H. Coburn, James Hector, John White, W. T. L.

Travers, and de Quatrefages; and there are many others. Others, again, assert that the bird became extinct in prehistoric times, and that the Maori was almost in utter ignorance of its existence. Among these we find the names of B. S. Booth, von Haast, J. W. Stack, Colenso, Sir George Grey, Alexander Mackay, and J. H. F. Wohlers. These writers assert that the Maori had no tradition on the subject, no songs, and, with one exception, no proverbs. These gentlemen were for the most part studious Maori scholars; yet in this theory they were wrong, for the Maori certainly has references to the moa, but under different names. Every one knows that Maori songs bristle with untranslatable names of gods, men, and places connected with far-away Hawaiki, and it is therefore not surprising that such names as *kura-nui* and *manu-whakatau* were passed over without comment.

Briefly, the discussion has been on the following lines :—

The first collection of moa-bones was made by W. Colenso, W. Williams, and others about 1840.

In 1847 Mantell deduced from his discoveries that the moa had been eaten by man. In 1864 Buller published letters to the effect that the moa was extinct, but was contemporaneous with the Maori, as shown by the burnt and broken bones on the site of their feasts, and also by traditions still held by the Maori.

Next came the discovery of a skeleton with skin and ligaments attached. A discussion followed, arising from a paper read by Alles, when the general opinion was that the bird had probably been living within ten years. Then, in 1868, E. Newman concluded that the last moa died about 1800, or even later.

The same year Mantell pointed out that the extermination of the bird must have taken place shortly after the appearance of man, as the allusions to the moa by the Maori were so extremely rare. He also pointed out that nephrite appeared to have been discovered at a later date than the extinction of the moa, as it was never found in conjunction with bones at a Maori cooking-place.

In 1871–74 von Haast published papers on moas and moa-hunters, in which he denied the existence of Maori tradition, and sought to prove that the extermination had taken place by a race of people prior to the Maori—a race who were unacquainted with greenstone, and had not even acquired the art of grinding stone. Later von Haast modified his ideas somewhat, and stated that the moa-hunters had reached a certain stage of civilization. This arose from the finding of some polished instruments with moa-bones.

Colenso, in a very valuable paper on the subject, recognizes that the ancestors of the Maori knew the moa, but in a very vague sort of way, as it was extinct long before the genealogical descent of the tribes, which extends back some twenty-five generations. He says allusions are to be found to the bird in Maori poetry, but that these allusions are largely mythical. He mentions that there is a tradition among the Maoris of the East Cape district that the moas were exterminated by the fire of Tamatea, captain of the "Takitimu" canoe. As will be noticed later, the song evidence that I have collected bears out Colenso's arguments to some extent; indeed, according to the evidence now to hand, his deductions are probably nearer the truth than the theory held by present-day writers.

Mantell says, "The extermination of the moas must have taken place shortly after the Maoris reached New Zealand, as allusions to the bird in their most ancient traditions are very slight and obscure."

W. G. Mair, a thorough Maori scholar, says, "In all these thousands
of pages of Maori lore that I have written from the mouths of [Maori]
witnesses there is not one word about the moa." J. W. Stack says practically the same thing. On the other hand, a great number of writers
affirm that the Maori knew the moa well, and has plenty of traditions
touching upon its appearance, food, methods of hunting it, &c., some of
which are purely romances, others probably true; but the weight of evidence by the bulk of writers is, to my mind, in favour of the ancient period.

In this paper it is not my intention of joining in a discussion to which
there seems no finality; what I want to point out is a fact that was
unknown or overlooked by both the Maori students and those gentlemen
who gave their opinion in regard to Maori songs. The fact that they have
overlooked is this: In several of the very old songs and proverbs, probably
the oldest, the bird is known by several names—first *kura-nui*, second *manu-
whakatau*, and (probably) at a later date *moa*.

I have carefully looked through between eight hundred and a thousand
Maori *waiata, karakia, oriori,* and *whakatauki* (songs, incantations, lullabies,
and proverbs), and have found a fair number of references, from which I
have selected from three periods the following. The first and oldest is the
last section of a very fine *karakia* composed by a *tohunga* named Tuhoto-
ariki, who belonged to a period about one generation after the coming of
the fleet. This song was composed on the birth of Tutere-moana, whose
mother was in prolonged labour. As it almost entirely deals with the Maori
ideas of conception, a full translation might be considered out of place; it is,
moreover, apart from the present purpose. The last verse, therefore, will
alone be given, also a table showing the period of the composer. As this
song has been handed down from father to son right from the period of the
migration, it proves to my mind that the moa was extinct, and recognized
by the Maori as such, in Tuhoto-ariki's time—that is, a few years after the
arrival of the canoes. Indeed, it is probable that he was a boy when the
fleet arrived.

The word "*manu-whakatau*" probably means a bird superior to all others,
as indicated by the following ancient proverbs: "*Hopara makau rangi
manu-whakatau*" (The finest chieftain is like the manu-whakatau—moa).
"*Ko wai tou tamaiti; kapa ko te manu-whakatau?*" (Is your child noble
like the manu-whakatau—moa?).

Kura-nui had a somewhat similar meaning. The usual interpretation
of the word "*kura-nui*" would be "great-red." ("Red" to the Maori includes
"brown.") "*Kuranga o te ao*" (The redness of the morning). Also, *kura*
is applied to anything highly prized. A second meaning is a valued feather
used for head-dress, such as a huia or amokura feather. *Kura-nui* was
likewise used as a term of address to a high-born chieftainess. Tahurangi
(the highest peak on Ruapehu) was another word of address representing
the same high idea. *Kura* was also a term applied to sacred things connected with the *whare wananga*, or house of learning, in far-away Hawaiki.

I mentioned my discovery to the Rev. T. G. Hammond, of Hawera,
who was exceedingly interested, and the day after our conversation sent
me the following note, which adds additional interest to the words: "Since
our talk about those *waiata*, I have seen an old Maori who confirms the
names *kura-nui* and *manu-whakatau* as names of the moa. I have also
learned that these names indicate varieties of the moa." He was aware
that there were great differences as shown by the remains, and this may
account for the word *momo* (race) in this connection.

A point that should be noticed is that in these, as in most of the songs and proverbs relating to the moa, the word " *huna* " is used, meaning " hidden," never " *ngaro*," which we would think more applicable, as it means " lost." This seems to imply that the habit of hiding the head when pursued, characteristic of all wingless birds, had been noted by the Maori and the idea incorporated into his references. It is also worthy of mention that as early as Tuhoto-ariki's time the natives had noticed that the swamps held quantities of bones, and their attempt at an explanation of the fact is certainly noteworthy.

·Table I.

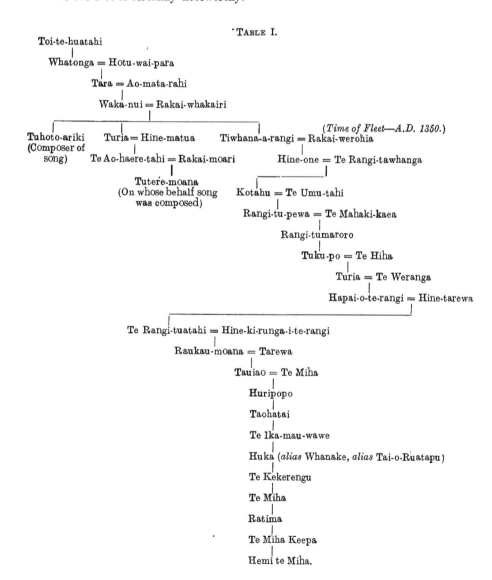

TABLE II.

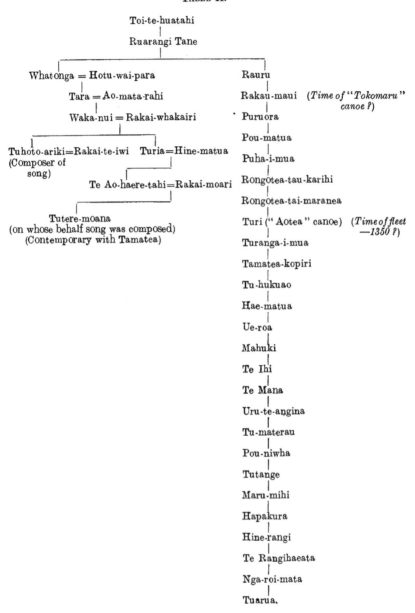

```
                            Toi-te-huatahi
                                 |
                            Ruarangi Tane
                                 |
        ┌────────────────────────┴──────────────────┐
  Whatonga = Hotu-wai-para                         Rauru
            |                                        |
          Tara = Ao-mata-rahi                    Rakau-maui   (Time of "Tokomaru"
                |                                                    canoe ?)
            Waka-nui = Rakai-whakairi             Puruora
                |                                    |
   ┌────────────┴──────────────┐                 Pou-matua
 Tuhoto-ariki=Rakai-te-iwi   Turia=Hine-matua        |
 (Composer of                       |             Puha-i-mua
   song)                            |                |
              Te Ao-haere-tahi=Rakai-moari       Rongotea-tau-karihi
          ┌──────────────────┘                      |
      Tutere-moana                              Rongotea-tai-maranea
 (on whose behalf song was composed)                |
   (Contemporary with Tamatea)                   Turi ("Aotea" canoe)   (Time of fleet
                                                     |                      —1350 ?)
                                                 Turanga-i-mua
                                                     |
                                                 Tamatea-kopiri
                                                     |
                                                 Tu-hukuao
                                                     |
                                                 Hae-matua
                                                     |
                                                 Ue-roa
                                                     |
                                                 Mahuki
                                                     |
                                                 Te Ihi
                                                     |
                                                 Te Mana
                                                     |
                                                 Uru-te-angina
                                                     |
                                                 Tu-materau
                                                     |
                                                 Pou-niwha
                                                     |
                                                 Tutange
                                                     |
                                                 Maru-mihi
                                                     |
                                                 Hapakura
                                                     |
                                                 Hine-rangi
                                                     |
                                                 Te Rangihaeata
                                                     |
                                                 Nga-roi-mata
                                                     |
                                                 Tuarua.
```

TABLE III.

Toi
|
Oho
|
Tipua-ki-rua-rangi
|
Puhi
|
Rere
|
Tata
|
Maika
|
Ira-manawa-piko
|
Tamatea-nui
|
Tamatea-roa
|
Tamatea-mai-tawhiti
|
Muri-whenua
|
Tamatea-pokai-whenua
(of "Takitimu" canoe—1350).

Owing to tribal variations in many of these ancient genealogies it is somewhat difficult to arrive at a satisfactory solution of Tuhoto-ariki's period. In Table I this well-known (to the Maori) *tohunga* is situated about one generation after Tamatea, or the period usually spoken of as "the time of the fleet" (A.D. 1350). It is now generally agreed that Whatonga and Tara flourished some considerable time before the "coming of the fleet"; and, if that is so, Tuhoto-ariki may have been earlier, but he could not well have been later.

In Table II we have Tuhoto-ariki placed about four generations before the "fleet" period, but in considering this we have the difficulty of Tamatea being mentioned in the song. Now, it is Tamatea-pokai-whenua, the navigator, who is usually credited with the destruction of the moa by his sacred fire, but if we accept this table we must look to one of the other Tamateas of that period (of which there were many) as the hero. In the life of Tamatea of the "Takitimu" canoe, as received by myself, and recorded in the history of Ngati-Kahungunu ("Journal of the Polynesian Society"), and also in the exploits of this man as recounted in Kauwae-raro, not a word about the fire myth is mentioned; consequently it is possible that an earlier Tamatea was the hero.

Table III gives an idea of the frequency with which the name can be met with, but in this particular genealogy it is probable that there was originally only one man bearing several names, which later came to be recognized as different men. However, as these Tamateas could scarcely have been aborigines, it is improbable that they could have been associated with the sacred fire.

Whatever way we accept the evidence of genealogy, it is clear to my mind that Tuhoto-ariki lived about the time of the fleet, and, on the evidence of the song, that the moa was destroyed by those who came hither with the migrations prior to the time of the fleet; and, further, that myth surrounded the memory of the bird as early as Tuhoto-ariki's time.

This is the first song I have selected :—

Haramai, E tama !
E huri to aroaro ki Turanga-nui-a-Rua, ki Whangara.
E hara i konei, he ingoa whakahua no Hawaiki-nui-a-Rua-matua.
Ka waiho nei hei papa mo te kakano korau a Iranui,
Hei papa mo te kumara i maua mai e Tiunga-rangi, e Haronga-rangi ;
Ka waiho nei hei mana mo Maahu ki Marae-atea.

Tenei, E tama ! Te whakarongo ake nei ki te hau mai o te korero,
Na Tu-wahi-awa te *manu-whakatau* i mau mai i runga i a Tokomaru
Parea ake ki muri i a koe, he atua korero ahiahi.
Kotahi tonu, E tama ! Te tiaki whenua, ko te *kura-nui*,
Te manu a Rua-kapanga, i tahuna e to tipuna, e Tamatea,
Ki te ahi tawhito, ki te ahi tipua, ki te ahi na Mahuika,
Na Maui i whakaputa ki te ao ;
Ka mate i whare huhi o Reporoa, te rere te momo
E tama—e—i !

[TRANSLATION.]

Come hither, son, and turn thyself
With face towards the east.
There thou wilt behold
Turanga-nui-a-Rua,* and likewise Whangara,†
Names that originated in far Hawaiki,
In Hawaiki-nui-a-Rua-matua,
Now left as a korau plantation for Iranui,
And also as a field for the kumara,
Brought hither by Tiunga-rangi and Haronga-rangi,
Now left as a renowned *mana* for Maahu.

Listen, my son, for I hear rumours spoken
That the *manu-whakatau* was brought here
By Tu-wahi-awa on the "Tokomaru" (canoe).
Reject this story as an idle tale.
One guardian only, O son, had this land,
The *kura-nui*, the bird of Rua-kapanga.
Destroyed by your ancestor, by Tamatea, with subterranean
    and supernatural fire,
The fire of Mahuika, brought to this world by Maui.
Thus were they driven to the swamps and perished ;
Thus was the species lost, O son !

[TRANSLATION OF THE LATTER PORTION OF THE SONG, BY REV. T. G. HAMMOND.]

O son, the only one who took care of the land
Was Te Kura-nui, the bird of Rua-kapanga.
Your ancestor Tamatea lit the sacred fire,
The fire obtained by Maui from Mahuika and brought to the world,
And they (the birds) were destroyed,
Even the breed of them in the swamps of Reporoa.

The second song selected is of a later period, and is a lament composed by Hau-te-horo for his ancestors and relatives who were killed by Ngati-Ira at Pueru-maku (Tolaga Bay). Tawhipare gave a feast of crayfish to his intended victims, and this food apparently causes thirst. Tawhipare prepared for this by placing an ambush round the only available water, so that those who went to drink would be slain. Many were killed, but in the struggle for water some of the warriors jumped into the pond with their heavy flax *pueru* cloaks, and, saturating their garments, they ran back to the *pa* and gave relief to some of their people. This incident gave the name to the affair—*pueru*, a garment ; *maku*, wet.

---

* Gisborne.   † A place north of Gisborne.

DESCENT OF HAU-TE-HORO FROM TAMATEA, AN IMMIGRANT FROM EASTERN POLYNESIA
(about Ten more Generations to Present Time).

Tamatea
|
Rangi-nui
|
Taotu (?)
|
Te Tohia
|
Kuri-nui
|
Maru-houa
|
Te Whakaroro-a-te-po
|
Tu-hokai-rangi
|
Murupara
|
Te Hauerangi (?)
|
Hau-te-horo.

The lament is as follows :—

Tera ia te ao pukohu te huripoki ra i runga o Hikurangi
He tohu aitua tenei ka tata mai ki ahau,
E noho wairangi noa nei i raro,
Whakarongo iho ai ki roto i ahau,
E haruru ana me he tai e whati ana,
Haere ra, e hine ma, e tama ma
E koro ma e, ki roto o' Hawaiki-rangi
I runga o Irihia i Tawhiti-nui
I Tawhiti-pamamao ki te Hono-i-wairua,
Ki te huna i a te *Kura-nui* e ngaro nei
I te ahi a te tipua nana i huna ; takoto kau Aotearoa,
Ko ia te ngaro ia koutou, e Tama ma—e—i.
Whakamau atu ra ki roto o Hawaiki-rangi
Ka whai e koutou ki te ara tiatia
Ki te Toi huarewa i kake ai Tane,
Ki te pumotomoto o Tikitiki o Rangi,
Kia urutomo atu koutou ki roto te Rauroha,
Ma tini o te Marei-kura koutou e powhiri mai
Ki roto o Rangi-atea, ka whakaoti te manako
Ki taiao nei, e Tama ma—e—i.

[PARAPHRASE.]

Afar off the fleecy mists enshroud the summit of Hikurangi ; an evil omen this that hovers near me, dwelling restless and apprehensive below, hearkening to hidden dangers resounding like unto breaking seas. Farewell, O maids and lads. Farewell, O elders, to Hawaiki-rangi, on Irihia, at Tawhiti-nui, at Tawhiti-pamamao—to the Hono-i-wairua ; there to be lost to me even as the *Kura-nui* is lost, destroyed by mysterious fires, and lone lies Aotearoa. Thus are ye lost to me, O sons. Fare on to Hawaiki-rangi, follow the whirling way by which Tane ascended to the entrance of the uppermost heaven. That you may enter within the Rauroha, where multitudes of celestial maids will welcome you within Rangi-atea, where all desire for this world shall cease, O sons.

The third period that has been chosen is some twenty years before the first discovery of bones by Europeans—that is, about 1820.

The accompanying *tangi* was part of a lament composed on the death of Te Momo, a great chief allied to Ngati-Tuwharetoa, who was killed by Ngati-Kahungunu at Kahotea, near Roto-a-Tara, about the year 1820.

From a second lament of the same period, composed by Nuku, who died about 1840, the word "moa" occurs in a similar phrase; but one example is enough to show that the bird was known under the name of "moa" about that time.

TE TANGI MO TE MOMO, I MATE I A NGATI-KAHUNGUNU, I TE ROTO A-TARA (OTIRA I KAHOTEA).

Tera te whetu, kamokamo ana mai,
Ka tangi te whatitiri,
Ka rapa te uira,
Te tohu o Hoturoa
I maunu atu ai.

Kaitoa kia mate,
Nau i rere mua,
He waewae tapeka
Ki te ara ripeka,
He pukainga pakake
Ki Te Roto-a-Tara.

Ma wai e huaki
Te umu ki Kahotea ?
Ma Te Rauparaha,
Ma Tohe-a-Pare,
Mana e tamoe,
Te awa kei Ahuriri.
Kia riro ana mai
Taku kai ko Te Wera,

Me horo mata tonu
Te roro o Pare-ihe,
Hei poupou ake
Mo roto i a au.

Iri mai E Pa !
I runga te turuturu ;
To uru mawhai
Ka piua e te tai.
To kiri rauwhero
Ka whara kei muri,
Kia koa noa mai ra
Te wahine 'Ati-Puhi.

Tahuri mai o mata
Te tihi ki Tirau
Mowai rokiroki.
Ko te huna o te moa
I makere iho ai,
Te tara o te marama.

[TRANSLATION.]

A LAMENT FOR TE MOMO, WHO WAS KILLED BY NGATI-KAHUNGUNU AT KAHOTEA (NEAR TE ROTO-A-TARA).

Yonder the star glittereth (winks),
Crashes the thunder,
Flashes forth the lightning,
The sign of Hoturoa*
(Whose descendant has departed).

'Tis well you died;
Forward you ran,
With hurrying feet
That led to insecurity.
By the cross-roads you sped,
Where fell the chiefs at Tara's Lake.

Who now will be the first at Kahotea ?
The ovens to uncover
(Who shall avenge the death-ovens at Kahotea ?)
'Twill be for Te Rauparaha,
Or perhaps Tohe-a-Pare,†

Who shall cause the sleep of death
To fall on the waters of Ahuriri,
That I may obtain Te Wera as my food
And swallow raw the brains of Pare-ihe
As an all-sustaining food within.

Alas, O father ! thy head now rests
On the (accustomed) stake,‡
Thy curly hair
Blown by the sea-breezes,
Thy warm-tinted flesh,
Now a thing of the past,
Giving delight to the 'Ati-Puhi women.

Turns now thy glance (in vain)
To the summit at Tirau,§
That place of Ocean-like calm ;
Hidden art thou like the extinct moa ;
The horns of the moon have fallen.

* The thunder crashed and lightning flashed at the death of a great chief, as it did at the death of Hoturoa, captain of the "Tainui," an ancestor of Te Momo.
† Tohe-a-Pare was the second name of Te Whata-nui, head chief of Ngati-Raukawa at that time.
‡ "*Turuturu*" here is the stake on which preserved heads were stuck, and often exposed in the houses where women were engaged in weaving, and these women used to jeer at the head, as the Nga-Puhi women are said to do in the above.
§ Tirau is a well-known place (now), east of Cambridge, in the Ngati-Raukawa territories, and presumably was the home of Te Momo.

ART. XLVI.—*Maori and Maruiwi: Notes on the Original Inhabitants of New Zealand and their Culture; on the Question of how that Culture affected the Later-coming Maori; and on the Existence in these Isles of Customs, Arts, and Artifacts not traceable to Polynesia.*

By ELSDON BEST, Hector Memorial Medallist.

[*Read before the Wellington Philosophical Society, 27th October, 1915.*]

THE student of the history, customs, and arts of the Maori of New Zealand must recognize the existence of some interesting problems connected with those subjects. He knows the Maori to be a member of the far-spread Polynesian race, speaking a dialect of the racial tongue. He knows that the Maori came to these isles from Polynesia in past times, and that he kept up communication with the northern isles apparently for some centuries. He also knows that the natives of those far-scattered groups and lone islands were a fairly homogeneous folk in regard to their various arts and customs. But when examining those of the natives of New Zealand he must necessarily be impressed by the fact that in these islands there existed certain customs, implements, and arts not traceable, apparently, to the kindred peoples of Polynesia. It is the desire of the writer to draw attention to some of these discrepant features, and to throw thereon such small rays of light as may be gathered from observation and native tradition.

In the first place, it may be stated that no attempt will be made in this brief paper to uphold any special theories as to origins, or to make arbitrary remarks on any of the debatable subjects discussed herein. There is by no means sufficient evidence available to justify any person in assuming such an attitude. The small amount of such evidence here brought forward may enhance to some extent the interest of these matters, and serve to direct attention to some hitherto unexplored fields of inquiry.

THE ORIGINAL INHABITANTS OF NEW ZEALAND : THEIR ORIGIN, PHYSICAL PECULIARITIES, AND CULTURE.

The amount of information available under the above heading is, unfortunately, very limited, and soon quoted.

According to Maori tradition, the first inhabitants of New Zealand were a people of unknown origin, whose racial or tribal name, if any, has not been preserved. The Maori knows them as Maruiwi, which name is said to have been not a tribal one, but merely that of one of their chiefs at the time when the Maori from eastern Polynesia arrived on these shores. The first of these Maori settlers are shown in tradition to have reached New Zealand twenty-eight to thirty generations ago. At that time the Maruiwi folk were occupying many portions of the North Island. They were the descendants of castaways who had reached these shores in past times, and landed on the Taranaki coast. They had been driven from their own land by a westerly storm. Their home-land, according to the accounts given by their descendants, was a hot country—a much warmer land than this. In appearance these folk are said to have been tall and slim-built, dark-skinned, having big or protuberant bones, flat-faced and flat-nosed, with upturned nostrils. Their eyes were curiously restless, and they had a habit of glancing sideways without turning the head. Their hair in some cases stood upright, in others it was bushy.

When, during the fighting on the East Coast in the "sixties," native prisoners were sent down to Chatham Isles it was noted that some of the women of No. 4 batch, who came from Tarawera and Te Whaiti, much resembled Moriori women in physique, and more particularly in their frizzy hair of Fijian appearance. A member of the Ngati-Awa Tribe there remarked, "They are exactly like Moriori women." A few of the Tuhoe hill tribe seen by the writer at Te Reinga in 1877 had the same Fijian-like heads of hair.

The culture plane of these Maruiwi seems to have been lower than that of the Maori of Polynesia, so far as we can gather from tradition. They are said to have been ignorant of their own lineage, a sure mark of an inferior people in Maori eyes: "They were an indolent and chilly folk (*kiriahi*), fond of sitting round a fire. They slept anyhow, and in summer-time went almost naked, wearing merely some leaves. In the winter season they wore rough capes made of the fibrous leaves of *toi* (*Cordyline indivisa*), of *kiekie* (*Freycinetia Banksii*), and *harakeke* (*Phormium*). They were improvident in the matter of food-supplies, and did not construct good houses, but merely rude sheds that our ancestors called *tawharau*. On account of these peculiarities of those people our ancestors called them, in contempt, *kiri whakapapa* and *pakiwhara*. It is also said that Maruiwi had overhanging or projecting eyebrows, and were thin-shanked: an unpleasant and treacherous folk. Our ancestors from Hawaiki and Rarotonga were given some of these women as wives when they first arrived. Latercomers asked for them; in yet later days they took them, enslaving women and young men. They always selected the best-looking women as wives; and those women approved of it, for the Maori men were much betterlooking than their own, and more industrious. Now, as time rolled on and generations went by, the mixed folk became numerous in the land, the result of the Maori taking Maruiwi wives. Then troubles between the two peoples became frequent, Maruiwi stealing from our folk and murdering them. At last it was resolved to exterminate them, and they were attacked in all parts. War raged all over the island—a war of extermination against all of Maruiwi not connected with the Maori. Thus were they slain at Te Wairoa, Mohaka, Taupo, Rotorua, Maketu, Tauranga, Tamaki, Hauraki, Hokianga, Mokau, Urenui, and all other places where they lived. Thus originated the famed saying 'Te Heke o Maruiwi,' as meaning death. But ever were spared those living with the Maori people. Some of the survivors of Maruiwi are said to have fled to forest ranges in the interior. Some fled to Arapaoa from Taranaki and Te Whanganui-a-Tara (Port Nicholson). These were attacked by the party of Tama-ahua that was going south to seek for greenstone. The survivors of Maruiwi fled to Rangitoto (D'Urville Island), where they were again attacked, and many women captured. The last seen of the remnants of these folk was the passing of six canoes through Raukawa (Cook Strait) on the way to Whare-kauri (Chatham Isles). Such is the story of the folk to whom this land belonged, and it is known that all of us are descended from Maruiwi—from those women taken by our Maori ancestors."

Such is the account of Maruiwi, though much abbreviated, preserved by oral tradition. We here have, if reliable, a description of a people much inferior to the Maori in appearance and general culture. We are also told that the thick projecting lips, the bushy frizzy hair, dark skin, and flat nose often seen among the Maori are derived from Maruiwi. The writer has seen many natives showing these peculiarities among the Tuhoe Tribe; and we know from the traditions of that tribe that some of the Maruiwi folk at one time lived at Te Waimana, in their territory. Another tradition of much interest has been preserved by the Ngati-Awa people of the Bay of Plenty, to the effect that about five hundred or more years ago a canoe reached Whakatane with a number of black-skinned people on board. Presumably these were waifs from some island of Melanesia—possibly Fiji or the New Hebrides, or even New Caledonia, where the natives used double canoes. Forster's description of the natives of Malekula, as seen during Cook's second voyage, reminds us of the Maruiwi of Maori tradition. He remarks, " They were all remarkably slender, and in general did not exceed 5 ft. 4 in. in height. Their limbs were often indifferently proportioned, their legs and arms long and slim, their colour a blackish-brown, and their hair black, frizzled, and woolly. . . . They had the flat broad nose and projecting cheek-bones of a negro, and a very short forehead. . . . All went stark naked. . . . Their ugly features and their black colour often provoked us to make an ill-natured comparison between them and monkeys."

Forster remarks on the superiority of the natives of the adjacent isles of Tana, &c. One of these, Futuna, we know to be inhabited by people speaking a dialect of the far-spread Maori tongue.

The evidence of language has now to be considered, as also that of a less direct nature regarding the culture of the aborigines. Of the language of the aborigines we know very little. We have some place, tribal, and personal names preserved in tradition which are said to have pertained to the aborigines. These names are undoubtedly Maori, or, at least, Polynesian; and if preserved in their correct form, then these Maruiwi must have spoken a tongue closely allied to the Maori dialect of the Polynesian language. Among these names are the following :—

Te Tini o Tai-tawaro ⎫
Te Pananehu ⎪
Te Tini o Rua-tamore ⎬ Names of tribes.
Te Tini o Te Wiwini ⎭

Maruiwi ⎫
Pohokura ⎪
Matakana ⎪
Reretua ⎬ Names of persons.
Orotu ⎪
Poa-tau-tahanga ⎭

It is plainly seen that these names are Maori in form and sound; and if original personal names, then the bearers thereof must have been a Maori-speaking people. Here we have something approaching a paradox, for if the physical appearance and culture of Maruiwi were such as described in tradition it is most improbable that they spoke a purely Polynesian tongue, no Polynesians answering to such a description.

In addition to the names given above we have a few words of the Maruiwi tongue also preserved :—

Maruiwi.	Maori.	English.
Kohi mai	Haere mai	Come hither.
Hakana	Tangata	Person, man.
Mahau	Wahine	Woman.

These three words are said to have been from the vocabulary of the Mamoe Tribe of aborigines, of the Napier district. The following Maruiwi words have also been preserved :—

Maruiwi.	Maori.	English.
Waihi	Wahine	Woman.
Kana	Tangata ..	Person.
Punui o kana ..	Tangata nui ..	Big person *or* important person.
Nakua ..	Tena koe ..	(A salutation).
Kohai rahu ? ..	Ko wai koe ? ..	Who are you ?
Papau aka ..	Ka pai koe ..	You are agreeable.
Hine a waihi (?) ..	Kotiro *or* hine ..	Girl.
Pakaraka mai ..	Oma mai ..	Run hither.

Here we have words simulating Maori in sound and form; also a few resemble some Maori or Polynesian equivalents. *Kohi mai*, carrying the same meaning of "Come hither," is an expression of the Moriori folk of the Chatham Islands, as noted in Deighton's Vocabulary. *Kana* and *hakana* bear some resemblance to *kanaka*, the Hawaiian form of *tangata*—person. *Punui* contains the Maori *nui*—big; *kohai* is near to Maori *ko wai*—who. *Hine* is Maori for "girl," while *mai* (hither) is a common Maori form. The other terms do not resemble any known Maori forms. The two words for "woman" are peculiar; and *waihi* might perhaps be compared with Maori *wahine*, but *mahau*, as meaning "woman," is quite unknown to us. The question of the authenticity of these terms is one that can scarcely be settled, unless they are encountered in the vocabularies of one of the islands of the Pacific.

There is one point of view that should not be neglected in connection with this subject. We have noted that these so-called Maruiwi words resemble Maori in form and phonology. This seems an important point, until one remembers that any isolated folk of the culture stage of the Maori would, necessarily and inevitably, so treat any foreign word that it would conform to their own usages of sound and pronunciation. Thus they would convert any sound foreign to their own tongue into the nearest equivalent they might possess. Hence the sound of "l" would be replaced by "r," "s" by "t" or "h," "b" by "p," and so on. Should any word end in a consonant, then a vowel would be placed after the consonant. Had we left these islands after having introduced new objects and ideas, the Maori would not now be able to transmit our names and words so borrowed in their correct form. A horse would be *hoiho*, scriptures would be *karaipiture*, measure would be *mehua*, and mantelpiece would have become *manatarapihi*. Thus the so-called Maruiwi words that have been preserved may or may not be the original forms used by the aborigines. The evidence of place-names is on the same footing; indeed, there is considerable doubt as to which were original names. It is quite possible that some unusual forms, such as Nuhaka and Mohaka, are "Maorized" forms of Maruiwi names. At least we can say this much : that, taking the circumstances into consideration, the evidence of language, in the matter of the origin of Maruiwi, is not to be relied on.

THE PA MAORI, OR NATIVE FORT.

There is one matter in connection with the Maruiwi aborigines that seems to show that in one direction at least they may have exhibited intelligence of a fairly high order. Tradition states that they constructed hill forts, and mentions those of Okoki, Pohokura, and Urenui, in northern Taranaki, as having been occupied by them. The writer has carefully examined those forts, and found them to be of a type common on the Taranaki coast—small hills of which the sides have been excavated into terraces. Those terraces were protected by lines of stockading along their outer edges. Fosses and ramparts formed only a small part of the defences of this type of fort. As these places have been occupied by the Maori for centuries, down to the nineteenth century, they may not now present the same features that they did when occupied by Maruiwi: the style of defence may have been altered since that time.

This brings us to the question of the origin of the *pa maori*, or native fort. In the North Island are the remains of thousands of old-time fortified places, mostly hill forts, exhibiting an advanced knowledge of the science of fortification on the part of those who formed them. Some are of great size, and must have accommodated thousands of persons; some are very small; the greater number are of medium size. The terraced hills, the fosses and ramparts (presenting scarps in some cases of 20 ft.), the double and treble systems of circumvallation, the ingeniously contrived earthwork defences for weak places and entrance passages—all these are of much interest, and well worthy of study. Where or how did they originate?

We know that the Maori who settled in New Zealand came from the eastern Pacific area; we know that no such remains of fortified places are found in that area. A few stone-walled refuges exist on the lone isle of Rapa. The Tongan fortified places were based on those of Fiji, but the Polynesian was not a fort-builder. Apparently the only place outside the North Island of New Zealand where hill forts, the defensive works of which were fosses, ramparts, stockades, and fighting-stages, were numerous is the Island of Viti Levu, in the Fiji Group.

Did the local type of forts originate here? If so, was it Maruiwi or Maori who was responsible for them? We know that the Maori was a fighter before he came to these isles; that he fought in the open in Polynesia, as he always preferred to do here down to our own time. The first Polynesian that brought a party of settlers to this island was Toi, who lived at the Ka-pu-te-rangi Fort, at Whakatane, according to all traditions. Did the Polynesian become a fort-builder as soon as he stepped ashore here? Did he evolve the idea of an earthwork fort out of his inner consciousness, or did he adopt a Maruiwi custom? The origin of the *pa maori* is a field for inquiry.

CANNIBALISM.

There is another subject that carries an element of interest. Though cannibalism was practised in some isles, yet it was no universal Polynesian custom. In the Society Group, whence the Maori of New Zealand came, it was rare, and it horrified several Tahitians who sailed on Cook's vessels in the Pacific. How is it that our Maori has become such a pronounced cannibal in these islands? No such a condition of general cannibalism— of its becoming such a common practice—is known among Polynesians of the south-eastern area. In order to find the eastern limit of this custom as a common habit we must turn to Fiji, in the Melanesian area. It is fairly clear that the Maori did not bring this shocking custom in any excessive form with him to New Zealand. Did he borrow it from Maruiwi? Tradition shows that the aborigines were of a lower plane of culture than that on which the Maori stood. The Maori immigrants took large numbers of Maruiwi women, first as gifts, afterwards by force: such a wholesale system of intermarriage must have had some effect on the culture and customs of the intruding people. Knowing as we do the effect of such a crossing of peoples, does it not appear probable that some of the Maruiwi customs were followed by the mixed folk that succeeded them? Was cannibalism as a common custom so acquired by the Maori? The dreadful Maori custom—or, at least, occasional habit—of *kai pirau* was also a Fijian custom—the exhuming and eating of buried human bodies.

HUMAN SACRIFICE.

We are aware that the practice of human sacrifice was followed in eastern Polynesia, and probably the Maori brought it with him to New Zealand. There is, however, some evidence to show that in former times two singular examples of this custom obtained here that we cannot trace to the former home of the Maori: these were the burial of human beings at the bases of the main forts of the stockade of a *pa*, or fortified village, and also at the bases of posts supporting a house. There are several allusions to the latter custom in Maori tradition; and, curiously enough, there is proof that in many cases some other object—such as a bird, a lizard, or a stone—was so buried, the human sacrifice being omitted. It would be interesting to know whether or not the depositing of a stone, &c., was the more modern custom, such objects serving as substitutes for a human sacrifice. Or were both forms of the ceremony practised during the same period? There is a certain amount of evidence to show that such sacrifices at the completion of a new fort or superior house, and perhaps also of a new canoe of the larger type, were practised at one time, but that in later times they became much less frequent, if, indeed, they did not entirely cease in some districts. Again, the custom of human sacrifice, or at least of slaying a person, at a certain ceremonial performed over the first-born child of a family of high rank does not seem to have been practised by the Takitumu tribes, as it was among some others.

The allusions in tradition to the burial of a human being at the base of a house-post are but few, and there is no record, so far as the writer is aware, of such an occurrence in late generations. One case, in which the mother of a child so sacrificed was a Maruiwi woman, hence probably a slave wife, occurred about two hundred and fifty years ago. Although Maori tradition says little about this custom, we do know that in Fiji the burial of human beings at the bases of house-posts was a custom of the natives.

In regard to the burial of human beings at the bases of stockade-posts, we know of no tradition concerning this custom, and no old natives questioned on the subject know anything about it. We have, however, some very direct evidence in the fact that the remains of such sacrifices have been found in one locality. The Tawhiti-nui *pa*, or fort, at Opotiki is said by natives of the district to be a very old one. It was occupied by members of the Toi tribes (a mixed Maori-Maruiwi folk) when the last Maori immigrants arrived here from Polynesia some twenty generations—or, say, five hundred years—ago. All signs of stockades have long since disappeared from Tawhiti-nui, leaving only the earthworks. When, some years ago, these earthworks were being levelled in order to facilitate farming operations, the workmen found remains of the butts of the main posts of the old stockade within the ramparts. At the base of each of these post-butts were the remains of a human skeleton.

Now, this is the only case in which such remains have been discovered, so far as we are aware ; but it must be borne in mind that such earthworks are not often removed. Most of such fortified places are situated on hill-tops ; such earthworks are not likely to be removed for any purpose. Tawhiti-nui is situated on the brink of a ploughable terrace. Maori tradition tells us nothing of this wholesale sacrifice of human life at the building of a new fort. It was evidently a ceremonial practice, connected with some idea of securing good luck for the fort and its inhabitants. Such an offering to gods or demons is quite a different thing from the slaying of a single person in order to give *éclat* to a function, as not infrequently occurred among the Maori. It is most improbable that Tawhiti-nui represents an isolated example of such a singular ritual performance ; such offerings must have been a customary procedure among former inhabitants of New Zealand. Were the folk who made such a wholesale sacrifice of human beings Maori or Maruiwi ? If Maori, then presumably he did not bring the custom with him from eastern Polynesia, for he did not employ stockades there. Again, if this custom was universal at one time in New Zealand, it certainly was not practised in late generations, not even in the Opotiki district. Why was it discontinued ? The following account of Fijian human sacrifice at the building of a new house is taken from " At Home in Fiji," by C. F. Gordon Cumming : " A series of large holes was dug to receive the main posts of the house ; and as soon as these were reared a number of wretched men were led to the spot, and one was compelled to descend into each hole, and therein stand upright with his arms clasped round it. The earth was then filled in, and the miserable victims were thus buried alive, deriving what comfort they might from the belief that the task thus assigned to them was one of much honour, as ensuring stability to the chief's house. The same idea prevailed with respect to. launching a chief's canoe, when the bodies of living men were substituted for ordinary rollers." (For " rollers " read " skids.")

On the death of a Fijian chief his wives were strangled and buried with him. Something similar obtained among the Maori, though here it seems to have been voluntary on the part of the widows—in fact, suicide. Was this a general Polynesian custom, or was it practised in eastern Polynesia ?

THE MAORI ADOPTS CERTAIN MARUIWI WEAPONS.

While engaged in discussing matters pertaining to war, let us inquire into the origin of three Maori weapons not employed by him in his former home in eastern Polynesia. It is distinctly stated in Maori tradition that the *huata*, the *hoeroa*, and the *kurutai* were Maruiwi weapons, and that they were adopted by the Maori. The first of these is a very long spear, in some cases 20 ft. in length, pointed at one end and having a knob at the other. It was used principally in defending and attacking fortified places. The *hoeroa* is the curiously curved weapon made from whale's bone that is said to have been sometimes thrown at an adversary and recovered by means of a cord. It is also known as a *tatu paraoa* and *paraoa-roa*. The *kurutai* is a short striking-weapon of stone, in form something like a *wahaika*, and of which specimens are seen in the Dominion Museum. These weapons appear to have been found at the Chatham Isles, and some are reported from the South Island. It must, however, be stated that natives do not agree as to which weapon was called a *kurutai.* Some seem to apply the name to the *patu onewa.*

Maruiwi are also said to have used throwing-spears, a form of fighting-implement but little favoured by the Maori; as also the *whiuwhiu,* or spear thrown with a whip. This latter weapon was adopted and used by the Maori, but not to a great extent. It was used principally in an attack on and defence of fortified places.

Was the Bow used by Maruiwi?

There are a few fragmentary items preserved in Maori tradition in reference to a weapon employed by the Maruiwi aborigines that are of much interest. An old Maori graduate of the *whare wananga,* or school of learning, in describing the Maruiwi folk and their habits and customs, at Wai-hinga in the year 1860 mentioned the weapons used by those people, concluding with the words, " *I wareware ake i a au tetahi o nga rakau a nga iwi nei, he pere, whakawhana ai te manuka hei pere* " (Overlooked by me was yet another weapon of those peoples, a *pere;* a piece of *manuka* was bent as a means of projecting it). Now, *pere* is a name applied to anything in the form of a dart or arrow. *Pere* and *kopere* are both applied to the dart or spear thrown with a whip. Both words are also used as verbs, meaning " to propel or cast, as a *pere.*" They seem to be used only when some instrument of propulsion is employed; the casting of a spear with the hand, minus any aid, is described by the word *whiu.*

Here, then, we have a statement that the aborigines bent a piece of the strong and tough wood of the *manuka* in order to gain a propelling force for an arrow-dart.

Another learned man of last century—Te Matorohanga, of Wai-rarapa—in describing the Maruiwi folk, made the following statement: " *Na taua iwi tenei hapai rakau te tarerarera i te tokotoko, te patu kurutai, me te kopere, he mea whakawhana ki te rakau, he kiri kuri te aho* " (Employed by that people was the custom of throwing spears—the *kurutai* striking-weapon, and the *kopere,* which was projected by means of a wooden implement, the cord being of dog-skin). Here we have a fairly clear statement that seems to refer to the bow and arrow, a dog-skin thong being used as a bow-string. The two usages of the word *whakawhana* call for close attention. Firstly, we have an allusion to the missile spear, mentioned as though it were a usage not commonly employed by the Maori. Now, in the casting of the whip-thrown spear no bent wooden implement was employed, nor were the means of propulsion acquired by a recoil or spring impulse; hence the above account cannot apply to this method. Moreover, the very next sentence spoken by Te Matorohanga dealt with the whip-thrown spear, as follows: " *Tetahi he whiuwhiu te ingoa, he mea here te aho ki te pito koi o te rakau, ka whakatakoto ai ki te whenua, ka takiri ai, ka rere taua rakau, ka kaha te rere me te tu ki te tangata*" (Another was called a *whiuwhiu;* the cord was tied to the pointed end of the weapon, which was laid on the ground and jerked suddenly, the weapon flying off and striking a person with great force).

In these sentences quoted above the double meaning of the word *rakau* has to be borne in mind. It implies, in the first place, any form of wood or timber, from a wand or small twig to a giant forest-tree, and as an adjective it means " wooden." It also means a weapon: all weapons are *rakau,* whatever the material may be.

Here we have two statements made by two different old men, acknowledged as being well versed in Maori tradition. Both seem to allude to the bow and arrow as having been known to, and employed as weapons by, the aborigines of New Zealand. One other item may here be mentioned —namely, the bow now in the Dominion Museum, having been deposited by Mr. Tregear. This bow was found by persons engaged in excavating a draining-ditch north of Auckland, and is said to have been found about 2 ft. below the surface. It closely resembles those from the New Hebrides in the Museum. How long has that bow been so buried, and to whom did it belong ? This query will never be answered ; if it were, then probably a new chapter of the story of man in these isles would be opened.

It may be asked, How is it that the Maori did not adopt the bow and arrow as a weapon, if it ever existed here, as they adopted other Maruiwi weapons ? Now, the answer to this query illustrates a very singular trait of Polynesian character. The bow has been known to the Polynesian for many centuries, and he has frequently come into contact with bow-using Melanesians, yet he has ever steadfastly refused to adopt it as a weapon. He has used it for killing game and in archery contests, northward to the Hawaiian Group and eastward to Tahiti, but never as a weapon. And that is the reason why he would not adopt it here—that is to say, if he really had the opportunity to do so. When the Maori fought, he loved to feel his weapon bite into the skull of his enemy ; he felt the keen joy of the fighting-man as he thrust his slim spear-head through the fish of Tu.

That is how the bow has been forgotten by the Maori people, and why the natives of Cook's time were ignorant of it. The knowledge their ancestors had of it was preserved only in old, old traditions handed down orally from one generation to another by the wise men of the *whare wananga*, the trained and close-lipped record-keepers of the Maori school of learning.

As an illustration of how a people may possess the knowledge of usages among a far-distant race, we may note a remark made by a native of the Marquesas Isles, away off in eastern Polynesia, to Porter, an American voyager of the "twenties" of last century. This was to the effect that far away across the ocean, in a southern land, dwelt a black folk who used the bow and arrow as a weapon.

STONE IMPLEMENTS OF UNKNOWN USE.

On the coast of the Bay of Plenty have been found some curious stone implements quite unknown to the present Maori inhabitants of the district. These objects are carefully fashioned flat stone discs, resembling a cheese in form, but much smaller. The only known objects in Polynesia which they resemble are certain stone discs formerly used by the natives of the Hawaiian Isles in a game called *maika*, resembling our game of bowls. These objects have not been found in any other part of New Zealand.

ARTIFACTS NOT TRACEABLE TO POLYNESIA.

Another singular stone instrument of unknown use has been found in the Bay of Plenty district, several specimens being known. In form this object may be compared to a flattened tipcat, a wooden item beloved by ungodly boys, who utilize it for the purpose of destroying windows. In cross-section it is almost diamond-shaped, and each end tapers to a point. This implement seems to be quite unknown to natives, and absolutely nothing is known as to its origin or use.

Yet another stone object, of which a number have been found on old village-sites, is what the writer usually refers to as a stone spool. It bears a resemblance to a couple of cotton-reels placed end to end. These implements are about 3 in. in length, and are very carefully fashioned and finished. A hole is bored axially through the middle, as though for the insertion of a cord, and one side is flat. The outstanding rims or ends and intermediate projections are notched on their edges. A fine specimen found by Captain Bollons is of black stone, and has a very fine finish; it has five projections adorned with notches. Another, at Whanga-nui, is of greenstone; another was found at the Chatham Isles. All have been made with much care, and at the expense of considerable time and labour. Their use is unknown, though some absurd guesses have been made in that direction.

The only object known to the writer as resembling this spool implement is an object of similar form worn by women (Mohammedan presumably) in Cairo, and probably elsewhere also. This is so worn as to cover the nose, and apparently has some connection with the veil worn by such women. The New Zealand object is so carefully finished that it can scarcely have been a tool, as some suppose, but may have been a pendant. Its form is a most singular one. The Maori can tell us nothing concerning it.

In addition to the above there are other manufactured objects of stone and bone in museums and private collections, the names and uses of which are unknown to the Maori. If these various objects were made and used by Maori folk it seems singular that all knowledge of them should have been lost. In this connection I do not refer to the younger generation, but to the old grey-heads who take pride in preserving knowledge of the customs of their ancestors.

Stone Adzes of New Zealand.

In common with all other branches of the Polynesian race, the Maori hafted his timber-working stone tools as adzes, not as axes. In connection with these implements there is a peculiar and unusual element to which attention does not appear to have been drawn. In the northern Pacific area we find at the Hawaiian Isles a well-defined type of stone adze possessing an angular tang, easily recognizable wherever seen. In the eastern Pacific we find at the Society Isles another well-marked type of peculiar form, marked by excessive thickness in comparison with its length. At the Cook Isles also we have a definite form of these tools. In the Fiji Group—Melanesian in name, but with a considerable mixture of Polynesian blood in its eastern area—we find two leading types, one of which is circular in cross-section, a form that found little favour among Polynesians. All of the above forms differ widely from the thin-bladed stone tools of the Solomon Isles and New Guinea.

Turning now to New Zealand with some expectation of finding one or two local types of stone adzes, it is somewhat surprising to find that our collections cannot be reduced to two or even four common types. We find here a considerable number of forms illustrating widely different types. We do not see a common form of cross-section among our specimens, as we do among those of the Cook, Society, Hawaiian, and other groups. In New Zealand we note numbers of implements in which the cross-section is rectangular, triangular, oval, ovoid, or subovoid, &c. We find a long narrow form, some thin, some remarkably thick; a flat, wide, comparatively thin type; a short form, thick and carrying an abrupt blade-angle; a form with angular tang; another carrying a curious shoulder-ridge across the upper end of the back of the blade ; as also others. We see specimens with parallel sides, others narrowing from cutting-edge backward to the poll, with many other forms. The one form lacking in adzes is the truly circular cross-section, though it is found in small stone tools of the gouge or chisel type. A curious and persistent form is one that presents an extremely narrow face, a narrow cutting-edge, and wide back, the cross-section being subtriangular, the use of which is by no means clear, and, indeed, most puzzling even to us old timber-workers.

This diversity of form in these stone implements of New Zealand is a subject of some interest, and worthy of study. It seems a pity that no effort has apparently been made to make collections of the stone implements of the various island groups, the possession of which would be of much value in the future, when a close study of such artifacts will assuredly be made. The variety of types among our stone adzes awaits an explanation.

WOODEN COFFINS, OR BURIAL-CHESTS.

The most interesting of late discoveries of Maori antiquities is assuredly that of the finding of a number of old wooden coffins in the North Auckland district. It seems strange that no specimens were found in earlier years, and that so many have come to light lately. Apparently they are confined to the northern part of the island. They were used not for containing the body, as with us, but merely as a receptacle for the bones after exhumation. They have been carefully fashioned out of durable timber, and show highly curious carved figures of archaic design—designs often differing from the Maori forms known to us, but presenting the well-known and far-spread three-fingered or three-clawed hands that have caused so much conjecture. Some of these coffins are large enough to contain the bones of an adult when the cleverly fitted lid at the back is in place ; others are so small that the receptacle would contain only very small objects. Possibly the latter were used for the preservation of some particular *tapu* bone, such as the *manu tu* (a small bone at base of skull), or the *iho* (umbilical cord) of a child of high rank. The fact that these coffins were fashioned with stone tools enhances their value to a marked degree.

Information as to the age of these coffins is by no means satisfactory, but the character of the carved designs upon them certainly denotes a considerable age. Any statements made by the younger generation of natives as to their being only a few generations old may be disregarded. The durable heart-wood of which they are composed might endure for centuries in a favourable situation, such as a dry cave. Hence it is possible that these coffins were made by some of the old tribes of the northern districts, of whose origin we have no definite knowledge, but who must have carried Maruiwi blood in their veins—descendants of the aborigines and the intrusive Maori.

DECORATIVE ART.

In three branches of decorative art we find the Maori utilizing designs that at once strike us as differing widely from those employed in Polynesia. The branches alluded to are the arts of wood-carving, painting, and tattooing. Professor Rivers has drawn attention to the fact that whereas Polynesian art is essentially rectilinear, that of the Maori of New Zealand is curvilinear. This dictum is borne out by the evidence of carved implements from Polynesia, and illustrations of similar objects to be found in many works. In Melanesia we encounter both of the above forms. A comparison of Maori tattoo-patterns with those of Polynesia, particularly of the Marquesas Group, serves to mark New Zealand forms as emanating from a different source. The writer has seen no series of illustrations of tattoo-patterns of Melanesia. Is there any series of designs in that region in any way resembling Maori forms? The *tara whakairo* was known in New Zealand and Fiji, but is not reported from Polynesia. In regard to the designs adopted by the Maori in his wood-carving, some of which are intricate and involved, we look in vain to Polynesia for archetypal forms. These designs bear not the impress of modern development; their general aspect is archaic, and often highly conventional. It seems probable that in some cases they are symbolical, but, unfortunately, no attempt was apparently made to gain an insight into this branch of Maori knowledge while the men who possessed such knowledge were living—a remark that may be equally applied to Maori star lore. It is certain that some of the grotesque semi-human figures, such as the Marakihau and Kekerepo, bear names found in Maori mythology. One outstanding fact is that the Maori did not attempt to represent his gods in his publicly exposed wood-carvings. Of the great number of carved figures in human form to be seen in the first-class house, not one of such figure represented a god, though heroes and mythical creatures were so shown. The carved figures on the slabs of house-walls represented ancestors. In two cases we can trace designs to Melanesia—those of the scroll and the *manaia*—while another resembling the *puhoro* is also to be found there. Professor Haddon, in his work "Evolution in Art," speaks of the occurrence of scrolls and spirals in New Guinea, and remarks, "I suspect that most of the Oceanic wood-carving is due to Melanesian influence." We can trace some of the wood-carving patterns of the Maori to Melanesia, but not, so far as the writer is aware, to Polynesia. In the textile art of the Maori we certainly encounter rectilinear designs, often largely made up of various dispositions of the triangle. Presumably this is owing to the difficulty of forming curved lines in the curious style of plaiting (not true weaving) employed by the natives of New Zealand. Wherever the Maori used chisel or brush he indulged in curved lines. A trained artist has suggested that the Maori was unwittingly influenced by his surroundings— that the rounded contours of foliage masses and other natural forms caused him to evolve in these isles those curvilinear designs for which his decorative art is remarkable. The writer is unable to discuss this subject, owing to his utter ignorance of this phase of culture; but if analogous conditions obtained in Polynesia, then the rectilinear art of that region would demand rectilinear contours in nature.

We know the curved lines of Maori patterns of painting, as seen on house-rafters, canoes, &c., many depicting graceful and pleasing designs of a superior type. We know the curved-line designs in his tattooing and carving. We also know that the Maori came from Polynesia, that he speaks the Polynesian language, and that he retains many Polynesian customs and myths. Did he, as he stepped ashore here, relinquish his artistic designs, and proceed to evolve others of a totally different type, or did he adopt them from a people already in possession of these isles?

Another interesting object not traceable to Polynesia is the *heitiki*, a highly prized pendant of singular form known to us all, usually fashioned from the intensely hard nephrite, or greenstone, a task demanding a great expenditure of time and labour. The curious form of this grotesque image is not without its meaning, and tradition states that it originated in very far-away times—in fact, in the days of the gods. Was this archaic form evolved here, together with decorative art-designs, weapons, forts, and other things mentioned above ?

This paper has now been carried far enough, intended as it was merely to draw attention to some interesting subjects for inquiry and discussion, most of which have received little attention, and present some curious discrepancies.

The field of inquiry is a wide one ; its exploration would call for many correspondents. There are many subjects that might repay research, in addition to those already given. For example : Did the excavated house-site obtain in Polynesia, as it did in New Zealand, and as it does in the Torres Group (where it could scarcely be made necessary by coldness of climate) ? Why does the Maori carry burdens strapped on his back, and why did he discard the balance-pole of his former home ? How comes it that his system of numeration is apparently a compound of two forms, and that he has several distinct series of month-names ? Why did the year commence among some tribes with the heliacal rising of Matariki, the Pleiades (as it also did in the Cook Group), and with that of Puanga, or Rigel, among others ? Whence the confusion in the number of the heavens ? And . . . But *kati noa iho*, lest weariness wait upon the answers. The queries put have been numerous, and followed by no intelligent explanation ; that portion of the task is calmly left for the consideration of others in the days that lie before.

" *Mo a muri mo a nehe.*"

ART. XLVII.—*Maori Voyagers and their Vessels : How the Maori explored the Pacific Ocean, and laid down the Sea Roads for all Time.*

By ELSDON. BEST, Hector Memorial Medallist.

[*Read before the Auckland Institute, 8th November, 1915.*]

FAR away across the dark waters of the Great Southern Ocean, within two thousand miles of the coast of South America, lies the lone Polynesian outpost of Easter Island. Away to the north-west, beyond many a far meridian, lies Nukuoro, south of the Carolines. A vast distance of something like seven thousand miles separates the two isles ; but the inhabitants of both speak the Maori tongue. In the southern extremity of New Zealand, about 48° S. latitude, and at Kauai, in the Hawaiian Group, about 22° N. latitude, early voyagers found peoples speaking the Maori tongue. Eastward to the Marquesas and westward to the Ellice Group they found the Maori in occupation. Over a great oceanic area of four thousand by five thousand miles in extent, flecked with many isles, the Maori alone held sway. Members of a common race, speaking dialects of a common tongue, these units in far-sundered lands not only held undisputed possession of the central and eastern Pacific, but also heard dim echoes of their racial tongue from their outposts in Melanesia and Micronesia. The Islands of Futuna (in the New Hebrides), Tikopia (north of that group), Nukuoro (in the Carolines), and some others, are held by Maori-speaking Polynesians.

How comes it that we find divisions of one uncultured race, ignorant of the use of metals, occupying so vast an area of Oceania, dwelling in archipelagoes and lones isles hundreds—even thousands—of miles apart ?

How came these scattered folk to possess common customs, myths, and, in some cases, genealogies to a certain point, to know the names of many lands they had not seen for long centuries ? How came the Hawaiian to speak of his old-time voyages to Tahiti, and relate the deeds of ancestors of the New Zealand Maori; the Samoan to relate his exploration of the Paumotus; the Tongarevan to maintain his descent from immigrants from New Zealand ? Why do Moriori and Hawaiian claim the same gods; the Tahitian describe voyages made to Aotearoa of the Maori; and the Maori of these isles recount his ocean wanderings from Tahiti, Samoa, and Raro- tonga to New Zealand ?

The answer to these queries is that all these widely separated peoples are descendants of common ancestors, of the Polynesian Vikings, of the Maori voyagers—the bold sea-rovers who broke through the hanging sky in times long past away, who fretted the heaving breast of Hine-moana with the wake of their swift canoes, who ranged over every quarter of the vast Pacific, and marked off the sea roads for all time.

For the Maori is truly a Polynesian, the Polynesians are essentially Maori, and no ethnological quibbles can separate them. This fact lightens our task of describing Maori vessels and Maori voyagers, though it increases the scope of the paper. It teaches us to look abroad for the origin of the Maori canoe as seen here; it compels us to follow the *ara moana,* or sea roads, traversed by the Maori voyager in the days when the Romans held Britain. In those voyages we shall cross the famed sea-ridge, the back- bone of Hine-moana, and look upon the wonders of the deep. We shall pass through great areas of the "many-isled sea," and range northward until strange stars rise above the sea horizon; we will seek the rising sun, even unto the land of strange gods. Southward will we go until we view frozen seas and drifting white islands, and the hand of Pārā-weranui lies heavy upon us, and westward to far-distant lands where strange black folk dwell.

For the Maori voyager was no fair-weather sailor, nor was he content to hug the shores of his home-land. He boldly crossed wide seas beneath changing skies, and rode out the fierce ocean gale; or went down to death in the embrace of Hine-moana. But when our voyager was following distant sea roads he was not a New-Zealander—he was a Polynesian of the Pacific isles. After he settled in New Zealand his voyages were apparently confined to expeditions to the Cook and Society Groups—say, from fifteen to eighteen hundred miles distant.

As late as the time of Toi, who flourished thirty generations ago, the Maori of New Zealand did not exist, for Polynesians had not yet settled in these isles. He made his voyages hither as a Polynesian of the northern isles, in the carvel-built Tahitian prototype of the Maori seagoing canoe known to us. All of which leads up to the statement that one cannot study the Maori canoe, or the Maori as a voyager, without including in one's purview the canoes and voyagers of Polynesia.

THE VESSELS OF THE VOYAGERS.

Two forms of vessels have been used by Polynesians in their deep-sea voyages—the double canoe and the single canoe provided with an outrigger. Both types were employed by voyagers to New Zealand, the latter being probably the most favoured. The double canoe, though apparently pos- sessing more stability than the outrigger, was not so handy in rough seas; it was somewhat cumbrous, and liable to meet disaster under such conditions.

This form of vessel needs no outrigger, the second canoe taking the place of that attachment. Ethnographers have derived both the double canoe and the outrigger from the primitive log raft.

Early European voyagers found the double canoe in use throughout Polynesia. They were specially numerous at Tahiti, where, in 1774, as related by Forster, 159 large double canoes, from 50 ft. to 90 ft. in length, were seen ranged in order off shore. These were war-canoes, with large platforms and fighting-stages. In addition were seventy smaller double canoes, each with a roof or cabin at the stern. The smallest district of Tahiti at that time possessed forty of the larger vessels.

In New Zealand all canoes seen by Tasman seem to have been double craft. Cook saw a number of such canoes on South Island coasts, but mentions only one in the North, seen in the Bay of Plenty. Our information concerning these vessels is meagre in the extreme, for no one of the early writers has left us any detailed description thereof, and the illustration given in Tasman's voyage is too grotesque to be taken seriously. The two canoes are said to have been connected by cross-spars, with from 1 ft. to 2½ ft. of space between the hulls, with a central platform. In the North Auckland district two forms seem to have been used. The *waka hourua* consisted of two vessels secured together side by side with cross-beams, while in the *mahanga* type the two canoes were about 30 in. apart. The cross-beams were the most important feature in a double canoe ; should these give way at sea in rough weather, disaster followed. Double canoes were employed on South Island coasts as late as the " thirties" of last century, long after their disuse in the North. As to the outrigger canoe, Cook does not seem to have seen one until he reached Queen Charlotte Sound.

The *pahi* of the Cook Group was a large double canoe furnished with masts and sails. This name was applied by the Moriori, or Mouriuri, folk of the Chatham Isles to a singular double-keeled vessel of a most uncommon type, between canoe and raft, constructed of timber and flax-stalks, and rendered buoyant with dried and reinflated bull-kelp. Curiously enough, these folk worked paddles as we do oars, using a thole-pin. Lack of timber led to the use of some very extraordinary craft among the Moriori, and effectually prevented any voyages to New Zealand.

The big double canoes of Paumotu, Samoan, and Fijian types did not go about in tacking, but the sheet of the sail was shifted from one end to the vessel to the other. In his single seagoing canoe the Maori of New Zealand employed two or four steersmen, but the big double canoe of Tahiti called for eight steersmen.

The double canoe, like the outrigger, can be traced across the Pacific from New Zealand to the Hawaiian Isles, and from eastern Polynesia to India. It was employed by Polynesians, Melanesians, Micronesians, Indonesians, and in northern Australia, Ceylon, Burmah, and India. There are two forms of this vessel—one in which the two canoes are of equal size, another in which one is much smaller than the other. The big double sea-going canoe of the Samoans, long discarded, was of the latter type ; the larger of the two being, in some cases, as much as 150 ft. in length. This was the style of vessel in which the natives of the Samoan and Cook Groups made their deep-sea voyages.

Cook reckoned that Polynesian canoes might sail forty leagues a day or more. Given favourable conditions, this would apparently be a moderate estimate. Morrell, a Pacific voyager of the early part of the nineteenth

15—Trans.

century, states that the outrigger canoes of the Carolines sail eight miles an hour within four points of the wind, and that, in running large, he reckoned they would sail twelve miles an hour. Dampier, who tested the sailing-powers of these craft, gives some astonishing results. If the sailing-rate of the outrigger employed by the Maori voyager be taken at seven miles an hour, and fair-weather conditions be granted, he might have made the run from Tahiti to New Zealand in eleven days, or from Rarotonga in nine days. He would undoubtedly carry sea stores for a considerably longer period, and thus be prepared for the buffetings of fate.

It has been said that the "Arawa," one of the vessels that reached these shores from Polynesia about five hundred years ago, was a double canoe, though evidence seems to be lacking. This statement appears to rest on a passage in Grey's "Polynesian Mythology," viz.: "I will climb upon the roof of the house which is built upon the platform joining the two canoes"; but this passage is not a translation of the original, which contains no reference to a platform and two canoes. In like manner, there is no evidence to show that "Tainui," "Matatua," "Tokomaru," "Horouta," "Mātā-hourua," and "Kura-hau-po" were double canoes, while "Takitumu" is distinctly described as an outrigger vessel. In one tradition only, so far as the writer is aware of, are double canoes distinctly mentioned as having made the voyage from Polynesia to New Zealand, and here is the story thereof:—

Voyage of Manaia and Nuku to New Zealand.

About twenty-eight generations ago two chieftains of eastern Polynesia quarrelled and fought in their island home. One of these, Manaia by name, having suffered grievously, resolved to migrate to New Zealand, here to dwell in peace. He therefore manned his vessel, called "Tokomaru," with such trained adept seafarers as were necessary in lifting the rolling sea roads of the Realm of Kiwa, and quietly left home between two days. His enemy, one Nuku, came to hear of his departure, and resolved to pursue and attack him. He therefore collected a number of warriors and started in pursuit. Tradition asserts that he was careful to select "sea-paddling braves," experts on the *ara moana*, and he also brought three *tohunga*, or priestly adepts, to assist him in overcoming the dangers of the deep. These folk came in three vessels, named "Te Houama," "Waimate," and "Tangi-apakura." Now, it is distinctly stated in the legend that the first-named was a single canoe (*waka marohi*), and the other two *waka unua*, or double canoes: "*Enei waka, e rua nga waka unua, kotahi te waka marohi, ko Te Houama*" (These canoes, two were double canoes, one was a single canoe, Te Houama).

Both these expeditions touched at Rarotonga, as was usual in making the voyage from the Society Isles to New Zealand. Nuku saw no sign of Manaia's vessel in the run down to Aotearoa, but when he entered Cook Strait and landed on D'Urville Island he found there the smouldering remains of the camp-fires of Manaia's party. He at once started in pursuit, and caught sight of his enemy off Pukerua, near Porirua Harbour. "Te Houama," the single canoe, being the swiftest craft, was the first to come up with "Tokomaru." Of the sea fight that occurred on the waters of Raukawa, and the later Homeric combat on the sands of Pae-kakariki, there is no space here to discuss details, but one statement in the tradition is of much interest. It is said that Nuku, when about to leave on his return to Polynesia, dismantled his two double canoes, and sailed them back across the Southern Ocean as single vessels, doubtless provided with outriggers. This was done in order to expedite his return passage. In the original we find: " *Ka tahuri a Nuku ki te mahi i ona waka; ka marohitia anake nga waka nei, kua kore e unuatia, kia māmā ai te hoki ki tona whenua.*"

Manaia pursued his way to Whaingaroa, thence to Kaipara, to Whaka-tane, to Tokomaru, a place named after his vessel, finally returning to Whaingaroa, where his career as a Maori voyager ends. His further adventures consisted of fighting with the aborigines of Taranaki, the feats of a landsman, which concern us not. Ngati-Awa, of Taranaki, claim him as an ancestor.

THE OUTRIGGER CANOE.

We have now to treat of the single canoe furnished with an outrigger. Concerning the small coastal outrigger seen by Cook on our shores we have no precise details. D'Urville, who left us the only diagrams drawn to scale that we possess of Maori canoes, affords us no help with the out-rigger or double canoe. Apparently he saw neither of these forms. We have, however, something of much interest in a description, preserved by oral tradition, of an outrigger canoe that arrived on these shores from Tahiti about five hundred years ago. This was "Takitumu," one of the old-time deep-sea craft of the ancestors of the Maori, and which brought hither the forbears of East Coast and South Island natives.

On a fair morning, nearly a hundred years before Columbus felt his way across the Western Ocean, a large concourse of brown-skinned folk gathered on the hill called Puke-hapopo, whence they could look down upon the waters of Pikopiko-i-whiti. Those waters were of calm appearance, being protected from the ocean by a rocky reef, and girdled the shores of an island known as Hawaiki. These folk had assembled in order to witness a canoe race, in which two vessels known as "Horouta" and "Te Pu-whenua" took part, and also others, as "Tainui," "Te Arawa," and "Matatua." In this contest "Te Puwhenua" distanced all others. As she sped over the placid waters Rua-wharo cried, " *Tena a Te Puwhenua te horo na i te whenua!* " (There is Te Puwhenua speeding past the land). And Te Rongo-patahi said, " *Koia ra ano he ingoa mo to waka, E Paoa!* " (O Paoa! now there is a name for your canoe). And that was how "Horouta" gained her name and "Te Puwhenua" received her per-manent name of "Takitumu."

Owing to severe intertribal wars, many people were at that time leaving the isles of eastern Polynesia, and the above vessels, with many others, brought a considerable number to New Zealand. These isles had already long been known to Polynesians, and a number of migrants and rovers had settled here, intermarrying with the aborigines. A number of voyagers had also visited these shores and returned to northern isles, as shown in the traditions of New Zealand, Mangaia, Rarotonga, and Manihiki. In some cases these voyagers called at Sunday Island, known to the Maori of Aotearoa and Rarotonga as Rangi-tahua.

Omitting a great amount of detail, we give some part of the story of "Takitumu," from the tree-stump to the Waiau River of our South Island: When the dugout hull had been roughly dubbed out, as also the *haumi*, or pieces to lengthen it, the top strakes, and other timbers, all these were placed in a huge trench and covered with earth, there to remain for months. This was a seasoning method, said to have the effect of expel-ling sap from green timber, without danger of warping or splitting. The timbers were then taken out of the pit, placed on a scaffold, and covered so as to be protected from the sun. When seasoned, the final adzing reduced them to the desired form and finish, and the construction of the canoe com-

15*

menced. The first task was to attach the pieces to lengthen the hull. Then
the side boards were lashed on with the butted join of carvel-built boats
There were four of them on either side; they were retained in position
and braced by means of lashing on the thwarts. The stem and bow pieces
were attached, the decking, or floor, below the thwarts laid down, the
korewa or outrigger was attached, the masts fitted, as also the stanchions,
cross-pieces, and battens of the awning. Sails, paddles, bailers, and awning-
mats were provided, and then, after the recital of certain ritual over her at
the *turuma*, a *tapu* spot, "Takitumu" was launched on the waters of
Pikopiko-i-whiti, at far Hawaiki.

We here see that "Takitumu" was provided with four side boards, or
strakes, on either side. She was apparently one of the well-known type
of Polynesian canoe in which the dugout hull is a shallow trough, the
sides being built up by attaching several tiers of plank placed one above
the other, carvel fashion. The single *rauawa*, or top strake, of the New
Zealand canoe would be due to the much greater size of our timber. Here,
and in most parts of Polynesia proper, these planks are lashed by means
of passing cords through holes bored near their edges, such lashings enclosing
battens that cover the joints. The Tongans and Samoans, however, employed a different method, borrowed from Fiji, in which the lashing-cords
were passed through cants formed on the inside edges of all planks when
hewn. Thus, such lashings did not appear on the outer sides of the
planks.

The *korewa*, or outrigger, was formed of a very light timber, and was
connected with the canoe by means of spars, termed *hokai*.

In order to render these vessels the more snug in rough weather or
broken seas, a series of splashboards, called *taupa karekare wai* and *pare
arai wai parati*, were secured along the sides. Then, again, the greater
part of the vessel was covered with a kind of awning. Stanchions (*tokotu*)
were lashed in upright positions along the sides, and to these were lashed
the *whiti-tu*, curved rods that extended across the vessel in the form of
an arch. Battens (*kaho*) were lashed horizontally to these, and then the
huripoki, or cover (awning) of mats (*tuwhara*), was stretched over this frame-
work, hauled taut, and lashed down along the sides of the vessel. These
covering-mats were in some cases made from the bark of the *aute*, apparently
a stout form of *tapa*.

When a storm was encountered at sea, where no haven was near, our
Maori voyager was compelled to face and ride it out, and the operations
entailed thereby called for the direction of the *amotawa*, or sea expert. All
ocean-going canoes carried two anchors; the *punga korewa*, or smaller one,
was used as a drift-anchor, while the big heavy *punga whakawhenua* was
the ground-anchor. Both, however, were often used in riding out a storm
in deep waters. The smaller one was lowered a certain depth in the ocean
at the prow; the heavy one was lowered at the stern. This kept the prow
well up, and served to steady the vessel. In addition to this, four steersmen were on duty. At the stern were two, one on either side, manipulating
the long steer-oars termed *hoe whakatere*. Near the bow were stationed
two others, wielding two long oars known as *hoe whakaara*, the manipula-
tion of which by experts lessened swaying and pitching of the bow. Much
depended on these four men in times of danger, for theirs was the task of
keeping the vessel in a proper position. At such times, also, two men
were stationed at each *puna wai*, or bailing-well.

And then, with his longboat covered and splashboards rigged, his sea-
anchors down and outrigger braced, with stalwart, half-naked steersmen

gripping their long steer-oars, and facing the driving storm with courageous hearts and a sublime faith in their gods, the Maori voyager calmly awaited the wrath of Hine-moana—the storm at sea.

Prior to leaving the home-land " Takitumu " had been solemnly placed under the protection of the gods Kahukura, Tama-i-waho, Tunui-a-te-ika, Hine-korako, Rongomai, and Ruamano. These were the protecting deities who brought " Takitumu " safely across the Great Ocean of Kiwa. For such are the beliefs of the Maori.

In accordance with a racial custom of applying proper names in manner most generous, each one of the twenty-six thwarts of this vessel had its special name. These names, as also those of the principal people who occupied them, have been preserved. In like manner, the outrigger timbers, anchors, cables, steer-oars, masts, sails, ropes, sprits, bailers, &c.—all had proper names assigned to them, to recite which would be tedious and unprofitable.

In ocean voyages of considerable length, when voyagers took their families with them, each family, as a rule, occupied the space between two thwarts, where the decking was covered with mats, on which the people sat and slept. Paddlers occupied the ends of the thwarts, each man having his appointed place; reliefs sat on the thwarts between the paddlers. The stern thwart of " Takitumu " (and its adjacent space) was occupied by the three wise men, or priestly experts, Te Rongo-patahi, Tupai, and Rua-wharo. Here also abode the spirit gods in whose care the vessel had been placed. The next thwart was occupied by the steersmen, the next by the principal chief of the party, Tamatea, father of the eponymic ancestor of the Ngati-Kahungunu Tribe of the East Coast.

Each man was provided with two paddles, though sails were always used as much as possible, hence the close study of wind-conditions by Polynesians. Sea stores consisted of dried food products, as fish and shell-fish, and some vegetable foods. Coconuts were carried in quantities, while water was conserved in gourd, seaweed, and bamboo vessels, as procurable.

" Takitumu " left the Society Isles after the' other vessels enumerated above, and all seem to have called at Rarotonga. Apparently, " Takitumu " did not call at Sunday Island; but there is a curious story of certain happenings at a place in mid-ocean called Te Tuahiwi o Hine-moana, where rough seas were encountered and some strange ceremonies were performed, ritual explaining the use of certain ceremonial stone adzes formerly possessed by the Maori, and in which figured Te Awhio-rangi, now preserved at Wai-totara.

Even so these old argonauts swung south from the summer isles of Eden, and sailed boldly out into the Great Southern Ocean. Happily ignorant of the fact that they possessed only frail canoes, and could not possibly make a deep-sea voyage (as we are told by some modern writers), they relied stoutly on their own sea-craft and the assistance of their gods. They traversed the water roads marked out by Kupe in past times, and watched the wheeling stars as they sought the land-head at Aotearoa. For these were feats of which it was written :—

> The sun sags down on Tama's path,
>   Across the changing sky;
> New stars do leap above the deep
>   To meet the wondering eye;
> New seas are spread on every side,
>   New skies are overhead;
> New lands await the sea-kings
>   In the vast grey seas ahead.

"Takitumu" made her landfall at Whanga-paraoa, on the East Coast. Here was found "Tainui," that, with others, had arrived before her. As this coast was already occupied by aborigines and former Maori immigrants, Tamatea took his vessel northward in search of lands whereon to settle. The voyagers called at Muri-whenua, in the far North, afterwards proceeding to Hokianga, where they dwelt for some time. Leaving here they returned down the East Coast to Tauranga, thence to Nuku-taurua, where some seem to have remained. The others proceeded to Te Whanga-nui-a-Tara (Port Nicholson), where they lived some time with the Ngai-Tara folk, descendants of Toi and Whatonga, of eastern Polynesia. From here they went to Waiau, in the South Island, where they settled, and assumed the tribal name of Waitaha. But Tamatea and a few others made another canoe, named it "Te Karaerae," and went to Kapiti, thence to Whanga-nui, where they met Turi and other members of the crew of "Aotea." Of the further adventures of Tamatea we need not speak, inasmuch as they were not those of a Maori voyager, but of a land traveller. Eventually Tamatea returned to Hokianga, where he died.

The Discovery of New Zealand.

Voyage of Kupe and Ngahue from Eastern Polynesia.

This is one of the old-time voyages of which the approximate date is not fixed, but it must have occurred long before the time of Toi, who flourished about seven hundred and fifty years ago. For Kupe is said to have found this island uninhabited by man, whereas Toi found a large population of the Maruiwi, or Mouriuri, folk occupying the North Island.

Kupe and Ngahue (*alias* Ngake) were natives of eastern Polynesia, of an island then known as Hawaiki, but which is almost assuredly Tahiti, as is shown by traditional accounts mentioning the relative position of the Islands of Maitea and Raiatea. The father of Kupe was a native of Hawaiki, his mother was a Rarotongan, while his maternal grandfather belonged to Raiatea (known to the Maori as Rangiatea), facts illustrating the free movements of island-folk in those far-off days.

Kupe made his voyage to New Zealand in a vessel named "Matahorua"; that of Ngahue, his companion, was "Tawiri-rangi." They came to land near the North Cape, then proceeded down the East Coast to Rangi-whaka-oma (Castle Point), thence to Te Kawakawa in Palliser Bay, thence to Port Nicholson, camping at Hataitai (Miramar Peninsula), and naming the two islands Matin (Somes) and Makaro (Ward), after the daughters of Kupe. At Porirua Kupe left one of his anchors, named Maungaroa, brought from a place named Maungaroa at Rarotonga, and took another stone in its place.

These voyagers sailed round the South Island, discovered nephrite (greenstone) at Arahura, recognized its value, and took blocks of it back to their homes. They returned northward through Cook Strait to Whanga-nui, Patea, and Hokianga. None of the crews remained here: all returned to Rarotonga, thence to Rangiatea and Hawaiki. On his arrival at the latter place Kupe recounted the story of his voyage, the discovery of Aotearoa (New Zealand), the aspect of these islands and their products, also explaining how the new lands might be reached by navigators. All these particulars were preserved by oral tradition, and, in later centuries, when voyagers wished to reach these isles they applied to the wise men, the record-keepers, who had retained the directions left by Kupe.

THE COMING OF MARUIWI.

Settlement of the North Island by an Unknown People.

Subsequent to the discovery of New Zealand by Kupe, the North Island was settled in many parts by a dark-skinned folk of inferior culture, whose origin is unknown. They are said to have been a people of spare build, thin-shanked, with flat noses, distended nostrils, and generally unpleasant appearance. Their eyes were peculiarly restless, their hair upstanding. They lived in rude huts, wore little clothing, and were an indolent people, fond of hugging the fireside. Their ancestors had come from a very warm far-away land—a much warmer land than New Zealand. They arrived here in three canoes, named " Kahu-tara," " Tai-koria," and " Okoki." These vessels had been driven from their home-land by a westerly wind, and, after a long drift, reached the Taranaki coast, where these folk settled. As time went on they occupied many parts of the North Island, and were most numerous at Taranaki, Tamaki, the Bay of Plenty, and Hawke's Bay, when the voyager Toi arrived.

A SEA-FOG BRINGS EASTERN POLYNESIANS TO NEW ZEALAND.

The Voyages of Toi and Whatonga to Aotearoa.

Sea-mists, ocean currents, and winds have caused many drift voyages in Pacific waters, have settled many lands, and sent many souls down to Rarohenga, the spirit world of the Maori. When the Polynesian voyager became enshrouded by a dense mist, such as occur during easterly winds in that region, he was compelled, lacking a compass, to trust to the regular roll of the waves in the guidance of his vessel. A change of wind under such circumstances often utterly confused him, as noted by Mariner when sailing with some Tongans. Mariner's native companions were actually sailing away from their island home when he induced them to trust to his despised compass.

It was a sea-fog that brought about the second settlement of New Zealand, this time by men from eastern Polynesia, the home of Kupe. This event occurred three centuries before Columbus saw the world of life.

On the waters of Pikopiko-i-whiti, on which in after-generations " Takitumu " was to float, a canoe-race was being held by the folk of the Isles of Hawaiki and Tuhua (after which Tuhua, or Mayor Isle, in the Bay of Plenty, was named). These competing canoes left the sheltered waters and went out to sea in their enthusiasm. Here they were caught in a storm, and some were carried away by it, while others regained the land. Among the drift canoes was that of Whatonga and Tu-rahui. When the storm died out these hapless folk found themselves enveloped in a mist, and unable to return home. Eventually they landed at Rangiatea, where they remained some time. Meanwhile Toi, the grandfather of Whatonga, had set forth in search of the ocean-waifs, proceeding westward. Some castaways were found at Samoa, but not his grandson. Hence Toi visited the islands as far south as Rarotonga, still without success. He then determined to sail across the Southern Ocean to the strange land—the great land —discovered by Kupe in past times, to see if the waifs had perchance been carried there. And so, ever seeking his grandson, the old sea-rover boldly sailed out into the vast trackless expanse that rolls between Raro-tonga and Aotearoa. And his final word to the folk of Rarotonga was, " I go to seek my child in strange lands, in the moist land discovered by

Kupe, and I will greet the land-head at Aotearoa or be engulfed in the stomach of Hine-moana."

How the gallant old voyager sailed his craft across the Southern Ocean, how he missed New Zealand and discovered the Chathams, how he ranged westward to this land, coasted the North Island, and settled at Whakatane, are matters of traditional history. Also how Whatonga, returning home after many adventures, found that Toi was absent in search of him, how he fitted and manned the famous vessel "Kura-hau-po," sailed forth in search of Toi, and followed him down the long sea roads to Rarotonga, heard of his voyage to Aotearoa, and lifted the rolling water trail of Te Ririno all across the dark ocean to these shores. How he made his landfall at Tonga-porutu, coasted round the North Cape, and finally joined Toi at Whakatane; there these Vikings settled down, never more to look upon the palm-clad isles of the sunny north, never again to listen to the thunder of far-driven seas on the guardian reef.

These were the first folk from eastern Polynesia to settle in New Zealand among the Maruiwi aborigines, many of whom were living at Maketu, known then as Moharuru. It was inland of that place that Rua-kapanga, brother-in-law of Toi, met with his surprising adventure with a flock of five moa. Soon other immigrants came from the eastern Pacific, including Manaia, and the return of Nuku to the islands seems to have induced others to come and settle here. So the new-comers remained here, took aboriginal wives, and became the progenitors of the mixed Tini o Toi tribes found here by the immigrants of "Tainui," "Aotea," "Te Arawa," "Takitumu," and other vessels, nearly two hundred years later.

Tradition relates that the Maruiwi women were attracted by the comparatively fair-skinned, good-looking, industrious Maori men. Their progeny lived as Maori, and even now we plainly see the aboriginal element in natives, in hair, and features, and skin-colour. But all this made for trouble, and, as time went on, quarrels took place between the domineering Maori and half-breeds on one side and the aborigines on the other. Fighting and wars followed, ceaseless harrying of the aborigines until none remained save the Toi tribes, the mixed breed. Now, it is recorded that seven vessels manned by survivors sailed from Cook Strait in search of the Chatham Isles, discovered by Toi, of which they had heard. Those vessels, or at least some of them, reached the Chathams, where the refugees settled, twenty-seven generations ago, and where their descendants were found by Lieutenant Broughton on the 29th November, 1791.

Now, one of these vessels, under a chief named Te Kahu, sailed from the mouth of the Rangitikei River. Her crew were unable to rig a deep-sea vessel, hence they obtained the services of a Maori expert from Whanga-nui, one Aka-roroa by name. Both this man and his sister accompanied the party to the Chathams, and his name was preserved in tradition by the Moriori, or Mouriuri, folk of the Chathams, as shown in the writings of the late Mr. Shand. Hau-te-horo, fourth in descent from Aka-roroa, returned to New Zealand in after-years, and his descendants are at Whanga-nui. And that is how the Maori came to know of the arrival of the refugees at the Chathams. This information was obtained from Hauauru and Takarangi, of Whanga-nui, in the year 1854. The description of the vessel of Te Kahu shows that it was a dugout single

Te Aka-roroa
|
Kauri
|
Waitaha
|
Rangi-tuataka
|
Hau-te-horo.

canoe, with *haumi* and top strake, and covered with a roof or awning, as already described. She crossed over to D'Urville Island, and there stayed some time; doubtless her crew were awaiting favourable weather-conditions. And then, on the Omutu night of the month of Akaaka-nui, these harassed folk launched their vessel, and, passing through the Strait, sailed forth upon the sullen seas in search of a new home.

THE PEOPLING OF THE PACIFIC.

Cook speaks of finding the Polynesian Maori located over an area extending twelve hundred leagues north and south by sixteen hundred leagues east and west, and even then he cut off some Maori communities to the westward. We will now inquire into the manner in which these far-spread isles were settled by the ancestors of our Maori folk, and quote a few more of their voyages.

The earliest voyagers of whom the Maori has preserved tradition were those who left the fatherland of the race. That home-land was known as Irihia; an extremely hot land, wherein grew the prized food called *ari*—a land inhabited by many dark-skinned peoples, a land of great extent. Here was situated the sacred place known as Hawaiki-nui, and on the summit of a mountain in that land, the ascent of which occupied two days, were performed all ritual performances connected with Io, the Supreme Being. After a long sojourn among the slim-built thin-shanked dark peoples, wars with them became numerous, and vast numbers of men were slain. Thus many left Irihia in order to seek new homes across the ocean.

These explorers steered toward the rising sun; by night their guides were the stars, moon, and the sea-breeze. In the tradition of this voyage it is distinctly said that outriggers were fixed and the vessel covered in on the approach of rough weather, hence, presumably, the outrigger timbers must have been carried inboard during calm weather. Also the vessels must have been of wide beam. The double outrigger also seems to be alluded to. These voyagers settled in a land far across the ocean, from which they, or their descendants, moved on to other lands, ever sailing toward the rising sun, until we find them located in Polynesia. How long this eastward movement lasted it is impossible to say.

As to voyages throughout Polynesia we have only time to give a few illustrations. About the seventh century, as recorded in Mr. Percy Smith's " Hawaiki," one Hui-te-rangiora sailed southward until he encountered icebergs and a frozen sea, marvellous sights to Polynesians. Traditions state that about that time many voyages were made, and many isles were visited by Polynesians, who were occupied in exploring the oceanic area, and in peopling its far-spread islands, or possibly in repeopling them. New Zealand, known to the natives of south central Polynesia as " Hawaiki-tahutahu," is said to have been first visited about the seventh century. The Society Isles were inhabited forty generations ago, and probably long before. It is fairly clear, as shown by many traditions of many isles, that for a period of at least eight centuries the Polynesians must have made many voyages in the Pacific, some of great length, traversing vast areas, peopling and repeopling many lands. In later times long sea voyages of set purpose to outlying lands were of much rarer occurrence, those to New Zealand and the Hawaiian Isles apparently ceasing altogether.

Quiros, who sailed with Mendana in 1595, and, later on, made another voyage across the Pacific in 1606, spent much time in wondering how the isles received their population. He maintained that, with no compass, the Polynesians could not voyage to any island not in sight from their own Hence he judged that the islands must be close together, or that a great mother-land existed in the south, from which the various islands had been settled, "as otherwise the islands could not have been populated without a miracle." Nearly three hundred years later Colenso wrote, "I note you seem to adhere to the myth of the Maoris coming to this land; I had thought I had fully exposed that many years ago." But neither Quiros nor Colenso could do that. Between these two comes James Cook, who saw clearly how the islands had become populated, and puts the case in clear, simple language.

A voyage made by one Uenga, of Samoa, about the twelfth century, extended to Tonga, Tongareva, Rimatara, the Austral Group, Tahiti, and the Paumotus, a jaunt of over three thousand five hundred miles. Tangihia, a voyager of the thirteenth century, made a yet longer one. Starting, apparently, from Samoa, he visited Niue, Keppel Isle, the Marquesas, Tahiti, Rapa, the Austral and Cook Groups, Rimatara, and other isles. Whiro took a party of settlers to Rarotonga, then sailed to the Marquesas, Tahiti, Rapa, and other places.

In 1616 Le Maire and Schouten encountered a double canoe under sail, out of sight of land, west of the Paumotu Group, with twenty-five men, women, and children on board. These folk had exhausted their water-supply, and were seen to drink sea-water. These natives being unarmed, the Dutch gentlemen had quite a pleasant time shooting them. The historian remarks on the enterprise of natives who "without compass, or any of the aids from science which enable the navigators of other countries to guide themselves with safety, ventured beyond the sight of land."

In former times the Tongans were in the habit of making frequent voyages to Fiji, which group was reached in three days' sail from Tonga-tapu. They also made voyages to the New Hebrides and New Caledonia. Futuna, in the New Hebrides, and Tikopia to the north of that group, are occupied by Polynesians. The Tongans have been the most daring and energetic of Polynesians voyagers in modern times.

Marquesan traditions tell us of voyages made in double canoes to lands to the westward. These vessels carried not only stocks of food and water, but also hogs, fowls, and food plants, and that is how these things were spread over the Pacific. These plants were yams, sweet potatoes, *taro*, gourd, also the breadfruit; banana, coconut, &c., while the orange was advancing eastward when Europeans began to traverse Pacific waters. Most of these are traced by Candolle to a western source. The animals introduced into New Zealand were the dog and rat; the other food products were the sweet potato, *taro*, and gourd, possibly the yam. The *aute* tree was also introduced.

The Maori voyager recognized the influence of ocean currents on navigation, and had his peculiar method of ascertaining their movements. Even as the Great Black River carried many Japanese vessels to the western coast of North America, and its reflux bore one such to Oahu, Hawaiian Isles, in 1833, so did the ocean streams farther south affect the Polynesian voyager. These currents flow in different directions, some for long distances. Thus the branch of the antarctic drift that swerves westward from the South American coast seems to coalesce with the westward - sweeping equatorial current, the southern branch of which, flowing south of the Tongan Group, passes, under the name of Rossel's Drift, the New Hebrides, on its way to Torres Strait. This helps to explain the arrival of drift canoes at and near the New Hebrides containing waifs from Polynesia. Several such occurrences are on record.

Even as winds assisted our Polynesian voyager in his navigation of Pacific waters, so also did they, in many cases, cause drift voyages, and send many souls down to Rarohenga, the spirit world of the Maori. A few of the many known cases of drift voyages are quoted as illustrating how many islands must have been discovered and settled by their agency. We have already seen that the first and second peoplings of New Zealand were owing to drift voyages—the first directly so, the second indirectly.

Ellis held the curious view that the Polynesians must have originally come from the east, as it would be impossible for them to come from the west against the prevailing winds. And yet he must have known of the fairly frequent communication between the Society and Paumotu Groups, as also other such movements. The south-east trades are by no means constant the year round, as shown by observers as far back as Cook's time. The strong north-west winds that strike the Samoan Group have carried canoes from there as far as the Austral Isles. At the Society Isles the prevailing wind blows from between east-south-east and east-north-east for the greater part of the year, but in December and January the winds are variable, frequently blowing from north-west and west-north-west. Cook tells us that this is the wind by which the natives of the isles to leeward come to Tahiti. Such a wind is often followed by one from the south-west or west-south-west. We have not space to give much data under these heads, but we do know that Polynesians carefully studied wind-conditions. Barstow writes of several weeks of westerly wind at Tahiti, and mentions the case of some Polynesian voyagers he encountered there. Their canoe, containing men, women, and children, had come from the Paumotu Group, to the eastward, in search of some ocean-waifs from that region. They had visited Huahine and other islands, and were compelled to wait over six months at Tahiti for a fair wind to take them home. The Polynesian voyager, indeed, passed much of his time in waiting for fair winds, though that fact would not disturb his equanimity. Possibly this was why he often took his family with him. If he did not live to reach his destination, why, then, his son or grandson might do so.

Barstow records a drift voyage from Chain Island, east of Tahiti, away west to Manua, in the Samoan Isles. This occurred in 1844, and the boat contained three natives and one white man, the latter being the sole survivor.

Colonel Gudgeon informs us that Polynesians are quite capable of navigating their vessels to any island they may desire to visit, always selecting a favourable season of the year. Also that they had well-known starting-places for each such voyage, and stopping-places at intermediate isles in long voyages. Thus voyagers from Tahiti to New Zealand first made the run to Rarotonga, leaving there in December for the run south-west to New Zealand, calling in some cases at Sunday Island. The return voyage was made in June. This is corroborated by Maori tradition, which states that voyagers left Rarotonga for these shores in the month Akaaka-nui, equivalent to our December. An old native of the Nga Rauru Tribe stated that Whanga-rei and Whanga-te-au were starting-places for canoes leaving New Zealand for Rarotonga.

Missionary Williams, the man of many voyages in Polynesia, remarks that westerly winds occur about every two months. He sailed from Rurutu to Tahiti, three hundred and fifty miles north-north-east, in forty-eight hours. On another occasion, from a point two hundred miles west of Niue, he sailed, with a fair wind, seventeen hundred miles to the eastward in fifteen days. In October, 1832, during a voyage from Rarotonga to Samoa, he sailed eight hundred miles in five days without once shifting a sail.

The trade-winds that pass northward of New Zealand would carry Tongan raiders to the New Hebrides, Loyalty Isles, and New Caledonia. In 1793 the expedition in search of La Pérouse saw a canoe on the coast of New Caledonia containing eight Polynesians—seven men and one woman—who spoke the Tongan dialect. They had come from Uvea, in the Loyalty Group, a day's sail distant. Pritchard, in his " Polynesian Reminiscences," mentions that, in his time, there were living at this Uvea, or Uea, the grandchildren of Tongan castaways who had, in a double canoe, drifted over eleven hundred miles to that isle.

In 1696 two canoes, containing thirty persons of both sexes, drifted nine hundred miles to the Philippines. In 1721 two canoes reached Guam, in the Ladrones, after a twenty-day drift. In 1817 Kotzebue found on one of the Radack Chain a native of the Carolines, one of a party that had made a fifteen-hundred-mile drift due east. Cook, when on his third voyage, found at Atiu some castaways from Tahiti, driven thither when trying to make Raiatea. Of this incident Cook remarked, " It will serve to explain, better than a thousand conjectures . . . how the islands of the South Seas may have been first peopled."

Kotzebue tells us of finding a Japanese vessel off the Californian coast in 1815 that had drifted for seventeen months across the Pacific. Only three of her crew of thirty-five were alive. Dillon speaks of a drift voyage of 465 miles made by four Rotuma men who were cast away on Tikopia, a small island north of the New Hebrides. As this island is peopled by Polynesians speaking a dialect closely resembling that of New Zealand, it was probably settled by drift voyagers from the east. The above drift occurred about the year 1800. Dillon states that other drift canoes from Rotuma have reached Tikopia, Fiji, and Samoa.

In 1832 Williams found at Manua, Samoa, a native of Tubuai, in the Austral Group, south of Tahiti. He was one of a party sailing from Tubuai to an adjacent isle. Their canoe, storm-caught, drifted for three months ere it reached Manua, when most of the crew had perished. In such cases the catching of rain-water, and of fish, usually sharks, preserved life in some of the waifs. Coconuts, usually carried in canoes, would presumably furnish some extra water-vessels.

Another recorded drift is that of some natives of Aitutaki, who thus reached Proby's Island, a thousand miles to the westward. Beechey found some natives of Anaa or Chain Island, at Bow Island, Paumotu Group. Three canoes had drifted six hundred miles eastward; two had been lost, while those in the third, owing to a series of accidents and bad luck, had been for three years trying to get home by working from island to island.

On the 8th March, 1821, a canoe reached Raiatea from Rurutu, Austral Isles, after being buffeted about the ocean for six weeks.

Easter Island was resettled by people from Rapa Isle, who are said to have found a strange "long-eared" folk in possession—possibly the authors of the strange script and the stone images of that lone isle.

But enough of drift voyages, for their number is legion. Cases of drift voyages in many directions across the Pacific Ocean are on record. Feckless writers have told us that no drift or other voyage in an easterly direction could have been made by Polynesians, on account of the trade-winds; that no Polynesian could have reached New Zealand; that no Polynesian canoe could carry sea stock for a lengthy voyage; that such canoes were too frail for deep-sea navigation. The hapless Polynesian could not sail out of sight of land because he possessed no compass; he could not traverse the open ocean because it provided no cabbage-trees to tie his canoe to at night! Pretty soon we shall hear that there never was a Polynesian canoe, or a Polynesian to use it if there had been one. The fact of natives occupying all groups and most isolated isles of Polynesia has apparently been viewed by the above writers as a personal injury, hence the evolving of the sunk-continent theory, the sudden disappearance of half a world, leaving a few continental folk clinging desperately to mountain-peaks, somewhat startled doubtless, but by no means downhearted.

For centuries the Maori voyager was crossing the Southern Ocean between New Zealand and Polynesia; for a very much longer period he was weaving innumerable sea roads across northern oceans. No timid coast paddler was he, but a bold navigator of great oceanic areas, who, ever listing to the lure of Hine-moana, broke through the hanging skies, and lifted every water trail of the Realm of Kiwa.

But we do not like it, and cannot grasp it. For we feared to do these things when in the same culture stage as the Maori, and for long after. Hence our search for lost continents and land bridges, and a special creation of man for Auckland and another for the Great Barrier. Our fears ran to the anger of the gods, ever averse to wild enterprises, and initiative, and a round earth, and other desirable things. The Polynesian voyager who pushed out into the unknown went down the changing centuries as a hero. We would probably have burnt him. We poled a log raft, with anxious hearts, across the raging Thames, but the Maori hewed him a dugout with a sharp stone, tied a top strake to it with a piece of string, dumped his wife and bunch of coconuts into it, and paddled forth to settle an isle beyond the red sunrise.

The voyages of Tama-ahua, Tu-moana, Tuwhiri-rau, Mou-te-rangi, and Pahiko from New Zealand to Polynesia we have no time to discuss— a remark that also applies to two traditions of drift canoes from New Zealand reaching those parts, and returning here.

Though the Maori has long ceased his voyages to Polynesia—for the last we know of took place ten generations ago—yet has much of the adventurous spirit been retained to our own time—the days of the white man. When Ngati-Awa seized the "Rodney" at Port Nicholson, in 1835, to raid and settle the Chatham Isles, they wrote the last chapter in the long, long history of the Maori buccaneers. And is it not on record that these daring Vikings had arranged with an American whaler to transport them to Samoa, when the arrival, cutting-off, and plunder of the "Jean Bart" marred the scheme, and saved Samoa some stirring times.

The Polynesian voyager left the so-called adventurous Turanian folk to longshore traffic, and the isles adjacent to their homes; he passed through the dark-skinned folk of Melanesia, despising them for their colour and lack of daring; he roamed far and wide over the vast Pacific Ocean, and carried his speech from Nukuoro to the Chathams, from Easter Island to Madagascar.

For the Maori as a voyager feared not the dangers of the deep, known or unknown. He harnessed his gods to the task of assisting him; he traced out the *ara moana*, the sea roads, over two great oceans for western folk to treasure, and western keels to furrow.

The scene changes. Our Maori voyager has boldly crossed the sullen seas and made his landfall under alien skies. Afar off on the rolling waves of Hine-moana his strained vessel cuts the sky-line. Strained and sea-weary is she, worn and battered from the passage of Te Moana nui a Kiwa. Her land-hungering crew gaze eagerly on green hills, and brown-skinned experts scan the surging surf. The coast swings in nearer, the roar of breakers strikes upon the ear. For this is no fair landing; it is the rolling *tai maranga*, the leaping surge of Hine-moana dashing wildly against Raka-hore, the iron bounds of her realm placed there by the gods of old.

They call upon the *amotawa*, the sea expert, wise with the wisdom of those who brave the wrath of the Ocean Maid. He takes command, and all await his orders. The sails are lowered, the paddles hold the tossing craft, or edge her in in search of Hine-tuakirikiri, the fair landing-beach. The steersmen and paddlers are all attention, for this is the *tai maranga*; a single error shall open the gates of death. The expert knows that, in this sea, eleven *ngaru wharau*, curling dangerous combers, are followed by the *mutu moana*, a smooth, rounded, crestless billow, the only one on which the canoe may ride safely to land. He awaits that wave. As it reaches the craft and lifts her, there comes the sharp order, " *Kia aronui te hoe!* " and instantly every paddle is held stationary in the water, blade broadside on to the sea run. So is the canoe held on the swell of the wave. The correct position is for the prow to project somewhat in front of the wave-crest; to allow it to forge ahead or drop behind is to court disaster; the dreaded *tai maranga* is following and preceding her. Hence the order to meet and hold her. Should the canoe show signs of slipping back off the wave, the command, " *Kia korewa te hoe!* " brings all paddles turned edgewise on to the sea, whereupon she forges ahead on the wave, and is there held with the paddles. Two steersmen at the stern wield long steer-oars (*hoe whakatere*), two more at the bow manipulate the *hoe whakaara*. These play an important part in the management of the vessel.

The canoe is now rushing shoreward, poised on the *mutu moana*, or rounded wave, while every man, vigilant, ready for instant action, watches the swift rush as she leaps to land, and awaits the quick commands of the expert. As the wave grounds, and begins to dissolve, there comes the quick cry, " *Kumea te hoe!* " and the long bow oars are taken in, while every paddle is plied with fierce energy to impart additional impulse that will carry the canoe well up the beach. As one man, all hands now drop their paddles inboard, leap out, and run her up beyond reach of the next wave—the Maori voyager has made his landing, and upheld the saying of yore, " *He ihu waka, he ihu whenua.*"

Thus the Maori voyager comes to land, and enters into his rest. But not as you would! He does not paddle ashore, make fast, and go into camp with careless mien and prosaic mind. He steps softly on the flanks of the land, and placates the demons thereof; he conducts solemn ritual and performs strange rites to introduce his gods, and to preserve his physical and spiritual welfare ; he forgets not those who have protected, guided, and succoured him. For the Maori was ever in sympathy with his surroundings, and ever he vivified them. He endowed them, for weal or woe, with strange powers ; he loved to personify the elements, the forces of nature, and inanimate objects ; to feel that he was in unison with them, that all possessed life in common, that all were the offspring of the first all-embracing parents—the Sky Father and the Earth Mother.

Impelled by Tawhiri-matea, and borne by Tane across the broad, heaving breast of Hine-moana ; guided by Hine-korako, and urged forward by Huru-moana ; succoured in time of stress by Te Ihorangi and Tangaroa, our voyager eludes iron-ribbed Rakahore, and is received by Hine-tua-kirikiri. Fair to his sea-weary eyes, Hine-rau-wharangi greets him ; while, sheltering within Tane-mahuta, Punaweko cries him welcome. Rolling down the rugged flanks of Hine-tu-maunga comes Para-whenua-mea to restore his waning energies, while Hine-pukohu-rangi casts her white mantle over him.

Even so does our Maori voyager return to the Primal Parent ; the Parent who brought man forth to the World of Life, and who takes him again to her sheltering breast, when, weary and wayworn, he returns from his journey ; the Parent to whom all voyagers and all men return at last —the first Mother Parent, Papa-tuanuku, Papa-matua-te-kore, the Parent and the Parentless—the old, old Earth Mother !

EXPLANATION OF SOME MAORI TERMS USED.

Aotearoa. Maori name of the North Island.
Ara moana. Sea roads ; sea paths.
Haumi. Piece fastened to the hull of a canoe to lengthen it.
He ihu waka, he ihu whenua. A canoe-nose (prow), a land nose. (Implies that the two shall meet, as noses do in the *hongi* salute.)
Hine-korako. Personified form of some celestial glow.
Hine-moana. Personified form of the ocean.
Hine-pukohu-rangi. Personified form of mist.
Hine-rau-wharangi. Personified form of vegetable growth.
Hine-tuakirikiri. Personified form of sand and gravel.
Hine-tu-maunga. Personified form of ranges.
Huru-moana. Personified form of sea-birds.
Kiwa. Presiding genius or guardian of the ocean.
Papa-matua-te-kore. Papa the Parentless. (Papa and Rangi had no parents, but were themselves the first parents.)
Papa-tuanuku. The Earth Mother.
Pārā-weranui. Personified form of south wind.
Para-whenua-mea. Personified form of waters of earth.
Punaweko. Personified form of land-birds.
Rakahore. Personified form of rock.
Rarohenga. The spirit world.
Tama-nui-te-ra. Honorific name for the sun.
Tane. Personified form of forests and trees.
Tangaroa. Personified form of fish.
Tawhiri-matea. Personified form of winds.
Te Ihorangi. Personified form of rain.
Te Moana nui a Kiwa. The Great Ocean of Kiwa.

ART. XLVIII.—*Investigation into the Resistance of Earth Connections.*

By L. BIRKS, B.Sc., M.Inst.C.E., M.I.E.E., and ERIC WEBB, Lieut. R.E
(A.I.F.).

[*Read before the Philosophical Institute of Canterbury, 7th July, 1915.*]

THE subject of this investigation is one which is being constantly referred
to by the electrical engineer, and called for by innumerable regulations
framed for the protection of life and plant ; but in practice it is often so
vague and uncertain, and so few engineers actually measure their earth-
resistances, that there is room for much investigation before we can claim
to understand the principles involved or their effect when applied to any
particular set of circumstances. The electrical engineer, in practice, early
admits the difficulty of the problem, and realizes the absence of any comprehensive treatise on the subject. There is such diversity in the data
extant, and so strange a variation between his own experience and the
stated results of others, that he is only too glad to escape further attention from official inspectors by obtaining permission to "earth" to a water-
pipe. And in this way the matter, for the most part, is quietly shelved.

The present paper aims primarily at the publication of data obtained
during an investigation of earthing devices, for use on the Canterbury
Plains (New Zealand), in connection with the high-voltage electric-power
transmission-line recently erected by the Public Works Department of New
Zealand for the supply of Christchurch and district from Lake Coleridge,
a distance of sixty-two miles. Some description of the country traversed
is essential to an intelligent apprehension either of the data itself or of the
difficulty experienced in attaining the objective.

The Lake Coleridge power-house is situated in the Rakaia River valley,
sixty-two miles west of Christchurch, and from this point two 66,000-volt
three-phase transmission-lines are run to the city on independent pole-lines.
For the first sixteen miles from the power-house the lines are located about
3 chains apart, and pass along the terraces and low hills of the river-valley,
in many places over screes or fans of shingle detritus. At Windwhistle
Point they divide. In the next fourteen miles the north line continues
over low hills and rolling downs to Glentunnel, while the south stretches
almost directly across the thirty-two miles of shingle-beds which constitute
the Canterbury Plains. At the city end the lines meet again at Bealey
Road, seven miles from the substation, and continue about 1 chain apart.
For this distance the soil is largely of a sandy to peaty nature, the remains
of a marsh formed by the sandhills of a former seashore and in the bed of
an ancient river. It is thus evident that, although the major part of the
line runs through dry shingle country, there is at the same time extensive
variety. In many places across the plains water in ordinary soakage wells
is almost unobtainable, not being met till below 60 ft., and in some cases
120 ft. to 140 ft. Evidently the permanent water-level may thus be at a
very great depth, leaving at the top a considerable thickness of dry shingle.
The shingle is, in the main, a hard, highly crystalline, bluish-grey sandstone,
commonly known as greywacke. In size it ranges from sand to boulders
18 in. in diameter.

The line conductors consist of 7/·135 bare aluminium wires, and are
carried on Thomas No. 4000 pin insulators, mounted on wooden (jarrah)

cross-arms, and carried on ironbark poles standing 36 ft. out of the ground and sunk 6 ft. into the ground. There are 864 poles in the northern line and 873 in the southern line. The standard pole-spacing is 6 chains (396 ft.) —*i.e.*, about thirteen poles to the mile. In order to obviate the burning of poles and cross-arms, it was decided to earth the whole of the high-tension insulator-pins; and owing to the large inductance with high-frequency current and considerable capacity of a ground wire, not to mention the greater cost, it was decided to earth each pole with an individual earth-strip. In addition to this, the apparatus in telephone depots along the line required to be thoroughly earthed, while the stations at either end, and the numerous transformers in the distribution system, all required very efficient earths. The station lightning-arresters in particular depend for their effectiveness on a low-resistance connection to ground.

The substance of the British Board of Trade specification in regulations concerning station earths and earths for tramways and general power purposes is that the resistance shall be less than 1 ohm between two plates of copper, cast iron, or galvanized iron, packed in coke, and 60 ft. apart.

Regulations under the Coal-mines Act (Great Britain), 1911, specifies that " All conductors of an earthing system shall have a conductivity at all parts and at all joints at least equal to 50 per cent. of that of the largest conductor used solely to supply the apparatus a part of which it is desired to earth." In practice this would involve an earth-resistance never greater than 10 to 15 ohms, and usually not more than 2 or 3 ohms. " Sparks " (journal of the Institution of Electrical Engineers), 15th March, 1915, recommended packing-plates placed vertically at 6 ft. or more beneath the surface, with a foot of coke on each side : " Tests made on the earth plates constructed on the lines recommended by the General Regulations, the plates being buried in excavations 4 ft. by 2 ft. by 8 ft. deep in clay, under favourable conditions as to moisture, vary from 1·8 to 2·2 ohms, while the resistance of similar earth plates in another district, in excavations 4 ft. by 2 ft. by 6 ft. deep in well-consolidated marl and clay, the earth plates resting on clay in damp positions, reached 2·6 and 2·7 ohms. . . . If the conditions are not favourable a much higher resistance to earth will be found, as the resistance to earth is directly affected by the nature and temperature of the surrounding strata and by the amount of moisture."

It is now fairly common knowledge that the Board of Trade test is seldom obtained, and that values such as those quoted above are decidedly the exception rather than the rule.

Some tests were made locally on plate earths of the Christchurch City Council. These plates were ⅛ in. copper, 2 ft. square, and were placed horizontally in a bed of coke 1 ft. thick above and below the plate, and at a depth of 6 ft. to 8 ft. These were tested against the city high-pressure water-main, and (assuming the water-main as of zero resistance to earth) these gave the following results : Beckenham transformer-house plate earth = 51 ohms (loam and shingle) ; Sydenham transformer-house plate earth = 17·2 ohms (clay) ; Montreal Street transformer-house plate earth = 13 ohms (clay).

The Christchurch Tramway Board also had some considerable difficulty with earth connections. Finally two large copper plates 8 ft. square were placed some 10 ft. below the surface (60 ft. apart), and well packed with coke. Tests on these plates failed to reach that required by the Board of Trade. To avoid further trouble and expense the ground wire was earthed to the city water-main. In July, 1915, tests of these earths by the authors gave the following results : Plate 1 to plate 2 = 6 ohms ; plate 1 to water-main = 4 ohms ; plate 2 to water-main = 5 ohms.

During the course of the present investigation plates were tried on the transmission-line. Three copper plates 12 in. square were buried at a depth of 5 ft. 6 in. to 6 ft., and bedded in clay. In one case the country was loam to shingle, while the others were clay to sand. The plates were placed at intervals of 6 chains, and tests of pairs in series gave the following values : Resistance of plates immediately after placing—No. 760 plate to No. 762 plate = 260 ohms ; No. 759 plate to No. 760 plate = 200 ohms ; No. 759 plate to No. 760 plate = 180 after 4 gallons water had been poured over each ; No. 760 plate to No. 762 plate = 130 after 3 gallons of brine had been poured over No. 762 plate. Values by A.C. bridge test.

The contrast of these values with those quoted is obvious, and these are from by no means in the worst districts. This chiefly indicates the unreliability of the earth connection by means of small copper plates. At the same time, since each earth plate in place costs from £1 to £1 10s., the cost for such inadequate result precludes the adoption of the plate earth on any extended scale.

At the erection of the transmission-line poles provision was made for earthing by carrying a strip of galvanized hoop iron ($1\frac{1}{4}$ in. by $\frac{1}{16}$ in.) down the pole and carried round the butt three times below ground-level, terminating underneath the bottom of the pole. The connection was thus buried at a depth of at least 6 ft. No precautions in the way of selecting earth to make the contact were taken when filling in, and at bends and in special cases the poles were concreted. Subsequently tests of these earths were made with an instrument consisting of an ordinary Wheatstone bridge and 4-volt dry cell.

To get the resistance of each separate earth connection, the strips were tested in groups of three throughout the 1,737 poles, and the values computed from the three simultaneous equations obtained. (Notes concerning both the test and the method of computation of individual values will be found later.)

The values obtained ranged from 10 ohms to 5,000 ohms for individual connections. In some districts the resistances were consistently low ; in others consistently high ; while still others gave resistances indiscriminately high and low. Of the 864 earths on the north line, more than 30 per cent. were over 1,000 ohms, and 83 per cent. over 100 ohms ; while on the south line, out of 873 poles, 18 per cent. were over 1,000 ohms, and 90 per cent. over 100 ohms. As earths of this order are quite useless for practical purposes, attention was turned to pipe earths.

By far the most useful data on the subject available consists of a paper by E. E. F. Creighton in the " General Electric Review," vol. 15, page 66, February, 1912. The most important and pertinent conclusions are here quoted :—

" 1. Resistance of a pipe earth varies inversely as the depth of the pipe after the pipe has reached a uniformly conducting stratum.

" 2. Practically all the resistance is in the earth in the immediate vicinity of the pipe. This resistance depends on the specific resistance of the material. The specific resistance depends on acids, salts, or alkalies in solution about the plate. To get the lowest possible resistance strong salt water should be poured around the pipe. The chemical action of salt on the iron of the pipe under average conditions is negligible.

" 3. If an iron pipe 1 in. in diameter is driven into normally moist earth to a depth of 6 ft. to 8 ft. it will usually have a resistance of about 15 ohms— 8 ohms may be considered unusually low—while in dry soils it may give a resistance of 50 ohms and upwards. When it is desired to lower the resistance to earth below that of a single-pipe earth, others should be driven at distances of not less than 6 ft. apart. Then the total conductances will be only slightly less than the sum of the conductances of the individual pipes.

" 4. Half the total resistance lies within 6 in. of the pipe. At a distance apart of 6 ft. the resistance reaches nearly a constant value.

" 5. Since the resistance of a pipe earth lies mostly in the immediate vicinity of the pipe, the greatest potential drop when the current flows will also be concentrated there. Heating and drying out of the soil will tend to magnify this effect. The more salt water placed round a pipe earth the less the potential gradient.

" 6. The resistance of a pipe earth does not decrease in direct proportion to the increase in diameter of the pipe. A pipe 2 in. in diameter has a resistance only about 6 per cent. to 12 per cent. less than a pipe 1 in. in diameter."

Mr. H. P. Liversidge, before the International Association of Municipal Electricians, 19th August, 1913, gives data recording resistances with pipe earths (2 in. in diameter and 6 ft. to 12 ft. deep) ranging from 0·08 ohms to 138 ohms. He states that the general average of some other twenty-five different readings, representing fairly average conditions in reference to character of soil, contained moisture, &c., give the following contact resistances : Clay, 13·60 ohms ; gravel, 6·01 ohms ; top soil, 1·80 ohms. There is no specific mention of the method of test employed, but apparently a voltmeter-ammeter method, with about 220 volts D.C., was used. All these results were obtained after a salt solution had been poured round the pipe, and probably these values are about one-half of the initial resistance. Mr. Liversidge suggests as desirable values for an earth for station work 1 to 2 ohms, and for other purposes 2 to 6 ohms.

As it was apparently necessary to investigate in the actual country, some preliminary pipe earths were then tried on the transmission-line at Bealey Corner, where the line passes into the heavy shingle of the old Waimakariri river-bed. Galvanized-iron pipes 1½ in. diameter were driven 4 ft. to 6 ft. deep beside line-poles, 6 chains apart. In sand and clay the average individual resistance of seven pipes 5 ft. deep was 126 ohms by the D.C. bridge method. In heavy shingle the average individual resistance of four pipes 4 ft. deep was 1,800 ohms (D.C. bridge method). These pipes were driven into the ground with a 14 lb. hammer, and no brine or other electrolyte was added.

Further pipes were then driven in the neighbourhood of three other poles. The ground about the three poles selected was loam for a foot at the top, to clay and sand or clay and light shingle below. The uncertainty of the substratum was demonstrated by the fact that, though one pipe was easily driven in sand and clay, another, 6 ft. away, when driven a little more than 2 ft. refused to go farther on account of heavy shingle.

The pipes were drilled with two ⅜ in. holes at intervals of 1 ft. along the length of the pipe, and after starting with a crowbar were driven 5 ft. to 8 ft. deep with a hammer. They were then tested as driven, and results are tabulated below. Pipes A and B were then " salted " by filling with salt, pouring water in, and filling again with salt. They were then tested again, and at intervals subsequently. Pipes C were salted by pouring in at intervals hot saturated solutions of brine up to 20 gallons, testing between successive charges, and also subsequently. In pipes D cold saturated solutions (up to 9 gallons) of copper sulphate were used.

TABLE I.—TEST PIPES, BEALEY CORNER.
(Average Individual Pipe-resistances to Earth.)

Particulars of Pipes, 6 Chains apart.	Pipes as driven.	Immediately after salting.	Seven Days after salting.	Twenty-one Days after salting.	Three Months after salting.	Seven Months after salting.
	Ohms.	Ohms.	Ohms.	Ohms.	Ohms.	Ohms.
A. Three 2 in. pipes, 7 ft. deep ..	115	95	84	48	38	45
B. Three 1¼ in. pipes, 9 ft. deep ..	138	105	92	56	51	70
C. Two 1¼ in. pipes, 9 ft. deep ..	175	85	63	..	36	46
D. Two 1½ in. pipes, 7 ft. 6 in. deep (salted with copper sulphate)	78	68	..	..	..	94

These are all individual pipe-resistances obtained with a low-voltage D.C. bridge method.

Some discussion is here necessary concerning the various methods of test that have been used. The low-voltage D.C. Wheatstone bridge method has been most commonly used, owing to its ease of application and the portable nature of the apparatus. This test cannot give the true resistance under working-conditions, when thousands of volts A.C. are concerned, but gives a proportional result, as indicated by the following observations: Some earths at Addington Substation were tested by three different methods —(a) By D.C. bridge; (b) by voltmeter-ammeter D.C. 110 volts; (c) by Wheatstone bridge, using surging A.C. voltage and telephone-receiver.

In reference to (a) and (b), in all tests two values were obtained with reversed polarity, to eliminate as far as possible electrolytic effect, and the tabulated values are the mean of the two. Test (c) was effected by placing an interrupter (a small buzzer) in the primary of a small transformer, with a 4-volt cell as source of e.m.f., and connecting the secondary to the "battery" terminals of the bridge. A telephone-receiver replaces the galvanometer for indicating a balance of the arms. The results are tabulated below for three earths marked A, D, and P.

TABLE II.—COMPARISON OF METHODS OF TESTING.

Earth-resistances (Two in Series).	(a.) D.C. Bridge.	(b.) D.C. Ammeter.	(c.) A.C. Bridge.
	Ohms.	Ohms.	Ohms.
A and P	31·3	23·6	22·8
D and P	37·3	31·2	28·8
A and D	15·3	13·0	12·8

Although the tests are not numerous, they are sufficient to indicate a large but fairly constant error in the ordinary low-pressure D.C. bridge test, and a fairly close approximation of the A.C. bridge test to the positive voltammeter determination.

In the same connection Mr. Creighton may be quoted—"that this method (A.C. bridge) gives values within 5 per cent. of those by D.C. or A.C. voltammeter."

Some measurements of earth-resistances by D.C. and A.C. voltmeter-ammeter method were made by Mr. Parry, Chief Electrical Engineer, in Wellington. These were made on hemispheres 3 in. and 6 in. in diameter, placed in soil without addition of any salt. Results are tabulated below, with percentage of the D.C. value greater than the A.C. :—

TABLE III.—COMPARISON OF D.C. AND A.C. VOLTAMMETER METHODS.
(Two Earths in Series.)

Number of Test Pair.				D.C.	A.C.	Ratio.
				Ohms.	Ohms.	
1 ..	..	..	..	213	177	1·21
2 ..	..	..	..	158	138	1·14
3 ..	..	..	..	161	138	1·16
4 ..	..	..	..	164	139	1·19
5 ..	..	..	..	115	101	1·14
6 ..	..	..	..	158	135	1·18
7 ..	..	..	..	112·5	100	1·12
	Average ..	..	..	..	..	1·16

This gives an average of D.C. values greater than the corresponding A.C. values by 16 per cent. The A.C. ammeter used in these tests had a very close scale, and results are not entirely satisfactory.

Some later comparative values were obtained at Addington Substation and on the transmission-line, with the following results :—

TABLE IV.—COMPARISON OF D.C. AND A.C. BRIDGE TESTS.
(Two Pipes in Series.)

Number of Pairs.			Distance apart.	By D.C. Bridge.	By A.C. Bridge.	Ratio.
Pipes unsalted—				Ohms.	Ohms.	
4	..	..	18 in. ..	26·3	16·9	1·55
3	..	..	3 ft. ..	29·7	20·2	1·47
2	..	..	4 ft. 6 in. ..	33·2	23·5	1·41
1	..	..	6 ft. ..	35·5	25·0	1·42
1	..	..	30 ft. ..	40·5	28·0	1·45
2	..	..	6 ch. ..	1,770	1,390	1·27
3	..	..	6 ch ..	287	238	1·21
1	..	..	100 ft. ..	41·8	32·0	1·31
Pipes salted—						
2	..	..	6 ch. ..	89	79	1·13
1	..	..	6 ch. ..	92	70	1·31
1	..	..	6 ch. ..	187	143	1·31
3	..	..	6 ch. ..	383	328	1·17
3	..	..	6 ch. ..	142	111	1·28
	Average	..	..	..	..	1·33

The D.C. bridge thus gives a resistance about 33 per cent. higher than the A.C. bridge. There is a considerable range in the percentage difference, showing that the D.C. test probably gives a rather erratic result. Although results are not sufficiently numerous to warrant a definite statement, there is an indication that the difference unsalted is in the region of 30 to 40 per cent., while salted it is from 20 to 30 per cent.

Thus resistances by D.C. measurement are all higher than those by A.C. As the resistance is—at least, very largely—the resistance of an electrolyte, this is to be expected by reason of the polarization effect with D.C. Obviously, values by voltmeter-ammeter methods must be accepted as correct, and whether D.C. or A.C. determinations should be considered in any specific case will depend on working-conditions.

The A.C. bridge was found both quicker and more convenient in operation than the D.C. bridge. A galvanometer, however well packed, will not stand continuous transport by motor bicycle without losing in sensitiveness, while a telephone-receiver is easily packed, and with reasonable care does not suffer from carriage.

A very compact form for the transformer and interrupter can be obtained by rewinding the electro-magnet of an ordinary 4-ohm buzzer, and using one leg as primary and for working the interrupter, and winding a secondary on the other leg. Eight layers of No. 36 copper on the primary, and eleven layers of same gauge on secondary, gave very good results.

Some comment is necessary relative to the practice of solving simultaneous equations to obtain the resistances of individual earths. The solutions obtained will be correct as long as the unknowns are independent quantities—*i.e.*, as long as one earth does not interfere with another. This depends mainly on the distance between the earths being greater than some critical value, which, again, is dependent on the class of soil. Where individual resistances are tabulated herein, unless otherwise stated, the earths are sufficiently far apart not to interfere with each other in any way.

It became evident at once that in the class of ground encountered earth-resistances below 100 ohms were going to be very difficult and expensive in practice.

From the tests made it appears that once a pipe reaches a depth of 6 ft. to 7 ft., increase in depth has little effect on the final value. Any improvement produced by increased depth is so small as to be lost in the effect of other factors. It is common experience that in considering ground connections the character of the soil is the main factor concerned. However, it is not the solids constituting the soil so much as the moisture and the salts in solution which command attention. Messrs. McCollum and Logan (Proc. A.I.E.E., June, 1913) publish some material bearing on the conductivity of soils with varying moisture-content (diagram No. 1). The curve shown was obtained from a sample of red-clay soil which had been dried out at 105° F., and water afterwards added. It will be noticed that the resistance is practically constant for a content greater than 20 per cent. In the same paper are recorded ninety-two values of specific resistances of "a wide variety of different kinds of soil." In 50 per cent. of the determinations the moisture-content, which is also recorded, lies between 19 and 30 per cent.—*i.e.*, they were moist soils—and at these values the resistance-variation with moisture-content is almost negligible. Yet the recorded specific resistances range from 41,490 to 470 ohms. In some cases—few rather than many—the quantity of moisture is distinctly responsible for the variation in resistance; but in general the main factor affecting the conductivity of the pipe earth is the electrolytic quality of this moisture.

Considering the exceedingly high resistances of both approximately dry soil and approximately pure water, we must conclude that the conductivity —at least, in the immediate neighbourhood of the earth connection—is almost solely due to the salts, alkalies, or acids in solution. The exceedingly large range of resistances is then explicable by electrolytes of variable specific conductivity, determined by the solutes or the degrees of concentration. That such is a feasible explanation is evident from the following specific conductivities of electrolytes (" Physical and Chemical Constants," by Kaye and Laby), which admit of combinations to provide almost infinite range of values (diagram No. 2).

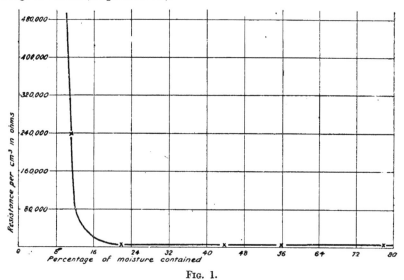

Fig. 1.

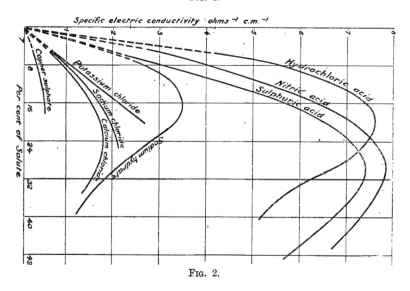

Fig. 2.

In the course of this investigation a few comparatively very low-resistance " earths " were obtained, ranging from 4 to 40 ohms—*e.g.*, at Windwhistle Point single pipes gave 20 ohms; at Acheron River single pipes gave 35 ohms; and see values in Table VIII : all by A.C. bridge method.

These invariably occurred either in marshy places or where there had been extensive accumulation of vegetable matter and a good depth of soil. It is probable that in these localities there is a liberal proportion of those decomposition products which may be included under "humus acids." Low-percentage solutions of acids have many times the conductivity of equivalent solutions of neutral salts or alkalies, and the existence of these in particular localities probably accounts for the comparatively low resistances encountered there.

Further valuable data were obtained from tests of artesian wells about Christchurch. Since these are from 70 ft. to several hundred feet deep, they might be expected to give resistances approaching zero.

Four wells consisting of 2 in. pipes, 70 ft. to 80 ft. deep, at Addington, gave, on the average, a resistance of 6·2 ohms (A.C. bridge and voltammeter determinations). A 2 in. well, 80 ft. deep, at the city waterworks, Beckenham, had a resistance of 10 ohms (A.C. bridge). The average of three 8 in. wells, 80 ft. deep, at the same place gave resistance of 8 ohms (A.C. bridge).

Thus these well pipes have resistance of the same order as earth pipes 6 ft. deep in the same districts. It would therefore appear that depth and area of pipe in contact affect resistance only to a limited extent. The pure artesian water in the vicinity of the pipe is probably responsible for this effect.

A trial was made in sand and shingle ground at Sockburn to ascertain the effect of treating earth pipes with a higher-conductivity electrolyte— viz., sulphuric acid. Three 1 in. pipes 7 ft. long were placed (well tamped) about 1 to 2½ chains apart in much the same class of soil. These were then tested against an independent pipe 1 to 1½ chains distant. These gave individual resistances as follows (A.C. bridge test): The earth-resistance of the first pipe, on treatment with 5-per-cent. sulphuric acid, fell from 460 to 170 ohms—*i.e.*, by 63 per cent.; that of the second, on treatment with 15-per-cent. sulphuric acid, fell from 910 to 350 ohms—*i.e.*, by 61·5 per cent.; that of the third, on treatment with cold concentrated sodium chloride, fell from 360 to 170 ohms—*i.e.*, by 52·8 per cent.

The higher-conductivity solution gives a bigger reduction of the initial resistance, but soil, though mechanically uniform, is apparently so variable chemically that a great many tests would be necessary to give a definite result. The degree of penetration into the surrounding soil and the composition of the soil are both very uncertain. There was, of course, considerable action of the acid on the pipe, and there is no suggestion whatever to use such a method commercially.

It has been previously stated that the resistance lies almost wholly within a radius of a few feet of the pipe. This is shown clearly in diagrams Nos. 3 and 3A of the resistances obtained in different localities between pipes at varying distances apart. The effect of saturating the ground round these pipes with an electrolyte (a strong solution of sodium chloride) is also shown by these curves. The Wellington pipes were 3 in. pipes, 6 ft. long, and were placed in reclaimed ground with a good admixture of surface soil. The Addington pipes were 1 in. in diameter and 10 ft. deep, driven in peaty to sandy soil. The pipes at Bealey Corner were 1¼ in. and 1½ in. pipes, 6 ft. deep. At Sockburn 1 in. pipes were placed in shingle and sand over 7 ft. deep. The pipes from which curve 3A was obtained were placed in line in the same class of soil. The pipes were all salted with 4 gallons of cold concentrated brine poured into each. When salting, the most distant pipe was salted and tested, then the next nearer salted and tested, and so on, to avoid any interference-effect due to salted soil between. The curve of retests after two or three days, when the soil between was more or less impregnated with salt, is also shown. The introduction of the salt has substantially reduced the resistance of the pipe earth, and has at the same time very considerably flattened the curve of potential drop.

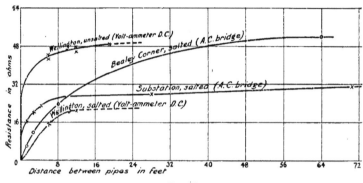

FIG. 3.

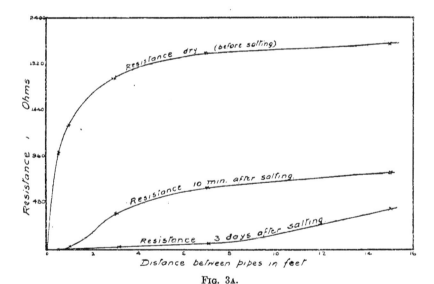

FIG. 3A.

Messrs. Liversidge and Creighton both record that salting had the effect of reducing the resistance to about 50 per cent. of the original value. Apparently the tests on which this statement is based were of earths of fairly low initial resistance—*i.e.*, below 150 ohms.

On the majority of the pipe earths tested during the course of our investigations, with initial resistances ranging from 400 to 4,000 ohms, we found much larger percentage reduction of the initial resistance as the result of salting. This is very clearly shown in diagram No. 4. This is based on the result of tests on over 100 pipes, each plotted point being the value from a group of three to eight 1 in. pipes, each 6 ft. to 7 ft. long. No two pipes were less than 6 ft. apart or more than 12 chains, and they were placed in groups of three to eight over the tract of country described, and therefore in all classes of soil. Salting was effected by pouring into each 4 gallons of saturated-brine solution, and then filling the pipe with salt. Tests were repeated within an hour or so of salting. Initial resistances up

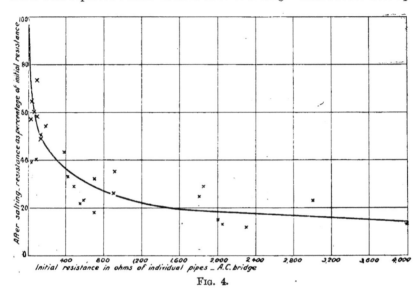

FIG. 4.

to 250 ohms are in loam, sand, or clay, and over 700 are in heavy shingle; others are in a mixture of earth and shingle. As the degree of penetration of the salt into the surrounding soil is a matter of considerable uncertainty, the percentage decreases in resistance are much more uniform than might be expected. The curve drawn through the plotted points suggests the probable average reduction. Thus it appears that the reduction of resistance by salting is largely dependent on the magnitude of the initial resistance, and the salting is highly efficacious in reducing high resistances.

A further interesting feature of salting is the improvement of the " earth " with time. Apparently the salt continues to work into the surrounding soil for a considerable time after the initial salting. In time the salt must be dissolved by successive rains, and finally the " earth " must deteriorate. Curves are given demonstrating this in diagram No. 5. Apparently the brine continues to percolate into the surrounding soil, partly by soakage and partly by capillarity. Eventually the solution is weakened by the addition of rain and subsoil water until the resistance again increases.

From a cursory inspection of these values it is evident that—at least, in many cases—single-pipe earths, however well salted, will not give the value of resistance desired. The only loophole remaining is. by way of placing a number of pipes in a group. The question immediately arises as to how close such pipes may be placed and still remain effective. A consideration of curves of the decrease in resistance with increase of distance indicates that the minimum useful distance between two pipes in good ground is about 6 ft. when the pipes are unsalted. After pipes have been salted the minimum distance must be increased.

In general, when pipes are salted, they must be placed at least 12 ft. apart to retain 90 per cent. of the efficiency of each pipe. The actual distance

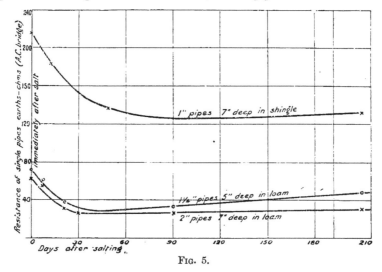

Fig. 5.

necessary in any particular case depends on the conductivity of the soil in which the pipes are placed, and the higher the conductivity the farther apart must the pipes be to retain their efficiency.

Groups of pipes were placed on the transmission-line for the purpose of earthing depot telephones and apparatus. The following table gives particulars and results obtained with several pipes in parallel. No two pipes were placed less than 12 ft. apart.

Table VI.—Pipes in Parallel. (All A.C. Bridge.)

Locality.	Number of Pipes.	Average Resistance of each Pipe.		Resistance of all Pipes in Parallel.		Class of Soil.
		Unsalted.	Salted.	Immediately after salting.	Five Months after salting.	
		Ohms.	Ohms.	Ohms.	Ohms.	
Greendale ..	5	1,900	244	47	32	Very dry clay and shingle.
Charing Cross	4	840	141	44	57	Heavy shingle cemented with clay.
Aylesbury ..	7	2,000	440	80	78	Heavy shingle.
West Melton	4	910	305	55	80	Sand and clay to heavy shingle.
Bealey ..	4	534	150	34	61	Loam to heavy shingle.
Glenroy ..	4	180	94	26	..	Loam and clay, heavy boulder, somewhat marshy.
Highfield ..	4	556	126	31	..	Loam and sand for 4 ft. to shingle.
Sandy knolls	5	2,334	417	90	..	Heavy running shingle.

It remains now to cite some actual methods employed, with the results attained.

Pipe earths, as well as the plate earths before mentioned, were used by the Christchurch City Council (diagram No. 6). These were placed in

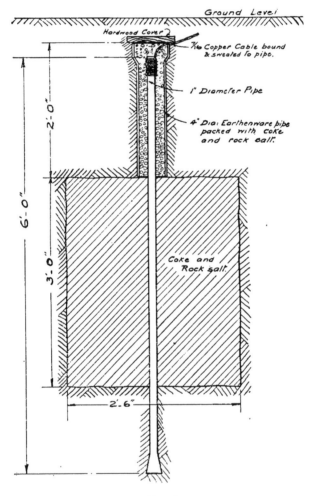

Fig. 6.

excavations 6 ft. to 7 ft. deep. The bottom of the pipe was packed with a mixture of rock salt (refuse salt from tanneries) and coke, while round the top was placed a drainpipe filled with rock salt and coke, and protected by a cover. Water was then poured in. The lead was $\frac{7}{16}$ S.W.G. braided copper. These tested against independent earths (either water or gas mains) gave the following results (A.C. bridge method) :—

TABLE VII.—CHRISTCHURCH CITY COUNCIL EARTHS.

Locality.		Resistance of Earth.	Computed Value of Gas or Water Main.
		Ohms.	
Corner Madras and Armagh Streets ..	(pipe)	22	Gas = 0.
N.E. corner Latimer Square	,,	19	
S.E. corner ,, ...	,,	21	Gas = minus 2.*
S.W. corner ,, ...	,,	20	
Corner Cashel and Madras Streets (N.W.)	,,	23	Gas = plus 3.
,, ,, (S.E.)..	,,	18	
Corner Peterborough and Madras Streets	,,	76·5	Gas = plus 1·5.
Corner Salisbury and Madras Streets . ..	,,	21·5	
Corner Chester and Madras Streets ..	,,	44	
Corner Victoria and Montreal Streets ..	,,	11·5	Gas = plus 0·5, assuming water-main = 1 ohm.
Beckenham transformer	,,	21	
,, ..	(plate)	50	
Sydenham transformer ..	(pipe)	15	
,, ..	(plate)	16	
Montreal Street	(pipe)	12	
,, ...	(plate)	12	
Madras Street, over Bealey Avenue ..	(pipe)	13·8	
,, ,, ..	(plate)	10·7	
Armagh Street – Hanmer Street ..	(pipe)	4	
,, ,, ..	(plate)	6	

* Since the assumption is that values are independent quantities, and in a city pipes run in unknown directions, some of the values will be slightly in error, as in this case, owing to the unknowns of the simultaneous equations being in some cases not entirely independent. Hence the above negative values.

At the transformers, the pipe earths and plate earths were placed within a few feet of each other, and tests show in favour of the pipe earth. As might be expected, in good conducting soil there is very little difference between plate and pipe earths, but in high-resistance soil the salted pipe usually has a considerable advantage.

The practice pursued by the Public Works Department (diagram No. 7) in the general transmission and distribution earths is to drive a 1 in. pipe, drilled with small holes every foot, to a depth of 6 ft. to 8 ft. in ground ; then to pour 4 gallons of concentrated brine down the pipe, and fill the pipe with rock salt, placing some rock salt round the top. The top of the pipe is stopped with a wooden plug, which can easily be removed for re-plenishing the supply of salt inside the pipe. The conductor consists of 1 in. galvanized-iron strip, riveted round the pipe and bolted through. In attaching the lead it is essential to make a good connection mechanically. Also any combination of metals which would yield to electrolysis should be avoided, or otherwise the joint must be thoroughly covered with some waterproof material. On this account, as well as for economy in cost, the joint is not sweated.

Resistances with this device, in soils similar to those of the previous tests (Table VII), are given in Table VIII. Values are of individual pipes, and are obtained by the A.C. bridge method.

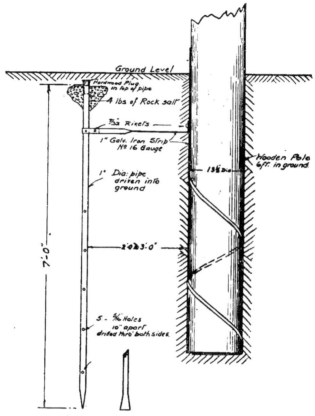

Fig. 7.

TABLE VIII.—PUBLIC WORKS DEPARTMENT EARTHS ABOUT CHRISTCHURCH.

Locality.	Resistance in Ohms.	Remarks.
Substation, Addington ..	16	1 in. pipe, 10 ft., unsalted; peaty to sand.
Christchurch Brick Company ..	6	1 in. pipe, 8 ft., unsalted; marshy loam.
,, ,, ..	4	1 in. pipe, 8 ft., salted; marshy loam.
Corner Lincoln Road and Barrington Street	17	1 in. pipe, 12 ft., salted; loam to peat.
Lincoln Road	15	1¼ in. pipe, 12 ft., salted; loam to peat.
Colombo Street	24	1 in. pipe, 12 ft., unsalted; loam to clay.
,, ...	34	1 in. pipe, 10 ft., unsalted; loam to clay.
Allen's Mill	15	1 in. pipe, 10 ft., unsalted; loam to clay.
Sunnyside transformer ..	14	1 in., unsalted; loam to clay.
Lincoln Road	20	1 in. pipe, 10 ft., unsalted; loam to clay.
Barrington Street	100	1 in. pipe, 7 ft., unsalted; loam to gravel.
,, ...	40	1 in. pipe, 7 ft., salted; loam to gravel.

A comparison of these two sets of values (Tables VII and VIII) indicates that the bare pipe salted is more efficient than the pipe packed with coke and costing twice as much to place. The more elaborate earth, however, will probably tend to retain its efficiency longer.

A number of sample values (A.C. bridge method) of the resistances of pipes in various parts of the transmission-line are indicated in curve No. 4. The initial resistance of individual pipes ranges from 12 ohms to 4,000 ohms, and averages several hundred ohms. Immediately after salting, the range is 7 ohms to 560 ohms.

Pipes are now being placed at each pole on the transmission-line. Where there is nothing heavier than ordinary gravel, pipes can easily be started with a crowbar, and then driven to 6 ft. or 7 ft. with a hammer. The lower end of the pipe is flattened and closed, while the top is protected for driving by a loosely fitting cap turned from mild steel. In heavy shingle ground it is sometimes necessary to make an excavation in depth almost equal to the length of the pipe, and this very considerably increases the cost of placing. The lead from the pipe, as before, is 1 in. galvanized-iron strip, riveted round and bolted through by $\frac{1}{4}$ in. bolt to the pipe, and riveted on to the earth-strip running up the pole. Bolting and riveting were resorted to after it was found that the sweated joint broke away while driving. Particulars of some samples of these earths are given below (sum of two resistances in series values equivalent to A.C. bridge test) :—

Pole Numbers.	Resistance Original Earth-strip.	Initial Resistance of Pipe.	Resistance immediately after salting.	Ten to Fourteen Days after salting.	Forty-four Days after salting.	Five Months after salting.	Character of Soil.
	Ohms.	Ohms.	Ohms.	Ohms.	Ohms.	Ohms.	
759 and 758	305	170	98	84	65	104	Loam to sand.
757 ,, 756	241	170	126	99	91	129	,,
755 ,, 754	209	216	118	93	71	100	Loam to sand and clay.
753 ,, 752	309	330	160	135	131	189	Sand to shingle.
751 ,, 750	890	1,600	493	387	290	420	Heavy shingle.
749 ,, 748	700	1,667	440	393	333	420	,,
747 ,, 746	667	1,200	360	327	267	374	Heavy shingle and sand.
745 ,, 744	680	1,093	467	360	267	327	,, ,,
743 ,, 742	660	1,067	393	363	360	360	Heavy shingle.
742 ,, 741	593	1,467	433	380	213	297	Heavy shingle to sand.
603 ,, 604	..	7,000	1,060	..	..	800	Heavy shingle.
605 ,, 606	..	7,000	1,000	..	..	800	,,
607 ,, 608	..	7,000	667	..	..	800	,,
609 ,, 610	..	4,667	847	..	..	780	Heavy shingle and loam.
611 ,, 612	..	2,667	380	..	..	400	,, ,,
482 ,, 483	..	500	..	..	..	360	Clay to sand and shingle.
483 ,, 484	..	560	253	..	..	277	Sand, loam, and shingle.
485 ,, 486	..	280	143	..	..	143	Sand, loam, and clay (scrub).
487 ,, 488	..	173	83	..	..	110	,, ,, ,, ,,
489 ,, 490	..	313	123	..	..	154	Shingle and sand to loam.
491 ,, 492	..	833	187	..	..	190	Clay to shingle.
493 ,, 494	..	713	145	..	..	176	,,
495 ,, 496	..	1,800	293	..	..	230	,,

For salting these pipes, ordinary coarse salt is best for the initial charge, as it readily dissolves, while rock salt is better for filling and leaving about the pipe, on account of its taking a considerable time to dissolve. At present refuse rock salt from tanneries is being used, and found equal to ordinary salt for the purpose, and much cheaper.

It still appears that in some districts single pipes will not provide a reasonably low-resistance earth connection. In such districts it will be necessary either to drive several pipes round each hole or to run a ground wire along that portion of the line and connect it with the single pipe at each pole.

This discussion would not be complete without some reference to the efficiency of the earth connection when called on to dissipate current.

Concerning the behaviour of pipe earths under working-conditions Mr. Creighton notes, "With an applied potential of 1,000 volts the current was so great that the earth round the pipe was quickly dried out, and 90 per cent. of the drop of potential took place within 1 ft. of the pipe (unsalted). The pipe had lost its effectiveness as a ground. With 900 volts drop in the immediate vicinity, it was a dangerous condition."

The effect of salting on the potential drop has been illustrated by previous curves (Nos. 3 and 3A), as the potential curve is of the same order as the curve of resistance. Mr. Creighton publishes curves of potential distribution which are very instructive, and further insists on the value of thorough salting. With a potential difference applied of 120 volts D.C. a drop of 70 volts took place in the first 6 in. unsalted, while the same drop was extended over 3 ft. 6 in. after the pipe was salted.

The capacity of an earth to discharge current over an extended period is also treated by Mr. Creighton : "The quantity of electricity that can be passed through a pipe earth without materially changing its resistance increases directly with the wetness of the earth in contact with the iron, and the area of the iron surface exposed to the passage of current; and decreases as the resistance of the earth in contact with the pipe increases. Certain critical values of the current may be carried continuously by a pipe earth without varying the resistance. The higher the current above this critical value, the more rapid the drying-out. To increase the ampere-hour capacity, keep the pipe earth wet with salt water."

Curves are also published showing the advantage of the pipe earth over a solid rod in both states, and particularly its value when salted. A pipe earth thoroughly salted carried a current of 56 amperes for a period of forty hours without appreciable variation, while the same pipe earth initially carried only 30 amperes, and was baked out to 4 amperes in four hours and a half.

Mr. Liversidge also made some tests, passing a current of 25 amperes D.C. : "In most of the tests it was found that the pipe connection or other grounding devices were able to carry a current of 25 amperes continuously without any marked increase in contact resistance. The results which were obtained indicated in most cases that the limiting feature of the current carrying-capacity of a ground connection is the ampere-discharge per unit area of contact surface. If the current-density was high enough to drive off the electrolytic moisture as a consequence of excessive heating of the earth immediately surrounding the grounding device, then the contact resistance would gradually increase."

With alternating current it appears that the ampere-hour capacity of a given earth is, in general, considerably higher than with D.C. This is consequent on the excessive formation of films in the interspaces, consequent on electrolysis, these films forming to a very much lesser degree, and almost negligible, when the current is alternating. With a high frequency the inductance of the earth and lead is a proportionally large factor for consideration, and this must be reduced as far as possible by using shortest possible leads. This question will be the subject of further investigation as opportunity offers.

To conclude, the observations here set out demonstrate the uniformly high resistance of earth in the country investigated. Comparison with published data accentuates the contrast with the average values found elsewhere. It is not suggested that this is the only place of the kind, but it is evident that some circumstance is responsible for the abnormal high resistances.

A further item which points to a high resistance of soil in the district is the fact that very little electrolysis of waterpipes, &c., has been noted.

Although the dryness of the shingle plain may be responsible for some of the high resistances, it cannot account for their persistent occurrence. Christchurch and environs is extensively supplied with artesian water which itself is almost pure, containing the very lowest percentage of salts or impurities of any kind. Since pure water has such a high resistance, this may account for the high " earth " resistances encountered.

ART. XLIX.—*Resistance to the Flow of Fluids through Pipes.*

By E. PARRY, M.I.E.E., Assoc.M.Inst.C.E.

[*Read before the Technological Section of the Wellington Philosophical Society, 10th November, 1915.*]

THE present work is the result of an effort to express the resistance offered by rough pipes, as distinct from smooth pipes, to the flow of fluids in terms of $\dfrac{vd}{\nu}$ where v is the mean velocity of flow, d the diameter of the pipes, and ν the kinematic viscosity of the fluid.

The investigation was prompted by the publication in the " Philosophical Transactions of the Royal Society A," vol. 214 (1914), of Stanton and Pannell's experiments upon smooth pipes, and more particularly by the publication in the " Proceedings of the Royal Society A," vol. 91 (1914), of Professor Lees's discussion of those experiments. The latter work contains an admirable historical *résumé* of the subject, so there is no need to dwell further upon that aspect than to state that the resistance offered by a viscous fluid of any kind flowing in a circular pipe is proved by Stoker, Helmholtz, Rayleigh, and Reynolds from dynamical considerations to be a function of the expression vd/ν.

Below a certain critical value of the velocity the resistance is purely viscous, and the function mentioned is a simple one, viz.,—

$$\frac{R}{\rho v^2} = a \left(\frac{\nu}{vd} \right) \quad \dots\dots\dots\dots\dots\dots\dots\dots\dots\dots(1)$$

where R is the resistance per unit of surface, ρ is the density of the fluid, a is a coefficient, the other values having the same significance as before.

Above the critical value of the velocity the resistance is apparently partly viscous and partly an inertia effect, and the relation between the elements a complex one; so much so that the probability is that it cannot be exactly expressed by any formula; neither is a formula an absolute necessity, though an approximate formula, if obtainable, is undoubtedly a convenience.

16—Trans.

In this work the resistance curve is regarded somewhat in the same manner as a stress strain or magnetic force and magnetization curve is regarded—that is to say, it is a complicated function obtained by experiment and expressed by means of a diagram which shows the relation between two chosen functions.

Professor Lees, on examination of Stanton's experimental results, finds that within the limits of experimental error the law of resistance above the critical value of a smooth pipe can be expressed very approximately in the form

$$\frac{R}{\rho v^2} = a \left(\frac{\nu}{vd}\right)^{\chi} + b \dots \dots \dots (2)$$

where a, b, and χ are experimental coefficients. When the equation is expressed in absolute units he gives the following values to the coefficients, viz. :—

$$a = \cdot 0765 \; ; \; b = \cdot 0009 \; ; \; \chi = 0 \cdot 35,$$

so that the equation becomes

$$\frac{R}{\rho v^2} = \cdot 0765 \left(\frac{\nu}{vd}\right)^{0 \cdot 35} + \cdot 0009 \dots \dots \dots (3)$$

Converting to foot pound units, and also transforming from resistance per unit area and per unit of mass to energy per pound of fluid per foot of length or "head" per foot of length, we have

$$i \frac{d}{4} \bigg/ v^2 = \cdot 00801 \left(\frac{\nu}{vd}\right)^{0 \cdot 35} + \cdot 000028 \dots \dots \dots (4)$$

where i is the hydraulic gradient or slope, or, in other words, the resistance head per foot of pipe. This form is more convenient for practical purpose, and conforms to engineering practice. The formula most commonly in use is that known as Chezy's formula, or, rather, a modification of the same, viz.,—

$$v = C\sqrt{ri} \dots \dots \dots (5)$$

where C is an experimental coefficient, i the hydraulic gradient, and r is the hydraulic mean radius which for round pipe is equal to $\frac{d}{4}$. It will be seen by comparing equation (4) and (5) that for smooth pipe

$$\frac{1}{C^2} = \cdot 00801 \left(\frac{\nu}{vd}\right)^{0 \cdot 35} + 000028 \dots \dots \dots (6)$$

Returning to equation (4), this is plotted in fig. 1, curve a, with $i\frac{d}{4}\big/v^2$ as ordinates and log vd/ν as abscissae. The reason why it is necessary to adopt logarithmic values is that the range is so great that a comprehensive diagram could not otherwise be drawn within the limits of ordinary sheet. The range covered by experiment and observation lies between the values 5·2 and 7·2 of the expression log vd/ν, and it will be seen later that certain observations on a wood stave pipe lie on an extension of the curve up to a value of 7·7, which tends to confirm the law as expressed by Professor Lees.

Accepting curve a, fig. 1, as a sort of datum-line from which to gauge experimental results, the sequel is an account of an examination of experiments carried out by different observers on the resistance to the flow of water in pipes with the object of determining whether it is possible in the light of present knowledge to systematize them and to deduce a law for their behaviour.

In order to comprehend the whole range satisfactorily both below and above the critical velocity it is necessary to take logs of both sides, and the standard curve is set out again in fig. 2, curve a, with $\log \dfrac{4v^2}{id}$ as ordinates and $\log vd/v$ as abscissae.

The diagram fig. 2 shows the first of stream-line stage as a straight line, and as the index χ is equal to unity in this region the line is inclined at an angle of 45°. The equation of the line is

$$i \frac{d}{4}\Big/ v^2 = a \left(\frac{v}{vd} \right)$$

where a has the value 8, an experimental value obtained from an examination of Stanton's experiments already referred to. Thus straight portion of the characteristic in fig. 2 intercepts the abscissa at log 8.

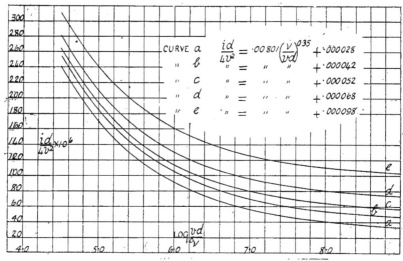

Fig. 1.

The curved portion of the characteristic marked a is Professor Lees's equation already referred to, which represents the behaviour of fluids in smooth pipes above the critical velocity in the eddy state of motion. Inasmuch as the change is abrupt from one state to the other, the exact nature of the connection cannot be determined, but the two portions are joined together on the principle that it requires a higher force to change a state of motion than to maintain either the previous or subsequent stage. This phenomenon was observed by Reynolds, and is similar in its behaviour to elastic and magnetic phenomena.

Having obtained a standard of reference in this way, the next step is to plot to the same scale all the observations available on some one class of pipe and see how they stand in reference to the standard. Comparing first riveted pipes in general, the dots on the sheet represent 129 observations made on riveted pipes, varying from 12 in. in diameter to 102 in. in diameter, recorded by Messrs. Marx, Wing, and Hoskins in the "Transactions of the American Society of Civil Engineers," vol. 40. This is a

16*

collection of all known experiments on riveted pipe up to the date of the record—viz., 1898—to which is added a series of thirty-one observations made by the same authors on a riveted pipe 72 in. in diameter and recorded in vol. 44 of the same society.

In these experiments the speeds vary from a fraction of a foot per second to 20 ft. per second.

The observations referred to are plotted in fig. 2, and it will be noticed that despite the number and range of observations the range is extremely limited from a law-determining point of view, and one is entitled to say at once on regarding them that no law of friction could possibly be deduced from them, and that every effort in that direction has been a waste effort. The range of $\log vd/\nu$ in these experiments is from 7 to 8, whilst Stanton's observations, before referred to, ranged from $\log vd/\nu = 5\cdot2$ to $\log vd/\nu = 7\cdot2$, the extent of which enabled Professor Lees to deduce a law for smooth pipes with some degree of certainty. Even this range could with advantage be extended.

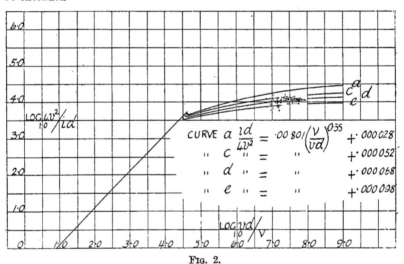

FIG. 2.

In the absence of a similar range for riveted pipe it is necessary to consider what degree of guidance Professor Lees's curve, as set out in fig. 2, curve a, affords towards deducing a resistance-curve for riveted pipe.

Regarding the plotted points by themselves, all that can be said regarding them is that they indicate an inclination towards the left, but whether a straight line or a curved line would best represent the law no definite answer can be given.

Most recent experimenters work on the assumption that the law is of the form given in equation (1), which would give a straight line when the logarithmic values are plotted. It will be seen that if the true law is of the form given in equation (2), this would result in a curved line, and inasmuch as any one series of experiments is, as a rule, of an extremely small range from the point of view under discussion, the points would fall very approximately upon a straight line, though in reality such lines are segments of a curve. It follows that as the inclination is different in different parts of the curve, and the range of each series being small, a different index would result according to the position of the series, and which might vary from

1·7 to nearly 2·0. An examination of experimental results confirm this view. Each experimenter has a different index, which varies, as a matter of fact, between the limits mentioned.

The diagram of fig. 2 affords a clear indication of the direction in which further experiments upon riveted pipe should take, and it is evident that no addition to our present knowledge can be made; neither can a law be for certain deduced unless the observations are extended so as to cover a greater range.

Regarding the results plotted in fig. 2, and taking curve *a* as a guide, it is reasonable to assume that the curve will be of the same form as curve *a* as expressed in equation (2); and, further, one might also reasonably assume that the viscosity element is constant, and that the inertia effect is greater because of the disturbing effect of the rivets and joints. Proceeding on this principle, three curves—viz., *c*, *d*, and *e*—are drawn through the points in fig. 2, *c* and *e* being drawn through what may be deemed to be the lower and upper limit after neglecting what are obviously random or stray shots,

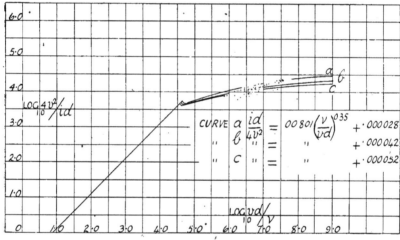

Fig. 3.

or errors in observation. Curve *d* is drawn through the thickest part of the cluster, and may be deemed to represent the commonest or most usual value.

It will be seen that these curves are as likely to represent the true form as any other, and there is ground to believe them to be the true representation of the law.

These curves have been transferred to fig. 1, curve *a* being the standard, whilst *c*, *d*, and *e* are the same as *c*, *d*, and *e* in fig. 2. These may be expressed respectively as follows:—

$$\text{Curve } c \quad \ldots \quad \ldots \quad i\frac{d}{4}\bigg/v^2 = \cdot00801 \left(\frac{v}{vd}\right)^{0.35} + \cdot000052$$

$$\text{,, } d \quad \ldots \quad \ldots \quad \text{,,} = \text{,,} \quad + \cdot000068$$

$$\text{,, } e \quad \ldots \quad \ldots \quad \text{,,} = \text{,,} \quad + \cdot000098$$

Large numbers of experiments have been carried out on wood stave pipes, and on this account it is instructive to plot observations in the same manner as for riveted steel pipes. This is done in fig. 3, where *a* is the curve for smooth pipes, as in figs. 1 and 2.

The dots represent a series of sixty-six observations made by Moritz on wood stave pipe recorded in the " Transactions of the American Society of Civil Engineers," vol. 74 (1911). The pipe-diameters range from 18 in. to 55¾ in., and velocities from a fraction of a foot per second to 5·874. These experiments are noteworthy on account of the extent of and care exercised in carrying out the experiments.

Regarding the results, it will be seen that quite a number of the dots are on the standard curve for smooth pipe, and otherwise by their disposition strengthen the evidence in favour of the curve *a* being the true law for smooth pipe.

Regarding these results, it may be truly said that under favourable circumstances a wood stave pipe may be treated as a smooth pipe, and the resistance truly represented by curve *a* in the figures. The other curves have been drawn in fig. 3—viz., *b* and *c*—the former a sort of mean curve, and the latter an outside or superior limit after neglecting what appear to be errors of observation. These curves have been transferred to fig. 1, and may be identified by the same lettering. The conclusion so far as regards wood stave pipe is that the value of $1/C^2$ lies between

$$\cdot 00801 \left(\frac{\nu}{vd} \right)^{0\cdot 35} + \cdot 000028$$
$$\text{and}$$
$$\cdot 00801 \left(\frac{\nu}{vd} \right)^{0\cdot 35} + \cdot 000052,$$

whilst the mean or commonest value may be taken as

$$\cdot 00801 \left(\frac{\nu}{vd} \right)^{0\cdot 35} + \cdot 000042.$$

It is reasonable to assume that asphalted pipes of any material will yield the same results as wood stave pipe, provided the joints are well made and even, and that there are no projections of any kind such as rivet-heads or straps, and that the asphalting lies smoothly and evenly. The common defect in asphalted pipes is that the pipe has been immersed in an asphalt mixture having too low a boiling-point, with the result that the asphalt coating is of uneven thickness and corrugated, thereby considerably increasing the resistance. Cast-iron pipe or solid-drawn or welded iron or steel pipes should, if this asphalting is properly done, yield results within the limits given for wood stave pipe.

Whilst no certain conclusion can be drawn as to the correct expression of the law of resistance of rough pipes, we can say the whole range or experience lies bounded by the curves *a* and a curve similar to and not far different in shape and position from curve *e*, and as an instance of the application of the curves of fig. 1. Suppose that the conditions are such that the value of log vd/ν is 7, and that we wish to know the value of $1/C^2$ for riveted pipe, we can be certain it is not less than ·000056, whilst the probability is that it is not less than ·00008 nor more much than ·000126, whether curves *c* and *e* are correctly drawn or not, or whether the equations given for these curves are right or wrong, whilst there is a preponderance of experience in favour of a value of ·000096, which is found on curve *d*.

Applying the same reasoning to a wood stave pipe, and assuming that the value of log vd/ν is 7 as before, the value of $1/C^2$ cannot be less than ·000056, and is not likely to exceed ·00008, whilst the probable value is ·00007.

It is claimed as a result of this investigation that fairly definite limits have been set within which the loss of head for almost any kind of pipe can be ascertained, but that the exact expression for any class of pipe is not

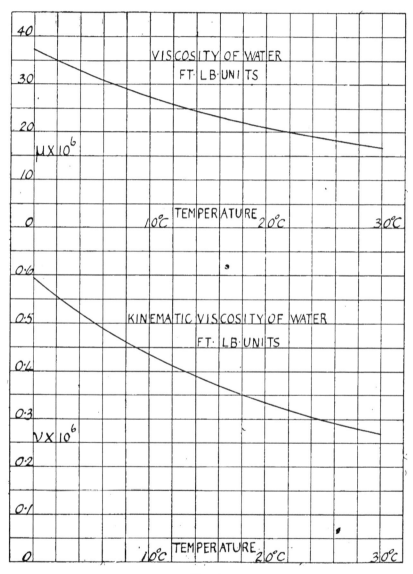

FIG.

at present ascertainable and cannot be ascertained unless the range of observations is considerably extended. Certain conclusions may also be drawn as to the value of $1/C^2$ for very large values of log vd/v, and that such values tend towards being constant, the value being that of the second term on the right-hand side of the equations. By modification of the diagrams these may be made to apply to open conduits and canals, and as the cross-sections, generally speaking, **are** very much larger compared with the largest pipe in use, the value of the expression which would correspond to log vd/v is large, and in consequence the variation of $1/C^2$ is small, and tends to become nearly constant, and will vary roughly with the surface and very little with the viscosity and consequently with temperature.

It should be noted that whilst the fluid under discussion has been water, the curves, and therefore the equations, are quite general and apply to any fluid, the link being the kinematic viscosity v. For convenience of reference two curves are shown in fig. 4, showing the viscosity and kinematic viscosity of water and its variation with temperature, which enables the value of the function vd/v to be obtained for any given temperature, whilst the value of log vd/v can be obtained for any value of v and d, and any temperature between 0° centigrade and 30° centigrade, by reference to the curves in fig. 5.

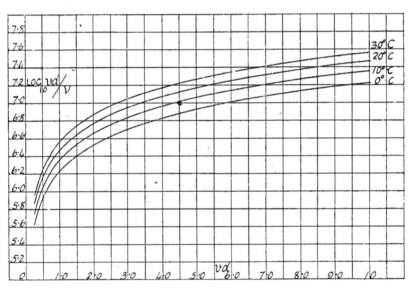

FIG. 5.

The following is a summary of the foregoing and of the present state of knowledge in respect to the friction of fluids in pipes :—

1. The resistance offered to the flow of fluids through round pipes with smooth surface is represented by curve *a*, fig. 1, which may be expressed, according to Professor Lees, by the equation

$$i\frac{d}{4}\Big/ v^2 = \cdot00801 \left(\frac{v}{vd}\right)^{0\cdot35} + \cdot000028.$$

2. The value of $1/C^2$ in Chezy's modified formula $v = C\sqrt{ri}$ is approximately expressed for smooth pipe by

$$\cdot00801 \left(\frac{v}{vd}\right)^{0\cdot35} + \cdot000028.$$

3. Wood stave pipe under some conditions comes under the definition of smooth pipe, for which $1/C^2$ has the value given in paragraph 2 of this summary.

4. In the present state of knowledge no certainty exists as to the correct expression of the law of resistance to the flow of fluids in rough pipes. Regarding, however, the observations made of the resistance or loss of head in relation to the curve of resistance for smooth pipe, an equation of the same form as for smooth pipe, in which the first expression on the right is constant and of the same value as Professor Lees's equation, viz.,—

$$\cdot 00801 \left(\frac{\nu}{vd}\right)^{0\cdot 35}$$

whilst the second term varies with the roughness of the surface, will fit the facts as well as any other, and there are some indications that this is the correct mode of expression.

5. On the assumption made in paragraph 4, the value of $1/C^2$ for wood stave pipes lies between

$$\cdot 00801 \left(\frac{\nu}{vd}\right)^{0\cdot 35} + \cdot 000028$$

and

$$\cdot 00801 \left(\frac{\nu}{vd}\right)^{0\cdot 35} + \cdot 000052,$$

whilst the more usual value may be taken as

$$\cdot 00801 \left(\frac{\nu}{vd}\right)^{0\cdot 35} + \cdot 000042.$$

6. The value of $1/C^2$ for riveted pipe lies between

$$\cdot 00801 \left(\frac{\nu}{vd}\right)^{0\cdot 35} + \cdot 000052$$

and

$$\cdot 00801 \left(\frac{\nu}{vd}\right)^{0\cdot 35} + \cdot 000098,$$

whilst the more usual value may be taken as

$$\cdot 00801 \left(\frac{\nu}{vd}\right)^{0\cdot 35} + \cdot 000068.$$

7. For large values of vd/ν the value of $1/C^2$ for any kind of pipe tends towards a constant value and independent of viscosity and therefore of temperature, but varies with the nature of the surface.

8. For large conduits or canals the same reasoning applies as in the case of large pipes referred to in paragraph 7.

ART. L.—*On the Inscribed Parabola.*

By E. G. HOGG, M.A., F.R.A.S., Christ's College, Christchurch.

[*Read before the Philosophical Institute of Canterbury, 1st September, 1915.*]

1. IF $l + m + n = 0$, the conic inscribed in the triangle of reference ABC, whose equation is

$$S \equiv \sqrt{lax} + \sqrt{mby} + \sqrt{ncz} = 0,$$

is a parabola whose focus F has trilinear ratios $\left(\dfrac{a}{l}, \dfrac{b}{m}, \dfrac{c}{n}\right)$.

It may be remarked that if a line cutting BC, CA, AB in D, E, F respectively move so that the ratio DE : DF $= p : q$, the envelope of the line is the parabola $\sqrt{(p-q)\,ax} + \sqrt{-pb\beta} + \sqrt{qcz} = 0$; *i.e.*, $l : m : n = p-q : -p : q$.

The equation of the directrix of the parabola is $l \cos Ax + m \cos By + n \cos Cz = 0$.

The trilinear polar of F and the directrix of S meet on the conic which is the isogonal transformation of Euler's line (*i.e.*, the line passing through the orthocentre, centroid and circumcentre).

Let S touch BC, CA, AB in A', B', C' respectively: the equations of B'C', C'A', A'B' are respectively

$$L \equiv -\, lax + mby + ncz = 0$$
$$M \equiv \quad lax - mby + ncz = 0$$
$$N \equiv \quad lax + mby - ncz = 0.$$

The lines L, M, N pass through the fixed points $\left(-\dfrac{1}{a}, \dfrac{1}{b}, \dfrac{1}{c}\right)$, $\left(\dfrac{1}{a}, -\dfrac{1}{b}, \dfrac{1}{c}\right)$, $\left(\dfrac{1}{a}, \dfrac{1}{b}, -\dfrac{1}{c}\right)$ respectively, and the triangles ABC, A'B'C' are in perspective, the trilinear ratios of P—the centre of perspective—being $\left(\dfrac{1}{la}, \dfrac{1}{mb}, \dfrac{1}{nc}\right)$.

The locus of P is the Steiner ellipse of the triangle ABC (*i.e.*, the circumscribed ellipse whose centre is at the centroid, or maximum circumscribed ellipse). The line FP passes through the fixed point,

$$\left[\dfrac{1}{a\,(b^2 - c^2)}, \dfrac{1}{b\,(c^2 - a^2)}, \dfrac{1}{c\,(a^2 - b^2)}\right],$$ which is a point of intersection of

the circumcircle and Steiner ellipse of the triangle ABC.

The equation of the maximum inscribed ellipse of the triangle A'B'C' is $l\sqrt{L} + m\sqrt{M} + n\sqrt{N} = 0$; hence P is also situated on this conic.

The bisector of the diagonals of the quadrilateral formed by the lines L, M, N and the trilinear polar of P—viz., $lax + mby + ncz = 0$—is the line $l^2 ax + m^2 by + n^2 cz = 0$, which touches its envelope—the Steiner ellipse—at the point P.

The triangle A'B'C' is self-conjugate with respect to the Steiner ellipse, whose equation may be written $l\mathrm{L}^2 + m\mathrm{M}^2 + n\mathrm{N}^2 = o$.

The polar of the focus F $\left(\dfrac{a}{l}, \dfrac{b}{m}, \dfrac{c}{n}\right)$ with respect to the triangle A'B'C' has for its equation

$$\frac{\mathrm{L}}{b^2 + c^2 - a^2} + \frac{\mathrm{M}}{c^2 + a^2 - b^2} + \frac{\mathrm{N}}{a^2 + b^2 - c^2} = o\ ;$$

i.e., $\qquad \tan A\mathrm{L} + \tan B\mathrm{M} + \tan C\mathrm{N} = o,$

which may be written in the form

$$m\{(\tan B + \tan C - \tan A)\ ax - (\tan C + \tan A - \tan B)\ by\}$$
$$+ n\{(\tan B + \tan C - \tan A)\ ax - (\tan A + \tan B - \tan C)\ cz\} = o,$$

showing that the polar passes through a fixed point.

The sides of the medial triangle of the triangle A'B'C' envelop parabolas as F moves on the circle ABC. The equation of the side of the medial triangle parallel to B'C' is $- l^2\mathrm{L} + m^2\mathrm{M} + n^2\mathrm{N} = o$, which reduces to

$$(m^2 + mn + n^2)\ ax + m^2by + n^2cz = o,$$

and the envelope of this line is the parabola

$$a^2x^2 = 4\ (ax + by)\ (ax + cz).$$

If S be expanded and $- l\ (m + n)$, $- m\ (n + l)$, $- n\ (l + m)$ be substituted for l^2, m^2, n^2 respectively, we have

$$\mathrm{S} \equiv mn\ (by + cz)^2 + nl\ (cz + ax)^2 + lm\ (ax + by)^2 = o,$$

showing that the triangle whose medial triangle is the triangle ABC is self-conjugate with respect to S.

2. The line at infinity expressed in terms of L, M, N is $l^2\mathrm{L} + m^2\mathrm{M} + n^2\mathrm{N} = o$: if $(x'y'z')$ be the trilinear ratios of the centroid G' of the triangle A'B'C', then the polar of G' with respect to the triangle—viz., $\dfrac{\mathrm{L}}{\mathrm{L}'} + \dfrac{\mathrm{M}}{\mathrm{M}'} + \dfrac{\mathrm{N}}{\mathrm{N}'} = o$—is identical with $l^2\mathrm{L} + m^2\mathrm{M} + n^2\mathrm{N} = o$. Hence $l^2\mathrm{L}'$

$= m^2\mathrm{M}' = n^2\mathrm{N}' = \kappa$, and therefore $\mathrm{M}' + \mathrm{N}' = 2lax' = \kappa \left(\dfrac{1}{m^2} + \dfrac{1}{n^2}\right)$, whence

$$\frac{ax'}{l\ (m^2 + n^2)} = \frac{by'}{m\ (n^2 + l^2)} = \frac{cz'}{n\ (l^2 + m^2)}.$$

The isogonal conjugate of F—viz., the point $\left(\dfrac{l}{a}, \dfrac{m}{b}, \dfrac{n}{c}\right)$—lies at infinity in a direction perpendicular to the pedal line of F, and therefore parallel to the axis of S. The line joining G' to the isogonal conjugate of F is $(m - n)\ ax + (n - l)\ by + (l - m)\ cz = o$, which passes through the centroid G $\left(\dfrac{1}{a}, \dfrac{1}{b}, \dfrac{1}{c}\right)$ of the triangle ABC. Hence GG' is parallel to the axis of the parabola.

From the above equations for the centroid G' $ax' = \kappa\ (l^3 - 2lmn)$, $by' = \kappa\ (m^3 - 2lmn)$, $cz' = \kappa\ (n^3 - 2lmn)$, and therefore $ax' + by' + cz' = - 3\kappa\ lmn$, and thence we obtain $ax' - 2by' - 2cz' = 3\kappa\ l^3$.

The locus of G' is therefore the cubic curve

$$(ax - 2by - 2cz)^{\frac{1}{3}} + (by - 2cz - 2ax)^{\frac{1}{3}} + (cz - 2ax - 2by)^{\frac{1}{3}} = o.$$

The equation of the pedal line of F is

$$lmn \; 2R \sin A \sin B \sin C \; (ax + by + cz)$$
$$= (mna^2 + nlb^2 + lmc^2)(l \cos Ax + m \cos By + n \cos Cz);$$

hence, eliminating l, m, n between this and the equations, $l + m + n = o$, $(m - n)\,ax + (n - l)\,by + (l - m)\,cz = o$, we obtain for the pedal of the envelope of the pedal lines of the triangle ABC, the centroid being the pole of the pedal, the equation

$$2R \sin A \sin B \sin C \; \Sigma \; (ax)\; UVW = \Sigma\;(a^2 VW)\; \Sigma\; (\cos A \; xU),$$

where $U \equiv by + cz - 2ax$, $V \equiv cz + ax - 2by$, $W \equiv ax + by - 2cz$.

3. A parabola is the isogonal transformation with respect to any inscribed triangle of a tangent to the circumcircle of the triangle. Hence the parabola $S = o$, whose equation may be written in the form $\dfrac{1}{L} + \dfrac{1}{M} + \dfrac{1}{N} = o$, may be regarded as the isogonal transformation with respect to the triangle LMN of a tangent to the circumcircle of that triangle.

The equation of this circle is

$$\frac{l^2 \Omega_1}{L} + \frac{m^2 \Omega_2}{M} + \frac{n^2 \Omega_3}{N} = o,$$

where $\Omega_1 \equiv a^2 l^2 + b^2 m^2 + c^2 n^2 - 2bcmn \cos A + 2canl \cos B + 2ablm \cos C$, with similar expressions for Ω_2 and Ω_3.[*]

The tangent to this circle at the point Q, determined on it by the equations $\dfrac{l'L}{l^2 \Omega_1} = \dfrac{m'M}{m^2 \Omega_2} = \dfrac{n'N}{n^2 \Omega_3}$, is $\dfrac{l'^2 L}{l^2 \Omega_1} + \dfrac{m'^2 M}{m^2 \Omega_2} + \dfrac{n'^2 N}{n^2 \Omega_3} = o,$ where $l' + m' + n' = o$.

The isogonal transformation of this line with respect to the triangle LMN is $\dfrac{l'^2}{l^2 L} + \dfrac{m'^2}{m^2 M} + \dfrac{n'^2}{n^2 N} = o$, which reduces to $\dfrac{1}{L} + \dfrac{1}{M} + \dfrac{1}{N} = o$ if $l'^2 : m'^2 : n'^2 = l^2 : m^2 : n^2$. Hence S is the isogonal transformation of the tangent $\dfrac{L}{\Omega_1} + \dfrac{M}{\Omega_2} + \dfrac{N}{\Omega_3} = o$, whose point of contact Q is given by the equations $\dfrac{L}{l\Omega_1} = \dfrac{M}{m\Omega_2} = \dfrac{N}{n\Omega_3}$.

To determine the locus of Q we have $\kappa\,(M + N) = m\Omega_2 + n\Omega_3$, whence $2\kappa\, lax = (m + n)(l^2 a^2 + m^2 b^2 + n^2 c^2 + 2mnbc \cos A)$
$$+ 2l\,(m^2 ab \cos C + n^2 ca \cos B) - 2lmna^2,$$
$$2\kappa\, ax = -(l^2 a^2 + m^2 b^2 + n^2 c^2 + 2mnbc \cos A)$$
$$+ 2\,(m^2 ab \cos C + n^2 ca \cos B) - 2mna^2$$
$$= l\,(m + n)\,a^2 + m\,(n + l)\,b^2 + n\,(l + m)\,c^2 - 2mnbc \cos A$$
$$- 2m\,(n + l)\,ab \cos C - 2n\,(l + m)\,ca \cos B - 2mna^2$$
$$= -3mna^2 + nlb^2 + lmc^2.$$

---

[*] See "Messenger of Mathematics," No. 502, February, 1913, p. 129.

Solving for mn, nl, lm from the equations

$$- 3mna^2 + nlb^2 + lmc^2 = 2\kappa\, ax$$
$$mna^2 - 3nlb^2 + lmc^2 = 2\kappa\, by$$
$$mna^2 + nlb^2 - 3lmc^2 = 2\kappa\, cz,$$

we obtain

$$a^2mn : b^2nl : c^2lm = 2ax + by + cz : ax + 2by + cz : ax + by + 2cz ;$$

hence the locus of the point of contact Q is

$$\frac{a^2}{2ax + by + cz} + \frac{b^2}{ax + 2by + cz} + \frac{c^2}{ax + by + 2cz} = 0,$$

which on expansion becomes

$$abc\, (ayz + bzx + cxy) + (ax + by + cz) \times$$
$$[(a^2 + 2b^2 + 2c^2)\, ax + (b^2 + 2c^2 + 2a^2)\, by + (c^2 + 2a^2 + 2b^2)\, cz] = 0,$$

a circle of radius 4R, having its centre at the point whose trilinear ratios are

$$[\cos A - \cos B \cos C, \cos B - \cos C \cos A, \cos C - \cos A \cos B].$$

The triangle formed by the lines

$$2ax + by + cz = 0, \; ax + 2by + cz = 0, \; ax + by + 2cz = 0$$

may be called the $\triangle_0$.

4. Let $S \equiv \sqrt{lax} + \sqrt{mby} + \sqrt{ncz} = 0$

$$S' \equiv \sqrt{l'ax} + \sqrt{m'by} + \sqrt{n'cz} = 0,$$

when $l' = m - n$, $m' = n - l$, $n' = l - m$, be two parabolas inscribed in the triangle ABC.

The equation of the line joining their foci $F\left(\dfrac{a}{l}, \dfrac{b}{m}, \dfrac{c}{n}\right)$ and $F'\left(\dfrac{a}{l'}, \dfrac{b}{m'}, \dfrac{c}{n'}\right)$

is

$$ll'\frac{x}{a} + mm'\frac{y}{b} + nn'\frac{z}{c} = 0,$$

showing that FF′ is a chord of the circle ABQ passing through the symmedian point (a, b, c) of the triangle ABC.

The centres of perspective P, P′ of the triangle ABC and the two triangles formed by joining the points of contact of S and S′ with the sides of the triangle ABC have trilinear ratios

$$\left(\frac{1}{la}, \frac{1}{mb}, \frac{1}{nc}\right), \; \left(\frac{1}{l'a}, \frac{1}{m'b}, \frac{1}{n'c}\right)$$

respectively : the equation of the line PP′ is $ll'ax + mm'by + nn'cz = 0$, which is satisfied by the trilinear ratios of the centroid $\left(\dfrac{1}{a}, \dfrac{1}{b}, \dfrac{1}{c}\right)$, showing that P and P′ are the extremities of a diameter of the Steiner ellipse of the triangle ABC. Hence the lines joining the fourth point of inter-section of the circle ABC and the Steiner ellipse of the triangle to the extremities of any chord of the circle, which passes through the symmedian point of the triangle, meet the Steiner ellipse again at the extremities of a diameter of the ellipse.

The polars of the centroid of the triangle ABC with respect to S and S′ form a pair of parallel tangents to the Steiner ellipse of the triangle ABC. The equations of the two polars are $l^2ax + m^2by + n^2cz = 0$, $l'^2ax + m'^2by + n'^2cz = 0$, and at their intersection $ax : by : cz = ll' : mm' : nn'$, showing that the tangents meet on the line at infinity.

The equations of the lines joining P to the isogonal conjugate of F and P' to the isogonal conjugate of F' are respectively

$$l^2l'ax + m^2m'by + n^2n'cz = 0$$
$$ll'^2ax + mm'^2by + nn'^2cz = 0.$$

These lines, which are parallel to the axes of S and S' respectively, intersect on the Steiner ellipse of the triangle ABC in the point $\left(\dfrac{1}{all'}, \dfrac{1}{bmm'}, \dfrac{1}{cnn'}\right)$, which is the centre of perspective of the triangle ABC and the "contact" triangle of the inscribed parabola which has its focus at the trilinear pole $\left(\dfrac{a}{ll'}, \dfrac{b}{mm'}, \dfrac{c}{nn'}\right)$ of the chord FF'.

The equation of the trilinear polar of the isogonal conjugate of F' is $\dfrac{ax}{l'} + \dfrac{by}{m'} + \dfrac{cz}{n'} = 0$: from its form it is seen that it touches S and also the maximum inscribed ellipse, $\sqrt{ax} + \sqrt{by} + \sqrt{cz} = 0$, of the triangle ABC. The corresponding fourth common tangent of this ellipse and the parabola S' is $\dfrac{ax}{l} + \dfrac{by}{m} + \dfrac{cz}{n} = 0$; the trilinear ratios of the point of intersection of these lines are $\left(\dfrac{ll'}{a}, \dfrac{mm'}{b}, \dfrac{nn'}{c}\right)$, which satisfy the equation of the line at infinity. Hence these common tangents are parallel; they touch the ellipse at the extremities of the diameter whose equation is $ll'ax + mm'by + nn'cz = 0$; *i.e.*, the diameter of the Steiner ellipse whose extremities are the centres of perspective P, P'. These common tangents to the maximum inscribed ellipse and the parabolas touch S, S' at the points $\left(\dfrac{ll'^2}{a}, \dfrac{mm'^2}{b}, \dfrac{nn'^2}{c}\right)$, $\left(\dfrac{l^2l'}{a}, \dfrac{m^2m'}{b}, \dfrac{n^2n'}{c}\right)$ respectively, and the line joining these points—viz., $\dfrac{ax}{ll'} + \dfrac{by}{mm'} + \dfrac{cz}{nn'} = 0$—is a tangent to the maximum inscribed ellipse.

5. The equation of the lines joining A to the points in which S and S' touch BC is

$$(mby - ncz)(m'by - n'cz) = 0 ;$$

i.e,
$$mm'b^2y^2 + nn'c^2z^2 - (mn' + m'n) \, bcyz = 0,$$

or
$$mm'b^2y^2 + nn'c^2z^2 - 2ll'bcyz = 0.$$

Hence the six points of contact of S and S' with the sides of the triangle ABC lie on the conic

$$S_1 \equiv ll'a^2x^2 + mm'b^2y^2 + nn'c^2z^2 - 2ll'bcyz - 2mm'cazx - 2nn'abxy = 0.$$

If in S_1 we substitute for (x, y, z) the trilinear ratios of the isogonal conjugates of the foci F, F'—viz., $\left(\dfrac{l}{a}, \dfrac{m}{b}, \dfrac{n}{c}\right)$, $\left(\dfrac{l'}{a}, \dfrac{m'}{b}, \dfrac{n'}{c}\right)$—we obtain the expressions

$$l^2l' + m^2m' + n^2n' - 2lmn \, (l' + m' + n'),$$
$$ll'^2 + mm'^2 + nn'^2 - 2l'm'n' \, (l + m + n),$$

both of which vanish; hence the asymptotes of S_1 are parallel to the axes of S and S'.

It is at once seen that the conic S_1 passes through the four fixed points

$$\left(\frac{1}{a}, \frac{1}{b}, \frac{1}{c}\right), \quad \left(-\frac{3}{a}, \frac{1}{b}, \frac{1}{c}\right), \quad \left(\frac{1}{a}, -\frac{3}{b}, \frac{1}{c}\right), \quad \left(\frac{1}{a}, \frac{1}{b}, -\frac{3}{c}\right).$$

Its equation may be written

$$\frac{ll'}{2ax + by + cz} + \frac{mm'}{ax + 2by + cz} + \frac{nn'}{ax + by + 2cz} = o.$$

Hence S_1, which is the harmonic conic of S and S', circumscribes the triangle Δ_0. It is the locus of points whose polars with respect to Δ_0 pass through the point $\left(\frac{ll'}{a}, \frac{mm'}{b}, \frac{nn'}{c}\right)$; *i.e.*, whose polars are perpendicular to the pedal line of the point which is the trilinear pole of the chord FF' of the circle ABC. Its centre lies on the conic

$$\frac{1}{by + cz} + \frac{1}{cz + ax} + \frac{1}{ax + by} = o.$$

6. The triangle A'B'C' is self-conjugate with respect to the conic $\frac{l'}{ax} + \frac{n_4'}{by} + \frac{n'}{cz} = o$, whose equation may be written in the form $ll'L^2 + mm'M^2 + nn'N^2 = o$. This conic passes through the points G $\left(\frac{1}{a}, \frac{1}{b}, \frac{1}{c}\right)$ and P $\left(\frac{1}{al}, \frac{1}{bm}, \frac{1}{cn}\right)$, and the tangents to the conic at these points are $l'ax + m'by + n'cz = o$ and $l^2l'ax + m^2m'by + n^2n'cz = o$ respectively; they are parallel to the axis of the parabola S. The locus of the centre of this conic as F moves round the circle ABC is the maximum inscribed ellipse of the triangle ABC.

7. Let three points F_1, F_2, F_3, having trilinear ratios $\left(\frac{a}{l}, \frac{b}{m}, \frac{c}{n}\right)$, $\left(\frac{a}{m}, \frac{b}{n}, \frac{c}{l}\right)$, $\left(\frac{a}{n}, \frac{b}{l}, \frac{c}{m}\right)$ respectively, be taken on the circle ABC. The equation of F_2F_3 is $\frac{x}{al} + \frac{y}{bm} + \frac{z}{cn} = o$; hence it touches the conic $\sqrt{\frac{x}{a}} + \sqrt{\frac{y}{b}} + \sqrt{\frac{z}{c}} = o$, which is the Brocard ellipse of the triangle ABC. Hence $F_1F_2F_3$ is a triangle inscribed in the circle ABC and circumscribed to the Brocard ellipse of that triangle.

Let inscribed parabolas S_1, S_2, S_3 have their foci at F_1, F_2, F_3 respectively, and let P_1, P_2, P_3 be the centres of perspective of the triangle ABC and the "contact" triangles of S_1, S_2, S_3 respectively with the sides of the triangle ABC. The trilinear ratios of P_1, P_2, P_3 are respectively $\left(\frac{1}{la}, \frac{1}{mb}, \frac{1}{nc}\right)$, $\left(\frac{1}{ma}, \frac{1}{nb}, \frac{1}{lc}\right)$, $\left(\frac{1}{na}, \frac{1}{lb}, \frac{1}{mc}\right)$, and the equation of P_2P_3 is $\frac{ax}{l} + \frac{by}{m} + \frac{cz}{n} = o$, which touches the ellipse $\sqrt{ax} + \sqrt{by} + \sqrt{cz} = o$;

hence the centres of perspective determine a triangle inscribed in the Steiner ellipse and circumscribed to the maximum inscribed ellipse of the triangle ABC.

The Steiner ellipse of the triangle $P_1P_2P_3$ is

$$\frac{1}{\dfrac{ax}{l}+\dfrac{by}{m}+\dfrac{cz}{n}}+\frac{1}{\dfrac{ax}{m}+\dfrac{by}{n}+\dfrac{cz}{l}}+\frac{1}{\dfrac{ax}{n}+\dfrac{by}{l}+\dfrac{cz}{m}}=0,$$

which reduces to $bcyz + cazx + abxy = 0$; *i.e.*, to the Steiner ellipse of the triangle ABC.

The equations of the parabolas having their foci at F_2, F_3 are, when expanded,
$$mnw^2 + nlu^2 + lmv^2 = 0$$
$$mnv^2 + nlw^2 + lmu^2 = 0,$$

where $\quad u \equiv by + cz, \; v \equiv cz + ax, \; w \equiv ax + by$;

hence $\quad mn : nl : lm = u^4 - v^2w^2 : v^4 - w^2u^2 : w^4 - u^2v^2.$

The locus of the intersections of inscribed parabolas which have their foci at the vertices of a triangle inscribed in the circumcircle and circumscribed to the Brocard ellipse of the triangle ABC is $\dfrac{1}{u}+\dfrac{1}{v}+\dfrac{1}{w}=0$,
where $u \equiv (by + cz)^4 - (cz + ax)^2 (ax + by)^2.$

The condition that an inscribed parabola

$$S \equiv \sqrt{lax} + \sqrt{mby} + \sqrt{ncz} = 0, \text{ i.e., } \frac{u^2}{l}+\frac{v^2}{m}+\frac{w^2}{n}=0,$$

may pass through the fixed point $(x_1y_1z_1)$ is $\dfrac{u_1^2}{l}+\dfrac{v_1^2}{m}+\dfrac{w_1^2}{n}=0$. Hence the equation of the two parabolas which may be drawn through the point $(x_1y_1z_1)$ is

$$\frac{1}{v^2w_1^2 - w^2v_1^2}+\frac{1}{w^2u_1^2 - u^2w_1^2}+\frac{1}{u^2v_1^2 - v^2u_1^2}=0,$$

and the equation of the two directrices is

$$\frac{u_1^2}{y \cos B - z \cos C}+\frac{v_1^2}{z \cos C - x \cos A}+\frac{w_1^2}{x \cos A - y \cos B}=0.$$

The line joining the foci of the two inscribed parabolas has for its equation $u_1^2\dfrac{x}{a} + v_1^2\dfrac{y}{b} + w_1^2\dfrac{z}{c} = 0.$

If H be any point on the line $fx + gy + hz = 0$, then writing this line in the form $pu + qv + rw = 0$, where $p \equiv -\dfrac{f}{a}+\dfrac{g}{b}+\dfrac{h}{c}$, $q \equiv \dfrac{f}{a}-\dfrac{g}{b}+\dfrac{h}{c}$, and $r \equiv \dfrac{f}{a}+\dfrac{g}{b}-\dfrac{h}{c}$, we easily find that the line joining the foci of the two inscribed parabolas which pass through the point H envelops the conic $\dfrac{ap^2}{x}+\dfrac{bq^2}{y}+\dfrac{cr^2}{z}=0.$

8. If $L = o$, $M = o$, $N = o$ be *any* three lines, the equation of the parabola inscribed in the triangle formed by them is

$$\sqrt{f l L} + \sqrt{g m M} + \sqrt{h n N} = o,$$

where $l + m + n = o$, and the equation of the line at infinity is

$$f L + g M + h N = o.$$

The equations which determine the focus are

$$\frac{l L}{f \Omega_1} = \frac{m M}{g \Omega_2} = \frac{n N}{h \Omega_3},$$

and the equation of the directrix is

$$f l L \left(- f^2 \Omega_1 + g^2 \Omega_2 + h^2 \Omega_3 \right) + g m M \left(f^2 \Omega_1 - g^2 \Omega_2 + h^2 \Omega_3 \right) + h n N$$
$$\left(f^2 \Omega_1 + g^2 \Omega_2 - h^2 \Omega_3 \right).$$

The trilinear ratios of the centre of perspective of the triangle LMN and the triangle formed by joining the points of contact of the parabola with the sides of that triangle are given by the equations

$$f l L = g m M = h n N,$$

and these ratios satisfy the equation of the Steiner ellipse of the triangle LMN—viz.,

$$\frac{1}{f L} + \frac{1}{g M} + \frac{1}{h N} = o.$$

The trilinear ratios of the isogonal conjugate of the focus is found from the equations

$$\frac{f L}{l} = \frac{g M}{m} = \frac{h N}{n},$$

and the equation of the axis of the parabola is

$$\frac{f}{l} \left(\frac{g^2 \Omega_2}{m^2} - \frac{h^2 \Omega_3}{n^2} \right) L + \frac{g}{m} \left(\frac{h^2 \Omega_3}{n^2} - \frac{f^2 \Omega_1}{l^2} \right) M + \frac{h}{n} \left(\frac{f^2 \Omega_1}{l^2} - \frac{g^2 \Omega_2}{m^2} \right) N = o.$$

The equations of the sides of the "contact" triangle are

$$L_1 \equiv - f l L + g m M + h n N = o$$
$$M_1 \equiv f l L - g m M + h n N = o$$
$$N_1 \equiv f l L + g m M - h n N = o,$$

and the equation of the parabola may be written

$$\frac{1}{L_1} + \frac{1}{M_1} + \frac{1}{N_1} = o.$$

ART. LI.—*The Star Test for Telescopic Mirrors.*

By T. ALLISON.

[*Read before the Wanganui Philosophical Society, 17th May, 1915.*]

IN the manufacture of mirrors for telescopic work the usual test is that known as Foucault's shadow test. This test used in connection with the zonal one is very convenient for workshop use, but it has its drawbacks. It will be found that the depth of the shadow varies according to the length of focus of the mirror, and it is affected also by the size of the hole through which the light shines and by the amount of light in the room. It is very easy, therefore, to be deceived by the shadow test: it takes, also, some experience to know what shadow should be seen. Theoretically, the zonal test should be satisfactory, but the necessary measurements are so small and delicate that I should not care to depend on it. There therefore remains one test for the final figure of a mirror, which I think is supreme: this is the star test. I will endeavour to describe it, premising that, as a mirror is specially made for viewing the stars, testing it on the stars seems quite a rational proceeding.

Focus the mirror on to a bright star and then rack it slightly out of focus both inside and outside the focal point. The image swells into a ring of light with a dark centre, the shadow of the flat. This should be exactly similar at equal distances on each side of the focal point. If, on the other hand, with the eye-piece slightly outside the focus the ring of light with the dark centre is seen, but with it slightly inside the focus there is a disc of light with a star-like point in the centre, the figure of the mirror is elliptic. If the reverse appears—that is, with the eye-piece inside the focal point there is a disc of light with a dark centre, and with it outside the focal point the telescope shows a disc of light with a bright star in the centre—the figure is a hyperbola. On throwing the eye-piece still farther out of focus it will be found, if the figure is elliptical, that inside the focal point the disc of light expands with a fairly large dark spot in the centre, while outside the focal point, at an equal distance, the disc of light will have a small black spot in the centre. If the mirror should unfortunately prove hyperbolic, outside the focal point the disc of light will have a large black spot, while inside the focal point, at a similar distance, the disc of light will have a small black spot. If the expanded disc is hairy or ragged when inside but well defined outside the focus, the edge of the mirror is turned down.

With a perfect mirror, throwing the eye-piece out of focus at equal distances on each side of the focal point, the mirror should show the outside edge of the disc slightly heavier than inside; and it should contain concentric rings of light each slightly fainter than the one immediately outside it, and in the centre there should be a black spot. Each image at an equal distance from the focal point should be an exact copy of the other. In making these experiments the eye-pieces should, if possible, be achromatic, and for the results to be critical a bright star, such as Sirius, Canopus, Arcturus, or either of the Pointers, should be used, and an eye-piece of fairly high power. I used in these experiments three achromatic eye-pieces—180, 260, and 360—and also a low-power negative one of about 100. The mirror can only be perfectly balanced with one power. Any power from 180 to 260 would be a good power to finally correct it with. If the series of rings do not shade down uniformly it shows that there are zones in the mirror. A good hint can be obtained from Cooke,* and after making the necessary alterations I have no hesitation in quoting from him :—

" If on racking towards the mirror it is found that the central rings look feeble while the edge rings, and especially the outer one, look massive and luminous, while on racking out of focus away from the mirror the central rings look relatively brighter and the outer rings look weak in comparison to what they appeared when within focus, then the inference is that the edge rays fall short or come to a focus at a point nearer to the mirror than the focus for the central rays, or, in other words, there is positive aberration.

" If, on the other hand, the central rings when inside focus look about as luminous or even more so than the outer ring which is thin and weak, while on racking outside focus the complementary effect of a massive and luminous outside ring enclosing comparatively feeble central rings is observed, then the inference is that the edge rays come to focus at a point farther from the mirror than the focus for the central rays, and this fault is negative aberration."

It will thus be seen that the general disc of light gives a general idea of the figure. The black spot in the centre being larger on one side or the other shows at once whether the mirror is under- or over-corrected; and the outside ring of light viewed at equal distances inside and outside the focus, with a succession of stops getting smaller and smaller, shows at once the figure of each successive zone, and whether polished too long or too short in focus; or, if absolutely equal, quite correct.

When finally silvered and tested this way it gives at once the general figure and the individual zones. If there are zones in the mirror they are easily seen in the expanded discs of light, as these zones appear brighter or darker than the normal disc.

If the edge is badly turned down there will probably be a ring-system outside the focus, with a heavy outside ring, while inside of the focal point the disc of light will be faint on its margin, with a hairy and confused edge. It is useful to have a number of stops made reducing the aperture, and then to compare the outside ring as the aperture gets smaller. This outside ring should be the best-defined ring on the disc, and should be exactly equally bright inside and outside of focus at equal distances. If this is not the case there is a zone at the edge of the circle to which the mirror is stopped down, and it can be ascertained in manner previously mentioned whether it is polished too deep or not deep enough.

Another defect easily detected by the out-of-focus method is astigmatism. Astigmatism may be caused either by the large mirror, the eye-piece, or the eye of the observer; or the small flat may, if not flat, show a similar defect. The appearance of the out-of-focus image will be an oval instead of a circular disc.

If it is caused by a defective eye-piece the long diameter of the oval image will rotate as the eye-piece rotates. If it is caused by the eye of the observer it is only necessary to move the head a quarter of a circle round the eye-piece to see if the oval rotates. There remains the chance of the flat being in fault. If the large mirror is turned 90° and the oval image does not rotate the flat is in fault. If the oval rotates with the large mirror the fault lies with the mirror—that is, it is astigmatic : one diameter is ground and polished slightly flatter than the other diameter. This method of testing is particularly suited to finally testing large mirrors after they

---

* " Telescopic Objectives." London : De Little and Sons, 1896.

have been mounted, as a few minutes' scrutiny of a large star will show at once whether the figure is good or bad without removing the mirror from its tube. A high power must be used for large mirrors. Very large mirrors seldom have absolutely first-class figures for dividing double stars, but are sufficiently good for photography, as the halation hides the minor faults which make the difference between a perfect mirror and a good one. Until mirrors are figured with the same care as achromatics they will never give satisfaction; and I may add that a perfect flat is as essential as a perfect mirror. If the flat is concave it will give, slightly out of focus, an oval disc of light instead of a circle, and quite spoil the definition.

In conclusion, I should like to acknowledge my debt to Nichol's Cyclopaedia, published in 1837, and to Cooke's book on " Telescopic Objectives."

ART. LII.—*Southern Variable Stars.*

By C. J. WESTLAND, F.R.A.S.

[*Read before the Astronomical Section of the Wellington Philosophical Society, 4th August, 1915.*]

IN choosing " Southern Variable Stars " for my title I wish to explain that I am referring to stars which are well situated for observation in New Zealand, but without excluding some which may be visible to a certain extent in Europe also. The southern stars are not nearly so well known as the northern ones, and this is true of the variable as well as the other stars. Still, there is much to be learnt about variable stars in both hemispheres, and with reference to this I may quote a statement made by Professor Pickering in addressing the British Astronomical Association in June, 1913. Out of 4,525 variable stars now known, 3,371 have been discovered at Harvard College Observatory; and out of all this number probably less than a hundred have had their periods and ranges of magnitude determined.

The explanation of this is that a discovery is made by comparing two photographs taken on different dates; but determination of a star period requires prolonged observations for which no professional observer has time to spare. Only amateurs can find the opportunities to collect the information, and, fortunately for them, the work does not require very elaborate instruments. The Variable Star Section of the British Astronomical Association has done much valuable work in this direction, and several of its most diligent members have only 3 in. telescopes. When a star reaches naked-eye magnitude a pair of field-glasses is more useful than a telescope, because it may be necessary to compare the variable with a star of known magnitude several degrees away. The field-glasses may be turned rapidly from one star to the other in a way that is impossible with the telescope.

A good example of the long-period variables is the star R. Hydrae. It may be called a southern variable, for, although it is on the working list of the B.A.A. Variable Star Section, the difficulties found in observing it made it impossible to determine either maximum or minimum until a year ago, when one member in South Africa and myself undertook observations of it. Our results showed that at maximum it rose to 4·4 and at minimum it fell to 9·3, so that at one period of its career it gives a hundred times as much light as at the opposite extreme.

My own attention was called to this star in an unexpected way. In 1911 a comet appeared which passed through Hydra in November, and was among the comparison stars of R. Hydrae for several days. I exposed a plate on it on the 25th November, and on examining the result afterwards I found the variable rather bright. It was really about the sixth magnitude at the time, and, as the exposure had been long enough to show ninth-magnitude stars, a star of sixth magnitude left a very distinct impression. I watched the star for several weeks, and saw it pass through a maximum in January, 1912.

This star follows a custom which is common among variable stars, in that it rises to maximum very rapidly, but falls slowly. The usual method

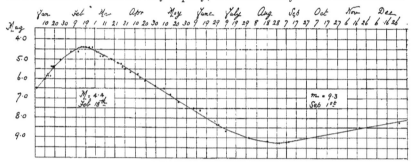

FIG. 1.—Curve of R. Hydrae in 1913.

of dealing with observations is to plot them on squared paper, making the horizontal scale represent time and the vertical scale show the magnitudes. Then the characteristic I have mentioned is expressed by saying that the rise is much steeper than the fall. The curve of this star in 1913 is shown in fig. 1.

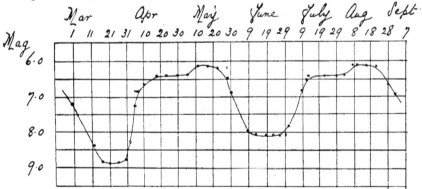

FIG. 2.—Curve of T. Centauri in 1913.

A peculiarity in the period of R. Hydrae is that it is growing decidedly shorter. The observations of last year show that only 409 days elapse between one maximum and the next, whereas all books of reference at present give the period as 425 days. Mr. Gore's book states that the period was 500 days in the year 1708, 487 days in 1785, 461 in 1857, and 437 in 1870.

There are a good many variables whose curves are similar to that of R. Hydrae. A few of these curves—Mira Ceti, R. Leonis, and others—are given herewith.* Fig. 2 is the curve of a southern star, T. Centauri,

---

. * Other diagrams and lantern-slides were shown at the meeting.

classified as a long-period variable, but its changes take much less time than the others just mentioned—apparently about thirteen weeks. This star has a larger measure of irregularity in its behaviour. One peculiarity I have noticed myself is that a week or two before the fall sets in, when the star is usually at a fairly flat maximum, it seems to rise quite suddenly about half a magnitude and then plunges rapidly down to its minimum. The maximum on this curve is more like a cairn on the top of a hill than'the hill-top itself.

Fig. 3 shows the curve of T. Gruis, in which the usual rapid rise and slow fall is not to be found. There is even a suspicion that the fall is the steeper part of the curve, and, if this is correct, it is a very unusual feature in variable stars.

The star R. Normae has the reputation of showing two maxima and minima. In my own experience, however, the range of variation has been so small that the curve is rather flat and shows no well-marked peaks of maximum.

All the stars I have mentioned so far are included in the second class of variables—namely, the long-period variable stars. The others of the five

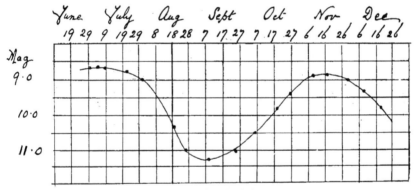

Fig. 3.—Curve of T. Gruis in 1913.

classes are—(1) Temporary stars ; (3) irregular variables ; (4) short-period stars ; (5) eclipsing stars. The fourth class has certain subdivisions, and sometimes two other classes are added—cluster variable and suspected variable.

The short-period stars are more difficult to observe, because they change so rapidly that unless the observations are frequent they teach us nothing. Of course, they require continuous good weather also. Three good specimens of their class are Kappa Pavonis, S. Trianguli, and S. Normae. I have watched these for several weeks while the weather was suitable, and found them wonderfully regular both as to period and range of magnitude. The period of S. Trianguli is six days; the other two take nine days. There are a few stars whose periods cannot be classified as either long or short, being about thirty or forty days, which is between the two classifications. The only one I know among the southern stars is l. Carinae, which is said to vary between 3·3 and 5·5 in thirty-five days. I have not seen it brighter than 3·8 or fainter than 4·5, and its period is apparently something just over a month, and may be perfectly regular.

The absolute regularity of the short-period stars gives us the first hint as to the reason of their fluctuations, for when we say their changes are as regular as the flow of time we say they are as regular as the earth's rotation. If we suppose that stars are sometimes brighter on one side than the other we have a ready explanation of a regular variable star. Stars of the Eta Aquilae type—that is, spectroscopic binaries which do not eclipse one another—can also be explained on this hypothesis, because if they always turn the same face towards one another, as the moon does to the earth, the period of variability must be equal to the period of revolution, just as the spectroscope shows. Notice also that if the periods of rotation were otherwise they would explain an apparently irregular variable. The two brightest sides turned earthwards would produce the extra-bright maxima, and the two darkest sides would in the same circumstances give the extra-dark minima. These would be the two extremes of brightness, and if the rotation periods were very unequal the intermediate values would be irregular both as regards time and brightness.

No theory of rotation will account for long-period stars which take months to complete their fluctuations. Mira Ceti and several others are known to undergo physical changes, and at the last maximum of this star I saw the hydrogen line of wave-length 4340 bright without any difficulty, although I have only a 4 in. telescope to collect light for my spectroscope.

The method of magnitude rating consists of comparing the variable with stars situated near it. Charts of the fields surrounding these stars are obtainable, so that identification of the stars is easy, and the magnitude of each comparison star is given. The observer finds it convenient to memorize the stars he makes use of, because the eye loses some of its sensitiveness if it has to leave the telescope and study a chart by lamplight. But after the observer has looked over the stars of known magnitude his eye is in tune. I use this metaphor purposely, because the conditions are similar to those of the ear when it has heard certain notes of the scale played: it is able to pick out the intermediate notes. Similarly, the eye can tell the magnitude of a variable whose brightness is intermediate between two stars of known magnitude, provided that care is taken to get the eye into that condition and to let it work under the best circumstances.

ART. LIII.—*The Distribution of Titanium, Phosphorus, and Vanadium in Taranaki Ironsand.*

By W. DONOVAN, M.Sc.

[*Read before the Technological Section of the Wellington Philosophical Society, 10th November, 1915.*]

THE present writer, in the course of an analysis of ironsand from Patea two years ago, found the amount of phosphorus present to be 0·11 per cent. The highest previously recorded percentage of phosphorus in Taranaki ironsand was 0·039 per cent., returned by Dr. J. S. Maclaurin in 1902 from a New Plymouth sand forwarded by Mr. E. M. Smith.* It seemed important to ascertain whether the high result was accidental, or abnormal in any way, or whether the ironsand deposits generally contain more phosphorus than has hitherto been supposed. In view, too, of the fact that the commercial utilization of the sands would seem to depend on the

---

* Thirty-sixth Annual Report, Colonial Laboratory, p. 7.

successful application of magnetic concentration, it appeared desirable to obtain some data as to the extent to which the percentage iron-content could be increased and objectionable impurities eliminated by such means.

Representative samples were obtained by Mr. W. Gibson, B.E., Assistant Geologist, from Patea, covering an area estimated to contain at least 5,000,000 tons of sand, and also from the neighbourhood of New Plymouth. These were analysed for iron, titanic oxide, phosphorus, and vanadium.

Iron.—This was estimated by reduction of 0·5 gram of the finely ground sample in a current of hydrogen gas at a red heat, followed by solution in semi-dilute sulphuric acid, and titration with deci-normal potassium permanganate.* The titrations were made at a temperature of 70° C., and corrected for vanadium by deducting 0·4 c.c. for every 0·001 gram of vanadium present.

Titanium.—Titanium was estimated colorimetrically by adding hydrogen peroxide to the solution after the permanganate titration, and comparing with a standard solution (Weller's method).

Phosphorus.—The determination of phosphorus in ironsand is rather difficult. In the presence of titanium there is a tendency, when eliminating silica, for a highly insoluble titano-phosphate to be formed. An excess of iron in solution would appear to retard the precipitation of traces of phosphoric acid. On the other hand, vanadium, when present, is partly carried down with the phosphorus, and contaminates, though only slightly, the final precipitate.

The method finally adopted was to fuse 1 gram of ironsand with 8 grams sodium carbonate, and extract with water. The insoluble residue was re-fused, and again extracted. A complete separation was thus obtained between the titanium and iron, which remained in the insoluble portion, and phosphorus and vanadium, which were entirely soluble.

The water-extract was acidified with nitric acid, evaporated to dryness, taken up with hot dilute nitric acid, filtered to remove silica, and again evaporated, this time to small bulk. The phosphorus was precipitated by the method of Stunkel, Wetzke, and Wagner (as described in Crooke's "Select Methods in Chemical Analysis," 3rd ed., p. 509), and weighed as magnesium pyrophosphate. Any vanadium in the precipitate was deter-mined colorimetrically, and deducted.

Blank determinations with known amounts of phosphorus, iron, and titanium gave excellent results.

Vanadium.—This was estimated colorimetrically with hydrogen peroxide in the presence of nitric acid. One gram of the sample was fused with 6 grams of sodium potassium carbonate, and extracted with water. The extract was acidified very slightly with nitric acid, and 1 c.c. of hydrogen peroxide (10 per cent. by volume) added. The brown tint developed was compared with that of a similar solution containing a known amount of vanadium.

Magnetic Separation.—The samples were also separated by the action of first a weak and then a strong magnet into three portions: No. 1, strongly magnetic, separated by a very weak magnet; No. 2, feebly magnetic, by the use of a triple horse-shoe magnet, after removal of portion 1; No. 3, non-magnetic residue. Portions 1 and 2 were then analysed for iron, titanium, phosphorus, and vanadium, and the results compared with the analyses of the original sands, as shown in Table I.

---

* Trans. N.Z. Inst., vol. 41 (1908), pp. 49–51.

TABLE I.—PATEA SANDS.

(1.) From sand-dune area west of Kakaramea and a quarter of a mile from coast-line.
(2.) Breakwater to 70 yards west, above neap-tide mark.
(3.) Same as No. 2, but neap tide to mean tide.
(4.) From 70 yards west to 220 yards west of breakwater, and above neap tide.
(5.) From 220 yards to 400 yards west, width 44 yards, and including sand-dune rising 12 ft.

(O. = Original sand; S.M. = Strongly magnetic portion; F.M. = feebly magnetic portion.)

	(1.)			(2.)			(3.)			(4.)			(5.)		
	O.	S.M.	F.M.	O.	S.M.	F.M.	O.	S.M.	F.M.	O.	S.M.	F.M.	O.	S.M.	F.M.
Iron (as metal)	21·03	54·26	9·97	57·12	60·42	39·62	35·42	50·93	13·50	37·47	56·55	6·00	33·60	52·29	4·90
Titanium dioxide	3·80	10·00	1·60	10·60	11·00	6·80	6·90	10·10	1·92	5·20	9·60	0·50	4·10	8·70	0·46
Phosphorus	0·16	0·27	0·21	0·16	0·18	..	0·25	0·31	0·19	0·19	0·22	0·18	0·20	0·26	0·18
Vanadium (as metal)	0·08	0·18	0·03	0·16	0·17	0·14	0·14	0·14	0·05	0·16	0·12	0·03	0·13	0·24	0·03
Phosphorus (as percentage of iron)	0·76	0·50	2·10	0·28	0·30	..	0·70	0·61	1·41	0·50	0·39	3·00	0·60	0·49	3·68
Titanium dioxide (as percentage of iron)	18·1	18·4	16·1	18·5	18·2	17·2	19·5	19·8	14·2	13·9	16·9	8·3	12·2	16·6	9·8

Sulphur : From nil to 0·03 per cent.

Degree of concentration for strongly magnetic portion : No. 1, 30 per cent. of original sand ; No. 2, 95 per cent. of original sand ; No. 3, 65 per cent. of original sand ; No. 4, 52 per cent. of original sand ; No. 5, 61 per cent. of original sand.

New Plymouth Sands.

Six representative samples were received from New Plymouth, and these, on analysis, gave the following results :—

No.			Iron (as Metal).	Titanium Dioxide.	Phosphorus.	Vanadium.
1	..	..	49·56	9·2	0·28	0·34
2	..	..	45·00	10·0	0·26	0·29
3	..	..	59·92	10·6	0·29	0·26
4	..	..	59·36	10·6	0·26	0·15
5	..	..	36·62	6·2	0·27	0·16
6	..	..	56·89	9·8	0·29	0·33

Of these six sands, No. 1, representing a fair average sample, and No. 5, the poorest in iron, were separated magnetically, as previously described, into three portions, and the analyses of the strongly magnetic portions, together with those of the original sands, are given in Table II.

TABLE II.

—	(1.)			(5.)		
	O.	S.M.	F.M.	O.	S.M.	F.M.
Iron (as metal)	49·22	56·22	47·0	36·46	49·30	41·0
Titanium dioxide	9·20	10·30	6·6	6·20	7·60	5·7
Phosphorus	0·28	0·32	..	0·28	0·32	..
Vanadium (as metal)	0·29	0·27	..	0·16	0·23	..
Phosphorus (as percentage of iron)	0·56	0·57	..	0·76	0·64	..
Titanium dioxide (as percentage of iron)	18·7	18·3	14·0	17·0	15·4	13·9

Degree of concentration for strongly magnetic portion: No. 1, 80 per cent. of original sand ; No. 5, 37 per cent. of original sand.

More complete analyses were made of two of the strongly magnetic concentrates: A, Patea sample No. 2, being highest in iron-content, Patea ; B, New Plymouth No. 1, being the richer of the New Plymouth samples treated.

	Analyses.			A.	B.
Silica (SiO_2) ..	..	..	..	1·10	6·20
Alumina (Al_2O_3)	..	..	..	2·60	2·00
Lime (CaO) ..	..	..	..	Nil	0·62
Magnesia (MgO)	..	..	..	1·90	2·80
Titanium dioxide (TiO_2)	..		..	11·00	10·30
*Phosphoric anhydride (P_2O_5)	..		..	0·39	0·74
Vanadium pentoxide (V_2O_5)	..		..	0·31	0·39
† {Ferrous oxide (FeO) ..	..		..	32·49	30·28
{Ferric oxide (Fe_2O_3) ..	..		..	50·21	46·67
				100·00	100·00
*Equivalent to phosphorus	..		..	0·18	0·32
†Equivalent to metallic iron	..		..	60·42	56·22

These may be taken as typical analyses of magnetically separated concentrate, and afford data for determining the nature and quantity of flux necessary for smelting.

SUMMARY.

The results of the investigation may be summarized as follows :—

(1.) The percentage of iron in the magnetically separated concentrates lies between 50 and 60 per cent., and under good working-conditions would approach the latter figure.

(2.) Vanadium varies with the iron-content—from 0·08 per cent. in the poorer samples to 0·25 and 0·34 in the richer ores. The greater portion of it remains with the iron when the sand is magnetically separated. Its presence in such small quantities confers no advantage, as, on smelting, it would pass into the slag, and not into the iron. But if means could be devised to completely recover all the vanadium, and convert it into ferro-alloy, 1 ton of sand containing 0·2 per cent. of vanadium would at the present prices produce alloy worth approximately £2.

(3.) The titanic oxide varies in the original sands from 6·2 to 10·6 per cent., and remains with the strongly magnetic portion, being slightly increased relatively to the iron. This confirms the results obtained by W. Skey (33rd Annual Report of the Colonial Laboratory, N.Z., p. 17).

(4.) The phosphorus in the untreated sands varies from 0·16 to 0·28 per cent. It is not appreciably decreased, relatively to the iron, by magnetic separation. Fine grinding prior to separation did not give any better result. The phosphorus still remained in such quantities in the concentrated ironsand as would necessitate the use of the basic process for the manufacture of steel from the ore.

The results from a commercial standpoint are disappointing. They indicate that the phosphorus and titanium are intimately associated with the iron in the sand, and cannot be readily eliminated, if at all, by magnetic means. It is possible, however, that a method will be found which will achieve this end.

In conclusion, I would acknowledge the ready assistance of Messrs. N. L. Wright and R. P. Wilson in much of the analytical work involved. I would also thank Dr. J. S. Maclaurin, Dominion Analyst. for permission to publish these results.

Art. LIV.—*The Manufacture of Iron and Steel in New Zealand.*

By S. H. Jenkinson.

[Read before the Technological Section of the Wellington Philosophical Society, 13th October, 1915.]

The manufacture of iron and steel is the most important industry a nation can possess, and it is hardly too much to assert that empire and power follow directly on and are measured by the production of iron and steel. Most English metallurgists recognize this so clearly that there must be a great (though as yet unexpressed) jubilation over the fact that Germany provoked this war a quarter of a century too early for her recent superiority in iron and steel production to assert itself unmistakably. Now that the engineering trades have clearly proved themselves the basis of all military and naval strength, we can hope that the tremendous and phenomenal strides of the German metallurgical industry within the last few years will be appraised by all our allied statesmen as the most desperate and inexorable menace to the peace and freedom of the world, and that definite, decisive steps will be taken to prevent the German Empire procuring that predominance over the European nations which would be the inevitable outcome of, say, twenty years more of her present superiority in the iron trade. Luckily, these steps are simple and obvious enough. The orefields on which Germany depends are in Lorraine and Luxemburg, and her fuel lies in Westphalia. The removal of these provinces from the domination of Prussia to the government of their racially congenial neighbours, France, Belgium, and Holland, would end at once the German Empire as an arbiter in the iron trade and a menace to peace and freedom in Europe. By careful effort and the sinking of some metallurgical fallacies Great Britain would again become the arbiter of European metallurgy, and her place as premier nation of the world would be assured once more. Every part of the Empire should strive for this one end—the supremacy of Great Britain in the iron and steel trade, and her consequent supremacy among the nations of the world in military and naval might and power. The effort should be to develop the cheapest Empire source of manufactured iron to the greatest possible extent, and to spoon-feed by tariff operations an uneconomic supply in any province at the expense of the cheapest source elsewhere in the Empire should be realized as a suicidal policy that will only tend to keep the Empire in the same desperate position it has been in for the last ten years. Iron and steel supremacy, which inevitably spells military and naval supremacy, can be attained only by concentrating all our iron-manufacture in the most economical position, where vast operations in the one province will induce economics impossible in smaller efforts, and where metallurgical skill and facility will become a racial trait of the inhabitants. A national outlook will warn us against the destructive fallacy of making each or any province of the Empire independent of the others in this industry, since it has been the selfish and disconnected policies of her colonies in regard to iron-importation that have endangered England's supremacy in the past; and no little part of Germany's might and power in August, 1914, can be traced directly to the want of foresight among English and colonial statesmen with regard to the fostering of the relatively declining national production of iron and steel.

The question of iron and steel manufacture in New Zealand must be considered in the light of these remarks.

So far as our present scanty knowledge goes, the potential iron-ore fields of the Dominion are—(1) the Parapara district, (2) the Taranaki ironsands, (3) Mount Royal, (4) Mount Peel, (5) the Cheviot district. Our knowledge, however, of all iron sources save Parapara is where it was left by Sir James Hector in 1873, and our crying need is for an organized survey of all the likely deposits. Until that is accomplished we can only consider the possibilities of Parapara and Taranaki.

PARAPARA.

Bulletin No. 3 (n.s.) of the Geological Survey Department establishes the fact that Parapara, Onakaka, and Tukurua Blocks comprise an extensive surface outcrop of at least 20,000,000 tons of ore, occurring in an easily mined and smelted form and in an accessible position; but the bulletin gravely errs on the optimistic side when reference is made to the metallurgical problem involved. The surface analysis given in the bulletin has since been supplemented by tunnel analyses published in the 1914 Laboratory Report, and beyond material error we may assert that the ore has the following composition :—

					Per Cent.
Iron	..	..	..	..	.. 48 to 50
Alumina	..	..	..	..	.. 3 to 6
Phosphorus	..	..	..	..	.. 0·15
Silica	..	..	..	..	.. 10 to 15
Water	..	..	..	..	.. 10
Sulphur	..	..	..	..	.. 0·08

Such ore would need calcining or roasting to expel the water and sulphur, and a product would result—

					Per Cent.
Iron ..	..	..	..	..	.. 50 to 55
Silica	..	..	..	..	.. 11 to 16
Alumina	..	..	..	..	.. $3\frac{1}{2}$ to 7
Phosphorus	..	..	..	..	.. 0·17
Sulphur	..	..	..	..	.. 0·03

This product contains too much silica to be economically dealt with in an electric smelting-furnace, and, while it could possibly be purified by magnetic concentration after fine crushing, the treatment adds expense, and the product could not compete for cheapness or purity with Taranaki ironsand. Small hot-blast furnaces using charcoal fuel are in use in America, Sweden, and Styria, but there they work with exceptionally pure ore, needing very little flux, and produce a very valuable steel-making pig in a costly manner. Such a furnace would burn all the timber off 1 acre of forest for every 20 tons of pig made, and is only practicable in heavily wooded districts within reach of navigable waters. With the impure ore of Parapara, and in a country of dear labour and difficult transport, a charcoal furnace is not economically possible.

The roasted ore would give a white steel pig running — Phosphorus, 0·35 per cent.; sulphur, 0·08 per cent.; silica, 1 per cent.; and worth £2 15s. a ton: or, making grey foundry pig, would give—Phosphorus, 0·35 per cent.; sulphur, 0·05 per cent.; and worth £3 per ton.*

---

* The prices given are based on those current in July, 1914, f.o.b. Glasgow.

To make pig of this low value economically cheap production is essential; hence large furnaces, hot blast, labour-saving appliances, and an output of at least 1,000 tons a week are required. The cheapest flux for the silica in the ore is limestone, of which (assuming slags running 40 per cent. silica, 16 per cent. alumina, and 39 per cent. lime for the white iron, and 35 per cent. silica, 14 per cent. alumina, and 46 per cent. lime for the grey) 12 cwt. would be required per ton of white iron and 15 cwt. per ton of grey iron. Such limestone must be over 95 per cent. calcium carbonate, and should not contain over 1 per cent. of magnesia. The presence of such pure limestone in accessible position and sufficient in quantity for an output of 40,000 tons per year is problematical, but geologists generally are reassuring on this point; frequently, however, speaking with an imperfect appreciation of the metallurgical requirements, as in Bulletin No. 3.

The only suitable fuel available at present in large quantities is the hard dense Westport coke, which would probably run from 0·7 per cent. to 1·2 per cent. sulphur. One ton would be required for making white iron, and 25 cwt. for grey, per ton of iron. Also per ton of iron 4–6 tons of highly preheated air would be required. The capital cost can now be assessed. Two furnaces (one stand-by) would be required, 80 ft. high, 20 ft. in diameter, and 20,000 cubic feet capacity. This would give 1,000 tons of pig iron per week from 2,200 tons of ore, 1,000–1,250 tons of coke, 400–600 tons of limestone, and 5,000 tons of air, and would cost £50,000 in England and about £90,000 in New Zealand. Estimating the cost of mining the ore at 4s. 6d., of calcining it at 1s. 3d., of limestone at 7s. 6d., and of coke at £1 15s. (all per ton), we would have—

TABLE I.—PARAPARA PIG.

White Pig for Steel		Cost per Ton Pig.	Grey Pig for Foundry.		Cost per Ton Pig.
		s. d.			s. d.
5,000 lb. roasted ore..	..	13 9	5,000 lb. roasted ore	..	13 9
1,350 lb. limestone	..	4 6	1,680 lb. limestone ..	..	5 6
2,200 lb. coke	..	35 0	2,800 lb. coke	..	43 9
Labour	..	7 0	Labour ..	..	8 0
Repairs	..	1 6	Repairs ..	..	1 9
Interest and depreciation	..	2 0	Interest and depreciation	..	2 0
5 tons air	..	Nil	5 tons air ..	..	Nil
At furnace	..	63 9	At furnace	..	74 9
Worth in Glasgow, July, 1914		55 0	Worth in Glasgow, July, 1914		60 0

The exportation of pig is clearly out of the question, and sole dependence must be placed on the New Zealand market, which would at present absorb about 4,000 tons per annum of grey foundry pig of Parapara quality, leaving 46,000 tons of white pig for steel-making. This pig is too impure to produce higher-quality steels for the foreign market, and there is little demand for such in New Zealand; hence the only outlook is for cheap steel for the bar and rod trade. This can only be made from Parapara pig by the basic open-hearth process, where we are at once confronted with the sulphur difficulty, both in the pig and in the coal for use in the gas-producers. The coal now mined in the Liverpool seam of the State mine runs under 1 per cent. sulphur, but is probably too caking in its nature for easy and cheap treatment in a gas-producer; however, it could certainly be used for this purpose, but would cost over £1 a ton.

The pig from Parapara ore offers a very difficult metallurgical problem when used for steel-making, and one which has prevented the use of such ores to any extent in other countries, as the solution involves a subsidiary process which increases the cost of steel-production. As given, the white pig contains 0·08 per cent. sulphur and 1 per cent. silica, both too high for the basic open-hearth process. Either could easily be reduced in the blast furnace, but only at the cost of increasing the other, so that the attempted solutions fall under two main headings : (1) Making a high Si, low S, pig, and desiliconizing by (*a*) blowing in a converter or (*b*) washing in a bath of molten basic slag ; (2) making a low Si, high S, pig, and de-sulphurizing by (*a*) using manganese-ores as a flux, (*b*) pouring the molten pig on to nitrate of soda (Heaton process), or (*c*) using $CaCl_2$ and CaF_2 as a flux (Saniter process). All these operations require skill and add to the cost. None of them have survived in practice, and all are of doubtful utility.

However, assuming this difficulty was overcome by one of the processes, an ingot steel would be obtainable costing £5 a ton in the ladle (against £3 5s. in England), and suitable only for bar and rod trade, which includes wire and gauge plates for galvanizing. New Zealand could absorb 50,000 tons a year of this material, but the selling-price of that manufactured in New Zealand would be at least £4 a ton above the imported. Parapara steel would be quite unsuitable for rails, boiler-plate, and good structural material, and the conclusion is inevitable that—(1) Parapara ore presents a very difficult problem to the steel-maker ; (2) its content of phosphorus and sulphur is such that it is economically unsuited for the production of high-class steel, such as boiler-plate ; (3) the cost of fuel and transport in New Zealand is at present prohibitive to the foundation of a large iron and steel industry ; (4) the value of the ore is too low to encourage the idea of exporting it.

It may be pointed out here that the analysis given in Bulletin No. 3 of the Geological Survey of what is stated to be excellent pig iron made from Parapara ore in Melbourne in 1873 proves that the metal in question was a commercially useless steely product, probably made in a blacksmith's fire.

TARANAKI SANDS.

This ore occurs as a very finely divided sand over a large stretch of sea-coast extending from Waitotara to the Awakino River, and forms workable deposits at Patea, New Plymouth, Waitara, and Mokau at least. While no useful estimate can yet be made of quantity, it is certain that millions of tons of iron could be obtained. Earlier analysis made this ore appear of remarkable purity, but lately this has been disproved, and it is virtually certain that the sand contains about 0·5 per cent. phosphorus, estimated on the iron. Whether this percentage can be lessened by magnetic concentration is a moot point, but Mr. Donovan's valuable paper (see p. 503 in this volume), goes far to disprove the probability. Accepting his figures, the average of the Patea and New Plymouth sands, after magnetic concentration, runs about 56 per cent. Fe, 10 per cent. TiO_2, 0·25 per cent. Va, which would give a pig running 0·5 per cent. phosphorus, suitable for heavy foundry-work and for steel by the basic process. Certainly this is a purer and (neglecting its form) more valuable ore than that of Parapara, but it cannot be compared with Swedish or Styrian, and, indeed, would not be accepted as a structural steel ore by the majority of English engineers.

The titanium present might give some trouble in the blast furnace, and neither it nor the vanadium add anything to the value of the ore, as neither would appear in the pig. However, the form of this ore is against any hope of successful utilization in the blast furnace, even if the purity made it worth while. Possibly the first suggestion to briquette such ores was made by Mr. Pharazyn before the Wellington Philosophical Society in 1869, and to this day those ignorant of metallurgical history vainly hope for success along this line. The problem of utilizing finely divided ore on the blast furnace arises wherever iron is smelted (flue-dust and fine stuff); but so far no method of briquetting or sintering has ever been successful, beyond making possible the use of the product to the extent of about 10 per cent. of the total charge. · The very nature of the briquetting process is against the product holding solid under the high temperature and great pressure of the modern furnace, and in the same way sintering (or fritting a mass of ore and coal together to form what some fancifully term ferro-carbon) cannot be efficacious beyond the temperature at which fritting takes place. Metallurgical experience throughout the world has made the fallacy of such ideas an axiom. At the most, a fritted product can be made which will stand quick passage through a low shaft furnace, and give a molten semi-steel containing under 2 per cent. carbon, which is quite useless commercially. The prospect of large works using Taranaki sand seems hopeless unless the ore is valuable enough to justify electrical treatment, or unless some cheaper electrical furnace is introduced. At present electrical furnaces use only expensive charcoal or coke as the source of the carbon chemically necessary for reduction, and the necessary heat is provided by electricity separately generated, while all the products of combustion are wasted. But if powdered coal in excess (thousands of tons of coaldust are destroyed every year in New Zealand as valueless) were fed in with ironsand to the top of a small shaft furnace, and the waste gases burnt under boilers to produce the small quantity of electricity needed to quickly carbonize and melt the already white-hot mass of mingled reduced iron and particles of coke, it is probable a much cheaper process would result. Assuming that the ore, fuel, and air are highly pre-heated by the waste gases, and remembering how efficient powdered coal is as a heating agent, it should be possible to smelt ironsand at the expense in fuel of 30 cwt. of coaldust, which presupposes an efficiency of about 20 per cent. Even in Taranaki this should not cost over £1 a ton, and if the low figure of 30 per cent. is taken as the concentration factor of the sand we would have—

TABLE II.—TARANAKI PIG.

(Magnetic concentrate, 30 per cent. of original sand ; iron, 56 per cent. ; TiO$_2$, 10 per cent. ; phosphorus, 0·3 per cent. ; sulphur, trace. Electric furnace, 60 tons per week. Cost, £25,000).

Grey Pig for Foundry.	Cost per Ton Pig.	
	s.	d.
Ore, 6 tons (mining, 2s. 6d. ; drying, 6d. ; concentrating, 1s. 6d.)	27	0
30 cwt. powdered coal	30	0
5 cwt. limestone	2	0
Labour	12	0
Repairs	8	0
Interest and depreciation	8	0
At furnace	87	0
Worth in Glasgow, July, 1914.	66	0

The resulting pig would at this price have a New Zealand market of about 4,000 tons per annum for foundry use, but the small cost of the plant—say, £25,000—would enable even this small output to be manufactured as an economic proposition. Another 1,000 tons could be refined while still molten to cast steel for steel castings, a good demand existing for these at remunerative prices. This, however, is the greatest extent to which the utilization of New Zealand iron-ores is likely to rise for many years, and all talk of plate and rail mills, shipbuilding, &c., and of New Zealand as a factor in the steel and iron business of the world, is quite ridiculous at present.

ART. LV.—*Studies on the Lime Requirements of certain Soils.*

By LEONARD J. WILD, M.A., F.G.S., Lecturer in Chemistry, Canterbury Agricultural College.

[Read before the Philosophical Institute of Canterbury, 1st December, 1915.]

I.

UNDER this heading Hutchinson and MacLennan, in the *Cambridge Journal of Agricultural Science* for March, 1915, describe a very rapid and simple method for determining the lime requirement of soils. Since so much is spoken and written, while so little is really known about the lime requirements of our New Zealand soils, it seemed desirable to test the validity of this method, which claims to give valuable and accurate information for very little trouble. The present article contains an account of such an inquiry, together with the results obtained from trials with certain Canterbury soils.

The method is based on the absorptive capacity of the soil for calcium carbonate presented to it in solution as bicarbonate. A known quantity of bicarbonate in solution is left in contact with a known weight of soil for a few hours, at the end of which time the amount of lime in the solution is again determined. The quantity of lime lost by the solution is the quantity required to satisfy the given weight of soil, and from these figures the percentage requirement of the soil, and hence the required weight of lime per acre, is calculated. To obtain this last result, the apparent density of the soil must also be determined.

The calcium-bicarbonate solution is easily prepared in an ordinary "sparklet" or refillable "soda-water" apparatus, for which bulbs of compressed carbon dioxide are used. Into this an excess of finely divided calcium carbonate (10 gr. to 12 gr. pure pptd. $CaCO_3$) is put before the gas is admitted. The solution of $CaH_2(CO_3)_2$ thus formed is siphoned off, and about one-third of its volume of water is added, after which it is filtered free from undissolved carbonate, standardized by titration with decinormal sulphuric acid, using methyl orange as indicator, and is ready for use. The strength of a solution prepared in this way is about one-fiftieth normal.

Ten grams of the soil under investigation is placed in a bottle or flask of about 500 c.c. capacity, and 200 c.c. to 400 c.c. of the solution is added. The air in the bottle is displaced by a current of CO_2 to prevent the precipitation of $CaCO_3$, and the bottle is placed in a shaking-machine for three

17—Trans.

hours, or shaken by hand at intervals of about twenty minutes for four hours, at the end of which time a portion of the solution is filtered, and its strength again determined by titration with decinormal acid.

II.

In the first trials of the method made in this laboratory the results shown in Table A were obtained.

TABLE A.

Soil.					Requirement of CaO indicated.	
					Percentage.	Pounds per Acre.
21*	..	..	..	..	0·18	2,450
4	..	..	..	..	0·16	2,178
16	..	..	..	..	0·18	2,450
29	..	..	..	..	0·14	1,906
Waipara Downs ..	..	..	..	0·13	1,770
Motueka	..	..	..	..	0·42	5,718

* Numbers in all cases refer to numbers of fields on the Lincoln College farm.

It will be noticed that the figures expressing the weight of lime required per acre are higher than the average farmer would care to believe. The Motueka soil is a very unpromising sample from the apple lands, and, as it is known from an analysis to be devoid of calcium carbonate, the extremely high figure appears not unreasonable. The question, however, presents itself, Suppose we give this or any other soil a preliminary dose of carbonate of lime, will the demand as indicated by the method under trial be correspondingly lowered? With a view to testing this, several experiments were made. To 10 grams of the Motueka soil was added 0·06 gr. $CaCO_3$ (that is, 0·336 per cent. CaO); 100 c.c. of water was poured on, and the bottle left for about an hour, with occasional shaking. The bicarbonate solution was then added, and the rest of the operation completed in the regular way. The expected lime demand was then 0·42 less 0·34—that is, 0·08 per cent.; the actual requirement indicated was 0·18 per cent. However, it was considered that the short preliminary exposure of the soil to the lime might account for this difference, and so a more comprehensive trial was arranged. Two soils were taken, and determinations were made in duplicate of their lime requirements, both in the natural state and after twenty-four hours' treatment, with varying quantities of $CaCO_3$ in 100 c.c. of water. The results thus obtained are shown in the following table.

TABLE B.

Soil.	Original Lime Requirement.		Equivalent of CaO added.		Lime Requirement finally expected.		Lime Requirement actually indicated.	
	Per-centage.	Pounds per Acre.	Per-centage.	Pounds per Acre.	Per-centage.	Pounds per Acre.	Per-centage.	Pounds per Acre.
21A ..	0·136	1,854	0·062	845	0·074	1,009	0·076	1,037
21A ..	0·136	1,854	0·118	1,609	0·018	245	0·019	260
21B ..	0·181	2,464	0·185	2,520	— 0·004	— 56	— 0·003	— 42
21B ..	0·181	2,464	0·213	2,900	— 0·032	— 436	— 0·037	—498

These results are very satisfactory; but it is obvious that a more practical test of the method would be to determine by it the lime requirements of two similar and adjacent soils, one of which had received a known dressing of lime at a sufficient length of time previously to allow of its being incorporated with the soil. Unfortunately, such conditions do not obtain on the College farm at the present time ; for, though a good deal of liming has been done on grass land during the past winter (1915), it is certainly not yet incorporated with the soil, owing to the abnormally low rainfall. However, the following trials were made with such material as was available :—

(1.) A sample was collected from field 21, consisting of thirty-three cores taken 4 in. deep, the assumption being that the 3 in. of rain received since the field was limed in June had distributed the lime through that depth of soil. Determinations were made on this sample, and, for comparison, on one collected from the same field before the application of the lime.

(2.) It is known that the east end of the College playing-field was heavily dressed with air-slaked lime about ten years ago, with a view to the eradication of rushes. Samples were therefore taken 6 in. deep from the east end, from the west end, and from among a patch of rushes in the north-west corner.

(3.) A determination was made on a soil from Weka Pass, formed from limestone rock, and obviously containing a high percentage of calcium carbonate. The results of this set of trials are shown in Table C.

TABLE C.

Soil.	Requirement of CaO indicated.	
	Percentage.	Pounds per Acre.
21 (before liming)	0·212	2,887
21 (after liming)	0·176	2,398
Play-field, east end	0·19	2,586
Play-field, west end	0·18	2,450
Play-field, north-west corner	0·24	3,267
Weka Pass	— 0·05	—680

These results presented several interesting features. First, the difference in lime requirement of the two samples from field 21 is about 4½ cwt., while the amount of lime actually applied was 5 cwt. or 6 cwt., so that the agreement is fairly close. There is practically no difference in the indicated requirements of the two ends of the football-field ; one must assume, therefore, that ten years is sufficient time to neutralize the effect of a dressing of lime in the top 6 in. As was expected, the corner of the ground covered with rushes showed a markedly high requirement. The limestone soil was able to give lime to the solution.

III.

Early in the course of the work it was noticed that the determinations made at different times on the same soil gave different results. Thus the following figures for the percentage lime requirement of field 21 have been obtained on different occasions : 0·14, 0·15, 0·17, and 0·18. Expressed in

17*

pounds per acre, the range is from 1,900 lb. to 2,400 lb.—that is, the highest figure is more than 25 per cent. greater than the lowest. It was recognized, however, that the determinations which failed to agree had been made under different conditions as to volume and strength of bicarbonate solution, and so it seemed necessary to make sets of determinations simultaneously on one soil under uniform conditions. Results of such are shown in Table D.

TABLE D.

| Soil. | Set. | No. | Requirement of CaO indicated. | |
			Percentage.	Pounds per Acre.
21 (before liming)	A	1	0·143	1,948
„	A	2	0·151	2,056
.. ..	A	3	0·143	1,948
„	B	1	0·179	2,440
„	B	2	0·179	2,440
21 (4 in. deep after liming) ..	C	1	0·156	2,126
„ ..	C	2	0·156	2,126

In sets B and C 400 c.c. of solution was used ; in set A only 300 c.c., the strength of which was also rather less than in the case of the other sets.

As a further test, three samples of soil from field 16 were treated simultaneously—A with 300 c.c. of solution of strength 0·02 N ; B with 400 c.c. of the same strength ; C with 400 c.c. of strength 0·014 N. The lime requirements indicated in these three cases were—A, 0·11 per cent. ; B, 0·12 per cent. ; C, 0·05 per cent. It seems, therefore, that the volume of solution may be varied within fairly wide limits without affecting the result, provided that the solution does not at any time fall below a certain concentration. This is an obviously suggestive point, which merits investigation.

IV.

In one experiment a positive, though small, lime requirement was indicated for a limestone-derived soil, and this suggested the possibility of a weakening of the solution (which corresponds to a lime requirement) from physical as well as from chemical causes. To test this idea a series of trials was made. Four soil-samples that had already been in contact with solution for twenty-four hours in connection with previous trials, and which were now presumably satisfied as regards their lime requirements, were filtered from their old solutions, and treated again with fresh solution. A sample of sand was prepared by treating alternately with concentrated HCl and strong ammonia solution, washing thoroughly, and separating a uniform sample by sedimentation. About 9 grams of this sand was treated in the same way as the soils. Another bottle contained a soil derived from limestone from Waikari ; and, lastly, a bottle of the bicarbonate solution without any soil at all was put through the same processes as the others samples of this series. The results are given in Table E, the first four soils being which, having been previously treated, were presumably already saturated.

TABLE E.

Soil.	Strength of Soil.	Requirement of CaO indicated.	
		Percentage.	Pounds per Acre.
Nelson*	0·024 N	+ 0·042	+ 572
Waikari*	0·024 N	+ 0·025	+ 342
21 (after liming)*	0·024 N	+ 0·042	+ 572
21 (before liming)*	0·024 N	+ 0·051	+ 695
Sand	0·019 N	Nil.	..
Weka Pass †	0·019 N	− 0·05	− 680
Check-bottle	0·024 N	Nil.	..
Waikari	0·025 N	+ 0·05	+ 680

* Second treatment of sample.
† Result given in Table C, and repeated here for comparison.
+ Means a positive requirement.
− Means that the solution has taken up lime from the soil.

These results indicate that all soils remove a certain quantity of lime from the bicarbonate solution independently of their actual lime requirements, provided that the solution is above a certain concentration initially. The Weka Pass soil can scarcely be in a different chemical condition as regards lime to that from Waikari, since both contain a large excess of calcium carbonate; and yet whereas the Waikari soil removed lime from solution, that from Weka Pass gave it up to its solution, the strength of which was increased from 0·019 N to 0·02 N.

No attempt is made in this paper to explain this phenomenon, nor will any attempt be made to give a definition in chemical language of the term "lime requirement." The aim of the work herein described is to find whether the method gives results for a given soil which agree with what is known from other sources of the lime requirements of that soil. Finding this to be sufficiently near the case for practical purposes, the writer is for the present prepared to accept the statement of Hutchinson that the amount so indicated is actually the optimum for plant-growth. It has been shown, however, that the result for any given soil varies with the strength of the solution, and that in practice it is necessary to make the determination under standard conditions. Either a solution of uniform strength must be employed for all determinations, or a correcting factor must be applied. As sufficient data has not yet been obtained to enable one to select a reliable correcting factor, the use of a solution of standard concentration is recommended, for by this means strictly comparable results are obtainable.

It may be added that this work has brought out many points of more theoretical interest, which will be discussed in another communication.

Art. LVI.—*Studies on the Chemistry of the New Zealand Flora.*

By T. H. Easterfield and J. C. McDowell, B.Sc.

[*Read before the Wellington Philosophical Society,* [*27th October, 1915.*]

PART V.—THE CHEMISTRY OF PODOCARPUS TOTARA AND PODOCARPUS SPICATUS.

(1.) Podocarpus totara.

In Part iv of this series (Trans. N.Z. Inst., vol. 43, 1911, p. 55) it was mentioned that a crystalline substance can be extracted by alcohol from totara sawdust.

In a preliminary experiment 150 grams of dry totara sawdust were boiled for two hours and a half with 90 per cent. alcohol, and yielded 18 grams of vacuum-dry resinous extract. About 70 per cent. of the extract was soluble in ether, yielding a light-coloured solution. The residue obtained by evaporating the ethereal solution was amorphous, but yielded beautiful crystals when its solution in light petroleum was allowed to evaporate spontaneously. Thirty kilograms of totara sawdust were then percolated with alcohol, and the extract treated in the same way as in the preliminary experiment, but, owing to the difficulty of treating such a large quantity without special appliances, a smaller percentage yield of dry extract was obtained.

The residue left on evaporating the light petroleum extract of the resin prepared by alcohol can be distilled in a partial vacuum without decomposition, and this is the quickest method of purifying the material. The distillate is a light-yellow liquid, which solidifies to a transparent glass if quickly cooled. The glass may be kept for months without showing any sign of crystallization, but, if it be moistened with light petroleum, crystals rapidly begin to form.

The name "totarol" is proposed for this crystalline substance, since, as will be shown, it is a tertiary alcohol.

Pure totarol melts at 127° C. (corrected); in a 4-per-cent. alcoholic solution the specific rotation $[A]_D = 42.08°$; in an 8-per-cent. solution the value is $42.18°$. It is insoluble in water and aqueous alkali, intensely soluble in alcohol and ether; 100 grams of light petroleum (b.p. 50°–80°) dissolve 18 grams of totarol at 15° C. A very small quantity of alcohol prevents the substance from crystallizing from light petroleum.

An alcoholic solution of totarol is neutral to phenol-phthalein, and does not react with alcoholic potash, hydrochloric acid, or ferric chloride.

The chemical formula for totarol is $C_{20}H_{30}O$, as shown by the following analyses and molecular-weight determinations :—

I. 0.1057 grams yielded 0.3248 gm. CO_2 and 0.0997 gm. H_2O.

II. 0.1003 ,, 0.3076 ,, 0.0949 ,,

III. 0.1162 ,, 0.3569 ,, 0.1099

Calculated for $C_{20}H_{30}O$.		Found.		
		I.	II.	III.
C = 83.9	.. .. ..	83.80	83.64	83.76
H = 10.49	.. .. ..	10.48	10.50	10.73

Unless the copper oxide was maintained at a very bright-red heat during the analyses the results found were invariably too low (*cf.* the formula $C_{18}H_{26}O$ assigned in the preliminary note, *loc. cit.*; also Dunstan and Henry, J.C.S., Proc., 1896, p. 48).

The molecular weight was determined by the cryoscopic method, using glacial acetic acid as the solvent. The values found were 285 and 288. The formula $C_{20}H_{30}O$ requires a molecular weight of 286. No compound of this formula is cited in chemical literature.

Acetyl Totarol.

Totarol can be recrystallized without change from acetic anhydride, but upon prolonged boiling an acetyl compound results. Acetylation takes place very rapidly and with considerable evolution of heat if a drop of concentrated sulphuric acid be added to acetic anhydride in which finely divided totarol is suspended. The compound is conveniently purified by crystallization from alcohol, and melts at 123°–123·5°. The specific rotation in light petroleum (4-per-cent. solution) $[A]_D = 44\cdot6°$.

Calculated for $C_{20}H_{29}O . C_2H_3O.$				Found.	
				I.	II.
C = 80·49	..	..	..	.. 80·14	80·15
H = 9·75	..	..	..	.. 9·96	9·92

The substance is perceptibly hygroscopic. The molecular weight determined by hydrolysis with alcoholic potash gave the values 331·8, 328·7, and 326·7. The calculated value for the above formula is 328.

The totarol regenerated during hydrolysis had the same melting-point and optical activity as the original totarol.

The slow rate at which totarol undergoes acetylation suggests that it is a tertiary alcohol. The same conclusion is to be drawn from the fact that it yields no acid phthalic ester when the benzene solution is boiled with phthalic anhydride, and that no acetyl ester results when totarol is heated with glacial acetic acid in a sealed tube.

Oxidation experiments also indicate that totarol is neither a primary nor secondary alcohol. Chromic acid in glacial acetic-acid solution attacks totarol very slowly at first; subsequently the action becomes vigorous, and much carbon dioxide is evolved. Unchanged totarol, but no oxidation product, can be isolated after the reaction has ceased. Similarly no acid or ketonic products could be isolated when totarol was oxidized with potassium bichromate and dilute sulphuric acid, though the experiment was tried at various temperatures.

A neutral crystalline oxidation product, but no acid, results when totarol dissolved in pure benzene or purified light petroleum is agitated for some hours with decinormal permanganate solution. The yield is about 20 per cent. of the weight of the totarol taken. The substance is readily purified by crystallization from ether, and when pure melts at 205°, at which temperature sublimation commences.

Calculated for $C_{40}H_{60}O_3.$				Found.	
				I.	II.
C = 81·63	..	..	..	.. 81·41	81·47
H = 10·21	..	..	..	.. 10·31	10·29

The formula suggests that during oxidation two molecules of totarol have condensed and an atom of oxygen has been added. Until further evidence as to the constitution of totarol has been obtained it would be useless to speculate on the relationship of the two compounds to each other.

Bromine and Iodine Absorption.

Totarol in chloroform solution absorbs bromine very rapidly. At the same time a little hydrobromic acid is evolved. Very slight temperature differences make such large differences in the percentage of bromine absorbed that it is impossible to state definitely the number of unsaturated " linkings " in the molecule. The iodine absorption by Wij's method indicates that six atoms of iodine are absorbed by one molecule of totarol. Totarol yielded no definite products when submitted to sulphonation and nitration.

(2.) PODOCARPUS SPICATUS.

The occurrence of matai-resinol in the heart-cracks of this species has already been reported (Easterfield and Bee, Trans. Chem. Soc., 1910, p. 1028). The heart-cracks of old matai-trees frequently contain a considerable quantity of a liquid known as matai-beer. The liquid can be tapped by means of an auger, and is said to be eagerly drunk by the bushmen. A sample of matai-beer was kindly procured by Mr. Phillips-Turner, Secretary to the Forestry Commission. The liquid was of a light-brown colour ; its smell suggested the presence of butyric or caproic acid, and its taste was styptic and sweetish, followed by a bitter after-taste. 10 c.c. of the liquid left on evaporation a brown sticky residue weighing 0·265 gram. 10 c.c. required 5 c.c. $\frac{N}{10}$ sodium hydrate for neutralization. Practically the whole of the acidity disappears during evaporation.

In a 20 cm. tube the liquid rotated the ray of polarized sodium light 0·64° ; this value was unchanged after the liquid had been heated with hydrochloric acid.

The liquid instantly reduced Fehling solution in the cold, and gave a silver mirror with ammoniacal silver solution. It reduced mercuric chloride slowly on warming. With phenyl hydrazine in acetic acid it yielded a scarlet crystalline precipitate, but the amount obtained was too small for further examination. Ferric chloride produced a greenish-brown tint.

For the identification of the volatile acid 250 c.c. were distilled with steam. The distillate exactly neutralized with ammonia and excess of silver nitrate added. Analysis of the micro-crystalline silver salt showed it to be silver caproate.

Calculated for $C_6H_{11}O_2Ag$.				Found.	
				I.	II.
C = 32·28	..	..	..	32·49	32·10
H = 4·93	..	..	..	4·81	4·83
Ag = 48·43	..	..	..	49·00	48·66

Caproic acid exists in *Goupia tomentosa* (*Celastraceae* family), a timber-tree used for boatbuilding in British Guiana (Dunstan and Henry, Trans. Chem. Soc., 1898, p. 228), and which is closely allied to the well-known garden shrub *Euonymus*. Its occurrence in other plants does not appear to have been recorded.

Victoria University College, Wellington.

PROCEEDINGS.

PROCEEDINGS

OF THE

NEW ZEALAND INSTITUTE,

1915.

THIRTEENTH ANNUAL MEETING.

WELLINGTON, 28TH JANUARY, 1916.

THE annual meeting of the Board of Governors was held in the Dominion Museum Library on Friday, the 28th January, 1916, at 10 a.m.

Present: Hon. Mr. Russell, Minister of Internal Affairs, Mr. D. Petrie, President (in the chair); Mr. Charles A. Ewen; Mr. A. H. Turnbull, Dr. J. Allan Thomson, Mr. B. C. Aston, Professor T. H. Easterfield, Professor H. B. Kirk, Professor H. W. Segar, Dr. Hilgendorf, Mr. A. M. Wright, Professor Marshall, Mr. G. M. Thomson, Mr. H. Hill, Dr. L. Cockayne, Mr. J. W. Poynton, and Dr. Hatherly.

Changes in the Representation.--The Secretary announced that the only changes in the representation were: Mr. B. C. Aston replaced Professor von Zedlitz as a Government representative; Dr. Hilgendorf and Mr. A. M. Wright replaced Dr. C. C. Farr and Mr. R. Speight for the Canterbury Philosophical Institute; Mr. J. W. Poynton replaced Mr. K. Wilson for the Manawatu Philosophical Society.

President's Address.—The President then read his annual address. (See p. 530.)

The Minister of Internal Affairs (Hon. Mr. Russell), by invitation of the Chairman, then addressed the meeting, and promised to meet the Board again later in the day. A hearty vote of thanks was unanimously accorded the Minister for his interesting and comprehensive address.

The Annual Reports of the Incorporated Societies for their last financial years were received, and ordered to lie on the table. Dr. Hatherly explained why the Wanganui Philosophical Society's report had not been received.

Hon. Treasurer's Statements.—The statement of assets and liabilities of the Institute at the 31st December, 1915, and the statement of receipts and expenditure for the year ending 31st December, 1915, supported by the Public Trustee's certificates of the state of the Carter Bequest, the Hutton Memorial Fund, and the Hector Memorial Fund at the 31st December, 1915, all duly audited and signed by the Auditor-General, were, on the motion of Professor Kirk, seconded by Dr. Marshall, adopted.

STATEMENT OF RECEIPTS AND EXPENDITURE FOR THE YEAR ENDING 31ST DECEMBER, 1915.

Receipts.	£	s.	d.	Expenditure.	£	s.	d.
Balance at credit in Bank of New Zealand	494	18	3	Governors' travelling - expenses	32	14	8
Government grant	500	0	0	Fire-insurance premium	5	0	0
Transactions sold locally	1	14	0	Secretary's salary	50	0	0
"Maori Art" sold	8	8	0	Compiling card index	10	0	0
Bulletins sold	0	4	0	Bank charge	0	10	0
Public Trustee, Hutton Memorial Fund	15	0	0	Award to Mr. Oliver, Hutton Memorial Fund	15	0	0
Authors' reprints	8	4	5	Hon. Editor, petty cash	3	0	0
Wesley and Son, London: publications sold	11	10	4	Postage on Transactions	20	5	8
Postage refunded by societies	17	5	9	Secretary, petty cash	10	0	0
Hector Memorial Award	45	0	0	Hector Award to Mr. Elsdon Best	45	0	0
Balance at Bank of New Zealand £17 7 2				Deposited in Post Office Savings-bank	500	0	0
Less unpresented cheque .. 25 0 0				Government Printer, Vol. 46	418	7	3
	7	12	10				
	£1,109	17	7		£1,109	17	7

		£	s.	d.
Balance at credit of Institute in Post Office Savings-bank		500	0	0
Balance at Bank of New Zealand .. £17 7 2				
Less unpresented cheque .. 25 0 0				
		7	12	10
Credit balance		£492	7	2

STATEMENT OF LIABILITIES AND ASSETS AT 31ST DECEMBER, 1915.

	Liabilities.			Assets.		
	£	s.	d.	£	s.	d.
Dec. 31. To Balance due Government Printer	624	11	10			
West Newman Account	1	1	0			
Unpresented cheque	25	0	0			
By Balance in Post Office Savings-bank				500	0	0
Accrued interest thereon at 3 per cent.				10	0	0
Transactions, &c., sold—proceeds to come				12	1	5
Petty cash in hands of Secretary				5	18	9
Balance in Bank of New Zealand				17	7	2
	650	12	10	545	7	4
Debit balance				105	5	6
	£650	12	10	£650	12	10

The Institute has a large stock of Transactions on hand, valued as an asset at £750. The only other property possessed by the Institute is a very valuable scientific library.

CARTER BEQUEST.—STATEMENT OF RECEIPTS AND EXPENDITURE FOR THE YEAR ENDING
31ST DECEMBER, 1915.

Receipts.	£	s.	d.	*Expenditure.*	£	s.	d.
Balance brought forward ..	3,582	5	1	Public Trust Office, commission, 2½ per cent. on 13s. 1d.	0	0	4
New Zealand Loan and Mercantile Agency Company (Limited)—				Petty expenses—Postages ..	0	1	0
Interest, 1st October, 1914, to 1st October, 1915 ..	0	9	11	Balance	3,744	0	5
Dividend (preference), 30th June, 1914, to 30th June, 1915.. ..	0	3	2				
Public Trust Office, interest, 31st December, 1914, to 31st December, 1915 ..	161	3	7				
	£3,744	1	9		£3,744	1	9

HUTTON MEMORIAL FUND.—STATEMENT OF RECEIPTS AND EXPENDITURE TO 31ST
DECEMBER, 1915.

Receipts.	£	s.	d.	*Expenditure.*	£	s.	d.
Balance brought forward ..	722	15	9	New Zealand Institute Account: Grant to W. R. B. Oliver for 1914–15 ..	15	0	0
Public Trust Office, interest, 31st December, 1914, to 31st December, 1915, at 4¼ per cent.	31	18	9	Public Trust Office, postages	0	1	0
				Balance	739	13	6
	£754	14	6		£754	14	6

HECTOR MEMORIAL FUND.—STATEMENT OF RECEIPTS AND EXPENDITURE FOR YEAR
ENDING 31ST DECEMBER, 1915.

Receipts.	£	s.	d.	*Expenditure.*	£	s.	d.
Balance brought forward ..	1,084	8	6	New Zealand Institute Account: Elsdon Best, Hector Prize for 1914 ..	45	0	0
Public Trust Office, interest, 31st December, 1914, to 31st December, 1915, at 4¼ per cent.	47	2	6	Public Trust Office, postages	0	1	0
				Balance	1,086	10	0
	£1,131	11	0		£1,131	11	0

A financial discussion followed, in which the propriety of asking for a contribution from each incorporated society was considered. On the motion of Dr. J. Allan Thomson, seconded by Dr. L. Cockayne, it was resolved, That for every copy of the Transactions received by incorporated societies a contribution of 2s. 6d. towards the cost of printing shall be made during the current year by such society.

Hutton Memorial Grants for Research.—A report from Mr. W. R. B. Oliver on the results of his research for the past year was received.

On the motion of Mr. G. M. Thomson, seconded by Professor P. Marshall, it was resolved, That the sum of £25 be voted to the Portobello Marine Fish-hatchery for the purpose of prosecuting research on the distribution of native marine food fishes.

Standing Committee's Report.—The President moved the adoption of the Standing Committee's report.—Carried.

Report of the Standing Committee.

Five meetings of the Standing Committee have been held during the past year, the attendance being as follows: Mr. Petrie, 2; Dr. Cockayne, 5; Professor Easterfield, 2; Mr. C. A. Ewen, 3; Mr. Hill, 1; Professor Kirk, 2; Dr. J. Allan Thomson, 5; Mr. G. M. Thomson, 2; Professor von Zedlitz, 1.

Hector Memorial Award.—The presentation of the medal to Professor P. Marshall, of Otago University, took place on the 4th October, 1915, at a meeting of the Otago Institute, Mr. G. M. Thomson, a former President of the Otago Institute, making the presentation.

Hector Memorial Fund Research Grant.—Of the two grants to research workers made at the last meeting, only that to Mr. Oliver was availed of. A progress report from him has been received. Dr. Cotton, having received a grant from another body, has surrendered the grant made to him by the Institute, for which he has been thanked by the Standing Committee.

Publications.—Volume 47 of the "Transactions and Proceedings of the New Zealand Institute" was issued on the 12th July, 1915, and on the 16th three copies were received, and one was laid on the table of the Legislative Council on the 21st July, and one on the table of the House of Representatives on the 20th July, 1915. On the 2nd August the Government Printer delivered 223 volumes, and they continued to arrive in small lots, and were parcelled up and addressed until enough had been received to fill the New Zealand demands, which was not until the 20th August, when they were posted to members. An alteration in the manner of binding the Transactions has been introduced with vol. 47, at the suggestion of the Secretary, who pointed out that a binding in what is known as quarter-bound boards, instead of the previous limp covers, could be obtained at very little additional expense. It is hoped that the innovation will meet with general approval.

Finances of the Institute.—In view of the many demands made on the Government on account of the war, after interviewing the Hon. the Minister of Internal Affairs on the matter of an additional vote it was decided not to proceed further at present with the motion passed at the last annual meeting. The Government Printer is willing to allow a portion of the debt for printing the publications to stand over, providing the greater portion be paid. The Standing Committee is of opinion that in view of the financial position of the Institute a levy on the incorporated societies should be made in the coming year. The motion regarding handing over the Institute library to the Board of Science and Art has been referred to that body by the Hon. the Minister of Internal Affairs. Other motions passed at last annual meeting involving increased expenditure by the Government were also allowed to stand over in the meantime.

Decisions of the Standing Committee.—(1.) The storage of illustration blocks having been brought up by the Government Printer, the difficulty was disposed of by Dr. Thomson offering to store them in the Museum, an offer which was thankfully accepted.

(2.) It has been decided to circularize all public libraries in New Zealand to ascertain whether they would accept partial sets of Transactions as a donation from the Institute.

(3.) The following have been placed on the list of those to whom the Transactions are yearly sent free :—

Southland Museum, Invercargill.
Royal Scottish Geographical Society, Synod Hall, Castle Terrace, Edinburgh.
National Academy of Sciences, Smithsonian Institution, Washington, U.S.

Other applicants for the exchange or the gift of isolated volumes have been also dealt with on their merits by this Committee.

(4.) The Standing Committee have accepted a suggestion by the Publication Committee of allowing a Government Department to contribute the entire cost of preparing the figures for a paper which was submitted by a Government officer, who desired to publish the illustrations subsequently in a Government report.

Annual Reports and Balance-sheets of the following societies have been received :—

Auckland Institute, to 18th February, 1915.
Manawatu Philosophical Society, to 21st October, 1915.
Otago Institute, to 30th November, 1915.
Nelson Institute, to 31st December, 1915.
Hawke's Bay Philosophical Institute, to 23rd November, 1915.
Wellington Philosophical Society, to 30th September 1915.
Canterbury Philosophical Institute, to 31st October, 1915.

The report and balance-sheet of the Wanganui Philosophical Society has not yet been received.

Memorial to the late Mr. Hamilton.—The Hamilton Memorial Committee reports that the fund collected for the memorial to the late Mr. Augustus Hamilton, amounting at present to £122 2s. 10d., is deposited in the Post Office Savings-bank. Arrangements are now in progress for the erection of a suitable monolith over the grave at Russell, Bay of Islands.

Stock of Transactions.—The Parliamentary Librarian having intimated that he can no longer store the immense stock of Transactions, some 15,000 volumes, the Standing Committee decided to circularize all public libraries in New Zealand asking if they would be willing to accept, as a donation, partial sets of those volumes of which the Institute possesses such a large supply. The Parliamentary Librarian has kindly consented to help in the clerical work of issuing the circulars, of which as many as four to five hundred may be needed. The only conditions it is at present contemplated to impose on the recipients are that the freight charges shall be paid by the receiving body, and that the volumes shall become the property of that body, and not be disposed of. According to the decision of the Standing Committee, inquiries were made, and as a result eighty replies have been received, seventy of them accepting and ten refusing the conditional offer of the Standing Committee. It is now for the annual meeting to decide what further steps shall be taken in the matter.

29th January, 1916.   D. PETRIE, President.

Hamilton Memorial Fund.—Professor Easterfield explained why it had been found necessary to depart from the original intended form of the memorial.

Hector Award for 1916.—Professor Segar proposed, and Mr. G. M. Thomson seconded, That the action of the Standing Committee in supplying information to the Hector Award Committee be confirmed.—Carried.

The President then opened the recommendation of the Committee of Award received from Professor Carslaw in a sealed envelope, and announced that the Committee had selected Sir Ernest Rutherford, F.R.S. The recommendation was adopted.

Publication Committee's Report.—The report of the Committee was adopted, as follows:—

REPORT OF THE PUBLICATION COMMITTEE.

The Publication Committee begs to submit the following report for the year :—

Sixty-four papers were sent in for the consideration of the Committee, and, of these, fifty-eight were published in the "Transactions of the New Zealand Institute," vol. 47, the others being withdrawn or declined. Of those finally accepted many were much condensed at the request of the Committee, and the number of illustrations accompanying them was very greatly reduced. In a few cases the preparation of the papers and the arrangement of the figures for publication gave far more than the usual amount of work and trouble.

Volume 47 of the Transactions was issued on the 12th July, 1915, being somewhat later than usual owing to the cause just mentioned. It contains 704 pages, 12 plates, and a large number of text figures.

Bulletin No. 1, Part IV, on New Zealand *Coleoptera*, by Major T. Broun, was published on the 17th February, and contains 79 pages of text. A further instalment of MS. on the same subject had been received early in the year, and at the last annual meeting of the Board of Governors was referred to the Publication Committee for inquiry as to cost of publication, &c. An estimate has been received from the Printer, but owing to the state of the finances of the Institute the question of publication is held over for decision at the annual meeting.

Two lots of MS. by Mr. H. N. Dixon for the continuation of Bulletin No. 3. on New Zealand Bryology, have been received, and estimates of the cost of publication obtained for consideration at the annual meeting. The MSS. by Major Broun and Mr. Dixon are in safe keeping with the Government Printer.

At the last annual meeting the Committee was authorized to make arrangements for the publication of Dr. Mortensen's proposed report on the New Zealand *Echinodermata*, but Dr. Mortensen only recently reached Copenhagen, and no report has yet been received from him.

The estimates obtained by the Secretary from London firms of the cost of producing coloured illustrations have been considered by the Committee, which recommends that the question be held over till the finances of the Institute are in a better condition.

Towards the end of the year the Hon. Editor resigned owing to the pressure of other work. Arrangements have been made for papers intended for the next volume of the Transactions to be sent to the Secretary pending the appointment of an Hon. Editor at the annual meeting.

For the Committee.

CHAS. CHILTON,
Retiring Hon. Editor.

It was proposed by Dr. Cockayne, seconded by Mr. Poynton, That a sum of £50 be voted out of the Hutton Memorial Fund towards the publication of the researches of Major Broun on New Zealand *Coleoptera.*—Carried.

On the motion of Dr. J. Allan Thomson, seconded by Dr. Hilgendorf, it was resolved, That the Publication Committee be directed to insert a notice in the Transactions stating the privileges of members in relation to the libraries of the Institute and of the incorporated societies.

Report of the Library Committee.—The report of the Committee was adopted, as follows :—

REPORT OF LIBRARY COMMITTEE.

The rearrangement of the library, forecasted in the report of last year, has been completed, and it is now possible to trace easily the publications of any given society or institution. Your Committee feels that the facilities offered by the library are not understood by the majority of members outside Wellington, and recommends that a notice should be inserted annually in the Transactions stating the privileges of members and the conditions under which books may be lent through the post. The Librarian is always willing to answer queries from individual members as to whether or not any given book is in the library.

Although a large number of valuable journals are received annually in exchange for the Transactions, the number of societies with whom exchanges are effected bears a very small proportion to the number of societies which publish papers likely to be of interest to scientific workers in New Zealand, while the absence of many societies of the first importance from the exchange list is a matter of surprise and regret. This has a double aspect, for it means also that the Transactions are not available to, and are therefore not read by, the majority of scientific workers outside New Zealand. For instance, only seventeen universities and colleges outside New Zealand appear in the list of free copies, while in 1914 only seventeen copies were sold outside the Dominion. Your Committee has therefore drawn up a list of over eight hundred societies and institutions with which exchanges might be effected, and recommends that proposals to two hundred of these should be made during the current year.

J. ALLAN THOMSON,
Hon. Librarian.

Dr. J. Allan Thomson moved, Dr. Cockayne seconded, and it was unanimously resolved, That a set of the publications as complete as possible be presented to the University of Louvain.

The Secretary explained the difficulty which had arisen with regard to the storage of the large excess of the back numbers of the Transactions, and detailed the steps taken by the Standing Committee in circularizing libraries. A letter from the Hon. the Minister of Internal Affairs, dated the 26th November, 1915, was read dealing with the same subject.

On the motion of Dr. J. Allan Thomson it was resolved, That the Standing Committee be authorized to dispose of the stock of Transactions for those years in which the number is in excess of 200 by gift to suitable institutions or by sale at reduced terms.

Professor Easterfield moved, Mr. Ewen seconded, and it was resolved, That a statement be printed on the back of the forthcoming volume of the Transactions intimating that certain volumes of the Transactions are available to members at 2s. each, inclusive of postage.

On the motion of Dr. J. Allan Thomson it was resolved, That the Standing Committee be authorized to increase the exchange list.

On the motion of Mr. J. W. Poynton, seconded by Dr. Cockayne, it was resolved, That the storage of the excess volumes of Transactions be dealt with by the Standing Committee as they find necessary.

On the motion of Mr. G. M. Thomson, seconded by Dr. Cockayne, it was resolved, That the Minister of Internal Affairs be asked to obtain, if possible, a grant to enable the Board of Governors of the Institute to distribute spare volumes of Transactions to public libraries, secondary and technical schools of the Dominion, branches of the Teachers' Institute, &c.; also to suitably bind and forward the set of Transactions voted by the Institute to the University of Louvain.

Correspondence.—Letters were read and received from the Under-Secretary of the Internal Affairs Department relating to the Nobel Peace Prize (11/1/16), Kidnappers Reserve (21/9/15), Science and Art Board (12/3/15); from the Secretary of the Marine Department (24/3/15) re Catalogue of Fishes; from Dr. J. Allan Thomson regarding a paper by Mr. S. S. Buckman; and from the Wellington Philosophical Society regarding the date for sending in papers for publication (7/10/15). It was decided to refer Dr. Thomson's application to the Publication Committee, and to take no action in regard to the Wellington Philosophical Society's application.

International Catalogue of Scientific Literature.—It was decided to refer the matter of compiling the annual catalogue for the New Zealand Regional Bureau to the Standing Committee to take such action as they deem suitable.

Election of Officers.—On the motion of the President, Professor Benham was unanimously elected *President* of the New Zealand Institute for the ensuing year. The following officers were also elected : *Hon. Treasurer*—Mr. C. A. Ewen; *Hon. Editor*—Dr. L. Cockayne; *Joint Hon. Editor*—Dr. C. A. Cotton (subject to his acceptance); *Hon. Librarian*—Dr. J. Allan Thomson; *Hon. Secretary*—Mr. B. C. Aston; *Publication Committee*—The Hon. Editors, Professors Easterfield and Kirk, Dr. J. Allan Thomson, and Mr. Aston; *Library Committee*—Dr. Cotton, Dr. Cockayne, and the Hon. Librarian were re-elected; *Hector Award Committee for 1917* (subject Zoology)—Professors Haswell, of Sydney, and Baldwin Spencer, of Melbourne.

Travelling-expenses.—Mr. C. A. Ewen moved, and Professor Easterfield seconded, That the travelling-expenses of this meeting be reimbursed by the Institute.—Carried.

Election of Honorary Member.—Professor Jean Massart, of the University of Brussels, was elected.

Date and Place of next Annual Meeting.—On the motion of Professor Segar, seconded by Professor P. Marshall, Tuesday, 30th January, 1917, at Wellington, were fixed as the date and place of the next annual meeting.

Vote of Thanks.—On the motion of Mr. G. M. Thomson, seconded by Professor Kirk, it was resolved, That the Board pass a very hearty vote of thanks to Dr. Chilton for his valuable services as Hon. Editor during the past six years.

Government Grants.—The Hon. the Minister of Internal Affairs attended the meeting again at the close, when Mr. G. M. Thomson's motion regarding a request for a grant was read to him. The Hon. Mr. Russell again addressed the meeting, and promised to grant the application for funds to distribute the excess of Transactions and to bind a set of the Institute's publications in half-calf for the University of Lovain; and also to donate to that university any scientific Government publications which were available. He further promised to recommend to Cabinet the authorization of an additional grant to the Institute of £250, to be specially spent in furthering one or more branches of research not hitherto provided for from the Institute's funds; the Minister to be informed of those subjects on which the vote is to be spent, the only proviso being that the subject should have as practical an object as possible. A vote of thanks to the Minister for his attendance and offer was unanimously carried.

A vote of thanks to the President and officers concluded the business.

Confirmed 29th January, 1916.

D. PETRIE, President.

PRESIDENTIAL ADDRESS.

The following is the presidential address delivered at the annual meeting of the Board of Governors of the New Zealand Institute, at Wellington, on the 28th January, 1916, by Mr. D. Petrie, M.A., Ph.D. :—

GENTLEMEN OF THE BOARD OF GOVERNORS OF THE NEW ZEALAND INSTITUTE,— We meet again with the cloud of war hanging over the Empire. Our own land has had its share of suffering and sorrow, but the daring and heroism of its sons have already created a glorious tradition that can never be forgotten or sullied by the dwellers in these Islands. Our isolation and the protection of the British War Fleet allow us to go about our usual affairs with little distraction other than the anxiety which the passing weeks bring along.

The 47th volume of the Transactions of the Institute was accordingly issued in July last. The volume is considerably larger than those of recent years, extending to over 700 pages exclusive of plates. Among its varied contents is the usual large number of important contributions to local natural history, geology, and cognate subjects. The original papers in the various branches of zoological research are specially numerous and valuable, and several of them, it may be noted with satisfaction, are by young and promising workers.

For a good many years past the annual volume of Transactions has been issued in flimsy paper covers. At the suggestion of our Secretary I authorized the binding of the last-issued volume in stiff paper boards with a back of thin binding-cloth. This will prove a great convenience to those who do not care to bind the book and have frequent occasion to consult it, and the increased cost is practically met by the saving in packing the books for transmission by post, as an ordinary paper cover is now sufficient to ensure safe carriage.

It is obvious that the increased size of the volume means increased outlay in printing and postage. The Printer's bill alone exceeds by more than £70 the statutory grant of £500, which is all we can reckon on to cover the cost of publications and the Board's management. For two years a supplementary grant of £250 was voted by Parliament, but this aid to our scanty funds was not renewed last session. The Board is consequently faced with a considerable deficiency on the year's operations, and I fear will be compelled to resort to a levy on the funds of the incorporated societies. Should Parliament decline to pass a supplementary vote hereafter, a result which all interested in scientific inquiry must deplore, this levy will no doubt become a permanent feature in the Board's finance. As to the amount of the levy, it is imperative that it be kept as low as possible, for the incorporated societies have numerous unavoidable obligations which their income from subscriptions does little more than meet.

Until our financial position is greatly improved, retrenchment in our outlay is unavoidable. I am of opinion that the bulk of the Transactions (and, by consequence, the amount of the Printer's bill) could be very considerably reduced without any serious impairment of the value of their contents. Is it really needful to print year by year the whole of the matter included in the Appendix? Or is there any great urgency about the publication of lists of new plant-habitats, accounts of the fauna or flora of single counties or other limited districts, and many of the papers on Maori culture, customs, and folk-lore submitted for publication? Such articles might well be passed over until the Board is once more in a position to print them without "outrunning the constable." It seems to me, further, that papers on abstruse mathematical subjects might be altogether excluded from our Transactions, in the interests of their writers if for no better reason, for such papers are simply buried in our publications, and would far more fitly see the light of day in some of the special journals devoted to this branch of inquiry. The Publication Committee, by sternly refusing to accept for publication diffuse and verbose papers unless condensed to their satisfaction, could do much to ease our periodical financial difficulties.

Owing to our limited funds, only one bulletin has been issued during the year. The manuscript material for two additional bulletins has been held over. One of these is a long and valuable paper by Major T. Broun on new New Zealand *Coleoptera*. Major Broun is naturally greatly disappointed at the delay in publication, and I trust that the Board will authorize its publication as a bulletin early in the present year. Other means of publication are indeed available, but it is most desirable that the paper should be published here, as it is merely a continuation of other papers we have published already. Prolonged delay in dealing with this paper may result in loss of priority for the new genera and species described, in which case its learned and enthusiastic author would be deprived of the well-merited and sole reward of his months and months of continuous labour and research.

It is now more than two years since the Science and Art Act was placed on the statute-book, but the special Board set up under its provisions has not yet come to the birth. The arrangements for a possible transfer of the library of the Institute, authorized at the Board's last meeting, are consequently in abeyance. It is understood that a meeting will be held immediately.

In the course of the year the Institute's library has been rearranged, so that the publications of any given society or institution can now be easily traced. The books and papers have been stamped on the outside of the cover, and can thus be readily distinguished from the other works located in the Dominion Museum library-room. It is a pity that the Institute's stamp has not been placed here and there in the body of the books, as they are mostly in thin paper covers; this can, however, be done hereafter without difficulty. The Honorary Librarian and the other gentlemen who assisted him in carrying out these improvements deserve the cordial thanks of the Board. My predecessor in the President's chair argued in favour of a division of our library among the four University centres of the Dominion. To this proposal I am very decidedly opposed, but there is no need for recording the reasons for my view, as the project seems unlikely to meet with general support.

I may use the present opportunity to point out a conspicuous and most regrettable defect in the museums of this Dominion. I refer to their failure to provide any worthy collection of the native and introduced plants that grow within its borders. The only fairly complete herbaria in the country are the property of some two or three private persons; no museum contains anything at all comparable with these. It is high time that steps were taken to remedy this anomaly. The Dominion Museum at least should be provided as soon as may be possible with a full and varied collection of the native and naturalized plants of our Islands. Such a collection should not be confined to flowering-plants and ferns, but should cover the whole of the flora. It would be a signal service to biological science if the Director of the Dominion Museum could take this branch of museum service in hand, and make the institution over which he presides more and more a centre of light and leading for all who are prosecuting plant studies. Photographs of specimen plants and trees, and of selected spots of wild nature showing the plant societies that adorn our mountain valleys and slopes and other stations of interest, should also be got together and placed on exhibition. The late Mr. H. J. Matthews in the course of his wanderings about New Zealand accumulated a large and splendid collection of photographs of the kind here referred to, and it is a matter for sincere regret that his fine series of photographs was not secured for the Dominion Museum.

Some small sums of public money have been expended on special botanical surveys, and reports of these have been published, but so far as I am aware the botanical material collected has not been used to enrich any public museum. In any further research of this kind that may be undertaken, it might easily be arranged that as full collections as possible of the plants observed should be made, with a view to their permanent preservation in the Dominion Museum. If photographs could also be secured, so much the better. I would suggest further that the help of survey parties employed in the Government service should be enlisted in this good cause. Many of the gentlemen who direct such parties are interested in native plants, and could with little trouble collect and dry numbers of specimens not easy to procure in flower and fruit by other agencies. The Inspector and local officers of our forest and scenic reserves, and the Superintendents of the State nurseries, could also give valuable help in getting together a worthy natural collection of native and naturalized plants.

Early in the past year the two volumes of "Illustrations of the New Zealand Flora," edited by Mr. T. F. Cheeseman and Dr. W. Botting Hemsley, issued from the press some little time before, became available for reference by those interested in botanical research. This fine work is in all respects worthy of the reputation of its distinguished editors. I was a member of the deputation from a conference of School Inspectors that waited on the late Mr. Seddon to urge him to authorize the preparation of a new Flora of New Zealand, the work to be accompanied by a volume of illustrative drawings. With his usual public spirit and regard for the interests of country settlers, he promised favourable consideration of the deputation's request, and expressed his desire that the drawings should be such as would enable miners and country dwellers generally to gain, if they so desired, a knowledge of the common plants growing in their neighbourhood. The suggestion offered by the deputation was that there should be a drawing of one species of each of our genera of flowering-plants and ferns, and one for each section of the larger genera in which well-marked sections are recognized. This design was evidently known to the late Mr. T. Kirk, to whom the production of the new flora was entrusted, though he did not live to complete more than half the task. The preface to the "Students' Flora of New Zealand," as Mr. Kirk's work was entitled, shows that it was the intention of the Education Department, which was charged with the production of the book, to include in the series of plates many previously published drawings of native plants, no doubt on a reduced scale. Arrangements were even made with Messrs. Reeve and Co., of London, by payment of a small royalty, to utilize many of the numerous plates of native plants contained in the classical works on the floras of New Zealand and Tasmania by the late Sir J. D. Hooker. I consider it most regrettable that the Education Department should have consented to the abandonment of the plan roughly sketched out in the above-mentioned preface, no doubt with the late Mr. Seddon's approval. What was wanted to foster a popular interest in botanical inquiry was a set of plant drawings somewhat on the lines of Bentham's Illustrations of the British Flora. A work of some such kind would, no doubt, have aroused among intelligent country residents a growing interest in the local vegetation, and opened up for them a pleasant recreation; it would have made the path of all beginners in plant studies easy and sure; and would have helped to bring to the front many who are now turned away from such pursuits by the unfamiliar technical language in which accurate botanical descriptions must be set forth. For botanists outside our Dominion who wish to gain a more extended acquaintance with the New Zealand flora than Sir Joseph Hooker's works made available, the new volumes of illustrations are entirely suitable; but residents in the Dominion will find the books costly and unwieldy, and deficient in figures of a great many of the most common and most widely diffused native plants. As it seems to me, a great opportunity for stimulating popular interest in plant studies, and for enriching the non-selfish life interests of the coming generation, has been turned to poor account. The excellent list of illustrations of New Zealand plants previously published is a valuable feature of the new volumes.

An important scientific publication of the year is the atlas of plates in illustration of the recently published "Manual of the New Zealand Mollusca" by Mr. Henry Suter. This work contains a very large number of figures of Recent shells, in general beautifully executed, and is well fitted to stimulate closer and more general study of this department of zoology. Together with the author's Manual this atlas will enable any one drawn to the subject to get abreast of the present state of our knowledge of the molluscous fauna, and entice beginners in its study to go forward under highly favourable conditions. The book, it may be noted. is reasonably cheap, is of convenient size, and contains a very ample representative series of figures. The Government Printer can be warmly congratulated on the production of this fine work.

Two years hence the New Zealand Institute will have reached its fiftieth year of activity. The New Zealand Institute Act was passed in 1867, the "Abstracts of Rules and Statutes" was gazetted on the 9th March, and the inaugural meeting was held on the 4th August, 1868. It may be desirable to hold some formal celebration of the semi-centenary of the Institute's foundation.

An important event in the development of science and its practical applications within our Dominion is the recent generous bequest of a very large sum for these purposes by the late Mr. Cawthron, of Nelson. When the Cawthron Institute has commenced its activities we may look for very considerable benefits to many of our prominent industries, and to a growing number of scientific workers who may there be trained for research without submitting to the shackles that university degrees too often impose on the courses of study and the training of students at our universities.

A month or two ago the Honorary Editor of the Transactions (Dr. Charles Chilton) communicated to me his desire to be relieved of this onerous office, owing to a sudden enlargement of his other work. For his gratuitous and laborious services as Editor for some years past Dr. Chilton deserves the warmest thanks of this Board. It would be in several ways convenient if the new Editor were resident in Wellington. The filling of the position will come before you in the course of this meeting.

Our experienced Secretary (Mr. B. C. Aston) has been appointed a member of the Board as one of the Government representatives. Mr. Aston has intimated to me that he will be prepared to act as Honorary Secretary for some time to come. The Board will be invited later to consider how the office can be best filled. Should Mr. Aston's generous offer be accepted it will help to lessen the cost of administration.

Arrangements for distributing the large stock of surplus copies of the Transactions which, through the kindness of the Librarian, are now stored, not without inconvenience, in the Parliamentary Buildings will be submitted for the consideration of the Board. It is proposed to offer as complete sets as possible to all public libraries and all secondary and technical schools free of cost, other than that of transmission to their destination.

The Standing Committee has considered the propriety of increasing the number of scientific societies and institutions to which our Transactions are presented annually, and proposals to give effect to their views will no doubt be submitted at the present meeting.

WELLINGTON PHILOSOPHICAL SOCIETY.

FIRST MEETING : *28th April, 1915.*

Mr. T. King, F.R.A.S., President, in the chair, and about forty members present.

New Members.—Professor D. M. Y. Sommerville, M.A., D.Sc., F.R.S.E.; Professor E. Marsden, D.Sc.; Mr. C. J. Westland, F.R.A.S.; Miss Ellen Pigott, M.A.; Miss Elizabeth Pigott, M.A.

New Section.—The formation of a Literary Section was approved.

Rearrangement of the Library.—Dr. J. A. Thomson reported on the rearrangement of books in the library.

Presidential Address. — Mr. T. King delivered his presidential address, dealing with the war and the duties of scientific societies during its continuance.

Paper.—" Maori Art in the Arawa Country," by Mr. James Cowan; communicated by the President.

SECOND MEETING : *26th May, 1915.*

Mr. T. King, F.R.A.S., President, in the chair, and about forty members and friends present.

New Members.—Mr. W. Donovan, M.Sc.; Mr. H. S. Tily; and Sir John Findlay, K.C.

Papers.—1. " The Botany of the Kaikoura Mountains," by Mr. B. C. Aston, F.C.S., F.I.C.

2. " Fault Coasts," by Dr. C. A. Cotton, F.G.S.

THIRD MEETING : *23rd June, 1915.*

Mr. T. King, F.R.A.S., President, in the chair, and about forty-five members and friends present.

Address.—" A Review of Maori Art," by Dr. A. K. Newman, M.P.

Paper.—" The Maori as a Voyager," by Mr. Elsdon Best.

FOURTH MEETING : *28th July, 1915.*

Mr. T. King, F.R.A.S., President, in the chair, and about a hundred members and friends present.

Address.—"A Few Minutes with Microbes," by Dr. J. M. Mason, F.C.S., D.P.H.

Kinematograph Exhibition.—Mr. J. McDonald exhibited a number of films of scientific interest.

FIFTH MEETING : *25th August, 1915.*

Mr. T. King, F.R.A.S., President, in the chair, and about twenty-five members and friends present.

New Member.—Mr. J. W. Burbidge, M.Sc.

Lecture.—"Indonesia to Hawaiki and the Land of Hiwa," by Mr. F. W. Christian, corresponding member of the Polynesian Society of New Zealand.

SIXTH MEETING : *22nd September, 1915.*

Mr. T. King, F.R.A.S., President, in the chair, and about twenty-five members and friends present.

Addresses.—"The Manufacture of Searchlight Carbons,' and " The Preparation of Morphia from Confiscated Opium," by Professor T. H. Easterfield, M.A., Ph.D.

Papers.— 1. "East Coast Earthquakes, September to November, 1914," by Mr. G. Hogben, C.M.G., F.G.S.

2. "Catalogue of Changes proposed in the New Zealand Flora (Vascular Plants only) since the Appearance of Cheeseman's Manual in 1906," by Dr. L. Cockayne, F.R.S.

3. " Notes on the Plant Ecology of the Awatere River Basin, together with a List of the Vascular Plants," by Dr. L. Cockayne, F.R.S.

4. "Notes on the Occurrence of the Genus *Trachipterus* in New Zealand," by Mr. H. Hamilton.

5. "Records of Unconformities from Late Cretaceous to Early Miocene in New Zealand," by Mr. P. G. Morgan, M.A.

6. " The Continental Shelf," by Dr. C. A. Cotton, F.G.S.

7. " Notes on some of the Coast Features of New Zealand," by Dr. C. A. Cotton, F.G.S.

8. " Notes on *Puccinea otagoensis* found on *Clematis*," by Miss H. Jenkins, M.A.

9. " Early Stages in the Development of *Dolichoglossus otagoensis*," by Professor H. B. Kirk, M.A.

10. "On the Gonoducts of the Porcupine-fish (*Dicotylichthys jaculiferus* Cuvier)," by Professor H. B. Kirk, M.A.

11. On Stage Names applicable to the Divisions of the Tertiary in New Zealand," by Dr. J. Allan Thomson, F.G.S.

ANNUAL GENERAL MEETING : *27th October, 1915.*

Mr. T. King, F.R.A.S., President, in the chair, and twenty-eight members and friends present.

Annual Reports.—The annual report and balance-sheet, the report of the Hamilton Memorial Committee, the report of the Library Committee, and the annual reports of the Astronomical, Technological, and Geological Sections were read and adopted

ABSTRACT OF THE ANNUAL REPORT.

During the year there have been eight general meetings of the society, seven meetings of the Astronomical, eight of the Geological, and seven of the Technological Section.

At the general meetings seven lectures or addresses have been delivered, and twenty-five papers presented, which may be classified as follows : Ethnology, 5 ; botany, 10 ; geology, 9 ; ichthyology, 1 ; zoology, 4 ; chemistry, 2 ; and 1 of a more general character. The average attendance at the general meetings was over fifty.

The Sections.—The Astronomical, Geological, and Technological Sections have been very active, and a number of important papers have been read before them.

Membership.—Two members of the society have died during the year, fifteen have resigned their membership, four have been struck off the roll. Twelve new members have been elected. The roll at present contains 169 names, including those of ten life members and of seven on active service. Members on active service retain the full privileges of membership (including the right to the annual volume) without payment of any subscription during the time they are on active service.

Hamilton Memorial.—The fund collected to provide a memorial to the late Mr. Augustus Hamilton, amounting now to £122 2s. 10d., is deposited in the Post Office Savings-bank. Arrangements are now in progress for the erection of a suitable monolith over the grave at Russell, in the Bay of Islands.

Finance.—The receipts during the year amounted to £189 2s. 9d., and the total payments to £172 18s. 2d., including £60 17s. 4d. spent on the library. The Life Subscription Fund, with accrued interest, amounts to £80 7s. 1d., and the Research Fund to £47 12s. 2d. These two funds are invested with the Public Trustee. A sum of £122 2s. 10d. has been deposited in the Post Office Savings-bank, leaving a balance of £43 8s. 8d. in the current account at the Bank of New Zealand.

The report of the Astronomical Section shows that the Proctor Library Fund in connection with the proposed Solar Physics Observatory now amounts to £80 11s. 10d.

Astronomical Section.—The meetings have been well attended, the average number of members present at each meeting being twenty-five. The total number of members of the Philosophical Society who are registered as members of the Astronomical Section is sixty. The observatory at Kelburn has been open to the general public on fine Tuesday evenings from 7.30 to 9.30 p.m. During the absence of Dr. C. E. Adams, Mr. C. J. Westland, Acting Government Astronomer, has acted as Director and Curator of Instruments. Some very fine photographs of star clusters, nebulae, &c., have been presented to the section by Mr. A. C. Gifford, and these are now hung on the walls of the ante-room. Arrangements have been made with the Dunedin Astronomical Society to exchange papers of interest, and two such papers have been read during the year. The credit balance of the section, as shown by the balance-sheet, stands at £32 5s. 7d.

Geological Section.—On two occasions since the last annual meeting the section has been favoured with addresses by visiting geologists of note—Professors W. M. Davis and J. P. Iddings. A considerable number of papers have been read by members, and many interesting exhibits have been made.

Technological Section.—The active membership of the section is about fifty, and an average attendance of about twenty-five has been the rule. During the year nine papers were read, all of high character and interesting nature. The question of a technical library in Wellington is under consideration, and there are good prospects that a definite result will ensue next year, as a practicable scheme has been formulated.

Revision of Rules.—Dr. J. A. Thomson presented the report of the Revision of Rules Committee, and gave notice that he would move the adoption of the rules suggested by the Committee.

Election of Officers for 1916.—*President*—·Mr. Thomas King; *Vice-Presidents*—Dr. C. M. Hector and Dr. J. A. Thomson; *Council*—Mr. C. G. G. Berry (Chairman, Astronomical Section), Mr. G. Hogben (Chairman, Geological Section), Mr. E. Parry (Chairman, Technological Section), Dr. L. Cockayne, Professor H. B. Kirk, Professor T. H. Easterfield, Mr. B. C. Aston, Mr. P. G. Morgan, Dr. C. A. Cotton, Mr. S. H. Jenkinson; *Secretary and Treasurer*—·Mr. A. C. Gifford; *Auditor*—Mr. E. R. Dymock.

Address.—"Block Mountains in New Zealand," by Dr. C. A. Cotton, F.G.S.

Papers.—1. "Records of Unconformities from Late Cretaceous to Early Miocene in New Zealand," by Mr. P. G. Morgan, M.A.

2. "Maori and Maruiwi," by Mr. Elsdon Best.

3. "On the Much-abbreviated Development of a Sand-star (*Ophionereis schayeri?*)," by Professor H. B. Kirk, M.A.

4. "The ' Red Rocks ' and Associated Beds of Wellington Peninsula," by Mr. F. K. Broadgate, M.Sc.

5. "Some Hitherto-unrecorded Plant-habitats (No. 10)," by Dr. L. Cockayne, F.R.S.

6. "Notes on New Zealand Floristic Botany, including Descriptions of New Species (No. 1)," by Dr. L. Cockayne, F.R.S.

7. "Preliminary List of *Mollusca* from Dredgings taken off the Northern Coasts of New Zealand," by Miss Marjorie K. Mestayer; communicated by Dr. J. Allan Thomson, F.G.S.

8. "Additions to the Knowledge of the Recent and Tertiary *Brachiopoda* of New Zealand and Australia," by Dr. J. Allan Thomson, F.G.S.

9. "On the Geology of the Neighbourhood of Kakanui, Otago," by Mr. G. Uttley, M.Sc.; communicated by Dr. J. Allan Thomson, F.G.S.

10. "On the Flint-beds associated with the Amuri Limestone of Marlborough," by Dr. J. Allan Thomson, F.G.S.

11. "Note on Matai Beer," by Professor Easterfield and Mr. J. C. McDowell, B.Sc.

12. "Studies in the Chemistry of the New Zealand Flora : Part V—The Chemistry of *Podocarpus totara* and *Podocarpus spicatus*," by Professor Easterfield and Mr. J. C. McDowell, B.Sc.

13. "Block Mountains and a ' Fossil ' Denudation Plain in Northern Nelson," by Dr. C. A. Cotton, F.G.S.

14. "List of *Foraminifera* dredged from 15' South of the Big King at 98 Fathoms Depth," by R. L. Mestayer.

SPECIAL GENERAL MEETING : *8th December, 1915.*

Mr. T. King, F.R.A.S., President. in the chair, and about fifty-five members and friends present.

Lecture.—" The Physiology of Scenery," by Dr. L. Cockayne, F.R.S.

Papers.—(1) " A Comparison of the Montane Floras of the North Island," (2) " Plant-habitats Hitherto Unrecorded," and (3) " Wellington Island Florulas," by Mr. B. C. Aston, F.C.S., F.I.C.

ASTRONOMICAL SECTION.

Seven meetings were held, and the following papers were read : (4th November, 1914) " Magnetism of the Sun," by Rev. I. von Gottfried; " Ball's Theory of the Great Ice Age," by Mr. R. Gilkison (by arrange-ment with the Dunedin Astronomical Society) : (5th May, 1915) " Note on Kappa Crucis," by Mr. A. C. Gifford, M.A., F.R.A.S.; " Recent Astronomy," by Rev. P. Fairclough (by arrangement with the Dunedin Astronomical Society) : (2nd June, 1915) " Solar Radiations," by Rev. I. von Gottfried; " Spectra of Helium and Hydrogen," by Professor E. Marsden, D.Sc. : (7th July, 1915) " The Seasonal Variations in the Duration of Twilight," by Professor D. M. Y. Sommerville, M.A., D.Sc., F.R.S.E. : (4th August, 1915) " Southern Variable Stars," by Mr. C. J. Westland, F.R.A.S. : (1st September, 1915) " Some Engineering Problems of Mars," by Mr. W. S. La Trobe, M.A. : (6th October, 1915) " Some Points in the Theory of Optical Instruments," by Professor E. Marsden, D.Sc.

At the annual general meeting (6th October, 1915) the following officers for 1916 were elected : *Chairman*—Mr. C. G. G. Berry; *Hon. Member of Section*—Miss Mary Proctor, F.R.A.S.; *Vice-Chairmen*—Mr. C. P. Powles, Professor E. Marsden, D.Sc., and Mr. C. J. Westland, F.R.A.S.; *Committee*—Mr. G. Hogben, C.M.G., M.A., F.G.S., Mr. E. Parry, B.Sc., M.I.E.E., A.M.Inst.C.E., Professor D. M. Y. Sommerville, M.A., D.Sc., F.R.S.E., Dr. C. M. Hector, B.Sc., Mr. A. C. Gifford, M.A., F.R.A.S., Mr. W. S. La Trobe, M.A., Captain G. Hooper; *Director and Curator of Instruments*—Mr. C. J. Westland, F.R.A.S.; *Hon. Treasurer*—Mr. C. P. Powles; *Hon. Secretary*—Mr. E. G. Jones, B.A.

GEOLOGICAL SECTION.

At a special meeting on the 31st March, 1915, Professor J. P. Iddings delivered an address entitled " The Mechanics of Igneous Intrusion." Seven ordinary meetings were held during the year, and the following papers were read : (21st April) " The ' Red Rocks ' and Associated Beds in the Wellington Peninsula," by Mr. F. K. Broadgate : (19th May) " Types of Folding in the *Terebratulacea*," by Dr. J. A. Thomson; " The Amuri Limestone and Flint Series of Marlborough," by Dr. J. A. Thomson : (16th June) " Fault Coasts—Examples from Marlborough and Wellington," by Dr. C. A. Cotton : (21st July) " The Weka Pass District," by Mr. P. G. Morgan; " Evolutionary Stocks in New Zealand Tertiary *Brachiopoda*," by Dr. J. A. Thomson : (18th August) " The Structure of the Paparoa Range," by Dr. J. Henderson : (15th September) " East Coast Earthquakes, September to November, 1914," by Mr. G. Hogben : (20th October) " High-water Rock-platforms : A Phase of Shore-line Erosion," by Mr. J. A. Bartrum; " Stage Names appli-cable to Divisions of the Tertiary in New Zealand," by Dr. J. A. Thomson; " The Continental Shelf," by Dr. C. A. Cotton; " An Artesian Trial Bore at the Westshore, Napier," by Mr. R. W. Holmes.

At the annual general meeting (18th August) the following officers were elected for 1916 : *Chairman*—Mr. G. Hogben; *Vice-Chairman*—Dr. C. A. Cotton; *Hon. Secretary*—Dr. J. A. Thomson; *Committee*—Mr. W. Gibson, Dr. J. Henderson, Mr. R. W. Holmes, Mr. P. G. Morgan, Mr. M. Ongley. It was resolved that the Government should be ap-proached with a view to continuing the seismological observations at the Samoa Seismological Station.

TECHNOLOGICAL SECTION.

The following papers were read during the year: (12th May) "Technical Education," by Mr. W. S. La Trobe, M.A.: (9th June) "Fluid Friction in Pipes," by Mr. E. Parry, B.Sc.: (14th July) "The Electron Theory of the Conduction of Electricity," by Professor Marsden: (11th August) "The Collection of Hydrographic Data for Engineering Problems, with Special Reference to the Upper Taieri Basin," by Mr. F. W. Furkert, A.M.Inst.C.E.: (8th September) "Some Tests of Heat and Electrical Insulators," by Mr. G. B. Dall; "The Regulation of Water Turbines," by Mr. A. D. Cook, M.Sc.: (13th October) "The Manufacture of Iron and Steel in New Zealand," by Mr. S. H. Jenkinson: (10th November) "An Extension of the Theory of Fluid Friction," by Mr. E. Parry, B.Sc.; "The Distribution of Titanium, Phosphorus, and Vanadium in Taranaki Ironsand," by Mr. W. Donovan, M.Sc.

The Committee for 1916 was elected as follows: *Chairman*—Mr. E. Parry, B.Sc.; *Vice-Chairmen*—Mr. J. Marchbanks, M.Inst.C.E., Mr. F. W. Furkert, A.M.Inst.C.E.; *Committee*—Mr. A. Atkins, F.R.I.B.A., Mr. W. Ferguson, M.Inst.C.E., Mr. R. W. Holmes, M.Inst.C.E, Professor Marsden, Mr. W. Morton, M.Inst.C.E., Mr. H. Sladden (of Surveyors' Board); *Hon. Secretary*—Mr. S. H. Jenkinson.

AUCKLAND INSTITUTE.

FIRST MEETING: *7th June, 1915.*

Professor H. W. Segar, Vice-President, in the chair.

New Members.—Messrs. H. Atkinson, T. Crook, T. S. Culling, J. P. Grossmann, J. O. Horning, G. Knight, J. L. McColl, G. S. Poole, P. Upton, F. Whittome.

Lecture.—"Europe, 1815 and 1915: a Survey and a Contrast," by Professor J. P. Grossmann, M.A.

SECOND MEETING: *5th July, 1915.*

Hon. E. Mitchelson, President, in the chair.

Lecture.—"Constantinople," by E. D. Mackellar, M.D.

THIRD MEETING: *2nd August, 1915.*

Hon. E. Mitchelson, President, in the chair.

New Members.—Messrs. R. Jacobson, M.A.; G. W. Murray; J. A. Warnock.
Lecture.—"Crystals," by Professor F. P. Worley, M.Sc.

FOURTH MEETING: *30th August, 1915.*

Hon. E. Mitchelson, President, in the chair.

Lecture.—"The Elizabethan Debt to Rome," by Professor H. S. Dettmann, M.A.

FIFTH MEETING: *27th September, 1915.*

Hon. E. Mitchelson, President, in the chair.

Lecture.—"Modern Views of Matter," by Professor G. Owen, D.Sc.

SIXTH MEETING: *25th October, 1915.*

Hon. E. Mitchelson, President, in the chair.

New Member.—Mr. A. F. Ellis.
Lecture.—"The Newer Physiology," by Dr. Kenneth Mackenzie, F.R.C.S.

SEVENTH MEETING : *8th November, 1915.*

Hon. E. Mitchelson, President, in the chair.

Lecture.—"Maori Voyagers and their Vessels," by Mr. Elsdon Best. In the absence of the author the lecture was read by Mr. G. A. Hansard.

EIGHTH MEETING: *8th December, 1915.*

Hon. E. Mitchelson, President, in the chair.

New Members.—Messrs. E. C. Blomfield, C. A. Whitney.

Papers.—1. "Further Additions to the Flora of the Mongonui County," by Mr. H. Carse.

2. "New Species of Plants," by Mr. T. F. Cheeseman.

3. "Descriptions of New Native Phanerogams," by Mr. D. Petrie.

4. "New Genera and Species of *Coleoptera*," by Major T. Broun.

ANNUAL MEETING : *28th February, 1916.*

Hon. E. Mitchelson, President, in the chair.

Annual Report.—The annual report and audited financial statement was read to the meeting, and ordered to be printed and distributed among the members.

ABSTRACT

Members.—The number of new members added to the roll during the year has been fourteen. Against this, thirty-five names have been withdrawn—six from death, twenty-two from resignation or removal from the district, and seven from non-payment of subscription for more than two consecutive years. The net loss has thus been twenty-one, reducing the number on the roll from 356 to 335.

Among the members removed by death it is the painful duty of the Council to mention the names of Mr. A. E. T. Devore, for many years a consistent supporter of the society; of Mr. W. C. C. Spencer, of Mr. A. Wiseman, and of Mr. W. Coleman. One member, Mr. S. B. Bowyer, has been killed in action in the Dardanelles while serving his King and country; and another, Dr. T. C. Savage, died from sickness in Egypt while engaged in a similar capacity in the medical service of the Army. At the present time no less than seventeen members of the Institute are serving in the Expeditionary Forces equipped and maintained by the Dominion.

Finance.—Balance-sheets showing the financial position of the Institute are appended to this report, but it may be convenient to present a brief synopsis here. The total revenue credited to the Working Account, excluding the balance in hand at the beginning of the year, and also omitting for the present the particulars of a temporary advance made from the Investment Account to cover the cost of fitting up the new Foreign Ethnographical Hall, has been £1,483 17s. 5d. The amount for the previous year was £1,530 17s. 7d., so that there is a deficiency of £47 0s. 2d. Examining the separate items, it will be found that the members' subscriptions have reached £308 14s., against £322 7s. obtained last year. The slight decrease in the membership already alluded to is sufficient explanation for this. The receipts from the Museum endowment have been £550 16s. 10d., last year's amount being £537 14s. 5d. The invested funds of the Costley Bequest have realized £441 15s., against £480 15s. credited last year. The difference is partly due to a temporary delay in the payment of two items of interest, and partly to the fact that one of the securities has been discharged, thus causing a loss of interest for a brief period. The total expenditure has been £1,420 7s., but this does not include payments to the amount of £561 5s. 4d. on account of the expense of fitting up the new Foreign Ethnographical Hall, which have been met by means of an advance from the Investment Account. The cash balance in hand at the present time amounts to £282.

The position of the invested funds of the society must be regarded as satisfactory. Such funds consist of those comprised under the headings Costley Bequest, Museum Endowment Account, Mackechnie Bequest, Campbell Bequest, and one or two minor divisions, and include all capital the annual income from which can alone be used for the purposes of the society. The total of these funds has been materially increased during the year by the sale of some endowments, and now amounts to £21,457 18s. 5d., almost the whole of which is invested in specially selected mortgages or Government debentures.

Meetings.—Eight meetings have been held during the year, at which eleven papers were read and discussed.

Museum.—The attendance of visitors has been satisfactory, showing a slight increase over the figures for last year.

Since the establishment of a Municipal Art Gallery, and the association with it of the Mackelvie Gallery, it had become evident that the growth and management of art collections in Auckland would be most satisfactorily conducted by the City Council. Similarly, the great expansion in recent years of the Maori collections in the Museum clearly indicated the desirability of concentrating within the Museum all the ethnographical collections of the city. It was therefore decided to advocate a proposal to place the Russell collection of statues in the Art Gallery, and the Grey Maori collection in the Museum. The matter was sympathetically received by the City Council, and it was decided that the Russell collection should be handed over to the city on deposit, and the Grey collection to the Museum, each body retaining the actual ownership of its articles, and preserving the right of withdrawal if circumstances should ever make such a course necessary. At a later date the Council decided to grant a request by the City Council to deposit for exhibition in the Old Colonists' Museum a series of 127 pictures, drawings, photographs, historical documents, sets of old newspapers, &c., bearing on the early history of Auckland.

As soon as the removal of the statues placed the hall at the disposal of the Council it was decided to utilize it for the reception, in the first place, of the Grey Maori collection, which under the agreement with the City Council must be kept separate from the Maori collections belonging to the Museum, and, secondly, for the display of the fine series of foreign ethnographical articles in the possession of the Museum, a large proportion of which has never been exhibited. Much care has been taken in designing the show-cases and other fittings required, and only the best material and the best workmanship has been admitted. The total cost has been about £600, which amount has been temporarily borrowed from the invested funds of the society, with the understanding that it shall be returned in instalments as rapidly as possible, regular interest being payable on the sum outstanding.

Many additions of importance have been made to the Museum during the year. In the zoological department a special group has been prepared illustrating the habits and mode of life of the North Island kiwi (*Apteryx mantelli*). It includes several excellent specimens of the adults of both sexes and of the young, together with the nest and eggs. It represents a little glade in the Waitakerei Forest at the base of a large rata-tree, around which the kiwis are arranged. Another conspicuous addition is a specimen of the round-snouted swordfish (*Histiophorus herschelli*), caught by Mr. Campbell off Cape Brett, and kindly presented by him to the Museum. A painted plaster cast has been prepared of the well-known frost-fish (*Lepidopus caudatus*), based upon a remarkably fine specimen forwarded from Mercury Bay by Mr. W. Bonella. Reference should also be made to a series of nineteen skins of Chatham Island birds, purchased from Mr. S. Dannefaerd.

The most important accession to the Maori collections is the huge carved gateway, over 21 ft. in height, of the ancient pa at Te Koutou, Lake Okataina, which has been purchased from the Maori owners. So far as can be ascertained, it was carved prior to 1820, and thus may possibly be well over a hundred years of age. It is known to have been standing at the time of Hongi's raid on the Rotorua district in 1824. It was observed when the first missionaries reached Rotorua a few years later, and a rough sketch of it is given in Terry's "New Zealand," published in 1843. Other interesting additions are the stern-piece of an ancient war-canoe, dug up near the bottom of a deep drainage-canal on the Hauraki Plains, and presented by Mr. G. A. Hodge; a bone manaia, the figurehead of a small river-canoe, and various other articles donated by Mr. G. Graham; a carved burial-chest of unique type, an elaborately carved stern-post of a war-canoe, several ancient albatross-hooks, and other specimens purchased from various individuals.

In foreign ethnography, Dr. Bucknill, of Tauranga, has presented a valuable and comprehensive series of seventy-nine selected specimens illustrating the development of English glassware from the time of the Roman occupancy to the eighteenth century. Another donation of special value has been received from Mr. Henry Shaw, well known in Auckland by his previous benefactions both to the Institute and the Free Library. It consists of a collection of 131 Japanese ivories, bronzes, and cloisonné work, and contains many examples of indisputable age and of much ethnographic and artistic excellence.

Among numerous other additions the following deserve special mention: Mr. A. F. Ellis has contributed several specimens from Ocean Island, of the Gilbert and Ellice Protectorate; Mr. J. L. Young has presented two stone carvings of considerable size from eastern Polynesia; Miss Morrisby has donated an excellent little series of thirty-three specimens from South Africa; Mrs. Reid, of Motutapu Island, has presented a Polynesian collection of over 250 specimens, as well as a large series of shells; and, finally, Mr. Cameron, the Resident at Aitutaki, has forwarded two chiefs' carved seats, said to be the last remaining of that type on any of the islands in the Cook Group.

Library.—An expenditure of £141 3s. 8d. has been incurred on the library during the year. A consignment of ninety volumes was received in July. An order of rather larger size was dispatched during November last, but no advice has yet been received of its shipment. Various books and memoirs have been received in exchange, and several donations have been made by private individuals. Among the latter the Council have pleasure in mentioning Shelley's slendid monograph of the sun-birds, with 121 coloured plates, presented by Mr. H. Shaw.

Election of Officers for 1916.—President—Hon. E. Mitchelson; *Vice-Presidents*—C. J. Parr, C.M.G., M.P., Professor H. W. Segar; *Council*—Professor C. W. Egerton, Mr. J. Kenderdine, Mr. E. V. Miller, Professor G. Owen, Mr. T. Peacock, Mr. D. Petrie, Mr. J. A. Pond, Professor A. P. W. Thomas, Mr. J. H. Upton, Professor F. P. Worley, Mr. H. E. Vaile; *Trustees*—Messrs. T. Peacock, J. Reid, J. H. Upton; *Auditor*—Mr. S. Gray.

PHILOSOPHICAL INSTITUTE OF CANTERBURY.

Fɪʀsᴛ Meeting : *5th May, 1915.*

Present: Dr. Chilton, in the chair, and seventy-five others.

Ex-Presidential Address.—" Nitrates and the War," by Dr. W. P. Evans.

Seᴄᴏɴᴅ Meeting : *2nd June, 1915.*

Present: Mr. A. D. Dobson, President, in the chair, and sixty others.

New Members.—Mrs. Humphreys, Messrs. G. E. Blanch, H. T. Ferrar, W. O. R. Gilling, W. Martin, P. S. Nelson, H. Rands, H. V. Rowe, and George Scott, Drs. H. T. Thacker, J. P. Whetter, and J. C. Pairman.

Address.—" Flight," by Professor R. J. Scott.

Tʜɪʀᴅ Meeting : *7th July, 1915.*

Present: Mr. A. D. Dobson, President, in the chair, and forty-five others.

New Members.—Messrs. W. Murray, C. MacIndoe, E. E. Stark, J. Stevenson, R. E. Alexander, H. A. Knight, A. V. Mountford, John W. Garton, H. D. Broadhead, and Rev. A. T. Thompson.

Papers.—1. " On an Exhibit of Acorns and Leaves of Oaks grown by the Author at Greendale, Canterbury, New Zealand," by Mr. T. W. Adams.

2. " The Norfolk Island Species of *Pteris,*" by Mr. R. M. Lang.

3. " Investigations into the Resistance of Earth Connections," by Messrs. L. Birks and Eric Webb.

Fᴏᴜʀᴛʜ Meeting : *4th August, 1915.*

Present: Mr. A. D. Dobson, President, in the chair, and forty others.

New Members.—Messrs. H. C. Brent and C. E. St. John.

Microscopical Evening, under the direction of Mr. C. B. Morris.

FIFTH MEETING : *1st September, 1915.*

Present : Mr. A. D. Dobson, President, in the chair, and forty others.

Papers.—1. " A Note on the Estimation of the Increase of Iron Loss with Load in a Direct-current Machine," by Mr. P. H. Powell.

2. " On the Inheritance of Wool," by Mr. H. T. Ferrar.

3. " On the Rate of Growth of certain English Trees," by Mr. E. F. Stead.

4. " The Orientation of the River-valleys of Canterbury," by Mr. R. Speight.

5. " On the Inscribed Parabola," by Mr. E. G. Hogg.

SIXTH MEETING : *6th October, 1915.*

Present : Mr. A. D. Dobson, President, in the chair, and sixty others.

Address. — " Biology and Economics of Bread," by Dr. F. W. Hilgendorf.

SEVENTH MEETING : *3rd November, 1915.*

Present : Mr. A. D. Dobson, President, in the chair, and thirty others.

New Member.—Miss Ferrar.

Papers.—1. " Notes on some *Coccidae* in the Canterbury Museum," by Mr. G. Brittin.

2. " New *Coccidae*," by Mr. G. Brittin.

3. " Studies in the New Zealand Species of the Genus *Lycopodium*, Part I," by Rev. J. E. Holloway.

4. " Observations on the Lianes of the Ancient Forest of the Canterbury Plains," by Mr. J. W. Bird.

5. " Notes on the Marine Crayfish of New Zealand," by Mr. G. E. Archey.

6. " A New Species of *Orchestia*," by Dr. Charles Chilton.

7. " Some Australian and New Zealand *Gammaridae*," by Dr. Charles Chilton.

8. " Physiography and Geological History of Banks Peninsula," by Mr. R. Speight.

ANNUAL MEETING : *1st December, 1915.*

Present : Mr. A. D. Dobson. President, in the chair, and forty others.

Annual Report.—The annual report and balance-sheet were adopted.

ABSTRACT.

The number of Council meetings held during the year was eleven. The Council nominated Dr. Charles Chilton to be representative of this Institute on the Board of Trustees of the Riccarton Bush. On the suggestion of the Council of the Institute, the Mayor accorded a civic welcome to the scientific members of the Magnetic Survey vessel " Carnegie."

Eight meetings of the Institute have been held during the year, at which the following addresses were delivered : "Nitrates and the War" (ex-presidential address), by Dr. W. P. Evans; "Flight," by Professor R. J. Scott; "Biology and Economics of Bread," by Dr. F. W. Hilgendorf; also a "Microscopical Evening" was held under the direction of Mr. C. B. Morris. In addition to these, twenty-one papers have been read, which may be classified as follow : Botany, 7; geology, 4; mathematics, 1; zoology, 6; engineering, 2; chemistry, 1.

During the year twenty-six new members have been elected, and eleven have either resigned or have been struck off the roll, so that the number now stands at 179.

The Council desires to place on record that the following members of the Institute are now on active service in various parts of the Empire : Hon. R. Heaton Rhodes, Drs. Acland, Irving, and Whetter, Messrs. L. S. Jennings, H. Lang, F. S. Wilding, A. Taylor, and Major A. A. Dorrien Smith, D.S.O.

Arthur's Pass Tunnel Investigation : The usual temperature observations have been continued, and specimens of rock were received for examination, thus keeping the series complete.

The Council at several of its meetings during the year had under consideration the question of the publication of an account of the natural history of Canterbury, but after due consideration, while approving of the scheme, the Council decided that, in view of the present war-conditions, further consideration of the matter should be postponed until 1916. The Council desires to express its appreciation to Mr. T. D. Burnett, who has offered to donate the sum of £10 towards a certain portion of this investigation.

The Institute's representative on the Board of Trustees of the Riccarton Bush reports that the control of the bush has been taken over by the trustees and a Ranger appointed. The bush has been securely fenced, elderberry and other introduced plants are being carefully removed, and a few narrow paths have been made. It is hoped, as soon as sufficient funds are available, to erect a cottage for the Ranger, and to allow the bush to be visited by members of the public under the conditions necessary for its due preservation.

The library has been maintained in an efficient condition during the past year, and has been considerably increased by various gifts and purchases.

The balance-sheet shows that during the year the receipts were £227 2s. 11d. This includes a balance of £22 11s. 5d. carried forward from last year, and a sum of £52 transferred to the ordinary account from deposit with the Permanent Investment and Loan Association. The expenditure amounted to £112 13s. 3d., of which £73 12s. was spent on the library. The balance to the credit of the Institute in the Bank of New Zealand stands at £114 9s. 8d. The Council has decided to accord all members who are on active service the privilege of membership without payment of subscriptions.

Election of Officers for 1916.—President—Mr. L. Birks; *Vice-Presidents*—Messrs. A. D. Dobson and R. Speight; *Secretary*—Mr. A. M. Wright; *Treasurer*—Dr. Charles Chilton; *Librarian*—Mr. E. G. Hogg; *Council*—Drs. F. J. Borrie, C. Coleridge Farr, and F. W. Hilgendorf, and Messrs. R. M. Laing, G. E. Archey, and L. P. Symes; *Representatives on the Board of Governors of the New Zealand Institute*—Dr. Hilgendorf and Mr. A. M. Wright; *Auditor*—Mr. G. E. Way.

Papers.—1. "Studies on the Lime Requirements of certain Soils." by Mr. L. J. Wild.

2. "The Succession of Tertiary Beds in the Pareora District." by Mr. M. C. Gudex.

3. "Notes from the Canterbury College Mountain Biological Station": "No. 2, The Physiography of the Cass District, by Mr. R. Speight; "No. 3, Some Economic Considerations concerning Montane Tussock Grassland," by Mr. A. H. Cockayne; "No. 4, The Principal Plant Associations in the Immediate Neighbourhood of the Station," by Dr. L. Cockayne and Mr. C. E. Foweraker.

OTAGO INSTITUTE.

First Meeting: *4th May, 1915.*

Present: Mr. R. Gilkison, President, in the chair, and about a hundred members and friends.

Address.—"Recent Climbs in the Southern Alps," by Mr. H. F. Wright.

Second Meeting: *1st June, 1915.*

Present: Mr. R. Gilkison, President, in the chair, and about forty members and friends.

New Members.—Professor W. P. Gowland, M.D., and Mr. O. J. W. Napier, M.A.

Presidential Address.—"The Rise and Fall of Nations," by Mr. R. Gilkison.

Third Meeting: *6th July, 1915.*

Present: Mr. R. Gilkison, President, in the chair, and about thirty members and friends.

New Members.—Messrs. J. C. Begg, W. T. Monkman, and S. P. Seymour, B.A.

Address.—"Modern Problems of Chemistry," by Dr. J. K. H. Inglis, F.I.C.

Fourth Meeting: *3rd August, 1915.*

Present: Mr. R. Gilkison, President, in the chair, and about twenty members.

Address.—"Friedrich Nietzsche," by Professor F. W. Dunlop, M.A., Ph.D.

Fifth Meeting: *7th September, 1915.*

Present: Mr. R. Gilkison, President, in the chair, and about twenty members.

Papers.—1. "Notes on the Plant-covering of Breaksea Islands," by Mr. D. L. Poppelwell.

2. "Notes on the Plant-covering of Pukeokaoka, Stewart Island," by Mr. D. L. Poppelwell.

3. "Descriptions of New Species of *Lepidoptera*," by Mr. A. Philpott; communicated by Dr. W. B. Benham, F.R.S.

4. "Notes on New Zealand *Polychaeta*, Part II." by Dr. W. B. Benham, F.R.S.

18*

SIXTH MEETING : *5th October, 1915.*

Present: Mr. R. Gilkison, President, in the chair, and about forty members and friends.

Hector Medal.—Mr. G. M. Thomson, a past President of the New Zealand Institute, presented to Dr. P. Marshall, F.G.S., the Hector Memorial Medal awarded to him by the New Zealand Institute for his researches in New Zealand geology.

Address.—"Work at the Marine Fish-hatchery," by Mr. G. M. Thomson, F.L.S.

SEVENTH MEETING: *7th December, 1915.*

Present: Dr. W. B. Benham, F.R.S., Vice-President, in the chair, and about fifteen members.

New Members.—Messrs. J. M. Lowry, J. McNair. and J. E. Wingfield, and Dr. J. T. Bowie.

Papers.—1. "Notes on the New Zealand Cuckoo," by Mr. W. W. Smith; communicated by Mr. G. M. Thomson.

2. "A List of the *Lepidoptera* of Otago," by Mr. A. Philpott; communicated by Dr. W. B. Benham, F.R.S.

3. "The Younger Limestones of New Zealand," by Professor P. Marshall, D.Sc., F.G.S.

4. "Some New Fossil Gastropods," by Professor P. Marshall, D.Sc., F.G.S.

5. "Relations between Cretaceous and Tertiary Rocks," by Professor P. Marshall, D.Sc., F.G.S.

6. "Notes on a Botanical Visit to Bold Peak, Humboldt Mountains," by Mr. D. L. Poppelwell.

7. "The Occurrence of a Striated Erratic Block of Andesite in the Rangitikei Valley," by Professor J. Park, F.G.S.

Annual Report.—The annual report and the balance-sheet for 1915 were read and adopted.

ABSTRACT.

During the year the Council has met six times for the transaction of the business of the Institute.

Early in the year the Council decided to vote from its accrued funds the sum of £100 to the Portobello Marine Fish-hatchery Board, for the purpose of enabling it to prosecute researches on the fauna of our New Zealand seas, provided that the Government would supplement this grant with a pound-for-pound subsidy. The Minister, on being approached, was in sympathy with the proposal, and the desired grant was finally placed on the supplementary estimates. The Council has since paid over its promised donation to the Hatchery Board.

Meetings.—During the year seven ordinary meetings of the Institute have been held, at which there have been read or received eleven papers, embodying the results of original research.

The following addresses have also been delivered during the past session : "The Rise and Fall of Nations" (presidential address), by Mr. R. Gilkison; "Recent Climbs in the Southern Alps," by Mr. H. F. Wright; "Modern Problems of Chemistry," by Professor J. K. H. Inglis; "Friedrich Nietzsche," by Professor F. W. Dunlop; and "Work at the Marine Fish-hatchery," by Mr. G. M. Thomson.

At the October meeting opportunity was taken to present Professor P. Marshall with the Hector Medal (1915) of the New Zealand Institute, the presentation being made by Mr. G. M. Thomson, as a Governor and past President of that Institute. The society, as a body, has been honoured by this recognition of the value of the scientific research so ably carried out in recent years by one of its most active members, and takes this opportunity of placing on record its congratulations to Professor Marshall on his well-merited distinction.

The attendances at the meetings, though somewhat better than they were last year, are still poorer than they should be, considering the society's membership roll. It is to be hoped that the many distractions connected both directly and indirectly with the war will have ceased by next winter's session, so that the meetings may be attended in the way that they deserve to be.

Technological Branch.—The year 1915 is the fifth year of the existence of the Technological Branch. An innovation adopted by the Committee has been to restrict the meetings to the winter months. There will be no meeting in November as formerly; while yet another evening is saved by combining the business of the annual meeting with the final lecture for the year. It has not been thought advisable to attempt to continue the annual dinner for the present. The short-paper evening in September was allowed to lapse. The Committee has had under consideration the question of obtaining better attendances at our meetings by securing a room more central and accessible than the University. The Technical School has been suggested.

Astronomical Branch.—During the year six meetings of the branch have been held, at which seven papers have been read. Although the attendance of the meetings has suffered somewhat as a result of the war, on the whole they have been entirely satisfactory.

The observatory at Tanna Hill has been open every fine Friday night for two hours, a member of the Committee being in charge. The Committee hopes that the time is not far distant when the society will be in a position to secure an up-to-date refracting telescope, with the necessary accessories, so that interested members may be able to do some useful work. On account of the removal of Tanna Hill the Committee has been faced with the problem of securing a fresh site for the observatory, and has decided to approach the City Council on the matter. The Committee recommends a site at the end of Clyde Street, on the rise above the Leith, near the University, and hopes to be able to obtain permission for the re-erection of the observatory on the site chosen by the Committee. The society is grateful to Mr. Skey for the use of his telescope, and to the Otago University Council for the use of the Beverley telescope.

Librarian's Report.—During the year eight new works have been purchased, and five have been presented to the Institute, a smaller number than during last year. The majority of these works are of a scientific character.

As a result apparently of the troublous times, the journal *Bedrock* has ceased publication.

During the year some forty volumes have been bound; some of these were periodicals, others reports of Government Departments, and yet others were monographs which were in paper covers.

Mr. G. M. Thomson has been good enough to arrange with the Linnean Society that the Journal and Transactions of that society, due to him as a Fellow, shall in future be sent to the Institute library. To Mr. Thomson for this and other donations to the library the society's best thanks are due.

Quite recently the Council has presented to the University the eight volumes of Murray's New English Dictionary that have hitherto been in our own library. In a measure, this donation may be regarded as a recognition of the many kindnesses shown to the Institute by the University authorities in recent years.

Membership.—During the year six new members have been elected. On the other hand, nineteen members have resigned their membership, and four members (Dr. F. C. Batchelor, and Messrs. T. W. Kempthorne, R. Price, and John Sidey) have been removed by death. The membership roll, therefore, has suffered a net decrease of seventeen, and now stands at 184.

In connection with the war, it is perhaps worthy of record that twelve members of the Institute have enrolled themselves in the Dominion or Imperial Forces. Drs. Batchelor, Barnett, and O'Neill were given commissions in the New Zealand Medical Corps; Dr. Buddle is serving in the Royal Navy; the Rev. Dutton is Chaplain-Major on the hospital ship "Maheno"; Major R. Price, Major F. H. Statham, Lieutenant W. D. Stewart, and W. P. Macdougall left either in the New Zealand Expeditionary Force or in its reinforcements; Captain D. B. Waters is to sail with the Engineering Tunnelling Corps; Lieutenant T. R. Overton with the 4th Maori Contingent; and Lieutenant E. F. Roberts is in the Ordnance Engineers' at Home Of these, it is with regret that we have to record that Major Price was killed in action at the Dardanelles, and Major Statham has been posted as missing since August; whilst Dr. Batchelor died shortly after his return from Egypt on furlough.

Balance-sheet.—The balance-sheet, presented by the Treasurer (Mr. R. N. Vanes), showed a credit balance of £62 2s. The gross receipts totalled £732, including subscriptions amounting to £147, deposits at call amounting to £445 7s.

Election of Officers.—The election of officers for the year 1916 resulted as follows: *President*—Professor P. Marshall; *Vice-Presidents*—Mr. R. Gilkison and Professor J. K. H. Inglis; *Hon. Secretary*—Mr. E. J. Parr; *Hon. Treasurer*—Mr. R. N. Vanes; *Hon. Auditor*—Mr. H. Brasch; *Hon. Librarian*—Professor Benham; *Council*—Professors W. B. Benham, R. Jack, J. Park, Dr. R. V. Fulton, Messrs. W. G. Howes, J. B. Mason, and G. M. Thomson.

Mr. G. M. Thomson and Professor P. Marshall were re-elected representatives of the Institute on the Board of Governors of the New Zealand Institute.

TECHNOLOGICAL BRANCH.

Five meetings were held during 1915, and the following papers and addresses were read: (18th May) "The Future of Otago Harbour," by Mr. J. B. Mason; (15th June) "The Forth Bridge," by Mr. J. B. Mason; (20th July) "The Economics of Agriculture," by Mr. H. Mandeno; (17th August) "The Silting of Waipori River," by Mr. R. T. Stewart; (19th October) "The Strength of Materials," by Professor James Park.

At the meeting of the 10th October the annual report was read and adopted, and the following officers for 1916 were elected: *Chairman*—Mr. J. B. Mason; *Vice-Chairmen*—Professors J. Park and D. B. Waters, and Mr. B. B. Hooper; *Hon. Secretary*—Mr. H. Brasch; *Committee*—Messrs. G. W. Davies, H. Mandeno, W. D. R. McCurdie, G. Simpson, and R. N. Vanes.

ASTRONOMICAL BRANCH.

Six meetings were held during 1915, and the following papers and addresses were read: (25th May) "Cosmological Hypotheses," by Mr. R. T. A. Innes (communicated by Mr. J. W. Milnes): (22nd June) "Everyday Phenomena of Astronomy," by Professor D. J. Richards, M.A.; "The Value of a Little Knowledge of the Sun," by Rev. A. M. Dalrymple, M.A.: (27th July) "The Beginning of Worlds," by Mr. R. Gilkison; (24th August) "An Explanation of some Meteorological Phenomena," by Professor R. Jack, D.Sc.: (21st September) "Some Astronomical Phenomena," by Professor D. J. Richards, M.A.: (26th October) "Atmospheric Refraction," by Mr. W. T. Neill.

At the meeting of the 26th October the annual report was read and adopted, and the following officers for 1916 were elected: *Chairman*—Mr. R. Gilkison; *Vice-Chairmen*—Professors J. Park, D. J. Richards, and R. Jack; *Hon. Secretary*—Mr. J. Bremner; *Committee*—Messrs. H. Brasch, J. W. Milnes, W. T. Neill, W. S. Wilson, and Dr. P. D. Cameron.

HAWKE'S BAY PHILOSOPHICAL INSTITUTE.

Four meetings were held during 1915, and the following papers were read: "Between Two Rivers," by F. Hutchinson, jun.; "The Puketitiri Hot Springs," by F. Hutchinson, jun.; "National Character," by W. Dinwiddie; "Chemistry of some Explosives and Noxious Gases used in Warfare," by J. Niven.

At the annual meeting, 4th December, 1915, the annual report was read and adopted.

ABSTRACT.

The membership at the end of the year numbers seventy-nine. The loss by death of three members is noted—Messrs. Taylor White, J. N. Williams, and E. W. Andrews.

Appreciation of the action of Mr. Gordon in conveying to the Crown the Kidnappers gannet nesting-place as a public reserve is placed on record.

Additions have been made as usual to the library, among which are specially noted the two volumes of "Illustrations of the New Zealand Flora," presented to the library by the Minister of Internal Affairs.

The Treasurer's statement shows a credit balance of £54 5s. 1d.

*Election of Officers for 1916.—President—*Mr. D. A. Strachan, M.A.; *Vice-President—*Mr. W. H. Skinner; *Council—*Messrs. W. Dinwiddie, H. Hill, B.A., F.G.S., F. Hutchinson, jun., W. Kerr, M.A., T. C. Moore, M.D., T. Hyde; *Hon. Secretary—*Mr. J. Niven, M.A., M.Sc. (Technical College, Napier); *Hon. Treasurer—*Mr. J. Wilson Craig (Coote Road, Napier); *Hon. Auditor—*Mr. J. S. Large; *Hon. Lanternist—*Mr. E. G. Loten; *Representative on Board of Governors—*Mr. H. Hill, B.A., F.G.S.

NELSON INSTITUTE.

During 1915 four ordinary general meetings were held, at which the following papers were read: (26th April), "Maori Implements in the Museum," by Mr. F. V. Knapp; (31st May) "Rusts," by Mr. F. Whitwell; (5th July) Miscellaneous Items; (6th September) "Maori Implements in the Museum, Part 2," by Mr. F. V. Knapp.

At the annual meeting, 20th December, 1915, the annual report and balance-sheet were read and adopted.

ABSTRACT.

Early in the year it was decided to open the Atkinson Observatory to the public every Tuesday evening, weather permitting, and, though weather-conditions have been to a great extent unfavourable, the proposal met with fair success, particularly during the visit of Mellish's Comet in June.

It was with great regret that news was received of the death of the late Mr. Thomas Cawthron. During recent years Mr. Cawthron, whose many benefactions to Nelson are well known, had made several gifts of great value to the Museum. Among these may be mentioned a very fine set of show-cases in which to set out the Museum exhibits, and a splendid collection of curios purchased from Mr. Lukins, the chief value of which lay in its collection of Maori implements, some of which are unique.

*Election of Officers for 1916.—President—*Mr. G. J. Lancaster; *Committee—*Messrs. F. G. Gibbs, F. V. Knapp, W. F. Worley, T. A. H. Field, M.P., H. P. Washbourn, J. Strachan, G. R. Wise, and F. Whitwell; *Hon. Secretary and Treasurer—*Mr. E. L. Morley; *Hon. Auditor—*Mr. F. Whitwell.

MANAWATU PHILOSOPHICAL SOCIETY.

During 1915 eight general meetings (including the annual meeting) were held, at which the following papers were read: (11th February) "The Marine Biology of New Zealand," by Dr. Mortensen, of Copenhagen; "Continental Experiences during the War," by Mr. M. A. Eliott: (18th March) "The Effects of Recent Chemical and Physical Research on Astronomy," by Mr. C. T. Salmon: (15th April) "Luminiferous Ether, with Special Reference to Occult Forces, Telepathy, and Wireless Telegraphy," by Mr. J. W. Poynton, S.M.; "The Working of Wireless Telegraphy," by Mr. A. J. Colquhoun, M.Sc.: (21st May) "Two Notable Years, 1815 and 1915," by Mrs. J. H. Primmer: (19th August) "Radio-active Substances," by Mr. A. J. Colquhoun, M.Sc.: (16th September) "Wool in 1815 as compared with 1915," by Mr. M. A. Eliott; "X Rays," by Rev. H. M. Smyth, M.A.: (21st October) "A Survey of Bacteriology, Economic and Clinical," by Mr. J. W. Poynton, S.M.: (29th November) "Recent Progress in Chemical and Physical Research," by Mr. A. J. Colquhoun, M.Sc.; "The Use and Influence of Novel Machines in Warfare," by Mr. C. T. Salmon.

At the annual meeting, 29th November, 1915, the annual report was read and adopted.

ABSTRACT.

It was stated that the continuance of the war had prevented the Government from taking any further action in the matter of the preservation and improvement of the Tongariro Park and the reserves on Wharite.

In April last the Council had decided, on the recommendation of the Curator, to open the Museum on Sunday afternoons instead of Thursday, and the experiment had so far been a success, for, while the average daily attendance for the whole year had been twenty, for the Sundays it had been thirty-one.

During the year 140 fresh exhibits had been received, amongst the most notable of which were a collection of marine fossils from the Dominion Museum, received through the kindness of Dr. Allan Thomson, and one of rocks, presented by Mr. M. W. Walmsley, illustrating the geology of Otago Peninsula, and a valuable collection of native weapons from Fiji, presented by King Cacaban to the late Mr. Hawkins, of Palmerston.

During the year a limited advantage had been taken by the public schools of the free use of the telescope, and classes, accompanied by their teachers, had visited the Observatory.

Election of Officers for 1916.—President—Mr. J. W. Poynton, S.M.; *Vice-Presidents*—Messrs. M. A. Eliott and J. L. Barnicoat; *Officer in charge of the Observatory*—Mr. C. T. Salmon; *Secretary and Treasurer*—Mr. K. Wilson, M.A.; *Council*—Miss Ironside, M.A., and Messrs. R. Gardner, J. B. Gerrand, H. D. Skinner, B.A., W. Park, F.R.H.S., and J. E. Vernon, M.A.; *Auditor*—Mr. W. E. Bendall, F.P.A.

WANGANUI PHILOSOPHICAL SOCIETY.

Six ordinary meetings and two special meetings were held during the year 1915-16, at which the following lectures were delivered and papers read: (19th March, 1915) *Lecture*—" War Explosives," by Professor T. H. Easterfield, M.A., Ph.D. : (12th April, 1915) *Lecture*—" New Zealand Volcanoes," by Professor P. Marshall, D.Sc., F.G.S. : (17th May, 1915) *Papers*—1, "Easter Trip to Tongariro Group," by Mr. T. W. Downes; 2, "At the Observatory," by Mr. J. T. Ward; 3, "The Star Test for Telescopic Mirrors," by Mr. Thomas Allison : (21st June, 1915) *Paper*—" The Classic Architecture of Greece and Rome," by Mr. C. Reginald Ford, F.R.G.S. : (26th July, 1915) *Paper*—" Some Features of the Flora of the Swiss Alps," by Mr. J. A. Neame, B.A. : (29th September, 1915) *Lecture*—" The Architectural Monuments of Belgium," by Mr. S. Hurst Seager, F.R.I.B.A. : (1st November, 1915) *Papers*— 1, "The German Spirit," by Mr. H. E. Sturge, B.A. ; 2," Description of a New Species of *Melanchra*," by Mr. Morris N. Watt, F.E.S. ; 3, "Contributions to the Study of New Zealand Entomology, No. 8," by Mr. Morris N. Watt, F.E.S. : (26th January, 1916) *Papers*— 1, "Impressions of England in War-time," by Dr. Hatherly; 2, "New Light on the Period of the Extinction of the Moa (according to Maori Record)," by Mr. T. W. Downes.

At the annual meeting, 14th February, 1916, the annual report and balance-sheet were adopted.

ABSTRACT OF ANNUAL REPORT.

The two special meetings held were arranged to enable the society to help directly in the work of collecting for the patriotic funds, while simultaneously serving its own special aims. Both meetings were held in the Opera House, the public invited to attend, and a charge made for admission. In each case a considerable sum was handed over to the local Patriotic Committee. The lectures were—(1) "War Explosives," by Professor T. H. Easterfield, M.A. Ph.D., and (2) "The Architectural Monuments of Belgium," by Mr. S. Hurst Seager, F.R.I.B.A.

The roll of the society includes at date fifty-nine ordinary and thirty-six associate members. There is a slight falling-off in numbers, due to resignations, some of them unavoidable owing to members leaving the district.

The financial position is satisfactory, showing a balance on hand of £53 7s. 5d. after paying a subsidy to the Museum of £21 14s. 6d.

It was resolved to suggest to the Board of Governors the desirability of making a collection of interesting lantern-slides in connection with the library of the Institute, with the object of loaning them to the affiliated societies.

Dr. Hatherly was elected a life member.

Election of Officers for 1916.—*President*—Mr. H. R. Hatherly, M.R.C.S.; *Vice-Presidents*—Messrs. J. T. Ward, R. Murdoch; *Council*— Messrs. T. Allison, J. A. Neame, B.A., Morris N. Watt, F.E.S., C. P. Brown, M.A., LL.B., T. W. Downes, H. E. Sturge, B.A., Harry Drew; *Hon. Treasurer*—Mr. E. P. Talboys; *Hon. Secretary*—Mr. J. P. Williamson; *Representative on Board of Governors, New Zealand Institute*—Mr. H. R Hatherly, M.R.C.S.

APPENDIX.

NEW ZEALAND INSTITUTE ACT, 1908.

1908, No. 130.

AN ACT to consolidate certain Enactments of the General Assembly relating to the New Zealand Institute.

BE IT ENACTED by the General Assembly of New Zealand in Parliament assembled, and by the authority of the same, as follows :—

1. (1.) The Short Title of this Act is the New Zealand Institute Act, 1908.

(2.) This Act is a consolidation of the enactments mentioned in the Schedule hereto, and with respect to those enactments the following provisions shall apply :—

(a.) The Institute and Board respectively constituted under those enactments, and subsisting on the coming into operation of this Act, shall be deemed to be the same Institute and Board respectively constituted under this Act without any change of constitution or corporate entity or otherwise ; and the members thereof in office on the coming into operation of this Act shall continue in office until their successors under this Act come into office.

(b.) All Orders in Council, regulations, appointments, societies incorporated with the Institute, and generally all acts of authority which originated under the said enactments or any enactment thereby repealed, and are subsisting or in force on the coming into operation of this Act, shall enure for the purposes of this Act as fully and effectually as if they had originated under the corresponding provisions of this Act, and accordingly shall, where necessary, be deemed to have so originated.

(c.) All property vested in the Board constituted as aforesaid shall be deemed to be vested in the Board established and recognized by this Act.

(d.) All matters and proceedings commenced under the said enactments, and pending or in progress on the coming into operation of this Act, may be continued, completed, and enforced under this Act.

2. (1.) The body now known as the New Zealand Institute (hereinafter referred to as " the Institute ") shall consist of the Auckland Institute, the Wellington Philosophical Society, the Philosophical Institute of Canterbury, the Otago Institute, the Hawke's Bay Philosophical Institute, the Nelson Institute, the Westland Institute, the Southland Institute, and such others as heretofore have been or may hereafter be incorporated therewith in accordance with regulations heretofore made or hereafter to be made by the Board of Governors.

(2.) Members of the above-named incorporated societies shall be *ipso facto* members of the Institute.

3. The control and management of the Institute shall be vested in a Board of Governors (hereinafter referred to as " the Board "), constituted as follows :—

The Governor:

The Minister of Internal Affairs :

Four members to be appointed by the Governor in Council, of whom two shall be appointed during the month of December in every year:

Two members to be appointed by each of the incorporated societies at Auckland, Wellington, Christchurch, and Dunedin during the month of December in each alternate year; and the next year in which such an appointment shall be made is the year one thousand nine hundred and nine :

One member to be appointed by each of the other incorporated societies during the month of December in each alternate year ; and the next year in which such an appointment shall be made is the year one thousand nine hundred and nine.

4. (1.) Of the members appointed by the Governor in Council, the two members longest in office without reappointment shall retire annually on the appointment of their successors.

(2.) Subject to the last preceding subsection, the appointed members of the Board shall hold office until the appointment of their successors.

5. The Board shall be a body corporate by the name of the " New Zealand Institute," and by that name shall have perpetual succession and a common seal, and may sue and be sued, and shall have power and authority to take, purchase, and hold lands for the purposes hereinafter mentioned.

6. (1.) The Board shall have power to appoint a fit person, to be known as the " President," to superintend and carry out all necessary work in connection with the affairs of the Institute, and to provide him with such further assistance as may be required.

(2.) The Board shall also appoint the President or some other fit person to be editor of the Transactions of the Institute, and may appoint a committee to assist him in the work of editing the same.

(3.) The Board shall have power from time to time to make regulations under which societies may become incorporated with the Institute, and to declare that any incorporated society shall cease to be incorporated if such regulations are not complied with ; and such regulations on being published in the *Gazette* shall have the force of law.

(4.) The Board may receive any grants, bequests, or gifts of books or specimens of any kind whatsoever for the use of the Institute, and dispose of them as it thinks fit.

(5.) The Board shall have control of the property from time to time vested in it or acquired by it ; and shall make regulations for the management of the same, and for the encouragement of research by the members of the Institute ; and in all matters, specified or unspecified, shall have power to act for and on behalf of the Institute.

7. (1.) Any casual vacancy in the Board, howsoever caused, shall be filled within three months by the society or authority that appointed the member whose place has become vacant, and if not filled within that time the vacancy shall be filled by the Board.

(2.) Any person appointed to fill a casual vacancy shall only hold office for such period as his predecessor would have held office under this Act.

8. (1.) Annual meetings of the Board shall be held in the month of January in each year, the date and place of such annual meeting to be fixed at the previous annual meeting.

(2.) The Board may meet during the year at such other times and places as it deems necessary.

(3.) At each annual meeting the President shall present to the meeting a report of the work of the Institute for the year preceding, and a balance-sheet, duly audited, of all sums received and paid on behalf of the Institute.

9. The Board may from time to time, as it sees fit, make arrangements for the holding of general meetings of members of the Institute, at times and places to be arranged, for the reading of scientific papers, the delivery of lectures, and for the general promotion of science in New Zealand by any means that may appear desirable.

10. The Minister of Finance shall from time to time, without further appropriation than this Act, pay to the Board the sum of five hundred pounds in each financial year, to be applied in or towards payment of the general current expenses of the Institute.

11. Forthwith upon the making of any regulations or the publication of any Transactions, the Board shall transmit a copy thereof to the Minister of Internal Affairs, who shall lay the same before Parliament if sitting, or if not, then within twenty days after the commencement of the next ensuing session thereof.

SCHEDULE.

Enactments consolidated.

1903, No. 48.—The New Zealand Institute Act, 1903.

REGULATIONS.

THE following are the regulations of the New Zealand Institute under the Act of 1903 :—*

The word "Institute" used in the following regulations means the New Zealand Institute as constituted by the New Zealand Institute Act, 1903.

INCORPORATION OF SOCIETIES.

1. No society shall be incorporated with the Institute under the provisions of the New Zealand Institute Act, 1903, unless such society shall consist of not less than twenty-five members, subscribing in the aggregate a sum of not less than £25 sterling annually for the promotion of art, science, or such other branch of knowledge for which it is associated, to be from time to time certified to the satisfaction of the Board of Governors of the Institute by the President for the time being of the society.

2. Any society incorporated as aforesaid shall cease to be incorporated with the Institute in case the number of the members of the said society shall at any time become less than twenty-five, or the amount of money annually subscribed by such members shall at any time be less than £25.

---

* *New Zealand Gazette*, 14th July, 1904.

3. The by-laws of every society to be incorporated as aforesaid shall provide for the expenditure of not less than one-third of the annual revenue in or towards the formation or support of some local public museum or library, or otherwise shall provide for the contribution of not, less than one-sixth of its said revenue towards the extension and maintenance of the New Zealand Institute.

4. Any society incorporated as aforesaid which shall in any one year fail to expend the proportion of revenue specified in Regulation No. 3 aforesaid in manner provided shall from henceforth cease to be incorporated with the Institute.

PUBLICATIONS.

5. All papers read before any society for the time being incorporated with the Institute shall be deemed to be communications to the Institute, and then may be published as Proceedings or Transactions of the Institute, subject to the following regulations of the Board of the Institute regarding publications :—

(*a.*) The publications of the Institute shall consist of—

(1.) A current abstract of the proceedings of the societies for the time being incorporated with the Institute, to be intituled "Proceedings of the New Zealand Institute.";

(2.) And of transactions comprising papers read before the incorporated societies (subject, however, to selection as hereinafter mentioned), and of such other matter as the Board of Governors shall from time to time determine to publish, to be intituled "Transactions of the New Zealand Institute."

(*b.*) The Board of Governors shall determine what papers are to be published.

(*c.*) Papers not recommended for publication may be returned to their authors if so desired.

(*d.*) All papers sent in for publication must be legibly written, typewritten, or printed.

(*e.*) A proportional contribution may be required from each society towards the cost of publishing Proceedings and Transactions of the Institute.

(*f.*) Each incorporated society will be entitled to receive a proportional number of copies of the Transactions and Proceedings of the New Zealand Institute, to be from time to time fixed by the Board of Governors.

MANAGEMENT OF THE PROPERTY OF THE INSTITUTE.

6. All property accumulated by or with funds derived from incorporated societies, and placed in charge of the Institute, shall be vested in the Institute, and be used and applied at the discretion of the Board of Governors for public advantage, in like manner with any other of the property of the Institute.

7. All donations by societies, public Departments, or private individuals to the Institute shall be acknowledged by a printed form of receipt and shall be entered in the books of the Institute provided for that purpose, and shall then be dealt with as the Board of Governors may direct.

Honorary Members.

8. The Board of Governors shall have power to elect honorary members (being persons not residing in the Colony of New Zealand), provided that the total number of honorary members shall not exceed thirty.

9. In case of a vacancy in the list of honorary members, each incorporated society, after intimation from the Secretary of the Institute, may nominate for election as honorary member one person.

10. The names, descriptions, and addresses of persons so nominated, together with the grounds on which their election as honorary members is recommended, shall be forthwith forwarded to the President of the New Zealand Institute, and shall by him be submitted to the Governors at the next succeeding meeting.

General Regulations.

11. Subject to the New Zealand Institute Act, 1908, and to the foregoing rules, all societies incorporated with the Institute shall be entitled to retain or alter their own form of constitution and the by-laws for their own management, and shall conduct their own affairs.

12. Upon application signed by the President and countersigned by the Secretary of any society, accompanied by the certificate required under Regulation No. 1, a certificate of incorporation will be granted under the seal of the Institute, and will remain in force as long as the foregoing regulations of the Institute are complied with by the society.

13. In voting on any subject the President is to have a deliberate as well as a casting vote.

14. The President may at any time call a meeting of the Board, and shall do so on the requisition in writing of four Governors.

15. Twenty-one days' notice of every meeting of the Board shall be given by posting the same to each Governor at an address furnished by him to the Secretary.

16. In case of a vacancy in the office of President, a meeting of the Board shall be called by the Secretary within twenty-one days to elect a new President.

17. The Governors for the time being resident or present in Wellington shall be a Standing Committee for the purpose of transacting urgent business and assisting the officers.

18. The Standing Committee may appoint persons to perform the duties of any other office which may become vacant. Any such appointment shall hold good until the next meeting of the Board, when the vacancy shall be filled.

19. The foregoing regulations may be altered or amended at any annual meeting, provided that notice be given in writing to the Secretary of the Institute not later than the 30th November.

THE HUTTON MEMORIAL MEDAL AND RESEARCH FUND.

DECLARATION OF TRUST.

THIS deed, made the fifteenth day of February, one thousand nine hundred and nine (1909), between the New Zealand Institute of the one part, and the Public Trustee of the other part : Whereas the New Zealand Institute is possessed of a fund consisting now of the sum of five hundred and fifty-five pounds one shilling (£555 1s.), held for the purposes of the Hutton Memorial Medal and Research Fund on the terms of the rules and regulations made by the Governors of the said Institute, a copy whereof is hereto annexed : And whereas the said money has been transferred to the Public Trustee for the purposes of investment, and the Public Trustee now holds the same for such purposes, and it is expedient to declare the trusts upon which the same is held by the Public Trustee :

Now this deed witnesseth that the Public Trustee shall hold the said moneys and all other moneys which shall be handed to him by the said Governors for the same purposes upon trust from time to time to invest the same upon such securities as are lawful for the Public Trustee to invest on, and to hold the principal and income thereof for the purposes set out in the said rules hereto attached.

And it is hereby declared that it shall be lawful for the Public Trustee to pay all or any of the said moneys, both principal and interest, to the Treasurer of the said New Zealand Institute upon being directed so to do by a resolution of the Governors of the said Institute, and a letter signed by the Secretary of the said Institute enclosing a copy of such resolution certified by him and by the President as correct shall be sufficient evidence to the Public Trustee of the due passing of such resolution : And upon receipt of such letter and copy the receipt of the Treasurer for the time being of the said Institute shall be a sufficient discharge to the Public Trustee : And in no case shall the Public Trustee be concerned to inquire into the administration of the said moneys by the Governors of the said Institute.

As witness the seals of the said parties hereto, the day and year hereinbefore written.

RESOLUTIONS OF BOARD OF GOVERNORS.

RESOLVED by the Board of Governors of the New Zealand Institute that—

1. The funds placed in the hands of the Board by the committee of subscribers to the Hutton Memorial Fund be called "The Hutton Memorial Research Fund," in memory of the late Captain Frederick Wollaston Hutton, F.R.S. Such fund shall consist of the moneys subscribed and granted for the purpose of the Hutton Memorial, and all other funds which may be given or granted for the same purpose.

2. The funds shall be vested in the Institute. The Board of Governors of the Institute shall have the control of the said moneys, and may invest the same upon any securities proper for trust-moneys.

3. A sum not exceeding £100 shall be expended in procuring a bronze medal to be known as "The Hutton Memorial Medal."

4. The fund, or such part thereof as shall not be used as aforesaid, shall be invested in such securities as aforesaid as may be approved of by the Board of Governors, and the interest arising from such investment shall be used for the furtherance of the objects of the fund.

5. The Hutton Memorial Medal shall be awarded from time to time by the Board of Governors, in accordance with these regulations, to persons who have made some noticeable contribution in connection with the zoology, botany, or geology of New Zealand.

6. The Board shall make regulations setting out the manner in which the funds shall be administered. Such regulations shall conform to the terms of the trust.

7. The Board of Governors may, in the manner prescribed in the regulations, make grants from time to time from the accrued interest to persons or committees who require assistance in prosecuting researches in the zoology, botany, or geology of New Zealand.

8. There shall be published annually in the "Transactions of the New Zealand Institute" the regulations adopted by the Board as aforesaid, a list of the recipients of the Hutton Memorial Medal, a list of the persons to whom grants have been made during the previous year, and also, where possible, an abstract of researches made by them.

REGULATIONS UNDER WHICH THE HUTTON MEMORIAL MEDAL SHALL BE AWARDED AND THE RESEARCH FUND ADMINISTERED.

1. Unless in exceptional circumstances, the Hutton Memorial Medal shall be awarded not oftener than once in every three years ; and in no case shall any medal be awarded unless, in the opinion of the Board, some contribution really deserving of the honour has been made.

2. The medal shall not be awarded for any research published previous to the 31st December, 1906.

3. The research for which the medal is awarded must have a distinct bearing on New Zealand zoology, botany, or geology.

4. The medal shall be awarded only to those who have received the greater part of their education in New Zealand or who have resided in New Zealand for not less than ten years.

5. Whenever possible, the medal shall be presented in some public manner.

6. The Board of Governors may, at an annual meeting, make grants from the accrued interest of the fund to any person, society, or committee for the encouragement of research in New Zealand zoology, botany, or geology.

7. Applications for such grants shall be made to the Board before the 30th September.

8. In making such grants the Board of Governors shall give preference to such persons as are defined in regulation 4.

· 9. The recipients of such grants shall report to the Board before the 31st December in the year following, showing in a general way how the grant has been expended and what progress has been made with the research.

10· The results of researches aided by grants from the fund shall, where possible, be published in New Zealand.

11. The Board of Governors may from time to time amend or alter the regulations, such amendments or alterations being in all cases in conformity with resolutions 1 to 4.

AWARD OF THE HUTTON MEMORIAL MEDAL.

1911. Professor W. B. Benham, D.Sc., F.R.S., University of Otago—
For researches in New Zealand zoology.

1914. Dr. L. Cockayne, F.L S., F.R.S. — For researches on the
ecology of New Zealand plants.

GRANTS FROM THE HUTTON MEMORIAL RESEARCH FUND.

1916. (1.) To the Portobello Marine Fish - hatchery — £25 for the
purpose of prosecuting research on the distribution of native marine food
fishes.

(2.) To Major Broun—£50 towards the publication of researches on
New Zealand *Coleoptera* as a bulletin.

HECTOR MEMORIAL RESEARCH FUND.

DECLARATION OF TRUST.

THIS deed, made the thirty-first day of July, one thousand nine hundred
and fourteen, between the New Zealand Institute, a body corporate
duly incorporated by the New Zealand Institute Act, 1908, of the one
part, and the Public Trustee of the other part: Whereas by a declara-
tion of trust dated the twenty-seventh day of January, one thousand
nine hundred and twelve, after reciting that the New Zealand Institute
was possessed of a fund consisting of the sum of £1,045 10s. 2d., held
for the purposes of the Hector Memorial Research Fund on the terms of
the rules and regulations therein mentioned, which said moneys had been
handed to the Public Trustee for investment, it was declared (*inter alia*)
that the Public Trustee should hold the said moneys and all other moneys
which should be handed to him by the said Governors of the Institute
for the same purpose upon trust from time to time, to invest the same
in the common fund of the Public Trust Office, and to hold the principal
and income thereof for the purposes set out in the said rules and regulations in the said deed set forth: And whereas the said rules and regulations have been amended by the Governors of the New Zealand Institute,
and as amended are hereinafter set forth: And whereas it is expedient
to declare that the said moneys are held by the Public Trustee upon the
trusts declared by the said deed of trust and for the purposes set forth
in the said rules and regulations as amended as aforesaid:

Now this deed witnesseth and it is hereby declared that the Public
Trustee shall hold the said moneys and all other moneys which shall be
handed to him by the said Governors for the same purpose upon trust
from time to time to invest the same in the common fund of the Public
Trust Office, and to hold the principal and income thereof for the purposes set out in the said rules and regulations hereinafter set forth:

And it is hereby declared that it shall be lawful for the Public
Trustee to pay, and he shall pay, all or any of the said moneys, both
principal and interest, to the Treasurer of the said New Zealand Institute upon being directed to do so by a resolution of the Governors of
the said Institute, and a letter signed by the Secretary of the said Institute enclosing a copy of such resolution certified by him and by the

President as correct shall be sufficient evidence to the Public Trustee of the due passing of such resolution : And upon receipt of such letter and copy the receipt of the Treasurer for the time being of the said Institute shall be a sufficient discharge to the Public Trustee : And in no case shall the Public Trustee be concerned to inquire into the administration of the said moneys by the Governors of the said Institute.

As witness the seals of the said parties hereto, the day and year first hereinbefore written.

Rules and Regulations made by the Governors of the New Zealand Institute in relation to the Hector Memorial Research Fund.

1. The funds placed in the hands of the Board by the Wellington Hector Memorial Committee be called "The Hector Memorial Research Fund," in memory of the late Sir James Hector, K.C.M.G., F.R.S. The object of such fund shall be the encouragement of scientific research in New Zealand, and such fund shall consist of the moneys subscribed and granted for the purpose of the memorial and all other funds which may be given or granted for the same purpose.

2. The funds shall be vested in the Institute. The Board of Governors of the Institute shall have the control of the said moneys, and may invest the same upon any securities proper for trust-moneys.

3. A sum not exceeding one hundred pounds (£100) shall be expended in procuring a bronze medal, to be known as the Hector Memorial Medal.

4. The fund, or such part thereof as shall not be used as aforesaid, shall be invested in such securities as may be approved by the Board of Governors, and the interest arising from such investment shall be used for the furtherance of the objects of the fund by providing thereout a prize for the encouragement of such scientific research in New Zealand of such amount as the Board of Governors shall from time to time determine.

5. The Hector Memorial Medal and Prize shall be awarded annually by the Board of Governors.

6. The prize and medal shall be awarded by rotation for the following subjects, namely—(1) Botany, (2) chemistry, (3) ethnology, (4) geology, (5) physics (including mathematics and astronomy), (6) zoology (including animal physiology).

In each year the medal and prize shall be awarded to that investigator who, working within the Dominion of New Zealand, shall in the opinion of the Board of Governors have done most towards the advancement of that branch of science to which the medal and prize are in such year allotted.

7. Whenever possible the medal shall be presented in some public manner.

AWARD OF THE HECTOR MEMORIAL RESEARCH FUND.

1912. L. Cockayne, Ph.D., F.L.S., F.R.S.—For researches in New Zealand botany.
1913. T. H. Easterfield, M.A., Ph.D.—For researches in chemistry.
1914. Elsdon Best—For researches in New Zealand ethnology.
1915. P. Marshall, M.A., D.Sc., F.G.S.—For researches in New Zealand geology.
1916. Sir Ernest Rutherford, F.R.S.—For researches in physics.

THE CARTER BEQUEST.

Extracts from the Will of Charles Rooking Carter.

This is the last will and testament of me, Charles Rooking Carter, of Wellington, in the Colony of New Zealand, gentleman.

I revoke all wills and testamentary dispositions heretofore made by me, and declare this to be my last will and testament.

                *

I give to the Colonial Museum in Wellington the large framed photographs of the members of the General Assembly in the House of Representatives in the year 1860, and the framed pencil sketch of the old House of Commons, and the framed invitation-card to the Lord Mayor's dinner.

                *

As regards the following books, of which I am the author, and which are now stored in three boxes—namely, (1) "The Life and Recollections of a New Zealand Colonist," (2) "A Historical Sketch of New Zealand Loans," and (3) "Round the World Leisurely"—I direct that my executor shall retain possession of the same for a period of seven years, commencing from the date of my death, and that at the end of such period my executor shall place the same in the hands of Messrs. Whitcombe and Tombs (Limited) or some other capable and responsible booksellers in the City of Wellington, for sale, and so that the same shall be sold at such a price as will yield to my estate not less than six shillings per volume in respect of the first-named and second-named, and two shillings and sixpence in respect of the last-named works; and I further authorize my executor to sell and dispose of the copyright or right to reprint such works; and I direct that the moneys to be derived from the sale of such works and the privileges connected therewith shall be added to the sum provided for the purchase of a telescope as hereinafter mentioned.

                *

I direct my executor to subscribe the sum of fifty pounds towards the erection of a suitable brick room in which to house the priceless collection of books on New Zealand some time since given by me to the Colonial Museum and the New Zealand Institute.

                *

I give and devise unto the Public Trustee appointed under and in pursuance of an Act of the General Assembly of New Zealand intituled the Public Trust Office Act, 1894 (hereinafter called "my trustee"), all the rest, residue, and remainder of my property whatsoever and wheresoever situate, both real and personal, and whether in possession, reversion, expectancy, or remainder, upon trust, as to my freehold property at East Taratahi, containing by admeasurement two thousand one hundred and seventy-two acres, and being and comprising the whole of the land included in certificate of title, volume 51, folio 79, of the books of the District Land Registrar for the Registration District of Wellington (save and except such part of the said land, being portion of the section numbered 117 in the Taratahi Plain Block, as is hereinafter devised to my trustee for the purposes hereinafter appearing), and direct that my Trustee shall stand possessed of the same lands upon trust, to let and manage the same, and to pay and apply the rents and annual income in manner following, namely :—

      *   *   *   *   *   *

And as to all the residue and remainder (if any) of the said net proceeds of the sale, conversion, and getting-in of my estate as aforesaid, my trustee shall transfer the same to the Governors for the time being of the New Zealand Institute at Wellington, to form the nucleus of a fund for the erection in or near Wellington aforesaid, and the endowment of a Professor and staff, of an Astronomic Observatory fitted with telescope and other suitable instruments for the public use and benefit of the colony, and in the hope that such fund may be augmented by gifts from private donors, and that the Observatory may be subsidized by the Colonial Government; and without imposing any duty or obligation in regard thereto I would indicate my wish that the telescope may be obtained from the factory of Sir H. Grubb, in Dublin, Ireland.

NEW ZEALAND INSTITUTE.

BOARD OF GOVERNORS.

EX OFFICIO.

His Excellency the Governor.
The Hon. the Minister of Internal Affairs.

NOMINATED BY THE GOVERNMENT.

Mr. Charles A. Ewen (appointed December, 1914); Dr. J. Allan Thomson, F.G.S. (appointed December, 1915); Mr. B. C. Aston, F.I.C. (appointed December, 1915).

ELECTED BY AFFILIATED SOCIETIES (DECEMBER, 1915).

Wellington Philosophical Society ...	Professor T. H. Easterfield, M.A., Ph.D. Professor H. B. Kirk, M.A.
Auckland Institute	Mr. D. Petrie, M.A., Ph.D. Professor H. W. Segar, M.A.
Philosophical Institute of Canterbury...	Dr. Hilgendorf, M.A. Mr. A. M. Wright, F.C.S.
Otago Institute	Professor Marshall, M.A., D.Sc., F.G.S. Mr. G. M. Thomson, F.C.S., F.L.S.
Hawke's Bay Philosophical Institute ...	Mr. H. Hill, B.A., F.G.S.
Nelson Institute	Dr. L. Cockayne, F.L.S., F.R.S.
Manawatu Philosophical Society ...	Mr. J. W. Poynton, S.M.
Wanganui Philosophical Society ...	Dr. H. R. Hatherly, M.R.C.S.

OFFICERS FOR THE YEAR 1916.

PRESIDENT: Professor Benham, M.A., D.Sc., F.Z.S., F.R.S.
HON. TREASURER: Mr. C. A. Ewen.
JOINT HON. EDITORS: Dr. L. Cockayne, F.R.S., Dr. C. A. Cotton, F.G.S.
HON. LIBRARIAN: Dr. J. Allan Thomson, F.G.S.
HON. SECRETARY: Mr. B. C. Aston, F.I.C., F.C.S.
(Box 40, Post-office, Wellington).

AFFILIATED SOCIETIES.

Name of Society.	Secretary's Name and Address.	Date of Affiliation.
Wellington Philosophical Society	C. E. Adams, Hector Observatory, Wellington	10th June, 1868.
Auckland Institute ..	T. F. Cheeseman, Museum ..	10th June, 1868.
Philosophical Institute of Canterbury	A. M. Wright, Box 617, Christchurch	22nd October, 1868.
Otago Institute .. ·..	E. J. Parr, Boys' High School ..	18th October, 1869.
Hawke's Bay Philosophical Institute	James Niven, Technical College	31st March, 1875.
Nelson Institute	E. L. Morley, Waimea Street ..	20th December, 1883.
Manawatu Philosophical Society	K. Wilson, Palmerston North ..	6th January, 1905.
Wanganui Philosophical Society	J. P. Williamson	2nd December, 1911.

FORMER HONORARY MEMBERS.

1870.

Agassiz, Professor Louis.
Drury, Captain Byron, R.N.
Flower, Professor W. H., F.R.S.
Hochstetter, Dr. Ferdinand von.
Hooker, Sir J. D., G.C.S.I., C.B., M.D.,
F.R.S., O.M.

Mueller, Ferdinand von, M.D., F.R.S.,
C.M.G.
Owen, Professor Richard, F.R.S.
Richards, Rear-Admiral G. H.

1871.

Darwin, Charles. M.A., F.R.S.
Gray, J. E., Ph.D., F.R.S.

Lindsay, W. Lauder, M.D., F.R.S.E.

1872.

Grey, Sir George, K.C.B.
Huxley, Thomas H., LL.D., F.R.S.

Stokes, Vice-Admiral J. L.

1873.

Bowen, Sir George Ferguson, G.C.M.G.
Günther, A., M.D., M.A., Ph.D., F.R.S.

Lyell, Sir Charles, Bart., D.C.L., F.R.S.

1874.

McLachlan, Robert, F.L.S.
Newton, Alfred, F.R.S.

Thomson, Professor Wyville, F.R.S.

1875.

Filhol, Dr. H.
Rolleston, Professor G., M.D., F.R.S.

Sclater, P. L., M.A., Ph.D., F.R.S.

1876.

Clarke, Rev. W. B., M.A., F.R.S.

Etheridge, Professor R., F.R.S.

1877.

Baird, Professor Spencer F.

Weld, Frederick A., C.M.G.

1878.

Garrod, Professor A. H., F.R.S.
Müller, Professor Max, F.R.S.

Tenison-Woods, Rev. J. E., F.L.S.

1880.

The Most Noble the Marquis of Normanby, G.C.M.G.

1883.

Carpenter, Dr. W. B., C.B., F.R.S.
Ellery, Robert L. J., F.R.S.

Thomson, Sir William, F.R.S.

1885.

Gray, Professor Asa.
Sharp, Richard Bowdler, M.A., F.R.S.

Wallace, A. R., F.R.S., O.M.

1888.

Beneden, Professor J. P. van.
Ettingshausen, Baron von.

McCoy, Professor F., D.Sc., C.M.G.,
F.R.S.

1890.

Riley, Professor C. V.

1891.

Davis, J. W., F.G.S., F.L.S.

1895.

Mitten, William, F.R.S.

1896.

Langley, S. P.

1900.

Agardh, Dr. J. G.　　　　| Avebury, Lord, P.C., F.R.S.

1901.

Eve, H. W., M.A.　　　　· | Howes, G. B., LL.D.. F.R.S.

1906.

Milne, J., F.R.S.

1909.

Darwin, Sir George, F.R.S.

———

FORMER MANAGER AND EDITOR.
[UNDER THE NEW ZEALAND INSTITUTE ACT, 1867.]

1867–1903.

Hector, Sir James, M.D., K.C.M.G., F.R.S.

———

PAST PRESIDENTS.

1903–4.

Hutton, Captain Frederick Wollaston, F.R.S.

1905–6.

Hector, Sir James, M.D., K.C.M.G., F.R.S.

1907–8.

Thomson, George Malcolm, F.L.S., F.C.S.

1909–10.

Hamilton, A.

1911–12.

Cheeseman, T. F., F.L.S., F.Z.S.

1913–14.

Chilton, C., M.A., D.Sc., LL.D., F.L.S.

———

HONORARY MEMBERS.

1870.

FINSCH, Professor OTTO, Ph.D., Braunschweig. Germany.

1873.

PICKARD-CAMBRIDGE, The Rev. O., M.A., C.M.Z.S.

1876

BERGGREN, Dr. S., Lund, Sweden.

1877.
SHARP, Dr. D., University Museum, Cambridge.

1890.
LIVERSIDGE, Professor A., M.A., F.R.S., Fieldhead, Coombe Warren, Kingston Hill, England.

NORDSTEDT, Professor OTTO, Ph.D., University of Lund, Sweden.

1891.
GOODALE, Professor G. L., M.D., LL.D., Harvard University, Massachusetts, U.S.A.

1894.
CODRINGTON, Rev. R. H., D.D., Wadhurst Rectory, Sussex, England.

THISELTON - DYER, Sir W. T., K.C.M.G., C.I.E., LL.D., M.A., F.R.S., Witcombe, Gloucester, England.

1896.
LYDEKKER, RICHARD, B.A., F.R.S., British Museum, South Kensington.

1900.
MASSEE, GEORGE, F.R.M.S., Royal Botanic Gardens, Kew.

1901.
GOEBEL, Professor Dr. CARL VON, University of Munich.

1902.
SARS, Professor G. O., University of Christiania, Norway.

1903.
KLOTZ, Professor OTTO J., 437 Albert Street, Ottawa, Canada.

1904.
RUTHERFORD, Professor Sir E., D.Sc., F.R.S., Nobel Laureate, University of Manchester.

DAVID, Professor T. EDGEWORTH, F.R.S., C.M.G., Sydney University, N.S.W.

1906.
BEDDARD, F. E., D.Sc., F.R.S., Zoological Society, London.

BRADY, G. S., D.Sc., F.R.S., University of Durham, England.

1907.
DENDY, Dr. A., F.R.S., King's College, University of London, England.
DIELS, Professor L., Ph.D., University of Marburg.

MEYRICK, E., B.A., F.R.S., Marlborough College, England.
STEBBING, Rev. T. R. R., F.R.S., Tunbridge Wells, England.

1910.
BRUCE, Dr. W. S., Edinburgh.

1913.
DAVIS, Professor W. MORRIS, Harvard University.

HEMSLEY, Dr. W. BOTTING, F.R.S., Strawberry Hill, London, England.

1914.
ARBER, Dr. E. NEWELL, Cambridge, England.
BALFOUR, Professor I. BAYLEY, F.R.S., Royal Botanic Gardens, Edinburgh.

HASWELL, Professor W. A., F.R.S, University, Sydney.

1915.
BATESON, Professor W., F.R.S., Merton, Surrey, England.

1916.
MASSART, Professor JEAN, University of Brussels, Belgium.

ORDINARY MEMBERS.

WELLINGTON PHILOSOPHICAL SOCIETY.

[* Life members.   † On active service.]

Acland, E. W., P.O. Box 928, Wellington

Adams, C. E., D.Sc., A.I.A. (London), F.R.A.S., Hector Observatory, Wellington

Adams, C. W., Bellevue Road, Lower Hutt

Adamson, Professor J., MA., LL.B., Victoria University College, Wellington

Adkin, G. L., Queen Street, Levin

Alabaster, A. H., Head Office, Railways, Wellington

Anderson, W. J., M.A., LL.D., Education Department, Wellington

Aston, B. C., F.I.C., F.C.S., Dominion Laboratory, Wellington

Atkins, A., F.R.I.B.A., A.M. Inst.C.E., Grey Street, Wellington

Atkinson, E. H., Agricultural Department, Wellington

Bagley, G., care of Young's Chemical Company, 14 Egmont Street, Wellington

Bakewell, F. H., M.A., Education Board, Mercer Street, Wellington

Baldwin, E. S., 215 Lambton Quay, Wellington

Bates, Rev. D.C., F.R.G.S., F.R.Met. Soc., Weather Office, Wellington.

Beere, Wyn O., 155 Featherston Street, Wellington

Beetham, W. H., Masterton

Begg, Dr. C. M., 164 Willis Street, Wellington†

Bell, E. D., Panama Street, Wellington

Bell, Hon. Sir Francis H. D., K.C., M.L.C., Panama Street, Wellington

Berry, C. G. G., 35 Bolton Street, Wellington

Blair, David K., M.I.Mech.E., 9 Grey Street, Wellington

Blake, V., District Lands and Survey Office, Wellington†

Blow, H. J. H., Public Works Department, Wellington

Brandon, A. de B., B.A., Featherston Street, Wellington

Bridges, G. G., 2 Wesley Road, Wellington

Broadgate, F. L. K., M.Sc., Dominion Museum, Wellington†

Browne, M. H., Education Department, Wellington

Burbidge, P. W., M.Sc., Victoria University College, Wellington

Burnett, J., M.Inst.C.E., Chief Engineer, Head Office, Railways

Campbell, J., F.R.I.B.A., Government Architect, Public Works Department, Wellington

Carter, F. J., M.A., Diocesan Office, Wellington

Carter, W. H., care of Dr. Henry, The Terrace, Wellington

Chapman, Martin, K.C., Brandon Street, Wellington

Chudleigh, E. R., Orongomairoa, Waihou

Cockayne, L., Ph.D., F.L.S., F.R.S., 13 Colombo Street, Wellington

Cook, H. D., M.Sc., B.E. (Elect.), Bank Chambers, Lambton Quay, Wellington

Cotton, C. A., D.Sc., F.G.S., Victoria University College, Wellington.

Crawford, A. D., Box 126, G.P.O., Wellington

Crewes, Rev. J., 90 Owen Street, Wellington

Cull, J. E. L., B.Sc. in Eng. (Mech.), Public Works Department, Wellington

Donovan, W., M.Sc., Dominion Laboratory, Wellington

Dougall, Archibald, 34 Austin Street, Wellington

Dymock, E. R., A.I.A.N.Z., A.I.A.V., Woodward Street, Wellington

Earnshaw, Hon. W., M.L.C., 4 Watson Street, Wellington

Easterfield, Professor T. H., M.A., Ph.D., Victoria College, Wellington

Ewen, C. A., 126 The Terrace, Wellington

Ferguson, William, M.A., M.Inst. C.E., M.Inst.Mech.E., 131 Coromandel Street, Wellington

Field, W. H., M.P., 160 Featherston Street, Wellington

Findlay, Sir John G., K.C., LL.D., 197 Lambton Quay, Wellington

FitzGerald, Gerald, A.M.Inst.C.E., Box 461, Wellington

Fleming, T. R., M.A., LL.B., Education Board Office, Wellington

Fletcher, Rev. H. J., The Manse, Taupo

Fox, Thomas O., Borough Engineer, Miramar, Wellington

Fraser, G. V. R., Head Office, Railways, Wellington

Freeman, C. J., 95 Webb Street, Wellington*

Freyberg, C., Macdonald Crescent, Wellington

Fulton, J., 14 North Terrace, Kelburne†

Furkert, F. W., M.Inst.C.E., Inspecting Engineer, Public Works Department, Wellington

Garrow, Professor J. M. E., B.A., LL.B., Victoria College, Wellington*

Gavin, W. H., Inspecting Engineer, Public Works Department, Wellington

Gibbs, A., Electrical Branch, Telegraph Department, Wellington

Gibbs, Dr. H. E., 240 Willis Street, Wellington

Gibson, Miss G. F., M.A., 10 Hill Street, Wellington

Gibson, W., B.E. (Mining), Geological Survey Department, Wellington

Gifford, A. C., M.A., F.R.A.S., 6 Shannon Street, Wellington*

Girdlestone, H. E., F.R.G.S., Lands and Survey Department, Wellington

Gottfried, Rev. I. von, Boulcott Street, Wellington

Goudie, H. A., Whakarewarewa

Haast, H. F. von, M.A., LL.B., 41 Salamanca Road, Wellington

Hamilton, H., A.O.S.M., 58 Bowen Street, Wellington

Hanify, H. P., 18 Panama Street, Wellington

Harding, R. Coupland, 63 Matai Road, Hataitai*

Hastie, Miss J. A., care of Street and Co., 30 Cornhill, London, E.C.*

Hayward, Captain J. A., 113 Tasman Street, Wellington

Hector, C. Monro., M.D., B.Sc., F.R.A.S., Lower Hutt

Helyer, Miss E., 13 Tonks Grove, Wellington

Henderson, J., M.A., D.Sc., B.Sc. in Engineering (Metall.), Geological Survey Department, Wellington

Hislop, J., Under-Secretary for Internal Affairs, Wellington

Hodson, W. H., 40 Pirie Street, Wellington

Hogben, G., C.M.G., M.A., F.G.S., 32 Crescent Road, Khandallah

Holmes, R. L., F.R.Met.Soc., Kai Ora, Fern Street, Randwick, Sydney*

Holmes, R. W., M.Inst.C.E., Chief Engineer, Public Works Department, Wellington

Hooper, Captain G. S., Grant Road North, Wellington

Hudson, G. V., F.E.S., Inspector's Office, G.P.O., Wellington

Humphries, Thomas, F.R.G.S., Laery Street, Lower Hutt

Hursthouse, E. W., Routh's Buildings, Wellington

Isaac, E. C., Education Department, Wellington

Jack, J. W., Featherston Street, Wellington

James, L. G., Box 94 (Hunter Street), Wellington

Jenkinson, S. H., Railway Buildings, Wellington

Johnston, C. G., Head Office, Railways, Wellington†

Johnston, Hon. G. Randall, care of Martin Chapman, K.C., Wellington*

Jones, E. G., B.A., Wellington College, Wellington

Joseph, Joseph, Box 443 (Grant Road), Wellington

Kennedy, Rev. Dr. D., F.R.A.S., St. Patrick's College, Wellington

Kilroe, Miss F. C., B.Sc., 7 Hay Street, Wellington

King, G. W., B.E., care of A. H. King, Box 116, Christchurch†

King, Thomas, F.R.A.S., 58 Ellice Street, Wellington*

Kirk, Professor H. B., M.A., Victoria College, Wellington

La Trobe, W. S., M.A., Technical College, Wellington

Levi, P., M.A., care of Wilford and Levi, 15 Stout Street, Wellington

Lomax, Major H. A., Araruhe, Aramoho, Wanganui

Luke, J. P., Hiropi Street, Wellington

McCabe, Ultan F., care of Richardson and McCabe, 11 Grey Street, Wellington

McDonald, J., Dominion Museum, Wellington

Mackay, J., Government Printer, Wellington

McKenzie, Donald, care of Mrs. Elizabeth McKenzie, Marton.†

Mackenzie, Professor H., M.A., Victoria University College, Wellington

Mackenzie, James, Under - Secretary, Lands and Survey Department, Wellington

Maclaurin, J. S., D.Sc., F.C.S., Dominion Laboratory, Wellington

MacLean, F. W., M.Inst.C.E., Inspecting Engineer, Head Office, Railways, Wellington

Marchbanks, J., M.Inst.C.E., Harbour Board, Wellington

Marsden, Professor E., D.Sc., Victoria University College, Wellington†

Mason, J. Malcolm, M.D., F.C.S., D.P.H., Lower Hutt

Maxwell, J. P., M.Inst.C.E., 145 Dixon Street, Wellington

Mestayer, R. L., M.Inst.C.E., 139 Sydney Street, Wellington

Miles, P. H. R., Telegraph Department, G.P.O., Wellington

Miller, H. M., Public Works Department, Wellington

Moore, G., Eparaima, *via* Masterton (during session, Legislative Council)

Moore, W. Lancelot, care of H. D. Cook, Bank Chambers, Lambton Quay, Wellington†

Moorhouse, W. H. Sefton, 134 Dixon Street, Wellington

Morgan, P. G., M.A., F.G.S., Director Geological Survey, Routh's Buildings, Wellington

Morice, J. M., B.Sc., Assistant City Engineer, Town Hall, Wellington

Morison, C. B., 180 Featherston Street, Wellington

Morrison, J. C., Box 8, P.O., Eltham

Morton, W. H., M.Inst.C.E., City Engineer, Wellington

Myers, Miss P., B.A., 26 Fitzherbert Terrace, Wellington

Newman, A. K., M.B., M.R.C.P., M.P., 56 Hobson Street, Wellington

Nicol, John, 23, Cuba Street, Wellington

Ongley, M., M.A., Geological Survey Department, Wellington

Orchiston, J., M.I.E.E., Chief Telegraph Engineer, Telegraph Office, Wellington

Orr, Robert, 176 Featherston Street, Wellington

Patterson, Hugh, Assistant Engineer, Public Works Office, Ngatapu

Parry, Evan, B.Sc., M.I.E.E., A.M.Inst.C.E., Public Works Department, Wellington

Pearce, Arthur E., care of Levin and Co. (Limited), Wellington

Phillips, Coleman, Carterton*

Phipson, P. B., F.C.S., care of J. Staples and Co. (Limited), Wellington

Pigott, Miss Elizabeth, M.A., Victoria University College, Wellington

Pigott, Miss Ellen, M.A., Victoria University College, Wellington.

Pomare, Hon. Dr. M., M.P., Wellington

Porteous, J. S., 9 Brandon Street, Wellington

Powles, C. P., 219 Lambton Quay, Wellington

Reakes, C. J., D.V.Sc., M.R.C.V.S., Agricultural Department, Wellington

Reid, W. S., 189 The Terrace, Wellington

Richardson, C. E., Box 863 (11 Grey Street), Wellington

Richardson, J. M., 132 The Terrace, Wellington

Robertson, J. B., Public Works Department, Wellington

Robertson, R. E., 41 Grove Road, Kelburne (resigned 30th June, 1915)

Roy, R. B., Taita, Wellington*

Salmond, J. W., M.A., K.C., LL.B., Crown Law Office, Wellington

Shrimpton, E. A., Telegraph Department, Wellington

Sladden, H., Lower Hutt, Wellington

Smith, J. R., Telegraph Department, Wellington

Smith, M. Compton, Lands and Survey Department, Wellington

Sommerville, Professor D. M. Y., M.A., D.Sc., F.R.S.E., Victoria University College, Wellington

Spencer, W. E., M.A., M.Sc., Education Department, Wellington

Stewardson, G., 58 Taranaki Street, Wellington

Stout, T. Duncan M., M.B., M.S., F.R.C.S., 164 Willis Street, Wellington†

Stuckey, F. G. A., M.A., 21 Hobson Crescent, Wellington

Sunley, R. M., Karori

Swinburne, C. G., Town Hall, Wellington

Tennant, J. S., M.A., B.Sc., Training College, Wellington

Thomson, J. Allan, M.A., D.Sc., F.G.S., Dominion Museum, Wellington

Thomson, John, B.E., M.Inst.C.E., 17 Dorking Road, Brooklyn, Wellington

Tily, H. S., B.Sc., H.M. Customs, Wellington

Tolley, H. R., 34 Wright Street, Wellington

Tombs, H. H., Burnell Avenue, Wellington

Turnbull, A. H., care of W. and G. Turnbull and Co., Wellington

Turner, E. Phillips, F.R.G.S., Lands and Survey Department, Wellington

Vickerman, H., M.Sc., A.M.Inst. C.E., Public Works Department, Wellington†

Ward, Thomas, A.M.Inst.C.E., Grey Street, Wellington

Welch, J. S., 52 Wright Street, Wellington

Westland, C. J., F.R.A.S., Hector Observatory, Wellington

Widdop, F. C., District Railway Engineer, Thorndon Office, Wellington

Wilmot, E. H., Surveyor-General, Wellington

Wilson, Sir James G., Bulls

Wynne, H. J., Railway Buildings, Wellington

Zedlitz, Professor G. W. von, M.A., Belmont Road, Lower Hutt

AUCKLAND INSTITUTE.

(Members are requested to advise the Secretary of any change of address.)

[* Honorary and life members.]

Abbott, R. H., Elliott Street, Auckland

Abel, R. S., care of Abel, Dykes, and Co., Shortland Street, Auckland

Adlington, Miss H., District High School, Whangarei

Aickin, G., Queen Street, Auckland

Aldis, M., Empire Buildings, Swanson Street, Auckland

Alexander, L. M., Beauvoir, Hurstmere Road, Takapuna

Alison, A., Devonport Ferry Company, Auckland

Alison, E. W., Devonport Ferry Company, Auckland

Allen, John, Cheltenham, Devonport

Ardern, P. S., Remuera

Arey, W. E., Victoria Arcade, Auckland

Armitage, F. L., Gleeson's Buildings, High Street, Auckland

Arnold, C., Swanson Street, Auckland

Arnoldson, L., Quay Street, Auckland

Atkinson, H., Grafton Road, Auckland

Bagnall, L. J., Wynyard Street, Auckland

Baker, G. H., Commerce Street, Auckland

Ball, W. T., Sylvan Avenue, Mount Eden

Bamford, H. D., LL.D., Bank of New Zealand Buildings, Auckland

Bankart, A. S., Shortland Street, Auckland

Barr, J., Public Library, Wellesley Street, Auckland

Bartrum, J. A., M.Sc., University College, Auckland

Bates, T. L., Brookfield, Alfred Street, Waratah, Newcastle, New South Wales*

Batger, J., Mount Eden Road, Auckland

Bell, T., care of Union Oil, Soap, and Candle Company, Albert Street, Auckland

Benjamin, E. D., care of L. D. Nathan and Co., Shortland Street, Auckland

Biss, N. L. H., Shortland Street, Auckland

Blair, J. M., Market Road, Epsom

Bloomfield, J. L. N. R., St. Stephen's Avenue, Parnell

Bond, Elon, Commerce Street, Auckland

Bradley, S., Onehunga

Brett, H., *Star* Office, Shortland Street, Auckland

Briffault, R., M.B., Mount Eden Road, Auckland

Brooke-Smith, E., Manukau Road, Parnell

Broun, Major T., F.E.S., Mount Albert, Auckland

Brown, E. A., Cleave's Buildings, High Street, Auckland

Brown, Professor F. D., care of Auckland Institute, Auckland

Brown, John, Beresford Street, Bayswater, Devonport

Bruce, W. W., Williamson Chambers, Shortland Street

Buchanan, A. V., Victoria Avenue, Remuera*

Buddle, C. V., Wyndham Street, Auckland

Buddle, T., Wyndham Street, Auckland

Burns, R., Customs Street, Auckland

Burnside, W., Education Offices, Auckland

Burt, A., care of A. and T. Burt and Co., Customs Street, Auckland

Burton, Colonel, The Grove, Branksome Park, Bournemouth, England* .

Bush, W. E., City Engineer's Office, Town Hall, Auckland

Butler, Miss, Girls' Grammar School, Auckland

Buttle, J., New Zealand Insurance Company, Queen Street, Auckland

Campbell, J. P., Arney Road, Remuera

Casey, Maurice, Hamilton Road, Ponsonby

Cheal, P. E., Upper Queen Street, Auckland

Cheeseman, T. F., F.L.S., F.Z.S., Museum, Auckland

Choyce, H. C., Remuera Road, Auckland

Clark, A., Wellesley Street, Auckland

Clark, H. C., care of A. Clark and Sons, P.O. Box 35, Auckland

Clark, M. A., Wellesley Street, Auckland

Clay, T. B., care of S. Vaile and Sons, Queen Street, Auckland

Coates, T., Orakei, Auckland

Coe, James, Mount Eden Road, Auckland

Cole, Rev. R. H., Walford, Gladstone Road, Parnell

Cole, W., Mount Eden Road, Auckland

Coleman, J. W., Lower Queen Street, Auckland

Combes, F. H., Victoria Avenue, Remuera

Cooper, C., Bourne Street, Mount Eden

Cooper Mr. Justice, Judges' Chambers, Auckland

Cory-Wright, S., B.Sc., University College, Auckland

Court, G., Karangahape Road, Auckland

Court, J., Ponsonby, Auckland

Cousins, H. G., Normal School, Wellesley Street, Auckland

Craig, J. J., Queen Street, Auckland

Cranwell, R., Crescent Road, Parnell

Crook, T., 10 Prospect Terrace, Mount Eden

Cuff, J. C., Emerald Hill, Epsom, Auckland

Culling, T. F., Victoria Avenue, Remuera

Currie, J. C., care of J. Currie, 202 Queen Street, Auckland

Davis, Elliot R., care of Hancock and Co., Customs Street, Auckland

Davis, Ernest, care of Hancock and Co., Customs Street, Auckland

Dearsly, H., P.O. Box 466, Auckland

De Clive Lowe, Dr., Symonds Street, Auckland

Dempsey, J., Newmarket

Dennin, John, care of Hon. E. Mitchelson, Waimauku

Dettmann, Professor H. S., University College, Auckland

Devereux, H. B., Glen Road, Stanley Bay, Devonport

Devore, A. E. T., Wyndham Street, Auckland

Dickenson, J. C., Public School, Ponsonby

Downard, F. N. R., Public School, Kuaotunu

Duthie, D. W., National Bank of New Zealand, Wellington

Earl, F., Swanson Street, Auckland

Edgerley, Miss K., M.A., Girls' Grammar School, Auckland

Edson, J., Waimarama, Tudor Street, Devonport

Egerton, Professor C. W., University College, Auckland

Ellingham, W. B., Customs Street East, Auckland

Elliott, G., Bank of New Zealand Buildings, Queen Street, Auckland

Ellis, A. F., Argyle Street, Ponsonby

Ellis, J. W., Hamilton, Waikato

Ellison, T., Ellison's Buildings, Queen Street, Auckland

Endean, J., Jermyn Street, Auckland

Entrican, A. J., Customs Street East, Auckland

Ewington, F. G., Durham Street, Auckland

Fairclough, Dr. W. A., Watson's Buildings, Queen Street, Auckland

Farrell, R., Anglesea Street, Auckland

Fenwick, Dr. G., Premier Buildings, Queen Street, Auckland

Fleming, J., 142 Grafton Road, Auckland

Florance, R. S., Stipendiary Magistrate, Gisborne

Fowlds, Hon. G., Queen Street, Auckland*

Frater, J. W., care of R. Frater, Stock Exchange, Auckland

Garrard, C. W., M.A., Education Office, Auckland

George, G., Technical College, Auckland

George, Hon. S. T., St. Stephen's Avenue, Parnell

Gerard, E., Customs Street East, Auckland

Gilfillan, H., St. Stephen's Avenue, Parnell

Girdler, Dr., Khyber Pass Road, Auckland

Gleeson, J. C., Selwyn Lodge, Parnell

Gleeson, P., Selwyn Lodge, Parnell

Goldie, D., Breakwater Road, Parnell

Gordon, H. A., Ranfurly Road, Epsom

Gorrie, H. T., care of A. Buckland and Sons, Albert Street, Auckland

Gould, A. M., care of Parr and Blomfield, Shortland Street, Auckland

Graham, G., Tudor Street, Devonport

Grant, Miss J., M.A., Devonport

Gray, A., Smeeton's Buildings, Queen Street, Auckland*

Gray, S., Mount Eden Borough Council Offices, Mount Eden

Gribbin, G., Nicholson and Gribbin, Queen Street, Auckland

Griffin, L. T., Museum, Auckland

Grossman, Professor J. P., M.A., University College, Auckland

Gunson, J. H., Customs Street West, Auckland

Gunson, R. W., Customs Street West, Auckland

Haddow, J. G., Wyndham Street, Auckland

Haines, H., F.R.C.S., Shortland Street, Auckland

Hall, Edwin, Onehunga

Hamer, W. H., C.E., Harbour Board Offices, Auckland

Hansard, G. A., care of Parr and Blomfield, Shortland Street, Auckland

Hansen, P. M., Queen Street, Auckland

Hardie, J. C., care of Hardie Bros., Queen Street, Auckland

Hawkins, Ven. Archdeacon H. A., Remuera

Hay, Douglas, Stock Exchange, Queen Street, Auckland

Hay, D. A., Montpellier Nurseries, Remuera

Hazard, W. H., Queen Street, Auckland

Heather, H. D., Customs Street East, Auckland

Herbert, T., Shortland Street, Auckland

Hesketh, H. R., Wyndham Street, Auckland

Hesketh, S., Wyndham Street, Auckland

Holderness, D., Harbour Board Offices, Auckland

Horning, J. O., 78 Upper Pitt Street, Newton

Horton, E., *Herald* Office, Auckland

Horton, H., *Herald* Office, Auckland

Houghton, C. V., New Zealand Shipping Company, Quay Street, Auckland

Hutchinson, W. E., Nelson Street, Auckland

Inglis, Dr. T. R., Maroondah, Ponsonby Road, Auckland

Jackson, J. H., Customs Street East, Auckland

Jacobsen, N. R., M.A., Hamilton, Waikato

James, J. W., Mount Albert

Johnson, H. Dunbar, Mangauhenga, Te Aroha

Johnson, Professor J. C., M.A., University College, Auckland*

Johnstone, Halliburton, Howick

Jones, H. W., Public School, Papakura

Kenderdine, J., Sale Street, Auckland

Kent, B., Lower Symonds Street, Auckland

Kent, G. S., St. Stephen's Avenue, Parnell

Keyes, A., Birkenhead

Knight, G., Asquith Avenue, Mount Albert

Kronfeld, G., Eden Crescent, Auckland

Lamb, R. E., 55 Pitt Street, Sydney, New South Wales

Lamb, S. E., B.Sc., University College, Auckland

Lancaster, T. L., University College, Auckland

Lang, Sir F. W., M.P., Onehunga

Langguth, E., Customs Street, Auckland

Larkin, H., Harbour Board Offices, Auckland

Leighton, F. W., High Street, Auckland

Lennox, N. G., care of Auckland Institute, Auckland*

Lewis, Dr. T. H., Remuera

Leyland, W. B., care of Leyland and O'Brien, Customs Street West, Auckland

Leys, Cecil, *Star* Office, Shortland Street, Auckland

Leys, T. W., *Star* Office, Shortland Street, Auckland

Lindsay, Dr. P. A., O'Rorke Street, Auckland

Lunn, A. C., care of Collins Bros., Wyndham Street, Auckland

McColl, J. L., Newmarket

McDowell, Dr. W. C., Remuera

McFarlane, T., C.E., Victoria Arcade, Queen Street, Auckland

McGregor, W. G., Shortland Street, Auckland

Mackay, J. G. H., Ellison Chambers, Queen Street, Auckland

McKenzie, Captain G., Devonport

McKenzie, Dr. Kenneth, O'Rorke Street, Auckland

Maclean, Murdoch, Mount Albert

Macmillan, C. C., care of Gibson

Macmillan, Waingaro*

McMurray, Rev. G., St. Mary's Vicarage, Parnell

McVeagh, R., care of Russell and Campbell, High Street, Auckland

Mahoney, E., Shortland Street, Auckland

Mahoney, T., Swanson Street, Auckland

Mair, Captain G., Waiotapu, Rotorua

Mair, S. A. R., Hunterville, Wellington

Major, C. T., King's College, Remuera

Makgill, Dr. R. H., Public Health Office, Auckland

Marriner, H. A., New Zealand Insurance Company, Auckland

Marsack, Dr., care of Auckland Institute, Auckland

Mennie, J. M., Albert Street, Auckland

Metcalfe, H. H., C.E., Palmerston Buildings, Queen Street, Auckland

Miller, E. V., Chelsea, Auckland

Milnes, H. A. E., Training College, Wellesley Street, Auckland

Milroy, S., Kauri Timber Company, Customs Street West, Auckland

Milsom, Dr. E. H. B., 18 Waterloo Quadrant, Auckland

Mitchelson, Hon. E., Waitaramoa, Remuera

Moore, J. E., Esplanade Road, Mount Eden

Morrison, A. R., Palmerston Buildings, Queen Street, Auckland

Morrison, W. B., Hobson Buildings, Shortland Street, Auckland

Morton, E., Customs Street, Auckland

Morton, H. B., One Tree Hill, Epsom

Moss, E. G. B., Swanson Street, Auckland

Mulgan, E. K., M.A., Education Offices, Auckland

Murray, G. W., Omahu Road, Remuera

Myers, Hon. A. M., M.P., Campbell-Ehrenfried Company, Queen Street, Auckland

Napier, W. J., A.M.P. Buildings, Queen Street, Auckland

Nathan, D. L., care of L. D. Nathan and Co., Shortland Street, Auckland

Nathan, N. A., care of L. D. Nathan and Co., Shortland Street, Auckland*

Nathan, S. J., 5 Carlton Gore Road, Auckland

Neve, F., M.A., B.Sc., Technical College, Wellesley Street, Auckland

Newton, G. M., *Herald* Buildings, Queen Street, Auckland

Niccol, G., Customs Street West. Auckland

Nicholson, O., Imperial Buildings, Queen Street, Auckland

Oliphant, P., 24 Symonds Street, Auckland

Oliver, W. R. B., H.M. Customs, Auckland*

Osmond, G. B., Royal Insurance Buildings, Queen Street, Auckland

Owen, Professor G., M.A., University Buildings, Auckland

Pabst, Dr., Hobson's Buildings, Shortland Street, Auckland

Parr, C. J., C.M.G., M.P., Shortland Street, Auckland

Partridge, H. E., Queen Street, Auckland

Patterson, G. W. S., Shortland Street, Auckland

Peacock, T., Queen Street, Auckland

Peak, A., care of Walker and Peak, Wyndham Street, Auckland

Petrie, D., M.A., Ph.D., Rosemead, Epsom

Philson, W. W., care of Colonial Sugar Company, Quay Street, Auckland

Pond, J. A., F.C.S., Queen Street, Auckland

Poole, G. S., Harbour Board Offices, Auckland

Porter, A., care of E. Porter and Co., Queen Street, Auckland

Porter, Dr. A. H., Dominion Road, Mount Eden

Pountney, W. H., Commerce Street, Auckland

Powell, F. E., C.E., Ferry Buildings, Queen Street, Auckland

Price, E. A., care of Buchanan and Price, Albert Street, Auckland

Pryor, S. H., 36 Pencarrow Avenue, Mount Eden

Purchas, Dr. A. C., Carlton Gore Road, Auckland

Pycroft, A. T., Railway Offices, Auckland

Ralph, W. J., Princes Street, Auckland

Rangihiroa, Dr., care of Auckland Institute, Auckland

Rawnsley, S., Quay Street, Auckland

Reid, J., 45 Fort Street, Auckland

Renshaw, F., care of Sharland and Co., Lorne Street, Auckland

Rhodes, C., Ronaki, Remuera

Robb, J., Victoria Avenue, Mount Eden

Roberton, A. B., Customs Street, Auckland

Roberton, Dr. E., Market Road, Remuera

Robertson, Dr. Carrick, Alfred Street, Auckland

Roche, H., Horahora, Cambridge, Waikato

Rolfe, W., care of Sharland and Co., Lorne Street, Auckland

Rountree, S. G., Auckland Savings-bank, Auckland.

Scott, Rev. D., The Manse, One-hunga

Seegner, C., St. Stephen's Avenue, Parnell

Segar, Professor H. W., M.A., Manukau Road, Parnell

Shakespear, Mrs. R. H., Whanga-paroa, near Auckland.

Shaw, F., Vermont Street, Ponsonby

Shaw, H., Vermont Street, Ponsonby

Shepherd, H. M., Birkenhead

Simmonds, Rev. J. H., Wesley College, Epsom

Simson, T., Mount St. John Avenue, Epsom

Sinclair, A., Kuranui, Symonds Street, Auckland

Sinclair, G., care of Pilkington and Co., Queen Street, Auckland

Skelton, Hall, Watson's Buildings, Queen Street, Auckland

Smeeton, H. M., Binswood, View Road, Mount Eden

Smith, Captain James, Franklin Road, Ponsonby

Smith, H. G. Seth, Russell, Bay of Islands

Smith, S. Percy, F.R.G.S., New Plymouth*

Smith, W. Todd, Brooklands, Alfred Street, Auckland

Somerville, Dr. J., Alfred Street, Auckland

Somerville, J. M., Chelsea, Auckland

Spencer, Percy, Imperial Buildings, Queen Street, Auckland

Spragg, Wesley, Mount Albert

Stewart, James, care of Stewart Bros., Helensville

Stewart, John A., Kainga-tonu, Ranfurly Road, Epsom

Stewart, J. W., Wyndham Street, Auckland

Stewart, R. Leslie, care of Brown and Stewart, Swanson Street, Auckland

Streeter, S. C., Eden Street, Mount Eden

Strevens, J. L., care of Auckland Institute, Auckland

Swan, H. C., Henderson

Swanson, W., Church Street, Auckland

Talbot, Dr. A. G., A.M.P. Buildings, Queen Street, Auckland

Taylor, W., care of Kempthorne, Prosser, and Co., Albert Street, Auckland

Thomas, Professor A. P. W., M.A., F.L.S., Mountain Road, Epsom

Thornes, J., Queen Street, Auckland

Tibbs, J. W., M.A., Grammar School, Auckland

Tinne, H., Union Club, Trafalgar Square, London*

Tole, Hon. J. A., Queen Street, Auckland*

Trounson, J., Northcote, Auckland

Tunks, C. J., care of Jackson and Russell, Shortland Street, Auckland

Upton, J. H., Queen Street, Auckland

Upton, P., South British Insurance Company, Queen Street, Auckland

Urquhart, A. T., Karaka, Drury

Vaile, E. E., Broadlands, Walotapu

Vaile, H. E., Queen Street, Auckland

Wade, H. L., Queen Street, Auckland

Wake, F. W., Wyndham Street, Auckland

Walker, Professor Maxwell, University College, Auckland

Walklate, J. J., Electric Tramway Company, Auckland

Wallace, T. J., Waihi Gold-mining Company, Shortland Street, Auckland

Walters, J. R., Onslow Road, Kingsland, Auckland

Ward, Percy, Mountain Road, Remuera

Ware, W., Portland Road, Remuera

Warnock, J., 2 King Street, Grey Lynn

Watson, Rev. C. H. B., St. Paul's Vicarage, Auckland

Wells, T. U., Westbourne Road, Remuera

White, R. W., Wellington Street, Auckland

Whitley, W. S., Albert Street, Auckland

Whitney, C. A., Victoria Avenue, Remuera

Whittome, F., Carlton Road, Newmarket

Williams, Right Rev. W. L., D.D., Napier

Wilson, Albert, St. Stephen's School, Parnell

Wilson, A. P., Victoria Avenue, Remuera

Wilson, F. W., *Herald* Buildings, Queen Street, Auckland

Wilson, G. W., Cuba Street, Wellington

Wilson, H. M., C.E., Palmerston Buildings, Queen Street, Auckland

Wilson, John, Portland Cement Company, Shortland Street, Auckland

Wilson, J. M., Beach Road, Remuera

Wilson, Liston, Mountain Road, Remuera

Wilson, Martin, Rozelle, Lower Remuera

Wilson, R. M., Russell Road, Remuera

Wilson, W. R., *Herald* Office, Queen Street, Auckland

Winkelmann, H., Victoria Arcade, Queen Street, Auckland

Winstone, F. B., Customs Street, Auckland

Winstone, G., Customs Street, Auckland

Wiseman, J. W., Albert Street, Auckland

Withy, E., care of Auckland Institute, Auckland*

Woodward, W. E., Union Bank of Australia, Queen Street, Auckland

Woolcott, A. C., South British Insurance Company, Queen Street, Auckland

Worley, Professor F. P., D.Sc., University College, Auckland

Wright, R., care of A. B. Wright and Co., Commerce Street, Auckland

Wyllie, A., C.E., Electrical Power Office, Breakwater Road, Auckland

Yates, E., Albert Street, Auckland

Young, J. L., care of Henderson and Macfarlane, Fort Street, Auckland

PHILOSOPHICAL INSTITUTE OF CANTERBURY.
[* Life members.]

Acland, Dr. H. T. D., 381 Montreal Street, Christchurch

Acland, H. D., 42 Park Terrace, Christchurch

Adams, T. W., Greendale

Aldridge, W. G., M.A., Technical College, 13 Barbadoes Street, Christchurch

Alexander, R. E., Canterbury Agricultural College, Lincoln

Allison, H., care of Harman and Stevens, Christchurch

Anderson, Dr. C. Morton, 142 Worcester Street, Christchurch

Anderson, Gilbert, care of Bank of New Zealand, 1 Queen Victoria Street, London*

Archey, G. E., M.A., Canterbury Museum, Christchurch

Baker, T. N., care of Baker Bros., Manchester Street, Christchurch

Baughan, Miss B. E., Sumner

Beaven, A. W., care of Andrews and Beaven, Moorhouse Avenue, Christchurch

Bell, N. M., M.A., Muspratt Laboratory, Liverpool, England

Bevan-Brown, C. E., M.A., Boys' High School, Christchurch

Bird, J. W., M.A., Nelson College, Nelson

Birks, L., B.Sc., care of Public Works Department, Christchurch

Bishop, F. C. B., 101 Armagh Street, Christchurch

Bishop, R. C., Gas Office, 77 Worcester Street, Christchurch

Blackburne, S. L., Rolleston Avenue, Christchurch

Blanch, G. E., M.A., Christ's College, Christchurch

Boag, T. D., Webb's Road, Bryndwyr

Booth, G. T., Cashmere Hills

Borrie, Dr. F. J., 236 Hereford Street, Christchurch

Bowen, Sir Charles C., M.L.C., F.R.G.S., Middleton

Bradley, Orton, Charteris Bay

Brent, H. C., " Aberdare House," Christchurch

Brittin, Guy, 146 Armagh Street, Christchurch

Broadhead, H. D., 20 Eversleigh Street, St. Albans, Christchurch

Brock, W., M.A., Education Office, Chistchurch

Brown, Professor Macmillan, M.A., LL.D., Holmbank, Cashmere Hills*

Buddo, Hon. D., M.P., Rangiora

Bullen, Miss Gertrude, care of Mrs. Nixon, Harakeke Street, Christchurch

Burnett, T. D., Mount Cook, Fairlie

Chilton, Professor C., D.Sc., M.A., LL.D., M.B., F.L.S., Canterbury College*

Christensen, C. E., Hanmer Springs

Clark, W. H., 100 Bealey Avenue, Christchurch

Cocks, Rev. P. J., B.A., St. John's Vicarage, Christchurch

Cocks, Miss, Colombo Road South, Christchurch

Colee, W. C., M.A., Schoolhouse, Opawa

Coles, W. R., 256 Wilson's Road, Christchurch

Corkill, F. M., B.Sc., Canterbury College

Dash, Charles, 155 Norwood Street, Beckenham, Christchurch.

Deans, J., Kirkstyle, Coalgate

Denniston, Mr. Justice, 58 Rolleston Avenue, Christchurch

Dobson, A. Dudley, M.Inst.C.E., City Council Office, Christchurch

Dorrien-Smith, Major A. A., D.S.O., Tresco Abbey, Scilly, England

Drummond, James, F.L.S., F.Z.S., *Lyttelton Times*, Christchurch

English, R., F.C.S., M.I.M.E., Gas Office, 77 Worcester Street, Christchurch

Evans, Professor W. P., M.A., Ph.D., Canterbury College, Christchurch

Fairbairn, A., 53 Fendalton Road, Christchurch

Farr, Professor C. Coleridge, D.Sc., A.M.Inst.C.E., Canterbury College, Christchurch

Ferrar, Miss, 450 Armagh Street, Christchurch

Ferrar, H. T., M.A., F.G.S., " Merchiston," Opawa, Christchurch

Finlayson, Miss, M.A., West Christchurch School, Lincoln Road, Christchurch

Firman, Henry, 95 River Road, Beckenham

Fletcher, T., Public School, Sydenham

Flower, A. E., M.A., M.Sc., Christ's College, Christchurch

Foster, T. S., M.A., 19 Cashel Street, Christchurch

Foweraker, C. E., M.A., Canterbury College, Christchurch

Gabbatt, Professor J. P., M.A., M.Sc., Canterbury College

Garton, John W., 61 Richardson Street, Woolston, Christchurch

Garton, W. W., M.A., Elmwood School, Christchurch

Gibson, Dr. F. Goulburn, 121 Papanui Road

Gilling, W. O. R., B.A., Canterbury College, Christchurch

Godby, M. H., Hereford Street, Christchurch

Goss, W., Rossal Street, Fendalton

Gray, G., F.C.S., Lincoln

Grigg, J. C. N., Longbeach

Gudex, M. C., M.A., B.Sc., High School, Hamilton

Hall, J. D., Middleton

Hall, Miss, Gloucester Street West, Christchurch

Hansford, G. D., Parliament Buildings, Wellington

Haynes, E. J., Canterbury Museum, Christchurch

Herring, E., 46 Paparoa Street, Papanui

Hight, Professor J., M.A., Litt.D., Canterbury College, Christchurch

Hilgendorf, F. W., M.A., D.Sc., Canterbury Agricultural College, Lincoln

Hill, Mrs. Carey, 84 Papanui Road, Christchurch

Hitchings, F., 69 Durham Street, Sydenham

Hodgson, T. V., F.L.S., Science and Art Museum, Plymouth, England

Hogg, E. G., M.A., F.R.A.S., Christ's College, Christchurch

Hogg, H. R., M.A., F.Z.S., 13 St. Helen's Place, London, E.C.

Holland, H., 108 St. Asaph Street, Christchurch

Holloway, Rev. J., M.Sc., Hokitika

Hughes, T., B.A., Geraldine

Humphreys, G., Fendalton Road, Fendalton

Humphreys, Mrs. G., "Norholme," Bealey Avenue, Christchurch

Hutton, Mrs., Gloucester Street, Christchurch

Ingram, John, 39 Mansfield Avenue, St. Albans, Christchurch.

Irving, Dr. W., 56 Armagh Street, Christchurch

Jackson, T. H., B.A., Boys' High School, Christchurch

Jameson, J. O., Hereford Street, Christchurch

Jamieson, J., Hereford Street, Christchurch

Jennings, L. S., M.Sc., B.A., Waitaki Boys' High School, Oamaru

Jennings, Mrs. L. S., M.A., care of Girls' High School, New Plymouth

Kaye, A., 429 Durham Street, Christchurch

Kidson, E. R., M.Sc., Department of Terrestrial Magnetism, Washington, D.C., U.S.A.*

Kirkpatrick, W. D., Redcliffs, Sumner

Kitchingman, Miss, Hackthorne Road, Cashmere

Knight, H. A., Racecourse Hill

Laing, R. M., M.A., B.Sc., Boys' High School, Christchurch

Lang, H., B.A., Christ's College, Christchurch

Lester, Dr. G., 2 Cranmer Square, Christchurch

Louisson, Hon. C., M.L.C., 71 Gloucester Street, Christchurch

Macbeth, N. L., Canterbury Frozen Meat Company, Hereford Street, Christchurch

McBride, T. J., 15 St. Albans Street, Christchurch

MacGibbon, Dr. T. A., Royal Exchange Buildings, Christchurch

MacIndoe, George, B.E., Canterbury College, Christchurch

Macleod, D. B., M.A., B.Sc., Canterbury College, Christchurch

Marsh, H. E., Cashmere

Marshall, Mrs., New Brighton

Martin, William, 56 Carlton Street, Merivale, Christchurch

Meares, H. O. D., Fendalton

Mills, Miss C. B., M.A., B.Sc., Technical College, Christchurch

Morris, C. Barham, F.R.M.S., 78 Andover Street, Christchurch

Morrison, W. G., Hanmer

Mountford, A. V., 8 Park Street, Woolston, Christchurch

Murray, W., " Balgownie," Opawa, Christchurch

Murray-Aynsley, H. P., Clyde Road, Riccarton

Nairn, R., Lincoln Road, Spreydon

Nelson, P. S., M.Sc., 10 Flockton Street, St. Albans, Christchurch

Oliver, F. S., care of A. E. Craddock, Manchester Street, Christchurch

Olliver, Miss F. M., M.A., M.Sc., Waimate

Page, S., B.Sc., Canterbury College, Christchurch

Pairman, Dr. T. W, Governor's Bay

Pairman, Dr. J. C., Dominion Buildings, Christchurch

Pannett, J. A., Cashmere Hills

Powell, P. H., M.Sc., Canterbury College, Christchurch

Purchas, Rev. A. C., M.A., Geraldine

Purnell, C. W., Ashburton

Rands, Henry, M.A., Waitaki High School, Oamaru

Reece, W., Colombo Street, Christchurch

Relph, E. W., " Chilcombe," Fendalton Road, Christchurch

Rhodes, A. E. G., B.A., Fendalton

Rhodes, Colonel the Hon. R. Heaton, M.P., Tai Tapu

Robinson, W. F., F.R.G.S., Canterbury College

Ross, R. G., Telegraph Department, Wellington

Rowe, H. V., M.A., Canterbury College, Christchurch

Rowe, T. W., M.A., LL.B., 77 Hereford Street, Christchurch

Scott, Professor R. J., M.Inst.C.E., F.A.I.E.E., Canterbury College, Christchurch

Scott, G., Manchester Street, Christchurch

Seager, S. Hurst, F.R.I.B.A., Cathedral Square, Christchurch

Seth-Smith, B., 25 Stratford Street, Fendalton

Sheard, Miss F., M.A., B.Sc., Girls' High School, Christchurch

Sims, A., M.A., care of Sims, Cooper, and Co., Hereford Street, Christchurch

Skey, H. F., B.Sc., Magnetic Observatory, Christchurch

Sloman, C. J., Crown Brewery, 38 St. Asaph Street, Christchurch

Snow, Colonel, Holmwood Road, Christchurch

Speight, R., M.A., M.Sc., F.G.S., Canterbury College, Christchurch

Stark, E. E., B.Sc., P.O. Box 526, Christchurch

Stead, E. F., Ilam, Riccarton

Stevenson, Dr. J., Fendalton

Stevenson, James, Flaxton

St. John, Charles E., 745 Colombo Street, Christchurch

Stone, T., *Lyttelton Times* Office, Christchurch

Suter, Henry, 559 Hereford Street, Linwood

Symes, Dr. W. H., 176 Worcester Street, Christchurch*

Symes, Langford P., 20 May's Road, St. Albans

Tabart, Miss Rose, 97 Papanui Road, Christchurch

Taylor, A., M.A., M.R.C.V.S., Canterbury Agricultural College, Lincoln

Taylor, G. J., 440 Madras Street, St. Albans

Thacker, Dr. H. T. J., M.P., 25 Latimer Square, Christchurch

Thomas, Dr. W., 579 Colombo Street, Christchurch*

Thompson, Rev. A. T., M.A., B.D., St. Andrew's Manse, Christchurch

Tripp, C. H., M.A., Timaru*

Waller, F. D., B.A., West Christchurch District High School

Wallich, M. G., Manchester Street, Christchurch

Waymouth, Mrs., care of Mrs. R. M. Hughes, St. Buryan, S.O., Cornwall, England

Weston, G. T., B.A., LL.B., 173 Cashel Street, Christchurch

Whetter, Dr. J. P., 211 Gloucester Street, Christchurch

Whitaker, C. Godfrey, care of Booth, Macdonald, and Co., Carlyle Street, Christchurch

Whitehead, L. G., B.A., Boys' High School, Christchurch

Wigram, Hon. F., M.L.C., 1 Armagh Street, Christchurch

Wild, L. J., M.A., F.G.S., Canterbury Agricultural College, Lincoln

Wilding, Frank S., Hereford Street, Christchurch

Wilkins, C., Public School, Jerrold Street, Addington

Williams, C. J., M.Inst.C.E., Knowles Street, St. Albans

Wright, A. M., F.C.S., Box 617, Post-office, Christchurch

OTAGO INSTITUTE.

[* Life members.]

Allan, Dr. W., Mosgiel

Allen, Hon. James, M.P., Clyde Street

Allen, Dr. S. C., 220 High Street

Anscombe, E., 171 Princes Street

Balk, O., Driver Street, Maori Hill

Barnett, Dr. L. E., Stafford Street

Barr, Peter, 3 Montpellier Street

Bathgate, Alexander, Neidpath Road, Mornington*

Beal, L. O., Stock Exchange Buildings

Begg, J. C., 2 Elder Street

Bell, A. Dillon, Shag Valley*

Benham, Professor W. B., M.A., D.Sc., F.Z.S., F.R.S., Museum

Black, Alexander, 82 Clyde Street*

Black, James, care of Cossens and Black

Boys-Smith, Professor, University

Braithwaite, Joseph, 36 Princes Street

Brasch, H., 55 London Street

Bremner, James, Fifield Street, Roslyn

Brent, D., M.A., 19 New Street, Musselburgh*

Brown, W., 99 Clyde Street

Browne, Robert, Technical School, Hawera

Buchanan, N. L., 44 Bronte Street, Nelson*

Buckland, Mrs., Waikouaiti

Buddle, Dr. Roger, care of Buddle and Button, Wyndham Street, Auckland

Burt, Ross, care of A. and T. Burt (Limited)

Cameron, Dr. P. D., 145 Leith Street

Chamberlain, C. W., 6 Regent Road

Champtaloup, Dr. S. T., 4 Elder Street

Chapman, C. R., 135 Town Belt, Roslyn

Chapman, Mr. Justice, Supreme Court, Christchurch

Church, Dr. R., High Street

Clarke, C. E., 51 King Edward Road

Clarke, E. S., Woodhaugh

Colquhoun, Dr. D., High Street

Coombs, L. D., A.R.I.B.A. Stuart Street and Octagon

Crawford, W. J., 179 Walker Street

Dalrymple, Rev. A. M., M.A., 65 District Road, Mornington

Davidson, R. E., Hawthorn Road, Mornington

Davies, G. W., 9 Gladstone Street, Belleknowes

Davies, O. V., 109 Princes Street

Davis, A., Test Room, Cumberland Street

De Beer, I. S., London Street

Duncan, P., "Tolcarne," Maori Hill

Dunlop, Professor F. W., M.A., Ph.D., University

Dutton, Rev. D., F.G.S., F.R.A.S., Caversham

Edgar, G. C., Market Street

Edgar, James, 144 York Place

Fairclough, Rev. P. W., F.R.A.S., Kaiapoi

Farquharson, R.A., M.Sc., Geological Survey, Perth, W.A.

Fels, W., 48 London Street*

Fenwick, Cuthbert, Stock Exchange

Fenwick, G., *Otago Daily Times*

Ferguson, Dr. H. L., 434 High Street

Fisher, T. R., Alexandra Street, St. Clair

Fitchett, Dr. F. W. B., Pitt Street

Frye, Charles, Gasworks, Caversham

Fulton, H. V., A. and P. Society, Crawford Street

Fulton, Dr. R. V., Pitt Street

Fulton, S. W., The Exchange, Collins Street, Melbourne*

Garrow, Professor J. M. E., LL.B., Victoria College, Wellington*

Gibson, G. W., Silverton, Anderson's Bay

Gilkison, R., 14 Main Road, Northeast Valley*

Glasgow, W. T., Dunblane Street, Roslyn

Gowland, Professor W. P., M.D., University

Goyen, P., F.L.S., 136 Highgate, Roslyn

Green, E. S., Education Office

Gully, G. S., care of J. Rattray and Sons

Guthrie, H. J., 426 Moray Place E.

Hall, Dr. A. J., Stuart Street

Hamilton, T. B., M.A., B.Sc., University

Hanlon, A. C., Pitt Street

Hart, H. E., Royal Terrace

Henderson, M. C., Electrical Engineer's Office, Market Street

Henton, J. W., 140 York Place

Hercus, G. R., 20 Albert Street

Hooper, B. B., A.R.I.B.A., A.M.P. Buildings

Hosking, Mr. Justice, Supreme Corut, Wellington

Howes, Miss Edith, School, Gore*

Howes, W. G., F.E.S., 432 George Street

Hungerford, J. T., Gasworks

Inglis, Professor J. K. H., M.A., D.Sc., F.I.C., University

Jack, Professor R., D.Sc., University

Jeffrey, J., Anderson's Bay

Joachim, G., Randall Street, Mornington*

Johnson, J. T., Littlebourne Road, Roslyn

Johnstone, J. A., Driver Street, Maori Hill

Jones, R. C., Dunblane Street, Roslyn

Kerr, W. J., National Bank

King, Dr. F. Truby, Seacliff

Laing, John, 86 Queen Street

Lee, G. A., Bluff Harbour Board

Lee, Robert, P.O. Box 363

Loudon, John, 43 Crawford Street

Lough, F. J., 12 Queen's Drive, Musselburgh

Lusk, T. H., Black's Road, Opoho

McCurdie, W. D. R., Town Hall*

Macdougall, W. P., jun., 642 George Street

McEnnis, J. E., Public Works Office

McEvoy, W. L., Grove Street, Musselburgh

McGeorge, J. C., Eglinton Road, Mornington

McKellar, Dr. T. G., Pitt Street

McKerrow, James, F.R.A.S., 142 Ghuznee Street, Wellington

Mackie, A., Test Room, Cumberland Street

McRae, H., London Street

Malcolm, Professor J., M.D., University

Mandeno, H., New Zealand Express Company's Buildings

Marshall, Angus, B.A., Technical School

Marshall, Professor P., M.A., D.Sc., F.G.S., University

Mason, J. B., Otago Harbour Board

Massey, Horatio, Invercargill

Melland, E., Arthog Road, Hale, Cheshire, England*

Miller, David, 25 City Road, Roslyn

Milnes, J. W., 39 Lees Street*

Mitchell, W. J., U.S.S. Company, Port Chalmers

Morrell, W. J., M.A., Boys' High School

Monkman, W. T., 71 Bond Street

Morris, J. Fairly, Port Chalmers

Napier, O. J. W., M.A., University

Neill, W. T., Survey Office

Nevill, Rt. Rev. S.T., D.D., Bishopsgrove

Newlands, Dr. W., 12 London Street

Oakden, F., N.S.W. Cement, Lime, and Coal Company, Mutual Life Buildings, Sydney

O'Neill, Dr. E. J., 219 High Street

Orchiston, G. J., Test Room, Cumberland Street

20—Trans.

Overton, T. R., Test Room, Cumberland Street

Park, Professor J., F.G.S., University

Parr, E. J., M.A., B.Sc., Boys' High School

Payne, F. W., 90 Princes Street

Petrie, D., M.A., Epsom, Auckland*

Pickerill, Professor H. P., M.B., B.D.S., University.

Poppelwell, D. L., Gore

Price, W. H., 55 Stuart Street*

Reid, Donald, jun., Dowling Street

Richards, Professor D. J., M.A., University

Richardson, C. R. D., B.A., Education Office

Riley, Dr. F. R., Pitt Street

Ritchie, Dr. Russell, 400 George Street

Roberts, E. F., 128 Highgate, Roslyn

Roberts, John, C.M.G., Littlebourne

Ross, H. I. M., Willis Street

Ross, T. C., care of Ross and Glendining (Limited)

Rutherford, R. W., Playfair Street, Caversham

Salmond, J. L., National Bank Buildings

Sandle, Major S. G., Onslow House, St. Kilda

Sargood, Percy, " Marinoto," Newington

Scott, J. H., Converter Station, Cumberland Street

Seymour, S. P., B.A., Mossburn, Southland

Shacklock, J. B., Bayfield, Anderson's Bay

Shennan, Watson, 367 High Street

Shepherd. F. R., P.O. Box 361

Shortt, F. M., care of John Chambers and Sons

Sim, Mr. Justice, Musselburgh

Simpson, F. A., care of John Chambers and Sons

Simpson, George, 98 Russell Street

Simpson, George, jun., 9 Gamma Street, Roslyn

Smith, C. S., *Star* Office

Smith, J. C., 196 Tay Street, Invercargill

Solomon, S., K.C., 114 Princess Street

Somerville, W. G., 18 Leven Street, Roslyn

Stark, James, care of Kempthorne, Prosser, and Co.

Statham, F. H., A.O.S.M., 26 Dowling Street

Stewart, R. T., 21 Gamma Street, Roslyn

Stewart, W. D., M.P., LL.B., 62 Heriot Row

Stout, Sir Robert, K.C.M.G., Wellington

Symes, H., Hill Street, Mornington

Tannock, D., Botanical Gardens

Theomin, D., 42 Royal Terrace

Thompson, G. E., M.A., University

Thomson, G. M., F.L.S., 5 Sheen Street, Roslyn*

Thomson, T., Ngaruahia, Mangere

Thomson, W. A., A.M.P. Buildings

Uttley, G. H., M.A., M.Sc., Scots College, Wellington

Vanes, R. N., A.R.I.B.A., National Bank Chambers

Walden, E. W., 12 Dowling Street

Wales, P. Y., 2 Crawford Street

Walker, A., Lloyd's Surveyor, Wellington

Waters, Professor D. B., A.O.S.M., University

White, Professor D. R., M.A., 83 St. David Street

Whitson, T. W., 584 George Street

Whyte, W. J., 76 Highgate, Roslyn

Williams, Sir Joshua S., M.A., LL.D., K.C.M.G., Supreme Court*

Williamson, W. J., 27 Octagon

Wilson, W. S., 132 Bond Street

Wingfield, J. E., 663 Castle Street

Young, Dr. James, Don Street, Invercargill

HAWKE'S BAY PHILOSOPHICAL INSTITUTE.

[* Life members.]

Antill, H. W., Kumeroa
Armstrong, C. E., Gisborne
Asher, Rev. J. A., Napier
Bennett, A., Patoka
Bernau, Dr. H. F., Napier
Bull, Harry, Gisborne
Burnett, H., Woodville
Chambers, Bernard, Te Mata
Chambers, J., Mokopeka, Hastings
Chisholm, H. M., Napier .
Clark, Gilbert, Taradale
Clark, Thomas, Eskdale
Coe, J. W., Napier
Cooper, S. E., Napier
Cornford, Cecil, Napier
Craig, J. Wilson, Napier*
Dinwiddie, P., Napier
Dinwiddie, W., Napier
Donnelly, G. P., Napier
Duncan, Russell, Napier
Edgar, Dr. J. J., Napier
Edmundson, J. H., Napier
Fitzgerald, J., Napier
Fossey, W., Napier
Grant, M. R., Napier
Guthrie-Smith, H., Tutira
Harding, J. W., Mount Vernon, Waipukurau
Heaton, F., M.A., B.Sc., Dannevirke
Henderson, E. H., Te Araroa
Henley, Dr. E. A. W., Napier
Hill, H., B.A., F.G.S., Napier
Hislop, J., Napier*
Holdsworth, J., Havelock North
Humphrey, E. J., Pakipaki
Hutchinson, F., jun., Rissington
Hyde, Thomas, Napier
Kennedy, C. D., Napier
Kerr, W., M.A., Napier
Large, J. S., Napier*
Large, Miss, Napier
Leahy, Dr. J. P., Napier
Loten, E. G., Napier
Lowry, T. H., Okawa
McLean, R. D. D., Napier
Matthews, E. G., Gisborne
Mayne, Rev. Canon, Napier
Metcalfe, W. F., Kiritaki, Port Awanui
Moore, Dr. T. C., Napier
Morris, W., Hastings
Murphy, W., Woodlands Road, Woodville
Niven, J., M.A., M.Sc., Napier
Northcroft, E., Napier
Oates, William, J.P., Tokomaru Bay
Ormond, Hon. J. D., M.L.C., Napier
Ormond, G., Mahia
O'Ryan, W., Waipiro Bay
Piper, G. M., Pakowhai
Poole, P. Loftus, Tupuroa Bay, East Coast
Scott, William, Napier
Sheath, J. H., Napier
Sinclair, G. K., Hastings
Skinner, W. H., Napier
Smart, D. L., Napier
Smith, J. H., Olrig*
Smith, W., B.A., Gisborne
Stevenson, A., Woodville
Strachan, D. A., M.A., Napier
Tanner, T., Havelock North
Thomson, J. P., Napier
Tiffen, G. W., Cambridge
Townley, J., Gisborne
Townson, W., Pukekohe, Auckland
Wheeler, E. G., Havelock North
Whyte, D., Hastings
Williams, F. W., Napier
Williams, G. T., Mokoiwi, Tuparoa East Coast
Williams, Rev. H., Gisborne

NELSON INSTITUTE.

Bett, Dr. F. A., Trafalgar Square
Boor, Dr. L., Alton Street
Cooke, Miss M., Examiner Street
Crawford, A. C., Scotland Street
Curtis, W. S., Bronte Street
Fell, C. R., Brougham Street
Field, T. A. H., M.P., Ngatitama Street
Gibbs, F. G., M.A., Collingwood Street
Glasgow, J., Trafalgar Street
Graham, Mrs. C., Bridge Street
Harley, Charles J., Milton Street
Healy, F., Bridge Street
Hornsby, J. P., Weka Street
Jackson, R. B., Examiner Street

Knapp, F. V., Alfred Street
Lancaster, G. J., M.A., Nelson College
Morley, E. L., Waimea Street
Mules, Bishop, Trafalgar Square
Redgrave, A. J., Hardy Street
Short, William, Examiner Street
Snodgrass, W. W., Hardy Street
Strachan, J., care of Land Transfer Office, Government Buildings

Thompson, F. A., Commissioner of Crown Lands, Government Buildings
Ward, W. T., Chief Postmaster, Christchurch
Washbourn, H. P., Port Nelson
Whitwell, F., Drumduan, Wakapuaka
Wise, G. R., care of Pitt and Moore, Solicitors, Nelson
Worley, W. F., Trafalgar Street South

MANAWATU PHILOSOPHICAL SOCIETY.

[* Life members.]

Akers, H., Duke Street
Armstrong, E. J., C.E., The Square
Barnicoat, J. L., Union Bank
Batchelar, J. O., Willow Bank
Baylis, G. de S., Department of Agriculture
Bayly, Mrs., Palmerston North
Beale, J. Bruce, The Square
Bendall, W. E., College Street
Bett, D. H. B., M.B., Ch.B., M.R.C.S., L.R.C.P., Broad Street
Buick, D., M.P., Cloverlea
Clausen, C. N., Rangitikei Street
Cohen, M., Broad Street
Colquhoun, A.J., M.Sc., High School
Cooke, F. H., The Square
Doull, Rev. A., M.A., Church Street
Drake, A., Manakau
Eliott, M. A., The Square
Gardner, R., Terrace End
Gerrand, J. B., The Square
Graham, A. J., The Square
Greer, S., Broad Street
Hankins, J. H., Main Street
Hannay, A., care of Manson and Barr
Hodder, T. R., Rangitikei Street
Ironside, Miss, M.A., High School
Johnston, J. Goring, Oakhurst
Low, D. W., College Street
McNab, Hon. R., M.P., Litt.D., F.R.G.S., Palmerston North
Manson, T., Rangitikei Street
Martin, A. A., M.D., Ch.M., Fitzherbert Street
Morgan, A. H., M.A., High School

Mounsey, J., Rangitikei Street
Murray, J., M.A., High School
Nathan, F. J., Palmerston North
Natusch, C. A., Church Street
Park, W., F.R.H.S., The Square
Peach, C. W., M.B., C.M., Broad Street
Poynton, J. W., S.M., Featherston Street
Primmer, Captain J. H., North Street
Rainforth, J., Balt Street
Robertson, J., Rangitikei Street
Roth, C., Rongotea
Salmon, C. T., Assoc. in Eng., Rangitikei Street
Scott, G. J., The Square
Sinclair, D., C.E., Terrace End
Skinner, H. D., B.A., Palmerston N.
Smith, W. H., The Square
Smith, W. W., F.E.S., New Plymouth
Stevens, J., Church Street
Stowe, W. R., M.R.C.S., M.R.C.P., Linton Street
Sutherland, A., Boundary Road
Taplin, C. N., George Street
Vernon, J. E., M.A., B.Sc., High School
Waldegrave, C. E., Broad Street
Welch, W., F.R.G.S., Mosman's Bay, N.S.W.*
Williams, G., Farmer, Sandon
Wilson, K., M.A., Rangitikei Street*
Wollerman, H., Fitzherbert Street
Young, H. L., Cuba Street

WANGANUI PHILOSOPHICAL SOCIETY.

[* Life members.]

Allison, Alexander, No. 1 Line, Wanganui

Allison, Thomas, Ridgway Street, Wanganui

Amess, A. H. R., M.A., Collegiate School, Wanganui

Atkinson, W. E., Hurworth, Wanganui

Ball, J., *Chronicle*, Wanganui

Battle, T. H., Architect, Wanganui

Basset, W. G., St. John's Hill, Wanganui

Bourne, F., F.I.A.N.Z., Ridgway Street, Wanganui

Brown, C. P., M.A., LL.B., College Street, Wanganui

Burnet, J. H., St. John's Hill, Wanganui

Cave, Norman, Brunswick Line, Wanganui

Clark, E. H., College Street, Wanganui

Cowper, A. E., Victoria Avenue, Wanganui

Crow, E., Technical College, Wanganui

Cruickshank, Miss, M.A., M.Sc., Girls' College, Wanganui

D'Arcy, W. A., 11 Campbell Street, Wanganui

Downes, T. W., Victoria Avenue, Wanganui

Drew, Harry, Victoria Avenue, Wanganui

Duignan, Herbert, Ridgway Street, Wanganui

Dunkley, R., F.I.A.N.Z., Wickstead Place, Wanganui

Dunn, Richmond, St. John's Hill, Wanganui

Enderby, H. H., Dentist, Wanganui

Ford, C. R., F.R.G.S., College Street, Wanganui

Gibbons, Hope, Wanganui

Hall, William, Campbell Street, Wanganui

Hatherly, Henry R., M.R.C.S., Gonville, Wanganui*

Helm, H. M., Victoria Avenue, Wanganui

Hutchison, George, Wickstead Place, Wanganui

Hutton, C. C., M.A., St. John's Hill, Wanganui

Jack, J. B., Native Land Court, Wanganui

Jones, Lloyd, Victoria Avenue, Wanganui

Kerr, William, S.M., Magistrate's Court, Wanganui

Liffiton, E. N., J.P., Ridgway Street, Wanganui

McFarlane, D., Ridgway Street, Wanganui

Mackay, C. E., Mayor of Wanganui

Muir, A. G., B.Sc., Surveyor, Wanganui

Murdoch, R., Campbell Place, Wanganui

Murray, J. B., St. John's Hill, Wanganui

Neame, J. A., M.A., Collegiate School, Wanganui

Payne, H. M., 32 Bell Street, Wanganui

Peck, F. Leslie, Liverpool Street, Wanganui

Polson, D. G., St. John's Hill, Wanganui

Saywell, T. R., Public Trust Office, Wanganui

Staveley, N. C., C.E., Borough Engineer, Wanganui

Sturge, H. E., M.A., Collegiate School, Wanganui

Talboys, F. P., Tramways Manager, Wanganui

Thompson, H. H., Victoria Avenue, Wanganui

Ward, J. T., Victoria Avenue, Wanganui

Watt, J. P., B.A., LL.B., Ridgway Street, Wanganui

Watt, M. N., St. John's Hill, Wanganui

Williamson, J. P., College Street, Wanganui

Wilson, Alexander, M.D., Wickstead Street, Wanganui

LIST OF INSTITUTIONS

TO WHICH

THE PUBLICATIONS OF THE INSTITUTE ARE PRESENTED BY THE
GOVERNORS OF THE NEW ZEALAND INSTITUTE.

———

Honorary Members of the New Zealand Institute, 30.

New Zealand.

Cabinet, The Members of, Wellington.
Executive Library, Wellington.
Free Public Library, Auckland.
        ,,   Christchurch.
        ,,   Dunedin.
        ,,   Wellington.
Government Printer and publishing staff (6 copies).
Library, Auckland Institute, Auckland.
   ,,   Auckland Museum, Auckland.
   ,,   Biological Laboratory, Canterbury College, Christchurch.
   ,,   Biological Laboratory, University College, Auckland.
   ,,   Biological Laboratory, University of Otago, Dunedin.
   ,,   Biological Laboratory, Victoria College, Wellington.
   ,,   Canterbury College, Christchurch.
   ,,   Canterbury Museum, Christchurch.
   ,,   Dunedin Athenæum.
   ,,   General Assembly, Wellington (2 copies).
   ,,   Hawke's Bay Philosophical Institute, Napier.
   ,,   Manawatu Philosophical Society, Palmerston North.
   ,,   Nelson College.
   ,,   Nelson Institute, Nelson.
   ,,   New Zealand Geological Survey.
   ,,   New Zealand Institute of Surveyors.
   ,,   New Zealand Institute, Wellington.
   ,,   Otago Institute, Dunedin.
   ,,   Otago Museum, Dunedin.
   ,,   Otago School of Mines, Dunedin.
   ,,   Philosophical Institute of Canterbury, Christchurch.
   ,,   Polynesian Society, New Plymouth.
   ,,   Portobello Fish-hatchery, Dunedin.
   ,,   Reefton School of Mines.
   ,,   Southland Museum, Invercargill.
   ,,   Thames School of Mines.
   ,,   University College, Auckland.
   ,,   University of Otago, Dunedin.
   ,,   Victoria College, Wellington.
   ,,   Waihi School of Mines, Waihi.
   ,,   Wanganui Museum.
   ,,   Wellington Philosophical Society.

Great Britain.

Bodleian Library, Oxford University.
British Association for the Advancement of Science, London.
British Museum Library, London.
    „   Natural History Department, South Kensington, London S.W.
Cambridge Philosophical Society, Cambridge University.
Colonial Office, London.
Clifton College, Bristol, England.
Geological Magazine, London.
Geological Society, Edinburgh.
    „   London.
Geological Survey of the United Kingdom, London.
High Commissioner for New Zealand, London.
Imperial Institute, London.
Institution of Civil Engineers, London.
International Catalogue of Scientific Literature, 34 Southampton Street, Strand, London.
Leeds Geological Association, Sunnyside, Crossgate, Leeds.
Linnean Society, London.
Literary and Philosophical Society, Liverpool.
Liverpool Biological Society.
Marine Biological Association of the United Kingdom, Plymouth.
Natural History Society, Glasgow.
Nature, The Editor of, London.
Norfolk and Norwich Naturalist Society, Norwich.
North of England Institute of Mining and Mechanical Engineers, Newcastle-upon-Tyne.
Patent Office Library, 25 Southampton Street, London W.C.
Philosophical Society of Glasgow.
Royal Anthropological Institute of Great Britain and Ireland, 59 Great Russell Street, London W.C.
Royal Botanic Garden Library, Edinburgh.
Royal Colonial Institute, London.
Royal Geographical Society, 1 Savile Row, London W.
Royal Institution, Liverpool.
Royal Irish Academy, Dublin.
Royal Physical Society, Edinburgh.
Royal Scottish Geographical Society, Synod Hall, Castle Terrace, Edinburgh.
Royal Society, Dublin.
    „   Edinburgh.
    „   London.
Royal Society of Literature of the United Kingdom, London.
Royal Statistical Society, London.
University Library, Cambridge, England.
    „   Edinburgh.
Victoria College, Manchester.
Victoria Institute, London.
William Wesley and Son, London (Agents).
Zoological Society, London.

British North America.

Geological and Natural History Survey of Canada, Ottawa.
Hamilton Scientific Association, Hamilton, Canada.
Institute of Jamaica, Kingston.
Natural History Society of New Brunswick, St. John's.
Nova-Scotian Institute of Natural Science, Halifax.
Royal Canadian Institute, Toronto.

South Africa.

Free Public Library, Cape Town.
South African Association for the Advancement of Science, Cape Town.
South African Museum, Cape Town.
Rhodesia Museum, Bulawayo, South Africa.

India.

Asiatic Society of Bengal, Calcutta.
Colombo Museum, Ceylon.
Geological Survey of India, Calcutta.
Natural History Society, Bombay.
Raffles Museum, Singapore.

Queensland.

Geological Survey Office, Brisbane.
Queensland Museum, Brisbane.
Royal Society of Queensland, Brisbane.

New South Wales.

Agricultural Department, Sydney.
Australasian Association for the Advancement of Science, Sydney.
Australian Museum Library, Sydney.
Department of Mines, Sydney.
Engineering Association of New South Wales, Sydney.
Engineering Institute of New South Wales, Watt Street, Newcastle.
Library, Botanic Gardens, Sydney.
Linnean Society of New South Wales, Sydney.
Public Library, Sydney.
Royal Society of New South Wales, Sydney.
University Library, Sydney.

Victoria.

Australian Institute of Mining Engineers, Melbourne.
Field Naturalists' Club, Melbourne.
Geological Survey of Victoria, Melbourne.
Legislative Library, Melbourne.
Public Library, Melbourne.
Royal Society of Victoria, Melbourne.
University Library, Melbourne.

Tasmania.

Public Library of Tasmania, Hobart.
Royal Society of Tasmania, Hobart.

South Australia.

Public Museum and Art Gallery of South Australia, Adelaide.
Royal Society of South Australia, Adelaide.
University Library, Adelaide.

Western Australia.

Government Geologist, Perth.

Russia.

Finskoie Uchonoie Obshchestvo (Finnish Scientific Society), Helsingfors.
Imper. Moskofskoie Obshchestvo Iestestvo‑Ispytatelei (Imperial Moscow Society of Naturalists). -
Kiefskoie Obshchestvo Iestestvo‑Ispytatelei (Kief Society of Naturalists).

Norway.

Bergens Museum, Bergen.
University of Christiania.

Sweden.

Geological Survey of Sweden, Stockholm.
Royal Academy of Science, Stockholm.

Denmark.

Natural History Society of Copenhagen.
Royal Danish Academy of Sciences and Literature of Copenhagen.

Germany.

Botanischer Verein der Provinz Brandenburg, Berlin.
Königliche Bibliothek, Berlin.
Königliche Physikalisch-Oekonomische Gesellschaft, Königsberg, E. Prussia.
Königliches Zoologisches und Anthropologisch‑Ethnographisches Museum, Dresden.
Naturhistorischer Verein, Bonn.
Naturhistorischer Museum, Hamburg.
Naturwissenschaftlicher Verein, Bremen.
Naturwissenschaftlicher Verein, Frankfort-an-der-Oder.
Rautenstrauch-Joest-Museum (Städtisches Museum für Völkerkunde), Cologne.
Redaction des Biologischen Centralblatts, Erlangen.
Senckenbergische Naturforschende Gesellschaft, Frankfort-am-Main.
Verein für Vaterländische Naturkunde in Württemburg, Stuttgart.
Zoological Society, Berlin.

Austria.

K.K. Central-Anstalt für Meteorologie und Erdmagnetismus, Vienna.
K.K. Geologische Reichsanstalt, Vienna.

Belgium and the Netherlands.

Musée Teyler, Haarlem.
Académie Royal des Sciences, des Lettres, et des Beaux-Arts de
    Belgique, Brussels.
La Société Royale de Botanique de Belgique, Brussels.
Netherlands Entomological Society, Rotterdam.

Switzerland.

Naturforschende Gesellschaft (Société des Sciences Naturelles), Bern.

France.

Bibliothèque Nationale, Paris.
Musée d'Histoire Naturelle, Paris.
Société Zoologique de France, Paris.

Italy.

Biblioteca ed Archivio Tecnico, Rome.
Museo di Zoologia e di Anatomia Comparata della R. Universita,
    Turin.
Orto e Museo Botanico (R. Instituto di Studi Superiori), Florence.
R. Accademia di Scienze, Lettre, ed Arti, Modena.
R. Accademia dei Lincei, Rome.
Stazione Zoologica di Napoli, Naples.
Società Africana d'Italia, Naples.
Società Geografica Italiana, Rome.
Società Toscana di Scienze Naturali, Pisa.

United States of America.

Academy of Natural Sciences, Buffalo, State of New York.
        ,   Davenport, Iowa.
        ,   Library, Philadelphia.
        ,   San Francisco.
American Geographical Society, New York.
American Institute of Mining Engineers, Philadelphia.
American Philosophical Society, Philadelphia.
Boston Society of Natural History.
Connecticut Academy, New Haven.
Department of Agriculture, Washington, D.C.
Field Museum of Natural History, Chicago.
Franklin Institute, Philadelphia.
Johns Hopkins University, Baltimore.
Leland Stanford University, California.
Missouri Botanical Gardens, St. Louis, Mo.
Museum of Comparative Zoology, Cambridge, Mass.
National Academy of Sciences, Smithsonian Institution, Washington,
    D.C., U.S.A.
National Geographical Society, Washington, D.C.
New York Academy of Sciences.

Philippine Museum, Manila.
Rochester Academy of Sciences.
Smithsonian Institution, Washington, D.C.
Stanford University, California.
Tufts College, Massachusetts.
United States Geological Survey, Washington, D.C.
University of Minnesota, Minneapolis.
University of Montana, Missoula.
Wagner Free Institute of Science of Philadelphia.
Washington Academy of Sciences.

Brazil.

Museo Paulista, Sao Paulo.
Escola de Minas, Rio de Janeiro.

Argentine Republic.

Sociedad Cientifica Argentina, Buenos Ayres.

Uruguay.

Museo Nacional, Monte Video.

Japan.

College of Science, Imperial University of Japan, Tokyo.

Hawaii.

Bernice Pauahi Bishop Museum, Honolulu.
National Library, Honolulu.

Java.

Society of Natural Science, Batavia.

597

INDEX.

AUTHORS OF PAPERS.

MARCUS F. MARKS, Government Printer, Wellington.—1916.

Lightning Source UK Ltd.
Milton Keynes UK
UKHW011343100219
336964UK00010B/716/P